Wireless Network Evolution

2G to 3G

ISBN 0-13-028077-1

90000

9 780130 280770

Prentice Hall Communications Engineering and Emerging Technologies Series

Theodore S. Rappaport, *Series Editor*

GARG *Wireless Network Evolution: 2G to 3G*

GARG *IS-95 CDMA and cdma2000: Cellular/PCS Systems Implementation*

GARG & WILKES *Principles and Applications of GSM*

HAĆ *Multimedia Applications Support for Wireless ATM Networks*

KIM *Handbook of CDMA System Design, Engineering, and Optimization*

LIBERTI & RAPPAPORT *Smart Antennas for Wireless Communications: IS-95 and Third Generation CDMA Applications*

RAPPAPORT *Wireless Communications: Principles and Practice*

RAZAVI *RF Microelectronics*

STARR, CIOFFI & SILVERMAN *Understanding Digital Subscriber Line Technology*

Wireless Network Evolution

2G to 3G

Vijay K. Garg, Ph.D., P.E.
Department of Electrical and Computer Engineering,
University of Illinois, Chicago

PH
PTR

Prentice Hall PTR
Upper Saddle River, NJ 07458
www.phptr.com

Library of Congress Cataloging-in-Publication Data

Garg, Vijay Kumar, 1938-
Wireless network evolution : 2G to 3G / Vijay K. Garg
p. cm.
Includes bibliographical references and index.
ISBN 0-13-028077-1
1. Wireless communication systems. I. Title.

TK5103.2.G38 2001
621.382--dc21

2001021968

Acquisitions Editor: *Bernard Goodwin*
Editorial Assistant: *Michelle Vincenti*
Production Supervisor and Compositor: *Vanessa Moore*
Cover Design Director: *Jerry Votta*
Cover Designer: *Nina Scuderi*
Manufacturing Manager: *Alexis R. Heydt*
Marketing Manager: *Dan DePasquale*
Project Coordinator: *Anne Trowbridge*

PH PTR

Prentice-Hall, Inc.
Upper Saddle River, New Jersey 07458

Prentice Hall books are widely used by corporations and government agencies for training, marketing, and resale.
The publisher offers discounts on this book when ordered in bulk quantities.
For more information, contact Corporate Sales Department, Phone: 800-382-3419;
Fax: 201-236-7141; E-mail: corpsales@prenhall.com
Or write: Prentice Hall PTR, Corp. Sales Dept., One Lake Street, Upper Saddle River, NJ 07458.

Printed in the United States of America
10 9 8 7 6 5 4 3 2 1

ISBN: 0-13-028077-1

Pearson Education Ltd.
Pearson Education Australia PTY, Limited
Pearson Education Singapore, Pte. Ltd
Pearson Education North Asia Ltd.
Pearson Education Canada, Ltd.
Pearson Educación de Mexico, S.A. de C.V.
Pearson Education — Japan
Pearson Education Malaysia, Pte. Ltd.
Pearson Education, Upper Saddle River, New Jersey

I dedicate this book to my father,
the late Shri Reoti Saran, and my father-in-law,
Shri Anandi Lal Bansal. I also dedicate this book to my wife, Pushpa Garg.
Without her, it would not have been possible to complete this book.

V. K. G.

Contents

Chapter 4
Speech and Channel Coding, Spreading Codes, and Modulation 87

Chapter 5
Physical and Logical Channels of IS-95 141

Chapter 14
Third-Generation European Standards 449

Chapter 17
Wireless Application Protocol, Bluetooth, and Wireless Local Area Network 577

Chapter 19
CDMA RF Optimization 661

Preface

Although wireless data has not yet generated a huge market, it will become a primary force as service providers run out of cost-effective means to compete on price, coverage, and packaging of voice airtime. With the Internet and corporate intranets becoming essential parts to daily business activities, it is becoming increasingly advantageous to have wireless offices that can easily connect mobile users to their enterprises. The potential for technologies that deliver news and other business-related information directly to wireless handsets could also develop entirely new revenue streams for service providers. Since about 40 percent of mobile subscribers already carry cellular phones, these devices are uniquely positioned to integrate the incoming generation of mobile data applications. Integrated voice and data networks are the future for the fixed and wireless network operators. GSM and cdmaOne systems offer operators the opportunity to provide fully integrated networks.

GSM data services were planned from the beginning of the digital wireless communication system and offer many options similar to those offered over Integrated Service of Digital Network (ISDN). GSM data services include circuit-switched and packet-switched data. Circuit-switched data can be sent to an analog modem, an ISDN connection, or fax equipment. Packet-switched data connects to a packet network.

GSM short message service (SMS) allows the exchange of short alphanumeric messages between a mobile and the GSM system or between the cellular system and an external device capable of transmitting and optionally receiving short messages. GSM supports two types of short message services. The first type is the point-to-point service, in which a user sending a short message interacts with a short message entity (SME) in order to enter the message into system. The user can interact with the SME via e-mail, a phone call into an operator console that takes the message and enters it into the system, or by calling into a voice response system that accepts dual-tone multifrequency (DTMF) tones. The second type of service is the SMS call broadcast (SMSCB). In this service, multiple mobile stations are sending the same message via a broadcast message. The purpose of this service is to provide the same information to many mobile stations in the system at the same time. The information could consist of news or sports reports, stock data, and so forth.

Presently, data traffic on most of the GSM networks is modest, at less than 3 percent of total GSM traffic; some operators do not even bother to implement data facilities. However, with new initiatives planned during the course of next two to three years, exponential growth in data traffic is forecasted. Message-based applications may reach a penetration of up to 25 percent in developed markets by 2001, and 70 percent by 2003. GSM data transmission using high-speed circuit-switched data (HSCSD) and general packet radio service (GPRS) may reach a penetration of about 10 percent by 2001 and 25 percent by 2003.

Ushering faster data speeds into the mainstream will be 14.4 kbps and HSCSD protocols that approach wireline access speeds of up to 57.6 kbps using multiple 14.4-kbps time slots. The increase from 9.6 to 14.4 kbps is due to a nominal reduction in the error-correction overhead of GSM radio link protocol (RLP), allowing the freed bandwidth to be used for higher data streams. Implementation of v.42bis compression could double throughput.

Migration to HSCSD will bring data into the mainstream of GSM, and will be enabled in many cases by relatively standard software upgrades in base station and mobile switching center equipment. HSCSD will enable faster Web browsing, file downloads, mobile video conferencing and navigation, vertical applications, telemetry, and bandwidth-secure LAN access. Firmware on most current GSM PC cards will have to be upgraded.

The next phase in the high-speed roadmap will be the evolution of current SMS, such as smart messaging and unstructured supplementary service data (USSD), toward the new GPRS. The GPRS is a new radio service that uses TCP/IP or X.25 to offer data rates of up to 115 kbps. The GPRS extends the packet capabilities of GSM to higher data rates and longer messages. The service supports sending point-to-point and point-to-multipoint messages. The key factors in making GPRS a success are

- Cost-effective network deployment
- Billing and service support
- Application support
- User-friendly terminals

Two new nodes have been added to the GSM network to support GPRS. The serving GPRS support node (SGSN) communicates with mobile stations within its service area. The gateway GPRS support node (GGSN) communicates with packet networks that are external to the GSM network. Call setup via GPRS will be instantaneous and users will be charged only for actual packet data transmitted rather than for online time. GPRS will make it practical to transmit bursty e-mail and demanding multimedia applications over the GSM network through its very efficient use of the available radio spectrum and channels. GPRS terminals linked to laptops via a cable or Bluetooth device registered with a GSM network can simultaneously send and receive data packets, such as e-mail, and make and receive voice calls. This is because even during the busy hours, GSM time slots are free between the ongoing calls, which means that short GPRS packets will be able to get through when normal circuit-switched traffic would be blocked.

Although, GPRS can produce 115 kbps by using up to 8 GSM 14.4-kbps time slots, most phones will probably provide up to 56 kbps. The GPRS standard also defines a mechanism by which a mobile can request the amount of bandwidth it desires at the time of establishing a data session. The direct IP nature of the GPRS will in large measure also make wireless middleware unnecessary, as users contact directly into the IP of a carrier.

The next generation of data heading toward the third-generation (3G) and personal multimedia environments is known as Enhanced Data Rate for GSM Evolution (EDGE) or GSM++. It builds on GPRS and will allow GSM operators to use existing GSM radio bands to offer wireless multimedia IP-based services and applications at theoretical maximum speeds of 384 kbps, with a bit rate of 48 kbps per time slot, and up to 69.2 kbps per time slot with good radio conditions. EDGE will be relatively painless to implement and will require relatively small changes to network hardware and software since it uses the same time division multiple access (TDMA) frame structure, logical channel, and 200-kHz carrier bandwidth as GSM networks. As EDGE progresses towards coexistence with 3G W-CDMA, data rates of up to 2 Mbps could be available.

Although cdmaOne (15-95–based technology) started offering SMS and asynchronous data later than GSM networks, the two technologies are on the same timeline when it comes to high-speed data and packet data development. The faster speed is crucial to luring data users. From a marketing perspective of operators, high-speed data is important to customers. Although cdmaOne networks were not the first to offer data access, they are uniquely designed to accommodate data. To start with, cdmaOne networks handle data and voice transmission in much the same way. The cdmaOne's variable-rate transmission capability allows data rate determination to accommodate the amount of information being sent, so system resources are used only as needed. Since cdmaOne systems employ a packetized backbone for voice, packet data capabilities are already inherent in the hardware. Circuit-switched data capability up to 14.4 kbps was the first data communication technology available over cdmaOne, but considerable work has been done to offer packet data capability at over 14.4 kbps. The cdmaOne packet data uses a TCP/IP-compliant cellular digital packet data (CDPD) protocol stack to provide seamless connectivity with enterprise networks and expedite third-party application development. Data to cdmaOne networks allows an operator to continue using its existing radios, backhaul facilities, infrastructure, and handsets while merely implementing a software upgrade with interworking functionality.

The recent IS-95B upgrade allows for code or channel aggregation to provide data rates of 64–115 kbps, as well as offering improvements in soft handoffs and interfrequency hard handoffs. To achieve speeds of 115 kbps, up to eight CDMA traffic channels offering 14.4 kbps need to be aggregated. It is expected that operators will initially support data rates between 28.8 kbps and 57.6 kbps on the forward link to mobiles, and 14.4 kbps on the reverse link since mobile users generally receive more data than they send over the air. Looking further ahead, IS-95 (cdma2000 lX) may double the capacity of IS-95A, as well as provide a basic data rate of 24.4 kbps. IS-95C will provide ultrafast data of more than 1 Mbps using a dedicated data channel and separate base stations. This step will be close to 3G data speed. cdmaOne equipment manufactures have announced IS-707 packet data, circuit-switched data, and digital fax capabilities on their cdmaOne infrastructure equipment. Some equipment manufactures are also adding mobile IP, the proposed Internet standard for mobility, as an enhancement to basic packet data services. Mobile IP allows users to maintain a continuous data connection and retain a single IP address while traveling between base stations or roaming on other CDMA networks.

The International Telecommunications Union (ITU) began studies on globalization of personal communications in 1986 and identified the long-term spectrum requirements for future 3G mobile wireless telecommunications systems. In 1992, the ITU identified 230 MHz of spectrum in the 2-GHz band needed in order to implement the International Mobile Telecommunications 2000 (IMT-2000) system on a worldwide basis for satellite and terrestrial components. IMT-2000 capabilities include a wide range of voice, data, and multimedia services with quality equivalent or better than the fixed telecommunications networks in different radio environments. The aim of IMT-2000 is to provide a universal coverage that enables terminals to have seamless roaming across multiple networks. The ITU accepted the overall standardization responsibility of IMT-2000 with the aim to define a family of radio interfaces applied in different radio environments including indoor, outdoor, terrestrial, and satellite.

The 3G mobile telecommunications systems are expected to provide worldwide access and global roaming for a wide range of services. Standards bodies in Europe, Japan, and North America are trying to achieve harmonization on key and interrelated issues. These include radio interfaces, system evolution and backward compatibility, user's migration and global roaming, and phased introduction of mobile services and capabilities to support terminal mobility. The European Telecommunications Standards Institute's (ETSI) Universal Mobile Telecommunications System (UMTS) studies were carried out in parallel with IMT-2000 to harmonize its efforts with the ITU. In Japan and North America, the ARIB and TIA committee TR45 are carrying forward standardization efforts for 3G, respectively. Two partnership projects—3GPP and 3GPP2—are underway to achieve harmonization between various 3G regional standards.

In Europe, 3G systems will support a substantially wider and enhanced range of services as compared with 2G (GSM) systems. This enhancement will include high-speed data and multimedia services. These will be achieved through an evolutionary path using the GPRS and EDGE to capitalize on the investments for GSM systems in Europe.

In North America, the 3G wireless telecommunications system, cdma2000, has been proposed to the ITU. It meets most of the IMT-2000 requirements in indoor office, indoor-to-outdoor/pedestrian, and vehicular environments. In addition, the cdma2000 also satisfies the requirements for 3G evolution of 2G cdmaOne (TIA/EIA 95 family of standards).

In Japan, evolution of the GSM platform is planned for the IMT-2000 (3G) core network due to its flexibility and widespread use around the world. Smooth migration from GSM to IMT-2000 is planned. The service area of the 3G system will overlay with the existing 2G (PDC) system. The 3G system will connect and interwork with 2G systems through interworking functionality (IWF). IMT-2000 PDC dual-mode terminals as well as IMT-2000 single-mode terminals will be deployed.

Mobile telephony services and their tariff structures are likely to change substantially over next few years. Mobile tariffs will decline to the point where they are priced only slightly above fixed wireline services, and mobile communications will be the norm. The conventional view is that mobile operators will not be able to increase revenue per customer for simple voice services. However, additional spending for new applications will counterbalance the decline in average monthly bills. These new applications will include data communications and other related services to divert spending from fixed-network services.

Today's mobile use relates almost exclusively to voice telephony. Nonvoice communications will increase substantially and account for an ever-increasing share of total traffic. An increasing number of wireless smart telephones and other wireless data connections will facilitate this condition. The UMTS Forum predicts that by 2005 nonvoice usage will exceed voice usage over mobile communications devices in Europe. Analysis by Nokia shows that a GSM operator pursuing an aggressive strategy on wireless data would be able to generate about 20 to 30 percent of its revenue from new services by 2001. High-speed circuit-switched data over cellular telephony will play a significant role; packet-oriented transport may be far more important, meaning that pricing will have to incorporate aspects of bandwidth and quality of service (QoS). Other components will be usage related to the volume of data and QoS. Bundled-up tariffs with a low marginal per-minute cost are closer to this concept than traditional mobile tariffs.

Since the deployment of wireless communications in early 1990s, about three dozen books have been written to describe wireless technologies. Unfortunately, none of these books address wireless data technologies in great detail. Some of these books provide a high-level view of wireless data. Moreover, these books assume that the reader is exposed to the data communications field and possesses adequate knowledge of fundamental disciplines such as the open system interconnect (OSI) model, channel capacity, data link, and other higher level protocols.

Soon, wireless data communications will see a worldwide explosion with the introduction of 3G systems. The motivation for writing this book is to present a comprehensive treatment of the subject. Starting at ground zero, the reader is first introduced to basic principles that are essential to an understanding of the 2G wireless systems. Mobile data communication services that have been used for TDMA/CDMA-based air interfaces are then discussed. Details of network architecture; logical channel structure; framing; channel coding; radio link operations; physical, data link, and network layers; and messages flow between the various layers are also provided. Management of a wireless network and highlights of the 3G data services for the UTRA/W-CDMA, cdma2000, GPRS, and EDGE networks are also discussed. I also focus on the network aspects and present evolutionary paths for 2G networks to 3G.

This book has been divided into three parts. Part I addresses the basics of 2G wireless technologies (particularly cdmaOne). Part II focuses on 3G air interfaces (UTRA/W-CDMA, cdma2000) and provides details for the evolution of TDMA-based GSM and IS-136 networks. Part III concentrates on the network aspects of wireless networking, including network management, network planning, and network optimization. This book can be used by practicing telecommunications engineers involved in the design of 3G networks, as well as by senior or graduate students in electrical engineering, telecommunication engineering, computer engineering, and computer science curricula.

This book can be adapted for two courses in the CDMA and 3G technologies. For the first semester course in the CDMA technology, I recommend using Chapters 1–10, 18, and 19. The second semester course in the 3G technologies should include Chapters 11–17 and 20.

Acknowledgments

I would like to thank the many people who helped me prepare the material in this book. Bernard Goodwin provided his encouragement in motivating me to write the book. Professor Theodore Rappaport of Virginia Tech took me under the banner of his series, *Communications Engineering and Emerging Technologies*. Vanessa Moore provided me her assistance in preparing the manuscript.

Vijay Garg
August 2001

PART I

Fundamentals of CDMA and Its Applications to 2G Systems

The first part of the book discusses the fundamental concepts of code division multiple access (CDMA) and the applications of CDMA technology to both second-generation (2G) cellular and personal commmunication services (PCS) systems. In the first part, we focus on the concepts of CDMA for wireless communications and the underlying network needed to support these applications for voice and data communications. The primary emphasis is placed on the CDMA system standardized by the Telecommunications Industry Association (TIA) and the Alliance for Telecommunications Industry Solutions (ATIS) as IS-95 and IS-665 standards. The chapters in the first part sequentially present the CDMA technology.

- They focus on the fundamental concepts, provide a foundation for understanding the underlying mathematics of spread spectrum, and highlight CDMA wireless standards.
- They provide sufficient details so that a reader can understand the related third-generation (3G) wireless standards and they allow the reader to apply the concepts to practical wireless systems.

In this first part, we assume that the reader has some basic understanding of wireless communications including network architecture, speech processing, modulation, etc.

CHAPTER 1

An Overview of Second- and Third-Generation (3G) Air Interfaces

1.1 Introduction

During the last two decades, the world has experienced phenomenal changes in the telecommunications industry. Communications that were formerly carried on wires are now supplied over radio (wireless). Thus, wireless communication, that uncouples the telephone from its wires to the local telephone exchange, has exploded. During the early 1980s, six incompatible analog systems were operational in Western Europe. The deployment of these incompatible systems resulted in mobile phones designed for one system that could not be used with another system, and roaming between many countries of Europe was not possible. With the growth of the European Common Market, roaming between the countries of Europe became important.

In 1982, the main governing body of the European PTTs, Conference Européenne des Administration des Postes et des Télécommunications (CEPT), set up a committee known as Groupe Special Mobile (GSM), under the auspices of its Committee on Harmonization, to define a mobile system that could be introduced across Europe by the 1990s. The CEPT allocated the necessary duplex radio frequency bands in the 900 MHz region.

The GSM (renamed Global System for Mobile Communications) initiative gave the European mobile communications industry a home market of about 300 million subscribers, while at the same time providing it with a significant technical challenge. The early years of the GSM were devoted mainly to the selection of radio techniques for the air interface. In 1986, field trials of different candidate systems proposed for the GSM air interface were conducted in Paris. A set of criteria ranked in order of importance was established to assess these candidates.

The interfaces, protocols, and protocol stacks in GSM are aligned with the open system interconnect (OSI) principles. The GSM architecture is an open architecture that provides maximum independence between network elements such as the base station controller (BSC), the mobile switching center (MSC), the home location register (HLR), etc. This approach simplifies the design, testing, and implementation of the system. It also favors an evolutionary growth path, since network element independence implies that modification to one network element can be made with minimum or no impact on the others. Also, a system operator has a choice of using network elements from different manufacturers.

The GSM900 has been adopted in many countries, including the major parts of Europe, North Africa, the Middle East, many East Asian countries, and Australia. In most of these cases, roaming agreements exist to make it possible for subscribers to travel in different parts of the world and enjoy continuity of their telecommunications services with a single number and a single bill. The adaptation of GSM at 1,800 MHz (GSM1800) is also spreading outside of Europe to East Asian and some South American countries. GSM1900, a derivative of GSM for North America, is planned to cover a substantial area of the United States. All these systems also enjoy a form of roaming, referred to as subscriber identity module (SIM) roaming, between them and with all other GSM-based systems. A subscriber from any of these systems could access telecommunication services by using his or her personal SIM card in a handset suitable to the network from which coverage is provided. If the subscriber has a multiband phone, then one phone could be used worldwide. This globalization is making GSM and its derivatives one of the leading contenders to offer digital cellular and PCS worldwide. A three-band handset (900, 1,800, and 1,900 MHz) is available, and true worldwide seamless roaming is possible.

Two digital technologies, time division multiple access (TDMA) and code division multiple access (CDMA) have emerged as clear choices for the newer PCS systems. TDMA is a narrowband technology in which communication channels on a carrier frequency are apportioned by time slots. For TDMA technology, there are two prevalent systems: North America TIA/EIA/IS-136 [1,2] and European Telecommunications Standards Institute (ETSI) Digital Cellular System 1800 (GSM1800) [3], a derivative of GSM. CDMA (TIA/EIA/IS-95A) is a direct-sequence (DS) spread-spectrum system in which the entire bandwidth of the carrier channel is simultaneously made available to each user. The bandwidth is many times larger than the bandwidth required to transmit the basic information. CDMA systems are limited by interference produced by the DS signals of other users transmitting within the same bandwidth.

In this chapter, we describe the major attributes of the IS-95A/J-STD-008 CDMA systems, PCS standardization activities in North America, and the future of CDMA technology. We provide brief descriptions of the seven air interfaces that have been approved by the Joint Technical Committee (JTC) and have been adopted as American National Standards Institute (ANSI) standards for deploying PCS systems in North America.

1.2 PCS Standardization Activities in North America

The standardization activities for PCS in North America were carried out by the JTC on wireless access, consisting of appropriate groups within the T1 committee, a unit of the Alliance for Telecommunications Industry Solutions (ATIS), and the engineering committee TR46, a unit of the Telecommunications Industry Association (TIA). The JTC was formed in November 1992, and its first assignment was to develop a set of criteria for PCS air interfaces. The JTC established seven Technical Adhoc Groups (TAGs) in March 1994, one for each selected air interface proposal. The TAGs drafted the specifications document for the respective air interface technologies and conducted validation and verification to ensure consistency with the criteria established by the JTC. This was followed by balloting on each of the standards. After the balloting process, four of the proposed standards were adopted as ANSI standards: IS-136–based PCS, IS-95–based PCS, GSM1900 (based on GSM), and personal access communication system (PACS). Two of the proposed standards—hybrid CDMA/TDMA and wideband CDMA (by OKI)—were adopted as trial use standards by ATIS and interim standards by TIA. The personal wireless telecommunications-enhanced (PWT-E) standard was moved from JTC to TR46.1 which, after a ballot process, was adopted in March 1996. Table 1.1 provides comparisons of the seven technologies using a set of parameters that include access methods, duplex methods, bandwidth per channel, throughput per channel, maximum power output per subscriber unit, vocoder, and minimum and maximum cell ranges.

Table 1.1 Technical Characteristics of North American PCS Standards

	TAG-1	TAG-2	TAG-3	TAG-4	TAG-5	TAG-6	TAG-7
Standard	Hybrid CDMA/ TDMA	IS-95–based PCS	PACS	IS-136–based PCS	GSM 1900	PWT-E	Wideband CDMA (OKI)
Access	CDMA/ TDMA/ FDMA	CDMA	TDMA	TDMA	TDMA	TDMA	CDMA
Duplex method	TDD	FDD	FDD	FDD	FDD	TDD	FDD
Frequency reuse	3	1	16×1	7×3	7×1, 3×3	Portable selected	1
Bandwidth/channel	2.5/5 MHz	1.25 MHz	300 kHz	30 kHz	200 kHz	1 kHz	5, 10, 15 MHz
Throughput/channel (kbps)	8	8.55/13.3	32	8	13	32	32
Maximum power/ subscriber unit	600 mW	200 mW	200 mW	600 mW	0.5 W, 2.0 W	500 mW	500 mW
Vocoder	PHS HCA	CELP	ADPCM	VCELP/ ACELP	RPE-LTE ACELP	ADPCM	ADPCM
Maximum cell range (km)	10.0	50.0	1.6	20.0	35.0	0.15	5.0
Minimum cell range (km)	0.1	0.05	0–1	0.5	0.5	0.01	0.05

1.2.1 TAG-1: Hybrid CDMA/TDMA-Based PCS

The trial use standard J-STD-017/interim standard IS-661 is based on the composite CDMA/TDMA system design. It was introduced by Omnipoint Corporation. The proposed standard was balloted in November 1994 and was approved as a trial use standard by committee T1 and an interim standard by TIA in 1995.

The standard supports 5-MHz and 2.5-MHz channels. The 2.5-MHz channels were developed for 10-MHz PCS bands but may also be implemented in the 30-MHz bands. The standard uses a combination of TDMA, FDMA, and CDMA for multiuser access to the PCS network. Within a cell, TDMA is employed to separate users. To provide a large area coverage or to provide greater capacity in densely populated regions, multiple cells or sectorized cells are deployed using frequency division multiple access (FDMA) to separate cells by frequency.

To allow multicell deployment in a given region, CDMA is used. CDMA is used for each radio frequency (RF) link to reduce co-channel interference between cells by reusing the same RF carrier. CDMA also improves system response to RF channel impairments caused by multipath propagation.

The speech codecs are pulse code modulation (PCM) and adaptive differential pulse code modulation (ADPCM), as well as personal communication system high compression algorithm (PCS HCA), which is an 8-kbps vocoder for high spectral efficiency and high-quality voice in the presence of channel errors. Time division duplex (TDD) is used to allow 32 simultaneous 8-kbps full-duplex mobile users in a 5-MHz channel and 16 users in a 2.5-MHz channel. For data transmission, the standard supports a flexible structure of 32 8-kbps time slots capable of supporting up to 256-kbps full-duplex or 512-kbps half-duplex rates through time slot aggregation.

The standard is designed around an object-based software architecture to allow for flexibility in interconnection to the public switching telephone network (PSTN), advanced intelligent network (AIN), GSM1900 networks, and IS-41 networks. The standard supports a variety of AIN architectures for full integrated service of digital network (ISDN) connectivity. It provides interconnection to GSM1900 networks via the "A" interface, which is between the BSC and MSC. This gives a PCS provider flexibility to deploy the appropriate network infrastructure to meet service requirements. The standard supports features and services typically found in wireline environments, such as voicemail, call holding, call forwarding, call waiting, caller ID, and three-way calling. It also supports short message service, smart cards, and over-the-air provisioning.

The standard allows two types of handoffs:

- *Make-before-break* soft handoffs to provide seamless coverage.
- *Break-before-make* hard handoffs for use in emergency situations. In this situation, the terminal establishes contact with a new base station on a new frequency and a new time slot.

One of the strengths of this standard is its large spread-spectrum bandwidth. Since the power is spread out over a larger bandwidth, the average power is very low, reducing the chance of interference with other technologies. The system is more mobile-centric than most of the other systems. The handset has the intelligence to sense when it should go for handoff and does not need to be directed by the BSC. The standard is designed to keep the cost of technology deployment and implementation low. It does not offer exotic error correction schemes. The standard currently does not support unlicensed band frequencies. However, protocols are being developed to support the spectrum etiquette of the unlicensed band. The standard also supports a variety of backhaul facilities. For most applications, T1, fractional T1, or ISDN interfaces are likely to be used. Other backhaul interfaces include HDSL, microwave links, and the cable plant. Local exchange carriers' dedicated digital facilities can also be used.

1.2.2 TAG-2: CDMA IS-95–Based PCS

The ANSI J-STD-008 standard is an upbanded version of IS-95, the North American wideband digital cellular standard for operation in PCS licensed bands. The IS-95 standard was adopted by TIA in July 1993, and an upbanded version of the standard to operate at 1.9 GHz was subsequently submitted to the JTC. The PCS version of the standard was developed by TAG-2.

The standard is interoperable with the IS-95 standard as well as the advanced mobile phone system (AMPS). Thus, a dual-band, dual-mode handset can operate in both PCS and cellular bands. Handoffs are supported from the PCS band to both IS-95 and AMPS base stations. Thus, if a handset moves out of the range of a PCS base station, it will be handed over to an IS-95 base station, or in its absence, to an AMPS base station. The standard is not intended to be used in the unlicensed bands.

Two transmission rate sets are supported, each using a different speech codec. Within each rate set, transmission rate is variable and depends on speech activity of the user. Rate Set 1 (RS1) supports data rates of 9.6 kbps, 4.8 kbps, 2.4 kbps, and 1.2 kbps; Rate Set 2 (RS2) uses rates of 14.4 kbps, 7.2 kbps, 3.6 kbps, and 1.8 kbps. Every 20 ms, speech codec generates a different rate, dependent on speech activity. The variable transmission rate increases the channel capacity. RS1 and RS2 support user information at 8.55 kbps and 13.3 kbps, respectively, for full rate. In addition to supporting both 8.55 kbps and 13.3 kbps voice services, the standard also supports a range of data services. These are:

- **Asynchronous (async) data and facsimile:** IS-99 provides procedures for async data and Group 3 facsimile communications. For async data, the handset interface emulates a wireline modem that processes standard AT modem control commands. For fax, the handset emulates a Group 3, Class 2.0 digital fax modem. A notebook computer, a PDA, or other terminal device can connect to the handset via a standard EIA-232 connection or can have CDMA transmission equipment embedded in it, such as in a PCMCIA card.

- **Packet data:** IS-657 provides specifications for the transport of packet services. Cellular digital packet data (CDPD) services are currently supported, and work has been completed on mobile Internet protocol (IP) services.
- **Point-to-point and broadcast short messaging:** IS-637 provides procedures for both broadcast and point-to-point short messaging. The maximum size of a message is limited to 255 octets, including overhead. Several different methods to broadcast short message delivery are supported to achieve efficiency and minimize power consumption of handsets. The standard provides the capability to deliver broadcast short messages to a handset in a call. Point-to-point multiple messages are also supported.

This standard provides for an evolution to an extended system to support higher data rate codecs. Transmission rates of 64 kbps are being offered in the first stage, followed by higher rates. These rates are in the 144-kbps or more range. Over-the-air service provisioning (OTASP) to allow the user to have the user identity information downloaded into the handset is also supported. The power output of the subscriber unit dynamically varies with the range from the cell so that the signal-to-noise (S/N) ratio remains the same. The maximum output of the handset is 200 mW. The power output of the base station depends on implementation and meets the FCC requirements of 1,640 watts EIRP. The system supports seamless soft handoff. It has a physical limit of 64 channels per carrier frequency, but an operational limit of around 20 channels. Each base station has up to seven paging channels. The standard is supported by the IS-41 network standard for automatic roaming, call delivery, automatic billing, authentication, voice privacy, and handoff between mobile service switching centers.

1.2.3 TAG-3: Personal Access Communication System (PACS)

PACS was developed by TAG-3 and is defined in the ANSI J-STD-014 personal access communication system (PACS) air interface standard. Its two annexes are J-STD-014A personal access communication system unlicensed-version A (PACS-UA) and J-STD-014 B personal access communication system unlicensed-version B (PACS-UB). Air interface standard PACS-UA is based on Japan's personal handyphone system (PHS), and PACS-UB is a derivative of Bellcore's wireless access communication system (WACS).

PACS is designed for low-mobility (speeds less than 40 km/hour) outdoor applications. PACS-UA and PACS-UB support indoor applications. PACS has an eight time slot TDMA air interface with a frequency division duplex (FDD) mode for small-cell, licensed applications. PACS-UA and PACS-UB use TTD mode. The technology is based on a low-cost interface to address small-cell, low-mobility applications. The system was designed to allow low-cost subscriber units, high-speech quality, and relatively high data rates. The standard uses error detection only. The standard is designed for pedestrian, residential, and business applications and primarily intended for outdoor applications, such as wireless local loop to replace the last 1,000 feet of copper. However, by incorporating packet mode and higher data rate capability, the system has been enhanced to support applications in office environments, such as wireless centrex. The other features of the PACS are as follows:

- Both PACS-UA and PACS-UB are designed for wireless PBX applications. The standard allows interoperability between outdoor and indoor versions.
- PACS was developed for slow fading environments. It supports moderate mobility up to 65 km/hr, such as urban traffic. For line of sight, the system can support speeds up to 100 km/hr.
- PACS also supports fixed wireless access in situations where a drop connection replaces a wireline connection. In such a case, the subscriber units provide a standard interface, such as an RJ-11 jack, to the user, allowing access from a standard telephone set.
- The standard is designed to support the ISDN environment: call forwarding, three-way calling, call waiting, and emergency calls without subscriber registration. It uses a number of backhaul facilities, including T1 and HDSL.
- PACS has a short frame structure and low end-to-end delay. With a 16-cell repeat pattern, the system provides 97 percent coverage, and with a 21-cell repeat pattern, it allows 99 percent coverage. The system uses three RF channels per cell site.
- Transmit and receive diversity are used at the cell site, as well as post-detection selection diversity in the portable. The voice quality is toll quality, and grade of service is around 1 percent.
- PACS offers a high level of privacy and security. Each subscriber unit has a unique identity (electronic serial number [ESN]).
- The base station has a maximum power output of 800 mW. The subscriber unit has a maximum power output of 200 mW. The system uses $\pi/4$-differential quadrature phase-shift keying (DQPSK) modulation scheme.
- The standard does not specifically support smart cards. However, the authentication and encryption algorithms in the standard can accommodate a smart card. Thus, the system can use information from a smart card, or it can use information from an embedded system.
- The transmission rate of the air interface is 384 kbps and the transmission rate per channel is 48 kbps. The PACS RF carrier provides eight full-rate traffic channels of 32 kbps, 16 half-rate channels of 16 kbps, 32 quarter-rate channels of 8 kbps, and 64 eighth-rate channels of 4 kbps.
- The circuit-mode data service is a nontransparent mode, low-latency data service in which data is enciphered for privacy. The 32-kbps channels can be aggregated to support higher data rates.
- The system allows packet data to coexist on the same RF channel with voice, and dynamically adapt to the requirements of the voice channel. The system can also be implemented so that the entire RF channel is dedicated to packet services.
- The packet channels allow subscriber units to operate on a single time slot per TDMA frame as well as achieve higher throughput by using multiple time slots per frame. The maximum data rate can be 256 kbps.

- The system also supports messaging services for a maximum data size of about 4 MB. The applications include text messages, Group 3 fax, encoded sound, imaging, and video.

1.2.4 TAG-4: IS-136–Based PCS

The IS-136–based PCS specification was adopted as a stand-alone standard. The standard is a frequency-shifted derivative of IS-54 for PCS bands. It is described in the following ANSI standards documents:

- J-STD-009 PCS: PCS IS-136-based mobile station minimum performance standard
- J-STD-010 PCS: PCS IS-136-based base station minimum performance standard
- J-STD-011 PCS: PCS IS-136-based air interface compatibility standard

The Telecommunications Industry Association (TIA) adopted the IS-54 standard based on TDMA to meet the growing need for increased cellular capacity in high-density areas. IS-54 retains the 30-kHz channel spacing of AMPS to facilitate evolution from analog to digital systems. Each frequency channel provides a raw RF bit rate of 48.6 kbps. This is achieved by using $\pi/4$ -DQPSK modulation at a 24.3-symbols/second channel rate. This capacity is divided among six time slots, two of which are assigned to each user in the current full-rate implementation. A 7.95-kbps vector sum excited linear prediction (VSELP) (replaced by a 7.45-kbps algebraic code excited linear prediction [ACELP]) speech codec is used. Each 30-kHz frequency pair serves three users simultaneously with the same reuse pattern as used in AMPS and provides three times the capacity of AMPS. The IS-54 standard uses an adaptive equalizer to mitigate the intersymbol interference caused by large delay spreads, but due to the relatively low channel rate of 24.3 symbols/second, the equalizer will be unnecessary in many situations.

Since IS-54 systems were to operate in the same spectrum used by the existing AMPS systems, it provides for both analog and digital operation. This is necessary to accommodate roaming subscribers, given a large embedded base of AMPS equipment. Initially, the IS-54 standard used the AMPS control channel with 10-kbps Manchester-encoded frequency shift keying (FSK). IS-136, the new version of IS-54 for PCS, includes a digital control channel (DCCH) which uses the 48.6-kbps modem. With increased signaling rate, the DCCH offers capabilities such as point-to-point short messaging, broadcast messaging, group addressing, and private user groups. IS-54 equipment has already been deployed and is operational in a majority of the top cellular markets in the United States. IS-136 equipment has been deployed for PCS in several major cities.

For subscriber equipment, the maximum power output is 600 mW. The power output for the base station is variable but is within the FCC limits of 1,640 watts EIRP. The technology can adapt to channel variations for vehicle speeds up to at least 110 km/hr. The standard contains forward error correction to enhance the reliability of user information.

Two types of circuit-switched data services are provided. These include async data and Group 3 fax. The async data service supports modem-based access to PSTN subscribers. It transports user data in digital form over the radio interface, with modems residing in the PCS system. Modems supported include V.32 (4.8 kbps and 9.6 kbps), V.32 bis (7.2 kbps, 12 kbps, and 14.4 kbps), and V.34 (28.8 kbps). The async data service can provide access to public packet-switched networks. The standard has been upgraded to support the cellular digital packet data (CDPD) standard that was originally developed to work with AMPS.

1.2.5 TAG-5: GSM1900

In the past few years, a number of North American companies have become GSM MoU signatories, including American Personal Communications/Sprint Spectrum, American Portable Telecom, BellSouth Mobility DCS, Omnipoint Communications Inc., Pacific Bell Mobile Services, Pocket Communications Inc., Powertel Inc., Microcell Telecommunications Inc., and Western Wireless Corporation. These companies participate in the meetings of the MoU Association and ETSI TC SMG. During the September 1996 plenary meeting of the GSM MoU, the North American Interest group decided that GSM1900 should be named "GSM North America." By the fall of 1997, there were about 100,000 customers in 19 U.S. markets that used GSM digital services. GSM networks are operational in nine U.S. states, plus some border areas of neighboring states. Standardized roaming agreements between U.S. service providers have been created. Operational roaming is targeted for the near future. An SS7 gateway is installed for international roaming. GSM North America supports the recommendations of the SMG speech strategy experts group to focus on a multirate codec to avoid unnecessary proliferation of voice codecs. A multirate codec is now commercially available.

In 1991, ETSI defined the parameters of GSM1800, an upbanded version of GSM for the 1,800 MHz frequencies. The standard was taken up by TAG-5 for modification to the U.S. market as GSM1900. The proposed standard was balloted in November 1994, and its technical approval was completed in January 1995 for adoption by ANSI as J-STD-007.

GSM1900 is an advancement on the original GSM standard because it is based on Phase II of GSM. The standard offers a very rich feature set. It supports call forwarding, emergency calls, short message service, video text, and facsimile. Transparent and non-transparent data service up to 9.6 kbps is supported, as well as digital bearer service with a net rate of 12 kbps. The standard has the potential of adding V.42 bis and time slot aggregation to support higher data rates. The backhaul facility supported is T1.

The data rate over the radio channel is 270 kbps. Gaussian minimum shift keying (GMSK) modulation is used with a bandwidth (B) multiplied by bit period (T) equal to 0.3 and channel spacing of 200 kHz. Frequency hopping is an optional network capability on GSM. Hopping occurs at the TDMA frame rate of about 217 hops/second with the hop sequence being communicated to the mobile at call setup and handoff times. Frequency hopping provides an ability to further counteract mutipath fading over and above that already achieved with channel coding, interleaving, and antenna diversity. Frequency hopping also provides a better statistical distribution of interference, and its use is anticipated to enable very efficient frequency reuse within the cellular environment.

Two vocoders have been defined for GSM1900, and both use 13 kbps. The RPE-LTE is the conventional European vocoder. ACELP (12.2 kbps) is a more advanced vocoder which offers high-quality voice. The standard provides a high level of security, both to protect access to the services and to ensure the privacy of user-related information. The security functions for authentication of subscriber-related information and all processes involving the authentication key are contained in a removable part of the mobile station called the subscriber identity module (SIM), usually in form of a smart card.

The cells can have a range of up to 35 km in rural areas and a radius of up to 1 km in urban areas. Extended cell operation with a range of up to 120 km can also be achieved. The mobility speed is about 125 km/hr. The gross transmission rate per traffic channel is 22.8 kbps. For circuit-switched data, 9.6-kbps sync and async are supported. For packet-switched data transmission, two proposals were considered. One proposal uses a single time slot with the effective throughput averaging 9.6 kbps. The other proposal is based on using the entire RF channel, for a gross data rate of 270.8 kbps.

The power output for subscriber unit is 1 W when used as a handheld, and 2 W when it is mounted on a vehicle. The base station power output is within the FCC limits of 1,640 Watts EIRP. All traffic channels may use discontinuous transmission, whereby the transmitter is silent most of the time when no relevant information is being transmitted. In the case of speech, this is achieved due to the specification of speech activity detectors. This feature, combined with frequency hopping, increases the system capacity and prolongs battery life in handheld portables.

There are two frequency reuse patterns. In one case, the system is omnidirectional, and there are seven cells in the reuse group pattern, permitting 11 duplex frequencies per cell within a 30-MHz allocation. In the other case, there are three 120° sectors, and there are three cells grouped to form the complete reuse pattern, allowing 25 duplex frequencies per cell within a 30-MHz allocation. The capacity of the system is about three times the capacity of AMPS. Power control, discontinuous transmission, and frequency hopping can support improved frequency reuse, offering higher system capacity.

Work is under way to provide interoperability between GSM1900 and personal wireless telecommunications (PWT), the U.S. version of digital enhanced cordless telecommunications (DECT). The GSM1900 is interoperable with the composite CDMA/TDMA standard at the "A" interface.

1.2.6 TAG-6: Personal Wireless Telecommunications-Enhanced (PWT-E)

The DECT standard was adopted as a European standard by ETSI in June 1992. Subsequently, work was started in the U.S. to adopt the standard for operation in the licensed and unlicensed bands. The development of the standard for the licensed band was taken up by TAG-6, and, for the unlicensed band, it was taken up by TR41.6. Later, it was decided to move the standardization activity for the licensed band from TAG-6 to TR41.6. After going through the ballot process, PWT-E was adopted for licensed band by TR41.6 as EIA/TIA 662 in 1995.

PWT-E is a 12 time slot TDMA air interface. The system offers time division duplex (TDD) operation. PWT-E is an interoperability standard that allows products manufactured by one manufacturer to work on another manufacturer's products if they are based on PWT-E. The major features of PWT-E are the following:

- **Spectrum etiquette:** The spectrum etiquette is not needed for the licensed band, and therefore, it is not included in PWT-E.
- **Frequency bands:** The transmit and receive frequency bands for PWT-E are specified.
- **Power levels:** In the licensed bands, PWT-E uses 250 mW and 500 mW power.

PWT-E is suitable for small-cell outdoor applications. It is basically designed for wireless centrex and wireless local loop applications. PWT-E supports mobility speeds up to 65 km/hr and can be used in moving vehicles in campus environments with speeds between 50 and 60 km/hr. The modulation scheme is $\pi/4$ -DQPSK. The system offers antenna diversity and has a bandwidth of 1 MHz. The system has been designed to allow access to other networks and can be adapted to backbone wireless networks and wireline networks. Interoperability is offered between PWT-E and GSM1900.

The system has a range of 300 ft indoors, but it can be optimized to 500 ft outdoors using directional antennas. It allows 12 simultaneous users per cell site. The frequency reuse factor is selected by the portable through dynamic channel selection. The system uses equalizers. The speech codec is ADPCM with a data rate of 32 kbps. Data transmission at 9.6 kbps is supported, and work is under way to support higher data rates by concatenating channels.

The system supports high-density traffic. For eight carriers (96 access channels), the system uses 8-MHz bandwidth for an average traffic of six Erlangs per base station. With an average number of 300 subscribers per base station, the traffic per square kilometer is about 77 Erlangs.

1.2.7 TAG-7: Wideband CDMA (W-CDMA)

The TIA/EIA IS 665 and ATIS J-STD-015 (trial use) standard was developed by TAG-7. It is based on wideband CDMA technology developed by OKI America in conjunction with InterDigital. The standard was finalized by TAG-7 in June 1995.

The standard uses direct-sequence spread-spectrum and employs FDD. RF modulation is QPSK. The system supports secure voice and data services using advanced authentication and privacy procedures. The system capacity is about 16 times the AMPS capacity. The standard is designed to support both large- and small-cell applications. It supports mobility up to 100 km/hr. The standard supports 5-MHz channel spacing that can be expanded to 10 MHz and 15 MHz. Each of the bandwidths supports data rates of 16 kbps, 32 kbps, and 64 kbps. A 64-kbps channel allows a service provider to offer ISDN service.

The standard supports flexible aggregation of traffic channels within an RF channel to support higher data rates: $N \times 64$, where N is the number of channels. The following data rates are supported:

- Voiceband data rate up to 14.4 kbps and Group 3 fax up to 9.6 kbps
- 64-kbps transparent mode with forward error correction feature
- 64-kbps nontransparent mode with automatic repeat request (ARQ) feature for error-free transmission

An advanced codec developed by OKI is used to provide toll-quality speech, even in severe radio environments. The codec supports 32 kbps and is superior in quality to PCM, ADPCM, and LD-CELP. The standard also supports 16-kbps LD-CELP.

Soft handoff is allowed. The base station supports two types of handoff procedures that may be selected by service providers. The frequency reuse factor is 1. The standard supports licensed band applications for cells with a radius of more than 5 km. The system capacity is 12 Erlangs per cell per MHz. The number of channels per cell can be increased to 64 by an optional interference canceller system located at the base station.

W-CDMA has a flexible architecture to accommodate various switching systems. It can interconnect with the switching system using signaling system 7 (SS7). The power output for the base station is 1,000 watts. The subscriber unit power output is 500 mW. W-CDMA uses forward error correction, interleaving, multipath combining, and interference cancellation to obtain higher system capacity compared to other PCS standards.

1.3 Major Attributes of CDMA Systems

The major attributes of IS-95A/J-STD-008 CDMA systems are as follows:

- **System capacity:** The projected capacity of CDMA systems is much higher than of the existing analog/digital systems. The increased system capacity is due to improved coding gain/modulation scheme, voice activity, three-sector sectorization, and reuse of the same spectrum in every cell and all sectors.
- **Quality of service:** CDMA improves the quality of service by providing robust operation in fading environments and transparent (soft) handoffs. CDMA takes advantage of multipath propagation to enhance communications and voice quality. By using RAKE receiver and other improved signal-processing techniques, each mobile station selects the three strongest multipath signals and coherently combines them to produce an enhanced signal. Thus, the multipath propagation of the radio channel is used to an advantage in CDMA. In narrowband systems, fading causes a substantial degradation of signal quality. By using soft handoff, CDMA eliminates the ping-pong effect that occurs when the mobile is close to the border between cells, and the call is rapidly switched between two cells. This effect results in handoff noise, increases the load on switching equipment, and increases the chance of a dropped call. In soft handoff, a connection is made to the target cell while maintaining the connection with the serving cell, all operating on the same carrier frequency. This procedure ensures a smooth transition between cells, one that is undetectable to the subscriber. In comparison, many analog and other

digital systems use a break-before-make connection and require a change in mobile frequency that increases handoff noise and the chance of a dropped call.

- **Economies:** CDMA is a cost-effective technology that requires fewer cell sites and no costly frequency reuse pattern. The average power transmitted by CDMA mobile stations averages 6 to 7 mW, which is significantly lower than the average power typically required by FM and TDMA phones. Transmitting less power means that average battery life will be longer.

1.4 Market Trends of Digital Wireless Technologies

Most of the European city areas are covered by GSM. The rural coverage often exceeded 90 percent as of the fall of 1997. It is estimated that, worldwide, around 57 million subscribers are connected to GSM, and new subscribers are signing up for GSM services at a rate of about 10,000 per day. GSM is a market success based on a standardization success. Specifying a mobile telecommunications platform was completed in Europe earlier and in a more comprehensive manner than in other regions of the world. The availability of a type approval procedure, agreed between relevant parties, has also been an important factor.

In Japan, a development study of digital cellular systems with a common air interface was initiated in 1989 under the auspices of the Ministry of Posts and Telecommunications (MPT). The new digital system, called personal digital cellular (PDC), was established in 1991. The PDC system is based on TDMA with three time slots multiplexed onto each carrier, similar to IS-54. The channel spacing is 25 kHz with interleaving to facilitate migration from analog to digital. The RF signaling rate is 42 kbps and modulation is $\pi/4$ -DQPSK. A key feature of PDC is the mobile-assisted handoff, which facilitates the use of small cells for efficient frequency usage. The full-rate VSELP speech codec operates at 6.7 kbps (11.2 kbps with error correction). A 5.6-kbps CELP half-rate codec was also standardized and introduced. A total of 80-MHz spectrum is allocated to PDC. The frequency bands are 810–826 MHz paired with 940–956 MHz, and 1,429–1,453 MHz paired with 1,477–1,501 MHz. With antenna diversity, the required S/I ratio is reduced, giving a reuse factor of 4. Group 3 fax (2.4 kbps), as well as 4.8-kbps modem transmission with MNP class 4, are supported using an adaptor to provide the required transmission quality.

The CDMA Development Group (CDG) reports that only 18 months after the first commercial deployment of a CDMA network in the U.S., service providers have signed up some 7.8 million subscribers worldwide, 1.5 million of whom are located in North America. This figure compares to one million for GSM systems and 4.2 million for TDMA, which have been in operation at least three years longer than CDMA systems in the U.S.

After 18 months of market rollouts of CDMA, digital cellular and PCS operators have amply demonstrated the power of CDMA technology to support a marketing strategy based on low prices and better performance in several key areas such as voice quality, system reliability, and battery life. Lower power consumption enables CDMA handsets to support up to four hours of talk time or 48 hours of standby time on a single charge. It has also been observed that soft handoff characteristics of CDMA lead to fewer dropped calls than with GSM and IS-136.

The IS-95B standard is quite flexible and enables service providers to allocate data in increments of 8 kbps within 1.25-MHz CDMA channel based on how the service providers configure software downloaded to already installed network controllers. This implies that service providers can implement return data speeds at rates much lower than 64 kbps, ensuring much lower power consumption in handsets than would be the case at full 64-kbps return rate.

While service providers in the GSM and IS-136 areas are taking steps to ensure they won't be left behind as data becomes a factor, CDMA appears to have a clear edge in its ability to support high data speeds over the existing infrastructure. Initially, GSM and IS-136 (TDMA) will offer data at 13 kbps. Then things become more complex with the ability to assign specific time slots to data, which raises the rate to anywhere from 28.8 to 56 kbps. It might be possible for the GSM and IS-136 service providers to evolve their systems from real-time circuit-switched implementations of data at 13 kbps to a general packet radio service (GPRS) at 171 kbps. This approach requires hardware upgrades along with improvements in the software, depending on how the standard evolves and how manufacturers configure current generation base stations.

To avoid upgrades later, the time division-based operators will have to begin deploying next-generation DSP (digital signal processor) technology now. With enhanced gear going into the field at an early stage, they can accommodate a jump to 171 kbps when the software is available without requiring further hardware changes.

Even if the GSM and IS-136 service providers begin using such technology immediately as they continue to expand infrastructure, they still have to retrofit the existing infrastructure. Inevitably, there will be a broader base of CDMA facilities that are software upgradeable to IS-95B and 3GIX. This suggests that 64 to 144 kbps will be the benchmark for data services in most markets until third-generation systems offering much higher data rates over all platforms come into play toward the middle of the year 2002.

CDMA has a clear advantage over TDMA in the data area. While data over mobile networks has been a nonstarter in the past, there are plenty of reasons to embrace a service that can be easily used at a low cost. One of the major trends is that companies are trying to integrate wireless communications into their overall operations. Data will be a big part of that integration, given the use of the Internet and other forms of data communications.

The opportunity to use the CDMA platform to add a fixed service feature represents a complementary advantage for operators on the consumer side of their marketing efforts. Technical advances have reached the point where it is now feasible for service providers to begin adding this component. With ample spectrum to provide a fixed service on top of mobile, the companies are exploring the use of terminals that would be able to shift the handset between fixed and mobile service, depending on where the user is. The universal handset would serve as a cordless phone in the home and a mobile handheld outside.

With the cornucopia of benefits surrounding CDMA, it's clear that operators using this platform have every opportunity to increase the business once the commodity-based strategy begins to lose its luster. The question is, when should they get serious about bringing these new capabilities to market? Operators have no choice but to begin differentiating themselves by offering enhanced services.

Based on the information available, the market share of different digital technologies is as follows in Table 1.2.

Table 1.2 Worldwide Market Shares for Digital Cellular Wireless Systems (1999)

Standards	Technologies	Users	Market Share
GSM (GSM900, GSM1800, GSM1900)	Advanced TDMA	57 M	71%
Public digital cellular (PDC) [Japan]	Basic TDMA	8 M	10%
IS-54/136 (800, 1,900 MHz)	Basic TDMA	5 M	6%
IS-95 (800, 1,900 MHz)	Narrowband CDMA	10 M	13%

1.5 Third-Generation (3G) Systems

Taking into account the limitations imposed by the finite amount of radio spectrum available, the focus of 3G mobile systems is on economy of network and radio transmission design to provide seamless service from the customer perspective. Third-generation mobile systems have to provide users with a seamless access to the fixed data network. Third-generation systems are perceived as the wireless extension of future fixed networks, as well as an integrated part of the fixed network infrastructure.

In Europe, three related network platforms are currently the subject of intensive research. These are future land public mobile telephone systems (FLPMTS, now known as IMT-2000), mobile broadband systems (MBS), and wireless local area networks (WLAN). One major distinction of IMT-2000 relative to 2G systems is the hierarchical cell structure designed to support a wide range of multimedia broadband services within the various cell layers by using advanced transmission and protocol technologies. Second-generation systems mainly use one-layer cell structure and employ frequency reuse within adjacent cells in such a way that each single cell manages its own radio zone and radio circuit control within the mobile network, including traffic management and handoff procedures. The traffic supported in each cell is fixed because of frequency limitations and little flexibility of radio transmission mainly optimized for voice and low data rate transmissions. Increasing traffic leads to costly cellular reconfiguration such as cell splitting and cell sectorization.

The multilayer cell structure in IMT-2000 aims to overcome these problems by overlaying, discontinuously, pico- and microcells over the macrocell structure with wide area coverage. Global/satellite cells can be used in the same sense by providing area coverage where macrocell constellations are not economical to deploy and/or support long distance traffic.

With low mobility and small delay spread profiles in picocells, high bit rates and high traffic densities can be supported with low complexity as opposed to low bit rates and low traffic load in macrocells that support high mobility. The user expectation will be for service selected in a uniform manner with consistent procedures, irrespective of whether the means of access to these services is fixed or mobile. Freedom of location and means of access will be facilitated by smart cards to allow customers to register on different terminals with varying capabilities (speech, multimedia, data, short messaging).

The choice of a radio interface parameter set corresponding to a multiple access scheme is a critical issue in terms of spectral efficiency, taking into account the ever increasing market demand for mobile communications and the fact that radio spectrum is a very expensive and scarce resource. A comparative assessment of several different schemes has been carried in the framework of the Research in Advanced Communications Equipments (RACE) program. One possible solution is to use a hybrid CDMA/TDMA/FDMA technique by integrating advantages of each and meeting the varying requirements on channel capacity, traffic load, and transmission quality in different cellular/PCS layouts. Disadvantages of such hybrid access schemes are the high complexity, difficulties in achieving simplified low-power, low-cost transceiver design, as well as inefficient flexibility management in the several cell layers.

CDMA is the selected approach for 3G systems, as evidenced by the proposals in ETSI, ARIB (Japan), and the TIA. In Europe and Japan, wideband CDMA (W-CDMA) [4] has been proposed to avoid IS-95 intellectual property rights (IPR). In North America, cdma2000 [5] will be used based on CDMA air interface of the existing IS-95 standard to provide wireline quality voice service and high-speed data services ranging from 144 kbps for mobile users to 2 Mbps for stationary users. The 64-kbps data capability of IS-95B will provide high-speed Internet access in a mobile environment, a capability that cannot be matched by other narrowband digital technologies.

Mobile data rates up to 144 kbps and fixed peak rates beyond 2 Mbps are within reach before the end of 2005 using wideband CDMA technologies. These services will be provided without degrading the systems' voice transmission capabilities or requiring additional spectrum. This will have tremendous implications for the majority of operators that are spectrum constrained.

In North America, the Universal Wireless communication Consortium (UWC) has proposed UWC 136 [6] to evolve IS-136 TDMA-based systems to satisfy IMT-2000 requirements.

1.6 Summary

In this chapter we discussed the development of the GSM system in Europe. We presented functions of the GSM MoU and traced its growth from 15 to 208 signatories. The market trends for various digital technologies was also presented. Seven 2G air interfaces that have been approved by the JTC and adopted as ANSI standards for deploying PCS systems in North America were briefly discussed. We concluded the chapter by presenting 3G cellular/PCS systems that are being explored to enhance 2G systems.

1.7 References

1. Garg, V. K., and Wilkes, J. E., *Wireless and Personal Communications Systems*, Prentice Hall: Upper Saddle River, NJ, 1996.

2. Balston, D. M., and Macario, R. C. V., *Cellular Radio Systems*, Artech House: Noorwood, MA, 1993.
3. Balston, D. M., *The Pan-European Cellular Technology*, IEEE Conference Publication, 1988.
4. TSG-RAN Working Group 2, "3GPP TS25.301, Radio Interface Protocol Architecture," Sophia Antipolis, France, V3.10, July 5–9, 1999.
5. TIA TR45.5, "The cdma2000 ITU-R RTT Candidate Submission," TR45.5/98.04.03.03, April 1998.
6. TIA TR45, "Proposed RTT Submission (UWC 136)," TR-45.3/98.03.03.19, March 1998.
7. Marley, N., *GSM and PCN Systems and Equipments*, JRC Conference, Harrogate, 1991.
8. Mouly, M., and Pautet, M. B., *The GSM System for Mobile Communications*, Palaiseau, France, 1992.
9. Dasilva, J. S., Ikonomou, D., and Erben. H., "European R&D Programs on Third-Generation Mobile Communications Systems," *IEEE Personal Communications*, Feb. 1997 [Vol. 4(1), pp. 46–52].
10. "The European Path Towards UMTS," *IEEE Personal Communications*, special issue, Feb. 1995.
11. Rapeli, J., "UMTS: Targets, System Concepts, and Standardization in a Global Framework," *IEEE Personal Communications*, Feb. 1995.

CHAPTER 2

Propagation and Path Loss Models

2.1 Introduction

The upsurge in mobile communications has revitalized many topics in propagation. Much effort is now being devoted to refining propagation path loss models for urban, suburban, and other environments together with substantiation by field data. Propagation in urban areas is quite complex because it often consists of reflected and diffracted waves produced by multipath propagation. Propagation in open areas free from obstacles is the simplest to treat, but, in general, propagation over earth and water invokes at least one reflected wave.

For closed areas such as indoors, tunnels, and underground passages, no established models have been developed as yet, since the environment has a complicated structure. However, when the environment structure is random, the Rayleigh model [1] used for urban area propagation may be applied. When the propagation path is line of sight, as in a tunnel or an underground passage, the environment may be treated either by Rician model [1] or waveguide theory. Direct wave models may be used for propagation in a corridor.

An analytical solution of the urban area propagation problem is almost impossible, and a statistical approach is often used. Okumura [2] developed useful curves based on test data collected in Tokyo for estimating propagation loss in both urban and suburban areas. These curves later were converted by Hata [3] into empirical formulae and is known as the Okumura-Hata model. Okumura's measurements are valid only for the building types found in Tokyo. Experience with comparable measurements in the United States has shown that a typical United States suburban situation is often somewhere between Okumura's suburban and open areas. Okumura's suburban definition is more representative of a residential metropolitan area with large groups of "row" houses.

The Okumura-Hata model has been widely used for cellular systems in the 800- to 900-MHz range. The model provides a reasonable estimate of mean path loss. It does not do well with rapid changes in radio path profile. The Okumura model needs considerable engineering judgment in the selection of appropriate environmental parameters. Field data is required to predict the environmental parameters from the physical properties of the buildings and other structures surrounding a mobile receiver. In addition to the appropriate environmental parameters, path-specific corrections are needed to convert Okumura's mean path loss predictions to those that apply to the specific path under study. Corrections for irregular terrain and other path-specific features require considerable engineering interpretations. For a PCS system in the 1.8- to 1.9-GHz range, the Okumura model shows significant discrepancies between the predicted and measured field results.

A new model has been developed by ETSI's COST 231 [4] committee for estimating the path loss in an urban environment of a PCN/PCS system. The model takes into consideration several parameters (e.g., average building height around the mobile receiver, street width, street orientation, and so on) that were ignored in the Okumura model.

In this chapter, we first focus on free space line-of-sight propagation in which there are no obstructions due to the earth's surface or other obstacles. We then discuss signal attenuation over a flat reflecting earth's surface and also present the effect of the earth's curvature on signal attenuation. We also provide brief descriptions of the Rician, Rayleigh, and lognormal distributions that are generally used for signal fading statistics. Several empirical models to determine path losses are also included. Discussions on frequency diversity and wideband signals are also given. We conclude the chapter with a discussion of link budget and cell coverage.

2.2 Free-Space Attenuation

The energy radiated from a transmitting antenna may reach the receiving antenna over several possible propagation paths. For many wireless applications in the 50- to 2,000-MHz range two components of the space wave are of primary concern: energy received by means of the direct wave, which travels a direct path from the transmitter to the receiver; and a ground-reflected wave, which arrives at the receiver after being reflected from the surface of the earth. Other propagation paths such as sky and surface waves are often neglected.

The most simple wave propagation case is that of a direct wave propagation in free space. In this special case of line-of-sight (LOS) propagation, there are no obstructions due to the earth's surface or other obstacles. We consider radiation from an isotropic antenna. This type of antenna is completely omnidirectional, radiating uniformly in all directions. While there is no such thing as a purely isotopic antenna in practice, it is a useful theoretical concept.

The received power, P_r, at the receiving antenna (mobile station) located at a distance, d, from the transmitter (base station) is given for free-space propagation as [5]

$$P_r = P_t\left(\frac{\lambda}{4\pi d}\right)^2 G_b G_m \tag{2.1}$$

If other losses are also present, we can rewrite Equation (2.1) as

$$\frac{P_r}{P_t} = \left(\frac{\lambda}{4\pi d}\right)^2 \cdot \frac{G_b G_m}{L_0} = \frac{G_b G_m}{L_p L_0} \tag{2.2}$$

where

P_r = received power

P_t = transmitted power

λ = wavelength = $\frac{c}{f} = \frac{c \cdot 2\pi}{\omega_c}$

ω_c = carrier frequency in rad/sec

c = speed of light (3×10^8 m/s)

G_b = gain of the base station antenna

G_m = gain of the mobile antenna

d = antenna separation distance

L_0 = other losses expressed as a relative attenuation factor

$L_p = \left[\frac{4\pi d}{\lambda}\right]^2$ = free-space path loss, often expressed as an attenuation in decibels (dB)

$= 20\log\left(\frac{4\pi d}{\lambda}\right)$ (dB)

It should be noted that the free-space attenuation increases by 6 dB whenever the length of the path is doubled. Similarly, as frequency is doubled, free-space attenuation also increases by 6 dB.

2.3 Attenuation over Reflecting Surfaces

Free-space propagation is encountered only in rare cases such as satellite-to-satellite paths. In typical terrestrial paths, the signal is partially blocked and attenuated due to urban clutter, trees, and other obstacles. Multipath propagation also occurs due to reflection from the ground. Signals reflected from the ground are fundamentally no different than any other reflected signal. However, they are considered separately because they contribute to the received signal power at virtually all terrestrial receiving locations.

We assume the base station antenna height, h_b, and the mobile station antenna height, h_m, are small as compared to their separation distance, d, and reflecting earth surface is flat (see Figure 2.1). The received power at the antenna located at a distance, d, from the transmitter, including other losses, L_0, is given as [5]

$$P_r = \left[\frac{h_b h_m}{d^2}\right]^2 \frac{(G_b G_m)}{L_0} P_t \tag{2.3}$$

where

$d > \bar{d}$ and $\bar{d} = \frac{4 h_b h_m}{\lambda}$

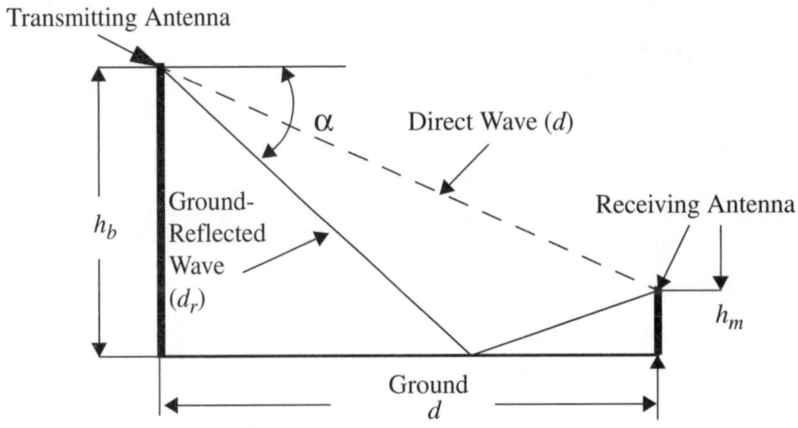

Figure 2.1 Geometry for Direct and Ground-Reflected Waves

Note that, under the assumed conditions, the received signal level is dependent only on the transmitted power, antenna heights, and separation distance; there is no frequency dependence. Furthermore, the total attentuation increases by 12 dB when the separation distance is doubled.

The expression for the effects of ground reflections from a flat or plane earth provides results that are approximately correct for $\Delta\alpha = \frac{2\pi}{\lambda(d-d_r)} \le \frac{\pi}{8}$ (see Figure 2.1). The results are not valid for $\Delta\alpha > \pi/8$. When $\Delta\alpha > \pi/8$, the attentuation factor will be

$$A_{gr} = \sqrt{1+\rho^2-2\rho\cos\Delta\alpha} \qquad \textbf{(2.4)}$$

where
ρ = reflection coefficient of ground (assumed to be –1)

EXAMPLE 2.1

With $h_b = 100$ ft, $h_m = 5$ ft, and a frequency of 881.52 MHz ($\lambda = 1.116$) ft, calculate signal attenuation at a distance equal to 5,000 ft. Assume antenna gains are 8 dB and 0 dB for the base station and mobile station, respectively. What is the free-space attenuation? Assume the earth's surface to be flat.

$$\bar{d} = \frac{4h_b h_m}{\lambda} = \frac{4 \times 100 \times 5}{1.116} = 1,792 \text{ ft}, \quad d > \bar{d}$$

$$G_b = 8 \text{ dB} = 6.3; \; G_m = 0 \text{ dB} = 1.0$$

Free-space attenuation:

$$\frac{P_t}{P_r} = \left(\frac{4\pi d}{\lambda}\right)^2 \cdot \frac{1}{G_b G_m} = \left(\frac{4\pi \times 5,000}{1.116}\right)^2 \cdot \frac{1}{6.3 \times 1} = 87 \text{ dB}$$

Attenuation on reflecting surface:

$$\frac{P_t}{P_r} = \left[\frac{d^2}{h_b h_m}\right]^2 \cdot \frac{1}{G_b G_m} = \left[\frac{5,000^2}{100 \times 5}\right]^2 \cdot \frac{1}{6.3 \times 1} = 86 \text{ dB}$$

2.4 Effect of the Earth's Curvature

We assumed a flat earth in the previous section. In reality, the earth is curved, preventing LOS propagation to great distances. Thus, Equations (2.2) and (2.3) are only approximately correct for distances less than the distance to radio horizon. The distance to radio horizon for terrestrial transmitters can be determined as in the following discussion.

We consider a circle representing the earth (see Figure 2.2). From the figure, we write

$$d^2 + R_e^2 = (R_e + h_b)^2$$

$$d^2 = 2R_e h_b + h_b^2$$

For $2R_e h_b \gg h_b^2$, then

$$d \approx \sqrt{2R_e h_b} \tag{2.5}$$

where

R_e = radius of earth

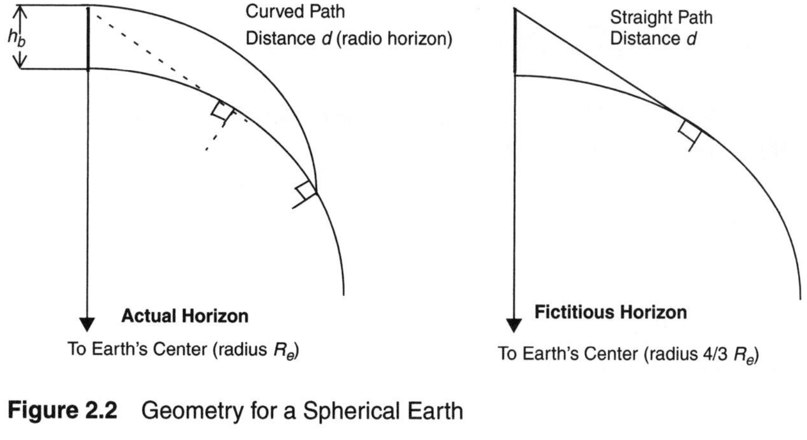

Figure 2.2 Geometry for a Spherical Earth

As an example with an antenna height of 600 m and earth's radius of about 6,400 km, the distance to the horizon is about 88 km. On average, the refractive index of the earth's atmosphere is such that the earth's radius appears to be about 33 percent more than the actual radius. Thus, the effective radius of the earth is often assumed to be 8,500 km. The curvature of the earth further affects the propagation of the space wave since the ground-reflected wave is reflected from a curved surface. Therefore, the energy diverges more than it does from a flat surface and a ground-reflected wave reaching the receiver is weaker than for a flat earth. The divergence factor D that describes this effect is less than unity and is given as [6]:

$$D = \frac{1}{\sqrt{1 + \frac{2}{[\{R_e(h_b + h_m)^3\}/(d^2 h_b h_m) - 0.5(h_b/h_m + h_m/h_b)]}}} \tag{2.6}$$

D in Equation (2.6) ranges from unity for a small value of d and approaches zero as d approaches the distance to the radio horizon. It can be combined with the ground reflection coefficient so that the attenuation due to ground reflections becomes

$$A_{gr} = \sqrt{1 + (\rho D)^2 - 2\rho D \cos(\Delta\alpha)} \tag{2.7}$$

where the reflection coefficient, ρ, has been modified to account for the divergence factor, D. The effect of the divergence factor is to reduce the effective reflection coefficient of the earth. The equation for calculating the received power is given as

$$P_r(dBm) = 60 + P_t + G_b + 20\log(A_{fs}) + 20\log(A_{gr}) \tag{2.8}$$

where

$A_{fs} = [\lambda/(4\pi d)]^2$, free-space attenuation

2.5 Signal Fading Statistics

The rapid variations in signal power caused by local multipath are represented by Rayleigh distribution. The long-term variations in the mean level are denoted by log-normal distribution. With LOS propagation path, the Rician distribution is often used. Thus, the fading characteristics of a mobile radio signal are described by the following statistical distributions:

- Rician distribution
- Rayleigh distribution
- Log-normal distribution

2.5.1 Rician Distribution

When there is a dominant stationary (nonfading) signal component present, such as an LOS propagation path, the small-scale fading envelope distribution is Rician. The Rician distribution has a probability density function (pdf) given by

$$p(r) = \frac{r}{\sigma^2} e^{-\left(\frac{r^2 + A^2}{2\sigma^2}\right)} I_0\left(\frac{Ar}{\sigma^2}\right) \qquad \text{for } A \geq 0, \quad r \geq 0 \tag{2.9}$$

where

A = peak amplitude of the dominant signal and $I_0 = (\ldots)$ modified Bessel Function of the first kind and zero order

$r^2/2$ = instantaneous power

σ = standard deviation of the local power

The Rician distribution is often described in terms of a parameter K, known as the Rician factor, and is expressed as

$$K = 10\log\frac{A^2}{2\sigma^2} \text{ dB} \tag{2.10}$$

As $A \rightarrow 0$, $K \rightarrow \infty$ dB and as the dominant path decreases in amplitude, the Rician distribution degenerates to a Rayleigh distribution (see Figure 2.3).

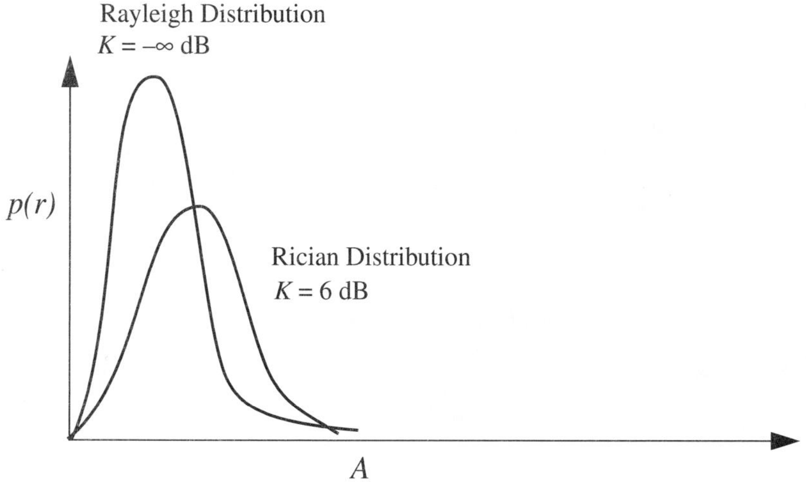

Figure 2.3 Rayleigh and Rician Distributions

2.5.2 Rayleigh Distribution

The Rayleigh distribution [1] is used to describe the statistical time-varying nature of the received envelope of a flat fading signal, or the envelope of an individual multipath component. The Rayleigh distribution is given as

$$p(r) = \frac{r}{\sigma^2} e^{-\left(\frac{r^2}{2\sigma^2}\right)} \qquad 0 \le r \le \infty \tag{2.11}$$

where
σ = rms value of the received signal
σ^2 = local average power of the received signal before envelope detection

Instead of the distribution of the received envelope, we can describe a Rayleigh fading signal in terms of the distribution function of its received normalized power.

Let

$$\Phi = \frac{r^2/2}{\sigma^2}$$

which is the instantaneous received power divided by the mean received power. Then

$$d\Phi = \frac{r}{\sigma^2} \cdot dr$$

and since $p(r)\,dr$ must be equal to $p(\Phi)d\Phi$, we get

$$p(\Phi) = [p(r)dr]/[(r/\sigma^2)dr] = \frac{(r/\sigma^2)e^{-(r^2/2\sigma^2)}}{(r/\sigma^2)} = e^{-\Phi},\ 0 \le \Phi \le \infty \tag{2.12}$$

Equation (2.12) represents a simple exponential density function. One can rightfully say that a flat fading signal is exponentially fading in power.

2.5.3 Log-Normal Distribution

The log-normal distribution [5] describes the random shadowing effects that occur over a large number of measurement locations that have the same transmitter and receiver separation, but have different levels of clutter on the propagation path. The signal, $s(t)$, typically follows the Rayleigh distribution but its mean square value or its local mean power is log-normal, i.e., $s = 10\log E\{s(t)\}$ is normal or Gaussian with variance equal to σ_s^2. Typically, the standard deviation, σ_s, equals 8 to 10 dB.

The log-normal distribution is given by (see Figure 2-4)

$$p(S) = \frac{1}{\sqrt{2\pi}\sigma_s} e^{-\left[\frac{(S-S_m)^2}{2\sigma_s^2}\right]} \tag{2.13}$$

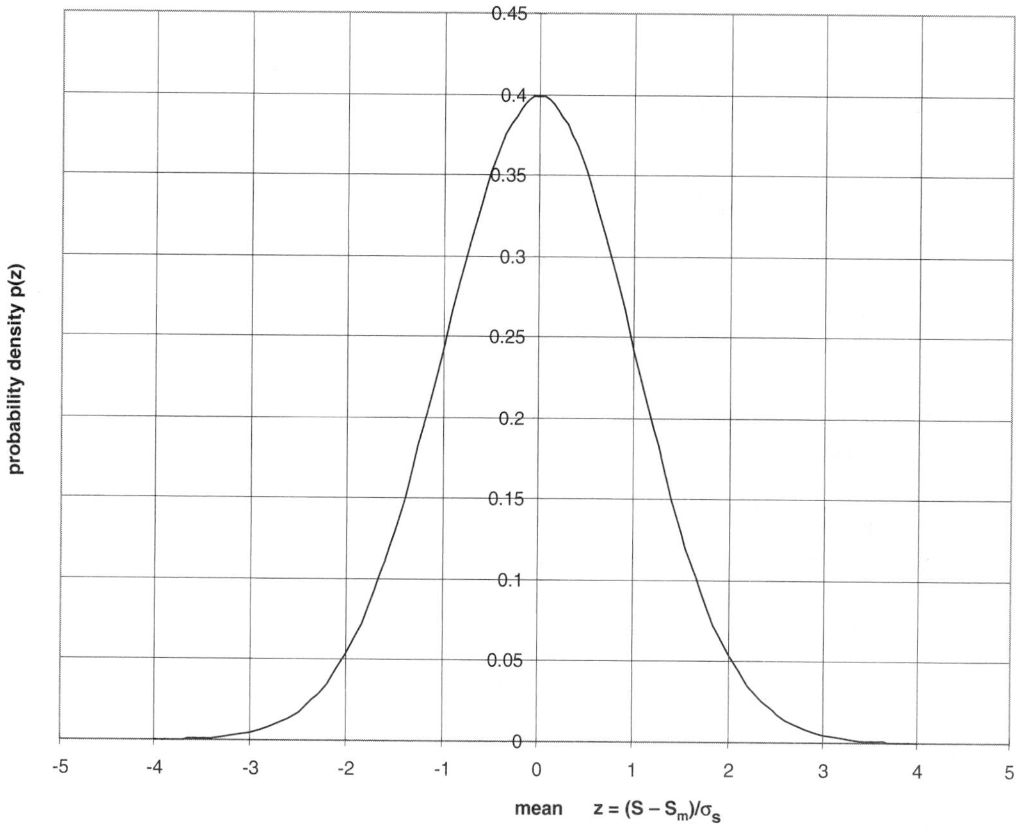

Figure 2.4 Log-Normal Distribution

where
S_m = mean value of S in dBm
σ_s = standard deviation of s in dB
$S = \log s$
s = signal power in dBm

2.6 Empirical Models for Path Loss

Here we discuss two widely used empirical models: Okumura-Hata and COST 231. The Okumura-Hata model has been used extensively in Europe and North America for cellular systems. The COST 231 model has been recommended by ETSI for use in PCN/PCS. In addition, we also present the empirical models proposed by IMT-2000 for the indoor office environment, outdoor-to-indoor pedestrian environment, and vehicular environment.

2.6.1 Okumura-Hata Model

Okumura et al. [2] analyzed path loss characteristics based on a large amount of experimental data collected around Tokyo, Japan. They selected propagation path conditions and obtained the average path loss curves under flat urban areas. Then, they applied several correction factors for other propagation conditions, such as

- Antenna height and carrier frequency
- Suburban, quasiopen space, open space, or hilly terrain areas
- Diffraction loss due to mountains
- Sea or lake areas
- Road slope

Hata [3] derived empirical formulae for the median path loss to fit Okumura curves. Hata's equations are classified into three models:

1. Typical urban:

$$L_{50} = 69.55 + 26.16\log f_c + (44.9 - 6.55\log h_b)\log d - 13.82\log h_b - a(h_m)\text{ dB} \tag{2.14}$$

where
$a(h_m)$ = correction factor for mobile antenna height and is given by:

- For large cities:

$$a(h_m) = 8.29[\log(1.54h_m)]^2 - 1.1 \qquad f_c \leq 200\text{ MHz} \tag{2.15}$$

$$a(h_m) = 3.2[\log(11.75h_m)]^2 - 4.97 \qquad f_c \geq 400\text{ MHz} \tag{2.16}$$

- For small- and medium-sized cities:

$$a(h_m) = [1.1\log(f_c) - 0.7]h_m - [1.56\log(f_c) - 0.8] \tag{2.17}$$

2. **Typical suburban:**

$$L_{50} = L_{50}(urban) - 2\left[\left(\log\left(\frac{f_c}{28}\right)^2\right) - 5.4\right] \text{ dB} \tag{2.18}$$

3. **Rural:**

$$L_{50} = L_{50}(urban) - 4.78(\log f_c)^2 + 18.33\log f_c - 40.94 \text{ dB} \tag{2.19}$$

where
f_c = carrier frequency (MHz)
d = distance between base station and mobile (km)
h_b = base station antenna height (m)
h_m = mobile antenna height (m)
The range of parameters for which the Hata model is valid is

$150 \le f_c \le 1,500$ MHz

$30 \le h_b \le 200$ m

$1 \le h_m \le 10$ m

$1 \le d \le 20$ km

2.6.2 COST 231 (Walfisch and Ikegami) Model

The COST 231 model [4,7] is a combination of empirical and deterministic models for estimating the path loss in an urban area over the frequency range of 800 MHz to 2,000 MHz. The model is used primarily in Europe for the GSM1800 system.

$$L_{50} = L_f + L_{rts} + L_{ms} \tag{2.20}$$

or

$$L_{50} = L_f \qquad \text{when } L_{rts} + L_{ms} \le 0 \tag{2.21}$$

where
L_f = free-space loss
L_{rts} = rooftop-to-street diffraction and scatter loss
L_{ms} = multiscreen loss

Free-space loss is given as

$$L_f = 32.4 + 20\log d + 20\log f_c \text{ dB} \tag{2.22}$$

The rooftop-to-street diffraction and scatter loss is given as

$$L_{rts} = -16.9 - 10\log W + 10\log f_c + 20\log \Delta h_m + L_0 \text{ dB} \tag{2.23}$$

where
W = street width (m)

$\Delta h_m = h_r - h_m$ (m)

$L_0 = -9.646$ dB $\qquad 0 \le \phi \le 35°$

$L_0 = 2.5 + 0.075(\phi - 35)$ dB $\qquad 35° \le \phi \le 55°$

$L_0 = 4 - 0.114(\phi - 55)$ dB $\qquad 55° \le \phi \le 90°$

ϕ = incident angle relative to the street

The multiscreen (multiscatter) loss is given as

$$L_{ms} = L_{bsh} + k_a + k_d \log d + k_f \log f_c - 9\log b \tag{2.24}$$

where b = distance between buildings along radio path (m)

$L_{bsh} = -18\log 11 + \Delta h_b \qquad h_b > h_r$

$L_{bsh} = 0 \qquad h_b < h_r$

$k_a = 54 \qquad h_b > h_r$

$k_a = 54 - 0.8h_b \qquad d \ge 500$ m; $h_b \le h_r$

$k_a = 54 - 1.6\Delta h_b d \qquad d < 500$ m; $h_b \le h_r$

Note: Both L_{bsh} and k_a increase path loss with lower base station antenna heights.

$k_d = 18 \qquad h_b < h_r$

$k_d = 18 - \dfrac{15\Delta h_b}{\Delta h_m} \qquad h_b \ge h_r$

$k_f = 4 + 0.7\left(\dfrac{f_c}{925} - 1\right)$ for midsized city and suburban areas with moderate tree density

$k_f = 4 + 1.5\left(\dfrac{f_c}{925} - 1\right)$ for metropolitan areas

The range of parameters for which the COST 231 model is valid is

$800 \le f_c \le 2{,}000$ MHz

$4 \le h_b \le 50$ m

$1 \le h_m \le 3$ m

$0.02 \le d \le 5$ km

The default values that may be used in the model are

$b = 20\text{–}50$ m

$W = b/2$

$\phi = 90°$

Roof = 3 m for pitched roof and 0 m for flat roof

$h_r = 3$ (number of floors) + roof

EXAMPLE 2.2

Using the Okumura and COST 231 models, calculate L_{50} path loss for a PCS system in an urban area at 1, 2, 3, 4, and 5 km distance. Assume $h_b = 150$ m, $h_m = 2$ m, and carrier frequency $f_c = 1.8$ GHz.

- **COST 231 model:**
  We assume the following data for the COST 231 model:

$W = 15$ m, $b = 30$ m, $\phi = 90°$, $h_r = 30$ m

$$L_{50} = L_f + L_{rts} + L_{ms}$$

$$L_f = 32.4 + 20\log d + 20\log f_c = 32.4 + 20\log d + 20\log 1800 \text{ dB}$$

$$L_f = 97.51 + 20\log d \text{ dB}$$

$$L_{rts} = -16.9 - 10\log W + 10\log f_c + 20\log \Delta h_m + L_0$$

$$\Delta h_m = h_r - h_m = 30 - 2 = 28 \text{ m}$$

$$L_0 = 4 - 0.114(\phi - 55) = 4 - 0.114(90 - 55) = 0$$

$$L_{rts} = -16.9 - 10\log 15 + 10\log 1800 + 20\log 28 + 0 = 32.83 \text{ dB}$$

$$L_{ms} = L_{bsh} + k_a + k_d \log d + k_f \log f_c - 9 \log b$$

$$k_a = 54 - 0.8 h_b = 54 - 0.8 \times 30 = 30$$

$$\Delta h_b = h_b - h_r = 30 - 30 = 0$$

$$L_{bsh} = -18 \log 11 + 0 = -18.75 \text{ dB}$$

$$k_d = 18 - \frac{15 \Delta h_b}{\Delta h_m} = 18 - \frac{15 \times 0}{28} = 18$$

$$k_f = 4 + 0.7\left(\frac{f_c}{925} - 1\right) = 4 + 0.7\left(\frac{1800}{925} - 1\right) = 4.66$$

$$L_{ms} = -18.75 + 30 + 18 \log d + 4.66 \log 1800 - 9 \log 30 = 13.14 + 18 \log d \text{ dB}$$

Table 2.1 Summary of Path Losses from the COST 231 Model

d (km)	L_f (dB)	L_{rts} (dB)	L_{ms} (dB)	L_{50} (dB)
1	97.51	32.83	13.14	143.48
2	103.53	32.83	18.56	154.92
3	107.05	32.83	21.73	161.61
4	109.55	32.83	23.98	166.36
5	111.49	32.83	25.72	170.04

- **Okumura model:**

$$L_{50} = 69.55 + 26.16 \log f_c + (44.9 - 6.55 h_b) \log d - 13.82 \log h_b - a(h_m) \text{ dB}$$

$$a(h_m) = (1.1 \log f_c - 0.7) h_m - (1.56 \log f_c - 0.8)$$
$$= (1.1 \log 1800 - 0.7)(2) - (1.56 \log 1800 - 0.8) = 1.48 \text{ dB}$$

$$L_{50} = 69.55 + 26.16 \log 1800 + (44.9 - 6.55 \log 150) \log d - 13.82 \log 150 - 1.48 \text{ dB}$$
$$= 123.16 + 30.647 \log d \text{ dB}$$

The results from the two models are shown in Figure 2.5.

Table 2.2 Summary of Path Losses from the Okumura Model

d (km)	L_{50} (dB)
1	123.16
2	132.39
3	137.78
4	141.61
5	144.58

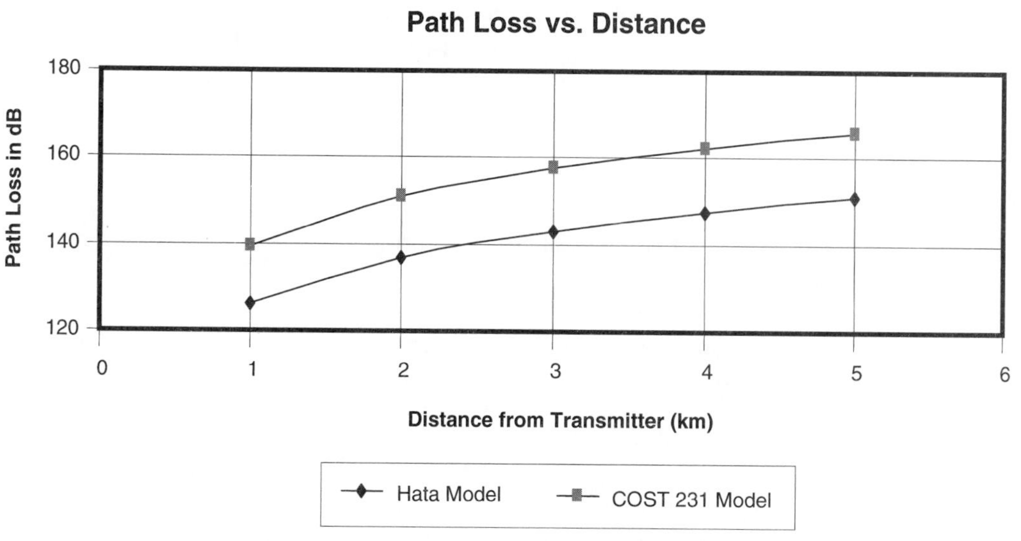

Figure 2.5 Comparison of COST 231 and Okumura-Hata Models

2.6.3 IMT-2000 (FPLMTS) Models

The operating environments are identified by appropriate subsets consisting of indoor office environment, outdoor-to-indoor and pedestrian environment, and vehicular environment. For narrowband technologies, delay spread may be characterized by its rms value alone. However, for wideband technologies, the number, strength, and relative time delay of the many signal components become important. In addition, for some technologies (e.g., those using power control) the path loss models must include the coupling between all co-channel propagation links to provide accurate predictions. Also, in some cases, the shadow fading temporal variations of the environment must be modeled. The key parameters of the propagation models are:

- Delay spread, its structure, and its statistical variation
- Geometrical path loss rule (e.g., $d^{-\gamma}$, $2 \le \gamma \le 5$)
- Shadow fading margin
- Multipath fading characteristics (e.g., Doppler spectrum, Rician versus Rayleigh) for envelope of channels
- Operating radio frequency

Indoor Office Environment. This environment is characterized by small cells and low transmit powers. Both base stations and pedestrian users are located indoors. Rms delay spread ranges from around 35 to 460 nsec. The path loss rule varies due to scatter and attenuation by walls, floors, and metallic structures such as partitions and filing cabinets. These objects also produce shadowing effects. A log-normal shadowing with a standard deviation of 12 dB can be expected. Fading characteristic ranges from Rician to Rayleigh with Doppler frequency offsets determined by walking speeds. Path loss model for this environment is

$$L_p = 37 + 30\log d + 18.3 \cdot n^{[(n+2)/(n+1) - 0.46]} \text{ dB} \tag{2.25}$$

where
d = separation between transmitter and receiver (m)
n = number of floors in the path

Outdoor-to-Indoor and Pedestrian Environment. This environment is characterized by small cells and low transmit power. Base stations with low antenna heights are located outdoors; pedestrian users are located on streets and inside buildings and residences. Coverage into building in high-power systems is included in the vehicular environment. Rms delay spread varies from 100 to 1,800 nsec. A geometrical path loss rule of d^{-4} is applicable. If the path is LOS on a canyonlike street, the path loss follows a rule of d^{-2}, where there is Fresnel zone clearance. For the region with longer Fresnel zone clearance, a path loss rule of d^{-4} is appropriate, but a range up to d^{-6} may be encountered due to trees and other obstructions along the path. Log-normal shadow fading with a standard deviation of 10 dB is reasonable for outdoors and 12 dB for indoors. Average building penetration loss of 18 dB with a standard deviation of 10 dB is appropriate. Rayleigh and/or Rician fading rates are generally set by walking speeds, but faster fading due to reflections from moving vehicles may occur some of the time. The following path loss model has been suggested for this environment:

$$L_{50} = 40\log d + 30\log f_c + 49 \text{ dB} \tag{2.26}$$

This model is valid for NLOS cases only and describes the worst-case propagation. Log-normal shadow fading with a standard deviation equal to 10 dB is assumed. The average building penetration loss is 18 dB with a standard deviation of 10 dB.

Vehicular Environment. This environment consists of larger cells and higher transmit power. Rms delay spread from 4 microseconds to about 12 microseconds on elevated roads in hilly or mountainous terrain may occur. A geometrical path loss rule of d^{-4} and log-normal shadow fading with a standard deviation of 10 dB are used in the urban and suburban areas. Building penetration loss averages 18 dB with a 10 dB standard deviation.

In rural areas with flat terrain, the path loss is lower than that of urban and suburban areas. In mountainous terrain, if path blockages are avoided by selecting base station locations, the path loss rule is closer to d^{-2}. Rayleigh fading rates are determined by vehicle speeds. Lower fading rates are appropriate for applications using stationary terminals. The following path loss model is used in this environment:

$$L_{50} = 40(1 - 4 \times 10^{-2}\Delta h_b)\log d - 18\log(\Delta h_b) + 21\log f_c + 80 \text{ dB} \quad \textbf{(2.27)}$$

where

Δh_b = base station antenna height measured from average rooftop level (m)

2.6.4 Delay Spread

A majority of the time, rms delay spreads are relatively small, but occasionally there are worst-case multipath characteristics that lead to much larger rms delay spreads. Measurements in outdoor environments show that rms delay spread can vary over an order of magnitude, within the same environment. Delay spreads can have a major impact on system performance. To accurately evaluate the relative performance of radio transmission technologies, it is important to model the variability of delay spread as well as the worst-case locations where delay spread is relatively large. Three multipath channels are defined by IMT-2000 for each environment. Channel A represents the low-delay spread case that occurs frequently; channel B corresponds to the medium-delay spread case that also occurs frequently; and channel C is the high-delay spread case that occurs only rarely. Table 2.3 provides the rms values of delay spread for each channel and for each environment.

Table 2.3 rms Delay Spread (IMT-2000)

	Channel A		Channel B		Channel C	
Environment	τ_{rms} (ns)	% Occurrence	τ_{rms} (ns)	% Occurrence	τ_{rms} (ns)	% Occurrence
Indoor office	35	50	100	45	460	5
Outdoor-to-indoor and pedestrian	100	40	750	55	1,800	5
Vehicular (high antenna)	400	40	4,000	55	12,000	5

2.7 Frequency Diversity and Wideband Signals

We denote the rms multipath spread in time due to differences in multipath delay by τ_d, and the maximum spread in frequency due to differences in multipath Doppler by f_m.

The coherence bandwidth (B_c) [8] between two frequency envelopes is given as

$$B_c \approx \frac{1}{2\pi\tau_d} \tag{2.28}$$

Frequency components of a signal separated by more than B_c will fade independently.

The coherence time, T_c, [8] over which the impulse response of the channel is nearly constant is equal to

$$T_c \approx \frac{1}{2\pi f_m} \tag{2.29}$$

where

$f_m = \frac{v}{\lambda}$

v = relative velocity

λ = wavelength

If the transmitted symbol interval exceeds T_c, then the channel will change during the symbol interval and symbol distortion will occur. In such cases, matched filter is impossible without equalization and correlator losses occur. A Rayleigh fading signal may change amplitude significantly in the interval T_c.

If the signal bandwidth $B_w \gg B_c$, the signal is called wideband signal and any fading will be frequency selective. This means, only a portion of the signal bandwidth will fade at any instant of time. If $B_w \ll B_c$, flat fading of the entire signal will occur.

If the signal symbol interval $T_s \gg T_c$, the channel changes or fades rapidly compared to the symbol rate. This case is called fast fading and frequency dispersion occurs, causing distortion. If $T_s \ll T_c$, the channel does not change during the symbol interval. This case is called slow fading.

EXAMPLE 2.3

Assume speed of a vehicle equal to 60 mph (88 ft/sec), carrier frequency of f_c = 860 MHz, and delay spread of τ_d = 2 μsec. Calculate coherence time and coherence bandwidth. At a coded symbol rate of 19.2 kbps (IS-95) what kind of symbol distortion will be experienced? What type of fading will be experienced by the IS-95 channel?

$v = 60$ mph (= 88 ft/sec)

$$\lambda = \frac{c}{f} = \frac{9.84 \times 10^8}{860 \times 10^6} = 1.1442 \text{ ft}$$

$$f_m = \frac{v}{\lambda} = \frac{88}{1.1442} = 77 \text{ Hz}$$

$$T_c = \frac{1}{2\pi f_m} = \frac{1}{2\pi \times 77} = 0.0021 \text{ sec}$$

$$T_s = \frac{10^6}{19,200} = 52\mu\text{sec}$$

The symbol interval is much smaller as compared to the channel coherence time. Symbol distortion is, therefore, minimal. In this case, fading is slow.

$$B_c \approx \frac{1}{2\pi\tau_d} = \frac{1}{2\pi \times 2 \times 10^{-6}} = 79.56 \text{ kHz}$$

This shows IS-95 is a wideband system in this multipath situation and experiences selective fading only over 6.5 percent (79.57 / 1228.8 = 0.0648) of its bandwidth.

2.8 Link Budget and Cell Coverage

The L_{50} path loss estimates provide the average signal strength at a given distance from the transmitter. The signal varies from that average by large amounts, both lower and higher. Field data show that, statistically, the signal strength at a given distance from the transmitter has a log-normal distribution with standard deviation equal to 8 to 10 dB. These variations are called *shadow losses*, and are caused due to obstructions between the transmitter and the receiver. Field data also indicate that an obstruction affects the path loss for an average distance of 500 m in the suburban environment and 50 m in the urban environment.

2.8.1 Link Margin for the Coverage

With log-normal shadowing, the propagation loss L_p in dB can be given as

$$L_p(d) = L_m(d) + \sigma_s X \tag{2.30}$$

where

$L_m(d)$ = mean path loss at distance d in dB

σ_s = standard deviation of propagation loss in dB

$X = G(0,1)$ = zero-mean, unit variance Gaussian random variable

The probability density function (pdf) for the path loss in dB will be

$$p(x) = \frac{1}{\sqrt{2\pi}\cdot\sigma_s} e^{-\left[\frac{(x-L_m)^2}{2\sigma_s^2}\right]} \tag{2.31}$$

The cumulative density function (CDF) for the path loss in dB will be

$$P(x) = \int_{-\infty}^{\frac{(x-L_m)}{\sigma_s}} \left(\frac{1}{\sqrt{2\pi}}\cdot e^{-u^2/2}\right)du \equiv P\left(\frac{x-L_m}{\sigma_s}\right) \tag{2.32}$$

The complementary CDF for the path loss will be

$$Q(x) = 1 - P\left(\frac{x-L_m}{\sigma_s}\right) \equiv Q\left(\frac{x-L_m}{\sigma_s}\right) \tag{2.33}$$

where Q is the complementary error function

The system design must include a margin, M, intended to compensate for the variation in the path loss. The margin is achieved by using more transmitter power than would be necessary if there were no variations in the path loss. The received E_b/N_0 in the coverage-limited case can be given as

$$\frac{E_b}{N_0}(d) = P_m + G_{net} - L_m(d) - \sigma_s G - N_T = \left(\frac{E_b}{N_0}\right)_{min} - \sigma_s G \tag{2.34}$$

where

P_m = mobile transmitted power in dBm

G_{net} = net gain of the various fixed gains and other losses in dB

N_T = total cell receiver noise in dBm

$\left(\frac{E_b}{N_0}\right)_{min} = P_m + G_{net} - L_m(d) - N_T$

The coverage area of the cell is the maximum distance $d_{max} = R_c$. For successful link operation, the coverage area is defined as

$$\left(\frac{E_b}{N_0}\right)(R_c) = \left[\left(\frac{E_b}{N_0}\right)_{min}(R_c) - \sigma_s G\right] > \left(\frac{E_b}{N_0}\right)_{reqd} \tag{2.35}$$

Since the path loss varies randomly, we define reliability as the probability of a successful operation.

$$Reliability = P_r\left\{\left[\left(\frac{E_b}{N_0}\right)_{min}(R_c) - \sigma_s G\right] > \left(\frac{E_b}{N_0}\right)_{reqd}\right\} \quad (2.36)$$

$$P_{rel} = P_r\left\{G < \frac{\left(\frac{E_b}{N_0}\right)_{min}(R_c) - \left(\frac{E_b}{N_0}\right)_{reqd}}{\sigma_s}\right\} = P_r\left\{G < \frac{M(R_c)}{\sigma_s}\right\}$$

$$P_{rel} = P_r\left\{\frac{M(R_c)}{\sigma_s}\right\} = 1 - Q\left\{\frac{M(R_c)}{\sigma_s}\right\} \quad (2.37)$$

$$\therefore Q\left\{\frac{M(R_c)}{\sigma_s}\right\} = 1 - P_{rel} \quad (2.38)$$

$$\frac{M(R_c)}{\sigma_s} = Q^{-1}(1 - P_{rel}) \quad (2.39)$$

where M = margin

If a link reliability of $P_{rel} = 90\%$ is required, then

$$M = \sigma_s P_{rel}^{-1}(0.90) = 1.28155\sigma_s \text{ dB}$$

Table 2.4 lists the required margins for different reliability percentages with $\sigma_s = 8$ dB.

Table 2.4 Link Reliability vs. Margin for $\sigma = 8$ dB

P_{rel}	Margin in dB (σ_s = 8 dB)
0.90	10.24
0.95	13.16
0.98	16.43

Figure 2.6 shows a typical plot of the path loss versus distance in meters with mean path loss (MPL) and MPL±10.24 dB. The path loss has a normal distribution with a standard deviation of ±8 dB. A normal distribution curve has ±32 percent of the values outside ±8 dB. This means 16 percent of the values will be more than 8 dB above the mean. We consider this variation when designing a system so that the received signal-to-interference ratio is greater than the minimum value required for an acceptable voice quality over most

of the coverage area. As discussed earlier, if the system objectives are to provide adequate voice quality over 90 percent of the coverage area, the system must be designed with at least a 10.24-dB margin.

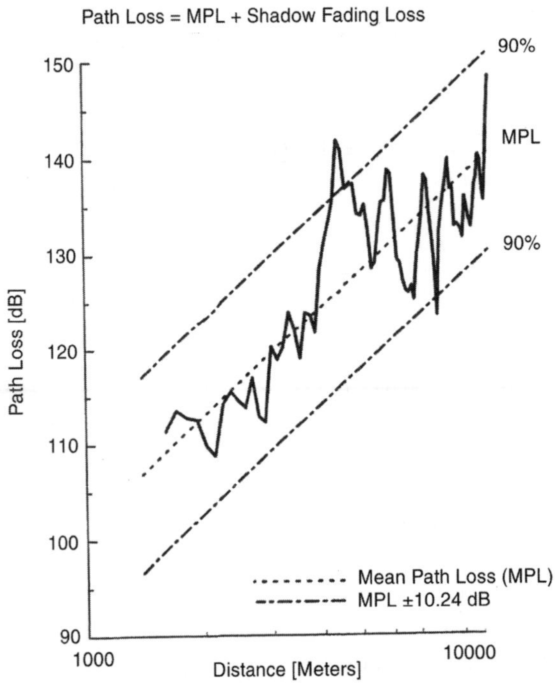

Figure 2.6 Shadow Margin

To determine the maximum base station range, the maximum allowable path loss to provide adequate signal strength at the cell boundary for acceptable voice quality over 90 percent of coverage area is calculated. The allowable path loss is the difference between the transmitter effective radiated power and minimum signal strength required at the receiver for acceptable voice quality. The components that determine the path loss are called the *link budget.*

Link budgets are used to calculate the coverage and performance for a base station and a mobile station. The components include propagation factors to calculate path loss and system parameters (transmitter power, receiver noise figure, antenna gains, receiver bandwidth, processing gain, and interference). Other losses such as power control errors, building penetrations, body/orientation losses, and interference from other sources are also included.

For a CDMA system, the link budget is used to

- Decide an appropriate network loading
- Allocate appropriate power to various forward link channels

The following procedure is used for link budget analysis:

- Identify parameters affecting the forward and reverse link.
  - Access technology-specific parameters
  - Access product-specific parameters
  - Access morphology-based parameters
- Determine the maximum allowable path loss to maintain communication on the forward and reverse link.
- Balance the forward and reverse link.

We next illustrate the procedure for calculating link budget and determining the range of a base station.

EXAMPLE 2.4

Refer to Figure 2.7 and use the following parameters to calculate the maximum allowable path loss:

- Information rate = 9,600 bps
- Mobile station effective radiated power (P_m) = 200 mW (23 dBm)
- Base station antenna gain (G_b) = 14 dBi
- Base station receiver antenna cable loss (L_c) = 2.5 dB
- PCS minicell receiver noise figure (F_b) = 5 dB
- Required margin (E_b/N_0) = 6.8 dB (with diversity antenna at base station)
- Base station noise floor (N_0) = –174 dBm/Hz
- Log-normal shadowing margin = 8 dB
- Body/orientation loss = 2 dB
- Building penetration loss = 10 dB

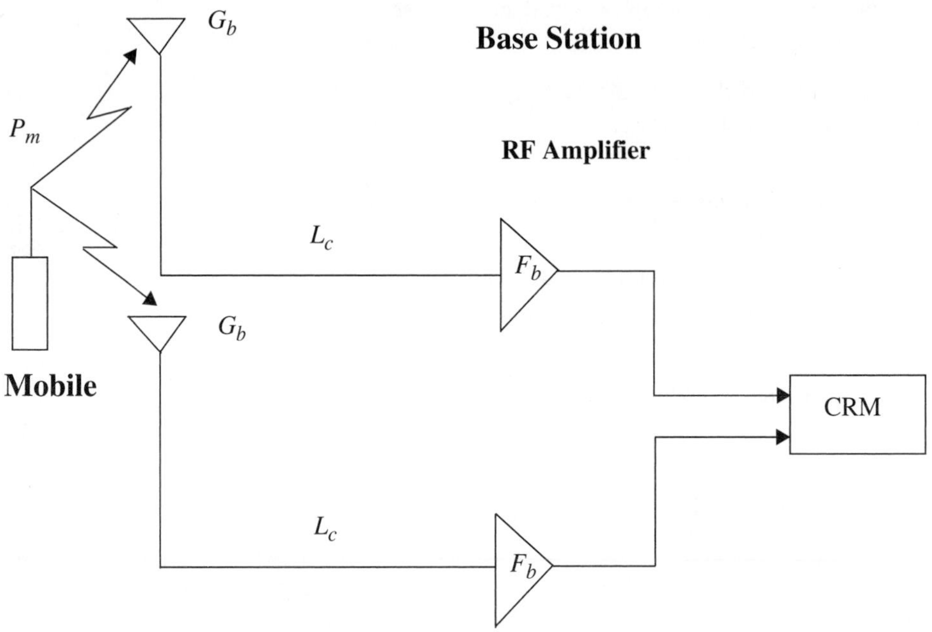

Figure 2.7 Transmission Between Mobile and Base Station

Base station noise floor:

$$N_T = N_0 + F_b = -174 + 5 = -169 \text{ dBm/Hz}$$

Minimum bit energy required for specified E_b / N_0:

$$(E_b)_{min} = N_T + (E_b/N_0)_{reqd} = -169 + 6.8 = -162.2 \text{ dBm/Hz}$$

Minimum signal strength required:

$$S_{min} = (E_b)_{min} + 10\log R = -162.2 + 10\log 9600 = -122.4 \text{ dBm}$$

Mean path loss (L_{50}) with S_{min}:

$$L_{50} = P_m - S_{min} + G_b - L_c = 23 + 122.4 + 14 - 2.5 = 156.9 \text{ dB}$$

To provide margin for shadowing:

$$\text{Path loss} = L_{50} - 10.2 = 156.9 - 10.2 = 146.7 \text{ dB}$$

To provide margin for body/orientation loss and building penetration loss:

$$\text{Allowable path loss} = 146.7 - 2.0 - 10 = 134.7 \text{ dB}$$

EXAMPLE 2.5

Using the allowable path loss value from Example 2.4, determine the coverage for the mini-PCS cell. Assume the following data:

- PCS frequency (f_c) = 1,800 MHz
- Street width (W) = 20 m
- Spacing between buildings (b) = 40 m
- Average roof height of building (h_r) = 40 m
- Mobile antenna height (h_m) = 2 m
- Base station antenna height (h_b) = 40 m
- Street orientation (ϕ) = 90 degrees

We use the COST 231 model:

$$\Delta h_m = h_b - h_m = 40 - 2 = 38\text{ m}$$

$$\Delta h_b = h_b - h_r = 40 - 40 = 0\text{ m}$$

$$L_0 = 4 - 0.114(\phi - 55) = 4 - 0.114(90 - 55) = 0$$

$$L_{bsh} = -18\log 11 + \Delta h_b = -18.75\ \text{dB}$$

$$k_a = 54,\ k_d = 18 - \frac{15\Delta h_b}{\Delta h_m} = 18,\ k_f = 4 + 1.5\left(\frac{f_c}{925} - 1\right) = 4 + 1.5\left(\frac{1800}{925} - 1\right) = 5.42$$

$$L_f = 32.4 + 20\log d + 20\log 1800 = 97.5 + 20\log d\ \text{dB}$$

$$L_{rts} = -16.9 - 10\log 20 + 10\log 1800 + 20\log 38 + 0 = 34.25\ \text{dB}$$

$$L_{ms} = -18\log 11 + 0 + 54 + 18\log d + 5.42\log 1800 - 9\log 40 = 38.47 + 18\log d\ \text{dB}$$

$$\therefore 134.7 = 97.5 + 20\log d + 34.25 + 38.47 + 18\log d$$

$$38\log d = -35.52$$

$$\log d = -0.935 = 0.116\ \text{km or } 116\text{ m}$$

2.9 Summary

In this chapter, we discussed signal attenuation in free space and noted that the free-space attenuation increases by 6 dB whenever the length of path or transmission frequency is doubled. With the discussion of signal propagation over the reflecting flat earth's surface, we found that the received signal level is dependent only on the transmitted power, antenna heights, and separation distance between the transmitter and receiver; there is no frequency dependence. The signal attenuation increases by 12 dB when the separation distance is doubled.

We also observed that the calculated path losses with the Okumura-Hata model are about 13 to 15 dB lower than the path losses obtained using the COST 231 model. This is because the Okumura-Hata model neglects several important parameters such as street width, street orientation, etc.

When the signal bandwidth is much larger than the coherence bandwidth, the signal is called wideband signal and the fading is frequency selective. This means that only a portion of the signal bandwidth fades at any instant of time. With signal bandwidth much smaller than the coherence bandwidth, flat fading of the entire signal occurs.

With the signal symbol interval much larger than the coherence time, the channel fades rapidly compared to the symbol rate. This is called fast fading relative to symbol time, and frequency dispersion occurs, causing signal distortion. With signal symbol interval much smaller than the coherence time, the channel does not change during the symbol interval. This is referred to as a slow fading channel, relative to the symbol time.

2.10 References

1. Rappaport, T. S., "*Wireless Communications,*" Prentice Hall: Upper Saddle River, NJ, 1996.
2. Okumura, Y., Aomori, T., Kano, T., and Fukuda, K., "Field Strength and its Variability in VHF and UHF Land-Mobile Radio Service," Rev. Elec. Communication Lab, Sept./Oct. 1968, [Vol. 16, pp. 825–43].
3. Hata, M., "Empirical Formula for Propagation Loss in Land Mobile Radio Service," *IEEE Transactions on Vehicular Technology*, August 1980, [Vol. 29(3), pp. 317–25].
4. COST 231 TD (91) 73, "Urban Transmission Loss Models for Mobile Radio in the 900 and 1800 MHZ Bands," Revision 2, COST 231 Working Group 2, UHF Propagation, The Hague, September 1991.
5. Garg, V. K., and Wilkes, J. E., *Wireless and Personal Communications Systems*, Prentice Hall: Upper Saddle River, NJ, 1996.
6. Collins, G. W., "Wireless Wave Propagation," *Microwave Journal,* July 1998, [Vol. 42(7), pp. 78–86].
7. Walfisch, J., and Bertoni, H.L., "A Theoretical Model of UHF Propagation in Urban Environment," *IEEE Transactions on Antennas & Propagation*, to be published.

8. Sklar, B., "Rayleigh Fading Channels in Mobile Digital Communications Systems Part I: Characterization," *IEEE Communication Magazine,* Sept. 1997, [Vol. 35(9), pp. 136–46].
9. Sklar, B., "Rayleigh Fading Channels in Mobile Digital Communications Systems Part II: Mitigation," *IEEE Communication Magazine,* Sept. 1997, [Vol. 35(9), pp. 148–55].

2.11 Problems

1. Repeat Example 2.1 with a frequency of 1,800 MHz and separation between the transmitter and receiver equal to 8,000 ft. What is the divergence factor due to the earth's curvature?
2. Repeat Example 2.2 for a PCS system in a suburban area with separation between transmitter and receiver equal to 10 km.
3. Assuming vehicle speed equal to 100 mph, carrier frequency $f_c = 1.9$ GHz, and delay spread $\tau_d = 6$ µs, calculate coherence time and coherence bandwidth. At the coded symbol rate of 32 kbps, what kind of symbol distortion will be experienced? What type of fading will exist for an IS-95 channel of 1.25 MHz?
4. Find the maximum allowable path loss using the following data:
   - Information rate = 14,400 bps
   - Mobile station effective radiated power (P_m) = 200 mW (23 dBm)
   - Base station antenna gain (G_b) = 16 dBi
   - Base station receiver antenna cable loss (L_c) = 2.0 dB
   - Cell receiver noise figure (F_b) = 5 dB
   - Required margin (E_b/N_0) = 7 dB (with diversity antenna at base station)
   - Base station noise floor (N_0) = –174 dBm/Hz
   - Log-normal shadowing margin = 8 dB
   - Body/orientation loss = 2 dB
   - Building penetration loss = 10 dB

CHAPTER 3

Direct-Sequence Spread-Spectrum (DSSS) and TIA IS-95 CDMA

3.1 Introduction

In this chapter, we first present concepts of direct-sequence spread-spectrum (DSSS) transmission. We show how signal spreading and despreading can be achieved using BPSK and QPSK modulation schemes. We then focus on multipath issues in wireless communications and show how CDMA takes advantage of multipath to improve system performance using a RAKE receiver. We conclude the chapter by presenting a summary of challenges in implementing a CDMA system and providing some highlights of the IS-95 CDMA system.

A wideband spread-spectrum (SS) signal involves spreading the bandwidth required to transmit information. The SS signal is generated from a data-modulated carrier. The data-modulated carrier is modulated a second time by using a wideband spreading signal. A spread-spectrum signal has advantages in the areas of security, resistance to narrowband jamming, resistance to multipath fading, and supporting multiple-access techniques.

The spreading modulation may be phase modulation or it may be a rapid change of the carrier frequency, or it may be a combination of these two schemes. When spectrum spreading is performed by phase modulation, we call the resultant signal a *direct-sequence spread-spectrum* (DSSS) signal. When the spectrum spreading is achieved by a rapid change of the carrier frequency, we refer to the resultant signal as a *frequency-hop spread-spectrum* (FHSS) signal. When both direct-sequence and frequency-hop techniques are employed, the resultant signal is called a hybrid DS-FH SS signal. Another way to also generate a spread-spectrum signal is the *time-hop spread-spectrum* (THSS) signal. In this case, the transmission time is divided into intervals called frames. Each frame is further divided into time slots. During each frame, one and only one time slot is modulated with a message. All the message bits accumulated in previous frames are transmitted.

The spreading signal is selected to have properties to facilitate demodulation of the transmitted signal by the intended receiver, and to make demodulation by an unintended receiver as difficult as possible. These same properties also make it possible for the intended receiver to differentiate between the communication signal and jamming. If the bandwidth of the spreading signal is large relative to the data bandwidth, the spread-spectrum transmission bandwidth is dominated by the spreading signal and is nearly independent of the data signal.

3.2 The Concept of a Spread-Spectrum System

We consider the channel capacity as given by Shannon [1]:

$$C = B_w \log_2 (1 + S/N) \tag{3.1}$$

where

B_w = channel bandwidth in Hertz (Hz)
C = channel capacity in bits per second (bps)
S = signal power
N = noise power

Equation (3.1) provides the relationship between the theoretical ability of a channel to transmit information without errors for a given signal-to-noise (S/N) ratio and a given bandwidth of the channel. The channel capacity can be increased by increasing the channel bandwidth, the transmitted power, or a combination of both. We rewrite Equation (3.1) as

$$\frac{C}{B_w} = 1.44\log_e(1 + S/N) \tag{3.2}$$

Since

$$\log_e(1 + S/N) = \frac{S}{N} - \frac{1}{2}\left(\frac{S}{N}\right)^2 + \frac{1}{3}\left(\frac{S}{N}\right)^3 - \frac{1}{4}\left(\frac{S}{N}\right)^4 + \ldots \tag{3.3}$$

Assuming that the S/N ratio is small (e.g., $S/N \leq 0.1$), we can neglect the higher-order terms and rewrite Equation (3.2) as

$$B_w \approx \frac{C}{1.44} \times \frac{1}{(S/N)} \tag{3.4}$$

For any given S/N ratio, we can have a low information error rate by increasing the bandwidth used to transmit the information. As an example, if we wish a system to operate on an RF link in which the data information rate is 20 kilobits per second (kbps) and the S/N ratio is 0.01, we should use a bandwidth of

$$B_w = \frac{20 \times 10^3}{1.44 \times 0.01} = 1.38 \times 10^6 \text{ Hz or } 1.38 \text{ MHz} \tag{3.5}$$

Information can be modulated into the SS signal by several methods. The most common method is to multiply the information to the spread-spectrum code before it is used to modulate the carrier frequency (Figure 3.1). This technique applies to any SS system that uses a code sequence to determine RF bandwidth. If the signal that is being sent is analog (voice, for example), the signal must be digitized before being modulated by the spreading code.

One of the major advantages of an SS system is the robustness to interference. The system processing gain G_p quantifies the degree of interference rejection. The system processing gain is the ratio of spreading bandwidth to the information rate, R, and is given as

$$G_p = \frac{B_w}{R} \tag{3.6}$$

Typical processing gains of an SS system lie between 20 and 60 dB. With an SS system, the noise level is determined both by thermal noise and by interference. For a given user, the interference is processed as noise. The input and output S/N ratios are related as

$$(S/N)_o = G_p(S/N)_i \tag{3.7}$$

We express input $(S/N)_i$ ratio as

$$\left(\frac{S}{N}\right)_i = \left(\frac{E_b R}{N_0 B_w}\right)_i = \left(\frac{E_b}{N_0}\right)_i \cdot \frac{1}{G_p} \tag{3.8}$$

where

E_b = bit energy

N_0 = noise density

Using Equation (3.8) we rewrite Equation (3.7) as

$$\left(\frac{S}{N}\right)_o = G_p \cdot \left(\frac{S}{N}\right)_i = \left(\frac{E_b}{N_0}\right)_i \tag{3.9}$$

but

$$\left(\frac{S}{N}\right)_0 = \left(\frac{E_b R}{N_0 B_w}\right)_0 = \left(\frac{E_b}{N_0}\right)_0 \cdot \frac{1}{G_p} \tag{3.10}$$

Therefore, using Equations (3.9) and (3.10),we get

$$\left(\frac{E_b}{N_0}\right)_o = G_p \cdot \left(\frac{E_b}{N_0}\right)_i \tag{3.11}$$

EXAMPLE 3.1

A DSSS system has a 1.2288-Mcps (megachips per second) code clock rate and 9.6-kbps information rate. Calculate the processing gain. How much improvement in information rate will be achieved if the code generation rate is changed to 5 Mcps and the processing gain is changed to 256?

$$G_p = \frac{1.2288 \times 10^6}{9.6 \times 10^3} = 128 = 21 \text{ dB}$$

$$R = \frac{5 \times 10^6}{256} = 19.53 \text{ kbps}$$

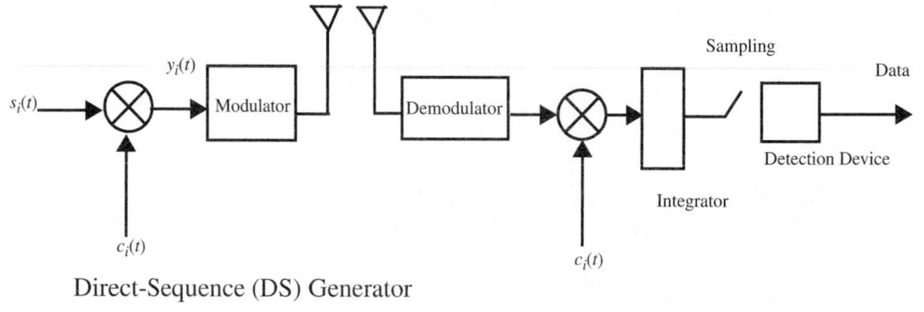

Figure 3.1 Spread-Spectrum System

In an SS system the information signal, $s_i(t)$, is multiplied by a wideband code signal, $c_i(t)$, which is the output signal of the direct-sequence (DS) generator (see Figure 3.1). The signal $s_i(t) \cdot c_i(t) \equiv y_i(t)$ is modulated and transmitted. This occupies a bandwidth far in excess of the minimum bandwidth required to transmit the information signal $s_i(t)$. We can observe from Figure 3.2 that the combined signal waveform has more high-frequency changes in the data information since $1/T_c \gg 1/T_b$, where T_b is the bit interval of the information stream and T_c is the bit interval of the DS stream. T_c is also called *chip interval.* The ratio of T_b to T_c is referred to as the processing gain G_p (since $R = 1/T_b$ and $B_w = 1/T_c$, $G_p = B_w/R = T_b/T_c$).

When $s_i(t)$ and $c_i(t)$ have the same rate, the product $y_i(t)$ contains all the information of $s_i(t)$ and has the same rate as $c_i(t)$. The spectrum of the signal is unchanged, and the incoming bit stream is said to be encrypted or *scrambled*. However, when $c_i(t)$ is faster than $s_i(t)$, $y_i(t)$ contains all the information of $s_i(t)$, has a faster bit rate compared with $s_i(t)$, and is said to have had its spectrum spread (refer to Figure 3.2).

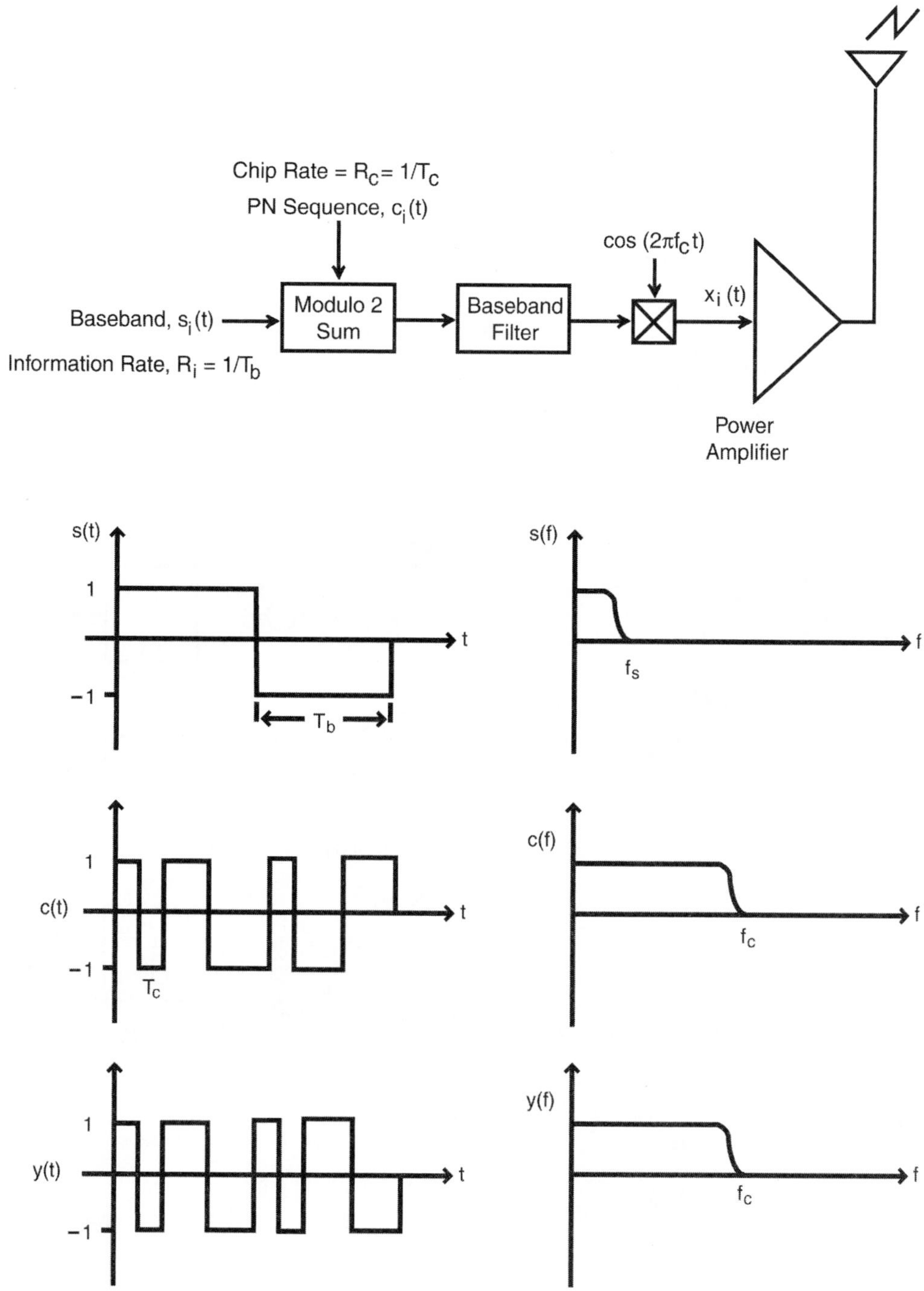

Figure 3.2 Direct-Sequence (DS) Spreading

We consider the downlink CDMA operation (see Figure 3.3) where the base station generates a data stream for mobiles 1, 2, and 3—$s_i(t)$—and multiplies the data stream by an appropriate DS code, $c_i(t)$. Next, we add the coded data streams.

$$x(t) = \sum_{j=1}^{3} y_j \tag{3.12}$$

We modulate the resultant baseband spread-spectrum signal, Equation (3.12), by a carrier to obtain

$$z_i(t) = A_i\left\{\sum_{j=1}^{3} c_j(t)\cdot s_j(t)\right\}\cos(\omega_c t) \tag{3.13}$$

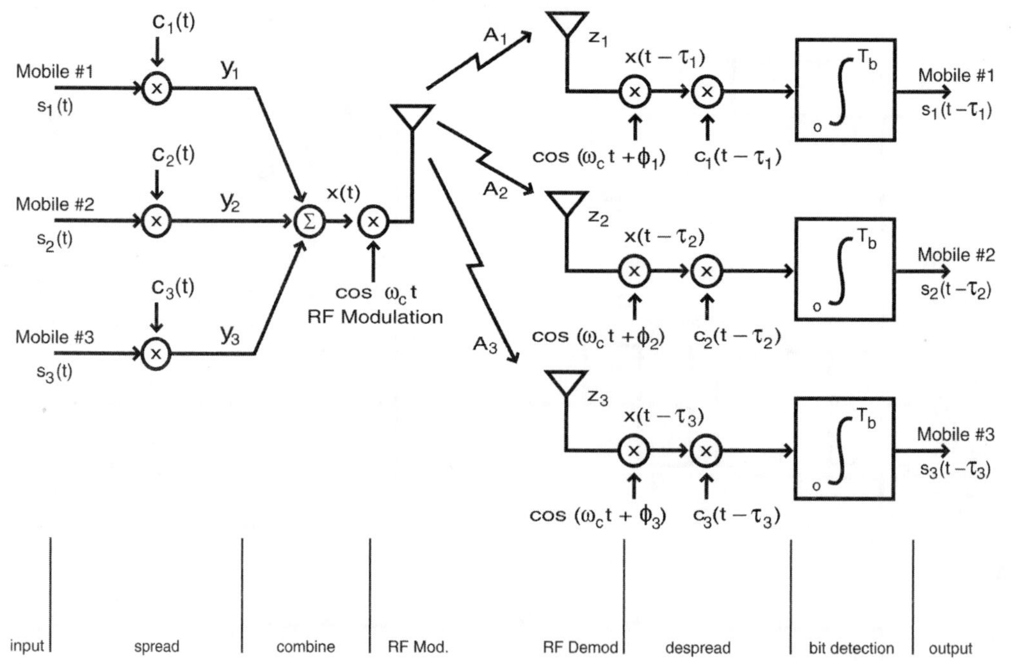

Figure 3.3 Multiuser DS CDMA—Downlink System

$z(t)$ is transmitted over the allocated bandwidth. The spread signal at the receiver for mobile, i, is $z_i(t)$ + noise, where

$$z_i(t) = A_i\left[\sum_{j=1}^{3} c_j(t-\tau_j)\cdot s_j(t-\tau_j)\right]\cos(\omega_c t+\phi_i) \quad (3.14)$$

Equation (3.14) contains the desired signal and other users' signals. After multiplication by a coherent carrier phase ϕ_i estimated by the receiver, a locally generated DS sequence that is an exact replica to the desired code transmitted multiplies the incoming signal—despreading. We assume that this DS sequence is in perfect synchronization (receiver estimates delay τ_j) with the transmitted version so that

$$c_j(t-\tau_j)\cdot c_j(t-\tau_j) = 1 \quad (3.15)$$

The multiplier output yields the desired data signal $s_i(t-\tau_i)$ plus interfering terms due to other users. Ideally, the integrator, an *integrate-and-dump* over T_b, should produce a cross-correlation between the desired signal and interferers that is 0. Hence, the output for mobile, i, is proportional to the transmitted data stream, $s_i(t-\tau_i)$.

EXAMPLE 3.2

We consider signals transmitted from a base station to mobiles 1, 2, and 3 (see Figure 3.3). The data streams $s_1(t)$, $s_2(t)$, and $s_3(t)$ are multiplied with codes $c_1(t)$, $c_2(t)$, and $c_3(t)$, respectively (see Figure 3.4). The resultant demodulated signal $z(t)$, sum of $y_1(t)$, $y_2(t)$, and $y_3(t)$ is given in Figure 3.5. Show that the transmitted signals to mobiles 1, 2, and 3 are recovered at the receivers by despreading the resultant signal $z(t)$. Neglect propagation delay. Note that the code length is $15T_c$ $[(2^4-1)\,T_c = 15T_c]$.

From Tables 3.1, 3.2, and 3.3, we observe that the transmitted signals to mobiles 1, 2, and 3 are recovered by despreading at the receivers.

Table 3.1 $z(t)\cdot c_1(t)$ [see Figure 3.6(a)]

	T_b	$2T_b$	$3T_b$	$4T_b$	$5T_b$
Value of integration at end of bit period	–6	4	6	–2	–12
Bit value	1	0	0	1	1

Table 3.2 $z(t)\cdot c_2(t)$ [see Figure 3.6(b)]

	T_b	$2T_b$	$3T_b$	$4T_b$	$5T_b$
Value of integration at end of bit period	–2	–6	8	–4	–8
Bit value	1	1	0	1	1

Table 3.3 $z(t) \cdot c_3(t)$ [see Figure 3.6(c)]

	T_b	$2T_b$	$3T_b$	$4T_b$	$5T_b$
Value of integration at end of bit period	–6	4	–8	–8	8
Bit value	1	0	1	1	0

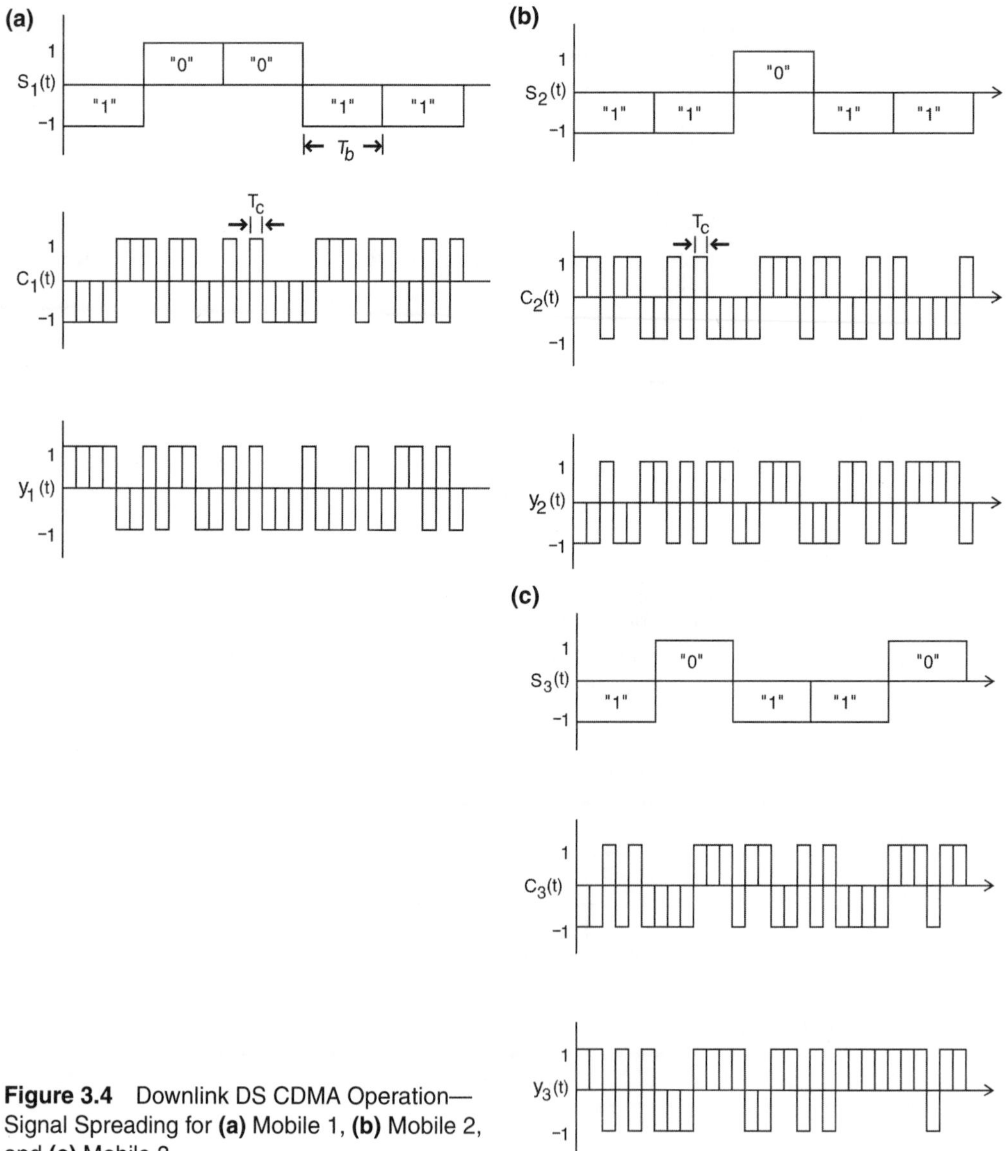

Figure 3.4 Downlink DS CDMA Operation—Signal Spreading for **(a)** Mobile 1, **(b)** Mobile 2, and **(c)** Mobile 3

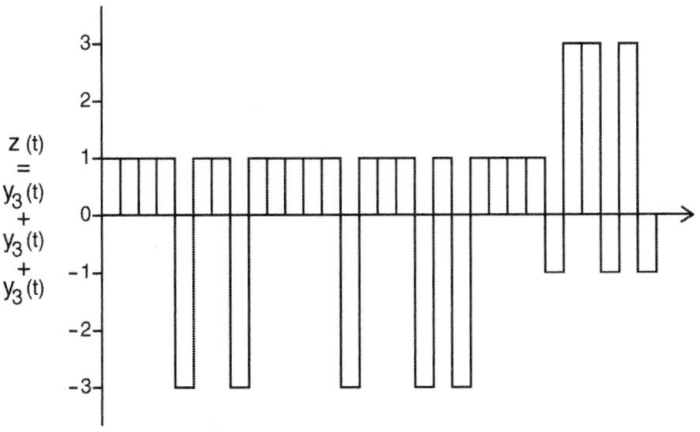

Figure 3.5 Resultant Demodulated Signal at a Mobile

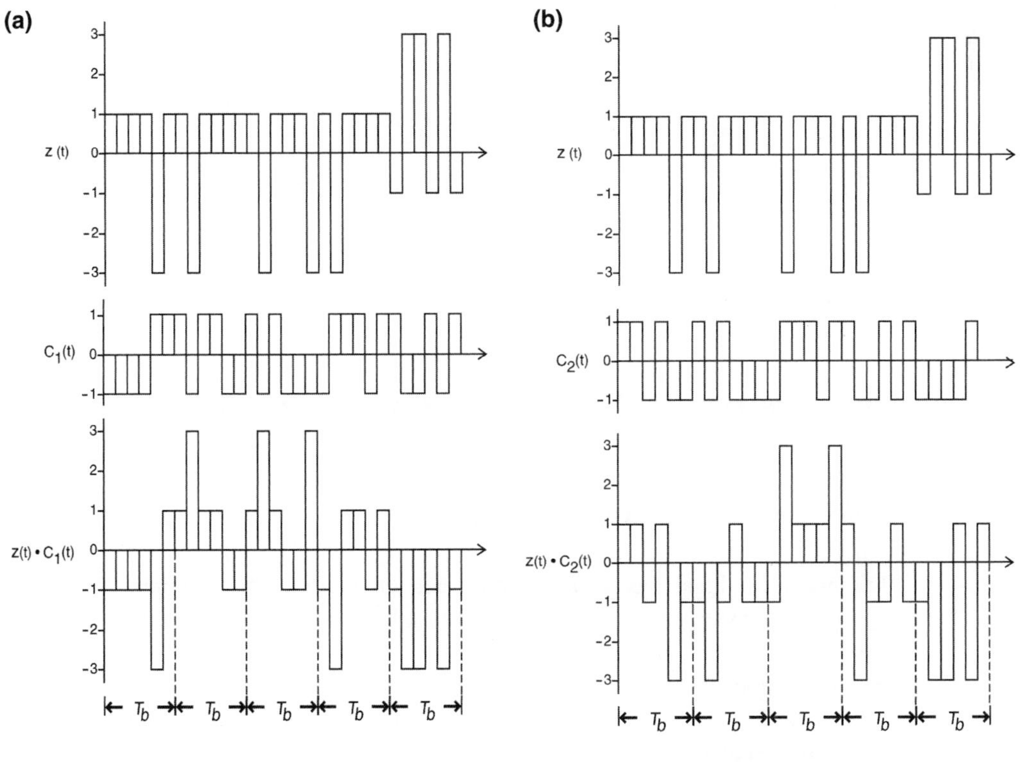

Figure 3.6 Despreading of Resultant Demodulated Signal at Mobile Receivers

(c)

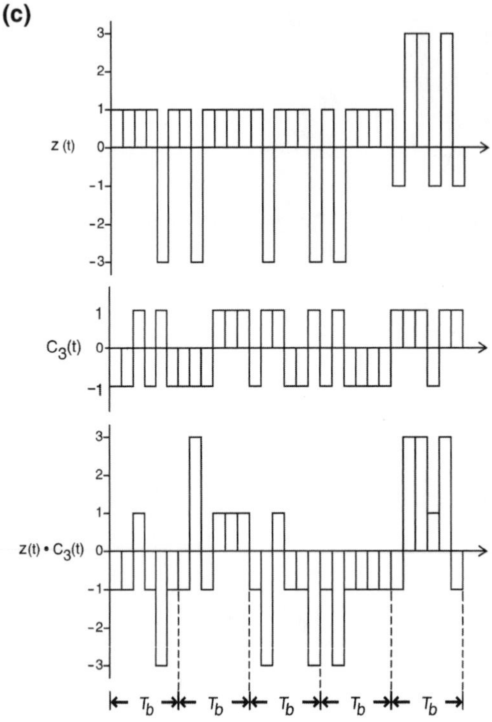

Figure 3.6 *(cont.)*

3.3 Requirements of Direct-Sequence Spread-Spectrum

In the DSSS system, the entire bandwidth of the RF carrier is made available to each user. The DSSS system satisfies the following requirements:

- The spreading signal has a bandwidth much larger than the minimum bandwidth required to transmit the desired information, which for a digital system is baseband data.
- The spreading of the information is performed by using a spreading signal, called the *code signal.* The code signal is independent of the data and is of a much higher bit rate than the data signal.
- At the intended receiver, despreading is accomplished by cross-correlation of the received spread signal with a synchronized replica of the same signal used to spread the data.

3.4 Coherent Binary Phase-Shift Keying DSSS

The simplest form of a DSSS communication system uses coherent binary phase-shift keying (BPSK) for both the data modulation and spreading modulation. But the most common form uses BPSK for the data modulation, and quadrature phase-shift keying (QPSK) for the spreading modulation.

We first consider the simplest case. The *i*th mobile station is assigned a spreading code signal $c_i(t)$, a periodic sequence with symbols (chips) of width T_C. Each mobile station has its own such code signal. Information bits are transmitted by superimposing the bits onto the code signal. If the *i*th mobile station transmits the binary data waveform $d_i(t)$ $[d_i(t) = \pm 1]$, it forms the binary sequence.

$$m_i(t) = d_i(t) \cdot c_i(t) \tag{3.16}$$

The transmitted signal from the *i*th mobile is

$$s_i(t) = d_i(t) \cdot c_i(t) \cdot \sqrt{2P} \cdot \cos(\omega t + \phi) \tag{3.17}$$

where

$d_i(t)$ = the baseband signal at the transmitter input and receiver output for the *i*th mobile
$c_i(t)$ = the spreading code for the *i*th mobile
ω = the carrier frequency
P = the signal power
ϕ = data phase modulation

If T_b is the bit period of $s_i(t)$, then T_b may correspond to either a full period for $c_i(t)$, or to a fraction of a period. If T_b is less than one code period, then the data bits are modulating the polarity of a portion of a code period. The code $c_i(t)$ serves as a subcarrier for the source data. Since each mobile station uses the entire channel bandwidth and since Equation (3.16) has a code chip rate of $1/T_c$ chips per second, each BPSK carrier uses an RF bandwidth of approximately

$$B_w = \frac{1}{T_c} \tag{3.18}$$

The available channel RF bandwidth determines the minimum chip width, and the code period determines its relation to the bit times. The number of code chips per bit is given by

$$G_p = \frac{T_b}{T_c} = \frac{B_w}{R} \tag{3.19}$$

The ratio B_w/R is the CDMA processing gain, G_p, or simply the spreading ratio of code modulation. This shows how much the RF bandwidth must be spread relative to the bit rate

R to accommodate a given spreading code length. Each mobile station uses the same RF carrier frequency and RF bandwidth, but with its own spreading code $c_i(t)$.

Equation (3.16) represents the modulo-2 addition of $c_i(t)$ and $d_i(t)$ as a multiplication because the binary 0 and 1 represent values of 1 and –1 into the modulator.

The signal in Equation (3.16) is transmitted using a distortionless path with transmission. The signal is received together with some type of interference and/or Gaussian noise. Demodulation is performed in part by remodulating with the spreading code appropriately delayed, as shown in Figure 3.7. This correlation of the received signal with the delayed spreading waveform is called *despreading*. This is a critical function in all spread-spectrum systems. The signal component of the output of the despreading mixer is

$$d_i(t-\tau_d)\cdot\sqrt{2P}\cdot c_i(t-\tau_d)\times c_i(t-\hat{\tau}_d)\cos(\omega(t-\tau_d)+\phi) \qquad \textbf{(3.20)}$$

where

$\hat{\tau}_d$ = the receiver's best estimate of the transmission delay

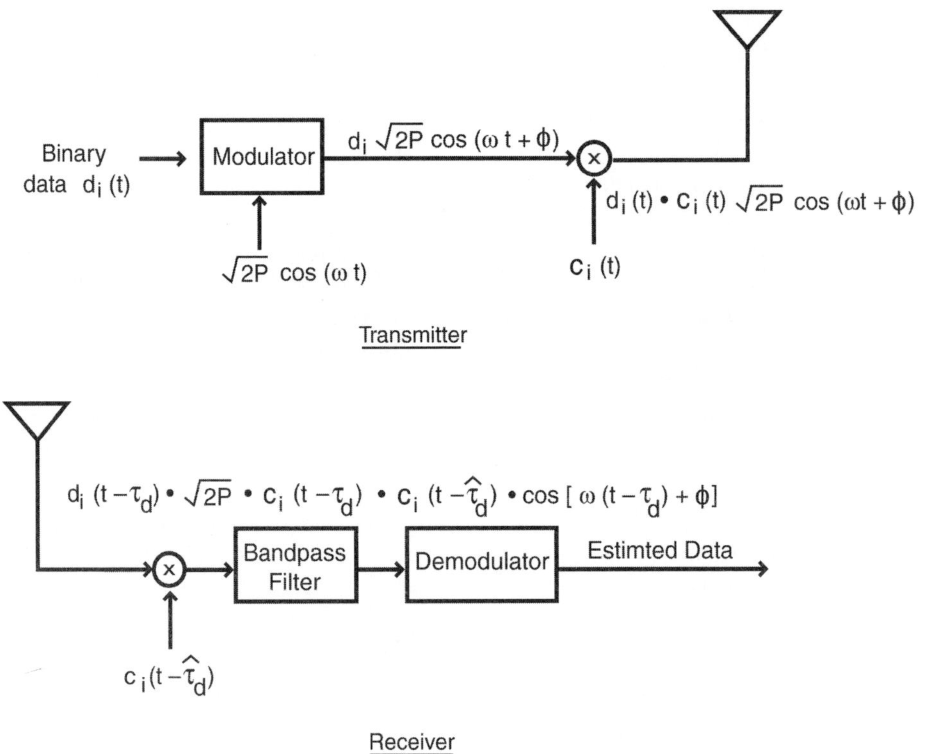

Figure 3.7 DSSS System with BPSK

Since $c_i(t) = \pm 1$, the product $c_i(t-\tau_d) \times c_i(t-\hat{\tau}_d)$ will be unity if $\tau_d = \hat{\tau}_d$, that is, if the spreading code at the receiver is synchronized with the spreading code at the transmitter.

When correctly synchronized, the signal component of the output of the receiver despreading mixer is equal to $\sqrt{2P} \cdot d_i(t-\tau_d)\cos(\omega(t-\tau_d)+\phi)$, which can be demodulated using a conventional coherent phase modulator.

The bit error probability P_b associated with the coherent BPSK spread-spectrum signal is the same as with the BPSK signal and is given as

$$P_b = \frac{1}{2}erfc\left[\sqrt{\left(\frac{E_b}{N_0}\right)_0}\right] = Q\left[\sqrt{2\left(\frac{E_b}{N_0}\right)_0}\right] = Q\left[\sqrt{2\left\{G_p \cdot \left(\frac{E_b}{N_0}\right)_i\right\}}\right] \tag{3.21}$$

where

$$Q(u) \approx \frac{e^{-u^2/2}}{(\sqrt{2\pi u})},\ u \gg 1 \quad \text{Gaussian Integral}$$

$E_b = P/R$ = energy per bit; R = bit rate

3.5 Quadrature Phase-Shift Keying DSSS

Sometimes it is advantageous to transmit simultaneously on two carriers which are in-phase quadrature. The main reason for this is to save spectrum, because for the same total transmitted power, we can achieve the same bit error probability, P_b, using one-half the transmission bandwidth. The quadrature modulations are more difficult to detect in low probability of detection applications. Also, the quadrature modulations are less sensitive to some types of jamming. We refer to Figure 3.8 and write

$$s(t) = \sqrt{P}c_I(t)\cos[\omega t+\phi] + \sqrt{P}c_Q(t)\sin[\omega t+\phi] \tag{3.22}$$

$$s(t) = a_I(t) + a_Q(t) \tag{3.23}$$

where $c_I(t)$ and $c_Q(t)$ are the in-phase and quadrature spreading codes. $a_I(t)$ and $a_Q(t)$ are orthogonal. This condition is satisfied in the present case since $c_I(t)$ and $c_Q(t)$ are independent code waveform.

The receiver for the transmitted signal is shown in Figure 3.8. The bandpass filter is centered at frequency ω_f and has a bandwidth sufficiently wide enough to pass the data-modulated carrier without distortion.

$$x(t) = \sqrt{P/2} \cdot c_I(t-\tau_d) \cdot c_I(t-\hat{\tau}_d) \cdot \cos[\omega t+\phi]$$

$$+ \sqrt{P/2} \cdot c_Q(t-\tau_d) \cdot c_I(t-\hat{\tau}_d) \cdot \sin[\omega t + \phi] \quad \textbf{(3.24)}$$

$$y(t) = \sqrt{P/2} \cdot c_I(t-\tau_d) \cdot c_Q(t-\hat{\tau}_d) \cdot \sin(\omega t + \phi)$$

$$+ \sqrt{P/2} \cdot c_Q(t-\tau_d) \cdot c_Q(t-\hat{\tau}_d) \cdot \cos[\omega t + \phi] \quad \textbf{(3.25)}$$

If the receiver-generated replicas of spreading codes are correctly phased, then

$$c_I(t-\tau_d) \cdot c_I(t-\hat{\tau}_d) = c_Q(t-\tau_d) \cdot c_Q(t-\hat{\tau}_d) = 1 \quad \textbf{(3.26)}$$

$$z(t) = \sqrt{2P} \cdot \cos[\omega t + \phi] \quad \textbf{(3.27)}$$

The signal $z(t)$ is the input to a conventional phase demodulator where data is recovered.

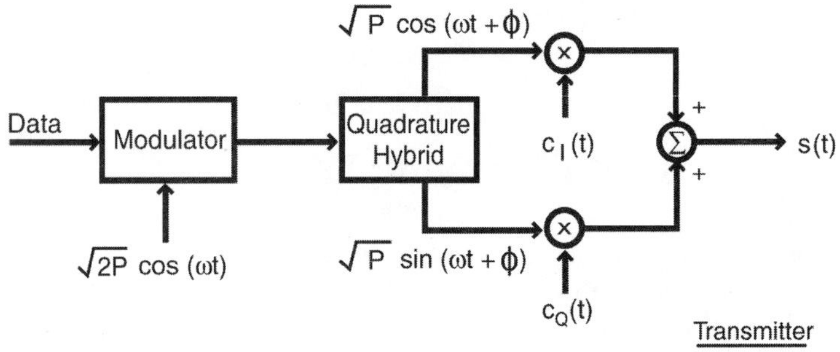

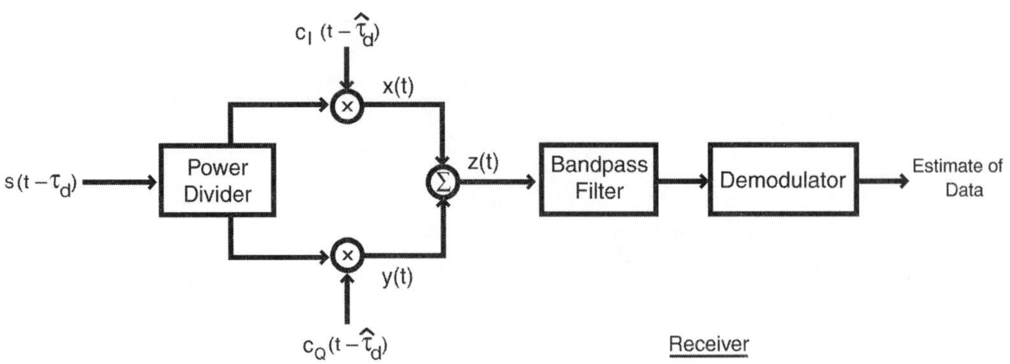

Figure 3.8 DSSS System with QPSK

When the spreading codes are staggered in one-half chip intervals with respect to each other, the QPSK is called OQPSK. In OQPSK, the phase changes every one-half chip interval, but it does not change more than $\pm 90°$. This limited phase change improves the uniformity of the signal envelope compared to BPSK and QPSK, since zero-crossings of the carrier envelope are avoided. Neither QPSK nor OQPSK modulation can be removed with a single stage of square-law detection. Two such detectors, and the associated loss of signal-to-noise ratio are required. QPSK and OQPSK offer some low probability of detection advantages over the BPSK method.

3.6 Spreading Codes

3.6.1 Bit Scrambling

Referring to Table 3.4, we consider the following activities at a given transmitter location (see Figure 3.9):

1. An arbitrary data sequence $s_i(t)$ is generated by a digital source.
2. An arbitrary code sequence $c_i(t)$ is generated by a DS generator.
3. Two sequences are modulo-2 added and transmitted to a distant receiver.
4. At the distant location, the resulting sequence (assuming no propagation delay) is picked up by the receiver (see Figure 3.10).
5. The code $c_i(t)$ used at the transmitter is also available at the receiver.
6. The original data sequence is recovered by modulo-2 adding the received sequence with the locally available code $c_i(t)$.

Table 3.4 Operations with Modulo-2 Addition

Transmitter	1	$s_i(t)$	1	1	0	1	0	0	1	1	1	1
	2	$c_i(t)$	1	0	0	1	1	1	0	1	0	0
	3	$s_i(t) \oplus c_i(t)$	0	1	0	0	1	1	1	0	1	1
Receiver	4	$s_i(t) \oplus c_i(t)$	0	1	0	0	1	1	1	0	1	1
	5	$c_i(t)$	1	0	0	1	1	1	0	1	0	0
	6	$s_i(t) \oplus c_i(t) \oplus c_i(t) = s_i(t)$	1	1	0	1	0	0	1	1	1	1

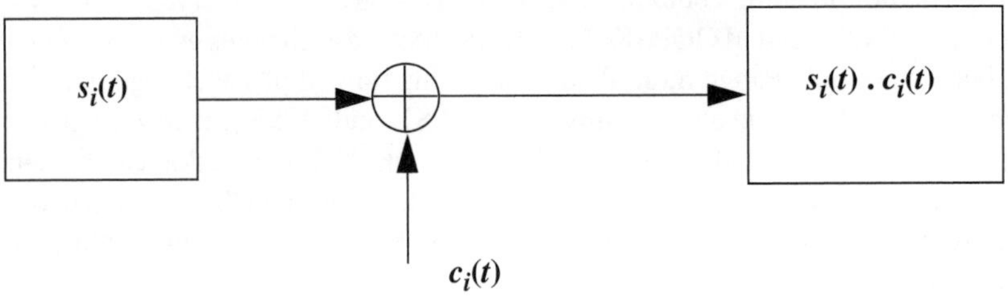

Figure 3.9 Mobile Transmitter

Next, referring to Table 3.5, we consider the following set of activities at the given transmitter location.

1. An arbitrary data sequence $s_i(t)$ is generated by a digital source. In this case, we use +1s and –1s to represent 0s and 1s as in the previous example (see Figure 3.9).
2. An arbitrary code sequence $c_i(t)$ is generated by a DS generator.
3. We multiply $s_i(t)$ and $c_i(t)$. The output of the multiplier is transmitted to a distant receiver.
4. At the distant location, the resulting sequence (again assuming no propagation delay) is picked up by the receiver (see Figure 3.10).
5. The code $c_i(t)$ used at the transmitting location is assumed to be available at the receiver.
6. The original data sequence is recovered by multiplying the received sequence by the locally available code $c_i(t)$.

Table 3.5 Operations without Modulo-2 Addition

Transmitter	1	$s_i(t)$	–1	–1	1	–1	1	1	–1	–1	–1	–1
	2	$c_i(t)$	–1	1	1	–1	–1	–1	1	–1	1	1
	3	$s_i(t) \bullet c_i(t)$	1	–1	1	1	–1	–1	–1	1	–1	–1
Receiver	4	$s_i(t) \bullet c_i(t)$	1	–1	1	1	–1	–1	–1	1	–1	–1
	5	$c_i(t)$	–1	1	1	–1	–1	–1	1	–1	1	1
	6	$s_i(t) \bullet c_i(t) \bullet c_i(t) = s_i(t)$	–1	–1	1	–1	1	1	–1	–1	–1	–1

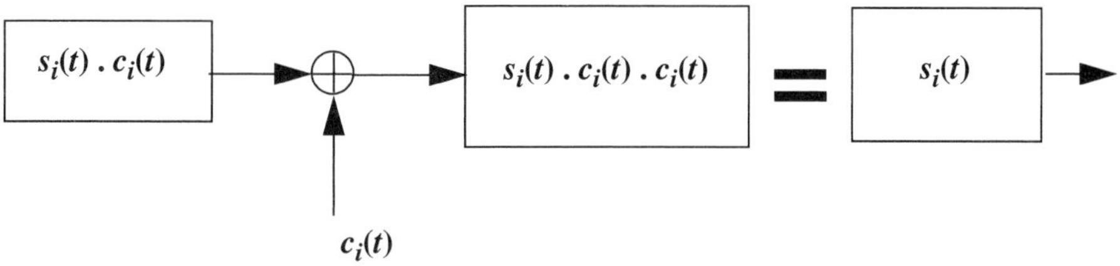

Figure 3.10 Mobile Transmitter

From Tables 3.4 and 3.5 we conclude that modulo-2 addition using 1 and 0 binary data is equivalent to multiplication using –1 and 1 binary data as long as we remain consistent in mapping 0s to +1s and 1s to –1s as shown in Table 3.5. For circuit implementation, modulo-2 addition is preferred since exclusive-OR gates are cheaper than multiplication circuits. However, for modeling purposes, the multiplication method is usually easier to formulate and understand than the modulo-2 approach.

We notice that for the output of the receiver to be identical to the original data, the following relationship must be satisfied:

$$s_i(t) \cdot c_i(t) \cdot c_i(t) = s_i(t) \quad \textbf{(3.28)}$$

In other words, $c_i(t) \cdot c_i(t)$ must be equal to 1. Note that $c_i(t)$ is a binary sequence made up of 1s and –1s, therefore

$$\text{If } c_i(t) \text{ is } 1,\ c_i(t) \cdot c_i(t) = 1 \quad \textbf{(3.29)}$$

$$\text{If } c_i(t) = -1,\ c_i(t) \cdot c_i(t) = 1 \quad \textbf{(3.30)}$$

In the previous discussion, we assume that there is no propagation delay and no other processing delay incurred between the transmitter and receiver input. Thus, the code copy used at the receiver is perfectly lined up with the initial code used at the transmitter. The two codes are said to be in phase or in synchronization (synch). In practice, however, a propagation delay and other processing delays (τ_i) incur between the transmitter and the receiver input. Therefore, the receiver may be time-shifted relative to the initial code at the transmitter. The two codes are no longer in synch. As a result, the output of the receiver will no longer be identical to the original data, $s_i(t)$.

In order to recover the original data $s_i(t)$, we must "tune" the receiver code sequence to that of the incoming code from the transmitter. In other words, we must time-shift the receiver code in order to line it up with the incoming code. It should be noted that by synchronizing or "tuning" the receiver code to the phase of the incoming code $c_i(t - \tau_i)$, the original data (shifted by propagation delay) can now be recovered at the output of the receiver. In these examples, the data sequence and code sequence have the same length (one code bit for each data bit) and are used for encrypting the data bits. This is referred to as *bit scrambling* and does not result in spectrum spreading.

3.6.2 Requirements of Spreading Codes

To spread the data sequence, the code sequence must:

- Be much faster than the data sequence
- Exhibit some random properties

By multiplying the data sequence with the faster code sequence, the resulting product yields a sequence with more transitions than the original data. Using suitable randomlike codes, the resulting sequence will have the same rate as the code sequence. It is desirable to use a set of orthogonal codes to provide good isolation between users. However, in practice, the codes used are not perfectly orthogonal, but they exhibit good isolation characteristics, i.e., they have low cross-correlation.

3.7 Multipath Signal Propagation and the RAKE Receiver

In the absence of a direct LOS signal from the base station to the mobile station, the received signal at the mobile station from the base station (and at the base station from mobile) is made up of the sum of many signals, each traveling over a separate path. Since these path lengths are not equal, the information carried on the radio link experiences a spread in delay as it travels between the base station and the mobile. In addition, to delay spread, some multipath environments cause severe local variations in signal strength as these multipath signals are added constructively and destructively at the receiving antenna. This type of variation is called *Rayleigh fading* and was discussed in Chapter 2. The movement of the mobile causes each received signal to be shifted in frequency as a function of the relative direction and speed of the mobile. This effect is called *Doppler shift*.

Multipath is treated as causing delayed versions of the signal to add to the system noise when the differential delay exceeds the chip time, T_c. Substantial performance improvement can occur by detecting each additional path separately, thereby enabling the signals to be combined coherently. A CDMA receiver can be implemented to resolve each individual path such that the paths can be combined to produce a net overall gain. This type of receiver is known as a *RAKE receiver.*

In the RAKE receiver (see Figure 3.11) for user #1 baseband demodulated signal $z(t)$ is the sum of N signals which arrive on N different paths. When we consider path 2, the multiplication of $z(t)$ by $c_1(t - \Delta_2)$, with integration starting at time Δ_2 and ending at T_b, yields the peak response. (The output of the integrator is the value of the correlation function of $c_1(t)$ for a particular delay. For path 2, this delay is zero, whereas for the other paths the delay exceeds the time duration of a chip.) The contributions from other paths average out to zero, since the differential delays exceed the chip duration T_c. The response from each path is summed to produce the stronger signal.

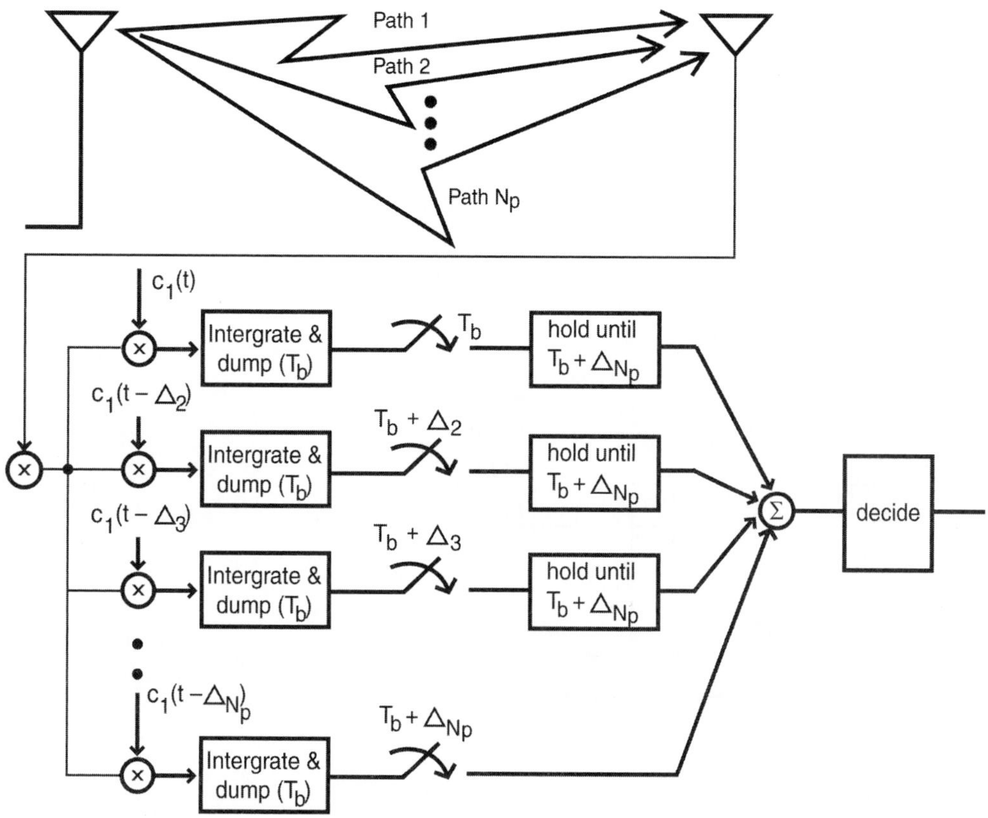

Figure 3.11 Simplified RAKE Receiver for Mobile 1—Equal Gain Combining

---

EXAMPLE 3.3

We consider the downlink in Example 3.2 where the demodulated signal at the mobile is $z(t) = y_1(t) + y_2(t) + y_3(t)$ (see Figure 3.5) for mobiles 1, 2, and 3. We assume two equal strength paths and write the demodulated signal as (see Figure 3.12):

$$\bar{z}(t) = z(t) + z(t - 2T_c)$$

The differential delay between paths is taken as $2T_c$ for simplicity. We will show how mobile 1 will detect its information using a two-path RAKE receiver. The results are shown in Figures 3.13 and 3.14.

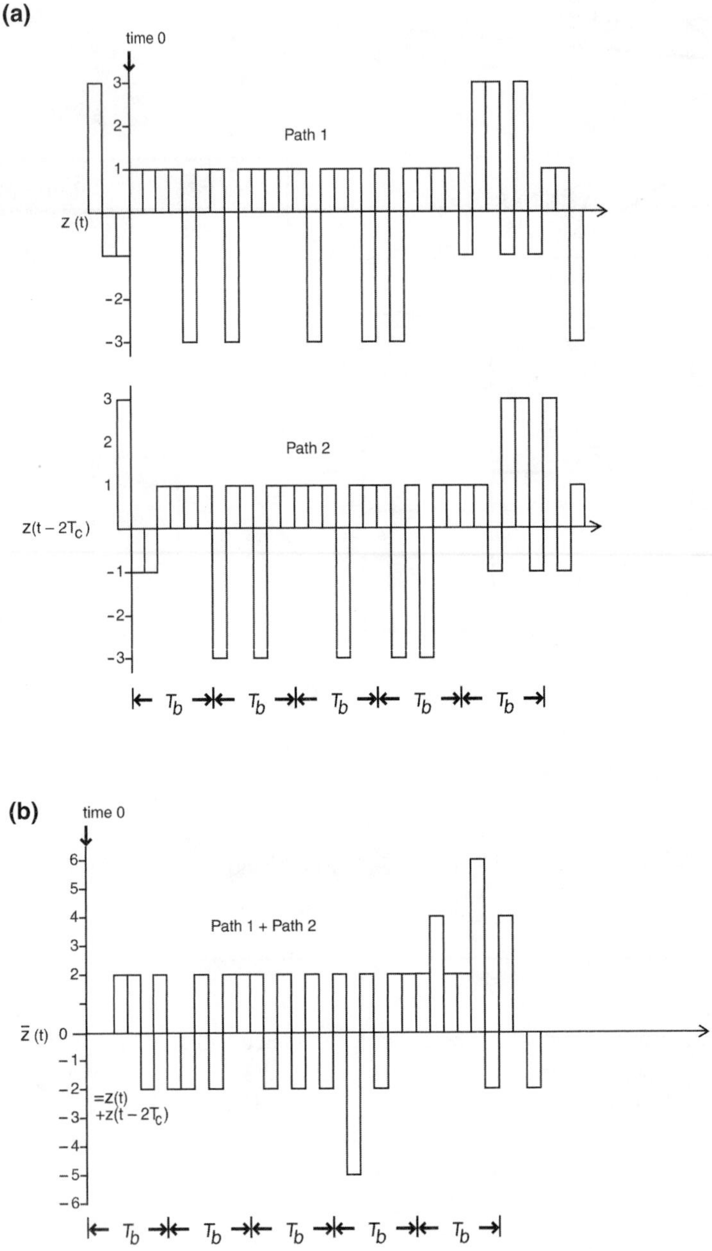

Figure 3.12 Two Equal Strength Paths for Mobile #1

Individual path outputs (Tables 3.6 and 3.7) yield an error in a particular bit position. The RAKE combining strengthens the signal and removes the error, as shown in Table 3.8.

Table 3.6 $\bar{z}(t) \cdot c_1(t)$ (see Figure 3.13)

	T_b	$2T_b$	$3T_b$	$4T_b$	$5T_b$
Value of integration at end of bit period	–4	–4	8	–8	–12
Detected bit value	1	1	0	1	1
Bit value	1	0	0	1	1

Table 3.7 $\bar{z}(t) \cdot c_1(t-2T_c)$ (see Figure 3.14)

	T_b	$2T_b$	$3T_b$	$4T_b$	$5T_b$
Value of integration at end of bit period	–8	8	12	4	–12
Detected bit value	1	0	0	0	1
Bit value	1	0	0	1	1

Table 3.8 Sum of Path 1 and Path 2 Integrator

	T_b	$2T_b$	$3T_b$	$4T_b$	$5T_b$
Path 1: integrator output	–4	–4	8	–8	–12
Path 2: integrator output	–8	8	12	4	–12
Sum of integrator outputs (RAKE receiver output)	–12	4	20	–4	–24
Detected bit value	1	0	0	1	1
Mobile 1 bits	1	0	0	1	1

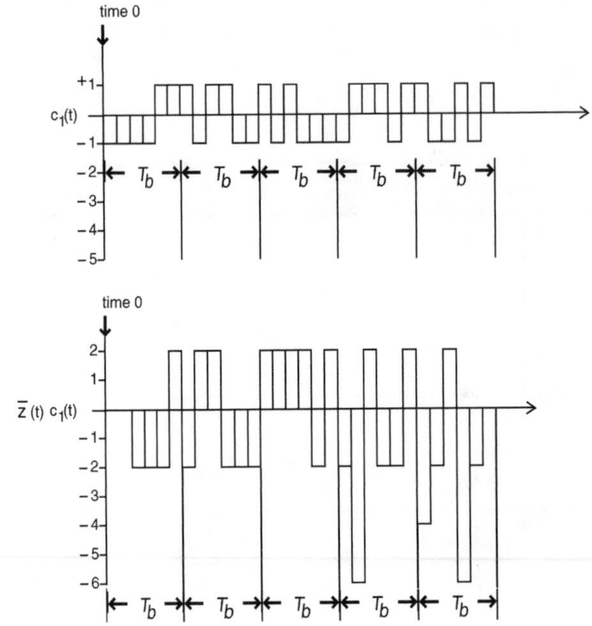

Figure 3.13 Mobile #1 Sequence Locked on Path 1

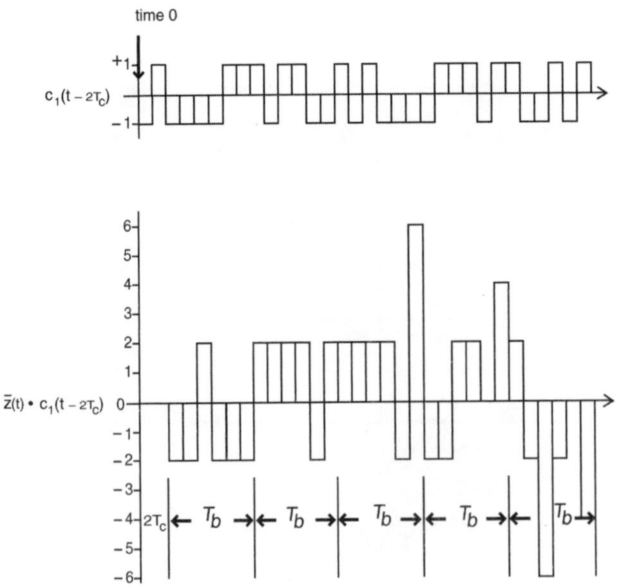

Figure 3.14 Mobile #1 Sequence Locked on Path 2

3.8 Resolution of Multipath

We show a transmitter, channel, and matched-filter receiver in Figure 3.15. The transmitted signal is

$$s(t) = \sqrt{2P} \cdot \cos(\omega t + \phi) \tag{3.31}$$

where
ω = carrier frequency in rad/sec = $2\pi f$

We assume a slowly fading channel having an impulse response function $h_c(t)$ that is given as

$$h_c(t) = \frac{1}{\sqrt{2A}} \sum_{i=1}^{N_p} \alpha_i \cdot \delta(t - \tau_i) \tag{3.32}$$

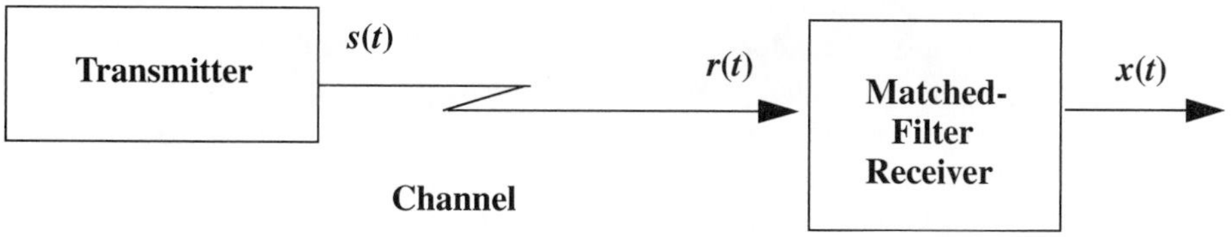

Figure 3.15 Transmitter, Channel, and Matched-Filter Receivers

where
α_i = contribution of the ith multipath component
N_p = significant multipath components
τ_i = delay of the ith multipath component
A = average power attentuation on a multipath component between the transmitter and the receiver

The received signal at the matched filter can be given as

$$r(t) = \sqrt{P_r} \cdot \sum_{i=1}^{N_p} \alpha_i \cdot s(t - \tau_i) \tag{3.33}$$

where
$P_r = P/A$

The receiver is matched to $s(t)$, i.e., the receiver impulse response function $h_r(t)$ will be

$$h_r(t) = s(T-t) \tag{3.34}$$

The output of the matched-filter receiver, denoted by $x(t)$, is the convolution of $r(t)$ with $h_r(t)$ yielding

$$x(t) = \sqrt{P_r} \cdot \sum_{i=1}^{N_p} \alpha_i \cdot \int_{-\infty}^{\infty} s(\xi - \tau_i)s(\xi + T - \tau_i)d\xi \tag{3.35}$$

But $\int_{-\infty}^{\infty} s(\xi)s(\xi + \tau)d\xi = R_x(\tau)$

where $R_x(\tau)$ is the autocorrelation function of the signal $s(t)$.

Thus from Equation (3.35), we will have

$$x(t) = \sqrt{P_r} \cdot \sum_{i=1}^{N_p} \alpha_i \cdot R_x(t + T - \tau_i) \tag{3.36}$$

Hence $x(t)$, the output of the matched-filter receiver, is a sequence of scaled and time-shifted replicas of the autocorrelation function of the signal to which the filter is matched. With approximately equal power, the scaled replicas are resolvable when the delays τ_i are separated by at least $1/B_w$, which is equal to PN (pseudorandom noise) chip interval. Therefore, the multipath components are separated by chip duration or more. For a chip rate equal to 1.2288 Mcps, the chip duration is $1/1.2288 \times 10^6 = 814$ nsec. Thus, nearly equal strength multipath components separated by more than about 800 ft ($9.84 \times 10^8 \times 814 \times 10^{-9} = 800$ ft) will be resolvable.

The power received on the ith multipath component will be

$$P_r = \left(\frac{P}{A}\right)\alpha_i^2 \equiv P_r\alpha_i^2 \tag{3.37}$$

where

$P_r = P/A$ = average received power on a multipath component

α_i^2 = the power on the ith multipath component relative to the average power of the several multipath components

Total received power P_{rTOT} will be

$$P_{rTOT} = \sum_{i=1}^{N_p} P_{ri} = P_r \cdot \sum_{i=1}^{N_p} \alpha_i^2 \tag{3.38}$$

α_i are ordered according to their magnitude, i.e., $\alpha_1 \geq \alpha_2 \geq \alpha_3$ and so on (see Figure 3.16).

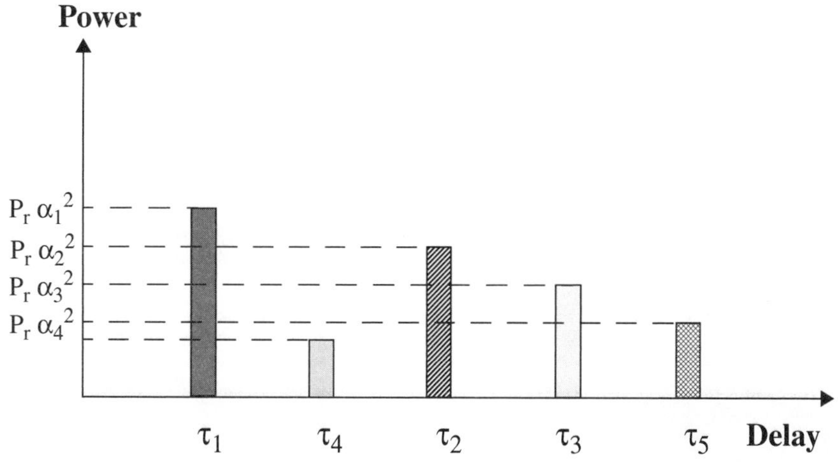

Figure 3.16 Resolution of Multipath

3.9 Frame Quality and Bit Error Rate (BER) Requirements

To establish the bit error rates on the forward and reverse links, we use the widely accepted values of frame error rate (FER) of about 1 percent for a 9.6-kbps vocoder and about 2 percent for a 14.4-kbps vocoder. From the BERs, we determine the required E_b/N_0, where N_0 includes the thermal noise power spectral density plus the interference power spectral density.

Frame quality indicator (FQI) bits, N_Q, are added to each full- and half-rate frame. Adding these FQI bits allows detection of one or more bit errors in the frame with a probability of approximately $(1 - 2^{-N_Q})$. The value of N_Q is 12 for full-rate and 8 for the half-rate frames with a 9.6-kbps vocoder.

The data stream with a 9.6-kbps variable-rate vocoder contains 172 bits (full-rate) and 80 bits (half-rate). The data stream with a 14.4-kbps variable-rate vocoder has 267 bits (full-rate) and 125 bits (half-rate). The FER on the reverse link is monitored and kept at about 1 percent or lower when the 9.6-kbps variable-rate vocoder is used.

$$P_R\{FQIError\} = P_R(1 \text{ or more bit errors in a frame}) = \text{FER} \quad (3.39)$$

$$= 1 - P_R(\text{no bit errors in a frame}) \quad (3.40)$$

$$= 1 - [1 - P_b]^{N_f} \approx N_f \cdot P_b \quad (3.41)$$

where

N_f = number of bits per frame

P_b = the probability of a bit error

Table 3.9 provides the bit error probability for FER 3, 2, 1, 0.5, and 0.25 percent with N_f equal to 172 and 267, respectively.

Table 3.9 Bit Error Probability to Achieve FERs

FER (%) P_R	Probability of Bit Error (P_b)	
	Full-Rate 9.6-kbps Vocoder	Full-Rate 14.4-kbps Vocoder
3	0.000174	0.0001124
2	0.000116	0.0000749
1	0.000058	0.0000375
0.5	0.000029	0.0000187
0.25	0.000015	0.0000094

We can make the following observations from Table 3.9:

- With FER between 1 percent and 2 percent for the 14.4-kbps vocoder, the required bit error probability is in between 3.8×10^{-5} and 7.5×10^{-5} or mean value of 5.7×10^{-5}.
- A 9.6-kbps vocoder with the same FER levels has a bit error rate between 5.8×10^{-5} and 1.2×10^{-4} or a mean of 8.9×10^{-5}.

In practice, the FER for Rate Set 1 (RS1) may be 1 percent or less and the FER for RS2 may be 2 percent or less. In these cases, the P_b for RS2 and RS1 would be about 5.7×10^{-5} and 8.9×10^{-5}, respectively.

3.9.1 E_b/I_t Requirements

Using the results of the previous section, we determine the required values of E_b / I_t for the forward and reverse links of a CDMA IS-95 system.

Forward-Link Requirements. The performance of the coherent BPSK with rate $r = 1/2$, $K = 9$ convolutional decoder can be upper-bounded for L equal mean-power Rayleigh fading paths. We use the procedures given in Viterbi's CDMA book [2] to relate P_b and required E_b/I_t. The results are shown in Figure 3.17.

Reverse-Link Requirements. The results for the reverse-link with 64-ary orthogonal noncoherent demodulation are given in Figure 3.18 for $r = 1/3$ used with the 8-kbps vocoder; and in Figure 3.19 for $r = 1/2$ used with the 13-kbps vocoder. In both cases, the results are given for $K = 9$, $L = 1$, and $L = 2$ equal mean-power Rayleigh fading paths.

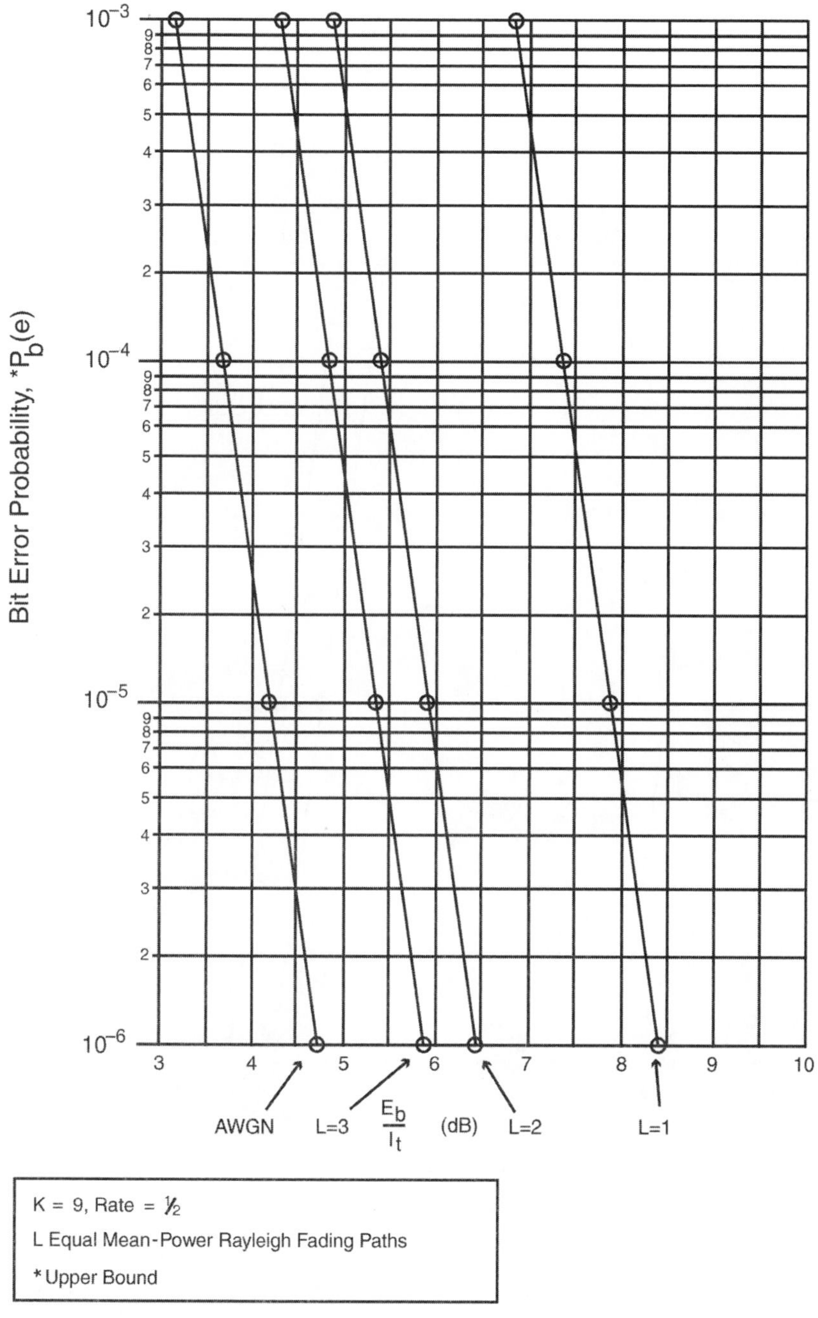

Figure 3.17 E_b/I_t Requirements for Forward Link, $r = 1/2$, $K = 9$

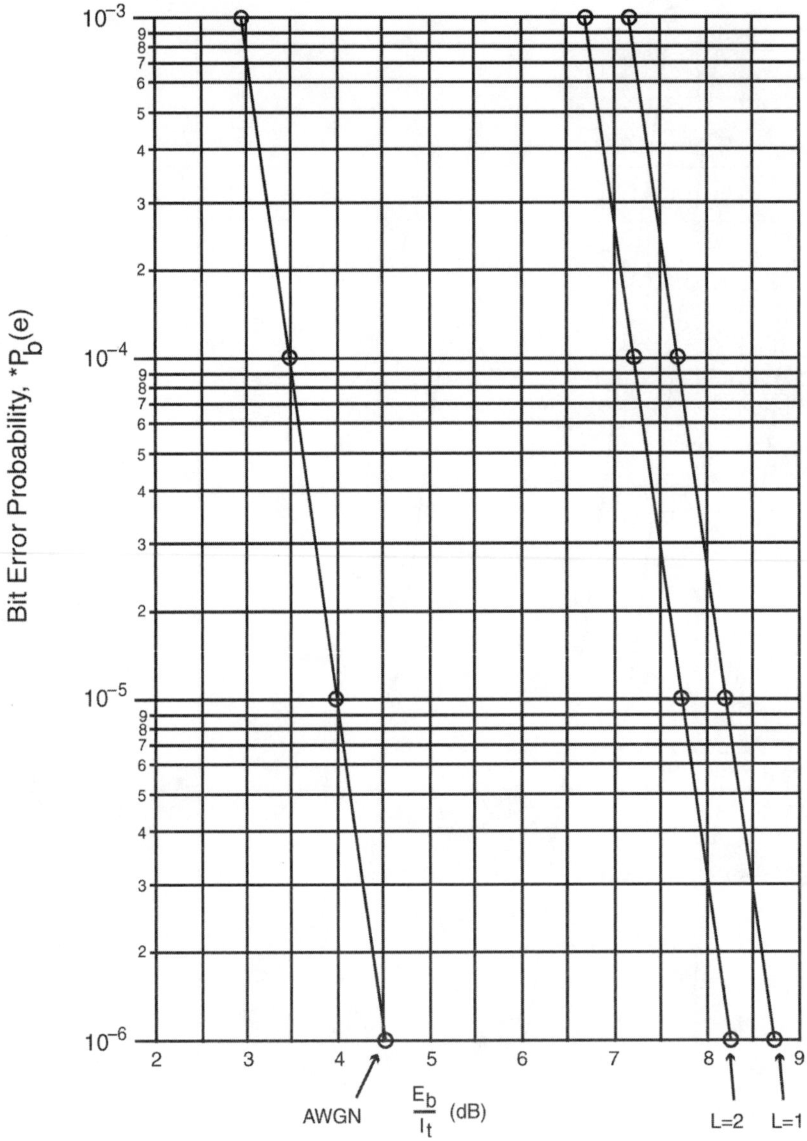

K = 9, Rate = 1/3

L Equal Mean-Power Rayleigh Fading Paths

* Upper Bound

Figure 3.18 E_b/I_t Requirements for Reverse Link, $r = 1/3$, $K = 9$ (8-kbps Vocoder)

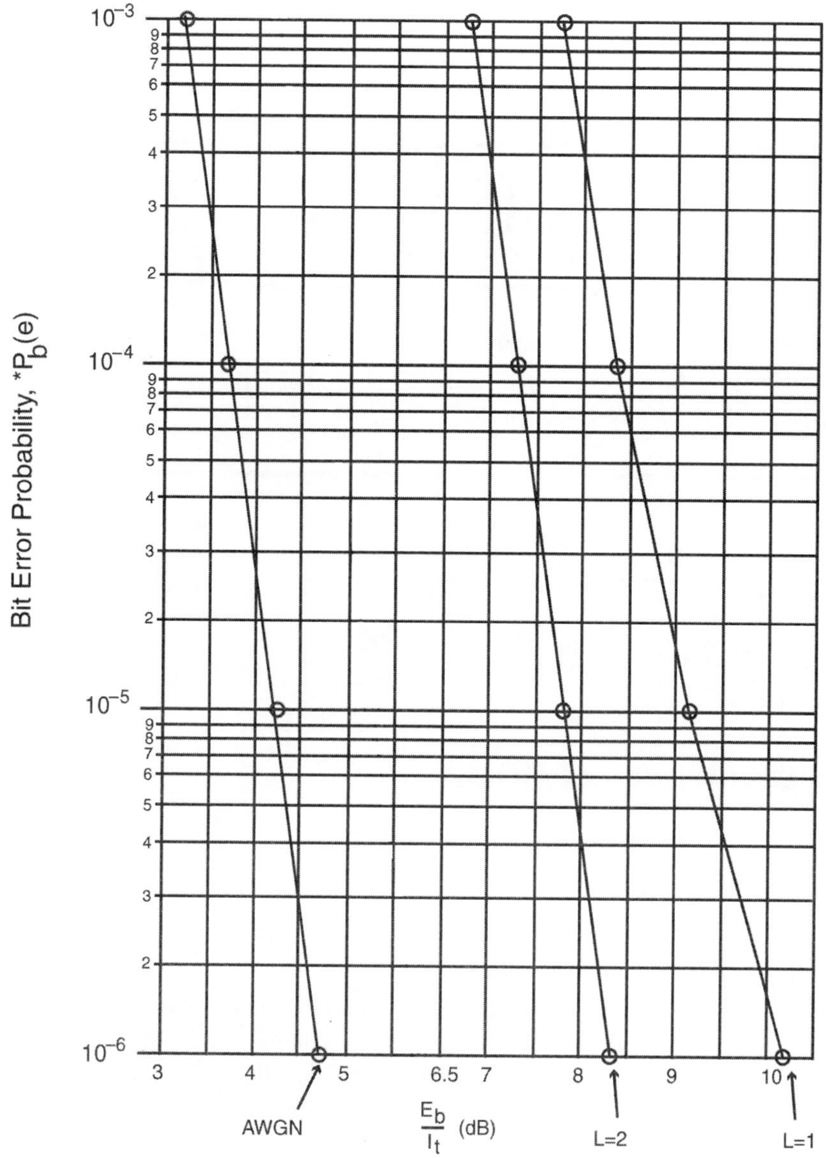

K = 9, Rate = ½
L Equal Mean-Power Rayleigh Fading Paths
* Upper Bound

Figure 3.19 E_b/I_t Requirements for Forward Link, $r = 1/2$, $K = 9$ (13-kbps Vocoder)

Next we estimate the required E_b/I_t values to achieve the required FER (refer to Table 3.10).

Table 3.10 E_b/I_t Required for Forward and Reverse Link

FER (%) P_R	P_b	Required E_b/I_t				
		Forward BPSK Link (r = 1/2, K = 9)			Reverse 64-ary Link (r = 1/3, K = 9)	
		$L = 3$	$L = 2$	$L = 1$	$L = 2$	$L = 1$
3	0.000174	4.40	4.90	6.40	7.10	7.40
2	0.000116	4.50	5.00	6.60	7.20	7.50
1	0.000058	4.70	5.20	6.80	7.25	7.70
0.5	0.000029	4.90	5.40	7.10	7.45	7.90
Required E_b / I_t at 2% FER		4.50	5.00	6.60	7.20	7.50

We conclude from the results in Table 3.10 that for the forward-link operation at 1 percent FER with RS1 [BPSK, $K = 9$, $r = 1/2$, and $L = 2$], we need $E_b / I_t = 5.2$ dB; whereas for the reverse-link operation at 1 percent FER with RS1 [64-ary, $K = 9$, $r = 1/3$, and $L = 2$], we need $E_b/I_t = 7.25$ dB.

3.10 Critical Challenges of CDMA

CDMA is more complex than other multiple-access technologies and as such poses several critical challenges:

- All users in a given cell transmit at the same time in the same frequency band. Can they be made not to interfere with each other?
- Will a user who is near the base station saturate the base station altogether so that it cannot receive users who are farther away (near-far problem)?
- CDMA uses a reuse factor of one. This means that the same frequencies are used in adjacent cells. Can the codes provide sufficient separation for this to work well in most real-life situations?
- CDMA uses soft handoffs where a moving user can receive and combine signals from two or more base stations at the same time. What is the impact on base station traffic handling ability?

3.11 TIA IS-95 CDMA System

Qualcomm proposed the IS-95 CDMA radio system for digital cellular phone applications. It was optimized under existing U.S. mobile cellular system constraints. The CDMA system uses the same frequency in all cells and all sectors. This system design has been standardized by the TIA and many equipment vendors sell CDMA equipment that meet the standard.

The IS-95 CDMA system operates in the same frequency band as the advanced mobile phone system (AMPS) using frequency-division duplex (FDD) with 25 MHz in each direction.[1] The uplink (mobile to base station) and downlink (base station to mobile) bands use frequencies from 869 to 894 MHz and from 824 to 849 MHz, respectively. The mobile station supports CDMA operations on AMPS channel numbers 1013 through 1023, 1 through 311, 356 through 644, 689 through 694, and 739 through 777 inclusive. The CDMA channels are defined in terms of an RF frequency and a code sequence. Sixty-four Walsh functions are used to identify the forward channels, whereas unique long PN code offsets are used for the identification of the reverse channels. The modulation and coding features of the CDMA system are listed in Table 3.11.

Table 3.11 Modulation and Coding Features of the IS-95 CDMA System

Modulation	Quadrature phase-shift keying (QPSK)
Chip rate	1.2288 Mcps
Nominal data rate	9,600 bps, full-rate with RS1
Filtered bandwidth	1.25 MHz
Coding	Convolutional with Viterbi decoding
Interleaving	With 20-msec span

Modulation and coding details for the forward and reverse channels differ. Pilot signals are transmitted by each cell to assist the mobile radio in acquiring and tracking the cell site downlink signals. The strong coding helps these radios to operate effectively at an E_b/I_t ratio in the 5 to 7 dB range.

The CDMA system uses power control and voice activation to minimize mutual interference. Voice activation is provided by using a variable-rate vocoder which for RS1 codec operates at a maximum rate of 8 kbps to a minimum rate of 1 kbps, depending on the level of voice activity. With the decreased data rate, the power control circuits reduce the trans-

1. The frequency spectrum for the A-System cellular service provider is split such that the spectrum is not divisible by 1.25 MHz. Thus, the A-System cellular provider cannot partition the spectrum into ten 1.25 CDMA channels. This restriction is not imposed for the B-System, however.

mitter power to achieve the same bit error rate. A precise power control, along with voice activation circuits are critical to avoid excessive transmitter signal power that is responsible for contributing to the overall interference in the system. The RS2 coding algorithms at 13 kbps are also supported.

A time interleaver with a 20-msec span is used with error-control coding to overcome rapid multipath fading and shadowing. The time span used is the same as the time frame of the voice compression algorithm.

A RAKE receiver is used in the CDMA radio to take advantage of a multipath delay greater than 1 μs, which is common in cellular/PCS networks in urban and suburban areas.

3.11.1 Forward Link (Downlink)

In this section, we summarize the operation of the downlink. The downlink channels include one pilot channel, one synchronization (synch) channel, and 62 other channels including up to seven paging channels. (If multiple carriers are implemented, paging channels and synch channels do not need to be duplicated.) The information on each channel is modulated by appropriate Walsh function and then modulated by a quadrature pair of PN sequences at a fixed chip rate of 1.2288 Mcps. The pilot channel is always assigned to code channel number zero. If the synch channel is present, it is given the code channel number 32. Whenever paging channels are present, they are assigned the code channel numbers 1 through 7 (inclusive) in sequence. The remaining code channels are used by forward traffic channels.

The synch channel operates at fixed data rate of 1,200 bps and is convolutionally encoded to 2,400 bps, repeated to 4,800 bps, and interleaved.

The forward traffic channels are grouped into sets. RS1 has four rates: 9,600, 4,800, 2,400, and 1,200 bps. RS2 contains four rates: 14,400, 7,200, 3,600, and 1,800 bps. All radio systems support RS1 on the forward traffic channels. RS2 is optionally supported on the forward traffic channels. When a radio system supports a rate set, all four rates of the set are supported.

The speech is encoded using a variable-rate vocoder to generate forward traffic channel data depending on voice activity. Since frame duration is fixed at 20 ms, the number of bits per frame varies according to the traffic rate. Half-rate convolutional encoding is used, which doubles the traffic rate to give rates from 2,400 to 19, 200 bits per second. Interleaving is performed over 20 ms. A long PN code of $2^{42} - 1$ ($= 4.4 \times 10^{12}$) is generated using the user's *electronic serial number (ESN)* embedded in the mobile station *long-code mask* (with voice privacy, the mobile station long-code mask does not use the ESN). The scrambled data is multiplexed with power control information which steals bits from the scrambled data. The multiplexed signal remains at 19,200 bps and is modulated at 1.2288 Mcps by the Walsh code W_i assigned to the ith user traffic channel. The signal is spread at 1.2288 Mcps by quadrature pseudorandom binary sequence signals, and the resulting quadrature signals are then weighted. The power level of the traffic channel depends on its data transmission rate.

The paging channels provide the mobile stations with system information and instructions, in addition to acknowledging messages following access requests on the mobile stations' access channels. The paging channel data is processed in a similar manner to the traffic channel data. However, there is no variation in the power level on a per frame basis. The 42-bit mask is used to generate the long code. The paging channel operates at a data rate of 9,600 or 4,800 bps.

All 64 channels are combined to give single I and Q channels. The signals are applied to quadrature modulators and resulting signals are summed to form a QPSK signal, which is linearly amplified.

The pilot CDMA signal transmitted by a base station provides a reference for all mobile stations. It is used in the demodulation process. The pilot signal level for all base stations is much higher (about 4 to 6 dB) than traffic channels with a constant value. The pilot signals are quadrature pseudorandom binary sequence signals with a period of 32,768 chips. Since the chip rate is 1.2288 Mcps, the pilot pseudorandom binary sequence corresponds to a period of 26.66 ms, which is equivalent to 75 pilot channel code repetitions every two seconds. The pilot signals from all base stations use the same pseudorandom binary sequence, but each base station is identified by a unique time offset of its pseudorandom binary sequence. These offsets are in increments of 64 chips, providing 512 unique offset codes. These large numbers of offsets ensure that unique base station identification can be obtained, even in dense microcellular environments.

A mobile station processes the pilot channel to find the strongest multipath signal components. The processed pilot signal provides an accurate estimation of time delay, phase, and magnitude of the multipath components. These components are tracked in the presence of fast fading, and coherent reception with combining is used. The chip rate on the pilot channel and on all frequency carriers is locked to precise system time, e.g., by using the global positioning system (GPS). Once the mobile station identifies the strongest pilot offset by processing the multipath components from the pilot channel correlator, it examines the signal on its synch channel, which is locked to the pseudorandom binary sequence signal on the pilot channel. Since the synch channel is time aligned with its base station's pilot channel, the mobile station finds the information pertinent to this particular base station. The synch channel message contains time-of-day and long-code synchronization to ensure that long-code generators at the base station and mobile station are aligned and identical. The mobile station now attempts to access the paging channel and listens for system information. The mobile station enters the idle state when it has completed acquisition and synchronization. It listens to the assigned paging channel and is able to receive and initiate calls.

3.11.2 Reverse Link (Uplink)

In this section, we summarize the operation of the uplink. The uplink channel is separated from the downlink channel by 45 MHz at cellular frequencies and 80 MHz at PCS frequencies. The uplink uses the same 32,768 chip code as is used on the downlink. The uplink channels are either access channels or reverse traffic channels. The access channel enables the mobile station to communicate nontraffic information, such as originating calls and responding to paging. The access rate is fixed at 4,800 bps. All mobile stations accessing a radio system share the same frequency assignment. Each access channel is identified by distinct access channel long-code sequence having an access number, a paging channel number associated with the access channel, and other system data. Each mobile station uses a different PN code; therefore the radio system can correctly decode the information from an individual mobile station. Data transmitted on the reverse channel is grouped into 20-ms frames. All data on the reverse channel is convolutionally encoded, block interleaved, and modulated by Walsh symbols transmitted for each six-bit symbol block. The symbols are from the set of the 64 mutually orthogonal waveforms.

The reverse traffic channel for RS1 may use either 9,600, 4,800, 2,400, or 1,200 bps data rates for transmission. The duty cycle for transmission varies proportionally with data rate, being 100 percent at 9,600 bps to 12.5 percent at 1,200 bps. An optional second rate set is also supported in the PCS version of CDMA and new versions of cellular CDMA. The actual burst transmission rate is fixed at 28,800 code symbols per second. Since six-code symbols are modulated as one of 64 modulation symbols for transmission, the modulation symbol transmission rate is fixed at 4,800 modulation symbols per second. This results in a fixed Walsh chip rate of 307.2 kcps. The rate of spreading PN sequence is fixed at 1.2288 Mcps, so that each Walsh chip is spread by four PN chips. Table 3.12 provides the signal rates and their relationship for the various transmission rates on the reverse traffic channel.

Following the orthogonal spreading, the reverse traffic channel and access channel are spread in quadrature. Zero-offset I and Q pilot PN sequences are used for spreading. These sequences are periodic with 2^{15} (32,768 PN chips in length) chips and are based on characteristic polynomials $g_I(x)$ and $g_Q(x)$ [see Equations (3.42) and (3.43)].

$$g_I(x) = x^{15} + x^{13} + x^9 + x^8 + x^7 + x^5 + 1 \tag{3.42}$$

$$g_Q(x) = x^{15} + x^{12} + x^{11} + x^{10} + x^6 + x^5 + x^4 + x^3 + 1 \tag{3.43}$$

The maximum-length linear feedback register sequences $I(n)$ and $Q(n)$, based on these polynomials, have a period $2^{15} - 1$ and are generated by using the following recursions:

$$I(n) = I(n-15) \oplus I(n-8) \oplus I(n-7) \oplus I(n-6) \oplus I(n-2) \tag{3.44}$$

based on $g_I(x)$ as the characteristic polynomial, and

$$\begin{aligned} q(n) = q(n-15) \oplus q(n-12) \oplus q(n-11) \oplus \\ q(n-10) \oplus q(n-9) \oplus q(n-5) \oplus \\ q(n-4) \oplus q(n-3) \end{aligned} \tag{3.45}$$

based on $q_Q(x)$ as the characteristic polynomial, where $I(n)$ and $q(n)$ are binary numbers (0 and 1) and the additions are modulo-2. To obtain the I and Q pilot sequences, a 0 is inserted in $I(n)$ and $q(n)$ after 14 consecutive 0 outputs (this occurs only once in each period). Therefore, the pilot PN sequences have one run of 15 consecutive 0 outputs instead of 14.

The chip rate for the pilot PN sequence is 1.2288 Mcps and its period is 26.666 ms. There are exactly 75 repetitions in every two seconds. The spreading modulation is offset quadrature phase-shift keying (OQPSK). The data spread by Q pilot PN sequence is delayed by half a chip time (406.901 ns) with respect to the data spread by I pilot PN sequence. Figure 3.20 and Table 3.13 describe the characteristics of OQPSK.

Table 3.12 CDMA Reverse Traffic Channel Modulation Parameters (Rate Set 1)

Parameter	9,600 bps	4,800 bps	2,400 bps	1,200 bps	units
PN chip rate	1.2288	1.2288	1.2288	1.2288	Mcps
Code rate	1/3	1/3	1/3	1/3	bits/code sym
Transmitting duty cycle	100	50	25	12.5	%
Code symbol rate	3 × 9600 = 28800	28,800	28,800	28,800	sps
Modulation	6	6	6	6	code symbol/ mod symbol
Modulation symbol rate	28,800/6 = 4,800	4,800	4,800	4,800	sps
Walsh chip rate	64 × 4800 = 307.2	307.2	307.2	307.2	kcps
Mod symbol duration	1/4800 = 208.33	208.33	208.33	208.33	μs
PN chips/code symbol	12288/288 = 42.67	42.67	42.67	42.67	PN chip/ code symbol
PN chips/mod symbol	1228800/4800 = 256	256	256	256	PN chip/ mod symbol
PN chips/Walsh chip	4	4	4	4	PN chips/ Walsh chip

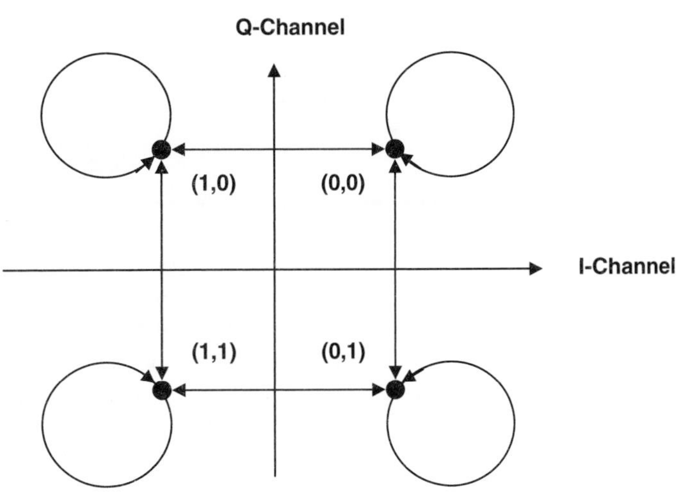

Figure 3.20 Signal Constellation and Phase Transition of Offset QPSK Use on Reverse CDMA Channel

Table 3.13 Reverse CDMA Channel I and Q Mapping

I	Q	Phase
0	0	$\pi/4$
1	0	$(3\pi)/4$
1	1	$-(3\pi)/4$
0	1	$-\pi/4$

Table 3.14 defines the signal rates and their relationship on the access channel.

Table 3.14 CDMA Access Channel Modulation Parameters

Parameter	4,800 bps	Units
PN chip rate	1.2288	Mcps
Code rate	1/3	bits/code symbol
Code symbol repetition	2	symbols/code symbol
Transmit duty cycle	100	%
Code symbol rate	28,800	sps
Modulation	6	code symbol/mod symbol
Modulation symbol rate	4,800	sps
Walsh chip rate	307.2	kcps
Mod symbol duration	208.33	μs
PN chips/code symbol	42.67	PN chip/code symbol
PN chips/mod symbol	256	PN chip/mod symbol
PN chips/Walsh chip	4	PN chips/Walsh chip

Each base station transmits a pilot signal of constant power on the same frequency. The received power level of the received pilot signal enables the mobile station to adjust its transmitted power such that the base station will receive the signal at the requisite power level. The base station measures the mobile station received power and informs the mobile station to make the necessary adjustment to its transmitter power. One command every 1.25

ms adjusts the transmitted power from the mobile station in steps of ±0.5 dB. The base station uses frame errors reported by the mobile station to increase or decrease its transmitted power.

In summary, a CDMA system operates with a low E_b/I_t ratio, exploits voice activity, and uses the same frequency in all sectors of all cells. Each sector has up to 64 CDMA channels on each carrier frequency. It is a synchronized system with RAKE receivers to provide path diversity at the mobile station and at the cell site.

More details about the forward and reverse links of the IS-95 CDMA system are given in Chapter 5.

3.12 Summary

In this chapter we first discussed the concept of spread-spectrum systems and provided the main features of the direct-sequence spread-spectrum system used in the IS-95 and IS-665 systems. A key component of spread-spectrum performance is the calculation of processing gain of the system, which is the relationship between the input and output signal-to-noise ratio of a spread-spectrum receiver.

Spread-spectrum systems trade bandwidth for processing gain and code division systems use a variety of orthogonal or almost orthogonal codes to allow multiple users in the same bandwidth. Thus, CDMA systems can have higher capacity than either analog or time division multiple access (TDMA) digital systems. However, because of practical constraints on CDMA systems, it is not possible to achieve the Shannon bound in system design. The upper bound of the capacity of a CDMA system is limited by the processing gain of the system, receiver modulation performance, power control accuracy, interference from other cells, voice activity, cell sectorization, and the ability to maintain accurate synchronization of the system. Practical CDMA systems are designed for a value of $E_b/I_t \approx 7$ dB.

We concluded the chapter by discussing the high-level features of the IS-95 system and listing the challenges in implementing CDMA system.

3.13 References

1. Shannon, C. E., "Communications in the Presence of Noise," *Proceedings of the IRE*, 1949, No. 37, pp. 10–21.
2. Viterbi, A. J., *CDMA*, Addison-Wesley Publishing Company: Reading, MA, 1995.
3. Bhargava, V., Haccoum, D., Matyas, R., and Nuspl, P., *Digital Communications by Satellite*, John Wiley & Sons: New York, 1981.
4. Dixon, R. C., *Spread Spectrum Systems*, 2nd ed., John Wiley & Sons: New York, 1984.
5. Feher, K., *Wireless Digital Communications Modulation and Spread Spectrum Applications*, Prentice Hall PTR: Upper Saddle River, NJ, 1995.

6. Garg, V. K., and Wilkes, J. E., *Wireless and Personal Communications Systems*, Prentice Hall: Upper Saddle River, NJ, 1996.
7. Lee, W. C. Y., *Mobile Cellular Telecommunication Systems*, McGraw-Hill: New York, 1989.
8. Pahlwan, K., and Levesque, A.H., *Wireless Information Networks,* John Wiley & Sons: New York, 1995.
9. Steele, R., *Mobile Radio Communications,* IEEE Press: New York, 1992.
10. Skalar, B., *Digital Communications: Fundamentals and Applications,* Prentice Hall: Englewood Cliffs, NJ, 1988.
11. TIA/EIA IS-95 "Mobile Station-Base Station Compatibility Standard for Dual-Mode Wideband Spread Spectrum Cellular System," PN-3422, 1994.
12. Torrien, D., *Principle of Secure Communication Systems,* Artech House: Boston, 1992.
13. Viterbi, A. J., and Padovani Roberto, "Implications of Mobile Cellular CDMA,"*IEEE Communication Magazine*, 1992, [Vol. 30(12), pp. 38–41].

3.14 Problems

1. Determine $(S/N)_o$ for a spread-spectrum system having bandwidth $B_w = 1.2288$ MHz, information rate $R = 9.6$ kbps, and $(S/N)_i = 15$ dB. Relate your result to E_b/N_0.
2. Use the data in Example 3.5 and show that the bit pattern at mobiles 2 and 3 can be determined from the demodulated signal using path 1 and path 2.
3. Consider a case where eight chips per bit are used to generate the Walsh codes. Mobile stations A, B, C, and D are assigned W_1, W_2, W_3, and W_4, respectively. The stations use the Walsh code to send a "1" binary bit and use negative Walsh to send a "0" binary bit. Assume all stations are synchronized in time; therefore chip sequences begin at the same instant. When two or more stations transmit simultaneously, their bipolar signals add linearly. Consider the following four different cases when one or more station transmit and show that the receiver recovers the bit stream of station B and C. (Note that the dash (-) means no transmission by the station.)

Station (A, B, C, D)	Transmitting Stations
- - 1 0	C + D
1 1 1 -	A + B + C
1 1 - -	A + B
1 1 0 0	A + B + C +D

CHAPTER 4

Speech and Channel Coding, Spreading Codes, and Modulation

4.1 Introduction

In this chapter, we first discuss speech coding methods and attributes of a speech codec. These are then followed by a brief discussion of linear-prediction-based analysis-by-synthesis (LPAS). We also focus on the QCELP and EVRC codecs. We then focus on channel coding, concentrating on three channel coding schemes: convolutional code, Reed-Solomon (R-S) code, and turbocode. Convolutional codes have been used in direct-sequence (DS) CDMA (IS-95) and R-S and turbocodes are the proposed channel coding schemes for cdma2000 and W-CDMA. We then present an overview of the spreading techniques that are used in DS CDMA systems. We briefly review the background for the sequences used in CDMA and wideband CDMA, and discuss important characteristics of maximal-length, Gold, and Kasami sequences, as well as variable- and fixed-length orthogonal codes. Finally, we present various modulation schemes used in DS CDMA and W-CDMA.

4.2 Speech Coding

The speech quality of codecs operating at a fixed bit rate is largely determined by the worst-case speech segments, i.e., those that are the most difficult to code at the given rate. Variable-rate coding can provide a given level of speech quality at an average bit rate that is substantially less than the bit rate that would be required by an equivalent quality fixed-rate codec.

The original CDMA codec known as QCELP, IS-96A [1] was an 8-kbps code-excited linear-prediction (CELP) variable-rate codec designed for use in the 900-MHz digital cellular band. The desire for improved voice quality spurred the TIA to begin working on a 13-kbps CELP variable-rate codec to provide higher quality voice transmissions. The 13-kbps CDMA codec takes advantage of the higher data rate (14.4 kbps as compared to 9.6 kbps for the 8-kbps codec) to improve speech quality. It has produced mean opinion scores (MOS) close to toll quality voice, the benchmark for comparison. Unfortunately, system capacity and cell site range are reduced by the higher data rate codec.

4.2.1 Speech Coding Methods

Speech coding is the process of reducing the bit rate of digital speech representation for transmission or storage, while maintaining a speech quality that is acceptable for the application. Speech coding methods can be classified as:

- Waveform coding
- Source coding
- Hybrid coding

In Figure 4.1, the bit rate is plotted on a logarithmic axis versus speech quality classes of poor to excellent corresponding to the five-point mean opinion score (MOS) scale values of 2 to 5, defined by the International Telecommunications Union (ITU). It may be noted, that for low complexity and low delay a bit rate of 32 to 64 kbps is required. This suggests the use of waveform codecs. However, for low bit rate between 4 and 8 kbps, hybrid codecs could be used. These types of codecs tend to be complex with high delay.

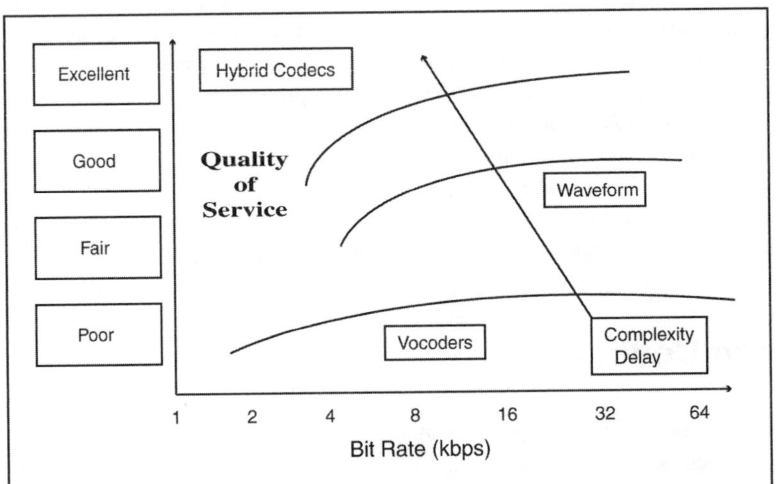

Figure 4.1 Bit Rate versus Quality of Service

4.2.2 Speech Codec Attributes

Speech quality as produced by a codec is a function of transmission bit rate, complexity, delay, and bandwidth. Therefore, when considering speech codecs it is essential to consider all of these attributes and their interactions. For example, low bit rate codecs, tend to have more delay as compared to the higher bit rate codecs. They are generally more complex to implement and often have lower speech quality than the higher bit rate codecs.

Transmission Bit Rate. Since the speech codec shares the communications channel with other data, the peak bit rate should be as low as possible so as not to use a disproportionate percentage of the channel. Codecs below 64 kbps are primarily developed to increase the capacity of circuit multiplication equipment used for narrow bandwidth links. For the most part, they are fixed bit rate codecs, meaning they operate at the same rate regardless of the input. In the variable bit rate codecs, network loading and voice activity determine the instantaneous rate assigned to a particular voice channel. Any of the fixed-rate speech codecs can be combined with a voice activity detector (VAD) and made into a simple two-state variable bit rate system. The lower rate could be either zero or some low rate needed to characterize slowly changing background noise characteristics. Either way, the bandwidth of the communications channel is used only for active speech.

Delay. The delay of a speech codec can have a great impact on its suitability for a particular application. For one-way delay of conversation greater than 300 ms, the conversation becomes more like a half-duplex or push-to-talk experience, rather than an ordinary conversation. The components of total system delay include frame size, look-ahead, multiplexing delay, processing delay for computations, and transmission delay.

Most low bit rate speech codecs process one frame of speech data at a time. The speech parameters are updated and transmitted for every frame. In addition, to analyze the data properly it is sometimes necessary to analyze data beyond the frame boundary. Hence, before the speech can be analyzed it is necessary to buffer a frame worth of data. The resulting delay is referred to as *algorithmic delay.* This delay component cannot be reduced by changing the implementation. All other delay components can be reduced by changing implementation. The second major contributor of delay comes from the time taken by the encoder to analyze the speech and the decoder to reconstruct the speech. This part of the delay is referred to as *processing delay.* It depends on the speed of the hardware used to implement the coder. The sum of the algorithmic and processing delays is called the *one-way codec delay.* The third component of delay is due to transmission. It is the time taken for entire frame of data to be transmitted from the encoder to the decoder. The total of the three delays is the *one-way system delay.* In addition, there is frame interleaving delay, which adds an additional frame delay to the total transmission delay. Frame interleaving is necessary to combat channel fading, and is part of the channel coding process.

Complexity. Speech codecs are implemented on special-purpose hardware, such as digital signal processor (DSP) chips. Their attributes are described as computing speed in millions of instructions per second (MIPS), random access memory (RAM), and read-only memory (ROM). For a speech codec, the system designer makes a choice about how much of these resources are to be allocated to the speech codec. Speech codecs using less than 15 MIPS are considered low complexity; those requiring 30 MIPS or more are thought of as high complexity. More complexity results in higher costs and greater power usage; for portable applications, greater power usage means reduced time between battery recharges or using larger batteries, which means more expense and weight.

Quality. Of all the attributes, quality has the most dimensions. In many applications there are large amounts of background noise (car noise, street noise, office noise, etc.). How well does the codec perform under these adverse conditions? What happens when there are channel errors during transmission? Are the errors detected or undetected? If undetected, the codec must perform even more robustly than when it is informed that entire frames are in error. How good does the codec sound when speech is encoded and decoded twice? All these questions must be carefully evaluated during the testing phase of a speech codec. The speech quality is often based on the five-point MOS scale as defined by ITU-T.

4.2.3 Linear-Prediction-Based Analysis-by-Synthesis (LPAS)

Linear-prediction-based analysis-by-synthesis (LPAS) methods [2,3] provide efficient speech coding at rates between 4 and 16 kbps. In LPAS speech codecs, the speech is divided into frame lengths of about 20 ms for which the coefficients of a linear predictor (LP) are computed. The resulting LP filter predicts each sample from a set of previous samples.

In analysis-by-synthesis codecs, the residual signal is quantized on a subframe-by-subframe basis (there are commonly two to eight subframes per frame). The resulting quantized signal forms the excitation signal for the LP synthesis filter. For each subframe, a criterion is used to select the best excitation signal from a set of trial excitation signals. The criterion compares the original speech signal with trial reconstructed speech signals. Because of the synthesis implicit in the evaluation criterion, the method is called analysis-by-synthesis coding. Various representations of excitation have been used [4]. For lower bit rates, the most efficient representation is achieved by using vector quantization. For each subframe, the excitation signal is selected from a multitude of vectors, which are stored in a codebook. The index of the best matching vector is transmitted. At the receiver this vector is retrieved from the same codebook. The resulting excitation signal is filtered through the LP synthesis filter to produce the reconstructed speech. Linear-prediction analysis-by-synthesis codecs using a codebook approach are commonly known as *code-excited linear-prediction* (CELP) codecs [5].

Parametric codecs are traditionally used at low bit rates. A proper understanding of speech signal and its perception is essential to obtain good speech quality with a parametric codec. Parametric coding is used for those aspects of the speech signal which are well understood, while the waveform matching procedure is employed for those aspects which are not well understood. Waveform matching constraints are relaxed for those aspects which can be replaced by parametric models without degrading the quality of the reconstructed speech.

A parameter which is well understood in parametric coding is the pitch of the speech signal. Satisfactory pitch estimation procedures are available [6]. Piecewise linear interpolation of the pitch does not degrade speech quality. Pitch period is typically determined once every 20 ms and linearly interpolated between the updates. The challenge is to generalize the LPAS method such that its matching accuracy becomes independent of the synthetic pitch-period contour used. This is done by determining a time wrap of speech signal such that its pitch-period contour matches the synthetic pitch-period contour. The time wraps are determined by comparing a multitude of time-wrapped original signals with a synthesized signal. This coding scheme is called the generalized analysis-by-synthesis method [7] and is referred to as *relaxed CELP* (RCELP). The generalization relaxes the waveform matching constraints without affecting speech quality.

4.2.4 Waveform Coding

In general, waveform codecs are designed to be independent of signal. They map the input waveform of the encoder into a facsimilelike replica of it at the output of the decoder. Coding efficiency is quite modest. However, the coding efficiency can be improved by exploiting some statistical signal properties, if the codec parameters are optimized for the most likely categories of input signals, while still maintaining good quality for other types of signals. The waveform codecs are further subdivided into: (1) time domain waveform codec and (2) frequency domain waveform codec.

Time Domain Waveform Coding. The well-known representation of speech signal using time domain waveform coding is the A-law (in Europe) or μ -law (in North America) companded pulse code modulation (PCM) at 64 kbps. Both use nonlinear companding characteristics to give near-constant signal-to-noise ratio (SNR) over the total input dynamic range.

The ITU G.721, 32-kbps adaptive differential PCM (ADPCM) codec is an example of a time domain waveform codec. More flexible counterparts of the G.721 are the G.726 and G.727 codecs. The G.726 codec is a variable-rate arrangement for bit rates between 16 and 40 kbps. This may be advantageous in various networking applications to allow speech quality and bit rate to be adjusted on the basis of the instantaneous requirement. The G.727 codec uses core bits and enhancement bits in its bit stream to allow the network to drop the enhancement bits under restricted channel capacity conditions, while benefiting from them when the network is lightly loaded.

In differential codecs, a linear combination of the last few samples is used to generate an estimate of the current one, which occurs in the adaptive predictor. The resultant difference signal (i.e., the prediction residual) is computed and encoded by the adaptive quantizer with a lower number of bits than the original signal, since it has a lower variance than the incoming signal. For a sampling rate of 8,000 samples per second, an 8-bit PCM sample is represented by a 4-bit ADPCM sample to give a transmission rate of 32 kbps.

Time domain waveform codecs encode the speech signal as a full-band signal and map it into as close a replica of the input as possible. The difference between various coding schemes is their way of using prediction to reduce the variance of the signal to be encoded in order to reduce the number of bits necessary to represent the encoded waveform.

Frequency Domain Waveform Coding. In frequency domain waveform codecs, the input signal undergoes short-time spectral analysis. The signal is split into a number of frequency domain subbands. The individual subband signals are then encoded by using different numbers of bits to fulfill the quality requirements of that band based on its prominence. The various schemes differ in their accuracies of spectral analysis and in the bit allocation principle (fixed, adaptive, semiadaptive). Two well-known representatives of this class are subband coding (SBC) and adaptive transform coding (ATC).

4.2.5 Vocoders

Vocoders are parametric digitizers that use certain properties of the human speech production mechanism. Human speech is produced by emitting sound pressure waves which are radiated primarily from lips, although significant energy can also emanate from the nostrils, throat, etc. In human speech, the air compressed by the lungs excites the vocal cords in two typical modes. When generating voice sounds, the vocal cords vibrate and generate quasi-periodic voice sounds. In the case of lower energy unvoiced sounds, the vocal cords do not participate in the voice production and the source acts like a noise generator. The excitation signal is then filtered through the vocal apparatus, which behaves like a spectral-shaping filter. This can be described adequately by an all-pole transfer function that is constituted by the spectral-shaping action of gloat, vocal tract, lip radian characteristics, etc.

In the case of vocoders, instead of producing a close replica of an input signal at the output, an appropriate set of source parameters is generated to characterize the input signal as close as possible for a given period of time. The following steps are used in this process:

1. Speech signal is segmented in quasistationary segments of 5 to 20 ms.
2. Speech segments are subjected to spectral analysis to produce the coefficients of the all-zero analysis filter to minimize the prediction residual energy. This process is based on the computation of the speech autocorrelation coefficients and uses either matrix inversion or iterative scheme.
3. The corresponding source parameters are specified. The excitation parameters as well as filter coefficients are quantized and transmitted to the decoder to synthesize a replica of the original signal by exciting the all-pole synthesis filter.

The quality of this type of scheme is predetermined by the accuracy of the source model rather than the accuracy of the quantization of the parameters. The speech quality is limited by the fidelity of the source model used. The main advantage of vocoders is their low bit rate, with the penalty of relatively low, synthetic speech quality. Vocoders can be classified into the frequency domain and time domain subclasses. However, frequency domain vocoders are generally more effective than time domain vocoders.

4.2.6 Hybrid Coding

Hybrid coding is an attractive trade-off between waveform coding and vocoders, both in terms of speech quality and transmission bit rate, although generally at the price of higher complexity. They are also referred to as analysis-by-synthesis (ABS) codecs.

Most recent international and regional speech coding standards belong to a class of LPAS codecs. This class of codecs includes ITU G723.1, G.728 (low-delay CELP, 16 kbps), and G.729 and all the current digital cellular standards including:

- Europe: Global System for Mobile communications (GSM), full-rate, half-rate, and enhanced full-rate (EFR)
- North America: Full-rate, half-rate, and enhanced full-rate for time division multiple access (TDMA) IS-136 and code division multiple access (CDMA) IS-95 systems
- Japan: Public digital cellular (PDC) full-rate and half-rate

In an LPAS coder, the decoded speech is produced by filtering the signal produced by the excitation generator through both a long-term (LT) predictor synthesis filter and a short-term (ST) predictor synthesis filter. The excitation signal is found by minimizing the mean-squared error over a block of samples. The error signal is the difference between the original and decoded signal. It is weighted by filtering it through a weighting filter. Both short-term and long-term predictors are adapted over time. Since the analysis procedure (encoder) includes the synthesis procedure (decoder), the description of the encoder defines the decoder. The short-term synthesis filter models the short-term correlations (spectral envelope) in the speech signal. This is an all-pole filter. The predictor coefficients are determined from the speech signal using linear-prediction (LP) techniques. The coefficients of the short-term predictor are adapted in time, with rates varying from 30 to as high as 400 times per second.

The long-term predictor filter models the long-term correlations (fine spectral structure) in the speech signal. Its parameters are a delay and gain coefficient. For periodic signal, delay corresponds to the pitch period (or possibly an integral number of pitch periods). The delay is random for nonperiodic signals. Typically, the long-term predictor coefficients are adapted at rates varying from 100 to 200 times per second.

An alternative structure for the pitch filter is the *adaptive codebook*. In this case, the long-term synthesis filter is replaced by a codebook that contains the previous excitation at different delays. The resulting vectors are searched, and the one that provides the best result is selected. In addition, an optimal scaling factor is determined for the selected vector. This representation simplifies the determination of the excitation for delays smaller than the length of excitation frames.

CELP coders use another approach to reduce the number of bits per sample. Both encoder and decoder store the same collection of codes (C) of possible length L in a codebook. The excitation for each frame is described completely by the index to an appropriate vector. This index is found by an exhaustive search over all possible codebook vectors, using the one that gives the smallest error between the original and decoded signals. To simplify the search it is common to use a gain-shape codebook in which the gain is searched and quantized separately. Further simplifications are obtained by populating the codebook vectors with a multipulse structure. By using only a few nonzero unit pulses in each codebook vector, efficient search procedures are derived. This partitioning of excitation space is referred to as *an Algebraic codebook*. The excitation method is known as *algebraic codebook-excited linear-prediction* (ACELP).

4.3 Speech Codecs in European Systems

4.3.1 GSM Enhanced Full-Rate

GSM enhanced full-rate (EFR) (same as a US1 vocoder with 8-PSK modulation) is a 12.2-kbps vocoder. It is used in GSM1900. The distribution of bits in a GSM EFR vocoder is given in Table 4.1.

Table 4.1 GSM Enhanced Full-Rate Vocoder (12.2 kbps)

	Bits per 20 ms
LP filter coefficients	38
Adaptive excitation	30
Fixed or algebraic excitation	$4 \times 35 = 140$
Gains	$4 \times 9 = 36$
Total	244 or 12.2 kbps

IS-136 uses the vector self-excited linear-predictor (VSELP) codec. The VSELP algorithm uses a codebook with a predefined structure to reduce the number of computations. The output of the VSELP codec for IS-136 is 7.95 kbps. It produces a speech frame every 20 ms containing 159 bits. Recently, the ACELP codec (IS-641 vocoder) was selected to replace the VSELP codec in IS-136. This codec has an output bit rate of 7.4 kbps.

ETSI adaptive multirate (AMR) speech coder design (which will be used for 3G UTRA) incorporates multiple submodes for use in full-rate or half-rate mode that are determined by the channel quality. AMR has submodes that incorporate bit-exact versions of both 12.2 kbps US1/GSM EFR and 7.4 kbps IS-641 full-rate speech coders. Tables 4.2 and 4.3 provide details of AMR speech coders.

Table 4.2 ETSI AMR Speech Coder Full-Rate Submodes

Submode	Speech Coder Source Rate (kbps)	Channel Coding Rate (kbps)
1	12.2 (US1/GSM EFR)	10.6
2	10.2	12.6
3	7.95	14.85
4	7.4 (IS-641 ACELP)	15.4

Table 4.2 ETSI AMR Speech Coder Full-Rate Submodes *(cont.)*

Submode	Speech Coder Source Rate (kbps)	Channel Coding Rate (kbps)
5	6.7	16.1
6	5.9	16.9
7	5.15	17.65
8	4.75	18.05

Table 4.3 ETSI AMR Half-Rate Submodes

Submodes	Speech Coder Source Rate (kbps)	Channel Coding Rate (kbps)
1	7.95	3.45
2	7.4 (IS-641 ACELP)	4.0
3	6.7	4.7
4	5.9	5.5
5	5.15	6.25
6	4.75	6.65

Comparisons of MOS and delay for IS-641, ITU LD-CELP, and GSM EFR are given in Tables 4.4 and 4.5.

Table 4.4 Comparison of MOS

Condition	Original	IS-641	ITU LD-CELP	GSM EFR
Clean speech	4.34	4.09	4.23	4.26
15-dB babble	3.75	3.49	3.81	3.70
20-dB car noise	3.72	3.61	3.64	3.75
15-dB office noise	3.70	3.40	3.61	3.58
15-dB music	3.99	3.82	3.98	3.99

Table 4.5 Comparison of Delay (ms)

Delay Cause	IS-641	GSM EFR	ITU LD-CELP
Look-ahead	5	0	0
Frame size	20	20	0.625
Processing	16	16	0.5
Bit stream buffer	0	0	19.375
Transmission	26.6	6.6	6.6
Total delay	67.6	42.6	27.1

Results show an advantage for ITU LD-CELP and GSM EFR when compared to IS-641 for every condition. The disadvantage of the LD-CELP is its higher bit rate. This means fewer bits are available for error protection. In weaker channel areas, GSM EFR will have an advantage over LD-CELP due to its lower bit rate, fewer sensitive bits to protect, and faster recovery from frame erasures. IS-641 also has the same advantages, and on an even weaker channel its performance would surpass that of GSM EFR.

ITU LD-CELP has a distinct advantage as far as total delay is concerned. The advantages of IS-641 over LD-CELP are in complexity and bit rate. LD-CELP is a low-delay coder and can produce better clear channel quality than IS-641 for a variety of conditions. GSM EFR has clear channel quality performance on par with LD-CELP, has a small delay advantage compared to IS-641, and seems like a good candidate as an upgrade to IS-641 for strong channel conditions.

4.4 QCELP Speech Codec

QCELP dynamically selects one of the four data rates every 20 ms, depending on the speech activity. The four rates are 8 kbps (full-rate), 4 kbps (half-rate), 2 kbps (quarter-rate), or approximately 1 kbps (eighth-rate). Typically, active speech is coded at the 8-kbps rate, while silence and background noise are coded at the lower rates. MOS testing has shown that QCELP provides speech quality equivalent to that of 8-kbps VSELP, while maintaining an average data rate under 4 kbps in a typical conversation. Figure 4.2 shows a block diagram of the decoder. The bit allocation for each data rate is given in Tables 4.6, 4.7, 4.8, and 4.9, respectively. A 10th-order LPC filter is used. Its coefficients are encoded using LSP frequencies due to the good quantization, interpolation, and stability properties of LSPs.

Table 4.6 Bit Allocation for QCELP at 8 kbps

LPC	40							
Pitch	10		10		10		10	
Codebook parameter	10	10	10	10	10	10	10	10

Table 4.7 Bit Allocation for QCELP at 4 kbps

LPC	20			
Pitch	10		10	
Codebook parameter	10	10	10	10

Table 4.8 Bit Allocation for QCELP at 2 kbps

LPC	10	
Pitch	10	
Codebook parameter	10	10

Table 4.9 Bit Allocation for QCELP at 1 kbps

LPC	10
Pitch	0
Codebook parameter	6

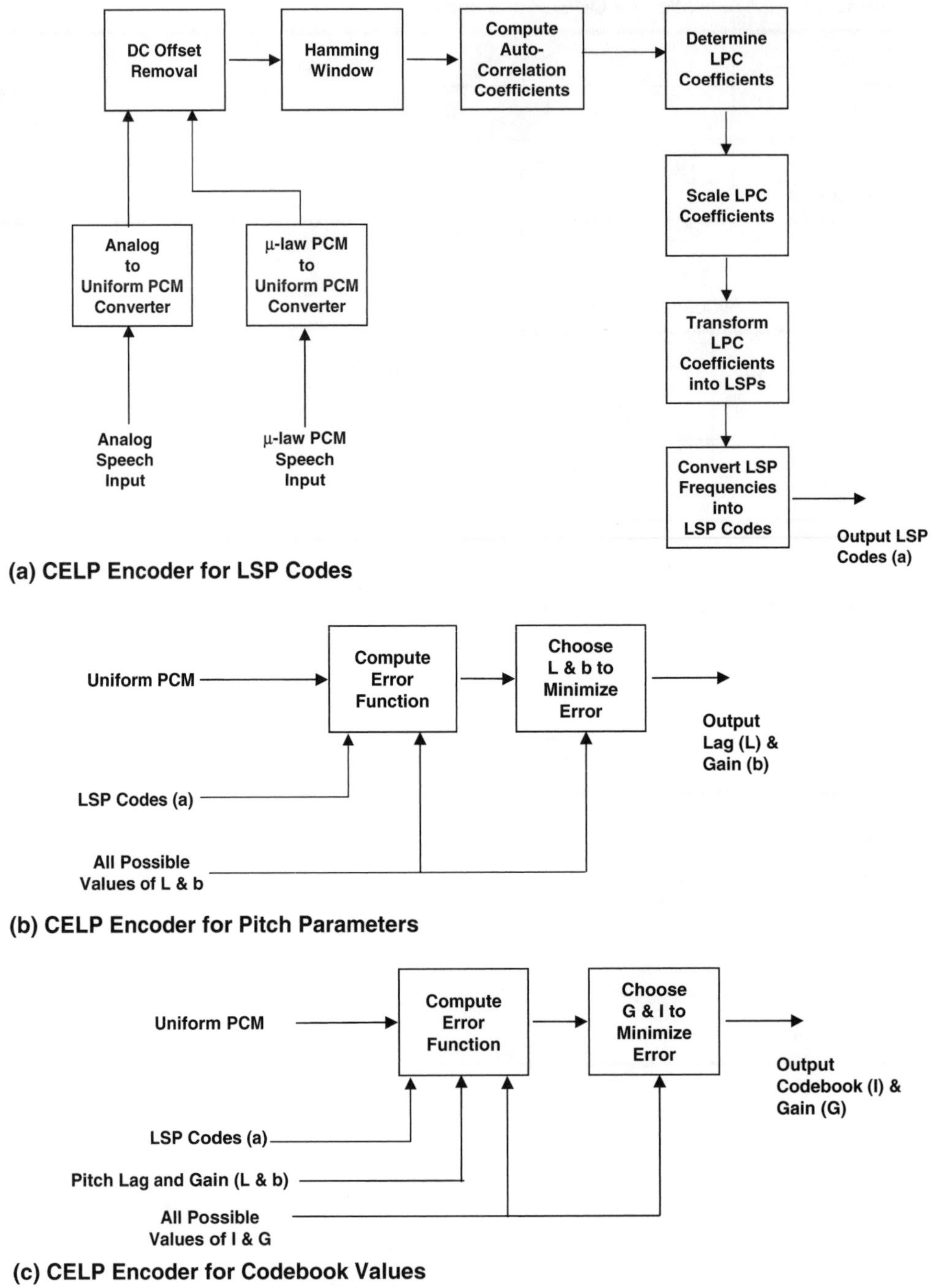

Figure 4.2 Block Diagram for QCELP Encoder and Decoder

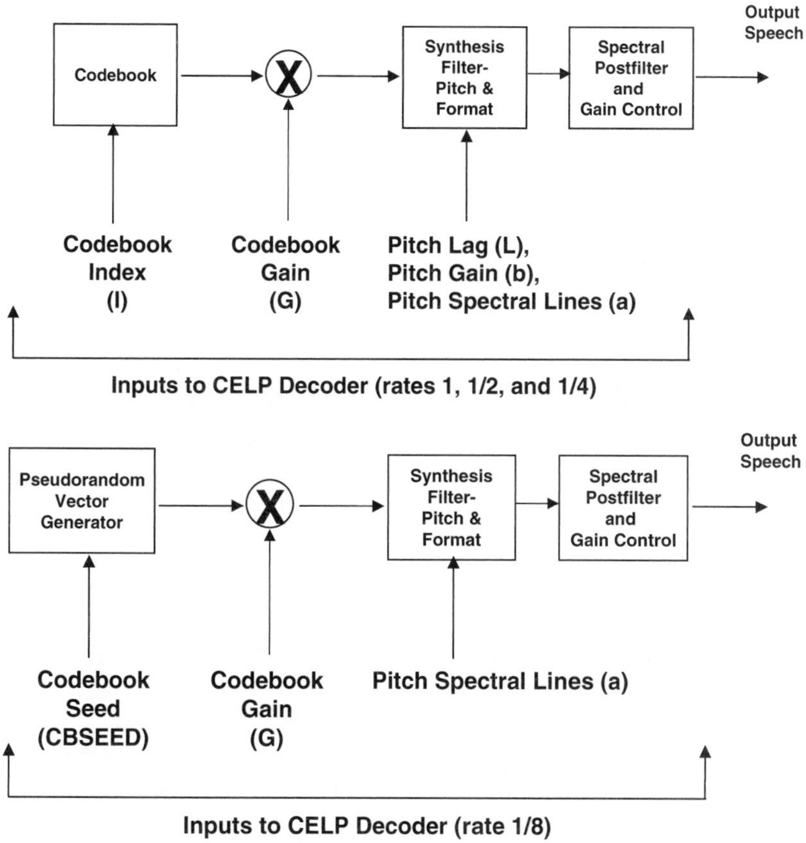

(d)

(e)

Figure 4.2 *(cont.)*

4.5 Enhanced Variable-Rate Codec (EVRC)

The CDMA Development Group's (CDG) enhanced variable-rate vocoder (EVRC) takes advantage of signal processing hardware and software techniques to provide 13-kbps voice quality at 8-kbps data rate, thereby maximizing both quality and system capacity. MOS data has shown that the 8-kbps EVRC (see Figure 4.3) compares favorably to the 16-kbps LD-CELP and ADPCM, the industry standards for comparison. More importantly, the tests have shown that EVRC maintains superior quality over 13-kbps CELP [8] as frame error rates (FERs) rise.

One of the important and unique aspects of CDMA wireless systems in that while there are limits to the number of mobile calls that can be handled by a given carrier at a given time, this capacity limit is not a fixed number. Rather, in CDMA, cell coverage depends on the way the system is designed and implemented. System capacity, voice quality, and coverage are all interrelated, enabling a service provider to trade off any one against the other two.

To maximize the number of simultaneous calls that can be handled at any given time in the allocated frequency spectrum, digital wireless systems utilize speech compression between the mobile and base station. A lower speech transmission rate enables a higher number of simultaneous calls that a system can handle at any given time within the allocated carrier frequency spectrum. The variable-rate vocoder in CDMA uses fewest bits to represent each call without sacrificing voice quality very much.

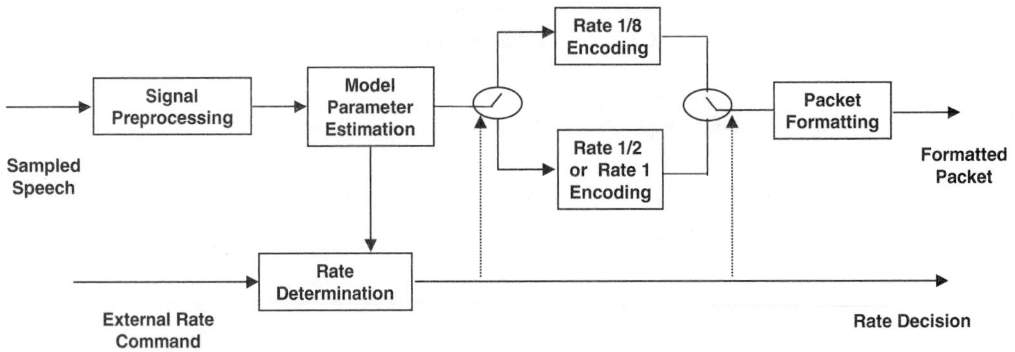

Figure 4.3 Block Diagram for Enhanced Variable-Rate Codec (EVRC)

The basis of the standard EVRC [9] algorithm is the relaxed code-excited linear-predictive (RCELP) coding. RCELP is a generalization of the CELP speech coding algorithm. It is particularly well-suited for variable-rate operation and robustness in the CDMA environment. CELP uses 20-ms speech frames for coding and decoding. In each 20-ms time interval, the encoder processes 160 samples of speech. With variable-rate coders, the encoder examines the contents of each speech frame to determine the necessary coding rate. Depending on the voice waveform (volume, pitch, rate, and so on), the coder represents speech at one of three bit rates: 8, 4, or 1 kbps. As a result, the average bit rate is less than 8 kbps. This differs from a half-rate or multirate coder, where the bit rate is determined once for each call. In addition, when no voice is detected, the vocoder drops its encoding rate and the effective bit rate goes to 1 kbps, further reducing the interference energy produced.

CELP codecs use three sets of bits to represent speech: linear-predictor filter coefficients, pitch parameters, and excitation waveform. For each 20-ms speech frame, the CELP algorithm examines the data and generates 10 linear-prediction coding filter coefficients. With EVRC, the coefficients are represented by a vector, which is a set of the most likely usable coefficients. Increased or decreased bit precision is applied as necessary.

CELP speech coders also perform long-term pitch analysis to generate a 7-bit pitch period and a 3-bit pitch gain. The analysis is based on a mathematical model of the human vocal track. At the different rates, the coder performs this pitch analysis on either four 5-ms subframes, two 10-ms subframes, or one 20-ms frame. The result is a variable number of bits per frame representing the pitch information. EVRC makes two pitch measurements and uses wrapped pitch delays, removing the requirement for a large number of bits and increased computations for fractional pitch delays.

The excitation waveform for every frame or subframe is selected from a codebook consisting of a large number of candidate waveform vectors. The codebook vector chosen to excite the speech coder filters minimizes the weighted error between the original and synthesized speech. The technique used by EVRC to reduce the number of bits required for linear-predictor coefficients and pitch synthesis enables algebraic codebooks to generate excitation. As a result, EVRC has higher voice quality.

Unlike conventional CELP encoders, EVRC does not attempt to match the original speech signal exactly. Instead, EVRC matches a time-wrapped version of the residual that conforms to a simplified pitch contour. The contour is obtained by estimating the pitch delay in each frame and linearly interpolating the pitch from frame to frame. While this adds to computational complexity, the result is higher voice quality per bit transmitted.

The simplified pitch representation also leaves more bits available in each packet for the stochastic excitation and channel impairment protection than would be possible if a traditional fractional-pitch approach were used. The result is enhanced error performance without degraded speech quality at the small cost of added processing requirements.

EVRC also enhances call quality by suppressing background noise. The IS-127 [9] standard recommends a noise suppressor algorithm, but allows system designers to define their own. This is an important factor in choosing a processing platform, making programmable DSPs a desirable choice.

The EVRC algorithm is based on the CELP algorithm. It uses the relaxed code-excited linear-prediction (RCELP) algorithm, and thus does not match the original residual signal but rather a time-wrapped version of the original residual signal that conforms to a simplified pitch contour. This approach reduces the number of bits per frame that are dedicated to pitch representation. This allows additional bits to be dedicated to stochastic excitation and to channel impairment protection. The EVRC algorithm categorizes speech into full-rate (8.55 kbps), half-rate (4 kbps), and eighth-rate (0.8 kbps) frames that are formed every 20 ms. The EVRC algorithm offers a significant performance improvement over the IS-96A speech codec. Table 4.10 shows the performance of the CDMA Development Group's 13-kbps (CDG-13 kbps) speech codec along with IS-96A and EVRC codecs. The CDG-13 kbps offers high voice quality but results in a decrease in channel capacity of about 40 percent over the 8-kbps codec.

Table 4.10 Comparisons of CDG-13 kbps, IS-96A, and EVRC Vocoders in MOS

FER %	CDG-13 kbps	IS-96A	EVRC
0	4.00	3.29	3.95
1	3.95	3.17	3.83
2	3.88	2.77	3.66
3	3.67	2.55	3.50

Table 4.11 provides the bit allocations by packet type.

Table 4.11 Bit Allocations by Packet Type in EVRC

Field	Packet Type			
	Rate 1	**Rate 1/2**	**Rate 1/8**	**Blank**
Spectral transition indicator	1			
Line spectral pair (LSP)*	28	22	8	
Pitch delay	7	7		
Delta delay	5			
Adaptive codebook gain (ACB)	9	9		
Fixed codebook shape (FCB)	105	30		
FCB gain	15	12		
Frame energy			8	
Unused	1			
Total encoded bits	171	80	16	
Mixed mode bit (MM)	1			
Frame quality indicator (CRC) (F)	12	8		
Encoder tail bits (T)	8	8	8	8
Total bits	192	96	24	8
Rate (kbps)	9.6	4.8	1.2	0.4

* A representation of digital filter coefficient in a pseudofrequency domain. This representation has good quantization and interpolation properties.

4.6 Channel Coding

High-speed digital communication systems development demands the optimization of

- Data transmission rate
- Data reliability
- Transmission energy

- Bandwidth
- System complexity
- Cost

Error correcting codes help to meet the requirements cost-effectivity. The higher the error correction code performance, the more flexibility a designer has to determine the required transmission energy, bandwidth, and system complexity. Signal-to-noise ratio (SNR) improvement of a communication channel depends on the error correction code used and the channel's characteristics.

Error control coding (ECC) is the process of adding redundant information to a message to be transmitted that can then be used at the receiving end to detect and possibly correct errors in the transmission. Since the redundant information adds overhead to the transmission, the type of coding must be chosen based upon how much error the system is expected to see, and whether the capability to request retransmission of data is available. There are two basic ECC classifications: *automatic repeat request* (ARQ) and *forward error correction* (FEC). ARQ is a detection-only type coding, where errors in a transmission can be detected by the receiver but not corrected. The receiver must ask for any data received and request that detected errors be retransmitted. FEC allows not only detection of errors at the receiving end, but correction of errors as well. In this book, we primarily focus on the FEC techniques and present the codes (i.e., block, convolutional, and turbo) that are generally used for FEC.

Reed-Solomon (RS), Viterbi (V) convolutional, and concatenated Reed-Solomon Viterbi (RSV) are the most common error correction codes implemented today. At a bit error rate (BER) of 10^{-6}, these codes are at least 2.5 to 3.0 dB short of the Shannon limit in an additive white Gaussian noise (AWGN) channel. Turbocodes have been shown to perform within 1 dB of the Shannon limit at a BER 10^{-6}. Turbocodes break a complex decoding problem into simple steps, where each step is repeated until a solution is reached.

4.6.1 Reed-Solomon (RS) Codes

RS [10] coding is a type of FEC. It has been widely used because of its relatively large error correction capability when weighed against its minimal added overhead. RS codes are also easily scaled up or down in error correction capability to match the error rates expected in a given system. It provides a robust error control method for many common types of data transfer mediums, particularly those that are one-way or noisy and sure to produce errors.

In *block codes* a sequence of K information symbols is encoded in a block of N symbols, $N > K$, to be transmitted over the channel. For a data source that delivers the information bits at the rate B bps, every T seconds the encoder receives a sequence of $K = BT$ bits which defines a message. After K information bits have entered the encoder, the encoder generates a sequence of coded symbols of length N to be transmitted over the channel. In this transmitted sequence or codeword, N must be greater or equal to K in order to guarantee a unique relationship between each codeword and each of the possible 2^K messages. Such a code which maps a block of K information symbols into a block of N coded symbols is called an *(N,K)* block code. The *code rate* is $r = K/N$ bits/symbol; N is called the *block length.*

RS codes are an example of a block coding technique. The data stream to be transmitted is broken up into blocks and redundant data is then added to each block. The size of these blocks and the amount of check data added to each block is either specified for a particular application or can be user-defined for a closed system. Within these blocks, the data is further subdivided into a number of symbols, which are generally from 6 to 10 bits in size. The redundant data then consists of additional symbols being added to the end of the transmission. The system-level block diagram for an RS codec is shown in Figure 4.4.

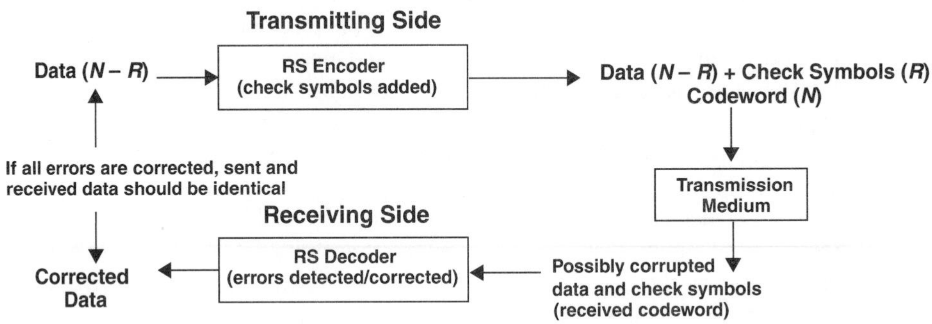

Figure 4.4 RS System-Level Block Diagram

The original data, which is a block consisting of $N - R$ symbols, is run through an RS encoder and R check symbols are added to form a codeword of length N. Since RS can be done on any message length and can add any number of check symbols, a particular RS code is expressed as RS (N, $N - R$) code. N is the total number of symbols per codeword; R is the number of check symbols per codeword; $N - R$ is the number of actual information symbols per codeword.

RS encoding consists of the generation of the check symbols from the original data. The process is based on finite field arithmetic. The variables that need to be known to generate a particular RS code include field polynomial and generator polynomial starting root. Field polynomial is used to determine the order of the elements in the finite field. Valid field polynomials are a function of the bit width to be operated on.

Another system-level characteristic of RS coding is whether the implementation is systematic or nonsystematic. A systematic implementation produces a codeword that contains the unaltered original input data stream in the first R symbols of the codeword. In contrast, in a nonsystematic implementation, the input data stream is altered during the encoding process. Most specifications require systematic coding, and the RS core implements systematic RS codes.

The simplified schematic representation for a systematic RS encoder is shown in Figure 4.5. The input data stream is immediately clocked back out of the function into the check symbol generation circuitry. The fact that the input data stream is clocked out immediately without being altered means that the implementation is systematic. A series of finite field adds and multiplies results in each register containing one check symbol after the entire input stream has been entered. At that point, the output select is switched over to the check symbol registers, and the check symbols are shifted out at the end of the original message.

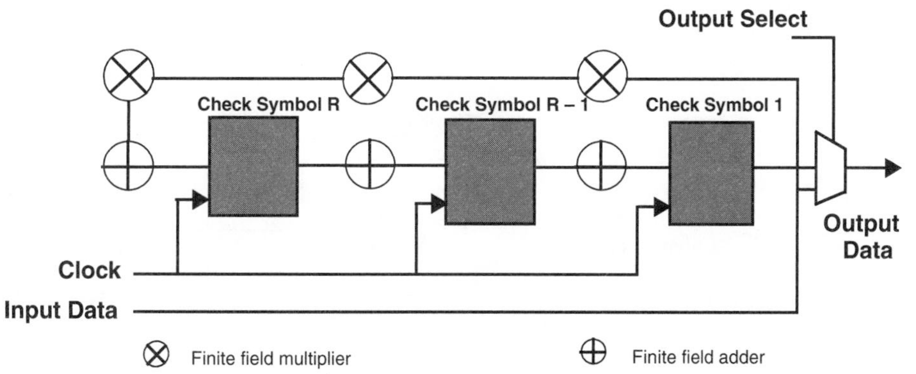

Figure 4.5 Systematic RS Encoder Schematic Representation

The size of the encoder is most heavily affected by the number of check symbols required for the target RS code. The total message length, as well as the field polynomial, and first root value do not have any appreciable effect on the device performance.

A typical RS decode algorithm consists of several major blocks. The first of these blocks is the syndrome calculation, where the incoming symbols are divided into the generator polynomial, which is known from the parameters of the decoder. The check symbols, which form the remainder in the encoder section will cause the syndrome calculation to be zero in case of no errors. If there are errors, the resulting polynomial is passed to the Euclid algorithm, where the factors of the remainder are found (see Figure 4.6). The result is evaluated for each of the incoming symbols over many iterations, and any errors are found and corrected. The corrected codeword is the output from the decoder. If there are more errors in the codeword than can be corrected by the RS code used, then the received codeword is output with no changes and a flag is set, stating that error correction has failed for that codeword.

The error correction capability of a given RS code is a function of the number of check bytes appended to the message. In general, it may be assumed that correcting an error requires one check symbol to find the location of the error, and a second check symbol to correct the error. In general then, a given RS code can correct $R/2$ symbol errors, where R is the number of check symbols in the given RS code. Since RS codes are generally described as an RS $(N, N - R)$ value, the number of errors correctable by this code is $[N - (N - R)]/2$. This error control capability can be enhanced by use of erasures, a technique that helps to determine the location of an error without using one of the check symbols. An RS implementation supporting erasures would then be able to correct up to R errors.

Since RS codes work on symbols (most commonly equal to one 8-bit byte) as opposed to individual data bits, the number of correctable errors refers to symbol errors. This means that a symbol with all of the bits corrupted is no different than a symbol with only one of its bits corrupted, and error control capability refers to the number of corrupted symbols that can be corrected. RS codes are more suitable to correct consecutive bits. RS codes are generally combined with other coding methods such as Viterbi, which is more suited to correcting evenly distributed errors.

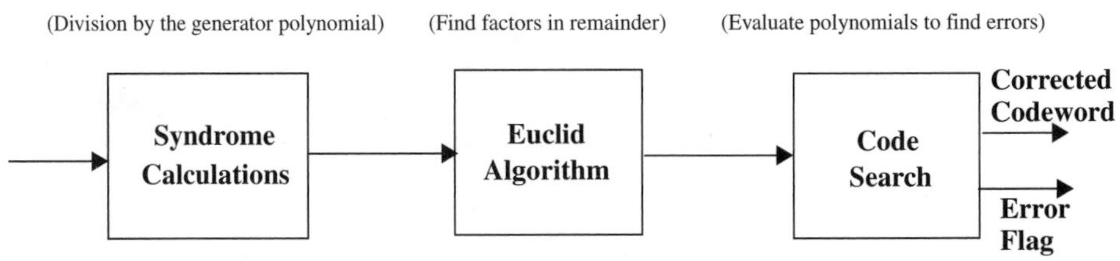

Figure 4.6 RS Decoder Block Diagram

The effective throughput of an RS decoder is a combination of the number of clock cycles required to locate and correct errors after the codeword has been received, and speed at which the design can be clocked. Knowing the latency and clock speed allows the user to determine how many symbols per second may be processed by the decoder. In the RS code, there are two RS decoder choices: a high-speed decoder and a low-speed decoder. The trade-off is that the low-speed decoder is usually approximately 20 percent smaller in device utilization. Note that both decoders operate at the same clock rate, but the low-speed decoder has a longer latency period, resulting in a slower effective symbol rate. As the number of check symbols decreases, the complexity of the decoder decreases, resulting in a smaller design and an increase in performance.

In a real-life RS coding implementation, functions that tend to reside on either side of the RS encoder or decoder are often implemented in programmable logic. One function that often resides after an RS encoder is an interleaver. The task of an interleaver is to scramble the symbols in several RS codewords before transmission, effectively spreading any burst error that occurs during transmission over several codewords. Spreading this burst error over several codewords increases the chance of each codeword being able to correct all of its induced errors. The interleaver scrambles the codewords and writes them into some type of memory prior to transmission. This function is easily and often implemented in programmable logic, even when a dedicated RS codec is being used.

4.6.2 Convolutional Code

The convolutional codes [11] are suitable when the information symbols to be transmitted arrive serially, in long sequences rather than in blocks. In convolutional codes, long sequences of information symbols are encoded continuously in a serial form. This is achieved by entering these symbols one at a time into the encoder, which has some finite memory capacity. The information symbols are sequentially shifted through a K-stage shift register, and following each shift some number v of coded symbols are generated and transmitted. These v coded symbols are obtained by parity checking, that is, by modulo-2 addition of the contents of various stages of the shift register according to the specific code. The length K of the shift register is called the *constraint length* of the code, and the *code rate* is $r = 1/v$ bit per transmitted symbol.

A binary convolutional code of rate $1/v$ bits per symbol can be generated by a linear finite-state machine consisting of a K-stage shift register, v modulo-2 adders connected to some of the shift registers, and a commutator that scans the output of the modulo-2 adders. The whole system is called a *convolutional encoder* (see Figure 4.7).

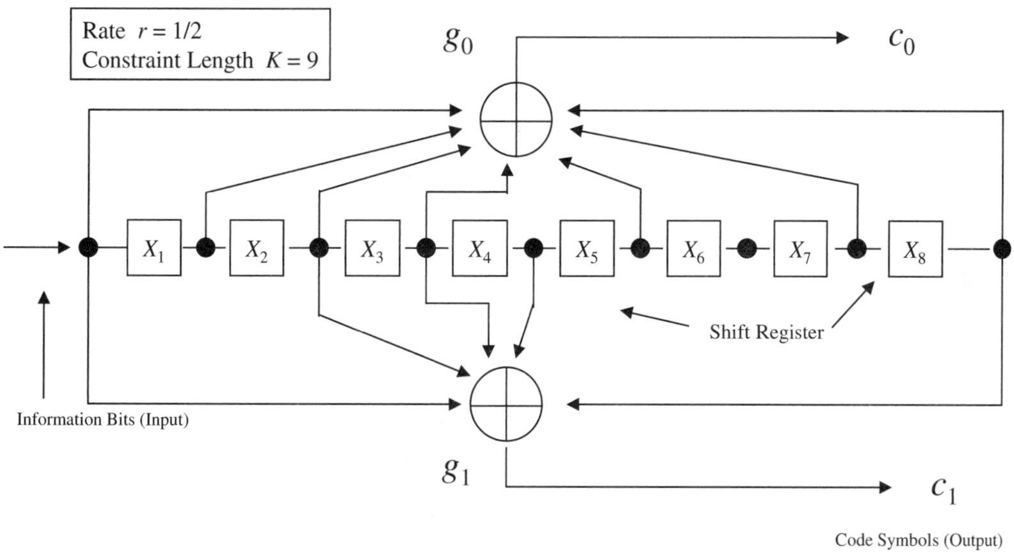

Figure 4.7 Convolutional Encoder

The error correcting capability of a convolution coding scheme increases as rate r decreases. However, the channel bandwidth and decoder complexity both increases with K. The advantage of lower code rates when using convolutional code with coherent PSK, is that the required E_b/N_0 is decreased, permitting the transmission of higher data rates for a given amount of power, or permitting reduced power for a given data rate. Simulation studies have indicated that for a fixed constraint length, a decrease in code rate from 1/2 to 1/3 results in reduction of the required E_b/N_0 of about 0.4 dB. However, the corresponding increase in decoder complexity is about 17 percent. For smaller values of code rate, the improvement in performance relative to the increased decoding complexity diminishes rapidly. Eventually, a point is reached where further decrease in code rate is characterized by reduction in coding gain.

The major drawback of the Viterbi algorithm is that while error probability decreases exponentially with constraint length, the number of code states, and consequently decoder complexity, grows exponentially with constraint length.

Figure 4.8 shows downlink channel coding that will be used with GSM EFR speech coder and 8-PSK modulation. The bits are combined such that every two class IA bits are used with one class II bit to form the first 89 triads. The remaining 44 triads are formed by using three class IB bits.

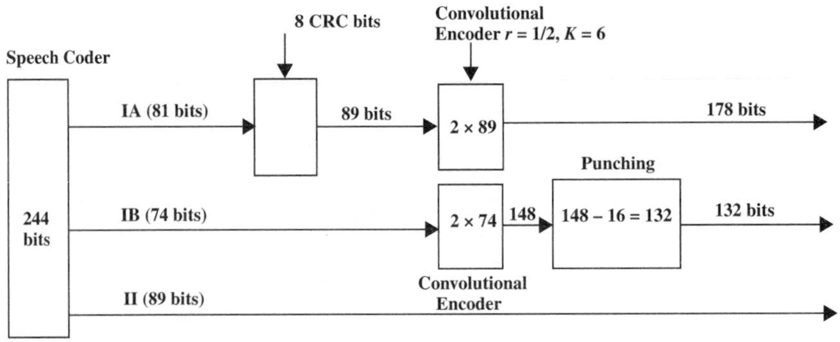

Figure 4.8 Channel Coding (Downlink) with GSM EFR Speech Coder

The bits are combined in a somewhat similar manner for the uplink (see Figure 4.9). Eighty-six triads consist of two class IA bits and one class II bit, three triads contain two class IB bits along with one class II bit, and 35 triads consist of three class IB bits.

The triads are reordered to provide additional interleaving gain, and then intraslot interleaving is applied. There are three primary interleaving options: one-slot, two-slot, and three-slot. In one-slot interleaving, there is no intraslot interleaving and the current triads are simply transmitted in the current slot.

In two-slot interleaving, certain triads (from the current triad vector) are transmitted in the current slot, and the remaining triads in the next slot. Thus, the current slot contains triads from the current and previous triad vectors.

In three-slot interleaving, the concept is extended to include another set of triads in each transmitted slot. Certain triads from the current triad vector are transmitted in the next slot, and the remaining triads are transmitted in the slot after the next slot. Thus, the current slot contains triads from the current triad vector, the previous triad vector, and one before the previous triad vector.

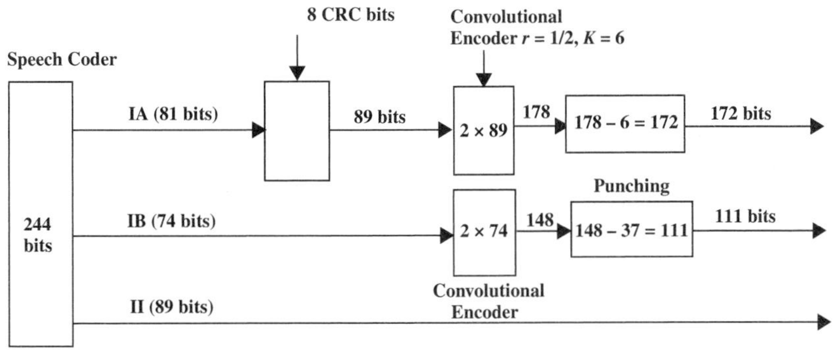

Figure 4.9 Channel Coding (Uplink) with GSM EFR Speech Coder

4.6.3 Soft and Hard Decision Coding

The decoding of a block code can be performed with hard or soft decision input, and the decoder may output hard or soft decision data. In hard decision antipodal decoding, each received channel bit is assigned a value of 1 or 0 at the demodulator, depending whether the received noisy data is higher or lower than a threshold. The decoder then uses the redundancy added by the encoder to determine if there are errors, and, if possible, correct them. The desired output of the decoder is a corrected codeword.

A soft decision decoder receives not only the binary value of 1 or 0, but also a confidence value associated with the given bit. If the demodulator is certain that the bit is a 1, it places a very high confidence on it. If it is less certain, it places a lower confidence value. A soft input decoder can output either hard decision data or soft decision data. For example, a Viterbi decoder receives soft information from the demodulator, and outputs hard decision data. The decoder can use the soft information to determine if a given bit is a solid 1 or solid 0, and it outputs this hard decision.

A soft input/soft output (SISO) decoder both receives soft decision data and generates soft decision output. For each bit in the codeword, the SISO decoder examines the confidence of the other bits in the codeword, and using the redundancy of the code generates an updated soft output for the given bit.

RS error correction codes are the "standard" algorithm for FEC. RS codes are block codes that are very efficient for error correction implemented in either hardware or software. RS codes are hard decision codes. RS codes followed with concatenated Viterbi codes (RSV) offer an improvement over the stand-alone RS codes in terms of BER performance.

The concept of SISO decoders has been applied to turbocodes. A turbocode feeds demodulated soft decision data into a SISO decoder. The output of this decoder is then fed into the same (or a different) SISO decoder. The output of this decoder, is then fed again. This iterative process continues until a confident solution is reached. The concept of feeding the output back into the input is similar to a turbocharger of an engine; therefore, the name turbocode is used.

For a turbocode to be effective, the given data must be encoded with two (or more) different codes. Then, when decoding, each of the codes will modify the confidence of each bit. With each iteration, all of the codes modify the confidence of the data, so each code sees slightly different data for each iteration. Each code pushes the confidence of a given bit higher or lower, and consequently changes the hard decision value of the bits in error. Eventually, the data will settle on an arrangement where all codes are pushing the confidence of all bits higher. The hard decision values at this point are closer to the transmitted data.

4.6.4 Turbocoding

A turbodecoder [12] consists of two concatenated decoders, each providing soft information and so-called intrinsic information. Two main classes of algorithms available for turbocodes are *soft-output Viterbi algorithms* (SOVA), and iterative soft-input and soft-output decoding algorithms such as the symbol-by-symbol *maximum a posterior* (MAP) algorithm [13], which is more complex but yields better performance. Since the capacity of a CDMA system is strongly dependent on the required E_b/N_0 value of the receiver, any improvement in E_b/N_0 value translates directly into capacity increase.

The inner receiver's task is to provide the best estimate for the outer receiver. It deals with the following signal impairments:

- The presence of multipath components
- The presence of multiuser interference (both inter- and intracell)
- The fading of each transmission path
- The near-far effect due to the relative position of all mobiles and the base station
- The time-varying nature of these impairments

These issues are tackled with a combination of the following techniques:

- Channel estimation and tracking
- A maximum ratio combination-based (coherent) RAKE receiver, to take advantage of multipath characteristics
- Multiuser detection (MUD) schemes, such as interference cancellation (IC) or decorrelating receivers
- Fast power control based on SNR estimation
- Antenna arrays (in the base station) to provide another form of diversity (space-diversity)

Turbocodes may use serial (concatenated) and/or parallel recursive convolution codes. Recursive means the output not only depends on the input sequence but also depends on the previous output. A turboencoder consists of two binary rate 1/2 convolution encoders separated by an N-bit random interleaver or permuter, together with an optional puncturing mechanism. The interleaver is used to spread coded symbols from the other encoder or from the input. The input to an encoder can be from an input sequence and/or from a coded output from another encoder. The encoders are configured in a manner reminiscent of classical concatenated codes. However, instead of cascading the encoders in the usual serial fashion, they are arranged in parallel concatenation (see Figure 4.10).

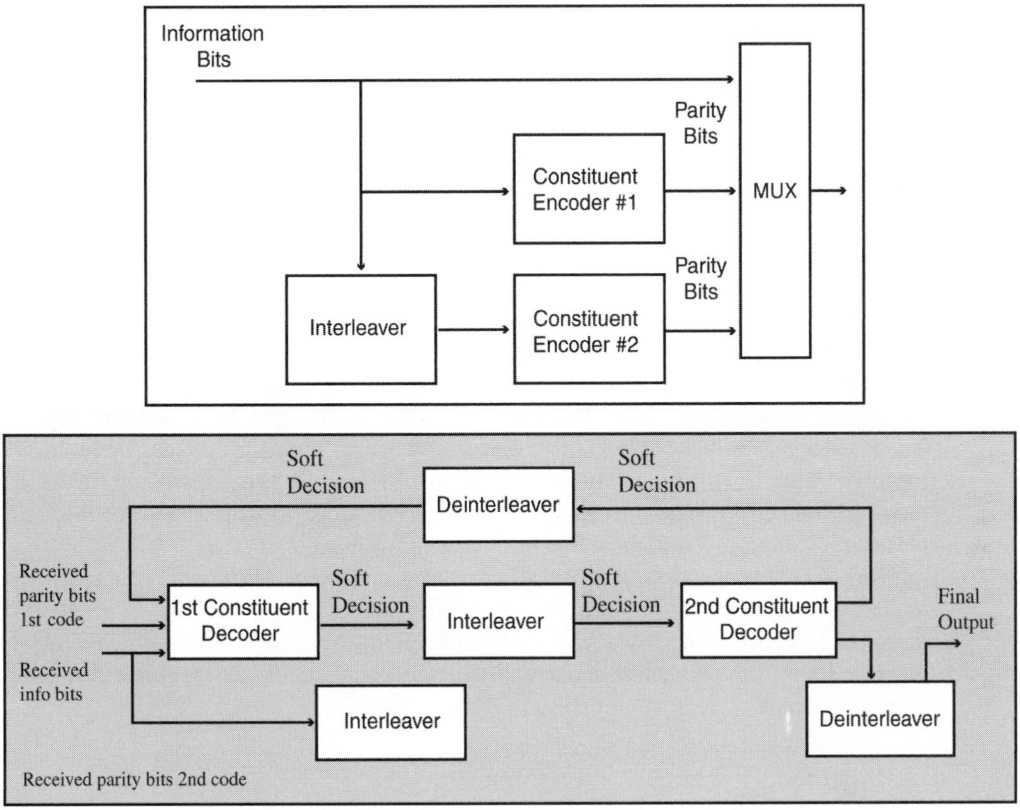

Figure 4.10 Turboencoding and Turbodecoding

The function of the permuter is to take each incoming block of N data bits and rearrange them in a pseudorandom fashion prior to encoding by the second encoder. Unlike the classical interleaver (e.g., block or convolution interleaver), which arranges the bits in some systematic manner, it is important that the permuter sorts the bits in a manner that lacks any apparent order. Also important is that N be selected quite large ($N > 1000$). The role of the turbocode puncture is identical to that of its convolutional code counterpart: to periodically delete selected bits to reduce coding overhead.

Each decoding process yields soft output decision for the next decoder. The key component is the soft input/soft output (SISO) decoder. To achieve benefits of turbocode, several iterations are provided. Thus, the turbocodes are processing intensive and are applied to less delay sensitive applications such as data.

Turbocodes have been implemented in both software and hardware as a single integrated circuit. Turbocodes don't suffer from the error floor at low BERs that have been attributed to other codes. Turbocodes are capable of providing high coding gain, even with high code rates.

Turbocodes provide about 1.5 to 3 dB bit energy-to-noise improvement over the current RS and RSV error correction codes. A 1.5 to 3.0 dB performance gain allows the system developer to reduce transmitter power and/or bandwidth and still maintain the same BER performance. Conversely, a 1.5 to 3 dB improvement can be used to improve overall system performance if transmitter power is not reduced. The use of turbocodes can allow the system engineer to reduce antenna size, lowering system cost. When used to provide improved BER performance, turbocodes can also lead to a clearer image transmission, improved audio, greater range, or better data integrity. A turbocode is applicable where digital data is transmitted over a noisy channel, because it spreads error uniformly over the interleaved duration. Turbocodes are much better than convolutional codes due to their strength in nonlinear coding/decoding and feedback property. Turbocodes can support the data rates achieved with other error correction codes, while offering improved correction capability.

Figure 4.11 gives a comparison of a turbocode having component codes of constraint length $k = 4$ decoded with four iterations with a convolutional code of constraint length $k = 9$ decoded with Viterbi decoder. Figure 4.11 shows the performance of a rate 1/3 turbocode compared with the corresponding rate 1/3 convolutional code. The results indicate that the turbocode outperforms the corresponding convolutional code with the same decoding complexity for data rates larger than 9.6 kbps; the improvement in performance increases data rate for a fixed frame length of 20 ms. The performance improvement with data rate is due to the fact that the number of bits in 20-ms frames increases with data rate and the performance of a turbocode increases with the number of bits in a frame. With a large number of bits in a frame, the interleaver separating the component codes can randomize the errors more effectively.

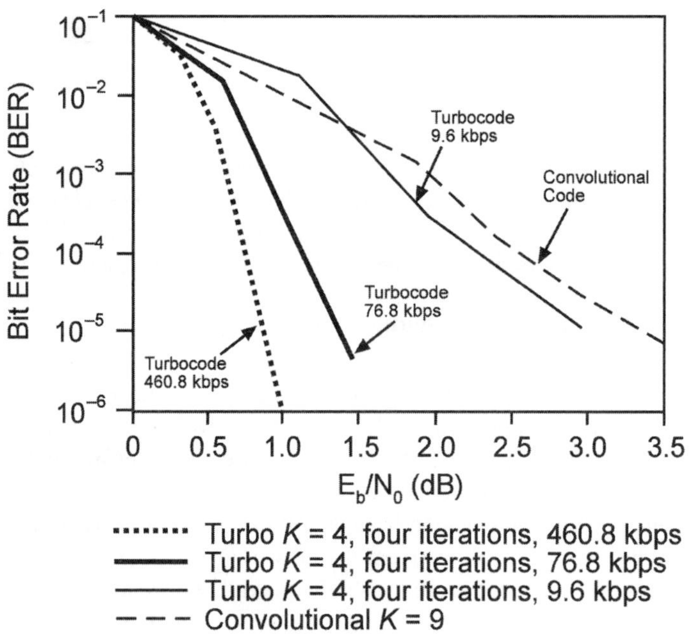

Figure 4.11 Performance of a Rate 1/3 Turbocode Compared with a Convolutional Code in a WGN Channel

4.7 Spreading Codes

CDMA uses a waveform that for all purposes appears random to anyone but the intended receiver of the transmitter waveform. For ease of both generation and synchronization by the receiver, the waveform is pseudorandom, implying that it can be generated by mathematically precise rules, but statistically it nearly satisfies the requirements of a truly random sequence.

In CDMA, direct-sequence spreading consists of multiplying the input data by a pseudorandom noise (PN) sequence having a bit rate much higher than the data bit rate. This increases the data rate while adding redundancy to the system. The ratio of PN sequence bit rate to data rate is called the *processing gain* or *spreading factor* (SF). The resulting waveform is wideband, noiselike, balanced in phase, and has a flexible timing structure. When there are two different I and Q branches, each one can be spread separately. To minimize the overall envelope variations and achieve high amplifier efficiency, a complex spreading technique will be used in third-generation systems (see Figure 4.12).

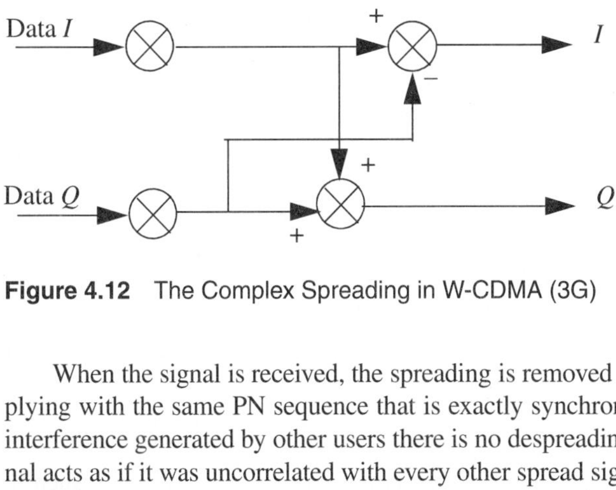

Figure 4.12 The Complex Spreading in W-CDMA (3G)

When the signal is received, the spreading is removed from the desired signal by multiplying with the same PN sequence that is exactly synchronized to the received PN. For the interference generated by other users there is no despreading, i.e., each spread-spectrum signal acts as if it was uncorrelated with every other spread signal using the same band. CDMA codes are designed to have very low cross-correlation. Orthogonal non-PN spreading codes with zero cross-correlation have been used in IS-95 to improve the detection reliability.

In this section we discuss the autocorrelation and cross-correlation of PN and orthogonal sequences. We briefly describe the maximal-length, Gold, and Kasami sequences along with fixed- and variable-length orthogonal codes.

4.7.1 Pseudorandom Noise (PN) Sequences

In CDMA systems, PN sequences are used to perform the following tasks:

- Spreading the bandwidth of the modulated signal to the larger transmission bandwidth
- Distinguishing among the different user signals using the same transmission bandwidth in multiple-access applications

PN sequences are not random. They are deterministic, periodic sequences. PN sequences are generated by combining the outputs of a feedback shift register. A feedback shift register consists of several consecutive two-state memory or storage stages and feedback logic. Binary sequences are shifted through the shift register in response to clock pulses. The contents of the stages are logically combined to produce the input to the first stage. The initial contents of the stages and feedback logic determine the successive contents of the stages. A feedback shift register and its output are called linear when the feedback logic consists entirely of modulo-2 adders.

The output sequences of a PN generator are classified as either maximal length or nonmaximal length. Maximal-length sequences are the longest sequences that can be generated by a given shift register of a given length. In binary shift register sequence generators, the maximal-length sequence is $2^n - 1$ chips, where n is the number of stages in the shift register. Maximal-length sequences have the property that for an n-stage linear feedback shift register, the sequence repetition period in clock pulses is $T_0 = 2^n - 1$. If a linear feedback shift register generates a maximal sequence, then all of its nonzero output sequences are maximal, regardless of the initial stage. When an n-stage shift register (see Figure 4.13) is configured to generate a maximal-length sequence, the sequence has the following properties:

1. The number of ones in a sequence equals the number of zeros within one chip. The number of ones is 2^{n-1} and number of zeros is $2^{n-1} - 1$.
2. The statistical distribution of ones and zeros is well defined and always the same. Relative positions of the runs vary from code sequence to code sequence but the number of each run length does not. A run is defined as a sequence of single-type binary digits, all ones or all zeros. The appearance of the alternate digit in a sequence starts a new run. The length of the run is the number of digits in the run.
3. A modulo-2 addition of a maximal linear PN sequence with a phase-shifted replica itself results in another replica with a phase shift different from either of the originals.
4. If a period of sequence is compared term by term with any cyclic shift itself, it is best if the number of agreements differ from the number of disagreement by not more than one count.
5. The normalized autocorrelation of a maximal linear PN sequence is such that for all values of phase shift the correlation value is $-1/(2^n - 1)$, except for the chip phase-shift area, in which correlation varies linearly from $-1/(2^n - 1)$ to 1 to $-1/ (2^n - 1)$. The autocorrelation function for a maximal-length PN sequence is triangular with a maximum value at shift, $\tau = 0$. With this property, two or more communicators can operate independently, if their codes are phase shifted more than one chip. For other code sequences, the autocorrelation properties may be markedly different than the properties of the maximal-length sequences.
6. Every possible state of a given n-stage generator exists at some time during the generation of a complete code cycle. Each state exists for one and only one clock interval. The exception is that the all-zeros state is not allowed to occur.

7. Cross-correlation of a maximal linear PN sequence is the measure of similarity between two different code sequences. (In CDMA systems, the cross-correlation is of interest because the receiver response to any other signal other than the proper addressing sequence is not allowed.)

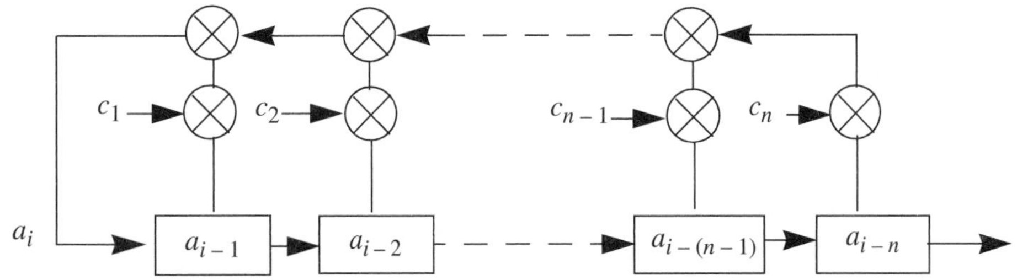

Figure 4.13 *M*-Sequence Generator Structure

Figure 4.13 shows the structure of the maximal-length linear feedback shift register sequences. Maximal-length sequences (*m*-sequences) are the largest codes that can be generated by a given shift register. Each clock time, the register shifts all contents to the right.

The generator function, $G(D)$, of the sequence can be expressed as a ratio of finite polynomials [14].

$$G(D) = \frac{g_0(D)}{f(D)} \tag{4.1}$$

$f(D)$ is the characteristics polynomial of the linear feedback shift register (LFSR) sequence generator. It depends on the connection vector $c_1, c_2, \ldots, c_n$ and determines the main characteristics of the generated sequence. The polynomial $g_0(D)$ depends on the initial condition vector $a_{i-n}, a_{i-(n-1)}, \ldots, a_{i-1}$ and determines the phase shift of the sequence. It has been shown in [15] that every LFSR sequence is periodic with period $N \le 2^n - 1$ for all nonzero initial vectors, where n is number of shift registers. A necessary condition for $G(D)$ to generate an *m*-sequence is that $f(D)$ be irreducible (nonfactorable). The irreducible polynomials that generate an *m*-sequence are called *primitive*. A primitive polynomial of degree n is simply one for which the period of the coefficients of $1/f(D)$ is $2^n - 1$. These polynomials exist for all degrees $n > 1$. The number of primitive polynomials of degree n is equal to [16]

$$N_p(n) = \frac{2^n - 1}{n} \cdot \prod_{i=1}^{k} \frac{P_i - 1}{P_i} \tag{4.2}$$

where $\{P_i, i = 1, 2, 3, \ldots\ldots\ldots, k\}$ is the prime decomposition of $2^n - 1$.

To demonstrate the properties of PN binary sequence, we consider the linear feedback shift register (see Figure 4.14) that has a four-stage register for storage and shifting, a modulo-2 adder, and a feedback path from adder to the input of the register. The operation of the shift register is controlled by a sequence of clock pulses. At each clock pulse the contents of each stage in the register is shifted by one stage to the right. Also, at each clock pulse the contents of stages X_3 and X_4 are modulo-2 added, and the result is fed back to stage X_1. The shift register sequence is defined to be the output of stage X_4. We assume that stage X_1 is initially filled with a 0 and the other remaining stages are filled with 0, 0, and 1; i.e., the initial state of the register is 0 0 0 1. Next, we perform the shifting, adding, and feeding operations, where the results after each cycle is given in Table 4.12.

Table 4.12 Distribution of Runs for a Chip Sequence

Shift	Stage X_1	Stage X_2	Stage X_3	Stage X_4	Output Sequence
0	0	0	0	1	1
1	1	0	0	0	0
2	0	1	0	0	0
3	0	0	1	0	0
4	1	0	0	1	1
5	1	1	0	0	0
6	0	1	1	0	0
7	1	0	1	1	1
8	0	1	0	1	1
9	1	0	1	0	0
10	1	1	0	1	1
11	1	1	1	0	0
12	1	1	1	1	1
13	0	1	1	1	1
14	0	0	1	1	1
15	0	0	0	1	1
16	1	0	0	0	0

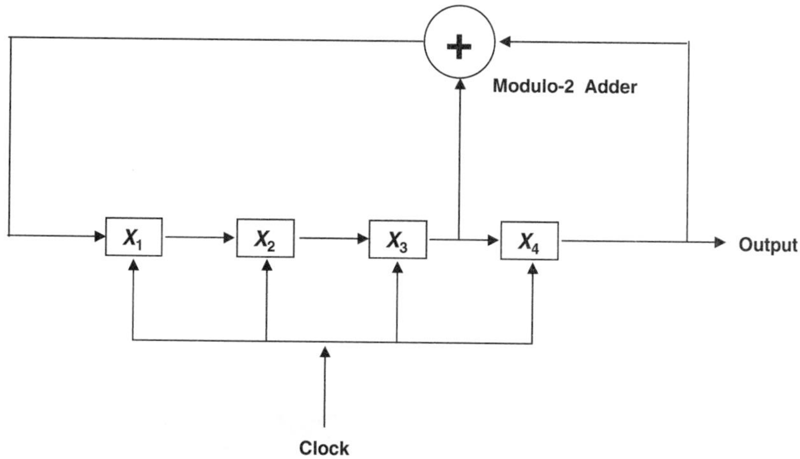

Figure 4.14 Four-Stage Linear Feedback Shift Register

We notice that the contents of the registers repeats after $2^4 - 1 = 15$ cycles. The output sequence is given as 0 0 0 1 0 0 1 1 0 1 0 1 1 1 1 (see Figure 4.15) where the left-most bit is the earliest bit. In the output sequence; the total number of zeros is seven and total number of ones is eight; the numbers differ by one.

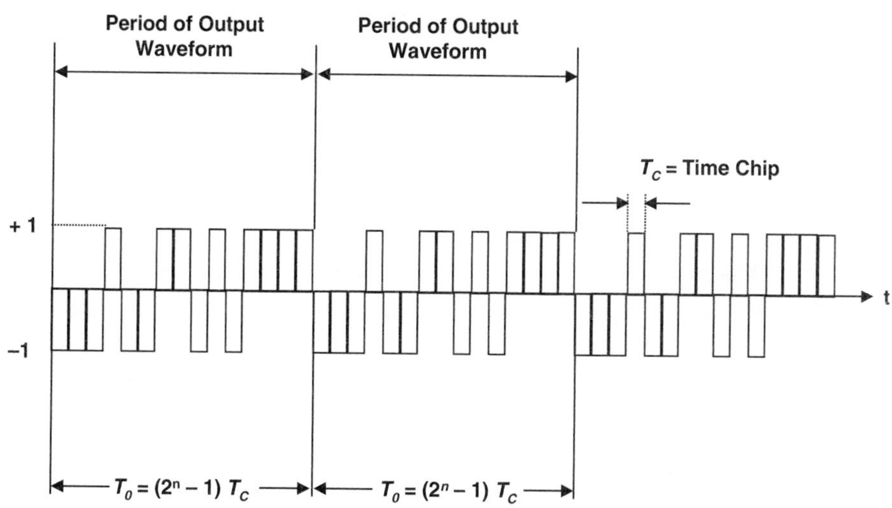

Figure 4.15 Output Waveform for Four-Stage Linear Feedback Shift Register

If a linear feedback shift register reached the zero state at some time, it would always remain in the zero state and the output sequence would subsequently be all zeros. Since there are exactly $2^n - 1$ nonzero states, the period of a linear n-stage shift register output sequence cannot exceed $2^n - 1$.

It has been shown [16] that there are exactly $2^{n-(p-2)}$ runs of length p for both ones and zeros in every maximal sequence [except that there is only one run containing n ones and one containing $(n - 1)$ zeros; there are no runs of zeros of length n or ones of length $(n - 1)$]. The distribution of runs for $2^4 - 1$ chip sequence is given in Table 4.13.

Table 4.13 Distribution of Runs for a $2^4 - 1$ Chip Sequence

Run Length	Ones	Zeros	Number of Chips Included
1	2	2	$1 \times 2 + 1 \times 2 = 4$
2	1	1	$1 \times 2 + 1 \times 2 = 4$
3	0	1	$0 \times 3 + 1 \times 3 = 3$
4	1	0	$1 \times 4 + 0 \times 4 = 4$
Total no. of chips			15

Whether an n-stage linear feedback shift register generates only one sequence with period $2^n - 1$ depends on its connection vector (see Figure 4.13). Let $h(x)$ be the nth-order polynomial given by:

$$h(x) = h_0 + h_1 x + h_2 x^2 + \ldots + h_n x^n \quad \textbf{(4.3)}$$

We refer to $h(x)$ as the associated polynomial of the shift register with feedback coefficients $(h_0, h_1, h_2, \ldots, h_n)$. Here $h_0 = h_n = 1$ and other feedback coefficients take values 0 or 1. Thus, the polynomial for the four-stage linear feedback shift register as shown in Figure 4.14 is given by

$$h(x) = 1 + x^3 + x^4 \quad \textbf{(4.4)}$$

Table 4.14 gives the number of maximal sequences available from register lengths 2 through 10 and provides an example of primitive polynomial of degree n.

Table 4.14 Number of Maximal Sequences available from Register Lengths 2 Through10

No. of Stage n	$2^n - 1$	Prime Decomposition of $2^n - 1$	No. of n-sequence $N_p(n)$	Example of Primitive Polynomial of degree n $h(x)$
2	3	3	$\frac{3}{2}\cdot\frac{2}{3} = 1$	$1+x+x^2$
3	7	7	$\frac{7}{3}\cdot\frac{6}{7} = 2$	$1+x+x^3$
4	15	3×5	$\frac{15}{4}\cdot\frac{2}{3}\cdot\frac{4}{5} = 2$	$1+x+x^4$
5	31	31	$\frac{31}{5}\cdot\frac{30}{31} = 6$	$1+x^2+x^5$
6	63	$3\times3\times7$	$\frac{63}{6}\cdot\frac{2}{3}\cdot\frac{6}{7} = 6$	$1+x+x^6$
7	127	127	$\frac{127}{7}\cdot\frac{126}{127} = 18$	$1+x^3+x^7$
8	255	$3\times5\times17$	$\frac{255}{8}\cdot\frac{2}{3}\cdot\frac{4}{5}\cdot\frac{16}{17} = 16$	$1+x^2+x^3+x^4+x^8$
9	511	7×73	$\frac{511}{9}\cdot\frac{6}{7}\cdot\frac{72}{73} = 48$	$1+x^4+x^9$
10	1023	$3\times11\times31$	$\frac{1023}{10}\cdot\frac{2}{3}\cdot\frac{10}{11}\cdot\frac{30}{31} = 60$	$1+x^3+x^{10}$

4.7.2 Autocorrelation and Cross-Correlation

Autocorrelation is shown in Figure 4.16. It refers to the degree of correspondence between a sequence and phase-shifted replica of itself. This characteristics of autocorrelation are used to great advantage in wireless communication. It is of most interest to select code sequences that give the least probability of false synchronization. With this characteristic, an m-sequence is indistinguishable from a pure random code when n is large.

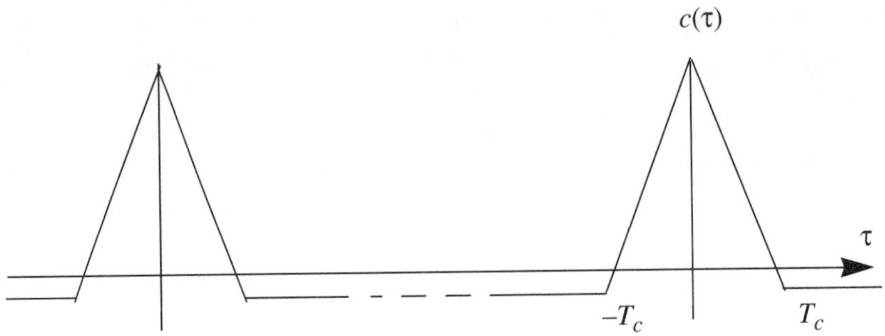

Figure 4.16 Autocorrelation of a PN Sequence

The autocorrelation function, $R_x(\tau)$, for a signal $x(t)$ is defined as

$$R_x(\tau) = \int_{-\infty}^{\infty} x(t)x(t+\tau)dt \tag{4.5}$$

An autocorrelation plot shows the number of agreements minus disagreements for the overall length of the two sequences being compared, as the sequences assume every shift in the field of interest. If $x(t)$ is a periodic pulse waveform representing PN sequence, we refer to each fundamental pulse as a PN sequence symbol or a chip. For such PN waveform of unit chip duration and period $T_0 = 2^n - 1$ chips, the normalized autocorrelation function is given as

$$R_x(\tau) = \frac{1}{T_0}\text{[number of agreements – number of disagreements in a comparison of one full period of sequence with a } \tau \text{ position cyclic shift of the sequence]}$$

The normalized autocorrelation function $R_x(\tau)$ of a periodic waveform $x(t)$ with T_0 period is given as

$$\frac{R_x(\tau)}{R_x(0)} = \frac{1}{T_0}\int_{-T_o/2}^{T_0/2} x(t)x(t+\tau)dt \qquad \text{for } -\infty < \tau < \infty \tag{4.6}$$

where $R_x(0) = \frac{1}{T_0}\int_{-T_0/2}^{T_0/2} x^2(t)dt$

4.7.3 Cross-Correlation

Cross-correlation is the measure of agreement between two different codes. The *m*-sequences are not immune to cross-correlation problems; they may have large cross-correlation values.

The cross-correlation function between two signals, $x(t)$ and $y(t)$, is defined as the correlation between two different signals $x(t)$ and $y(t)$ and is given as

$$R_c(\tau) = \int_{-T_0/2}^{T_0/2} x(t)y(t+\tau)dt \qquad \text{for} \qquad -\infty < \tau < \infty \tag{4.7}$$

The *m*-sequences are used as different codes with an excellent correlation property. In IS-95 networks, all base stations use GPS. Each base station is identified by a finite offset of its PN binary sequence in the forward direction.

EXAMPLE 4.1

Consider a three-stage shift register generator, generating a seven-chip maximal linear code. The reference sequence is 1 1 1 0 0 1 0. Sketch the autocorrelation function if the chip period is T_c.

Table 4.15 provides the sequence after each shift and shows the corresponding agreements (A) and disagreements (D) with the reference sequence.

Table 4.15 Agreements and Disagreements with Reference Sequence

Shift	Sequence	Agreement (A)	Disagreement (D)	$A - D$
1	0 1 1 1 0 0 1	3	4	–1
2	1 0 1 1 1 0 0	3	4	–1
3	0 1 0 1 1 1 0	3	4	–1
4	0 0 1 0 1 1 1	3	4	–1
5	1 0 0 1 0 1 1	3	4	–1
6	1 1 0 0 1 0 1	3	4	–1
0	1 1 1 0 0 1 0	7	0	7

It should be noted that the net correlation $A - D$ is –1 for all shifts except for the zero-shift or synchronous condition. This is typical of all n-sequences. In the region between zero and plus or minus one chip shift (T_c), the correlation increases linearly so that the autocorrelation function for an n-sequence is triangular, as shown in Figure 4.17. This characteristic of autocorrelation is very useful in communication systems. A channel can simultaneously support multiple users, if the corresponding codes are phase-shifted more than one chip.

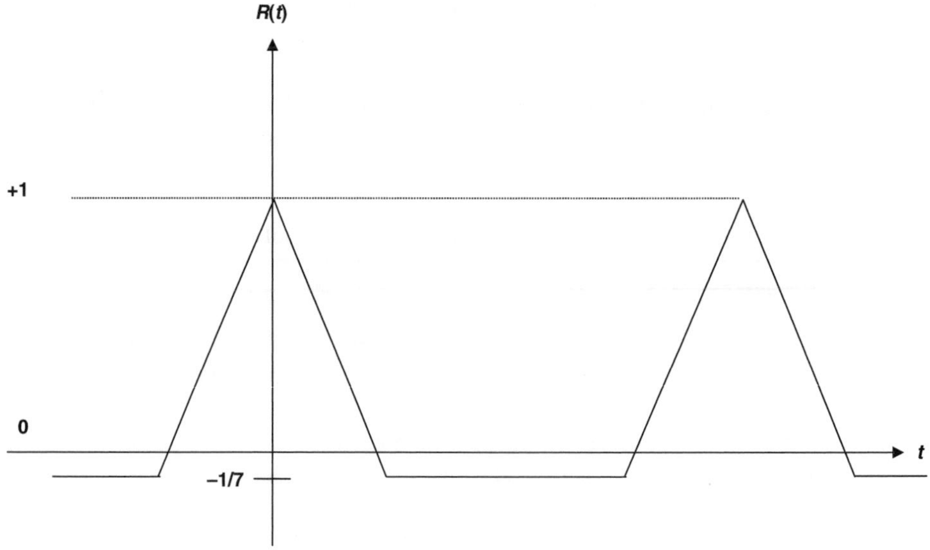

Figure 4.17 Autocorrelation of Three-Stage Linear Feedback Shift Register PN Sequence

4.7.4 Orthogonal Codes

Orthogonal codes are employed to improve the detection reliability of a spread-spectrum signal. Each mobile uses one member of a set of orthogonal codes representing the set of symbols used for transmission. On the downlink, orthogonal codes are used to create unique channels of information. While there are many different sequences that can be used to generate an orthogonal set of codes, the Walsh and Hadamard sequences make useful sets for CDMA.

Two different methods can be used to modulate the orthogonal codes into the information stream of the CDMA signal. The orthogonal set of codes can be used as a channel code or can be used to modulate the orthogonal functions into the information stream of the CDMA signal. These codes can form orthogonal modulation symbols.

With orthogonal modulation symbols, the information bit stream is divided into blocks so that each block represents a nonbinary information symbol associated with a particular transmitted code sequence. If there are b bits per block, one of the set of $K = 2^b$ functions is transmitted in each symbol interval. The signal at the receiver is correlated with a set of K matched filters each matched to the code function of one symbol. The outputs from correlators are compared, and the symbol with the largest output is taken as the transmitted symbol. In this technique the signal spectrum is spread by the factor $2^b/b$. Since all users use the same orthogonal functions, the signal of one user cannot be distinguished from another. Therefore, in IS-95 CDMA system each user's signal is also spread by a distinct PN sequence after orthogonal modulation.

Walsh code spreading can be used if the receiver is synchronized with the transmitter. In the forward link for IS-95, the base station transmits a pilot to enable the receiver to recover synchronization. In a IS-95 system, a pilot signal is not used in the reverse link. Therefore, Walsh symbol modulation is used from the mobile to the base station. In the W-CDMA system, the pilot will be sent in both directions, so the multiple-spreading technique will also be used in the reverse channels.

The TIA IS-95 CDMA system uses orthogonal functions as a channel code on the forward channel, and uses orthogonal functions for symbol modulation on the reverse channel. In the latter case, one of 64 possible modulation symbols is used for each hextet (a group of six transmitted bits). The modulation symbol is one member of the set of 64 mutually orthogonal functions. The orthogonal functions have the following characteristic:

$$\sum_{k=0}^{M-1} \phi_i(k\tau)\phi_j(k\tau) = 0 \qquad i \neq j \tag{4.8}$$

where

$\phi_i(k\tau)$ and $\phi_j(k\tau)$ are the ith and jth orthogonal members of an orthogonal set

M is the length of the set

τ is the symbol duration

Walsh functions are generated by codeword rows of special square matrices called *Hadamard matrices*. These matrices contain one row of all zeros, and remaining rows each with equal numbers of ones and zeros. Walsh functions can be constructed for block length j, where j is an integer.

The TIA IS-95 CDMA system uses a set of 64 orthogonal functions generated by using Walsh functions. The modulated symbols are numbered from zero through 63.

The 64×64 matrix is generated by using the following recursive procedure:

$$H_1 = \begin{bmatrix} 0 \end{bmatrix} \qquad H_2 = \begin{bmatrix} 0 & 0 \\ 0 & 1 \end{bmatrix} \tag{4.9}$$

$$H_4 = \begin{bmatrix} 0\ 0\ 0\ 0 \\ 0\ 1\ 0\ 1 \\ 0\ 0\ 1\ 1 \\ 0\ 1\ 1\ 0 \end{bmatrix} \qquad H_{2n} = \begin{bmatrix} H_N\ H_N \\ H_N\ \overline{H_N} \end{bmatrix} \tag{4.10}$$

where N is a power of 2 and $\overline{H_N}$ is the inverse of H_N.

The period of time needed to transmit a single modulation symbol is called a *Walsh symbol* interval and is equal to 1/4800 second (208.33 μs). The period of time associated with 1/64 of the modulation symbol is referred to as a *Walsh chip* and is equal to $1/307200$ second (3.255 μs). Within a Walsh symbol, Walsh chips are transmitted in the order 0, 1, 2, . . . , 63.

For the forward channel, Walsh functions (Figures 4.18 and 4.19) are used to eliminate multiple-access interference among users in the same cell. On the downlink, all Walsh functions are synchronized in the same cell and have zero correlation between each other. The following steps are used:

- The input user data (e.g., digital speech) is multiplied by an orthogonal Walsh function (TIA IS-95 standard uses the first 64 orthogonal Walsh functions).
- The user data is then spread by the base station pilot PN code and transmitted on the carrier.
- At the receiver, the mobile multiplies the detected signal by the synchronized PN code associated with the base station.
- The signal is then multiplied by the synchronized Walsh function for the *i*th user, which eliminates the interference from other users' signals on the transmission from the base station, and leaves the desired user information.

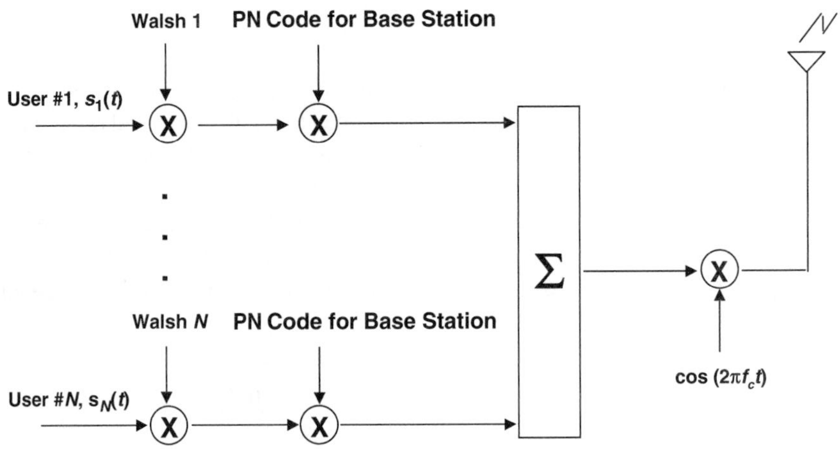

Figure 4.18 Applications of Walsh Functions and PN Code at the Base Station

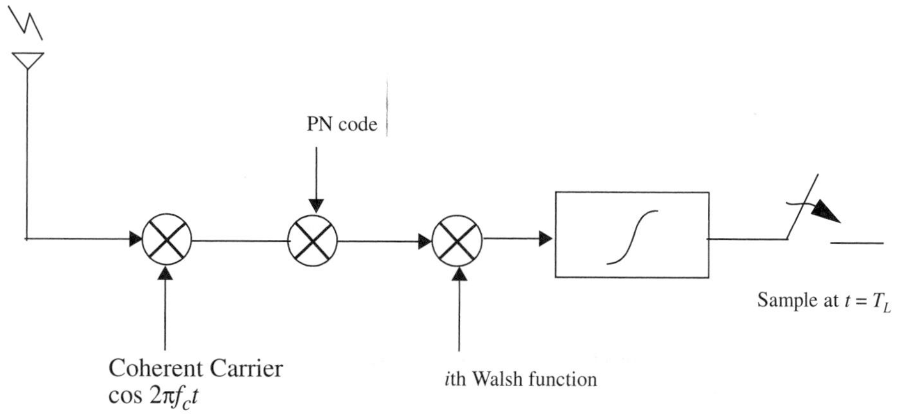

Figure 4.19 Application of Walsh Functions and PN Code in Mobile Station

The Walsh functions form an ordered set of rectangular waveform taking only two amplitudes of +1 and –1. They are defined over a limited time interval T_L, known as the *time base.* If ϕ_i represents the *i*th Walsh function and T_L is the time base, then

$$\frac{1}{T_L}\int_0^{T_L} \phi_i(t)\phi_j(t)dt = 0 \qquad \text{for } i \neq j \tag{4.11}$$

and

$$\frac{1}{T_L}\int_0^{T_L} \phi_i^2(t)dt = 1 \qquad \text{for all } is \tag{4.12}$$

(4.13)

The Walsh codes generated at the receiver must be synchronized with the Walsh codes generated at the transmitter. In the forward direction, the base station transmits a pilot signal to enable the receiver to recover synchronization.

EXAMPLE 4.2

We consider a case where eight chips are used per bit to generate the Walsh functions. Specify these functions, sketch them, and show that they are orthogonal to each other.

$$H_8 = \begin{bmatrix} H_4 & H_4 \\ H_4 & \overline{H_4} \end{bmatrix} = \begin{bmatrix} 0&0&0&0&0&0&0&0 \\ 0&1&0&1&0&1&0&1 \\ 0&0&1&1&0&0&1&1 \\ 0&1&1&0&0&1&1&0 \\ 0&0&0&0&1&1&1&1 \\ 0&1&0&1&1&0&1&0 \\ 0&0&1&1&1&1&0&0 \\ 0&1&1&0&1&0&0&1 \end{bmatrix} = \begin{bmatrix} \phi_1 \\ \phi_2 \\ \phi_3 \\ \phi_4 \\ \phi_5 \\ \phi_6 \\ \phi_7 \\ \phi_8 \end{bmatrix}$$

Figure 4.20 shows the sketches of the eight Walsh functions. We consider ϕ_2 and ϕ_4 to show orthogonality.

$$\frac{1}{T_L}\int \phi_2(t)\phi_4(t)dt = \frac{1}{T_L}[-1\times -1 + 1\times 1 + 1\times -1 + 1\times(-1) + (-1)\times(-1) + 1\times 1 + 1\times -1 + 1\times -1] = 0$$

Similarly, we can show that all eight Walsh functions are orthogonal to each other.

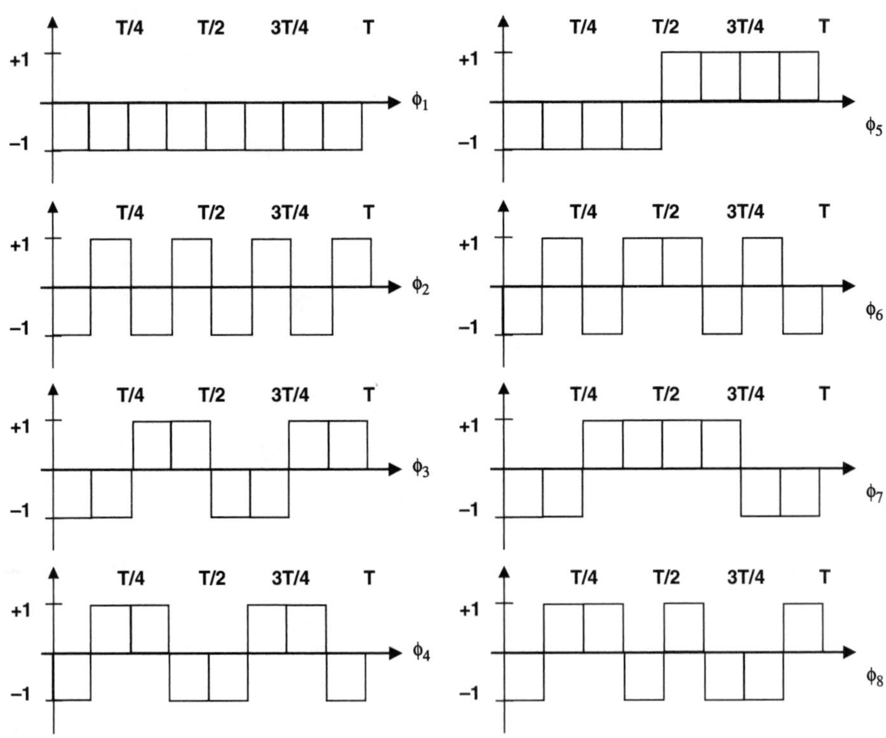

Figure 4.20 Walsh Functions $\phi_1, \ldots, \phi_8$

4.7.5 Variable-Length Orthogonal Codes

W-CDMA/cdma2000 is designed to support a variety of data services from low to very high bit rates. Since the spread signal bandwidth is the same for all users, multiple-rate transmission needs multiple spreading factor (SF) in the physical channels. A method to obtain variable-length orthogonal codes that preserve orthogonality between different rates and spreading factor is based on a modified Hadamard transformation.

If C_N is an $N \times N$ matrix that denotes the set of N binary spreading codes of N-chip length, $\{C_N(n)\}$; $n = 1,......, N$; where $C_N(n)$ is row vector of N elements and $N = 2^n$; it is generated from $C_{N/2}$ as

$$C_N = \begin{bmatrix} C_N(1) \\ C_N(2) \\ \vdots \\ C_N(N-1) \\ C_N(N) \end{bmatrix} = \begin{bmatrix} C_{N/2}(1)\ C_{N/2}(1) \\ C_{N/2}(1)\ \overline{C_{N/2}(1)} \\ \cdots \\ C_{N/2}(N/2)\ C_{N/2}(N/2) \\ C_{N/2}(N/2)\ \overline{C_{N/2}(N/2)} \end{bmatrix} \quad \text{(4.14)}$$

For $N = 2$: $C_2(1) = C_1(1)\ C_1(1)$; $C_2(2) = C_1(1)\ \overline{C_1(1)}$

For $N = 2^2 = 4$: $C_4(1) = C_2(1)\ C_2(1)$; $C_4(2) = C_2(1)\ \overline{C_2(1)}$; $C_4(3) = C_2(2)\ C_2(2)$;
$C_4(3) = C_2(2)\ \overline{C_2(2)}$

The variable-length orthogonal codes can be generated recursively using a tree structure (see Figure 4.21). The generated codes of the same layer constitute a set of Walsh functions that are orthogonal, although the rows of C_N are not in the same order of H_N. Any two codes of different layers are also orthogonal except for the case that one of the two codes is a mother code of the other. For example, all of $C_{16}(2)$, $C_8(1)$, $C_4(1)$, and $C_2(1)$ are mother codes of $C_{32}(3)$, and so are not orthogonal against $C_{32}(3)$. A code can be used in a channel if and only if no other code on the path from the specific code to the root of the tree or subtree produced by the specific code is used in the same channel. As an example, if $C_8(1)$ is assigned to a user, all codes $\{C_{16}(1), C_{16}(2), C_{32}(1),....., C_{32}(4), C_{64}(1),....., C_{64}(8), C_{128}(1),......, C_{128}(16)\}$ generated from this code cannot be assigned to other users requesting lower rates; in addition, mother codes $\{C_4(1), C_2(1)\}$ of $C_8(1)$ cannot be assigned to the users requesting higher rates. This means that the number of available codes is not fixed but depends on the rate and SF of each physical channel. These restrictions are needed to maintain orthogonality.

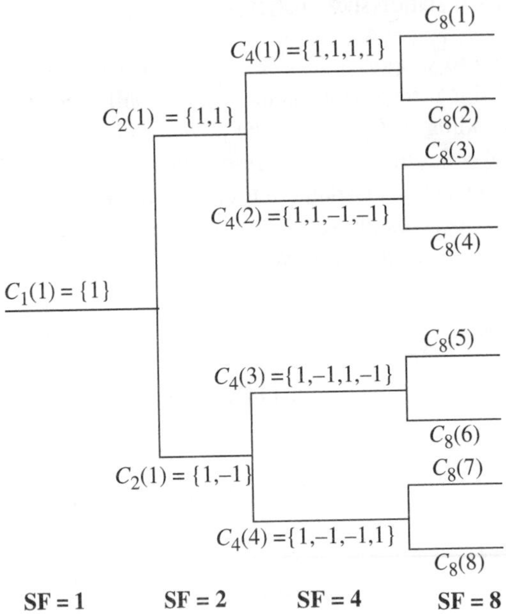

Figure 4.21 Spreading Factor (SF) Tree

4.7.6 Multiple Spreading

Multiple spreading, or two-layered spreading, code allocation provides flexible system deployment and operation. It is possible to provide waveform orthogonality among all users of the same cell while maintaining mutual randomness only between users of different cells. This is possible by the wide bandwidth of a spread-spectrum direct-sequence CDMA system that provides considerable waveform flexibility. Orthogonality can be attended by first multiplying each user's binary input by a short spread sequence which is orthogonal to that of every other user of the same cell. One class of such binary orthogonal sequences is the variable-length Walsh orthogonal set. The spread signal is followed by multiplication of a long PN sequence, which is cell-specific but common to all users of that cell in the forward link and user-specific in the reverse link. The short orthogonal codes are called *channelization codes*; long PN sequences are called *scrambling codes*. Therefore, each transmission *channel code* is distinguished by the combination of a channelization code and a scrambling code.

IS-95 standard uses synchronous systems, while European and Japanese systems employ the asynchronous technique wherein the reverse link mobile-unique scrambling code is allocated from the set of very large Kasami codes which are 256 chips long. Since this set of codes includes more than a million different codes, no extensive code planning is required. Each cell is allocated a suitable subset of these codes that are used by an active mobile within the cell. Also, a second scrambling code using Gold codes of length 2^{41} will be used for the reverse link. They will be truncated to form a cycle of 2^{15} bits (10-ms frame) and selected based on computer simulation such that cross-correlation is minimum. Since the asynchronous scheme does not require time synchronization between base stations, it makes system deployment from outdoor to indoor very flexible; however, it makes cell search and code synchronization somewhat more complex.

4.7.7 Gold Sequences

One of the goals of spread-spectrum designers for a multiple-access system is to find a set of spreading codes so that as many users as possible can utilize a band of frequencies with as little mutual interference as possible. Gold sequences are useful because of the large number of codes they supply. They can be chosen so that over a set of codes available from a given generator, the cross-correlation between the codes is uniform and bounded [17,18].

We consider an m-sequence represented by a binary vector a of length N and a second sequence a' obtained by sampling every qth symbol of a. The second sequence is said to be a *decimation* of the first, and the notation $a' = a[q]$ is used to indicate that a' is obtained by sampling every qth symbol of a. $a' = a[q]$ has a period N if and only if $gcd\ (N, q) = 1$ where gcd is the greatest common divisor. Any pair of m-sequences having the same period N can be related by $a' = a[q]$ for some q.

Two m-sequences a and a' are called the *preferred pair* if

- $n \neq 0$ (mod 4); i.e., n is odd or $n = 2$ (mode 4)
- $a' = a[q]$, where q is odd and either $q = 2^k + 1$ or $q = 2^{2k} - 2^k + 1$
- $gcd(n,k) = 1$ for n odd or 2 for $n = 2$ (mod 4)

The cross-correlation spectrum between a preferred pair is three-valued, where these three values are $-t(n)$, -1, and $t(n) - 2$, in which

$$t(n) = 1 + 2^{\frac{n+1}{2}} \quad \text{for } n \text{ odd}$$

$$= 1 + 2^{\frac{n+2}{2}} \quad \text{for } n \text{ even}$$

Finding preferred pairs of m-sequences is necessary in defining sets of Gold codes. If a and a' represent a preferred pair of m-sequences with period $N = 2^n - 1$, the family of codes defined by $\{a, a', a + a', a + Da', a + D^2a', \ldots\ldots\ldots, a + D^{N-1}a'\}$ (where D = delay element) is called the *set of Gold codes* for this preferred pair of m-sequences. With the exception of sequences a and a', the set of Gold sequences are not maximal sequences. Hence, their autocorrelation functions are not two-valued, and it takes the same three values as cross-correlation.

Maximal linear code sequences have many combining properties. The most interesting comes about when two maximal-length sequences of equal length are modulo-2 added. The resultant sequence is the new code of the same length. More important is that each phase shift produces another new code. The new codes are not maximal codes, but this is clearly a simple method of producing a generator with a large number of code options. When two code generators are used in this manner, the result is a Gold code (see Figure 4.22).

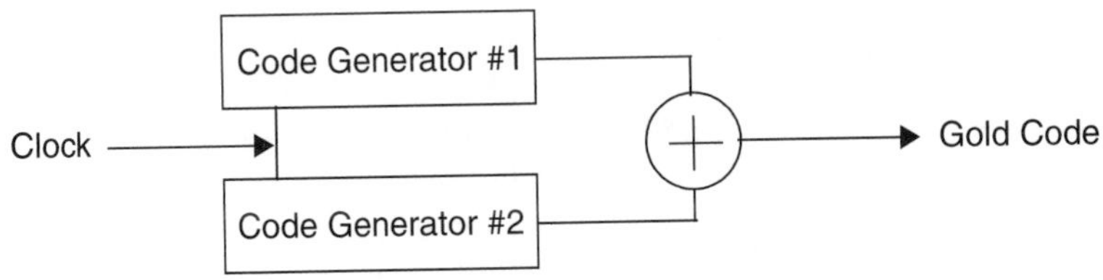

Figure 4.22 Generation of Gold Code

If length of a maximal-length PN sequence is $2^n - 1$, then the number of options is also $2^n - 1$. Gold codes offer better cross-correlation properties than *m*-sequence codes and, at the same time, can be long and offer good spectral characteristics.

4.7.8 Kasami Sequences

Kasami sequence sets are one of the important types of binary sequence sets because of their very low cross-correlation. There are two different sets of Kasami sequences. A procedure similar to that used for generating Gold sequence will generate the *small set* of Kasami sequence with $M = 2^{n/2}$ binary sequences of period $N = 2^n - 1$, where n is even. In this procedure, we begin with an *m*-sequence a and form the sequence a' by decimating a by $2^{n/2} + 1$. It can be shown [19] that the resulting a' is an m-sequence with period $2^{n/2} + 1$. For n = 10, the period of a is $N = 1023$ and period of a' is 31. Hence, if we observe 1023 bits of sequence a', we will see 33 repetitions of 31-bit sequence. By taking $N = 2^n - 1$ bits of sequences a and a' we form a new set of sequences by modulo-2 adding the bit from a and the bits from a' and all $2^{n/2} - 2$ cyclic shifts of the bits from a'. By including a in the set, we obtain a set of $2^{n/2}$ binary sequences of length $N = 2^n - 1$. The autocorrelation and cross-correlation function of the small set of Kasami sequences take on values from the set $\{-1, -(2^{n/2} + 1), (2^{n/2} - 1\}\}$.

The *large set* of Kasami sequences consists of sequences of period $2^n - 1$, for n even, and contains both Gold sequences and the small set of Kasami sequences as subsets. The *m*-sequence a' and a'' are formed by decimation of a by $2^{n/2} + 1$ and $2^{(n+2)/2} + 1$, and all sequences are formed by adding a, a', and a'' with different shifts of a' and a''. The number of such sequences is $M = 2^{3n/2}$ if $n = 0$ mod 4 and even larger, $M = 2^{3n/2} + 2^{n/2}$, if $n = 2$ mod 4. All values of autocorrelation and cross-correlation from members of this set are limited to five values $\{-1; -1 \pm 2^{n/2}; -1 \pm 2^{n/2+1}\}$. It can be deduced that all the sequences generated by $f(D)$, $f'(D)$, and $f''(D)$ form the large set of Kasami sequences, where $f'(D)$ and $f''(D)$ generate function of sequence a' and a'', respectively.

We discussed three main types of spreading codes: *PN codes, combinational code* (Gold codes), and *orthogonal code*. For channel separation, good cross-correlation properties are desired. For cell identification, several options are available for implementation. In general, the requirement is a good balance between autocorrelation and cross-correlation. For optimal spreading with an even distribution of power across the transmitted bandwidth, the requirement is usually a noiselike PN nature.

The receiver despreading operation is a correlation with the spreading code of the desired transmitter. Gold code sequence generators are useful because of the large number of codes they supply, although they require only one pair of feedback tap sets. The Gold codes are generated by modulo-2 addition of a pair of maximal linear (PN) sequences. The code sequences are added chip-by-chip by synchronous clocking. The codes themselves are the same length and the codes generated are the same length as the two base codes, which are added together. The generated codes are nonmaximal.

In cdmaOne and cdma2000, the GPS is used and all transmitters employ the same PN code for modulation. In GPS, each transmitter uses the same PN code but each transmitter's signal is offset in time. This prevents a receiver from synchronizing to more than one signal at once. The autocorrelation of the PN code is useful in this case to allow good effective signal orthogonality.

Orthogonal codes are only truly orthogonal when in the correct time alignment. They are only suited to channel separation in the forward link since there will not be time alignment between mobiles in the same cell. Orthogonal codes could be used to separate channels in the reverse link allocated simultaneously to a single mobile provided a second code is also used to separate mobiles. The second limitation is that these codes are not noiselike and pseudorandom. The result is that different codes within the code set will spread by different amounts even at the same chip rate. Thus when codes of this types are used, they need to be followed by a suitable spreading code.

EXAMPLE 4.3

Using two maximal-length PN sequences, 1 1 1 0 1 0 0 and 1 1 1 0 0 1 0, develop cross-correlation (Figure 4.23).

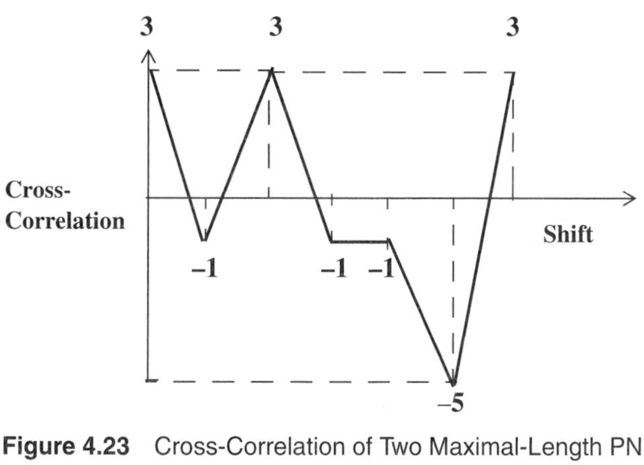

Figure 4.23 Cross-Correlation of Two Maximal-Length PN Sequences

A = Agreement; *D* = Disagreement

Shift 0

1 1 1 0 1 0 0
1 1 1 0 0 1 0

***A* – *D* = 5 – 2 = 3**

Shift 1

1 1 1 0 1 0 0
0 1 1 1 0 0 1

***A* – *D* = 3 – 4 = –1**

Shift 2

1 1 1 0 1 0 0
1 0 1 1 1 0 0

***A* – *D* = 5 – 2 = 3**

Shift 3

1 1 1 0 1 0 0
0 1 0 1 1 1 0

***A* – *D* = 3 – 4 = –1**

Shift 4

1 1 1 0 1 0 0
0 0 1 0 1 1 1

***A* – *D* = 3 – 4 = –1**

Shift 5

1 1 1 0 1 0 0
1 0 0 1 0 1 1

***A* – *D* = 1 – 6 = –5**

Shift 6

1 1 1 0 1 0 0
1 1 0 0 1 0 1

***A* – *D* = 5 – 2 = 3**

EXAMPLE 4.4

Using two maximal-length sequences, 1 1 1 0 1 0 0 and 1 1 1 0 0 1 0, generate a set of Gold codes with shift 0, 1, 2, 3, 4, 5 and 6.

Shift 0

```
1 1 1 0 1 0 0
1 1 1 0 0 1 0
---------------
0 0 0 0 1 1 0      Gold Code #0
```

Shift 1

```
1 1 1 0 1 0 0
0 1 1 1 0 0 1
---------------
1 0 0 1 1 0 1      Gold Code #1
```

Shift 2

```
1 1 1 0 1 0 0
1 0 1 1 1 0 0
---------------
0 1 0 1 0 0 0      Gold Code #2
```

Shift 3

```
1 1 1 0 1 0 0
0 1 0 1 1 1 0
---------------
1 0 1 1 0 1 0      Gold Code #3
```

Shift 4

```
1 1 1 0 1 0 0
0 0 1 0 1 1 1
---------------
1 1 0 0 0 1 1      Gold Code #4
```

Shift 5

```
1 1 1 0 1 0 0
1 0 0 1 0 1 1
---------------
0 1 1 1 1 1 1      Gold Code #5
```

Shift 6

```
1 1 1 0 1 0 0
1 1 0 0 1 0 1
---------------
0 0 1 0 0 0 1      Gold Code #6
```

EXAMPLE 4.5

Use Gold code #2 and #4 from Example 4.4 and show the cross-correlation properties.

Table 4.16 Cross-Correlation Calculations

Shift	Code Comparison	*A* – *D*
0	0 1 0 1 0 0 0 1 1 0 0 0 1 1	3 – 4 = –1
1	0 1 0 1 0 0 0 1 1 1 0 0 0 1	3 – 4 = –1
2	0 1 0 1 0 0 0 1 1 1 1 0 0 0	5 – 2 = 3
3	0 1 0 1 0 0 0 0 1 1 1 1 0 0	5 – 2 = 3
4	0 1 0 1 0 0 0 0 0 1 1 1 1 0	3 – 4 = –1
5	0 1 0 1 0 0 0 0 0 0 1 1 1 1	3 – 4 = –1
6	0 1 0 1 0 0 0 1 0 0 0 1 1 1	1 – 6 = –5

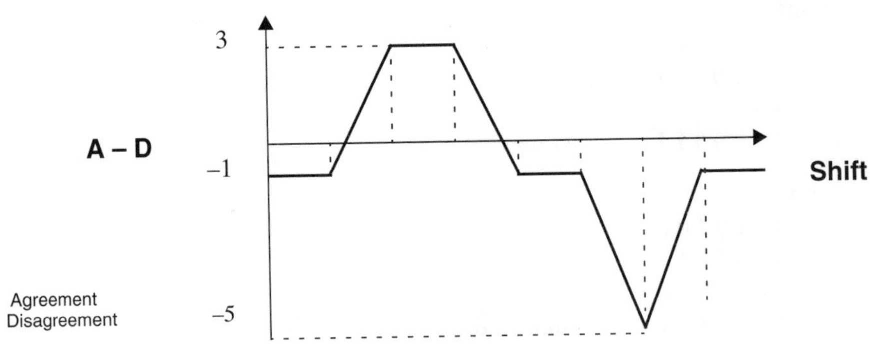

Figure 4.24 Cross-Correlation of Gold Codes

Note: The cross-correlation function has three values –1, –5, and 3 [as $t(m) = 2^{(m+1)/2} + 1$, e.g., $2^2 + 1 = 5$].

4.8 Modulation

Quadrature phase-shift keying (QPSK) [20] is characterized as two orthogonal binary phase-shift keyed (BPSK) channels. The QPSK bit stream is divided into even (I) and odd (Q) stream; each new stream modulates an orthogonal component of the carrier at half the bit rate of the original bit stream. The I stream modulates the $\cos\omega t$ term and the Q stream modulates the $\sin\omega t$ term. If the magnitude of the original QPSK vector has a value A, the magnitude of the I and Q component vectors each has a value of $0.707A$. Thus, each of the quadrature BPSK signals has half the average power of the original QPSK signal. Hence, if the original QPSK waveform has a bit rate of R bits/sec and an average power of S Watts, the quadrature partitioning results in each of the BPSK waveforms with a bit rate of $R/2$ bits/sec and an average power of $S/2$ Watts.

The E_b/N_0 characterizing each of the orthogonal BPSK channels, comprising the QPSK signal, is equivalent to E_b/N_0 since it can be written as:

$$\frac{E_b}{N_0} = \frac{S/2}{N_0} \cdot \left(\frac{1}{R/2}\right) = \frac{S}{N_0} \cdot \left(\frac{1}{R}\right) \tag{4.15}$$

Thus, each of the orthogonal BPSK channels, and hence the composite QPSK signal, is characterized by the same E_b/N_0 and hence the same bit probability error P_b as a BPSK signal. The natural orthogonality of the 90 degrees phase-shifts between adjacent QPSK symbols results in the bit error probabilities being equal for both BPSK and QPSK signals. It is important to note that the symbol error probabilities are not equal for the BPSK and QPSK signals.

On the $I - Q$ plane, QPSK is represented by four equally spaced points separated by $\pi/2$ (see Figure 4.25). Each of the four possible phases of carriers represents two bits of data. Thus, there are two bits per symbol. Since the symbol rate for QPSK is half of the bit rate, twice the information can be carried in the same amount of channel bandwidth as can be carried using BPSK. This is possible because the two signals I and Q are orthogonal to each other and, thus, transmitted without interfering with each other.

In QPSK, the carrier phase can change only once every $2T$ seconds. If from one $2T$ interval to the next, neither bit stream changes sign, the carrier phase remains the same. If one component $a_I(t)$ or $a_Q(t)$ changes sign, a phase-shift of $\pi/2$ occurs. However, if both components I and Q change sign, then a phase shift of π or 180 degrees occurs. When this 180 degree shift is filtered by the transmitter and receiver filters, it generates a change in amplitude of the detected signal and causes additional errors.

If the two bit streams, I and Q are offset by a half-bit interval, then the amplitude fluctuations are minimized since the phase never changes by 180 degrees. This type of modulation scheme, called offset quadrature phase-shift keying (OQPSK) [20], is obtained from conventional QPSK by delaying the odd bit stream Q by a half-bit interval with respect to the even bit stream, I. Thus, the range of phase transition is only 0 and 90 degrees and it occurs twice as often, but with half the intensity of the QPSK system. The I signal is the same for both QPSK and OQPSK, but the Q signal is delayed by half a bit. Thus, the 180 degree phase change at the end of the bit interval of the QPSK signal is replaced by a 90 degree phase change of the end of a bit interval of the OQPSK signal. While phase changes will still cause amplitude fluctuations in the transmitter and receiver, the fluctuations will have smaller magnitude. The bit error rate and bandwidth efficiency of QPSK and OQPSK are the same.

In theory, QPSK or OQPSK systems can improve the spectral efficiency of mobile communications. They do, however, require a coherent detector. In a multipath fading environment, the use of coherent detection is difficult and often results in poor performance over noncoherent-based systems. The coherent detection problem can be overcome by using a differential detector, but then OQPSK is subject to intersymbol interference (ISI) which results in poor system performance. The spectrum of offset QPSK is

$$P_{QPSK}(f) = T\left[\frac{\sin\pi fT}{\pi fT}\right]^2 \tag{4.16}$$

$\pi/4$-DQPSK is $\pi/4$-QPSK with differential encoding of symbol phases. The differential encoding mitigates loss of data due to phase slips. However, differential encoding results in loss of a pair of symbols when channel errors occur. This can be translated to about 3-dB loss in E_b/N_0 relative to coherent $\pi/4$-QPSK.

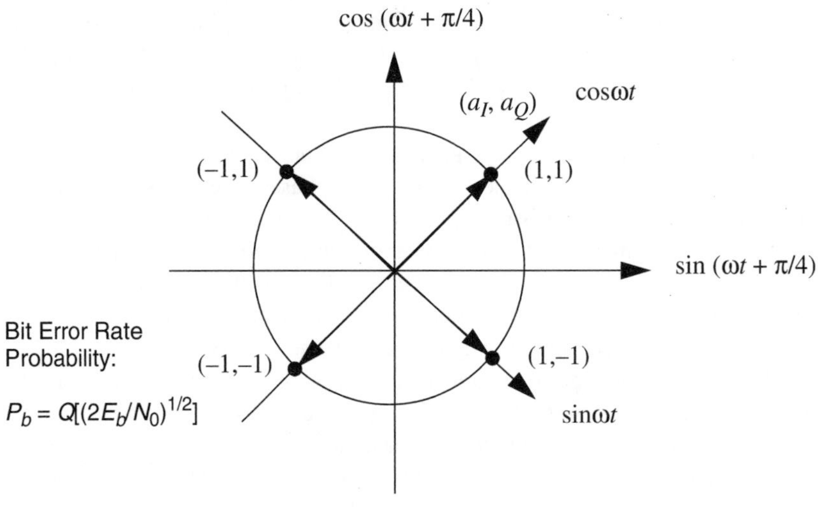

Figure 4.25 Signal Constellation for QPSK

4.9 Summary

In this chapter, we presented speech codec attributes and briefly discussed different coding schemes. We provided details of the EVRC which takes advantage of signal processing hardware and software technique to provide 13-kbps voice quality at 8-kbps data rate by maximizing both quality and system capacity. MOS data indicates that 8-kbps EVRC compares favorably to 16-kbps LD-CELP and ADPCM, the industry standards for comparison. The tests also show that EVRC maintains superior quality over 13-kbps CELP as frame error rates increase.

We focused on two types of channel coding schemes: convolutional codes and turbocodes. The error correcting capability of a convolutional coding scheme increases as rate decreases and constraint length increases. However, the number of code states and consequently, decoder complexity, grows exponentially with the constraint length. Turbocodes use serial and/or parallel recursive convolutional codes. They appear to approach Shannon bound with large enough iterations of decoding. Turbocodes spread error uniformly over the interleaved duration and, therefore, they are very useful for fading channels. Turbocodes provide additional coding gain of 2 to 3 dB over the convolutional codes.

We first reviewed the important aspects of the spreading codes used in the DS-CDMA technique. The spreading codes in CDMA are designed to have random behavior and very low cross-correlation. We also discussed the important characteristics of m-sequences and their generation by means of the LFSR structure. A technique to select m-sequence with good cross-correlation properties (preferred pairs) was presented which leads to the generation of Gold codes. Kasami sequence sets are also important because of the large number of codes they provide and their low cross-correlation properties. We also explained how the use of orthogonal codes and multiple spreading techniques provides flexible code allocation to the base station and mobile user.

Finally, we showed that QPSK is characterized as two orthogonal BPSK channels. Twice the amount of information can be carried for the QPSK in the same bandwidth as can be carried using BPSK. To reduce the phase change from 180 degrees to 90 degrees in QPSK, the two bit streams, I and Q, are offset by a half-bit interval. This type of modulation scheme is called OQPSK. The bit error rate and bandwidth efficiency of QPSK and OQPSK are the same.

4.10 References

1. TIA IS-96A, "Speech Service Option Standard for Wideband Spread Spectrum Digital Cellular System."
2. Chen, J., Cox, R., Lin, Y., Jayant, N., and Melchner, M., "Coder for the CCITT 16 kbps Speech Coder Standard," *IEEE Journal of Selected Areas of Communications*, 1988, [Vol. 6, pp. 353–63].
3. Atal, B. S., "Predictive Coding of Speech Signals at Low Bit Rates, "*IEEE Transactions on Communications*, 1982, [Vol. 30(4), pp. 600–14].
4. Kroon, P., and Deprettere, E. F., "A Class of Analysis-by-Synthesis Prediction Coders for High Quality Speech Coding at Rates between 4.8 and 16 kbps," *IEEE Journal of Selected Areas of Communications*, 1988, [Vol. 6, pp. 353–63].
5. Atal, B. S., and Schroeder, M. R., "Stochastic Coding of Speech at Very Low Bit Rate," *Proc. of International Conference on Communications*, Amsterdam, 1984, pp. 1610–13.

6. Hess, W., *Pitch Determination of Speech Signals*, Springer Verlag: Berlin, 1983.
7. Kleijn, W. B., Ramachandran, R. P., and Kroon, P., "Generalized Analysis-by-Synthesis Coding and Its Application to Pitch Prediction," *International Conference on Acoustics Speech, and Signal Processing*, San Francisco, 1992, pp. 1337–40.
8. TIA IS-733, "13 kbps Speech Coder."
9. TIA IS-127, "Enhanced Variable Rate Codec (EVRC) 8.5 kbps Speech Coder."
10. Ziemer, R. E., and Peterson, R. L., *Introduction to Digital Communication,* Macmillan Publishing Co.: New York, NY, 1992.
11. Skalar, B., *Digital Communication—Fundamental and Applications*, Prentice Hall: Englewood Cliffs, NJ, 1988.
12. TR45.5/98.04.03.03 "The cdma2000 ITU-R RTT Candidate Submission," April, 1998.
13. Bahl, L. R., Cocke, J., Jelinek, F., and Raviv, J., "Optimal Decoding of Linear Codes for Minimizing Symbol Error Rate," *IEEE Transactions on Information Theory,* [Vol. IT-20, No. 2].
14. Dinan, E. H., and Jabbri, B., "Spreading Codes for Direct Sequence CDMA and Wide CDMA Cellular Networks," *IEEE Communications Magazine*, Sept. 1998, [Vol. 36(9), pp. 48–55].
15. Viterbi, A. J., *CDMA Principle of Spread Spectrum Communication*, Addison-Wesley: 1995.
16. Golomb, S. W., *Shift Register Sequences*, Aegean Park Press, 1992.
17. Gold, R., "Optimal Binary Sequences for Spread Spectrum Multiplexing, "*IEEE Transactions on Information Theory,* Oct. 1967, [Vol. IT-B, pp. 619–21].
18. Gold, R., "Maximal Recursive Sequences with 3-Valued Recursive Cross-Correlation Functions," *IEEE Transactions on Information Theory,* Jan. 1968, [Vol. IT-4, pp. 154–56].
19. Kasami, T., "Weight Distribution Formula for Some Class of Cyclic Codes," *Coordinated Science Lab.*, University of Illinois, Urbana. Tech. Rep., R-285, Apr. 1966.
20. Garg, V. K., and Wilkes, J. E., *Principles & Applications Of GSM*, Prentice Hall PTR: Upper Saddle River, NJ, 1999.
21. Jarvinen, K., et al., "GSM Enhanced Full Rate Speech Codec," IEEE GLOBECOM '97.
22. Furuskar, A., et al., "System Performance of EDGE, a Proposal for Enhanced Data Rates in Existing Digital Cellular System," IEEE VTC '98, pp. 1284–89.
23. Van Nobelen, R., et al., "An Adaptive Radio Link Protocol with Enhanced Data Rates for GSM Evolution," *IEEE Personal Communications,* Feb. 1999, [Vol. 6(1), pp. 54–64].

4.11 Problems

1. Consider the four-stage linear shift register as shown in Figure 4.26. Calculate the output sequence. Assume the initial state of the registers as 1 1 0 1.

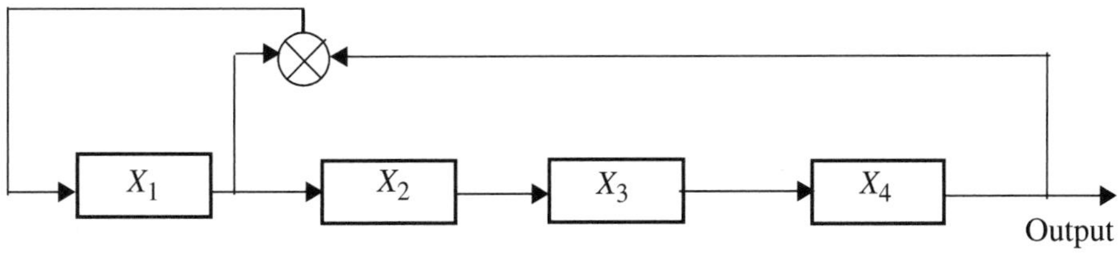

Figure 4.26 Four-Stage Register

2. Using the output sequence in Problem 1, demonstrate the properties of the maximal-length PN sequence.
3. Sketch the autocorrelation of the output sequence in Problem 1.
4. For a four-stage shift register generator, how many maximal-length PN sequences are generated? What is the location of the modulo-2 adder for each maximal-length sequence? What is the period of the maximal-length sequence?
5. Consider a case where 16 chips per bit are used to generate the Walsh codes, specify these codes, sketch W_0, W_8, W_{12}, and W_{16}. Show that these Walsh codes are orthogonal to each other.
6. Generate maximal-length PN sequences for a four-stage shift register generator. Sketch the cross-correlation of the output sequences.
7. Using the output sequences in Problem 6, develop Gold sequences.

CHAPTER 5

Physical and Logical Channels of IS-95

5.1 Introduction

In this chapter, we first discuss how to introduce a code division multiple access (CDMA) carrier in an existing advanced mobile phone system (AMPS) or time division multiple access (TDMA) system. We establish the number of AMPS or TDMA channels that should be removed in order to introduce the first and second CDMA carrier of 1.2288 MHz without interfering with the remaining AMPS or TDMA carriers. After this, we briefly describe modulation schemes, bit repetition, block interleaving, and channel coding, which are used in processing physical channels on the IS-95 CDMA forward and reverse links. Details about information processing, message types, and message framing are presented for the pilot, sync, paging, and traffic channels on the forward link. Similar details are also provided for the access and traffic channels on the reverse link.

5.2 Physical Channels

In IS-95, physical channels are defined in terms of a radio frequency (RF) and a code sequence. There are 64 Walsh codes available for the forward link (base station [BS] to mobile station [MS]), providing 64 physical channels. On the reverse link, channels are identified by long pseudorandom noise (PN) code sequences (see Chapter 4).

A CDMA system is implemented using N wideband frequency carriers, each capable of supporting M circuits that can be accessed by any user. A unique circuit is defined by a different code sequence for each user. Frequency assignment remains under control of the system in both downlink (BS to MS) and uplink (MS to BS) directions. Figure 5.1 shows N RF carriers for each direction. This is a *frequency division duplex* (FDD) arrangement and is called a *CDMA/FDD* system.

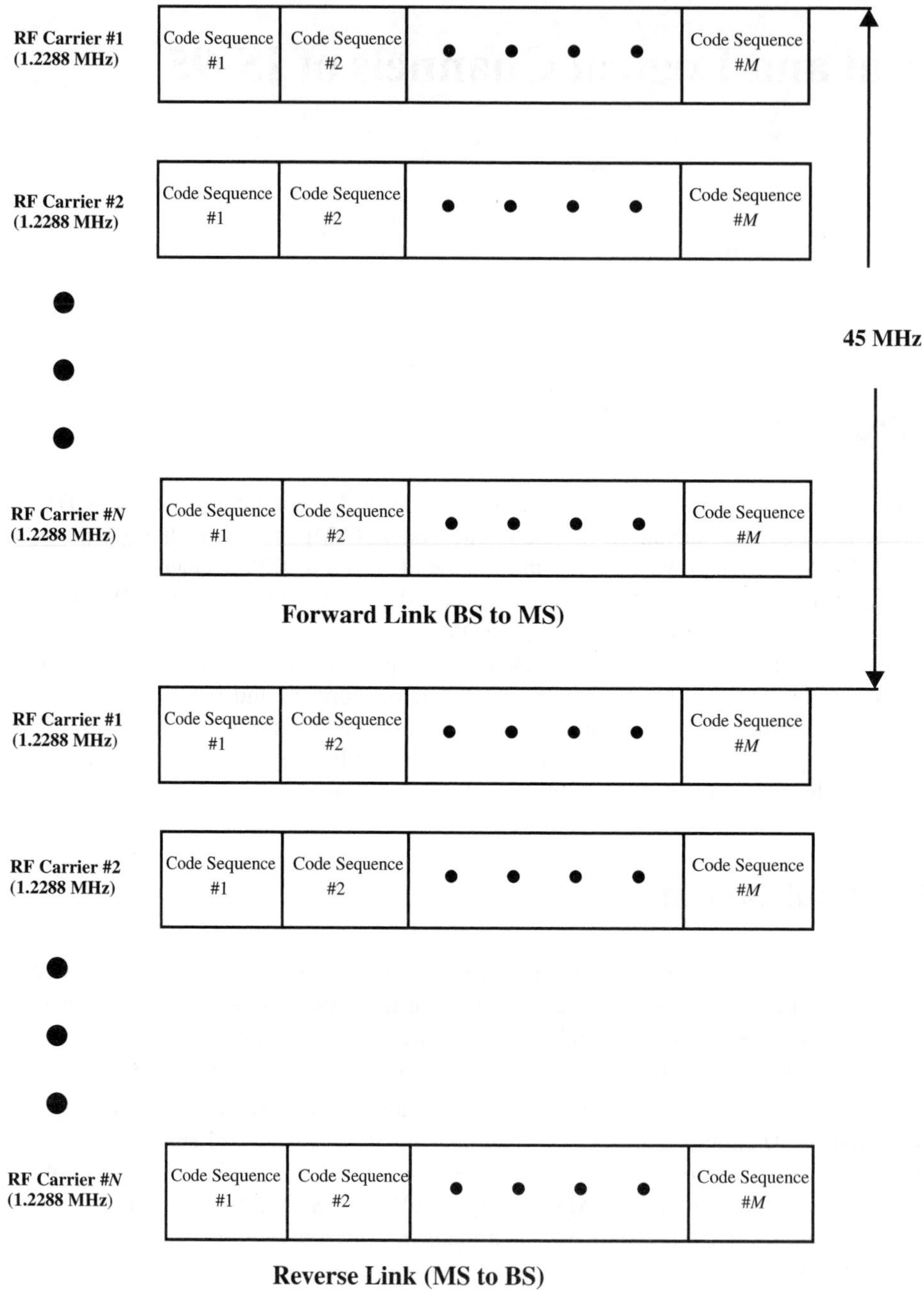

Figure 5.1 CDMA/FDD System

In IS-95, CDMA carrier band center frequencies are denoted in terms of AMPS channel numbers. Refer to Figure 5.2, in which AMPS channel number 283 is the center of a CDMA carrier band. To introduce one CDMA carrier, we need 41 30-kHz AMPS channels to provide a CDMA carrier bandwidth of 1.23 MHz. The recommended guard band between CDMA carrier band edge and an AMPS or a TDMA carrier is 0.27 MHz, which is equal to nine AMPS or TDMA channels. Thus, to introduce the first CDMA carrier without interfering with the remaining AMPS or TDMA channels, it is necessary that 59 AMPS channels be removed. In order to introduce the second CDMA carrier, we should remove only 41 additional AMPS channels. We see from Figure 5.3 that to introduce two CDMA carriers, we should remove 100 AMPS channels, a total of 3 MHz.

The *primary* and *secondary* CDMA carriers are the preassigned frequencies (AMPS channel numbers) that allow the mobile to acquire the CDMA system. A base station can support primary, secondary, or both types of channels.

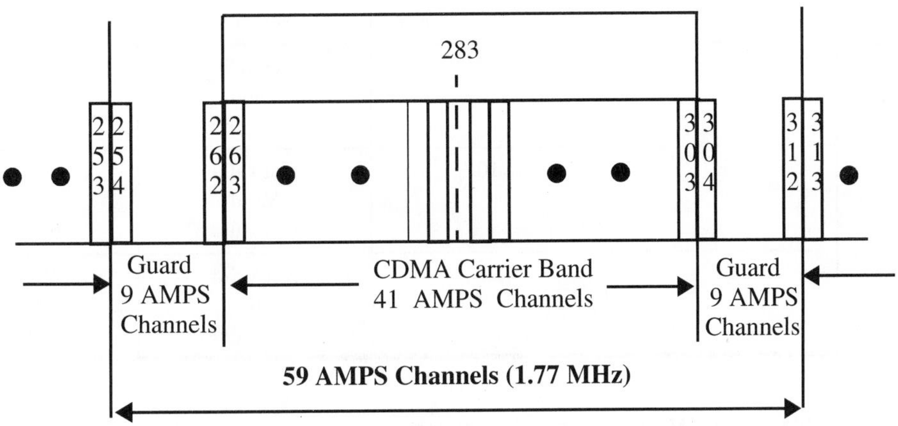

Figure 5.2 Introducing One CDMA Carrier

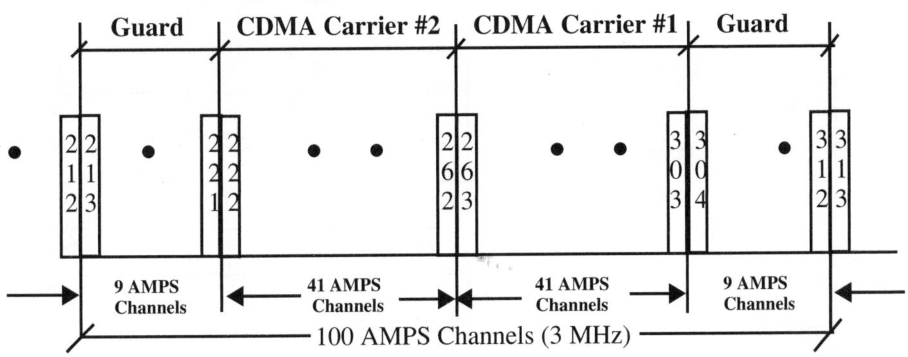

Figure 5.3 Introducing Two CDMA Carriers

The 1.23-MHz bandwidth for a CDMA carrier suggests that the minimum center frequency separation between two carrier frequencies is 1.23 MHz. The valid CDMA carrier frequencies are on AMPS channel numbers 1013–1023, 1–311, 356–644, 689–694, and 739–777 (see Table 5.1). Only the primary and secondary CDMA carrier center frequencies are specified in IS-95 standard. Other center frequencies are selected by each system operator.

Table 5.2 shows the CDMA center frequency (AMPS channel number) assignments for systems A and B with 41 AMPS channel separation.

Table 5.1 Definition of Valid Channel Numbers for CDMA at Cellular Frequencies

Frequency Band	Bandwidth (MHz)	Valid Regions	Channel Number
A'	1	Not valid Valid	991–1012 1013–1023
A	10	Valid Not valid	1–311 312–333
A"	1.5	Not valid Valid Not valid	667–688 689–694 695–716
B	10	Not valid Valid Not valid	334–355 356–644 645–666
B"	2.5	Not valid Valid Not valid	717–738 739–777 778–799

Table 5.2 CDMA Center Frequency Assignments for Systems A and B for Cellular

CDMA Channel Type	CDMA Frequency	AMPS Channels
	System A	**System B**
Primary	283	384
	242	425
	201	466
	160	507
	119	548
	78	589
	37	630
	1019	—
Secondary	691	777
Total number of CDMA channels	9	8

5.3 Modulation

The signals from each channel (pilot, sync, paging, and traffic) are modulo-2 added to *I* and *Q* PN short-code sequences. The *I* and *Q* spread signals are baseband filtered, and sent to a linear adder with gain control (see Figure 5.4). The gain control allows the individual channels to have different power levels assigned to them. The CDMA system assigns power levels to different channels depending on the quality of the received signal at a mobile station. The algorithms for determining the power levels are proprietary to each equipment manufacturer. The *I* and *Q* baseband signals are modulated by *I* and *Q* carrier signals, combined, amplified, and sent to the base station antenna. The net signal from the CDMA modulator is a complex quadrature signal that looks like noise.

The same PN short-code sequences are used on all channels (i.e., pilot, sync, paging, and traffic) of the forward link. All base stations in a system are synchronized using the Global Positioning System (GPS) satellite. Different base stations use time-shifted versions of these PN sequences to allow mobile stations to select the appropriate base station.

Unlike the forward direction, the CDMA system uses a different modulation scheme to generate the signal in the reverse direction. The net signal from modulator is a four-phase quadrature signal. The output from either the access channel or the traffic channel is sent to two modulo-2 adders, one for the in-phase and one for the quadrature channel. Two different PN short-code sequences are modulo-2 added to the data and filtered by a baseband filter. For a quadrature channel, a delay of half of a PN symbol (406.9 ns) is added before the filter. Thus, the reverse channel uses offset quadrature phase-shift keying (OQPSK). No pilot signal is used on the reverse link.

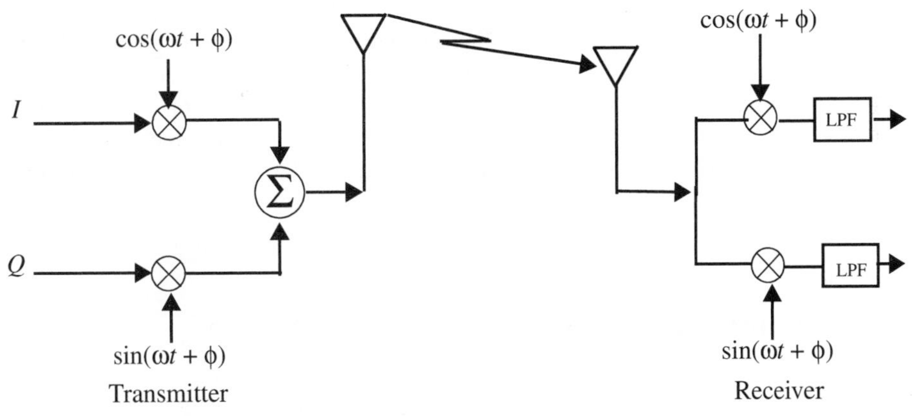

Figure 5.4 QPSK Modulator

5.4 Bit Repetition

The nominal Rate Set 1 data rate on the forward and reverse traffic channels is 9,600 bps. If the data is being transmitted at a lower rate (4,800, 2,400, or 1,200 bps), then the data bits are repeated n times to increase the rate to 9,600 bps.

5.5 Block Interleaving

The communications over a radio channel are characterized by deep fades that can cause large numbers of consecutive errors. Most coding schemes perform better on random data errors rather than on blocks of errors. By interleaving the data, no two adjacent bits are transmitted near each other, and the data errors are randomized.

The interleaver spans a 20-ms frame. In the reverse direction, the output of the interleaver is 28.8 kbps. If the data rate is 9.6 kbps, the resultant signal transmits with 100 percent duty cycle. If the data rate is lower (4,800, 2,400, or 1,200 bps), the interleaver plus randomizer deletes redundant bits and transmits with a lower duty cycle (50, 25, or 12.5 percent). Thus, bits are not repeated on the reverse CDMA traffic channel. On the access channel, the data bits are repeated. In the forward direction, the nominal data rate is 19.2 kbps, and lower data rates use a lower duty cycle.

On the reverse traffic channel, the output of the interleaver is processed by a data randomizer. The randomizer removes redundant data blocks generated by the code repetition. It uses a masking pattern determined by the data rate and the last 14 bits of the long code. For a 20-ms block (192 bits at 9,600 bps), the data randomizer segments the block into 16 blocks of 1.25 ms. At a data rate of 9,600 bps, all blocks are filled with data. At a data rate of 4,800 bps, 8 out of 16 blocks are filled with data in a random manner. Similarly, for 2,400 and 1,200 bps, 4 of the 16 and 2 of 16 blocks, respectively, are randomly filled with data. Thus, no redundant data are transmitted over the reverse channel.

5.6 Channel Coding

In IS-95, traffic data frames on uplink and downlink are fed to convolutional encoders. Both uplink and downlink encoders use an 8-bit shift register with a constraint length of 9. The rate of the uplink coder is 1/3; it outputs 3 bits for every input bit. At rates below 9.6 kbps, output bits are repeated to bring the number of bits in a 20-ms block to 576, for a gross rate of 28.8 kbps. The rate of the downlink encoder is 1/2; it outputs 2 bits for every input bit. At a rate below 9.6 kbps, output bits are repeated to bring the number of bits in a 20-ms block to 384, for a gross rate of 19.2 kbps.

5.7 Logical Channels

Logical channels in CDMA are the *control* and *traffic* channels. The control channels are the *pilot channel* (downlink), the *paging channels* (downlink), the *sync channels* (downlink), and the *access channels* (uplink) (see Figure 5.5).

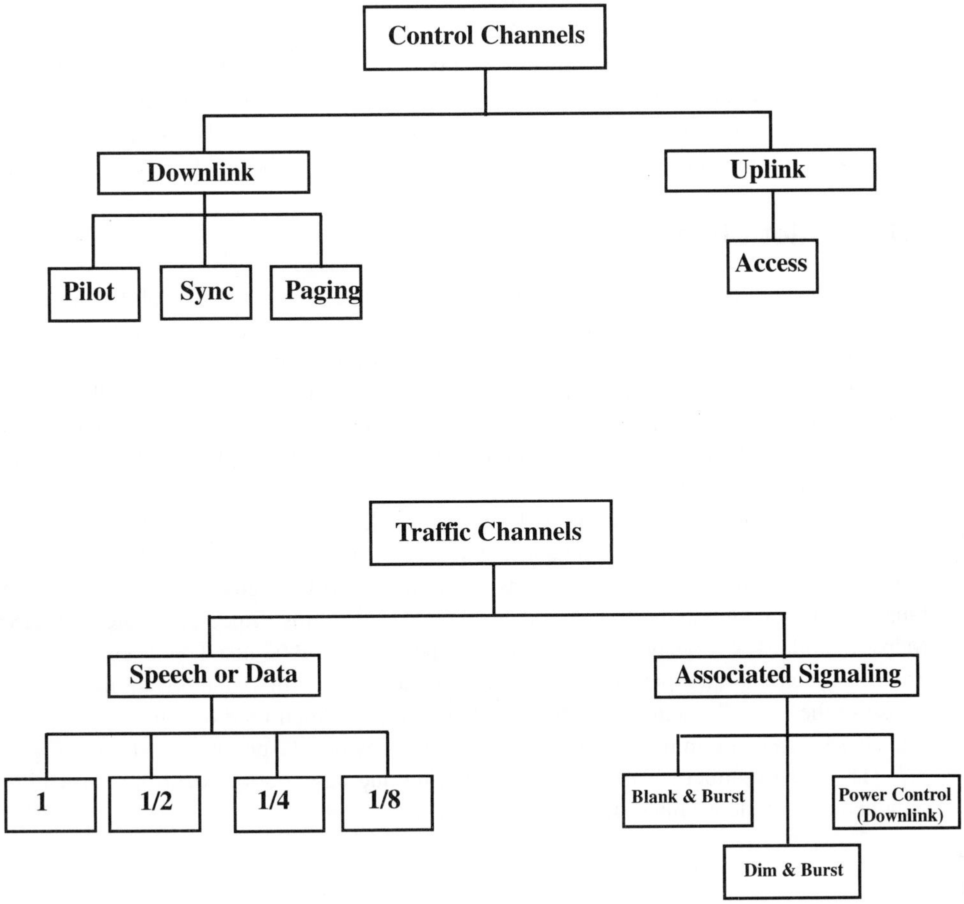

Figure 5.5 IS-95 Logical Channels

The traffic channels are used to carry user information (speech or data) between the base station and the mobile station, along with signaling traffic. Four different rates are used: Rate 1, Rate 1/2, Rate 1/4, and Rate 1/8. The downlink traffic channel is called the *forward link*, whereas uplink traffic channel is referred to as the *reverse link*.

When the user speech or data is replaced by the associated signaling, it is called *blank-and-burst*. When part of the speech is replaced by signaling information it is called *dim-and-burst*. All associated signaling is sent in Rate 1. In addition, on the downlink there is a power control subchannel that allows the mobile to adjust its transmitted power by ±1 dB every 1.25 ms (i.e., 16 times during a 20-ms speech frame).

In the forward direction there are the pilot channel, the sync channel, up to seven paging channels, and a number of forward traffic channels, all sharing the same center frequency. Out of the 64 Walsh coded channels available for use (W_0, W_1, W_2, . . . W_{63}), the pilot channel on W_0 is always required. There can be one sync channel (W_{32}), seven paging channels (W_1 to W_7, maximum allowed), and the remaining are traffic channels. The primary paging channel is always assigned the Walsh code W_1. The mobile examines the number of paging channel parameters in the *system parameter message*. If this value is not 1, a *hashing algorithm* is invoked to determine the correct paging channel number.

5.7.1 Pilot Channel

The pilot channel is used by a base station to provide a reference for all mobile stations. It provides a phase reference for coherent demodulation at the mobile receiver to enable coherent detection. It should be noted that the pilot channel does not carry any information and is assigned the Walsh code W_0 (see Figure 5.6). The pilot signal level for all base stations is kept about 4 to 6 dB higher than a traffic channel with a constant signal power. The pilot signal is used for signal strength comparisons between different base stations to decide when to perform handoff.

The pilot channel is needed to lock onto other channels on the same RF carrier. The pilot channel carries an unmodulated direct-sequence spread-spectrum (DSSS) signal that is transmitted continuously by each base station. The pilot signals are quadrature PN binary sequence signals with a period of 32,768 (2^{15}) chips. Since chip rate is 1.2288 Mcps, the pilot PN sequence corresponds to a period of 26.667 ms. This is equivalent to 75 pilot channel code repetitions every two seconds. The pilot signals from all base stations use the same PN sequences, but each base station is identified by a unique time offset. These offsets are in increments of 64 chips to provide 512 unique offsets (see Figure 5.7). The large number of offsets ensures that unique base station identification can be obtained even in dense microcellular environments.

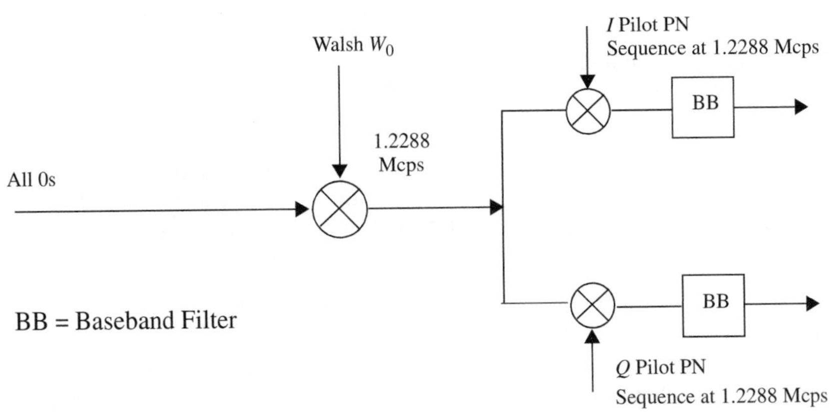

Figure 5.6 Pilot Channel Processing

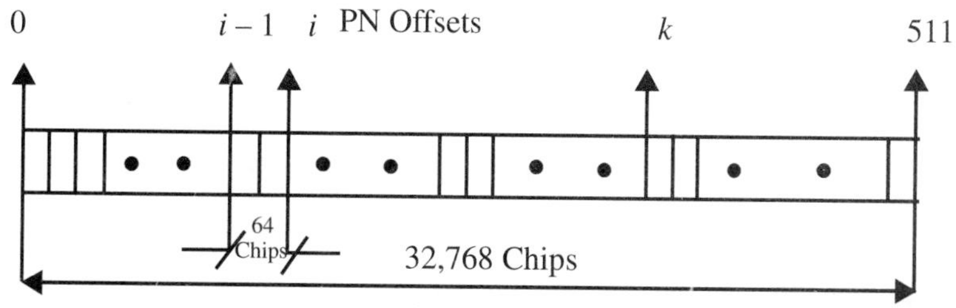

Figure 5.7 Short PN Offsets

5.7.2 Sync Channel

The sync channel is an encoded, interleaved, and modulated spread-spectrum signal that may be used with the pilot channel to acquire initial time synchronization. The sync channel is assigned the Walsh code W_{32}. The sync channel always operates at a fixed rate of 1,200 bps and is convolutionally encoded to 2,400 bps, repeated to 4,800 and interleaved (see Figure 5.8). The sync channel is used with the pilot channel to acquire initial time synchronization. Only the *sync message* is transmitted over this channel.

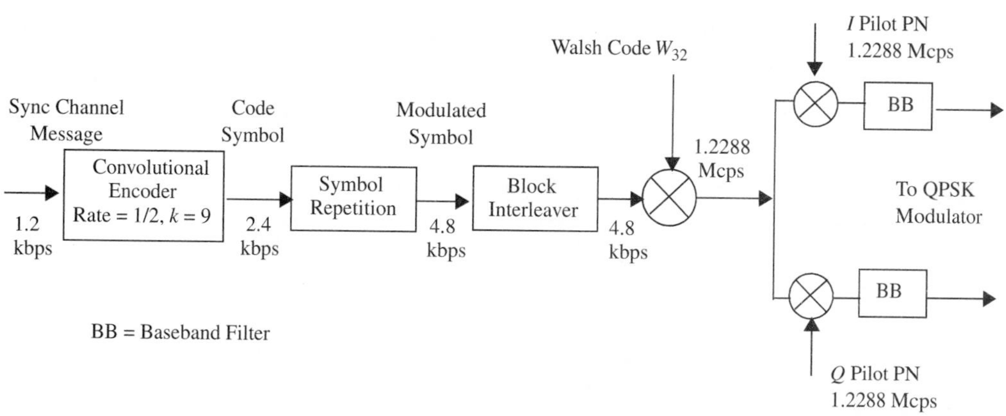

Figure 5.8 Sync Channel Processing

The sync message parameters are:

- **System identification (SID):** identifier number for the system
- **Network identification (NID):** identifier for the network

- **Pilot short PN sequence offset (PILOT_PN) index:** offset index, in units of 64 chips, for the base station or sector
- **Long code state (LC_STATE):** long code at the time specified in system time parameter
- **System time (SYS_TIME)**
- **Leap seconds (LP_SEC):** number of leap seconds that have occurred since the start of the system
- **Offset of local time:** offset from the system time
- **Daylight saving time indicator**
- **Paging channel rate (PRAT):** 4.8 or 9.6 kbps

The *sync channel message* itself is long and may occupy more than one *sync channel frame*. The sync channel message is organized in a *sync channel message capsule*. A sync channel message capsule contains the sync channel message and padding. When the sync channel message occupies more than one sync channel frame, padding is used to fill the bit positions up to the beginning of the next *sync channel superframe*, where the next sync message starts.

The sync channel message has an 8-bit message length header, a message body of a minimum of 2 bits and a maximum of 1,146 bits, and a cyclic redundancy check (CRC) code of 30 bits (see Figure 5.9). The message length includes the header, body, and CRC, but not the padding. The CRC is computed on the message length header and the message body using the following code:

$$g(x) = x^{30} + x^{29} + x^{21} + x^{20} + x^{15} + x^{13} + x^{12} + x^{11} + x^{8} + x^{7} + x^{6} + x^{2} + x + 1 \quad \textbf{(5.1)}$$

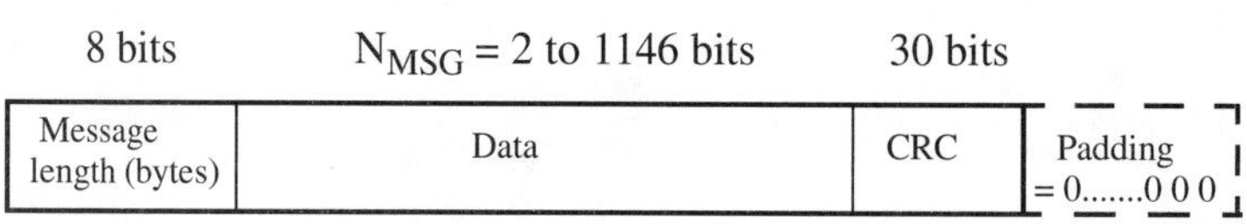

Figure 5.9 Message Framing on Sync Channel and Paging Channel

After a message is formed, it is segmented into 31-bit groups and sent in a sync frame (see Figure 5.10) consisting of a 1-bit start of message (SOM) field and 31 bits of the sync channel frame body. The value of 1 for SOM indicates that the frame is the start of the sync channel message, whereas a value of 0 for SOM indicates that the frame is a continuation of a sync channel message or padding. The sync channel frames are transmitted in groups of *sync channel superframes.* Three sync frames are combined to form a superframe. Each superframe carries 96 bits and lasts for 80 ms (see Figure 5.11). The entire sync channel message is then sent in N superframes. The padding bits are used so that the start message always starts 1 bit after the beginning bit of the superframe.

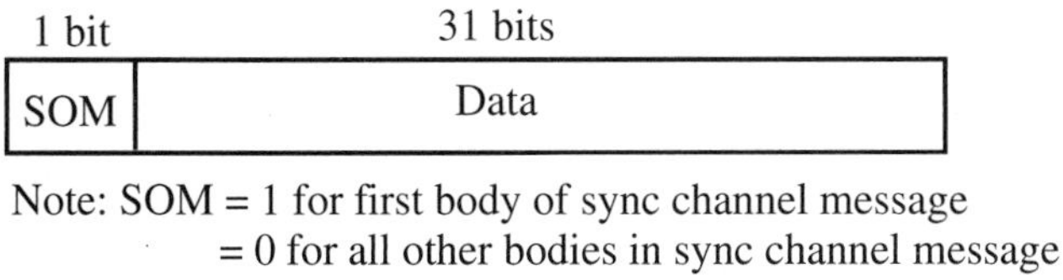

Figure 5.10 Sync Channel Frame

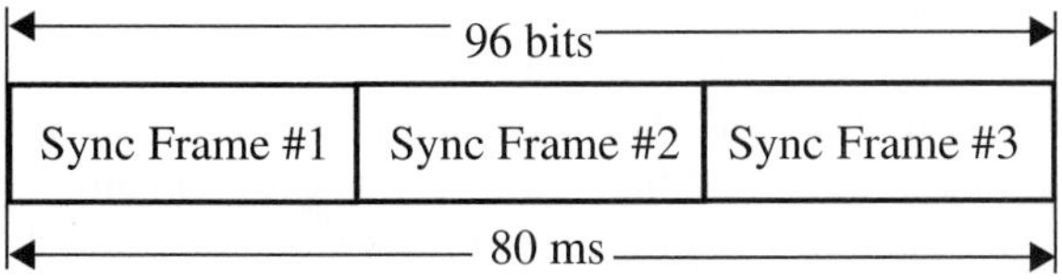

Figure 5.11 Sync Channel Superframe

5.7.3 Paging Channel

The paging channel is used to transmit control information to the mobile. When a mobile is to receive a call it will receive a "page" from the base station on an assigned paging channel. The paging channel provides the mobile stations with system information and instructions, in addition to acknowledging messages following access requests on the mobile station's access channels. The paging channel data are processed in a similar manner as the traffic channel data. However, there is no power control on a per-frame basis. The 42-bit mask is used to generate the long code (see Figure 5.12). The paging channel operates at a data rate of 4,800 or 9,600 bps.

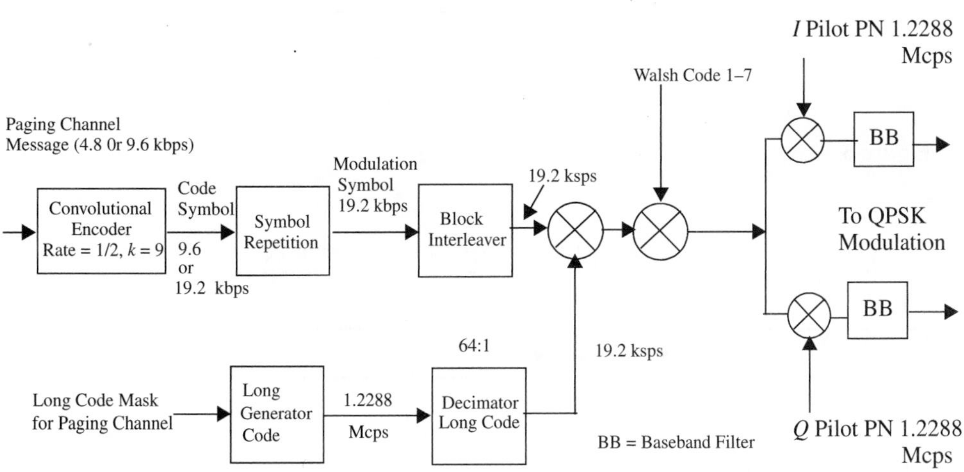

Figure 5.12 Paging Channel Processing

The paging channel message is similar in form to the sync channel message. It has an 8-bit message length header, a message body of a minimum of 2 bits and a maximum of 1,146 bits, and a CRC code of 30 bits. The message length includes the header, body, and CRC, but not the padding. The CRC is computed on the message length header and the message body using the same code as the sync channel.

Paging channel messages can use synchronized capsules that end on a half-frame boundary or unsynchronized capsules that can end anywhere within a half-frame. If synchronized paging channel messages are less than an integer multiple of 47 bits for 4,800-bps transmission (or 95 bits for 9,600-bps transmission), they are padded with 0 bits at the end of the message. Asynchronized messages are not padded.

After a message is formed, it is segmented into 45- or 95-bit chunks and sent in a paging channel half-frame (see Figure 5.13) that consists of a 1-bit synchronized capsule indicator (SCI) field and 47 or 95 bits of the paging channel frame body. A value of 1 for SCI indicates that the frame is the start of a paging channel message (either synchronized or asynchronized). Messages can also start in the middle of a frame and immediately after the end of an asynchronized message (with no padding bits). A value of 0 for SCI indicates that the frame is not the start of a message and can include a message (with or without padding), padding only, or the end of one message and start of another.

Eight paging channel half-frames are combined to form a *paging channel slot* (see Figure 5.14) of length 80 ms (384 bits at 4,800 bps and 768 bits at 9,600 bps). The entire paging channel message is then sent in N slots. The maximum number of slots that a message can use is 2,048. The base station always starts a slot with an asynchronized message capsule that starts 1 bit after the beginning of a slot. The first bit in a slot is SCI = 1.

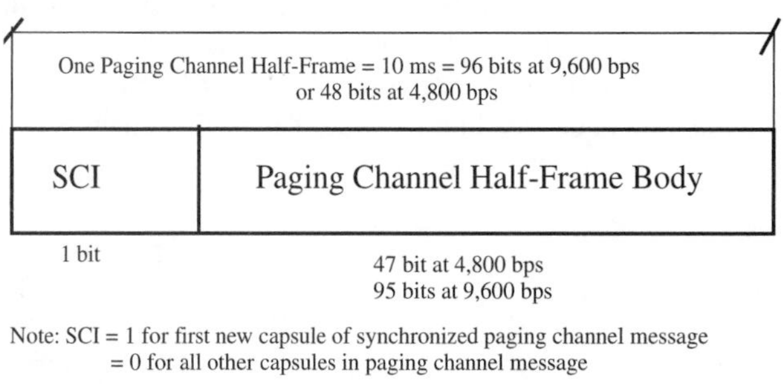

Figure 5.13 Paging Channel Half-Frame

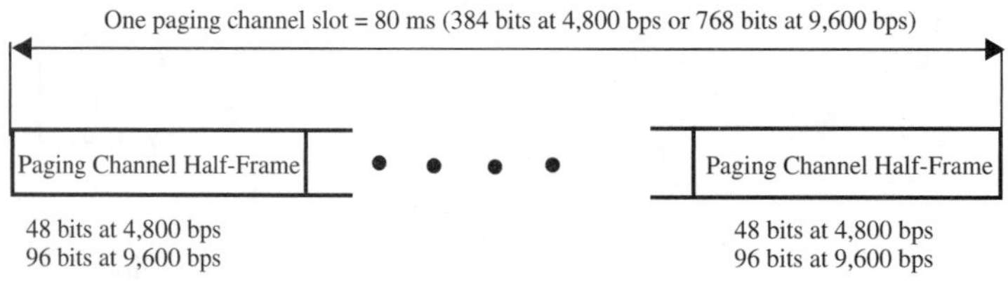

Figure 5.14 Paging Channel Slot Structure

The paging channel carries information to allow the network to

- Supply information to be displayed by the mobile station
- Identify the called party's number
- Identify the calling party's number
- Convey information to the mobile station by means of tones or other alerting signals
- Indicate the number of messages waiting

The paging channel carries the following messages:

1. **System parameter message**—provides overhead information such as pilot PN sequence offset index, *i*, in PN-I-i (t) and PN-Q-i (t), base station identifier, the number of paging channels, PAGE_CHAN, and other system information. PAGE_CHAN and a hashing algorithm (IS-95) are used by a mobile station to determine the correct paging channel number. Initially, the mobile station expects a PAGE_CHAN equal to 1, the primary paging channel. A different value invokes the hashing algorithm.
2. **Access parameters message**—defines parameters required by a mobile station to transmit on an access channel.
3. **Neighbor list message**—provides information about neighbor base station parameters, e.g., the neighbor pilot PN sequence offset index, *i*. If the neighbor does not have a page channel on the current CDMA carrier frequency, the ***CDMA channel list message*** contains this information.
4. **CDMA channel list message**—provides the list of CDMA carriers.
5. **Slotted page or page message**—provides data used to inform the mobile that it can receive a call. The mobile station monitors for its identification number in the MIN field. With the slotted page message, the mobile need only monitor specific time slots of the page channel message.
6. **Page message**—provides a page to the mobile station. The mobile monitors every time slot of the page channel message.
7. **Typical order message**—several order messages can be carried on the paging channel, such as abbreviated alert, base station challenge confirmation, reorder, audit, intercept, base station acknowledgment, lock until power-cycled, maintenance required, unlock, release (with or without reason), registration accepted, registration request, registration rejected, and local control.
8. **Channel assignment message**—message to inform the mobile station to tune to a new frequency.
9. **Data burst message**—data message sent by the base station to the mobile station.
10. **Authentication challenge message**—allows the base station to validate the mobile identity. The unique mobile authentication keys and/or shared secret data (SSD) for each mobile registered in the system will be used to perform the authentication calculations, which are then sent back to the base station in an authentication challenge response message.

11. **SSD update message**—request by the base station for the mobile station to update the SSD.
12. **Feature notification message**—contains information records to allow the network to supply information to be displayed by the mobile, to identify the called party's number, to identify the calling party's number, to convey information to the mobile by means of tones or other alerting signals, and to indicate the number of messages waiting.

The paging channel is divided into 80-ms slots called the *paging channel slots.* IS-95 allows two modes of paging, *slotted* and *nonslotted.* In the slotted mode, a mobile listens for pages only at certain times (i.e., during its page slot). This feature allows the mobile to turn off its receiver for most of the time, thereby saving battery power and increasing the time between battery charging. In the nonslotted mode of operation, the mobile is required to monitor all paging slots.

In the slotted mode, a mobile generally monitors the paging channel for one or two slots per slot cycle. The mobile can specify its preferred slot using the SLOT_CYCLE_INDEX field in the registration message, or in the origination message, or in the page response message, or traffic signaling to the base station. The length of the slot cycle, T, in units of 1.28 seconds is given by $T = 2^i$, where i is the selected slot cycle index.

There are $16T$ slots in a cycle for a particular mobile using some value of i, and four 20-ms full-frames in an 80-ms slot. A value of $i = 0$ means that the mobile listens to every 16th paging slot, $i = 1$ implies that the mobile monitors every 32nd slot. For $i = 2$, the mobile monitors every 64th slot. The $i = 0$ ensures that the pages are not missed by the mobile, but it is a drain on the mobile's battery power. The value of $i = 1$ is suggested. PGSLOT is a randomly calculated number that specifies the slot out of the $16T$ slots to be monitored by the mobile. This number is fixed for each mobile.

5.7.4 Access Channel

The *access channel* is used by the mobile to transmit control information to the base station. The access channel allows the mobile station to communicate nontraffic information (e.g., call origination and respond to page). The access rate is fixed at 4,800 bps. All mobile stations accessing a system share the same frequency assignment. Each access channel is identified by a distinct access channel long-code sequence having an access number, a paging channel number associated with the access channel, and other system data. There are many messages that can be carried on the access channel. When a mobile places a call it uses the access channel to inform the base station. This channel is also used to respond to a page.

The messages carried by an access channel are the following:

1. **Registration message**—The mobile station sends this message to inform the base station about its location status, identification, and other parameters required to register with the system. This is necessary so that the base station can page the mobile whenever a call is to be delivered to the mobile.
2. **Order message**—Typical order messages are base station challenge, SSD update confirmation, SSD update registration, mobile station acknowledgment, local control response, mobile station reject (with or without reason).

3. **Data burst message**—This is a user-generated data message sent by the mobile station to base station.
4. **Origination message**—This message allows the mobile station to place a call—sending dialed digits.
5. **Page response message**—This message is used by the mobile to respond to a page or slotted page in continuation of the process of receiving a call.
6. **Authentication challenge response message**—This message contains the necessary information to validate the mobile station's identity.

Figure 5.15 shows the processing of the access channel. The baseband information is error protected using a convolutional encoder of rate, $R = 1/3$. The lower encoding rate on the reverse link is used to make error protection more robust. The symbol repetition repeats the symbol, yielding a code symbol rate of 28.8 ksps. The data is then interleaved to combat fading. Following the interleaving, 64-ary orthogonal modulation is used. In this case, for each hextet (group of six symbols) sent to the modulator, one output Walsh function is generated. The reason for using orthogonal modulation of the symbol is the noncoherent nature of the reverse link. Since each group of six symbols is represented by a unique Walsh function, it is much easier for the base station to detect six symbols at a time by deciding which 64-bit Walsh function was transmitted during that period. A hextet of six coded symbols is used, as six binary symbols correspond to a decimal value between 1 and 64. The pattern of the six-symbol group determines the choice of the particular Walsh function transmitted.

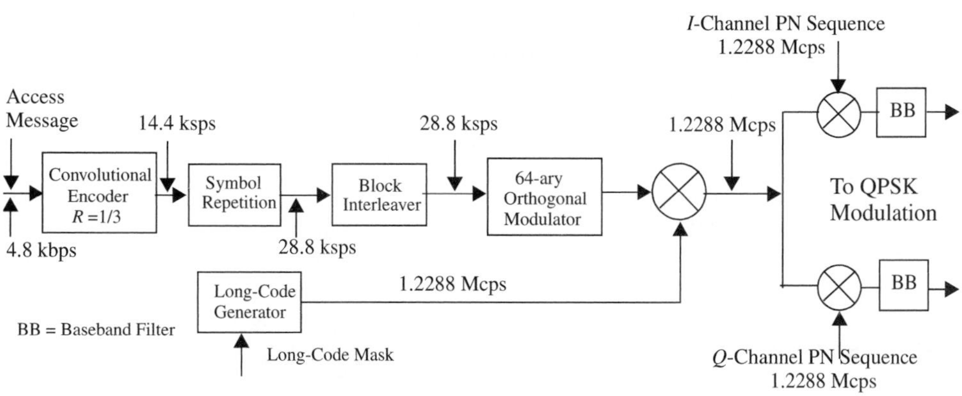

Figure 5.15 Access Channel Processing

The Walsh function is defined by

$$W_i = c_0 + 2c_1 + 4c_2 + 8c_3 + 16c_4 + 32c_5 \quad \textbf{(5.2)}$$

where

c_5 is the most recent and c_0 the oldest of the six symbols to be transmitted

W_i is selected from 0 of 63 orthogonal Walsh functions

As an example, suppose a hextet of six symbols (–1, 1, 1, –1, –1, –1) is an input. The corresponding bit values are (1 0 0 1 1 1). The output of the modulator in term of the Walsh function will be

$$W_{39} = 1 + 2 \times 1 + 4 \times 1 + 8 \times 0 + 16 \times 0 + 32 \times 1 = 39$$

The 64-ary modulated data at 4.8 ksps (modulation symbols) or at 307.2 ksps (code symbols) is spread using the long PN sequence at 1.2288 Mcps. The long code has a length of $2^{42} - 1$ chips and is generated by the following polynomial:

$$L(x) = x^{42} + x^{35} + x^{31} + x^{27} + x^{26} + x^{25} + x^{22} + x^{21} + x^{19} +$$

$$x^{18} + x^{17} + x^{16} + x^{10} + x^{7} + x^{6} + x^{5} + x^{3} + x^{2} + x^{2} + x + 1 \tag{5.3}$$

The output of the long-code generator is modulo-2 added. The long PN sequence is used to distinguish the access channel from all other channels occupying the reverse link. The data is scrambled in the *I* and *Q* paths by the short PN sequence. Since the reverse link uses OQPSK modulation, the data in the *Q* path are delayed by 1/2 a PN chip. The primary reason for this delay is to prevent collapse of QPSK signal envelope to zero. This is essential because the power amplifier of the mobile is typically small and limited in performance.

The message on the access channel consists of an access preamble of multiple frames of 96 zero bits with length of 1 + PAM_SZ frames (Figure 5.16), followed by an access message capsule of length 3 + MAX_CAP_SZ frames. The message capsule also consists of frames of length 96 bits. Since the data rate on the access channel is 4,800 bps, each frame has duration of 20 ms.

Access Channel Preamble = 00000......0000

96 × (1 + PAM_SZ) bits
(1 + PAM_SZ) frames

Figure 5.16 Access Channel Preamble

Thus, the entire access channel transmission occurs in an access channel slot that has a length of

$$4 + \text{MAX\_CAP\_SZ} + \text{PAM\_SZ} \text{ frames} \tag{5.4}$$

where the values of MAX_CAP_SZ and PAM_SZ are received on the paging channel.

An access channel slot nominally begins at a frame where

$$t \bmod (4 + \text{MAX\_CP\_SZ} + \text{PAM\_SZ}) = 0 \qquad (5.5)$$

where t is the system time in frames.

The actual start of transmission on the access channel is randomized to minimize collisions between multiple mobiles accessing the channel at the same time.

All access channels corresponding to a paging channel have the same slot length. Different base stations may have different slot lengths.

The *access channel message* (Figure 5.17) is similar in form to the sync channel message; it has an 8-bit message length header, a message body of a minimum of 2 bits and a maximum of 842 bits, and a CRC code of 30 bits. Following the message are padding bits to make the message end on a frame boundary. The message length includes the header, body, and CRC, but not the padding bits. The CRC is computed on the message length header and message body using the same code as the sync channel (Equation 5.1).

Each access channel frame contains either preamble bits (all zeros) or message bits. Frames containing message bits (Figure 5.18) have 88 message bits and 8 encoder tail bits (set to all zeros). Multiple frames are combined with an access channel preamble to form an *access channel slot* (Figure 5.19).

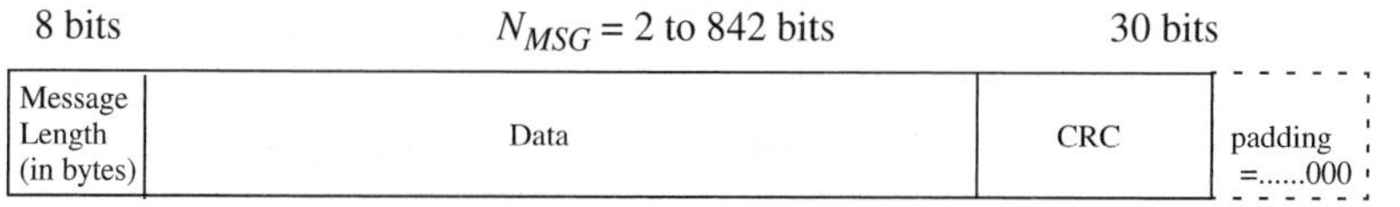

N_{MSG} = Message length in bits (including length field and CRC)

Figure 5.17 Message Framing on Access Channel

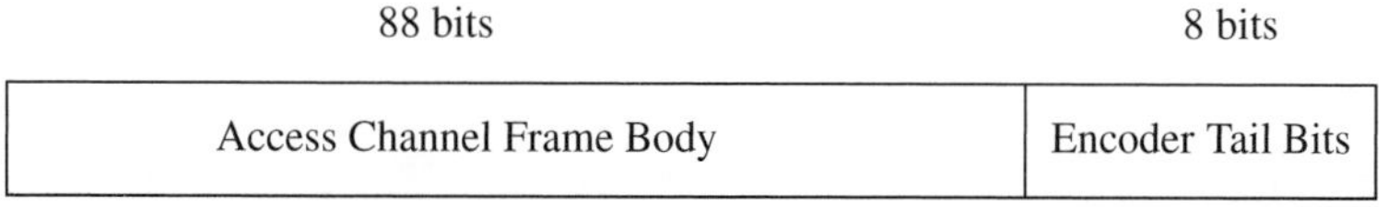

Figure 5.18 Access Channel Framing

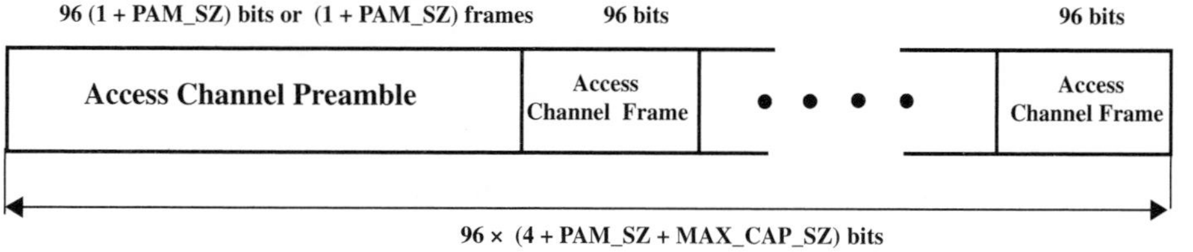

Figure 5.19 Access Channel Slot

5.7.5 Forward Traffic Channels

The forward traffic channels are grouped into rate sets. Rate Set 1 has four elements: 9,600, 4,800, 2,400, and 1,200 bps. Rate Set 2 uses four elements: 14,400, 7,200, 3,600, and 1,800 bps. All systems support Rate Set 1 on the forward traffic channels. Rate Set 2 is optionally supported on the forward traffic channels. When a system supports a rate set, all four elements are supported. Walsh codes that can be assigned to forward traffic channels are available at a cell or sector (W_2 through W_{31}, and W_{33} through W_{63}). After the first paging channel, each additional paging channel consumes one traffic channel Walsh code. So, if all seven paging channels are used, only 55 Walsh codes will be available for the forward traffic channels.

The speech is encoded using a variable rate vocoder to generate the forward traffic data depending on voice activity. Since frame duration is fixed at 20 ms, the number of bits per frame varies according to traffic rate. Since half-rate convolutional encoding is used, it doubles the traffic rate to provide rates from 2,400 to 19,200 symbols per second. Interleaving is performed over 20 ms. A long code of $2^{42} - 1 = 4.4 \times 10^{12}$ is generated containing a user's electronic serial number (ESN) embedded in the mobile station long-code mask (with voice privacy, the mobile station long-code mask does not use the ESN). The scrambled data are multiplexed with power control information that steals bits from the scrambled data. The multiplexed signal remains at 19,200 bps and is changed to 1.2288 Mcps by the Walsh code W_i assigned to the ith user traffic channel. The signal is spread at 1.2288 Mcps by quadrature PN binary sequence signals, and the resulting quadrature signals are then weighted. The power level of the traffic channel depends on its data transmission rate.

A *power control subchannel* is continuously transmitted on the forward traffic channel. A 0 specifies that the mobile increases its mean output power level by 1 dB (nominal) and a 1 indicates a decrease in mean output power level by 1 dB (nominal).

The power control bits puncture the modulated data symbols at a rate of 800 bps; a single power control bit replaces two data symbols. The location of a power control bit in a power control group is determined by bits 20–23 of the 1/64 long code in the previous 1.25-ms control group. The 19.2-ksps long code is decimated to 800 bps (see Figure 5.20) to establish the location of the bits in the power control subchannel.

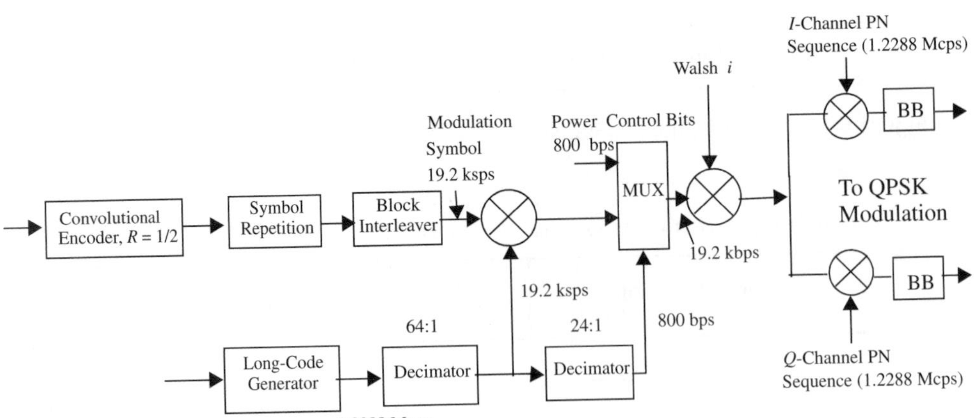

Figure 5.20 Forward Traffic Channel Processing

Each power control group contains 24 scrambled bits. The 24 bits are positioned within a 1.25-ms period and are numbered 0, 1, 2, . . . , 23 (see Figure 5.21). The position of the power control bits within the 1.25-ms power control group is determined from bits 20–23 of the 1/64 long code in the previous 1.25-ms control group period. This is done to randomize the location of the power control bits to avoid any spikes caused by to periodic repetition. The power control bits are always transmitted at full power.

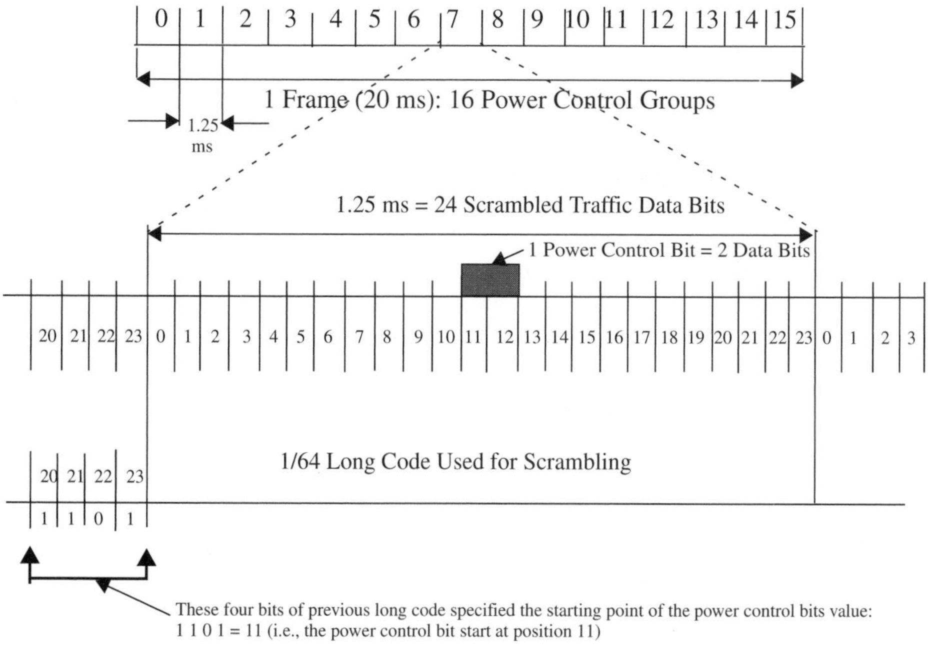

Figure 5.21 Position of the Power Control Bits

Channels not used for paging or sync can be used for traffic. Thus, the total number of traffic channels at a base station is 63 minus the number of paging and sync channels in operation at that base station. Information on the forward traffic channels includes the primary traffic (voice or data), secondary traffic (data), and signaling, in frames 20 ms long.

When the data rate on the forward traffic channel is 9,600 bps, each frame of 192 bits carries 172 information bits, 12 frame-quality bits, and 8 encoder tail bits (set of all zeros). At 4,800 bps, there are 80 information bits, 8 frame-quality bits, and 8 tail bits for a total of 96 bits. At 2,400 and 1,200 bps, there are 40 and 16 information bits and 8 tail bits, for a total of 48 and 24 bits, respectively. The base station can select the data transmission rate on a frame-by-frame basis. The data rate of 9,600 bps can support multiplexed traffic and signaling. Data rates of 1,200, 2,400, and 4,800 bps can support only primary traffic information.

The frame-quality indicator is a CRC on the information bits in the frame. At 9,600 bps, the generator polynomial is

$$g(x) = x^{12} + x^{11} + x^{10} + x^{9} + x^{8} + x^{4} + x + 1 \tag{5.6}$$

At 4,800 bps, the generator polynomial is

$$g(x) = x^{8} + x^{7} + x^{4} + x^{3} + x + 1 \tag{5.7}$$

At 9,600 bps, the 172 information bits consist of 1 or 4 format bits and 171 or 168 traffic bits. A variety of different multiplexing options are supported. The entire 171 information bits can be used for primary traffic, or the 168 bits can be used for 80 primary traffic bits and 88 signaling traffic bits or 88 secondary traffic bits. Other options use 40 and 128 or 16 and 152 bits for primary and signaling/secondary traffic. Alternatively, the entire 168 bits can be used for signaling or secondary traffic.

When the forward traffic channel is used for signaling, the message is similar in form to the paging channel (see Figure 5.9), with an 8-bit message length header, a message body of a minimum of 16 bits and a maximum of 1,160 bits, and a CRC code of 16 bits. Following the message are padding bits to make the message end on a frame boundary. The message length includes the header, body, and CRC, but not the padding. The CRC is computed on the message length header and the message body using the following:

$$g(x) = x^{16} + x^{12} + x^{5} + 1 \tag{5.8}$$

When the forward traffic channel is used for signaling, the typical messages that can be sent are the following:

1. **Order message**—This is similar to the order message on the paging channel.
2. **Authentication challenge message**—When the base station suspects the validity of the mobile, it can challenge the mobile to prove its identity.
3. **Alert with information message**—This allows the base station to validate the mobile identity.
4. **Data burst message**—This is a data message sent by the base station to the mobile.
5. **Handoff direction message**—This provides the mobile with information to begin the handoff process.
6. **Analog handoff direction message**—This tells the mobile to switch to the analog mode and begin the handoff process.
7. **In-traffic system parameters message**—This updates some of the parameters set by the system parameters message in the paging channel.
8. **Neighbor list update message**—This updates the neighbor base station parameters set by the neighbor list message on the paging channel.
9. **Power control message**—This tells the mobile how long the period is, or what threshold is to be used, in measuring frame error statistics that will be sent in the mobile's power measurement report message.

10. **Send burst dual-tone multifrequency (DTMF) message**—When the base station needs dialed digits, it can request them in this message. This message would be used for digits for a three-way call, for example.
11. **Retrieve parameters message**—This requests the mobile to report on any of the retrievable and settleable parameters (refer to IS-95 Appendix E).
12. **Set parameter message**—This informs the mobile to adjust any of the retrievable and settleable parameters (refer to IS-95 Appendix E).
13. **SSD update message**—This is a request from the base station for the mobile to update the shared secret data.
14. **Flash with information message**—This contains information that allows the network to supply display information to be displayed by the mobile, to identify the responding party's number (the connected number), to convey information to the mobile by means of tones or other alerting signals, and to indicate the number of messages waiting.
15. **Mobile registration message**—This message informs the mobile that it is registered and supplies the necessary system parameters.
16. **Extended handoff direction message**—This message is one of several handoff messages sent by the base station.

5.7.6 Reverse Traffic Channels

For the Rate Set 1, the reverse traffic channel may use either 9,600-, 4,800-, 2,400-, or 1,200-bps data rates for transmission. The duty cycle for transmission varies proportionally with the data rate, being 100 percent at 9,600 bps to 12.5 percent at 1,200 bps. An optional Rate Set 2 is also supported. The actual burst transmission rate is fixed at 28,800 code symbols per second. The reverse traffic channel processing is similar to the access channel. The major difference is that the reverse traffic channel uses a data burst randomizer (see Figure 5.22).

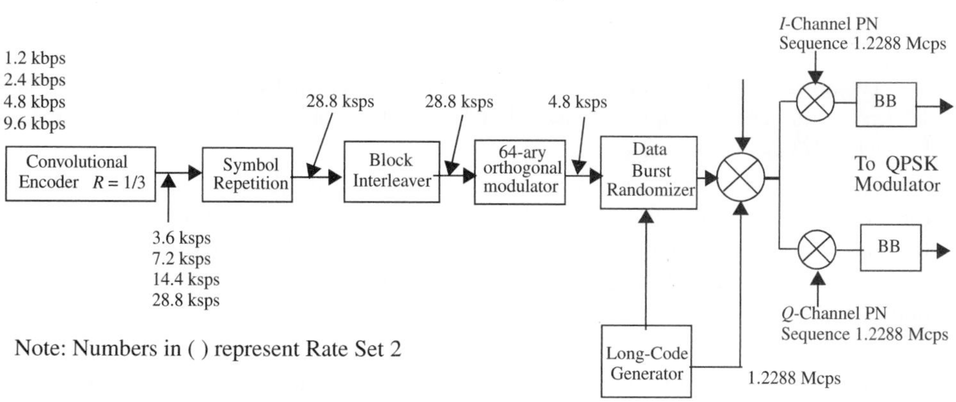

Figure 5.22 Reverse Traffic Channel Processing

Since hextets are modulated as one of 64-ary modulation symbols for transmission, the modulation symbol transmission rate is fixed at 4,800 modulation symbols per second. This results in a fixed Walsh chip rate of 307.2 kcps. The data from the 64-ary modulator is fed into the data burst randomizer. The data burst randomizer takes advantage of the voice activity on the reverse link. This is used to reduce reverse-link power during quieter periods of speech by pseudorandom masking out of redundant symbols produced by symbol repetition. This is achieved by the data burst randomizer. The data burst randomizer generates a masking pattern of 0s and 1s to randomly mask out redundant data. The masking pattern depends on vocoder rate. For a vocoder operating at 9.6 kbps, no data is masked out, whereas if the vocoder is operating at 1.2 kbps, then the symbols are repeated seven times and the data burst randomizer masks out seven out of eight groups of symbols.

Each 20-ms traffic channel frame is divided into 16 power control groups, each 1.25 ms (as discussed earlier). The data burst randomizer pseudorandomly masks out individual power control groups. With 9.6 kbps, no power control group (PCG) is masked out; with 4.8 kbps, 8 PCGs are masked out in a frame; with 2.4 kbps, 12 PCGs are masked out in a frame; and with 1.2 kbps, 14 PCGs are masked out in a frame. An example of this operation with a vocoder operating at 2.4 kbps is shown in Figure 5.23.

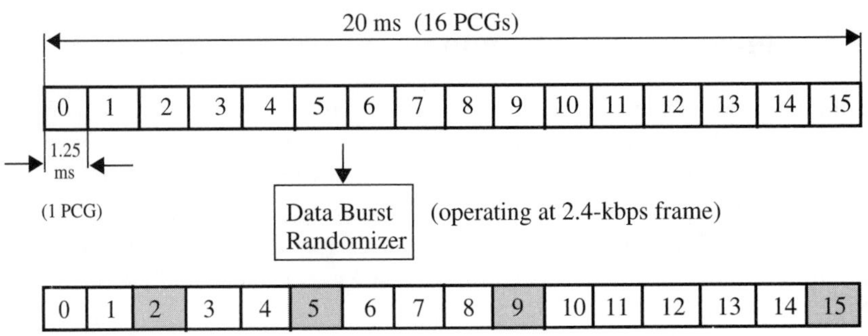

Figure 5.23 Data Burst Randomizer Operation at 2.4-kbps Frame

The reverse channel structure for Rate Set 2 is similar. The Rate Set 2 vocoder supports 14.4, 7.2, 3.6, and 1.8 kbps data rates. For Rate Set 2, the convolutional encoder is 1/2, instead of 1/3 as in Rate Set 1.

The system can multiplex primary (voice) and secondary (data) or signaling traffic on the same traffic channel. Multiplex Option 1 is used to transmit primary and secondary traffic. This option is also used to transmit primary (voice) and signaling (messaging) traffic. Multiplex Option 1 uses the following methods to simultaneously transmit primary and secondary traffic:

- **Blank-and-burst**—The entire traffic channel frame is used to send only secondary data. The entire traffic channel is also used to send signaling data. The secondary or signaling data blanks out the primary data (see Figure 5.24).

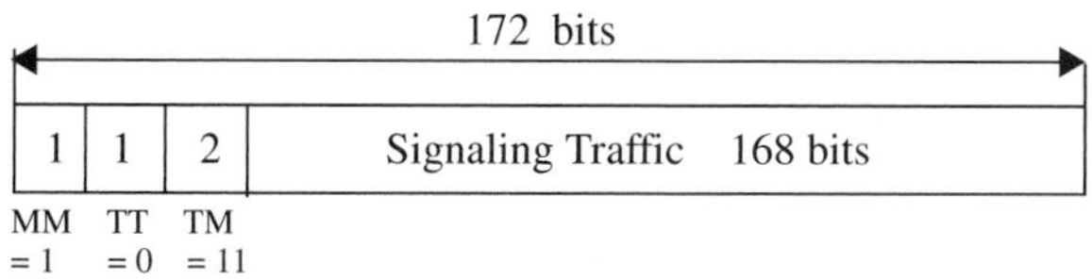

Figure 5.24 Blank and Burst for Speech (Rate Set 1)

- **Dim-and-burst**—The traffic frame is used to send both primary and secondary data. The traffic channel frame can be also used to send both primary and signaling data (see Figure 5.25). Figure 5.26 shows the traffic channel frame structure for the forward and reverse links.

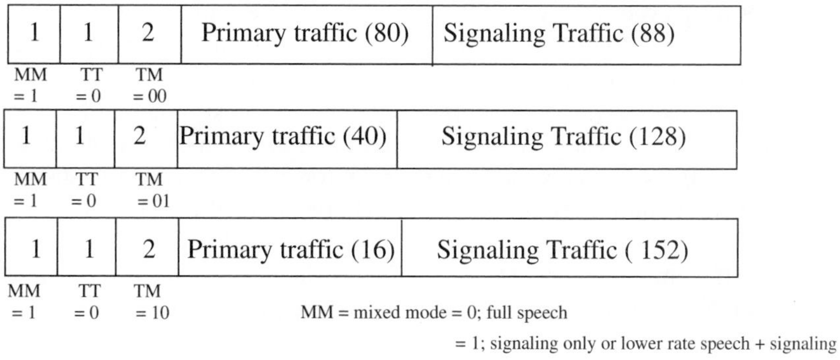

Figure 5.25 Dim-and-Burst Speech (Rate Set 1)

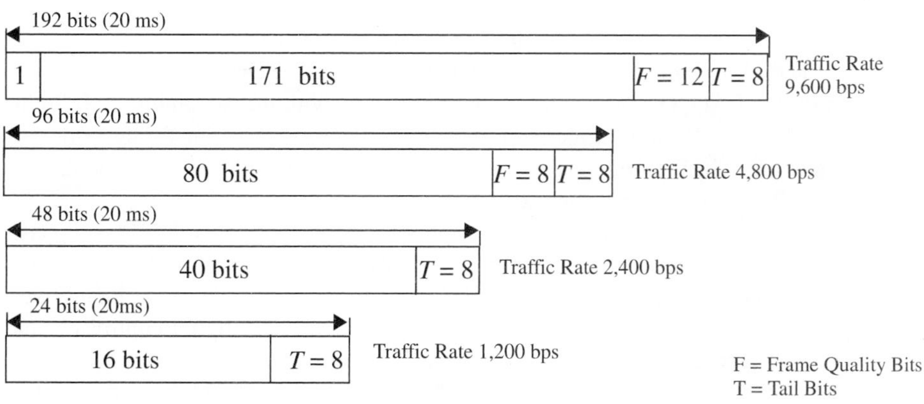

Figure 5.26 Traffic Channel Frame Structure for Forward and Reverse Links

The reverse traffic channel carries the following typical messages:

1. **Order** messages on the reverse traffic channel are base station challenge, SSD update confirmation, SSD update rejection, parameter update confirmation, mobile station acknowledgment, service option request, service option response, release (normal with power down indication), long-code transition request (public and private), connect, continuous DTMF tone (start and stop), service option control, mobile station reject (with and without a reason), and local control.
2. **Authentication challenge response** contains the information to validate the mobile's identity.
3. **Flash with information** contains information records from mobile concerning mobile features, mobile key pay facility, called party number, calling party number, and the connected number (i.e., the responding party).
4. **Data burst** is a user-generated data message sent by the mobile to the base station.
5. **Pilot strength measurement** sends information about the strength of other pilot signals that are not associated with serving base station.
6. **Power measurement report** message sends frame error rate statistics to the base station. The report is generated at specified intervals or when a threshold is reached.
7. **Send burst DTMF** message uses two tones, one low and one high frequency, to represent a dialed digit, and transmits dialed digits to the base station.
8. **Status** message contains information records from the mobile about mobile identification, mobile call mode, mobile terminal information, and security status.
9. **Origination continuation** message is a continuation of the origination message that was sent on the access channel if additional dialed digits need to be sent.
10. **Handoff completion** message is the mobile response to a *handoff direction message*.
11. **Parameter response** message is the mobile response to the base station for a *retrieve parameters message*.

5.8 Summary

In this chapter, we discussed implementation of IS-95 CDMA carriers in an existing AMPS or TDMA system. We established that 59 AMPS carrier channels must be removed to introduce the first CDMA carrier without interfering with the remaining AMPS channels. We provided details of information processing and message framing for the pilot, sync, paging, and traffic channels on the forward link. The pilot channel (W_0) is unmodulated; it consists of only short-code spreading sequences.

The pilot channel is used by all mobiles attached to a cell as a coherent phase reference and also provides a unique identifier for different base stations. The sync channel (W_{32}) transmits system timing information to allow mobiles to synchronize themselves with base stations. The paging channels (W_1 through W_7) are the digital control channels for the CDMA forward link. One base station can have up to seven paging channels. The first paging channel is always assigned W_1. The traffic channels (W_8 through W_{31} and W_{33} through W_{63}) carry digitized voice data and signaling from base station to mobile.

On the reverse link, an access channel is used by a mobile to register with the system, to access the system before assignment of a traffic channel, to originate a call, or to respond to a page. Several access channels per paging channel can be used. The reverse traffic channels are used to deliver encoded voice and reverse link signaling from mobile to base station.

We also presented typical messages that are carried over the channels of the forward and reverse links.

5.9 References

1. Garg, V. K., and Wilkes, J. E., *Wireless and Personal Communications Systems,* Prentice Hall PTR: Upper Saddle River, NJ, 1996.
2. Garg, V. K., Smolik, K., and Wilkes, J. E., *Applications of CDMA to Wireless/Personal Communications,* Prentice Hall PTR: Upper Saddle River, NJ, 1997.
3. Rappaport, T. S., *Wireless Communications*, Prentice Hall PTR: Upper Saddle River, NJ, 1996.
4. TIA/EIA/SP-3693, "Mobile Station-Base Station Compatibility Standard for Dual-Mode Wideband Spread Spectrum Cellular Systems," November, 1997.

CHAPTER 6

CDMA IS-95 Call Processing

6.1 Introduction

In this chapter, we discuss code division multiple access (CDMA) call processing states that a mobile station goes through in getting to a traffic channel. These include system initialization state, system idle state, system access state, and traffic channel state. Each of the call processing states has several substates that are also discussed. Idle handoff, slotted operation, CDMA registration, and authentication procedures are presented. Messages used to exchange data in different call processing states are also presented. We conclude the chapter by providing call flows for CDMA call origination, call termination, and call release.

6.2 CDMA Call Processing State

In getting to a traffic channel, a mobile station in CDMA goes through several states: system initialization state, system idle state, system access state, and traffic channel state (see Figures 6.1 and 6.2).

In the *system initialization state*, mobile acquires the pilot channel by searching all the PN-I and PN-Q possibilities and tuning to the strongest signal. Once the pilot is acquired, the synchronization (sync) is acquired using W_{32} Walsh function and the detected time offset of the pilot channel. The mobile obtains the system configuration and timing information.

Next, the mobile enters the *system idle state,* where it acquires the paging channel and monitors it. The mobile can then receive messages from the base station containing the necessary parameters to initiate or receive a call.

If a call is being placed (uplink or downlink) the mobile enters the *system access state.* The necessary parameters are exchanged from the mobile on the access channel and from the base station on the paging channel.

When the access attempts are successful, the mobile enters the *traffic channel state.* In this state, speech communication takes place with associated control messages replacing the digital speech by either of two methods:

- *Blank-and-burst:* The complete speech packet is replaced with signaling.
- *Dim-and-burst:* Part of speech packet is replaced with signaling.

Also, power control messages are sent by a method called *bit puncturing* on the downlink channel. In bit puncturing two gross-data bits are replaced by a single power control bit.

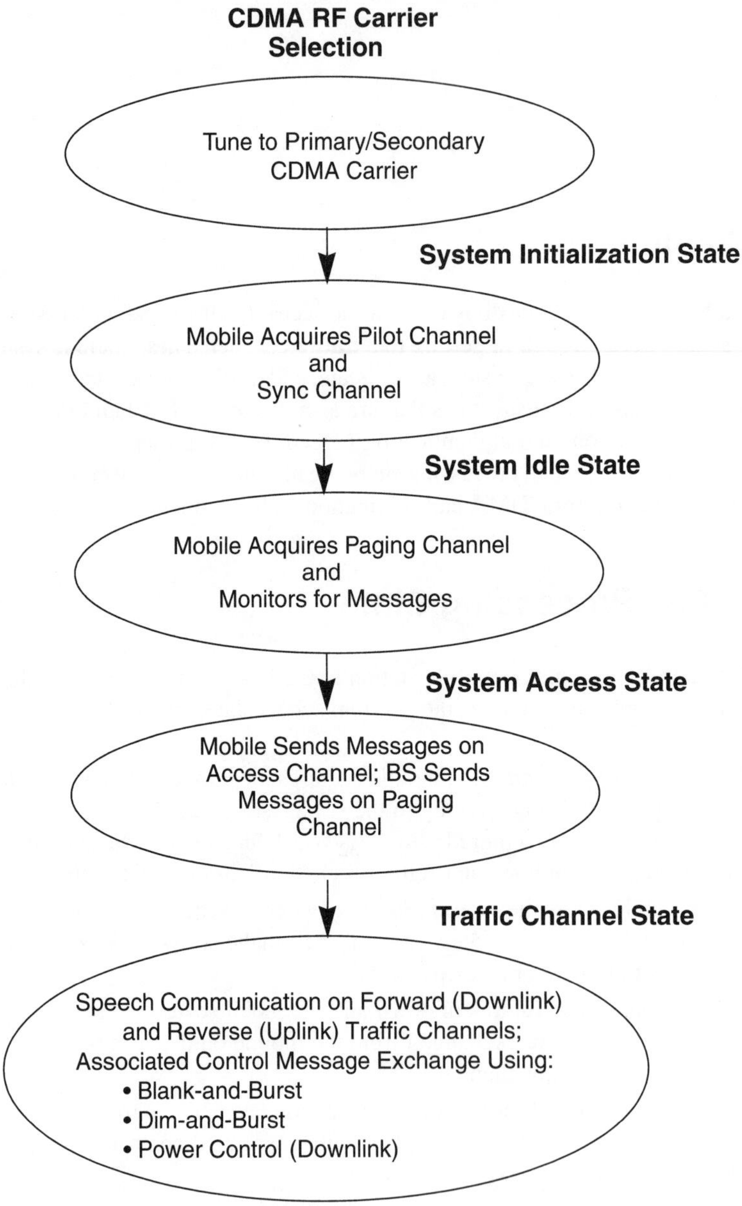

Figure 6.1 Getting to a Traffic Channel in CDMA

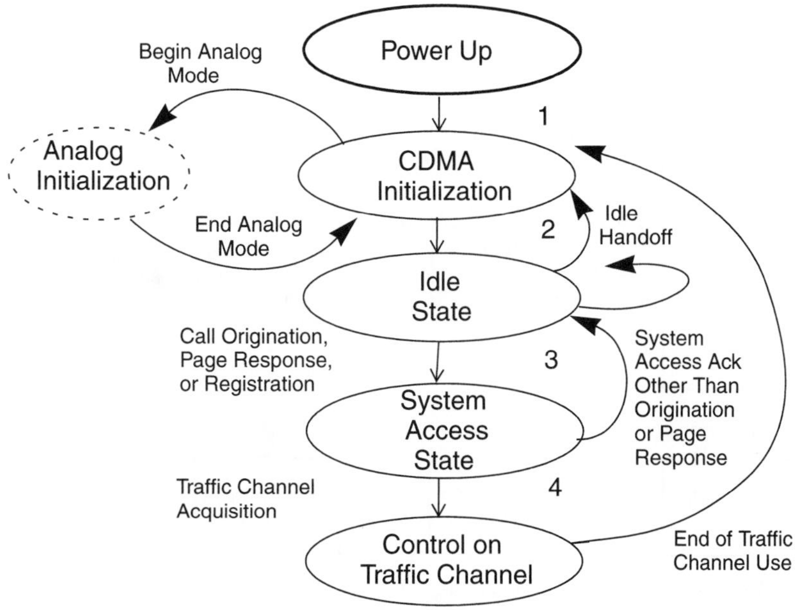

Figure 6.2 CDMA Call Processing States

6.2.1 System Initialization State

In the system initialization state, the mobile selects a system to use. If the selected system is a CDMA system, the mobile proceeds to acquire and synchronize to a CDMA carrier [1]. The system initialization state consists of the following substates:

- **System determination substate**—In this substate, the mobile selects the system to use (analog or CDMA). The mobile choices include service provider preference. If the CDMA is selected, the mobile sets the CDMA channel parameters (CDMA_CH) to N_i, where N_i is either a primary or secondary CDMA channel number.
- **Pilot channel acquisition substate**—In this substate, the mobile acquires the pilot channel of the CDMA system.
- **Synchronization (sync) channel acquisition substate**—In this substate, the mobile acquires the synchronization channel and obtains system configuration and timing information for the CDMA system.
- **Timing change substate**—In this substate, the mobile synchronizes its timing to the CDMA system after receiving and processing the synchronization message.

Figure 6.3 shows the system initialization state and the substates associated with it.

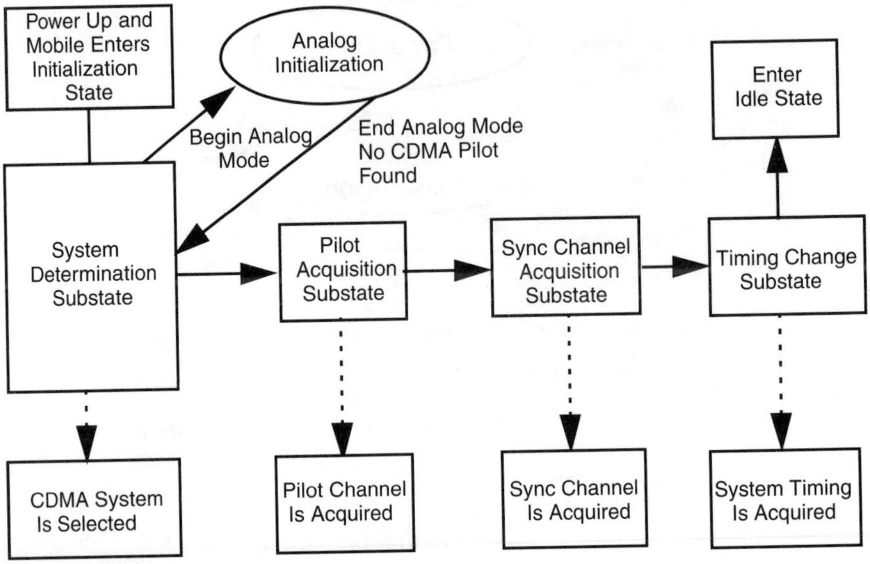

Figure 6.3 System Initialization State

Synchronization occurs when the phase $(t - T)$ of the locally generated pseudorandom (PN) code is equal to the phase $(t - T_i - T_p)$ of the incoming code.

$$(t-T) = (t-T_i-T_p) \quad \textbf{(6.1)}$$

$$\therefore T = T_i + T_p \quad \textbf{(6.2)}$$

where
T = phase of local PN code
T_P = propagation delay
T_i = pilot offset for ith pilot

At this stage of the system, we observe two points:

1. When the locally generated code phase matches with the incoming pilot $(T_i + T_p)$, pilot acquisition occurs. Thus, PN $(t - T_i - T_p)$ is known.
2. Although total phase $(T_i + T_p)$ is known, the pilot offset T_i is not known. We will get the pilot offset T_i from the sync message of the sync channel.

Once the sync channel is acquired and the sync message is received and processed, the mobile stores the following information from the sync message:

- System identification (SID)
- Network identification (NID)

- Pilot PN offset (T_i)
- System time (T_s)
- Long-code state at system time (LC_STATE)
- Paging channel data rate (PRAT)
- Number of leap seconds that have occurred since the start of system time (LP_SEC)
- Offset of local time from system time (LTM_OFF)
- Daylight saving time indicator (DAY_LT)

In the timing change substate, the mobile uses the pilot offset, the system time, and the long-code state information obtained from the sync message to synchronize its timing to the system time and synchronize its long-code phase to that of the system.

In IS-95, long code is generated with a 42-stage shift register. The mobile knows the generation polynomial. The problem is to get the correct code phase. The mobile obtains the long-code state from the sync message. The long code is a 42-bit sequence that corresponds to the contents of the shift registers at the system time, T_s. The mobile loads its shift registers with the 42-bit long-code state. The mobile waits and at system time T_s it starts shifting the contents of the shift registers at 1.2288 Mcps. At this point, long-code synchronization is achieved. The mobile may now tune to a paging channel in order to enter the idle state.

6.2.2 Idle State

In the idle state, the mobile monitors the paging channel [2]. In this state, the mobile can

- Receive messages and orders from the base station
- Receive an incoming call
- Initiate a registration process
- Initiate a call
- Initiate a message transmission

Figure 6.4 summarizes the activities in idle state. Upon entering the idle state, the mobile sets its Walsh code to the primary paging channel (W_1). The mobile sets its paging channel rate to the rate obtained from the sync message.

The paging channel is subdivided into 80-ms slots called *paging channel slots*. In the nonslotted mode, paging and control data for the mobile can be received in any of the paging channel slots. Therefore, the mobile monitors all slots on a continuous basis. In IS-95, the paging channel protocol also allows scheduling the transmission of messages for a given mobile in certain assigned slots. A mobile station that monitors the paging channel only during certain assigned slots is referred to as operating in the *slotted mode*. During the slots in which the paging channel is not monitored, the mobile can stop or reduce its processing activities to save battery power.

In the slotted mode operation, the mobile monitors the paging channel for one or two slots per cycle. Slotted page messages contain a field called MORE_PAGES. When this field is set to zero, it indicates that the remainder of the slots will contain no more messages addressed to the mobile. This allows the mobile to stop monitoring the paging channel as soon as possible. If a slotted page message with MORE_PAGES field set to zero is not received in the assigned slot, the mobile continues to monitor the paging channel for one additional slot. For each of its assigned slots, the mobile begins monitoring the paging channel in time to receive the first bit of the assigned slot. The mobile then continues to monitor the paging channel until one of the following conditions is satisfied:

1. The mobile receives a slotted page message with MORE_PAGES field set to zero.
2. The mobile monitors the assigned slot and the one following it, and receives at least one valid message.

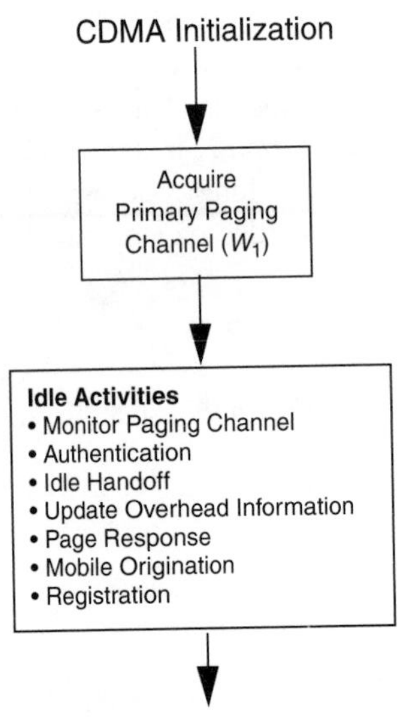

Figure 6.4 Idle State Activities in CDMA

The mobile can specify its preferred slot cycle using the SLOT_CYCLE_INDEX field in the registration message, origination message, or page response message. The mobile can also specify its preferred slot cycle using the SLOT_CYCLE_INDEX field of the terminal information record of the status message when in the mobile station control on the traffic channel state. The length of the slot cycle T, in units of 1.28 second, is given by

$$T = 2^i \quad (6.3)$$

where

i = selected slot cycle index

There are $16 \times T$ slots in a slot cycle for a particular mobile using some value of i, and four 20-ms full-frames in an 80-ms slot. SLOT_NUM is the paging slot number. To determine the assigned slots, the mobile uses a *hash algorithm* to select a slot number in the range 0 to 2,048. The minimum and maximum cycles are 16 slots (1.28 seconds) and 2,048 slots (163.84 seconds), respectively. The value of SLOT_NUM is given as

$$\text{SLOT\_NUM} = \left(\frac{t}{4}\right) mod 2048 \tag{6.4}$$

where
t = system time in frames

For each mobile station, the starting time of its slot is offset from the beginning of each of its slot cycles by a fixed, randomly selected number of slots, called PGSLOT. As an example, for $i = 0$, $T = 2^0 = 1$, so $16 \times T = 1.28$ seconds. Let the computed value of PGSLOT be equal to 6, then one of mobile station's slots begins when SLOT_NUM equals 6 (see Figure 6.5). The mobile begins monitoring the paging channel at the start of the slot in which SLOT_NUM equals 6. The next slot in which the mobile must begin monitoring the paging channel is 16 slots later, i.e., the slot in which SLOT_NUM is 22, since the slot cycle length is 16 for $T = 1$.

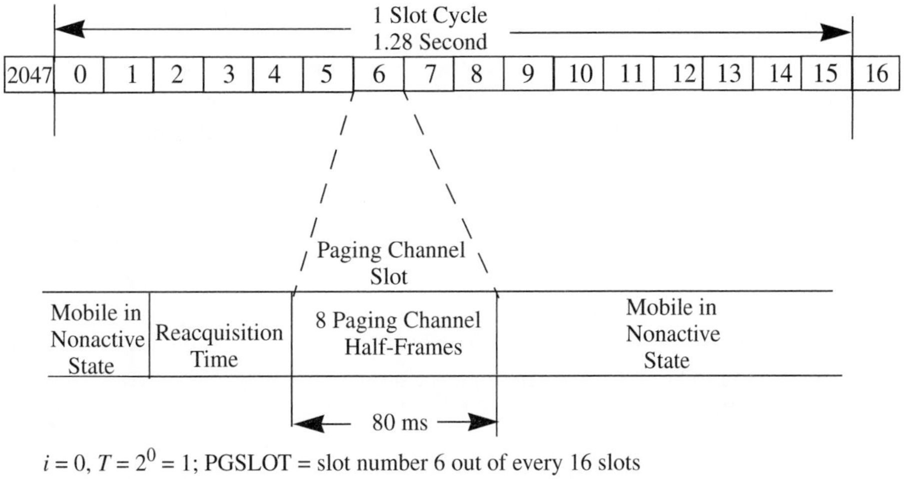

Figure 6.5 Slot Cycle with $i = 0$

Acknowledgment Procedure. We consider a message sent to the mobile by the base station on the paging channel. The message typically contains the address field, acknowledgment required sequence number, message sequence number, and other fields including data field (see Figure 6.6). Whether the mobile sends an acknowledgment to the base station depends on the value of the ACK_REQ field. Acknowledgment of messages received on the paging channel are sent on an access channel.

Other Field	ADDR_TYPE	ACK_REQ	MSG_SEQ

Figure 6.6 A Typical Message from Base Station to Mobile

If the ACK_REQ field of the message from the base station is 1, the mobile transmits an acknowledgment with the following parameters (see Figure 6.7):

- VALID_ACK field: 1
- ADDR_TYPE field: I (address type of message being acknowledged)
- ACK_SEQ field: I (message sequence number of the message being acknowledged)

Other Field	ADDR_TYPE	VALID_ACK	ACK_SEQ

Figure 6.7 A Typical Message with Acknowledgment from Mobile

When the page messages or slotted page messages addressed to a mobile do not have an ACK_REQ field, the mobile transmits a page response message including an acknowledgment in response to each record of a page message or slotted message.

When a message does not include an acknowledgment, the mobile sets the VALID_ACK field to 0. The ADDR_TYPE and ACK_SEQ fields are then set to the ADDR_TYPE and MSG_SEQ values of the last message that requires acknowledgment.

- VALID_ACK field: 0
- ADDR_TYPE field: I (address type of last message requiring acknowledgment)
- ACK_SEQ field: I (message sequence number of last message requiring acknowledgment)

When a message does not include an acknowledgment, the mobile sets the VALID_ACK field to zero. The ACK_TYPE and ACK_SEQ fields then set to 000 and 111, respectively, if there is no previous message that requires an acknowledgment.

Idle Handoff. An idle handoff occurs when a mobile has moved from the coverage area of one base station to the coverage area of another base station during the idle state. The mobile determines that a handoff should occur when it detects a new pilot that is sufficiently stronger than the current pilot.

Pilot channels are identified by the short PN offsets. They are grouped into sets describing their status with respect to pilot searching procedures. In the idle state, three pilot sets are maintained: active, neighbor, and remaining.

Using a strategy similar to a sliding correlator, it is possible to acquire a pilot if the correct phase of its short PN code is known. For each pilot set, a search window is specified. This allows the mobile to search for the direct path as well as multipath components of the pilot signal. The search window is centered on either the earliest arriving multipath or the short PN offset.

If the mobile determines that a neighbor set or remaining set pilot is sufficiently stronger than the active set pilot, idle handoff is performed. While performing an idle handoff, the mobile operates in nonslotted mode until at least one valid message is received from the new paging channel. On receiving a valid message from the new paging channel, the mobile may resume slotted mode operation. After performing an idle handoff, the mobile discards all unprocessed messages received on the old paging channel.

The paging channel used to transmit control information to the mobiles that have not been assigned to traffic channels carries two types of messages:

- The overhead messages that are broadcast messages for all mobiles
- The directed message addressed to a particular mobile or a specific group of mobiles

There are four overhead messages that are continuously broadcast on the paging channel. These are

- System parameter message
- Neighbor list message
- CDMA channel list message
- Access parameters message

The first three are called configuration messages. A configuration message sequence number (CONFIG_MSG_SEQ) is associated with a set of configuration messages sent on the paging channel. When the contents of one or more configuration messages change, the configuration message sequence number is incremented. The mobile stores the sequence number contained in each configuration message received.

Access parameters messages are independently sequence numbered by the access parameter message sequence number (ACC_MSG_SEQ). The mobile stores the most recently received access parameter message sequence number in the ACC_MSG_SEQ field.

Configuration and access parameters from one paging channel are not be used while monitoring a different paging channel. If the stored parameters are current, the mobile processes the messages on the paging channel. When a system parameter message SYS_PAR_MSG_SEQr is received, its associated configuration message sequence number CONFIG_MSG_SEQr is compared with the stored value of the system parameter message SYS_PAR_MSG_SEQs. If there is a match, the received system parameter message is ignored. If the comparison results in a mismatch, the mobile stores the configuration message sequence number, system identification (SID), network identification (NID), and base station identification (BASE_ID).

When an access parameters message is received, ACC_MSG_SEQr is compared with the stored value of the access parameters message sequence value ACC_MSG_SEQs. If there is a match, the received access parameters message is ignored. However, if the comparison results in a mismatch, the mobile stores the following parameters:

- Access parameter message sequence number (ACC_MSG_SEQ)
- Number of access channel (ACC_CHAN)
- Nominal transmit power (NOM_PWR)
- Initial power offset for access (INIT_PWR)
- Power increment or power step (PWR_STEP)

Neighbor List Message. When a neighbor list message (NGHBR_LST_MSG_SEQr) is received, its associated configuration message sequence number (CONFIG_MSG_SEQ) is compared with the stored value of the neighbor list message (NGHBR_LST_MSG_SEQs). If there is a match, the received neighbor list message is ignored. If the comparison results in a mismatch, the mobile stores the following parameters:

- Configuration message sequence number (CONFIG_MSG_SEQ)
- Short PN offsets of neighbor list members
- Short PN sequence offset increment (PILOT_INC)

The mobile updates the idle handoff neighbor set so that it contains only the pilot offsets listed in the neighbor list message.

CDMA Channel List Message. When a CDMA channel list message (CHAN_LST_MSG_SEQr) is received, its associated configuration message sequence number (CONFIG_MSG_SEQ) is compared with the stored value of the CDMA channel list message (CHAN_LST_MSG_SEQs). If there is a match, the received CDMA channel list message is ignored. If the comparison results in a mismatch, the mobile stores the following parameters:

- Configuration message sequence number (CONFIG_MSG_SEQ)
- CDMA channel list message sequence (CHAN_LST_MSG_SEQ)

6.2.3 System Access State

The system access state (see Figure 6.8) includes the following substates:

- **Update overhead information substate**—In this substate, the mobile monitors the paging channel until it has received a current set of configuration messages.

- **Mobile station origination attempt substate**—In this substate, the mobile station sends an origination message to the base station.
- **Page response substate**—In this substate, the mobile sends a page response message to the base station.
- **Registration access substate**—In this substate, the mobile station sends a registration message to the base station.
- **Mobile station order/message response substate**—In this substate, the mobile sends a response to a message received from the base station.
- **Mobile station message transmission substate**—In this substate, the mobile sends a data burst message to the base station.
- **PACA cancel substate**—In this substate, the mobile sends a priority access channel assignment (PACA) cancel message.

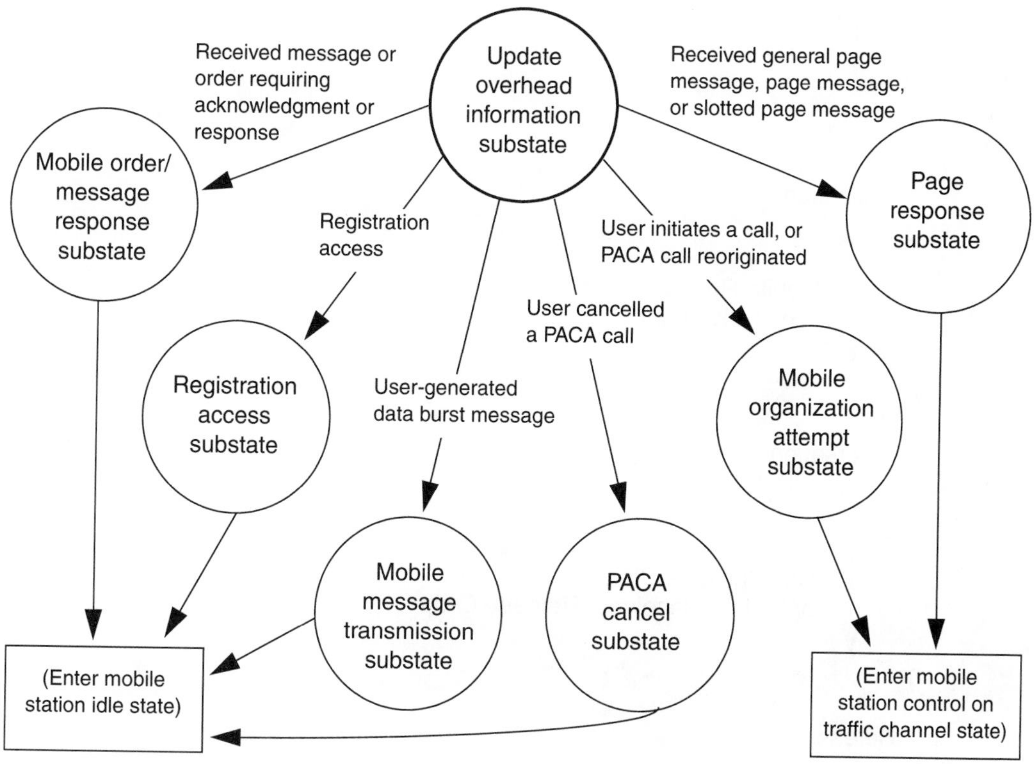

Figure 6.8 CDMA Mobile System Access State

6.2.4 Mobile Control on Traffic Channel State

Call Origination. The mobile station control on traffic channel state for call origination consists of the following substates (see Figures 6.9 and 6.10):

- **Traffic channel initialization substate**—In this substate, the mobile verifies that it can receive the forward traffic channel and begins to transmit on the reverse traffic channel.
- **Conversation substate**—In this substate, the mobile station exchanges primary traffic packets with the base station.
- **Release substate**—In this substate, the mobile station disconnects the call.

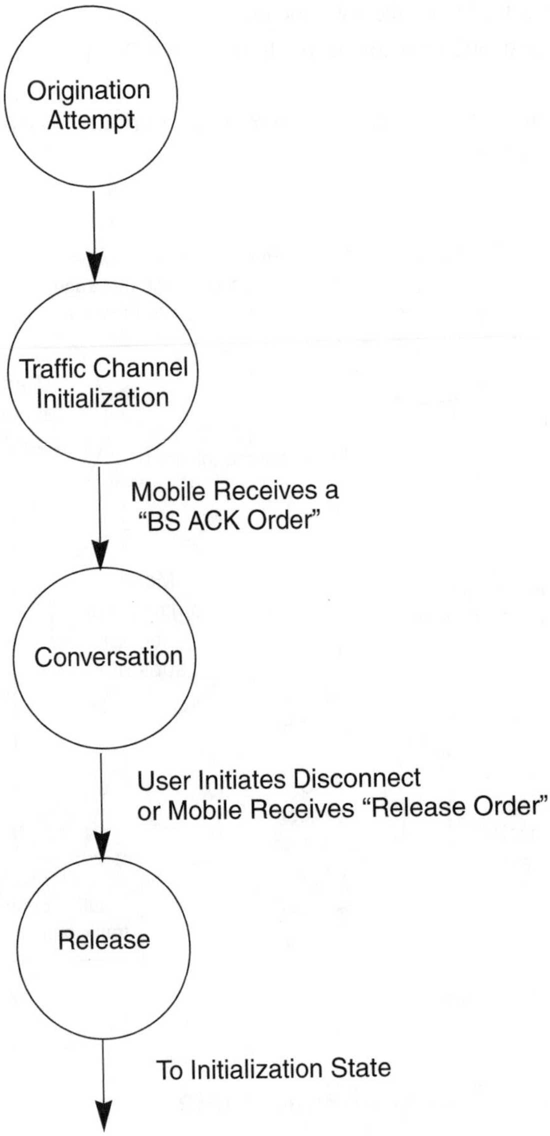

Figure 6.9 Mobile Control on Traffic Channel (Call Origination)

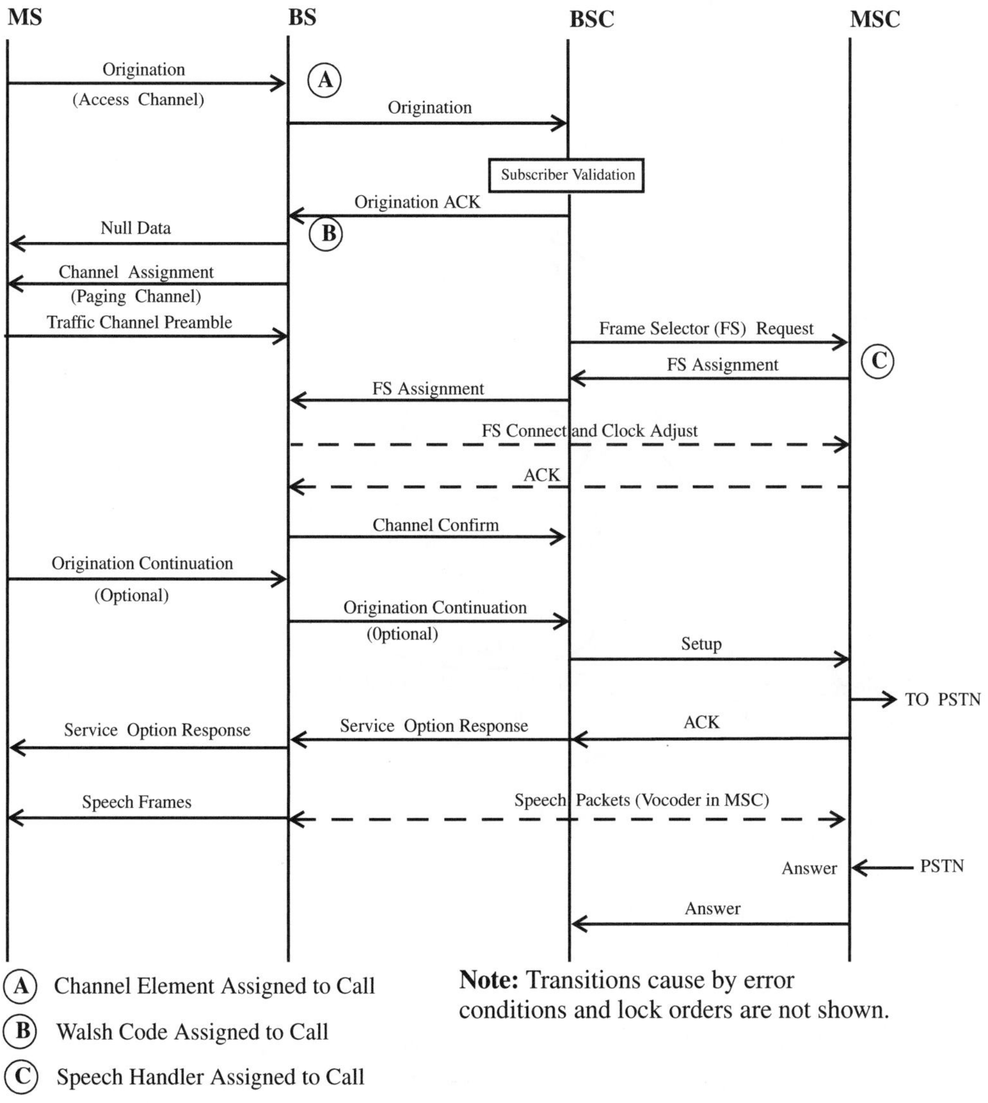

Figure 6.10 Flow Diagram for CDMA Call Origination

Call Termination. The mobile station's control of the traffic channel for call termination consists of the following substates (see Figures 6.11 and 6.12):

- **Traffic channel initialization substate**—same as in call origination.
- **Waiting for order substate**—In this substate, the mobile waits for an alert with information message. The information may be data such as calling party number, and/or voice.

- **Waiting for mobile station answer substate**—In this substate, the mobile waits for the user to answer the call.
- **Conversation substate**—In this substate, the mobile station's primary service option application exchanges primary traffic packets with the base station.
- **Release substate**—In this substate, the mobile disconnect the call.

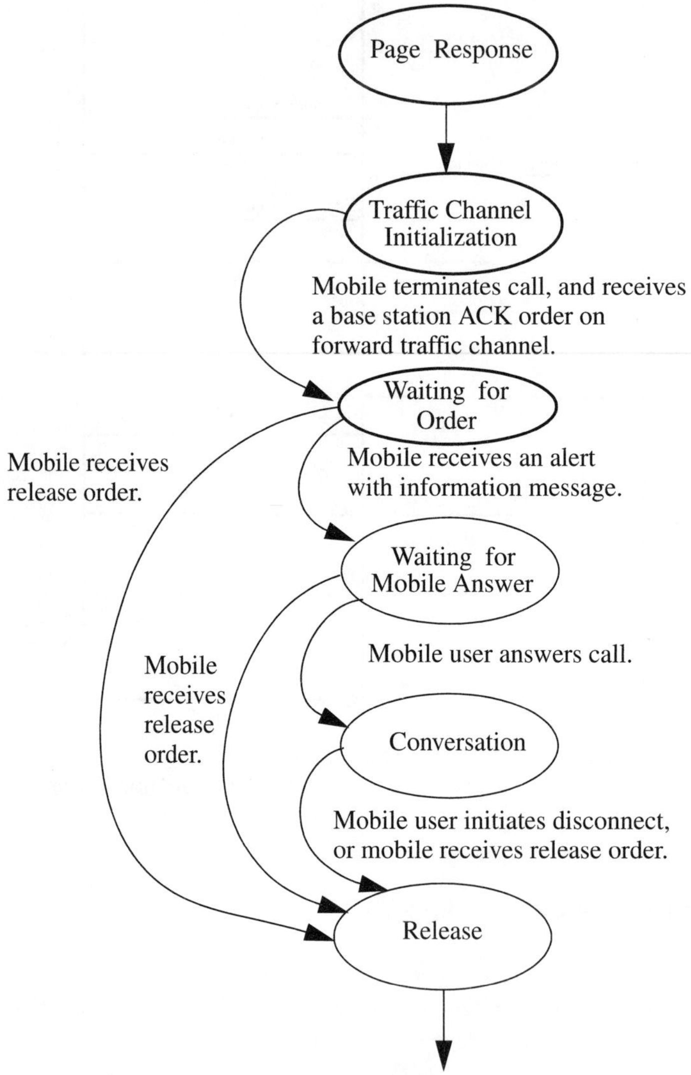

Figure 6.11 Mobile Control on Traffic Channel (Call Termination)

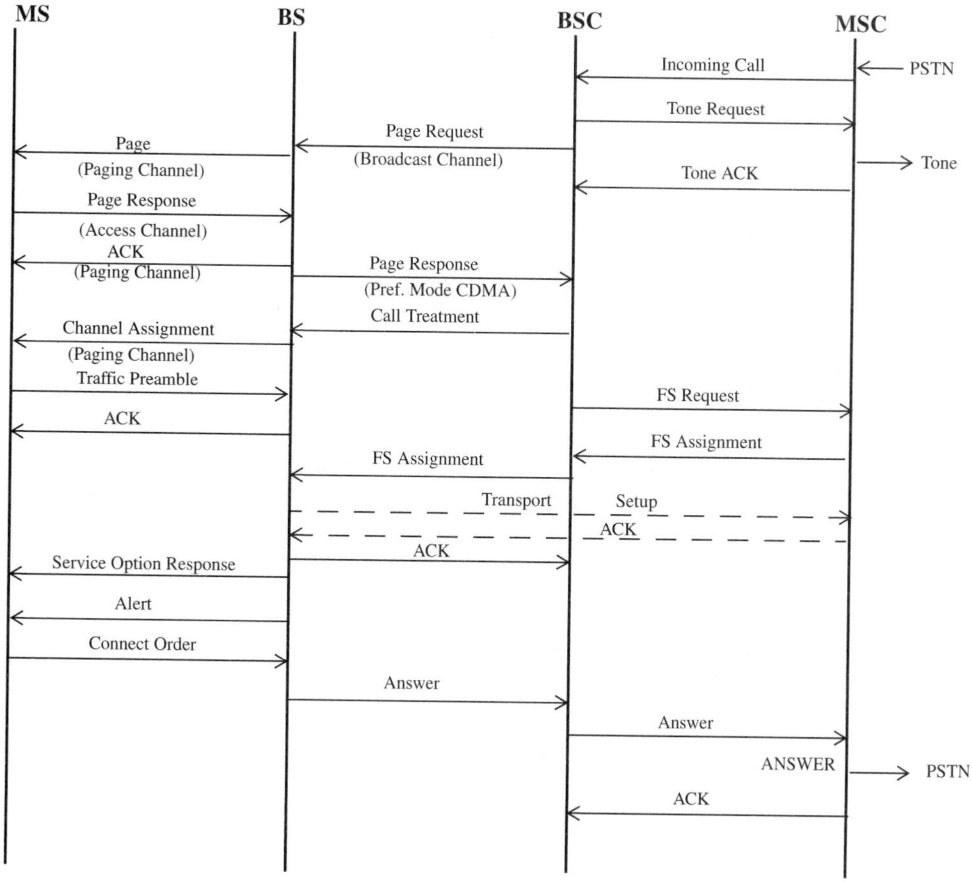

Figure 6.12 Flow Diagram for CDMA Call Termination

Flow diagrams for CDMA far-end initiated and mobile initiated call release are shown in Figures 6.13 and 6.14.

6.3 CDMA Registration

The mobile notifies the base station of its location, status, identification, slot cycle, and other characteristics by the registration process. The mobile informs the base station of its location and status so the base station can efficiently page the mobile station when establishing a mobile-terminated call. For operation in the slotted mode, the mobile supplies the SLOT_CYCLE_INDEX parameter so the base station can determine which slots the mobile is monitoring. The mobile supplies the station class mark and protocol revision number so the base station knows the capabilities of the mobile station.

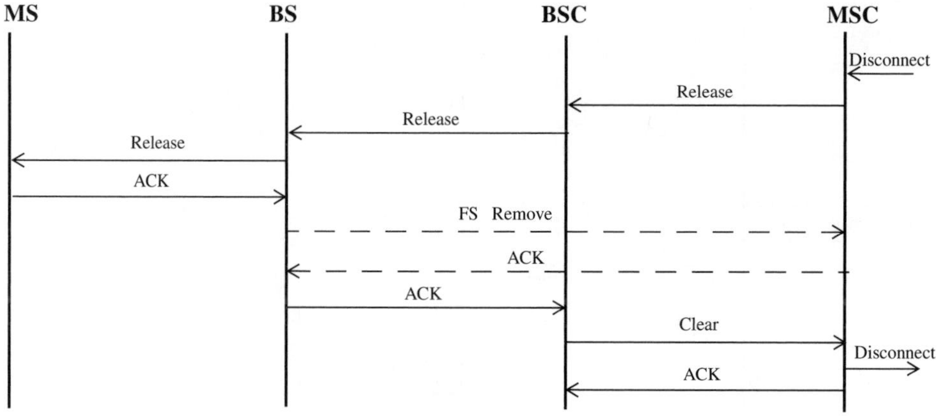

Figure 6.13 Flow Diagram for CDMA Call Release (Far-End Initiated)

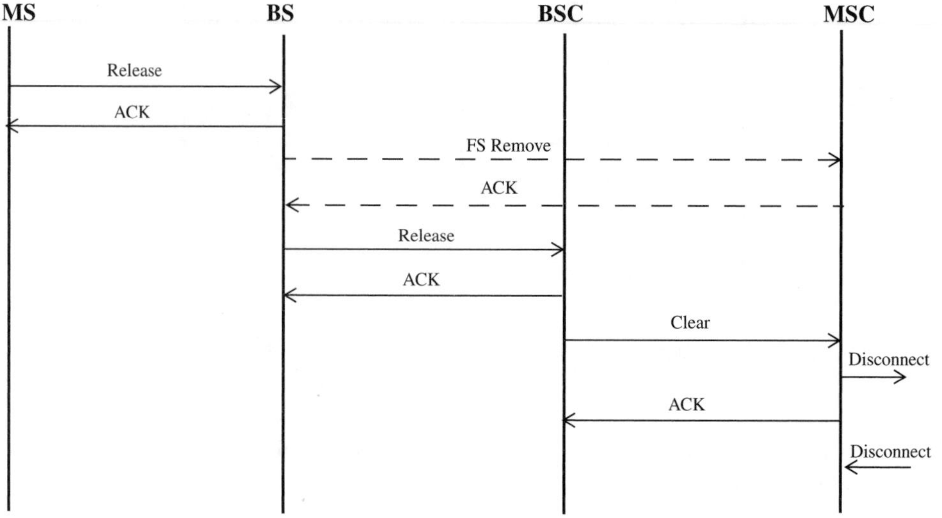

Figure 6.14 Flow Diagram for CDMA Call Release (Mobile Initiated)

IS-95 supports nine different forms of registration:

1. *Power-up registration*—The mobile registers when it powers on, switches from using an alternative serving system, or switches from using the analog system.
2. *Power-down registration*—The mobile registers with the current serving system when it powers down, informing the system that it is no longer active.

3. *Timer-based registration*—The mobile registers at regular intervals. Its use also allows the system to automatically deregister mobile stations that did not perform a successful power-down registration.
4. *Distance-based registration*—The mobile station registers when the distance between the current base station and the base station with which it last registered exceeds a threshold. The mobile determines that it has moved a certain distance by computing a distance based on the difference in latitude and longitude between the current and previous base stations where the mobile registered. If this distance exceeds the threshold value, the mobile station registers.
5. *Zone-based registration*—The mobile registers when it enters into a new zone. Zones are groups of base stations within a given system and network. A base station's zone assignment is identified by the REG_ZONE field of the system parameters message. Zone-based registration causes a mobile to register whenever it moves into a new zone not on its internally stored list of visited registration zones. A zone is added to the list whenever a registration (including implicit registration) occurs, and is deleted upon expiration of a timer. After a system access, timers are enabled for every zone except one.
6. *Parameter-change registration*—This is performed when a mobile station modifies any of the following stored parameters:
   - the preferred slot cycle index
   - the station class mark
   - the call termination enabled indicator
7. *Ordered registration*—The mobile registers when the base station requests it.
8. *Implicit registration*—When a mobile successfully sends an origination message or page response message, the base station can infer the mobile station's location. This is considered an implicit registration.
9. *Traffic channel registration*—Whenever the base station has registration information for a mobile that has been assigned to a traffic channel, the base station can notify the mobile that it is registered.

6.4 Authentication

The A-key is known only to the authentication center (AC) and mobile and is the most secure piece of secret data. The A-key and a special random number (RANDSSD) can be used by the AC and mobile to generate a new shared secret data (SSD). The AC may send SSD to the serving system, but it is never sent over the air link.

RAND is a 32-bit random number issued by the station in the system overhead data in two 16-bit segments: RAND_A and RAND_B. The mobile stores and uses the most recent version of RAND in the authentication process. The last RAND received by the mobile station is confirmed from the mobile with an 8-bit number RANDC, a part of RAND, since the current system RAND and the one used by the mobile station could differ when the base station receives mobile station results.

The *electronic serial number* (ESN) is a 32-bit binary number that uniquely identifies the mobile to any system. It is set by the factory and not readily alterable in the field. Modification of ESN requires a special facility not normally available to subscribers.

The *mobile identification number* (MIN) is derived from mobile station's 10-digit directory telephone number. The first three digits map into the 10 most significant bits, the second three digits map into the next 10 bits, and the last four digits map into the remaining 14 bits.

The SSD is a 128-bit pattern stored in the semipermanent memory of the mobile and is known by the base station. SSD is a concatenation of two 64-bit subsets: SSD_A and SSD_B. SSD_A is used to support the authentication procedure and SSD_B is used to support voice privacy and message confidentiality.

SSD is maintained during power off. It is generated using a 56-bit random number (RANDSSD created by the home AC), the mobile's A-key, and ESN. The A-key is a 64-bit secret pattern, assigned and stored in the mobile station's permanent security and identification memory; this eliminates the need to pass the A-key itself from system to system as the subscriber roams. SSD updates are carried out only in the mobile station and its associated home location register authentication center (HLR/AC), not in the serving system. The AC manages the encrypting keys associated with an individual subscriber, if such functions are provided within the network.

All mobiles are assigned an ESN at the time of manufacturing. They are also assigned a 15-digit international mobile subscriber identity (IMSI). When the mobile is turned on, it must register with the system. When it registers, it sends its IMSI and other data to the network. The visitor location register (VLR) in the visited system queries the home system's HLR for the security data and service profile information. The VLR then assigns a temporary mobile subscriber identity (TMSI) to the mobile station. The mobile station uses the TMSI for further accesses to that system. The TMSI provides anonymity of communications since only the mobile and the network know the identity of the mobile with a given TMSI. When a mobile roams into the new system, some air interfaces use TMSI to query the old VLR and then assign a new TMSI; other air interfaces request that the mobile station send its IMSI and then assign a new TMSI.

The network transmits a random number RAND, which is received by all mobile stations. When a mobile station accesses that system, it calculates AUTHR, an encrypted version of RAND using SSD_A. It then transmits the desired message to the network along with its authentication.

The network performs the same calculation and confirms the identity of the mobile. All communications between the mobile and network are encrypted to prevent decoding of the data and using the data to clone other mobile stations. Furthermore, each time a mobile places or receives a call, a call history count (CHCNT) is incremented. The counter is also used for clone detection since clones will not have a call history identical to the legitimate mobile.

Procedures have been designed to allow a system to challenge an individual mobile with a unique challenge and to update the SSD. All mobile stations accessing the network must respond to the global challenge as part of their access. The global challenge response is an integral part of the network access (call origination, page response, registration, etc.). The call flow for global challenge is given in Figure 6.15.

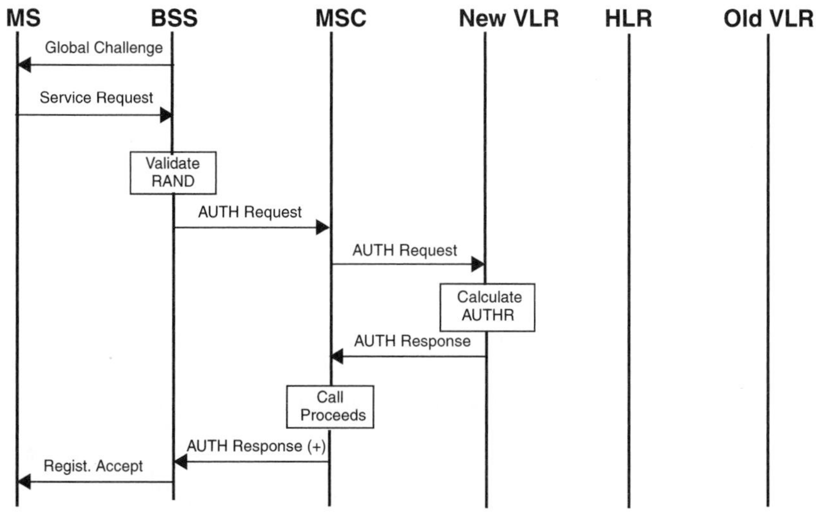

Figure 6.15 Call Flows for a Global Challenge

The unique challenge can be sent to a mobile station at any time. It is typically initiated by the mobile switching center (MSC) in response to some event (registration failure or after a successful handoff are the most typical cases). This is used to challenge the mobile to prove its identity. Figure 6.16 shows call flows for the unique challenge.

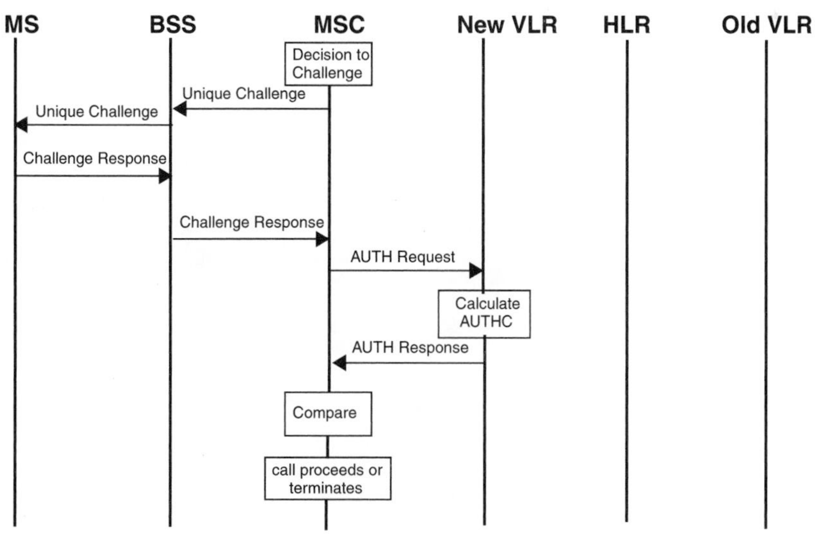

Figure 6.16 Call Flows for Unique Challenge

6.5 Summary

We discussed the CDMA call processing states and provided details of different messages that are used to exchange information in these states. We presented the registration and authentication procedures of CDMA. We concluded the chapter by discussing mobile substates and flow diagrams for CDMA call origination, call termination, and call release.

6.6 References

1. Garg, V. K., Smolik, K. F., and Wilkes, J. E., *Applications of CDMA In Wireless Communications,* Prentice Hall: Upper Saddle River, NJ, 1997.
2. TIA IS-95B, "Mobile Station and Radio Interface Specifications."

CHAPTER 7

Diversity, Combining, and Antennas

7.1 Introduction

In mobile radio systems, time variance and frequency selectivity of the fading radio channel and interference in the mobile environment have a degrading effect on the system performance. The signal-to-noise ratio (SNR) at the receiver varies with time and frequency. Mobile radio systems may be interference-limited where noise in the SNR is due to interfering signals of other users in the system while thermal noise is small and may be neglected. Since the interference power varies, the SNR at the receiver also varies depending on the actual interference situation.

In mobile radio, low SNRs that may result from deep fades of the radio channel or large interference power, can cause degradation in quality of service (QoS) or even an interruption of the radio link. The variation of the SNR at the receiver can be expressed as the normalized standard deviation [1]:

$$\sigma = \frac{\sqrt{var\{SNR\}}}{E\{SNR\}} \tag{7.1}$$

where

$var\{SNR\}$ = variance of SNR

$E\{SNR\}$ = expected value of SNR at the receiver

A large $var\{SNR\}$ may come from a large variance of the signal power as well as a large variance of the noise, i.e., interference in the SNR. The aim of *diversity* is to reduce the $var\{SNR\}$ or the value of σ in Equation (7.1) in order to improve the system performance. This is achieved by providing the receiver with two or more signals carrying the same information. These signals are combined at the receiver. The value of σ of the combined received signal is reduced as compared with the value of σ of individual signals owing to averaging over two or more different channel transmission conditions or interference situations.

In the case of averaging over different transmission conditions, the value of σ is decreased by reducing the variance of the signal in the SNR. In the case of averaging over different interference situations, the value of σ is decreased by reducing the variance of the noise in the SNR.

For combining the signals at the receiver, one of the combining methods—selection combining (SC), equal-gain combining (EGC), or maximal ratio combining (MRC)—can be used.

In this chapter, we discuss the concepts of diversity reception in which multiple signals are combined to improve the SNR of the system. Time diversity is used for IS-95 code division multiple access (CDMA) systems to improve system performance; therefore, we explore that system in more detail. We then describe various combining schemes used to combine the signals. Finally, we present some practical antennas used in cellular telephones today and also the intelligent or smart antennas that will be used in third-generation (3G) systems.

7.2 Diversity Reception

Buildings and other obstacles in built-up areas scatter the signal, and, because of the interaction between the several incoming waves, the resultant signal at the antenna is subject to rapid and deep fading. The average signal strength can be 30 to 40 dB below the free-space path loss. The fading is most severe in heavily built-up areas in an urban environment. In these areas, the signal envelope follows a Rayleigh distribution over short distances and a log-normal distribution over large distances [1].

Diversity reception techniques are used to reduce the effects of fading and improve the reliability of communication without increasing either the transmitter's power or the channel bandwidth. With increasing diversity, the cellular spectrum efficiency increases. A reduction of the required value of E_b/I_t in a single cell is equivalent in its effect on cellular spectrum efficiency to an increase of the value of S/I that can be achieved in the cellular environment.

The basic idea of a diversity reception is that if two or more independent samples of a signal are taken, these samples will fade in an uncorrelated manner. This means that the probability of all the samples being simultaneously below a given level is much lower than the probability of any individual sample being below that level. The probability of M samples all being simultaneously below a certain level is p^M, where p is the probability that a single sample is below the level. Thus, we can see that a signal composed of a suitable combination of the various samples will be much less likely to fade than will any individual sample alone.

The different signal contributions to be combined at the receiver can be provided in the three dimensions: frequency domain, time domain or domain of space, and direction. The frequency, time and space, and direction diversity aim at reducing the variance of the signal in the SNR. Interferer diversity aims at reducing the variance of the noise in the SNR by averaging over different interference situations. In real systems, combinations of the types of diversity given in Figure 7.1 are used since the higher the order of diversity in the system, the better the system performance.

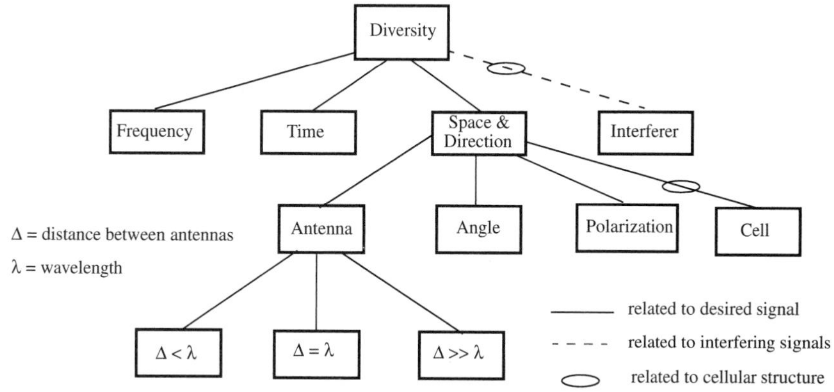

Figure 7.1 Classification of Different Types of Diversity

GSM with eight time slots per time division multiple access (TDMA) frame provides frequency diversity; coding and interleaving over eight time frames gives time diversity. Optionally, frequency hopping leads to both frequency and interferer diversity.

The gains achievable by different types of diversity depend on the

- Characteristics of the mobile radio channel
- Interference situation
- Combination of the types of diversity

Frequency diversity can be achieved either by applying a large user bandwidth or by applying a number of narrow frequency bands. In the case of the large user bandwidth, frequency diversity is equivalent to path diversity, since signal contributions received via a number of propagation paths can be resolved and subsequently combined. The gains achievable by frequency diversity increase with the multipath diversity potential, i.e., with increasing user bandwidth. User bandwidths larger than the coherence bandwidth of the channel are necessary to obtain independent signal contributions and considerable gains.

Time diversity improves the single cell performance. The faster the time variance of the channel and the larger the time period over which interleaving is performed, the larger the gains achievable by time diversity. Interleaving periods larger than the coherence time of the channel are necessary to obtain independent signal contributions and considerable gains. However, large interleaving periods entail large latency periods of the information, which some services cannot tolerate. On the one hand, fast time variance is favorable with respect to diversity gains, but on the other hand, fast time variance leads to deterioration of channel estimation and thus to performance degradation.

Antenna diversity with the distance Δ between the transmitter and receiver antennas in the order of wavelength λ, a larger $\Delta \geq \lambda$ improves the single cell performance. The gains achievable by antenna diversity with $\Delta \geq \lambda$ increase along with the variations of the channel characteristics, depending on the location and the larger the distance Δ of the antennas normalized to wavelength λ, Δ/λ. Distances Δ larger than the wavelength λ are necessary to obtain uncorrelated signal contributions at the antennas and considerable gains.

Antenna diversity with $\Delta > \lambda$ improves the single cell performance. This type of diversity and antenna designs are investigated mainly in Japan. Antenna diversity with $\Delta < \lambda$ can be usefully applied in mobile stations in the downlink where distances Δ of two or more antennas only much smaller than wavelength λ can be achieved. Antenna diversity with $\Delta >> \lambda$ improves the performance in the cellular environment. When using this type of diversity, two or more antennas are distributed over the cell region.

Cell diversity or *base station diversity*, which implies soft handoff, is used in the IS-95 system. In this system, which uses CDMA with single-user detection, cell diversity is necessary to obtain acceptable cellular spectrum efficiency. Interferer diversity is inherent due to the CDMA component.

Frequency diversity is useful if the radio channel is frequency selective. Time diversity is useful if the radio channel is time variant. In case of space and direction diversity, antenna, angle, polarization, and cell diversity can be distinguished. Antenna diversity is useful if the reception conditions depend on location, i.e., if the radio channel is time variant. With *angle diversity*, signals received with different arrival angles are resolved and combined at the receiver. In case of *polarization diversity*, signals with two different polarizations (in general orthogonal) are resolved and combined at the receiver. The speciality of the polarization diversity is that only two more-or-less independent signals can be made available for combining. *Cell diversity* means that signals transmitted by one mobile and received at different base stations in different sectors are combined. This type of diversity is uplink-specific.

In principle, diversity reception techniques can be applied either at the base station or at the mobile, although different problems must be addressed. Typically, the diversity receiver is used in the base station instead of the mobile station. The cost of the diversity combiner can be high, especially if multiple receivers are required. Also, the power output of the mobile station is limited by its battery life. The base station, however, can increase its power output or antenna height to improve coverage to a mobile station. Most diversity systems are implemented in the receiver instead of the transmitter since no extra transmitter power is needed to install the receiver diversity system. Since the path between the mobile station and the base station is assumed to be reciprocal, diversity systems implemented in a mobile station work similarly to those in a base station.

Depending on the type of diversity, either the single sector performance or the performance in the cellular environment is affected. If the single sector performance is affected, with the increasing diversity for a specified bit error rate (BER), the required E_b/I_t is reduced. With a large diversity, a smaller value of E_b/I_t is sufficient to achieve a certain BER. Thus, with increasing diversity affecting the single sector performance, the spectrum efficiency increases.

On the other hand, if the performance in the cellular environment is affected, with increasing diversity, the curves of S/I versus P_{out} become steeper. For small outage probability, the S/I curves move to larger values of S/I. With a large diversity, the outage probability for a given value of S/I becomes small. Therefore, with increasing diversity affecting the performance in the cellular environment, the spectrum efficiency also increases.

7.3 Types of Diversity

Diversity can be achieved using time, frequency, space, angle, multipath, and polarization [2]. In order to gain complete advantage of diversity, combining must be performed at the receiving end. Combiners are designed so that input signal levels, after phase and time delay corrections for the multipath effects, add vectorially while noise outputs are added randomly. Thus, on the average, the combined output SNR will be greater than that present at the input of a single receiver.

7.3.1 Macroscopic Diversity

Macroscopic diversity is used to reduce large-scale fading caused by shadowing. The local mean signal strength varies because of variations in terrain between the mobile station transmitter and the base station receiver. If only one antenna site is used, the traveling mobile unit may not be able to transmit a signal to the base station at certain geographical locations because of terrain variations such as hills or mountains. Therefore, two separate antenna sites can be used to receive two signals and to combine them to reduce long-term fading. The selective combining technique (see "Basic Combining Methods" on page 196) is recommended in the macroscopic diversity scheme since other methods require coherent combining that is difficult to achieve when the receivers are some distance apart. Macroscopic diversity is often used in short-wave radio systems to reduce the effects of fading from the ionosphere. Cellular and personal communication services (PCS) systems achieve the same effect by handoffs to nearby cell sites when the signal strength becomes weak.

With CDMA systems, the macroscopic diversity (i.e., soft handoff) is essential in order to achieve reasonable system performance because of the negative effects of frequency reuse and fast power control. If the mobile station is not connected to the base station to which the attenuation is the lowest, unnecessary interference is generated in adjacent cells. In the reverse direction (MS to BS), the macroscopic diversity is beneficial, since the more base stations try to detect the signals, the higher the probability is for at least one to succeed. In the reverse direction, the detection process itself does not utilize the information from the other base stations receiving the same signal, but the diversity is selection diversity, where the best frame is utilized in the network based on frame error rate (FER) indication from a cyclic redundancy check (CRC).

Macroscopic diversity is different in the forward direction as the transmission originates from several sources and diversity reception is handled by one receiver in the mobile station. All extra transmissions contribute to the interference. Capacity improvement is based on a similar principle as with a RAKE receiver in multipath channel, in which the received power level fluctuations tend to decrease as the number of separable paths increases. With forward link macroscopic diversity, the RAKE receiver's capability to gain from extra diversity depends also on the number of available RAKE fingers. If the RAKE receiver is not able to collect enough energy from the transmissions from two (or, in some cases, three) base stations owing to a limited number of RAKE fingers, the extra transmissions to the mobile station can have a negative effect on the total system capacity because of increased interference. This is most likely in the macroscopic cellular environment because the typical number of RAKE fingers considered adequate to capture the channel energy in most cases is four. If all connections offered that amount of diversity, then the receiver has only one or two branches to allocate for each connection.

7.3.2 Microscopic Diversity

Microscopic diversity uses two or more antennas that are at the same site (colocated) but designed to exploit differences in arriving signals from the receiver. Microscopic diversity techniques are used to prevent deep fades from occurring. Once the diversity branches are created, any of the combining schemes (e.g., selective, maximal-ratio, or equal-gain) can be used. The following methods are used to obtain uncorrelated signals for combining:

1. **Space diversity**—Several different transmission paths are used. Two antennas separated physically by a short distance d can provide two signals with low correlation between their fades. The separation d in general varies with antenna height h and with frequency. The higher the frequency, the closer the two antennas can be to each other. Typically a separation of a few wavelengths is enough to obtain uncorrelated signals.
2. **Frequency diversity**—Signals received on two frequencies, separated by the coherence bandwidth, B_c, are uncorrelated. To use frequency diversity in an urban or suburban environment for cellular and PCS frequencies, the frequency separation must be 300 kHz or more. The use of frequency hopping typical of TDMA systems (such as the Global System for Mobile Communications [GSM]) provides frequency diversity.

   The frequency diversity can be achieved either by using a user bandwidth, in general larger than the coherence bandwidth of the channel, or by applying a number of narrow frequency bands, whose center frequencies differ by at least the coherence bandwidth of the radio channel. In the case of the larger user bandwidth, frequency diversity is equivalent to path diversity, since signals received via a number of propagation paths can be resolved and subsequently combined. A larger bandwidth can be achieved by CDMA than by TDMA. In CDMA, spectrum spreading is performed by spreading sequences. The user bandwidth is increased over the symbol rate by a spreading factor. TDMA also performs spread spectrum. The user bandwidth is increased over the symbol rate by a factor of T if a user is only active in each time slot. Frequency diversity by a number of narrow frequency bands can be provided by frequency hopping (FH) together with coding and interleaving or by multicarrier transmission.
3. **Polarization diversity**—Horizontally or vertically polarized carrier waves are used. The horizontal and vertical polarization components, E_x and E_y, transmitted by two polarized antennas at the base station and received by two polarized antennas at the mobile unit, can provide two uncorrelated signal fadings. Polarization diversity results in a 3 dB power reduction at the transmitting site since the power must be split into two different polarized antennas. Polarization diversity requires different orientation of antennas.

4. **Angle diversity**—When the operating frequency is ≥ 10 GHz, the scattering of the signals from transmitter to receiver generates received signals from different directions that are uncorrelated with each other. Thus, two or more directional antennas can be pointed at different directions at the receiving site and provide signals for a combiner. This scheme is more effective at the mobile unit than at the base station since the scattering is from local buildings and vegetation and is more pronounced at street level than at the height of base station antennas. Angle diversity is based on directional antennas or antennas arrays.
5. **Time diversity**—The transmission of a symbol is spread out over time. If the identical signal is transmitted in different time slots, the received signals will be uncorrelated. This system will work for an environment where the fading occurs independent of the movement of the receiver. In a mobile radio environment, the mobile unit may be at a standstill at any location that has a weak local mean or is caught in deep fade. Although fading still occurs, even when the mobile is still, the time delayed signals are correlated and time diversity will not reduce the fades.

   Time diversity is achieved by coding, interleaving, and retransmissions. Channel coding is applied to achieve lower power levels and required signal quality in terms of BER/FER. Interleaving and channel coding processes are used to correct the errors caused by channel fades and interference peaks.
6. **Cell diversity**—Cell diversity implies soft handoff, which requires bringing together different signal contributions received at different base stations in different sectors via the fixed network.

7.3.3 RAKE Receiver

In 1958, Price and Green [3] proposed a method of resolving multipath using wideband pseudorandom (PN) sequences modulated onto a transmitter using other modulation methods (AM or FM). The PN sequence has the property that time-shifted versions of itself are almost uncorrelated. Thus, a signal that propagates from transmitter to receiver over multiple paths (hence, multiple different time delays) can be resolved into separately fading signals by cross-correlating the received signal with multiple time-shifted versions of the PN sequence. Figure 7.2 shows a block diagram of a typical system. In the receiver, the outputs are time shifted, and therefore must be sent through a delay line before entering the diversity combiner. The receiver is called a RAKE receiver because the block diagram looks like a garden rake.

When the CDMA systems were designed for cellular systems, the inherent wide-bandwidth signals with their orthogonal Walsh functions were natural for implementing a RAKE receiver. In addition, the RAKE receiver mitigates the effects of fading and is in part responsible for the claimed 10:1 spectral efficiency improvement of CDMA over analog cellular.

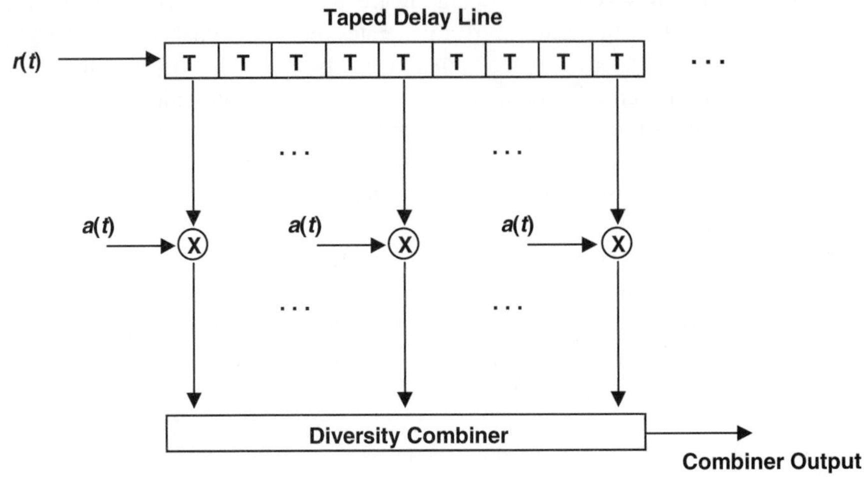

Figure 7.2 RAKE Receiver

In the CDMA system, the bandwidth (1.25 to 15 MHz) is wider than the coherence bandwidth of the cellular or PCS channel. Thus, when the multipath is resolved in the receiver, the signals from each tap on the delay line are uncorrelated with each other. The receiver can then combine them using any of the combining schemes. The CDMA system then uses the multipath characteristics of the channel to its advantage to improve the operation of the system.

The performance of the RAKE receiver will be governed by the combining scheme used. An important factor in the receiver design is obtaining synchronization of the signals in the receiver to that of the transmitted signal. Since adjacent cells are also on the same frequency with different time delays on the Walsh codes, the entire CDMA system must be tightly synchronized.

A RAKE receiver uses multiple correlators to separately detect the M strongest multipath components. The relative amplitudes and phases of the multipath components are found by correlating the received waveform with delayed versions of the signal or vice versa. The energy in the multipath can be recovered effectively by combining the (delay-compensated) multipath in proportion to their strengths. This combining is a form of diversity and can help to reduce fading. Multipath with relative delays less than $\Delta t = 1/B_w$ cannot be resolved, and if existing, contribute to fading; in such cases forward error-correction coding and power control schemes play the dominant role in mitigating the effects of fading.

The outputs of the M correlators are denoted as Z_1, Z_2, . . . and Z_M. The weights of the outputs are a_1, a_2, . . . and a_M, respectively [1] (see Figure 7.3). The weighting coefficients are based on the power or the SNR from each correlator output. If the power or SNR of a particular correlator is small, it is assigned a small weighing factor. The composite signal, $\overline{Z}$, is given by

$$\bar{Z} = \sum_{k=1}^{M} a_k \cdot Z_k \tag{7.2}$$

The weighting coefficients, a_k, are normalized to the output signal power of the correlator in such a way that the coefficients sum to unity, as shown in Equation (7.3).

$$a_k = \frac{Z_k^2}{\sum_{k=1}^{M} Z_k^2} \tag{7.3}$$

In CDMA cellular/PCS systems, the forward link (BS to MS) uses a *three-finger* RAKE receiver, and the reverse link (MS to BS) uses a *four-finger* RAKE receiver [4]. In the IS-95 CDMA system, the detection and measurement of multipath parameters are performed by a *searcher receiver*, which is programmed to compare incoming signals with portions of *I*- and *Q*-channel PN codes. Multipath arrivals at the receiver unit manifest themselves as correlation peaks that occur at different times. A peak's magnitude is proportional to the envelope of the path signal, and the time of each peak, relative to the first arrival, provides a measurement of the path's delay.

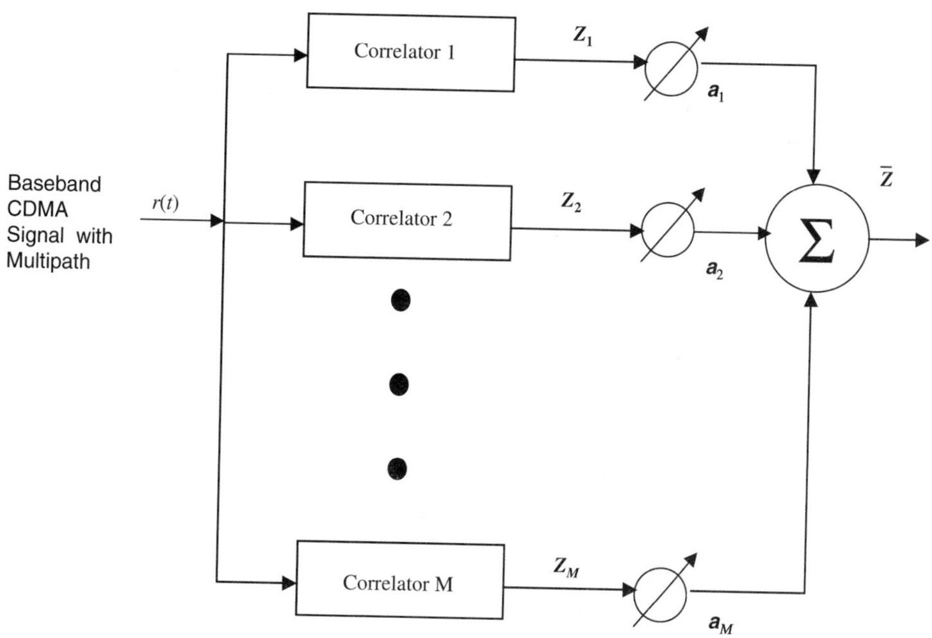

Figure 7.3 RAKE Receiver Correlator

The PN chip rate of 1.2288 Mcps allows for resolution of multipath at time intervals of 0.814 μs. Because all the base stations use the same *I* and *Q* PN codes, differing only in code phase offset, not only multipath but also other base stations are detected by correlation (in a different *search window* of arrival times) with the portion of the codes corresponding to the selected base stations. The searcher receiver maintains a table of the stronger multipath and/or base station signals for possible diversity combining or for handoff purposes. The table includes time of arrival, signal strength, and the corresponding PN code offset.

On the reverse link, the base station's receiver that is assigned to track a particular mobile transmitter uses the *I*- and *Q*-code times of arrival to identify mobile signals from users affiliated with that base station. Of the mobile signals using the same *I*- and *Q*-code offsets, the searcher receiver at the base station can distinguish the desired mobile signal by means of its unique scrambling long PN code offset, which is acquired using a special preamble before voice transmission begins on the link. As the call proceeds, the searcher receiver is able to monitor the strengths of multipath from the mobile unit to the base station and to use more than one path through diversity combining.

7.4 Basic Combining Methods

After obtaining the necessary signal samples, we need to consider the question of processing these samples to obtain the best results. For most communication systems, the process can be broadly classified as the *linear combination* of the samples. In the combining process, the various signal inputs are individually weighted and added together as [5]

$$r(t) = a_1 r_1(t) + a_2 r_2(t) + \ldots + a_M r_M(t) \tag{7.4a}$$

$$r(t) = \sum_{i=1}^{M} a_i r_i(t) \tag{7.4b}$$

where

$r_i(t)$ = the envelope of the *i*th signal

a_i = the weight factor applied to the *i*th signal

We make the following assumptions in the analysis of a combiner:

1. The noise in each branch is independent of the signal and is additive.
2. The signal amplitudes change because of fading, but the fading rate is much smaller than the lowest modulation frequency present in the signal.
3. The noise components are locally incoherent and have zero mean, with a constant local mean-square (i.e., constant noise power).
4. The local mean-square values (powers) of the signals are statistically independent.

Since the goal of the combiner is to improve the noise performance of the system, the analysis of combiners is generally performed in terms of SNR. We will examine several different types of combiners and compare their SNR improvements over no diversity.

7.4.1 Selection Combiner

The selection combiner is the simplest of all the diversity schemes. An ideal selection combiner chooses the signal with the highest instantaneous SNR, so the output SNR is equal to that of the best incoming signal. In practice, the system cannot function on an instantaneous basis; to be successful, it is essential that the internal time constants of a selection system are substantially shorter than the reciprocal of the signal fading rate.

We assume that the signal received by each diversity branch is statistically independent of the signals in other branches and is Rayleigh distributed with equal mean signal power P_0. The probability density function of the signal envelope on branch i is given by

$$p(r_i) = \frac{r_i}{P_0}e^{-r_i^2/2P_0} \tag{7.5}$$

where
$2P_0$ = mean-square signal power per branch = $<r_i>$
r_i^2 = instantaneous power in the ith branch

Let $\xi_i = r_i^2/2N_i$ and $\xi_0 = \frac{2P_0}{2N_i}$, where N_i is the noise power in the ith branch.

$$\therefore \frac{\xi_i}{\xi_0} = r_i^2/2P_0 \tag{7.6}$$

The probability density function for ξ_i is given by

$$p(\xi_i) = \frac{1}{\xi_0}e^{(-\xi_i/\xi_0)} \tag{7.7}$$

We assume that the signal in each branch has a constant mean; thus, the probability that the SNR on any one branch is less than or equal to any given value ξ_g is

$$P[\xi_i \le \xi_g] = \int_0^{\xi_g} p(\xi_i)d\xi_i = 1 - e^{(-\xi_g/\xi_0)} \tag{7.8}$$

Therefore, the probability that the SNRs in all branches are simultaneously less than or equal to ξ_g is given by

$$P_M(\xi_g) = P[\xi_1, \xi_2, \ldots, \xi_M \le \xi_g] = [1 - e^{(-\xi_g/\xi_0)}]^M \tag{7.9}$$

The probability that at least one branch will exceed the threshold SNR value of ξ_g is given by

$$P\ (\text{at least one branch} \geq \xi_g) = 1 - P_M(\xi_g) \tag{7.10}$$

The percentage of time the instantaneous output SNR ξ_M is below or equal to the threshold value, ξ_g, is equal to $P(\xi_M \leq \xi_g)$. We plot results for M = 1, 2, and 4 in Figure 7.4. Note that the largest gain occurs for the two-branch diversity combiner.

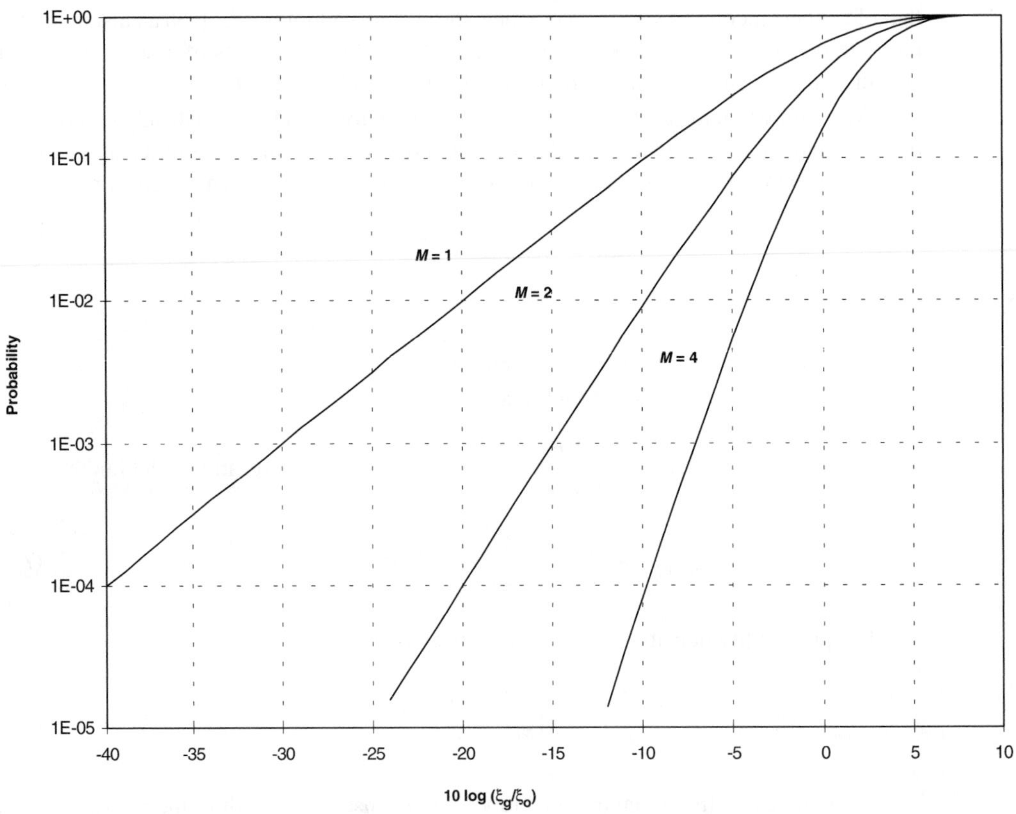

Figure 7.4 Probability for Different Values of *M*-Selection Combiner

By differentiating Equation (7.9) we get the probability density function

$$p_M(\xi_g) = (M/\xi_0)[1 - e^{(-\xi_g/\xi_0)}]^{M-1} e^{(-\xi_g/\xi_0)} \tag{7.11}$$

The mean value of the SNR can be given as

$$\overline{\xi_M} = \int_0^{\infty} M\left(\frac{\xi_g}{\xi_0}\right)[1 - e^{(-\xi_g/\xi_0)}]^{M-1} e^{(-\xi_g/\xi_0)} d\xi_g \tag{7.12}$$

Let $x = \frac{\xi_g}{\xi_0}$ and $dx = \frac{d\xi_g}{\xi_0}$

$$\therefore \frac{\overline{\xi_M}}{\xi_0} = M\int_0^{\infty} x[1 - e^{-x}]^{M-1} e^{-x} dx \tag{7.13}$$

Substituting $y = 1 - e^{-x}$ or $x = -\ln(1-y)$; then $dy = e^{-x}dx$

$$\therefore \frac{\overline{\xi_M}}{\xi_0} = M\int_0^1 [-\ln(1-y)]y^{M-1}dy = M\sum_{K=1}^{\infty} \int_0^1 \frac{1}{K} y^{M+K-1}dy \tag{7.14}$$

$$\therefore \frac{\overline{\xi_M}}{\xi_0} = \sum_{K=1}^{M} \frac{1}{K} \tag{7.15}$$

Table 7.1 shows that the mean SNR increases slowly with M.

Table 7.1 Number of Branches versus Mean SNR (dB)

M	$\frac{\overline{\xi_M}}{\xi_0}$	$10\log\frac{\overline{\xi_M}}{\xi_0}$
1	1.000	0.000
2	1.500	1.761
3	1.833	2.632
4	2.083	3.187
5	2.283	3.585
6	2.450	3.892

7.4.2 Maximal-Ratio Combining

Maximal-ratio combining was first proposed by Kahn [6]. The M signals are weighted proportionally to their signal voltage-to-noise power ratios and then summed.

$$r_M = \sum_{i=1}^{M} a_i r_i(t) \tag{7.16}$$

Since noise in each branch is weighted according to noise power,

$$\overline{n_i^2(t)} = \sum_{j=1}^{M} \sum_{i=1}^{M} a_i a_j \overline{n_i(t) n_j(t)} \tag{7.17}$$

$$\text{The Average Noise Power, } N_T = \sum_{i=1}^{M} a_i^2 \overline{n_i^2(t)} = 2 \sum_{i=1}^{M} |a_i|^2 N_i \tag{7.18}$$

where

$$\overline{n_i^2(t)} = 2N_i \tag{7.19}$$

The SNR at the output is given as

$$\xi_M = \frac{1}{2} \frac{\left| \sum_{i=1}^{M} a_i r_i(t) \right|^2}{\sum_{i=1}^{M} |a_i|^2 N_i} \tag{7.20}$$

We want to maximize ξ_M. This can be done by using the Schwartz inequality.

$$\left| \sum_{i=1}^{M} a_i r_i \right|^2 \le \left[\sum_{i=1}^{M} |r_i^2| \right] \left[\sum_{i=1}^{M} |a_i|^2 \right] \tag{7.21}$$

If $a_i = \dfrac{r_i}{\sqrt{N_i}}$ then

$$\xi_M = \frac{1}{2}\frac{\sum_{i=1}^{M} r_i^2 \sum_{i=1}^{M} \frac{r_i^2}{N_i}}{\sum_{i=1}^{M} r_i^2} \tag{7.22}$$

$$\therefore \xi_M = \frac{1}{2}\sum_{i=1}^{M} \frac{r_i^2}{N_i} = \sum_{i=1}^{M} \xi_i \tag{7.23}$$

Thus, the SNR at the combiner output equals the sum of the SNR of the branches.

$$\overline{\xi_M} = \sum_{i=1}^{M} \overline{\xi_i} = \sum_{i=1}^{M} \xi_0 = M\xi_0 \tag{7.24}$$

$$\therefore \frac{\overline{\xi_M}}{\xi_0} = M \tag{7.25}$$

The probability density function of the combiner output SNR is given by

$$p(\xi_M) = \frac{\xi_M^{M-1} e^{-\frac{\xi_M}{\xi_0}}}{\xi_0^M (M-1)!}\ , \quad \xi_M \geq 0 \tag{7.26}$$

The probability that $\xi_M \leq \xi_g$ is given by

$$P(\xi_M \leq \xi_g) = 1 - e^{-\frac{\xi_g}{\xi_0}} \sum_{K=1}^{M} \frac{\left(\frac{\xi_g}{\xi_0}\right)^{K-1}}{(K-1)!} \tag{7.27}$$

$$P(\xi_M > \xi_g) = e^{-\frac{\xi_g}{\xi_0}} \sum_{K=1}^{M} \frac{\left(\frac{\xi_g}{\xi_0}\right)^{K-1}}{(K-1)!} \tag{7.28}$$

The plot of P for $M = 1$, 2, and 4 is shown in Figure 7.5.

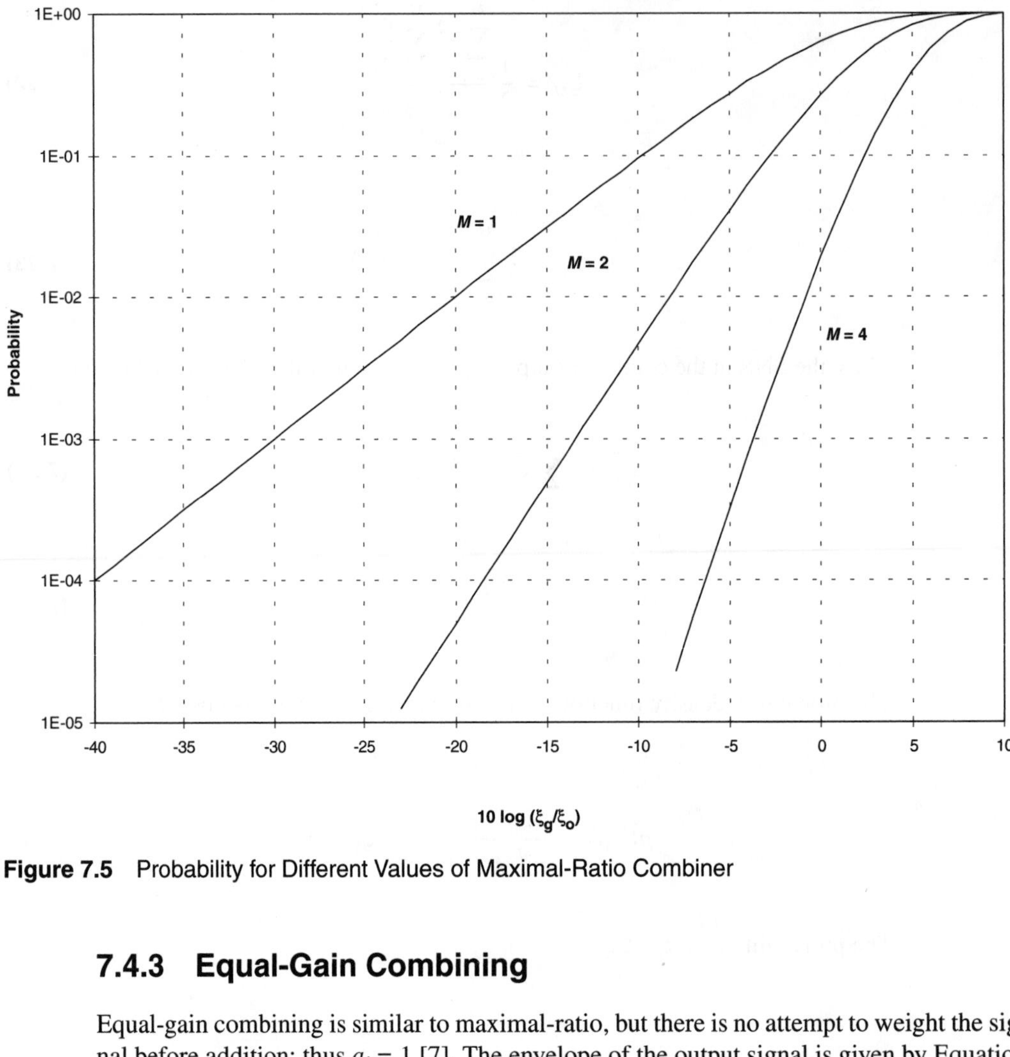

Figure 7.5 Probability for Different Values of Maximal-Ratio Combiner

7.4.3 Equal-Gain Combining

Equal-gain combining is similar to maximal-ratio, but there is no attempt to weight the signal before addition; thus $a_i = 1$ [7]. The envelope of the output signal is given by Equation (7.4b), with all $a_i = 1$,

$$r = \sum_{i=1}^{M} r_i \quad \textbf{(7.29)}$$

and the mean output SNR is given as

$$\overline{\xi_M} = \frac{1}{2}\frac{\overline{\left[\sum_{i=1}^{M} r_i\right]^2}}{\sum_{i=1}^{M} \overline{N_i}} \quad \textbf{(7.30)}$$

We assume that mean noise power in each branch is same (i.e., N), then Equation (7.30) becomes

$$\overline{\xi_M} = \frac{1}{2NM}\overline{\left[\sum_{i=1}^{M} r_i\right]^2} = \frac{1}{2NM}\sum_{j,\,i=1}^{M}\overline{r_j r_i} \tag{7.31}$$

but $\overline{r_i^2} = 2P_0$; and $\overline{r_i} = \sqrt{\frac{\pi P_0}{2}}$.

Since the various branch signals are uncorrelated, $\overline{r_j r_i} = \overline{r_i r_j} = \overline{r_i}\,\overline{r_j}$, for i not equal to j. Therefore, Equation (7.31) will be

$$\overline{\xi_M} = \frac{1}{2NM}\left[2MP_0 + M(M-1)\frac{\pi P_0}{2}\right] = \xi_0\left[1 + (M-1)\frac{\pi}{4}\right] \tag{7.32}$$

$$\frac{\overline{\xi_M}}{\xi_0} = 1 + (M-1)\frac{\pi}{4} \tag{7.33}$$

For $M = 2$, the probability P can be written in closed form as

$$P(\xi_M \le \xi_g) = 1 - e^{-\left(\frac{2\xi_g}{\xi_0}\right)} - \sqrt{\pi\left(\frac{\xi_g}{\xi_0}\right)}\, e^{-\frac{\xi_g}{\xi_0}} \cdot erf\sqrt{\frac{\xi_g}{\xi_0}} \tag{7.34}$$

For $M > 2$, the probability can be obtained by numerical integrations techniques. The plot of probability $P(\xi_M \le \xi_g)$ is given in Figure 7.6 for $M = 2$.

Table 7.2 shows M versus SNR at 1 percent probability for the selection, maximal-ratio, and equal-gain combiner. Table 7.3 shows SNR improvement for $M = 2$, 4, and 6 at 1 percent probability for the selection, maximal-ratio, and equal-gain combiner. It can be seen that selection diversity scheme has the poorest performance and maximal ratio the best. The performance of equal-gain combining is only marginally inferior to maximal ratio. The implementation complexity for equal-gain combining is significantly less than the maximal-ratio combining because of the requirement of correctly weighing factors. The data are compared in Figure 7.7.

Table 7.2 Signal-to-Noise Ratio (dB)

M	Selection	Maximal Ratio	Equal Gain
1	–20.0	–20.0	–20.0
2	–10.0	–8.5	–9.2
4	–4.0	–1.0	–2.0
6	–2.0	2.0	1.5

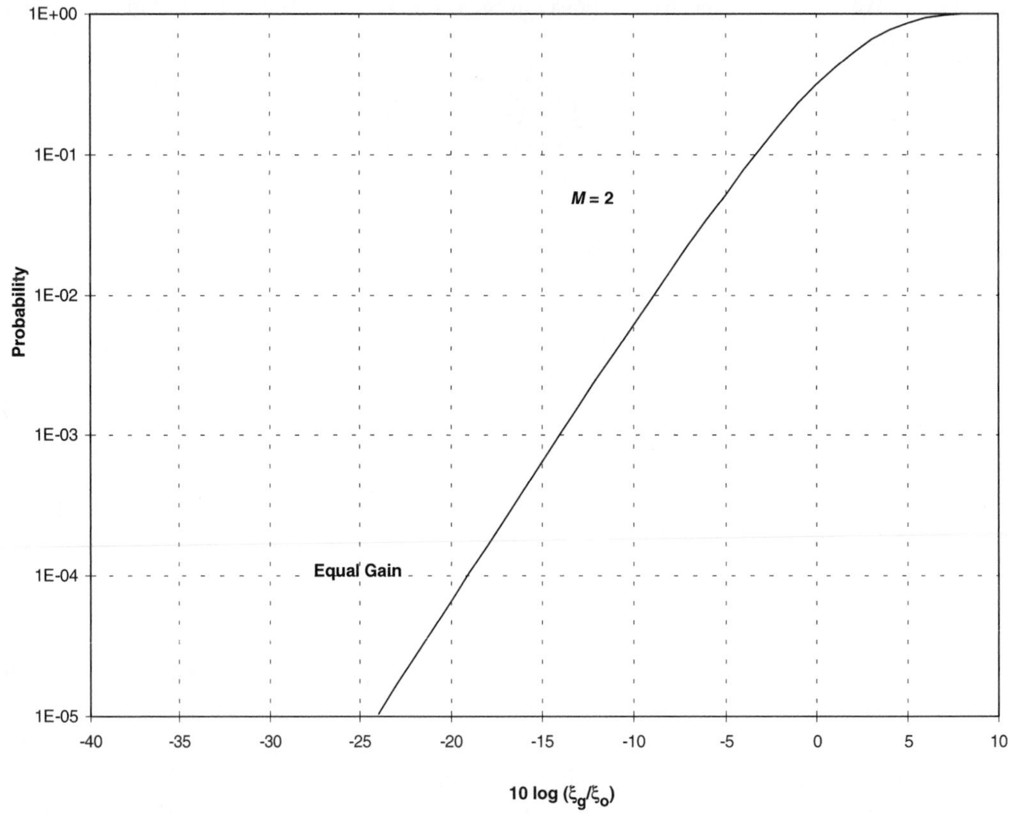

Figure 7.6 Probability for $M = 2$ Equal-Gain Combiner

Table 7.3 Signal-to-Noise Ratio Improvement (dB)

M	Selection	Maximal Ratio	Equal Gain
2	10.0	11.5	10.8
4	16.0	19.0	18.0
6	18.0	22.0	21.5

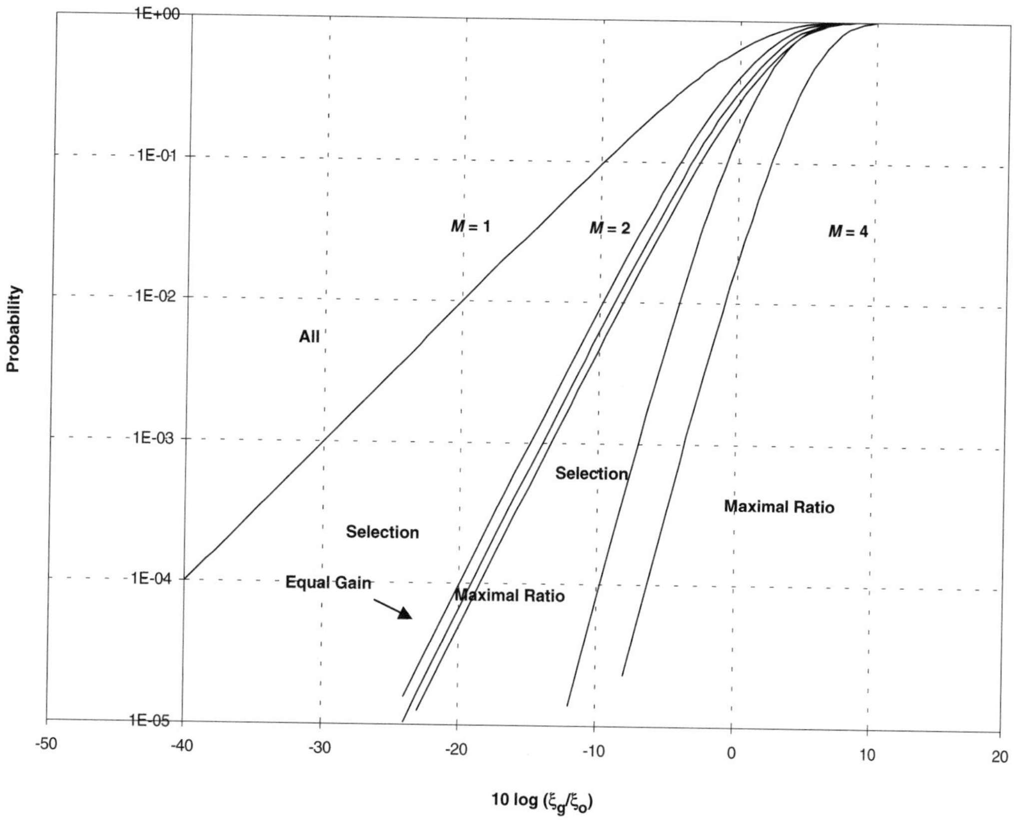

Figure 7.7 Performance Comparison Improvement of Various Combining Schemes

7.5 BPSK Modulation and Diversity

As we discussed earlier, to prevent deep fades from occurring, microscopic diversity techniques can exploit the rapidly changing signal. Macroscopic diversity can be used to reduce large-scale fading caused by the shadowing that arises from variations in the terrain between the transmitter and receiver. Macroscopic diversity is useful at the base station receiver. By using base station antennas sufficiently separated in space, the base station improves the reverse link performance choosing the antenna with strongest signal from the mobile.

For a binary phase-shift keying (BPSK) system, the bit error probability is given by

$$p_{BPSK} = Q\left(\sqrt{\frac{2 \cdot E_b}{N_0}}\right) \quad \textbf{(7.35)}$$

The bit error probabilities over M-branch diversity with selection, equal-gain, and maximal-ratio combining are given as [4]

- *Selection Combining*

$$p_{BPSK,M} = \frac{M}{2}\sum_{k=0}^{M-1}\binom{M-1}{K}\cdot\frac{(-1)^k}{k+1}\cdot\left[1-\sqrt{\frac{\rho_c}{k+1+\rho_c}}\right] \quad (7.36)$$

$$p_{\pi/4-DQPSK,M} = \frac{M}{2}\sum_{k=0}^{M-1}\binom{M-1}{k}\cdot\frac{(-1)^k}{k+1}\cdot\left[1-\sqrt{\frac{0.585\rho_c}{k+1+0.585\rho_c}}\right] \quad (7.37)$$

- *Equal-Gain Combining*

$$p_{BPSK,M} = (\bar{\rho}_{BPSK})^M\cdot\sum_{k=0}^{M-1}\binom{M+k-1}{k}\cdot(1-\bar{\rho}_{BPSK})^k \quad (7.38)$$

$$p_{\pi/4-DQPSK,M} = (\rho_{\pi/4})^M\cdot\sum_{k=0}^{M-1}\binom{M+k-1}{k}\cdot(1-\rho_{\pi/4})^k \quad (7.39)$$

where

$$\rho_{\pi/4} = \frac{1}{2}\left[1-\frac{\rho_c}{\sqrt{\rho_c^{\,2}+1/2+2\rho_c}}\right]$$

- *Maximal-Ratio Combining*

$$p_{BPSK,M} \approx \binom{2M-1}{M}\cdot\left(\frac{1}{4\rho_c}\right)^M \quad (7.40)$$

$$p_{\pi/4-DQPSK,M} = \frac{1}{2}\cdot(1-\mu)^M\cdot\sum_{k=0}^{M-1}\binom{M-1+k}{k}\cdot\left[\frac{1}{2}\cdot(1-\mu)\right]^k \quad (7.41)$$

where

$$\rho_c = \frac{E_b}{N_0\cdot M} = \text{average SNR per diversity branch}$$

$$\mu = \sqrt{\frac{\rho_c}{1+\rho_c}}$$

$$\bar{\rho}_{BPSK} = \frac{1}{2}\cdot\left(1-\sqrt{\frac{\rho_c}{1+\rho_c}}\right)$$

EXAMPLE 7.1

Compare the bit error performance of the BPSK modulation having SNR = 10 dB with two-branch diversity using selection, equal-gain, and maximal-ratio combining.

- *Selection Combining*

$$p_{BPSK,2} = \sum_{k=0}^{1}\binom{1}{k}\cdot\frac{(-1)^k}{k+1}\cdot\left[1-\sqrt{\frac{10}{k+1+10}}\right] = [1-0.95346]-\frac{1}{2}[1-0.91287]$$

$$= 0.002976$$

- *Equal-Gain Combining*

$$\bar{p}_{BPSK} = \frac{1}{2}\cdot\left(1-\sqrt{\frac{10}{11}}\right) = 0.02327$$

$$p_{BPSK,2} = (0.02327)^2\cdot\left[\sum_{k=0}^{1}\binom{1+k}{k}\cdot(1-0.02327)^k\right] = 0.002153$$

- *Maximal-Ratio Combining*

$$p_{BPSK,2} \approx \binom{3}{2}\cdot\left(\frac{1}{4\times 10}\right)^2 = 0.001875$$

Table 7.4 Bit Error Performance Comparison

Combining Type	$P_{BPSK,2}$	Performance with Respect to Maximal-Ratio Combining
Maximal-ratio combining	0.001875	1.0
Equal-gain combining	0.002153	1.148
Selection combining	0.002976	1.587

The bit error performance of the equal-gain and selection combining is about 15 percent and 59 percent worse than the maximal-ratio combining, respectively.

7.6 Examples of Base Station and Mobile Antennas

While the simple dipole antenna (Figure 7.8) is the reference example for antenna specifications, most practical antenna designs aim to improve on the gain of the dipole antenna. Practical antenna design must consider the following issues:

- **Antenna pattern**—Closely related to the gain of the antenna is the antenna pattern. As gain is increased, the beamwidth is decreased. This can be an advantage or a disadvantage depending on the antenna orientation and the needs of the system design.
- **Bandwidth**—The antenna must operate over the full range of frequencies in use for the cellular or PCS system. If the antenna bandwidth is small, channels at the edge of the band may not receive signals as well as those near the band center.
- **Gain**—The higher the gain of the antenna, the lower the power that is necessary at the transmitter. Since the antenna is purchased once and the transmitter power is purchased continuously, high-gain antennas are useful for saving money in electricity and help conserve the natural resources used to create the electricity.
- **Ground plane**—Some antennas require that they be mounted above a reflecting surface to function correctly. For example, a quarter-wave antenna is one-half of a dipole and requires that the other half of the dipole be developed by a mirror image below a ground plane. This can be used to advantage in designing antennas for vehicles but is a disadvantage when base station antennas (high above the earth) are designed.
- **Height**—The higher the antenna, the better the coverage of the system. However, if the coverage of the system is too good, interference from other cells may become troublesome. In an interference-limited system, all levels scale equally so at the first order, there will not be a problem. However, since radio wave propagation is statistical, there may be locations where good propagation exists from a point far removed from a base station. The higher the base station antenna, the more likely that these anomalous events will occur.

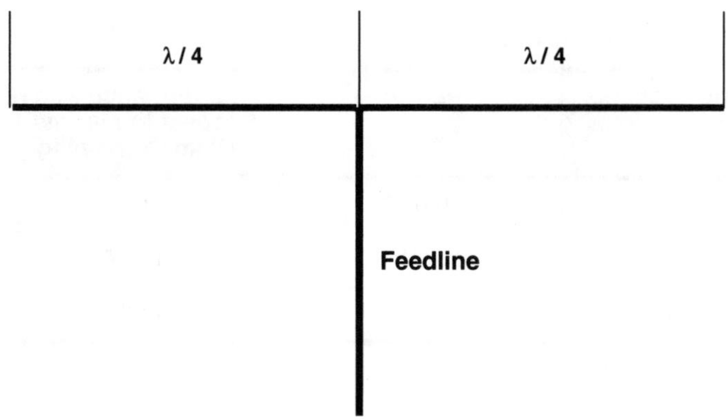

Figure 7.8 Dipole Antenna

- **Input impedance**—Most cables used as feedline from the transmitter/receiver to the antenna are either 50 ohms or 72/75 ohms. If the input impedance of the antenna is far removed from either of these values, it will be difficult to get the antenna to accept the power delivered to it and its efficiency η will be low.
- **Mechanical rigidity**—If the antenna flexes in the wind, it will introduce an additional fading component to the received signal. Ultimately, the continuous flexing will cause metal fatigue and mechanical failure of the antenna.
- **Polarization**—For wireless cellular and PCS communications, a vertical antenna is the easiest to mount on a vehicle; therefore, vertical polarization has been standardized. In general, horizontal or vertical polarization will work equally well.

7.6.1 Efficiency of a Sector Antenna

In a three-sector antenna there is always some overlap of the sector antenna patterns, so interference is not reduced exactly by a factor of three. The sectorization gain can be given as

$$\chi = N_s \times \left(\frac{1 + \beta_{omni}}{1 + \beta_{sector}}\right) \tag{7.42}$$

where

β_{omni} = interference factor for omnidirectional cell
β_{sector} = interference factor for three-sector cell
N_s = number of sectors

The sectorization efficiency is then defined as

$$\rho = \frac{\chi}{N_s} \tag{7.43}$$

EXAMPLE 7.2

Calculate the efficiency of a three-sector antenna. Assume β_{omni} and β_{sector} to be 0.60 and 0.85, respectively.

$$\chi = 3 \times \left(\frac{1 + 0.6}{1 + 0.85}\right) = 2.6$$

$$\rho = \frac{2.6}{3.0} = 0.865$$

7.6.2 Antenna Downtilt

Antenna downtilt is used to tilt the main beam to a certain angle to suppress the power level toward the reuse cell site and to reduce cochannel interference. The downtilt can be accomplished by mechanical or electrical means. Tilting the antenna vertical beam pattern reduces the interference to other cell sites because the received field intensity in these cell sites is weak.

The antenna downtilt angle ϕ is the function of antenna height, cell coverage radius, and antenna vertical beamwidth. In general, when the coverage radius of a service area is set to a specified value, the higher the antenna, the larger the downtilt angle, and larger the reduction in cochannel interference. On the other hand, when the height of the antenna at the base station is set, the smaller the coverage radius, the larger the downtilt angle. Equations (7.44) and (7.45) show two formulas of antenna downtilt that are used for different engineering considerations [8].

$$\phi_A = \tan{-1}\left(\frac{h_b}{2R}\right) + \frac{(HPBW)_{vert}}{2} \tag{7.44}$$

$$\phi_B = \pi - 2\tan{-1}\left(\frac{R}{h_b}\right) \tag{7.45}$$

where

$(HPBW)_{vert}$ = antenna vertical half-power bandwidth

h_b = antenna height at the base station

R = cell radius

Equation (7.44) represents the angle of antenna downtilt to reduce interference at the base of neighbor cell ($D = 2R$) by 3 dB. Equation (7.45) represents the angle of antenna downtilt that would be needed to preserve the coverage in the fringe of the cell ($D = R$). Table 7.5 gives the antenna downtilts obtained from the two equations for different environments.

Table 7.5 Antenna Downtilts in Different Environments

Environment	Height (m)	Coverage Radius (km)	R/h_b	Vertical Beamwidth (Degree)	ϕ_A	ϕ_B	ϕ_{av}
Large metropolitan	30	0.8	26.7	7	4.57	4.30	4.43
Urban	25	1.5	60	7	3.98	1.91	2.94
Suburban	20	3.0	150	7	3.69	0.76	2.23
Rural	50	10.0	200	7	3.64	0.57	2.1

With this background, we will examine some simple antennas that are used for base and mobile operation.

7.6.3 Quarter-Wave Vertical

The simplest antenna for a vehicle is the quarter-wave vertical (see Figure 7.9). A length of wire 1/4 wavelength long is mounted on the roof of the vehicle. With metal vehicles (most cars and trucks), the other half of the dipole is developed in the image in the ground plane. Since a vertical dipole antenna has an omnidirectional pattern [9], the quarter-wave vertical has an omnidirectional pattern. The gain of the antenna is the same as that of a dipole (0 dB dipole [dBd] or 2.1 isotropic dipole [dBi]). The impedance of a quarter-wave vertical antenna is 36.5 ohms and requires a matching transformer for proper feeding.

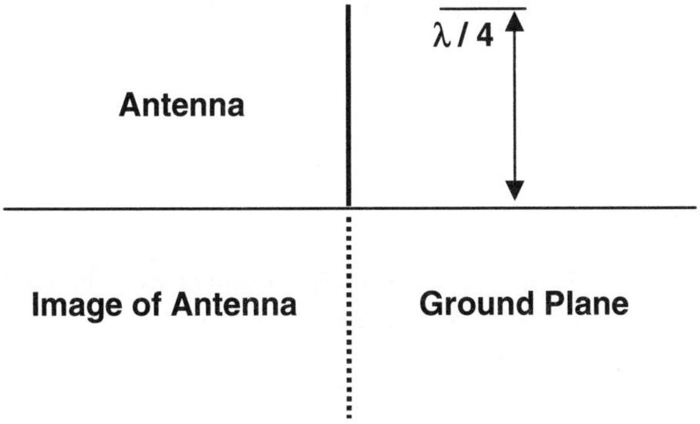

Figure 7.9 Quarter-Wave Vertical

7.6.4 Stacked Dipoles

Since a vertical dipole has an omnidirectional pattern, three or more dipoles can be stacked vertically to produce increased gain and maintain the omnipattern. A typical base station antenna will have four half-wave dipoles spaced by one wavelength vertically (see Figure 7.10). All the antenna elements are fed in phase with a signal from the transmitter. The resultant pattern is omnidirectional in the horizontal plane and has an 8.6 dBi gain on the horizon (0 degree elevation) with a vertical beamwidth of ±6.5 degrees. It has an impedance of 63 ohms; thus, a matching transformer is necessary, but this one is easier to build than the one for the quarter-wave vertical.

A variation of the base station antenna for use on a vehicle uses a half-wave dipole above a quarter-wave vertical (see Figure 7.11). The other half of the antenna is in the ground plane image. The two elements are decoupled from each other by a 1/4 wavelength long decoupling coil. This is the common cellular antenna seen on most vehicles. It has a gain of 7–10 dBi.

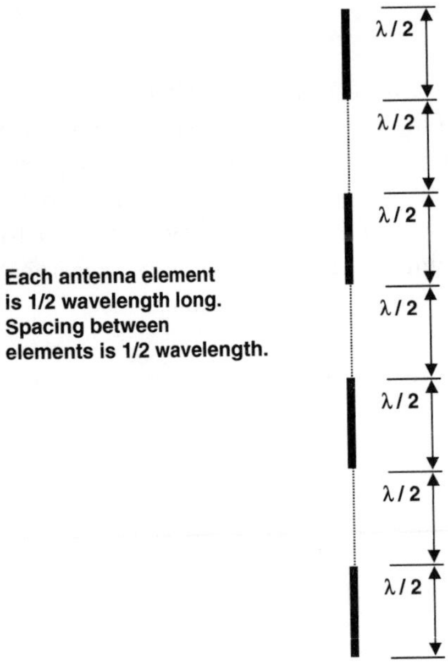

Figure 7.10 Stacked Vertical Dipoles

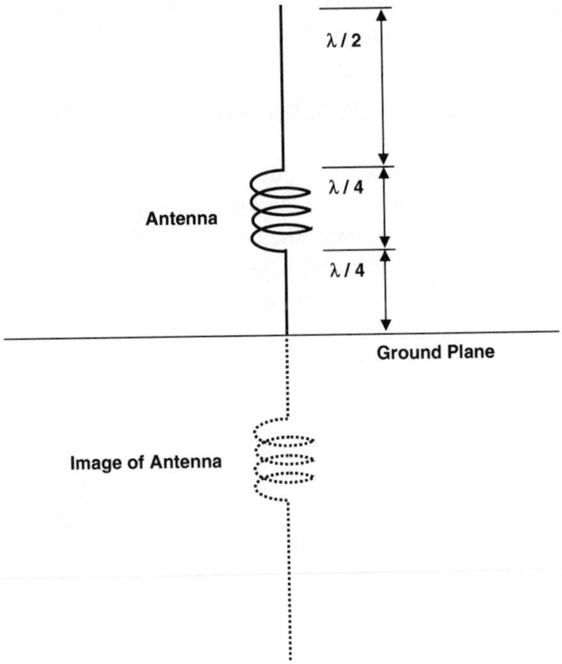

Figure 7.11 Half-Wave over Quarter-Wave Antenna (When Mounted on Vehicle)

7.6.5 Corner Reflectors

The previous antennas have omnidirectional patterns. A directional antenna at a base station can improve the SNR of the system and thus improve the spectral efficiency of the system. In 1939 Kraus [10] designed the corner reflector antenna consisting of a vertical dipole and two sheets of metal at a 45, 60, or 90 degree angles (see Figure 7.12).

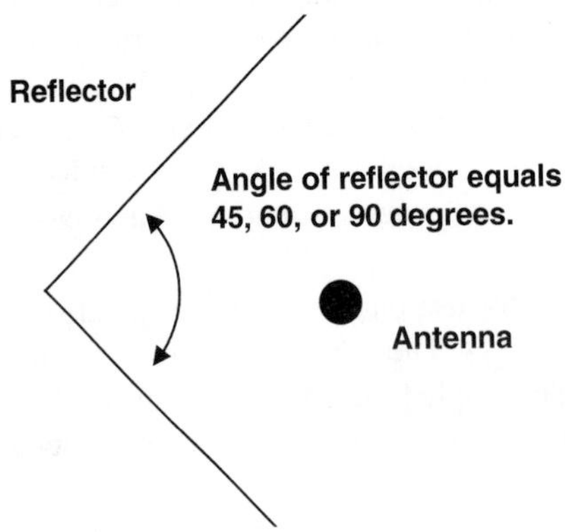

Figure 7.12 Corner Reflector

The impedance and gain of the antenna depends on the angle of the corner and the spacing from the corner to the dipole antenna. The gain of the antenna varies from 7 to 13 dBd and the impedance varies from 0 to 150 ohms. A colinear antenna can be used in place of the dipole for additional gain.

7.6.6 Smart Antenna

A smart or intelligent antenna refers to group of core radio frequency (RF) technologies that control directional antenna arrays by means of sophisticated digital signal processing (DSP) algorithms. A smart antenna evaluates signal conditions continuously for each signal that is transmitted or received. The smart antenna then uses this information to determine how to manipulate the incoming signals to maximize performance. The smart antenna constructs a composite signal from multiple antenna feeds by optimizing signal characteristics. This optimization is accomplished by assigning a specific weight to each incoming signal. A smart antenna functions autonomously and automatically, and makes complex decisions in real time.

There are two basic classes of smart antennas: switched beam and adaptive array antennas (see Figures 7.13, 7.14, and 7.15) [11]. These two classes differ according to the method by which they process incoming signal information. A switched beam antenna combines signals according to a fixed number of beam patterns. One of these patterns will be considered a best fit for the signal on an individual channel at a given instant. The system logic may select another pattern as conditions change, effectively switching between patterns as the signal is tracked. Although pattern characteristics may be selected, a beam may not be steered or swept on a continuous basis. For such a system to operate, signal processing occurs simultaneously over each of the hundreds of radio channels in a network. Such processing requires a powerful DSP engine, that must analyze the antenna signal across the entire frequency band occupied by the network, identify individual channels, and then apply appropriate processing. Switched beam antennas have been used to increase coverage and capacity of the first-generation frequency division multiple access (FDMA) systems and the second-generation TDMA systems. The switched beam antennas were not used for the second-generation CDMA systems because the process of switching from one pattern to another pattern destroys the system synchronization.

The adaptive antenna consists of a linear or rectangular array of M homogeneous radiating elements. These elements are coupled together via some type of amplitude control and phase-shifting mechanism to form a single output. The amplitude and phase control involves a set of complex weights.

Due to cost constraints, the adaptive array antennas are often used in the reverse direction or uplink (UL) (mobile to base station) only. In the UL, each signal arrives in a distinct path from an arbitrary direction. The received signal at each antenna element is composed of the desired signal, the sum of interfering signals, and thermal noise. The antenna reinforces the desired signal and suppresses the interfering signals and thermal noise by multiplying the total signal by a set of complex weights. The total array output in direction ϕ_k is given as

$$y_k(t) = \sum_{n=1}^{M} w_n \cdot e^{j(\omega t + \phi_{nk})} \tag{7.46}$$

where
w_n = complex weight applied to the output of the nth element
ω = frequency

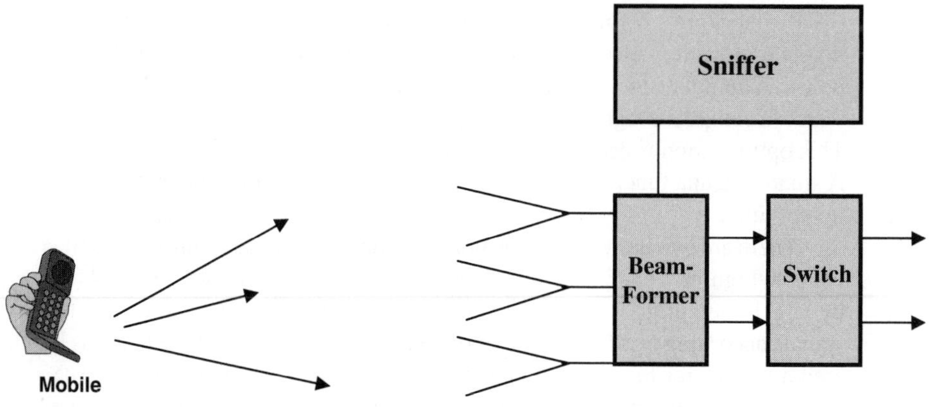

Figure 7.13 Switched Beam Smart Antennas

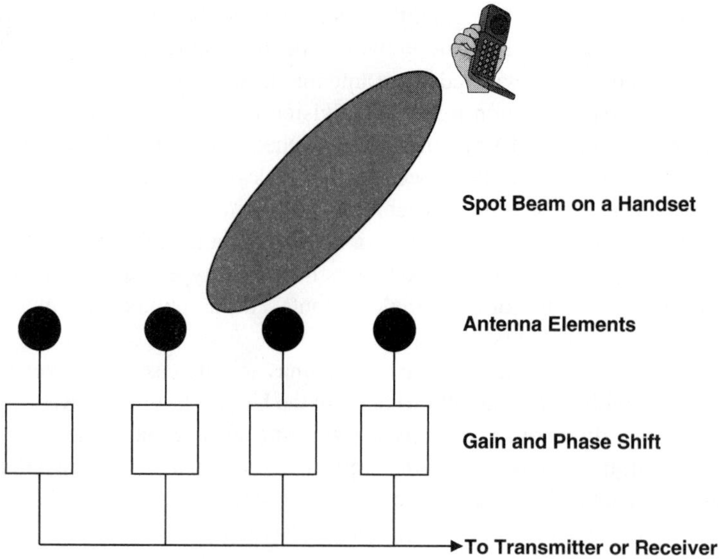

Figure 7.14 Adaptive Array Smart Antenna

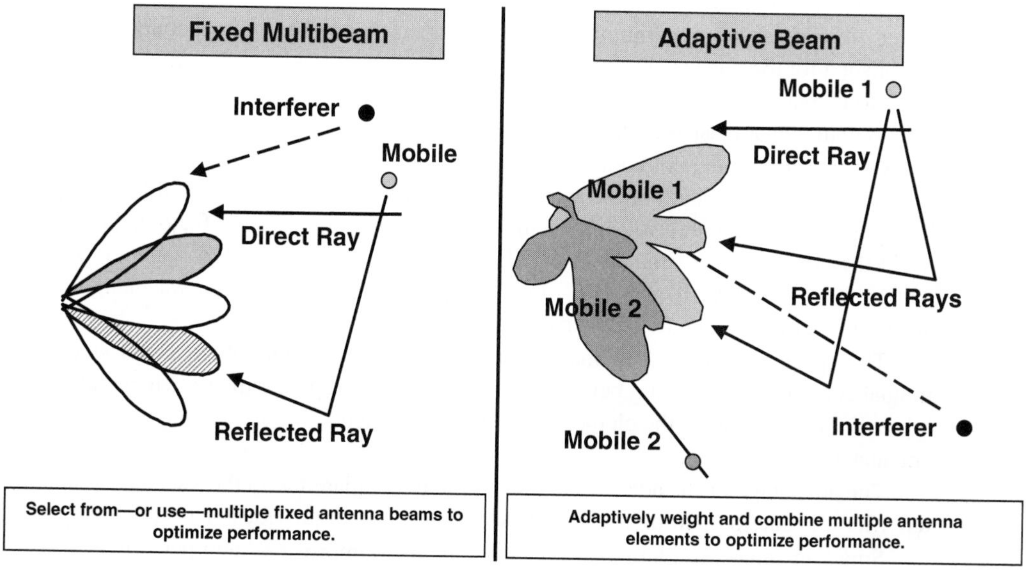

Figure 7.15 Switched Beam versus Adaptive Array Smart Antennas

With a suitable choice of weights, the array will accept a desired signal from direction ϕ_a and nullify interference signals originating from direction ϕ_k for $k \neq a$. The weighting mechanism is optimized to steer-beam in a specific direction or directions. This beam is essentially a radian pattern maxima of finite width. An M element array has $M - 1$ degrees of freedom, which yields a maximum of $M - 1$ independent pattern nulls. If the weights are controlled by a feedback loop to maximize the signal-to-interference ratio (SIR) at the array output, the system behaves as an adaptive spatial filter.

In an adaptive antenna system, the signals received by multiple antenna elements are combined to maximize the SNR using either the minimum mean-square error (MMSE) or least squares (LS) criterion. The antenna elements can be arranged in various geometries, with uniform line, circular, and planar arrays being quite common. The circular geometry provides complete coverage from a central base station as a beam can be steered through 360°. The spacing between antenna elements is very critical in the design of antenna arrays.

In a linear, equally spaced array, grating lobes can appear in an antenna pattern if elements are spaced more than $\lambda/2$ apart, where λ = wavelength. For an antenna array oriented along the x-axis, with every lobe that the array forms for $0 \le \phi \le \pi$, another lobe may also appear in $0 \le \phi \le \pi$. Similarly, for every null formed, a grating null may appear for antenna element spacing greater than $\lambda/2$. This effect is generally undesirable, so element spacing is typically kept less than or equal to $\lambda/2$.

Another limitation of the adaptive antenna array is that it may not be made arbitrarily small, since antenna elements spaced too closely will exhibit mutual coupling effects. It is difficult to generalize this effect since this depends on the type of antenna element and array geometry. The mutual coupling between two antenna elements has several undesirable effects. In general, the maximum element spacing, as dictated by mutual coupling, must be established for the particular antenna element type and array geometry. Mutual impedance tends to increase considerably for element spacing < $\lambda/2$. Therefore, it is generally advisable to maintain at least 1/2 wavelength spacing between array dipoles.

Over a certain range of element spacing, the maximum directivity of the array is proportional to the total length of the array divided by the wavelength. For a uniformly spaced linear array with identical isotropic elements with identical weighting factors, the directivity of the array is approximately $2d/\lambda$ when the phases are adjusted to form a broad side pattern for $\Delta z \leq 0.8\lambda$. If $d = (M-1)\Delta z$ is small compared with the wavelength λ, the array cannot exhibit a high directivity.

The spacing between antenna elements must be large enough to avoid significant mutual coupling. The spacing between elements must be kept less than $\lambda/2$ to avoid grating lobes. For these reasons, the element spacing is often kept close to $\lambda/2$ in a practical linear antenna array.

The advantage of the adaptive antenna system compared with the switched beam system (SBS) is that in addition to the M-fold antenna gain, it provides M-fold diversity gain.

Smart antennas offer a broad range of methods to improve system performance. They provide enhanced coverage through range expansion, hole filling, and better building penetration. Given the same transmitter power output at the base station and the mobile unit, smart antennas can increase range by increasing the gain of the base station antenna.

The M-fold antenna gain increases the range by a factor of $M^{1/\gamma}$ where γ is the path loss exponent and reduces the number of base stations needed to cover a given area by $M^{2/\gamma}$ [12]. An SBS with M beams can increase the system capacity by a factor of M by reducing the number of interferers. Adaptive antenna system can provide some additional gain by suppressing interferers further. However, since there are so many interferers the additional gain may not be worth the complexity.

7.7 Summary

In this chapter we discussed the role of antennas in the wireless system. We also presented the concepts of diversity reception where multiple signals are combined to improve the SNR of the system.

7.8 Problems

1. Compare the bit error performance of quadrature phase-shift keying (QPSK) modulation with SNR = 10 dB with three-branch diversity using selection, equal-gain, and maximal-ratio combining.
2. Calculate average antenna downtilt for the 40 m base station antenna in a large metropolitan area. The cell radius is 600 m and antenna vertical half-power bandwidth is 6 degrees.

7.9 References

1. Rapport, T. S., *Wireless Communications: Principle and Practice*, Prentice Hall: Upper Saddle River, NJ, 1996.
2. Garg, V. K., and Wilkes, J. E., *Wireless and Personal Communications Systems*, Prentice Hall: Upper Saddle River, NJ, 1996.
3. Price, R., and Green, Jr., P. E., "A Communication Technique for Multipath Channels," *Proceedings of the IRE,* March 1958 [Vol. 46, pp. 555–570].
4. Lee, J. S., and Miller, L. E., *CDMA Systems Engineering Handbook*, Artech House: Boston, 1998.
5. Mahrotra, A., *Cellular Radio Performance Engineering,* Artech House: Boston, 1994.
6. Kahn, L. R., "Radio Squarer," *Proceedings of the IRE*, 42, Nov. 1954 [Vol. 42, p. 1704].
7. Halpern, S. W., "The Theory of Operation of an Equal-Gain Predication Regenerative Diversity Combiner with Rayleigh Fading Channel," *IEEE Transaction on Communication Technology,* Aug. 1974, COM-22 (8), pp. 1099–1106.
8. Kim, K., *Handbook of CDMA System Design, Engineering and Optimization,* Apprentices Inc., 2000.
9. Lee, W. C. Y., "Antenna Spacing Requirements for a Mobile Radio Base Station Diversity," *Bell System Technical Journal*, July–Aug. 1971 [Vol. 50(6)].
10. Kraus J.D., *Antenna*, McGraw-Hill: New York, 1988.
11. Liberti, Jr., J. C., and Rappaport, T. S., *Smart Antennas for Wireless Communications*, Prentice Hall: Upper Saddle River, NJ, 1999.
12. Special Issue on Smart Antennas, *IEEE Personal Communications*, Feb. 1998 [Vol. 5(1)].

Soft Handoff and Power Control in CDMA

8.1 Introduction

Soft handoff distinguishes itself from the traditional hard handoff process used in time division multiple access (TDMA) and frequency division multiple access (FDMA) wireless systems. With hard handoff, a definite decision is made on whether to hand off. The handoff is initiated and executed without the user attempting to have simultaneous traffic channel communications with the two or more base stations. With soft handoff, a *conditional decision* is made on whether to hand off. Depending on the changes in pilot signal strength from the two or more base stations involved, a hard decision will eventually be made to communicate with only one. This normally happens after it is evident that the signal from one base station is considerably stronger than those from the others. In the interim period, the user has simultaneous traffic channel communications with all candidate base stations.

It is desirable to implement soft handoff in power-controlled code division multiple access (CDMA) systems because implementing hard handoff is potentially difficult in such systems. A system with power control attempts to dynamically adjust transmitter power while in operation. Power control is closely related to soft handoff. CDMA uses both power control and soft handoff as an interference-reduction mechanism. Power control is the main tool used in CDMA to combat the near-far problem. In theory, it is unnecessary to have power control if one can successfully implement a more intelligent receiver than that used in IS-95; this is the subject of the field of multiuser detection (MUD), a feature being proposed for the third-generation (3G) W-CDMA systems. Power control is necessary in order for a CDMA system to achieve a reasonable level of performance in practice. The use of power control in CDMA systems necessitates the use of soft handoff when the original and new channels occupy the same frequency band. For power control to work properly, the mobile must attempt to be linked at all times to the base station from which it receives the strongest signal. If this does not happen, a positive power control feedback loop could inadvertently occur, causing system problems. Soft handoff can guarantee that the mobile is indeed linked at all times to the base station from which it receives the strongest signal, whereas hard handoff cannot guarantee this.

The performance of CDMA systems is very sensitive to differences in received signal powers from various users on the reverse link. Owing to the nonorthogonality of the spreading codes used by different users, a strong interfering signal may mask out a weak desired signal, causing unreliable detection of the latter. This is called the *near-far problem.*

In this chapter, we first discuss handoff strategy used in CDMA and then focus on power control schemes for the reverse and forward links.

8.2 Types of Handoff

There are four types of handoffs:

1. **Intersector or softer handoff.** The mobile communicates with two sectors of the same cell (see Figure 8.1). A RAKE receiver at the base station combines the best versions of the voice frame from the diversity antennas of the two sectors into a single traffic frame.

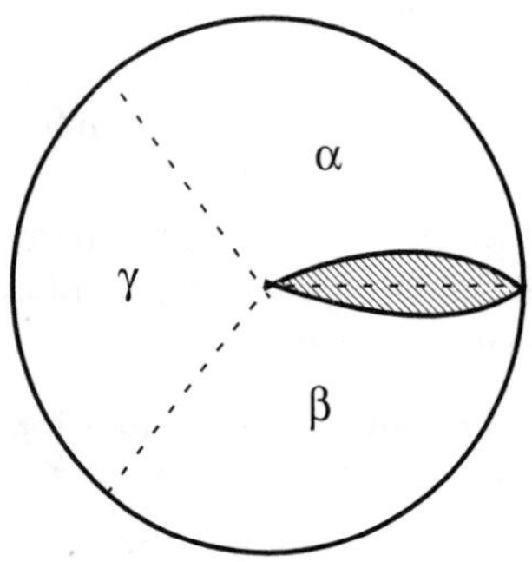

Figure 8.1 Softer Handoff

2. **Intercell or soft handoff.** The mobile communicates with two or three sectors of different cells (see Figure 8.2). The base station that has the direct control of call processing during handoff is referred to as the *primary base station*. The primary base station can initiate a forward control message. Other base stations that do not have control over call processing are called the *secondary base stations.* Soft handoff ends when either the primary or secondary base station is dropped. If the primary base station is dropped, the secondary base station becomes the new primary for this call. A three-way soft handoff may end by first dropping one of the base stations and becoming a two-way soft handoff.

The base stations involved coordinate handoff by exchanging information via signaling system 7 (SS7) links. Soft handoff uses considerably more network resources than the softer handoff.

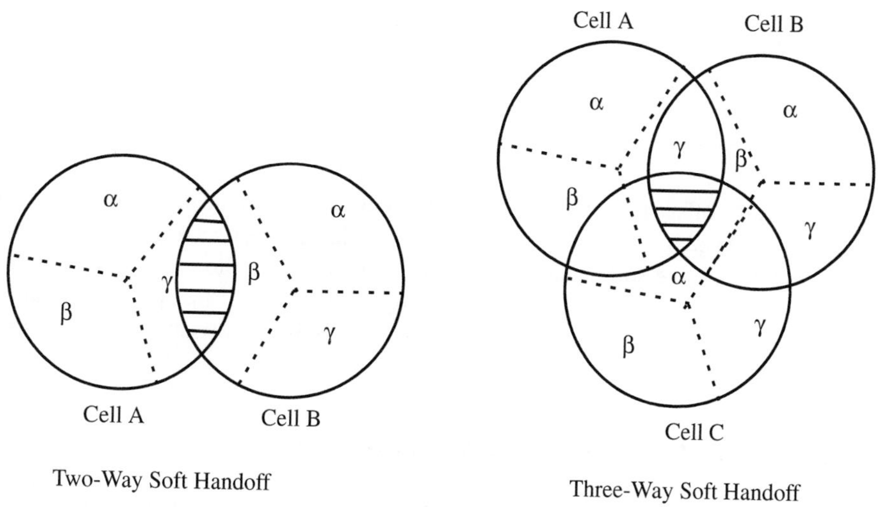

Figure 8.2 Soft Handoff

3. **Soft-softer handoff.** The mobile communicates with two sectors of one cell and one sector of another cell (see Figure 8.3). Network resources required for this type of handoff include the resources for a two-way soft handoff between cell A and B plus the resources for a softer handoff at cell B.

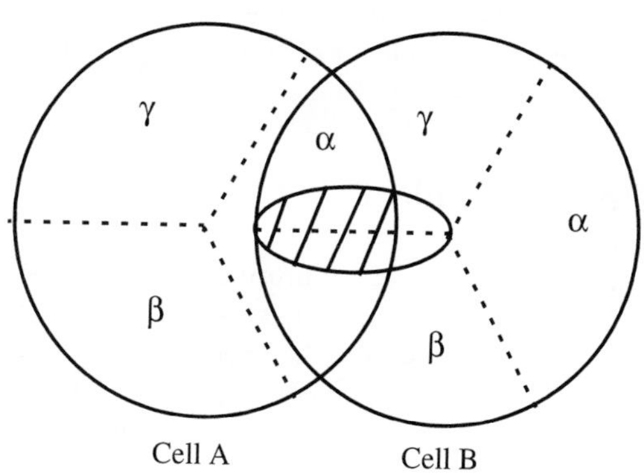

Figure 8.3 Soft-Softer Handoff

4. **Hard handoff.** Hard handoffs are characterized by the *break-before-make* strategy. The connection with the old traffic channel is broken before the connection with the new traffic channel is established. Scenarios for hard handoff include:
   - Handoff between base stations or sectors with different CDMA carriers
   - Change from one pilot to another pilot without first being in soft handoff with the new pilot (disjoint active sets)
   - Handoff from CDMA to analog system
   - Change of frame offset assignment (CDMA traffic frames are 20 ms long. The start of frames in a particular traffic channel can be at time 0 in reference to a system or it can be offset by up to 20 ms, as allowed in IS-95). This is known as the *frame offset.* CDMA traffic channels are assigned different frame offsets to avoid congestion. The frame offset for a particular traffic channel is communicated to the mobile. Both forward and reverse links use this offset. A change in offset assignment will disrupt the link. During soft handoff the new base station must allocate the same frame offset to the mobile as assigned by the primary base station. If that particular frame offset is not available, a hard handoff may be required. Frame offset is a network resource and can be used up.

8.2.1 Soft Handoff (Forward Link)

In this case all traffic channels assigned to the mobile are associated with pilots in the active set, and carry the same traffic information with the exception of the power control subchannel. When the active set contains more than one pilot, the mobile provides diversity by combining its associated forward traffic channels.

8.2.2 Soft Handoff (Reverse Link)

During intercell handoff, the mobile sends the same information to both base stations. Each base station receives the signal from the mobile with appropriate propagation delay. Each base station then transmits the received signal to the vocoder/selector. In other words, two copies of the same frame are sent to the vocoder/selector. The vocoder/selector selects the better frame and discards the other.

8.2.3 Softer Handoff (Reverse Link)

During intersector handoff, the mobile sends the same information to both sectors. The channel card/element at the cell site receives the signals from both sectors. The channel card combines both inputs and only one frame is sent to the vocoder/selector. It should be noted that extra channel cards are not required to support softer handoff, as is the case with soft handoffs. The diversity gain from soft handoff is more than the diversity gain from softer handoff because signals from distinct cells are less correlated than signals from sectors of the same cell.

8.2.4 Benefits of Soft Handoff

A key benefit of soft handoff is the path diversity on forward and reverse traffic channels. Diversity gain is obtained because less power is required on the forward and reverse links. This implies that the total system interference is reduced. As a result, the average system capacity is improved. Also, less transmit power from the mobile results in longer battery life and longer talk time.

In a soft handoff, if a mobile receives an *up* power control bit from one base station and a *down* control bit from the second base station, the mobile decreases its transmit power. The mobile obeys the *power down* command since a good communications link must have existed to warrant the command from the second base station.

8.3 Pilot Sets

The term *pilot* refers to a pilot channel identified by a pilot sequence offset and a frequency assignment. A pilot is associated with the forward traffic channels in the same forward CDMA link.

Each pilot is assigned a different offset of the same short pseudorandom (PN) code. The mobile search for pilots is facilitated by the fact that the offsets are the integer multiples of a known time delay (64 chips offset between adjacent pilots). All pilots in a pilot set have the same CDMA frequency assignment. The pilots identified by the mobile, as well as other pilots specified by the serving sectors (neighbors of the serving base stations/sectors), are continuously categorized by the mobile into the four following groups [1]:

- **Active set.** This set contains the pilots associated with the forward traffic channels (Walsh codes) assigned to the mobile. Because there are three fingers of the RAKE receiver in the mobile, the active set size is a maximum of three pilots. IS-95 allows up to six pilots in the active set, with two pilots sharing one RAKE finger. The base station informs the mobile about the contents of the active set by using the *channel assignment message* and/or the *handoff direction message* (HDM). An active pilot is a pilot whose paging or traffic channels are actually being monitored or used.
- **Candidate set.** This set contains the pilots that are not currently in the active set. However, these pilots have been received with sufficient signal strength to indicate that the associated forward traffic channels could be successfully demodulated. Maximum size of the candidate set is six pilots.
- **Neighbor set.** This set contains neighbor pilots that are not currently in the active or the candidate set and are likely candidates for handoff. Neighbors of a pilot are all the sectors/cells that are in its close vicinity. The initial neighbor list is sent to the mobile in the *system parameter message* on the paging channel. The maximum size of the neighbor set is 20 pilots.
- **Remaining set.** This set contains all possible pilots in the current system, excluding pilots in the active, candidate, or neighbor sets.

The mobile station assesses pilot signal strength and a set of (adjustable) thresholds to determine the movement of pilots among different sets. This activity is coordinated with the serving sector. The mobile station assesses pilots by comparing their strength and by comparing each pilot's power with the total received forward link power [see Equation (8.1)].

$$\left(\frac{E_c}{I_t}\right)_i = \frac{\left(\frac{\phi_p \cdot P_i}{B_w}\right)}{N_f \cdot N_0 + \sum_{all,\,k} P_k/B_w} \tag{8.1}$$

where
E_c = chip energy received from ith sector pilot
I_t = spectral density of total received interference
ϕ_p = fraction of sector power allotted to pilot signal
P_i = received power from ith sector
B_w = system bandwidth
N_f = base station noise figure
N_0 = thermal noise power density

While searching for a pilot, the mobile is not limited to the exact offset of the short PN code. The short PN offsets associated with various multipath components are located a few chips away from the direct path offset. In other words, the multipath components arrive a few chips later relative to the direct path component. The mobile uses the *search window* for each pilot of the active and candidate set, around the earliest arriving multipath component of the pilot. Search window sizes are defined in number of short PN chips. The mobile should center the search window for each pilot of the neighbor set and remaining set around the pilot's PN offset using the mobile time reference.

8.4 Search Windows

The mobile uses the following three search windows to track the received pilot signals:

- SRCH_WIN_A: search window size for the active and candidate sets
- SRCH_WIN_N: search window size for the neighbor set
- SRCH_WIN_R: search window size for remaining set

8.4.1 SRCH_WIN_A

SRCH_WIN_A is the search window that the mobile uses to track the active and candidate set pilots. This window is set according to the anticipated propagation environment. This window should be large enough to capture all usable multipath signal components of a base station, and at the same time it should be as small as possible in order to maximize searcher performance. As a rule of thumb, the search window should be large enough to span the maximum expected arrival time difference between the pilot's usable multipath components (i.e., the pilot's maximum delay spread, τ_{max}). The required delay budget $T_{d,active}$ will be

$$T_{d,\,active} > 2 \times \frac{\tau_{max}}{T_{chip}} \text{ chips} \tag{8.2}$$

where
T_{chip} = chip time (813.8 nanoseconds)

EXAMPLE 8.1

We consider the propagation environment of a CDMA network, where the signal with direct path travels 1 km to the mobile, whereas the multipath travels 5 km before reaching the mobile. What should be the size of SRCH_WIN_A?

Direct path travels a distance of $\frac{1000}{244} = 4.1 = \text{chips}$

Multipath travels a distance of $\frac{5000}{244^*} = 20.5 = \text{chips}$

The difference in distance travelled between the two paths = 20.5 – 4.1 = 16.4 chips

The window size chips $\geq 2 \times 16.4 = 32.8$

Use window size = 33 chips

*Note: $\frac{3 \times 10^8}{1.2288 \times 10^6} = 244 \text{ m}$

EXAMPLE 8.2

Consider cells A and B separated by a distance of 12 km. The mobile travels from cell A to cell B. The radio frequency (RF) engineer wishes to contain the soft handoff area between points *X* and *Y* located at distance 6 and 10 km from cell A (see Figure 8.4). What should be the search window size?

At point *X* the mobile is 6000/244 = 24.6 chips from cell A

At point *X* the mobile is 10000/244 = 41.0 chips from cell B

Path difference = 41.0 – 24.6 = 16.4 chips

At point *Y* the mobile is 10000/244 = 41.0 chips from cell A

At point *Y* the mobile is 6000/244 = 24.6 chips from cell B

Path difference = 41.0 – 26.4 = 16.4 chips

The SRCH_WIN_A > 2 × 16.4 > 32.8 chips

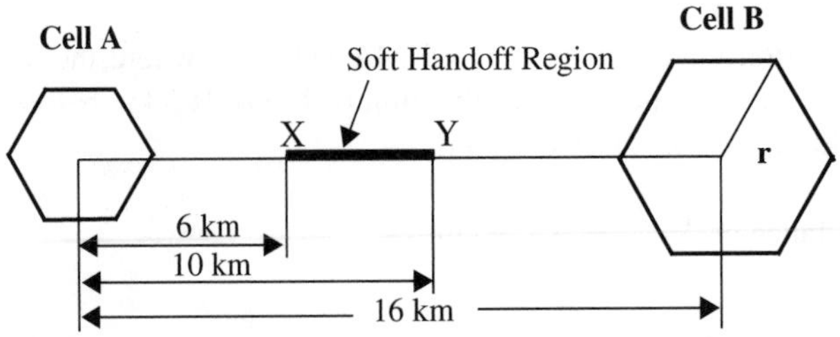

Figure 8.4 SRCH_WIN_A for Soft Handoff Between X and Y

This way, as the mobile travels from cell A to cell B, the mobile can ensure that beyond *Y* the pilot from cell A drops out of the search window.

8.4.2 SRCH_WIN_N

SRCH_WIN_N is the search window that the mobile uses to monitor the neighbor set pilots. The size of this window is typically larger than that of SRCH_WIN_A. The window needs to be large enough to not only capture all usable multipaths of the serving base station's signal, but also to capture the potential multipath of neighbors' signals. In this case, we need to take into account multipath and path differences between the serving base station and neighboring base stations. The maximum size of this search window is limited by the distance between two neighboring base stations. We consider two neighboring base stations located at a distance 6 km. The mobile is located right next to base station 1; therefore, the propagation delay from base station 1 to the mobile is negligible. The distance between base station 2 and the mobile is 6 km. The distance in chips is 6000/244 = 24.6 chips. The search window shows that the pilot from cell 2 arrives 24.6 chips later at the mobile. Thus, in order for a mobile (located within cells 1 and 2) to search pilots of potential neighbors, SRCH_WIN_N needs to be set according to the physical distances between the current base station and its neighboring base station. The actual size may not be this large, as this is an upper bound for SRCH_WIN_N.

The search window size for neighbor pilots must account not only for largest delay spread of the target pilot but also for the largest difference in propagation delays (i.e., difference in distance) between the reference pilot and target pilot. An equation to calculate the overall delay budget is

$$T_{d,N} > 2 \times \frac{d_{max}/v_c + \tau_{max}}{T_{chip}} \text{ chips} \tag{8.3}$$

where

d_{max} = maximum difference, in miles, (1) the mobile and cell transmitting active set pilot and (2) the mobile and the cell transmitting neighbor (or remaining) set pilot

v_c = speed of light (186,000 miles/sec)

8.4.3 SRCH_WIN_R

SRCH_WIN_R is the search window that the mobile uses to track the remaining set pilots. A typical requirement for the size of this window is that it is at least as large as SRCH_WIN_N.

The delay spread, and therefore the delay budgets, will depend on the propagation environment. In a large metropolitan urban environment more delay spread is experienced because of more multipath components. A typical value of the delay spread in an urban environment is 8–12 μs, whereas in a suburban environment, a typical value is about 4 μs. Generally, larger cells tend to have larger delay spread as compared with smaller cells. Table 8.1 provides the relationship between the delay budget, the search window size in PN chips, and the value of the search window parameter [2].

Table 8.1 Delay Budget and Search Window Size

Delay Budget (μs)	Window Size (PN Chips)	SRCH_WIN_A SRCH_WIN_N SRCH_WIN_R
$T_d \leq 1.64$	4	0
$1.64 < T_d \leq 2.45$	6	1
$2.45 < T_d \leq 3.27$	8	2
$3.27 < T_d \leq 4.09$	10	3
$4.09 < T_d \leq 5.72$	14	4
$5.72 < T_d \leq 8.17$	20	5
$8.17 < T_d \leq 11.44$	28	6
$11.44 < T_d \leq 16.34$	40	7
$16.34 < T_d \leq 24.51$	60	8

Table 8.1 Delay Budget and Search Window Size *(cont.)*

Delay Budget (μs)	Window Size (PN Chips)	SRCH_WIN_A SRCH_WIN_N SRCH_WIN_R
$24.51 < T_d \leq 32.68$	80	9
$32.68 < T_d \leq 40.85$	100	10
$40.85 < T_d \leq 53.11$	130	11
$53.11 < T_d \leq 65.36$	160	12
$65.36 < T_d \leq 92.32$	226	13
$92.32 < T_d \leq 130.72$	320	14
$130.72 < T_d \leq 184.42$	452	15

EXAMPLE 8.3

Calculate the search window size in PN chips of the active, neighbor, and remaining set. The maximum delay spread is 12 μs and spreading rate is 1.2288 Mcps. The maximum distance between (1) the mobile and cell transmitting active set pilot, and (2) the mobile and cell transmitting neighbor (remaining) set pilot is 4 miles.

$$\text{Chip duration} = \frac{1}{1.2288 \times 10^{6}} = 0.8138 \times 10^{-6}\ \text{s}$$

$$\text{Maximum delay spread} = 12 \times 10^{-6}\ \text{s}$$

$$T_{d,\,active} > \frac{2 \times 12 \times 10^{-6}}{0.8138 \times 10^{-6}} \approx 30\ \text{chips}$$

$$T_{d,\,R} = T_{d,\,N} > \frac{2 \times \dfrac{4}{186000} + 12 \times 10^{-6}}{0.8138 \times 10^{-6}} = \frac{43 + 12}{0.8138} \approx 68\ \text{chips}$$

8.5 Handoff Parameters

In IS-95A there are four handoff parameters. T_ADD, T_COMP, and T_DROP relate to the measurement of pilot E_c/I_t, and T_TDROP is a timer. Whenever the strength of a pilot in the active set falls below a value of T_DROP, a timer is started by the mobile. If the pilot strength goes back above T_DROP the timer is reset; otherwise the timer expires when a time T_TDROP has elapsed since the pilot strength falls below T_DROP. Mobile maintains a handoff drop timer for each pilot in the active set and the candidate set.

8.5.1 Pilot Detection Threshold (T_ADD)

The T_ADD parameter controls the movement of pilots from the neighboring/remaining sets to the active/candidate sets. Any pilot that is strong but is not in HDM is a source of interference. This pilot must be immediately moved to the active set for handoff to avoid voice degradation or a possible dropped call. Mobile stations move a neighbor or remaining set pilot with strength E_c/I_t greater than T_ADD to either the candidate or active set. This parameter is set on a per-sector basis. T_ADD affects the percentage of mobiles in handoff. It should be low enough to quickly add useful pilots, and should be high enough to avoid false alarms caused by noise.

8.5.2 Comparison Threshold (T_COMP)

The T_COMP parameter controls movement of pilots from the candidate set to the active set. The mobile station moves a candidate set pilot with strength E_c/I_t exceeding that of an active set pilot by T_COMP × 0.5 dB to the active set, replacing that pilot. The T_COMP has a similar effect on handoff percentage as T_ADD. It should be low enough to allow for faster handoffs and high enough to avoid false alarms. Raising T_COMP makes it more difficult for pilots to enter in the active set, but there is some risk of neglecting a strong pilot that should be included. This parameter is set on a per sector basis.

8.5.3 Pilot Drop Threshold (T_DROP) and Drop Timer Threshold (T_TDROP)

The T_DROP and T_TDROP parameters control movement of pilots out of the active/candidate sets. The mobile station starts a timer when strength E_c/I_t of a pilot in the active or candidate falls below T_DROP. When the timer exceeds T_TDROP, the pilot is moved out of the active or candidate set to the neighbor or remaining set. The T_DROP affects the percentage of mobiles in handoff. The T_DROP should be high enough not to quickly remove useful pilots from the active or candidate set or to drop a good pilot that goes into a short fade. The value of T_DROP should be carefully selected by considering the values of T_ADD and T_TDROP. The T_TDROP should be greater than the time required to establish handoff. T_TDROP should be high enough not to quickly remove useful pilots. A large value of T_TDROP may be used to force a mobile to continue in soft handoff in the weak coverage area. Both T_DROP and T_TDROP parameters are set on a per-sector basis.

8.5.4 NGHBR_MAX_AGE

This parameter controls the movement of pilots from the neighbor set to the remaining set. The mobile maintains an AGE counter for each pilot in the neighbor set and updates this counter under the direction of the serving site. The mobile moves a plot with AGE count exceeding NGHBR_MAX_AGE to the remaining set. This parameter is set on a per-sector basis.

8.5.5 SRCH_WIN_A, SRCH_WIN_N, and SRCH_WIN_R

These parameters govern the mobile's search for pilots in the active/candidate, neighbor, and remaining sets, respectively. As discussed earlier, these parameters specify the size of the search window to detect the pilot. Table 8.2 gives the range and recommended values for the search window parameters. Larger search window sizes require more mobile processing per search and thus reduce the overall number of pilots that can be searched in a fixed period of time.

Table 8.3 provides typical values of the handoff parameters.

Table 8.2 Search Window Parameters

Parameter	Range	Recommended Value
SRCH_WIN_A (active/candidate)	0–15	5–7
SRCH_WIN_N (neighbor)	0–15	7–13
SRCH_WIN_R (remaining)	0–15	7–13 during optimization 0 after optimization

Note: Search window parameter values are given in units that map into window size in PN chips (see Table 8.1).

Table 8.3 Handoff Parameter Values

Parameter	Range	Recommended Value
T_ADD	–31.5 to 0 dB	–13 dB
T_COMP	0 to 7.5 dB	2.5 dB
T_DROP	–31.5 to 0 dB	–15 dB
T_TDROP	0 to 15 seconds	2 seconds

8.6 Handoff Messages

Handoff messages in IS-95 are *pilot strength measurement message* (PSMM), *handoff direction message* (HDM), *handoff completion message* (HCM), and *neighbor list update message* (NLUM) [3].

The mobile detects pilot strength (E_c/I_t) and sends the PSMM to the base station. The base station allocates a forward traffic channel and sends the HDM to the mobile. On receiving the HDM, the mobile starts demodulation of the new traffic channel and sends the HCM to the base station.

The PSMM contains the following information for each of the pilot signals received by the mobile:

- Estimated E_c/I_t
- Arrival time
- Handoff drop timer

The HDM contains the following information:

- HDM sequence number
- CDMA channel frequency assignment
- Active set (now has old and new pilots [PN offsets])
- Walsh code associated with each pilot in active set
- Window size for active and candidate set
- Handoff parameters (T_ADD, T_DROP, T_COMP, T_TDROP)

The HCM contains the following information:

- A positive acknowledgment
- PN offset of each pilot in the active set

The NLUM is sent by the base station. It contains the latest composite neighbor list for the pilots in active set.

The mobile continuously tracks the signal strength for all pilots in the system. The signal strength of each pilot is compared with the various thresholds such as the T_ADD, T_DROP, T_COMP, and T_TDROP.

A pilot is moved from one set to another depending on its signal strength relative to the thresholds. Figure 8.5 shows a sequence on the threshold.

1. Pilot strength exceeds T_ADD. Mobile sends a PSMM and transfers pilot to the candidate set.
2. Base station sends an HDM to the mobile with the pilot to be added in active set.
3. Mobile receives an HDM and acquires the new traffic channel. Pilot goes into the active set and mobile sends an HCM to the base station.

4. Pilot strength drops below T_DROP; mobile starts the handoff drop timer.
5. Handoff drop timer expires. Mobile sends a PSMM to the base station.
6. Base station sends an HDM without related pilot to the mobile.
7. Mobile receives an HDM. Pilot goes into the neighbor set and mobile sends an HCM to the base station.
8. The mobile receives an NLUM that does not include the pilot. Pilot goes into the remaining set.

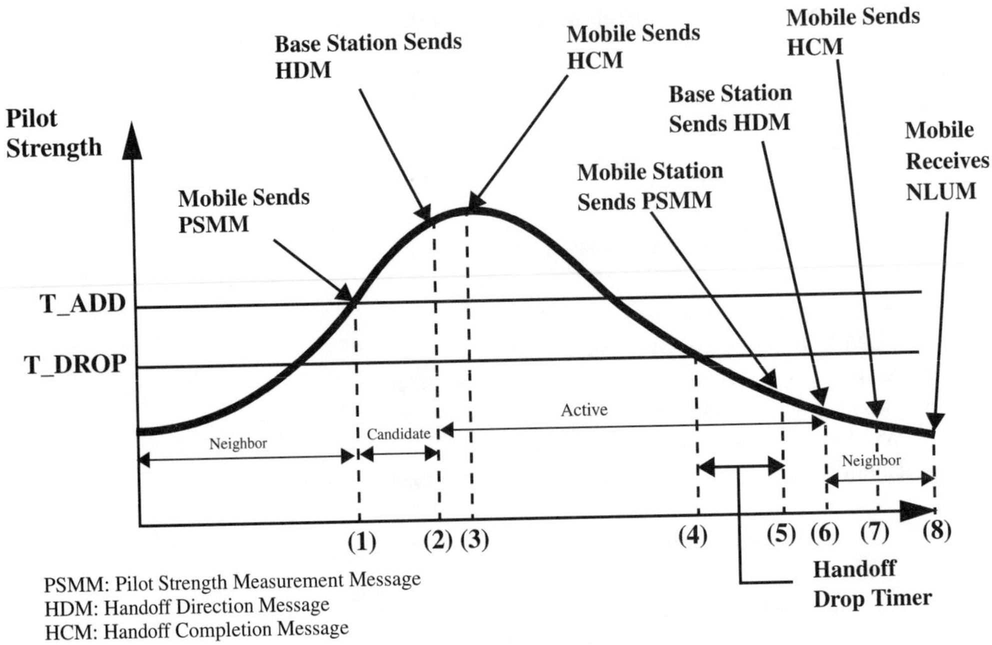

Figure 8.5 Handoff Threshold Example: Pilot Thresholds

The mobile maintains a T_TDROP for each pilot in the active set and candidate set. The mobile starts the timer whenever the strength of the corresponding pilot becomes less than a preset threshold. The mobile resets and disables the timer if the strength of the corresponding pilot exceeds the threshold.

When a member of the neighbor or remaining set exceeds T_ADD, the mobile moves the pilot to candidate set (Figure 8.6) and sends a PSMM to the base station. As the signal strength of candidate pilot P_c gradually increases, it rises above active set pilot, P_a. A PSMM is sent to the base station if only if

$$P_c - P_a > \text{T\_COMP} \times 0.5 \text{ dB}$$

where P_a and P_c are the strength of pilots in the active and candidate sets.

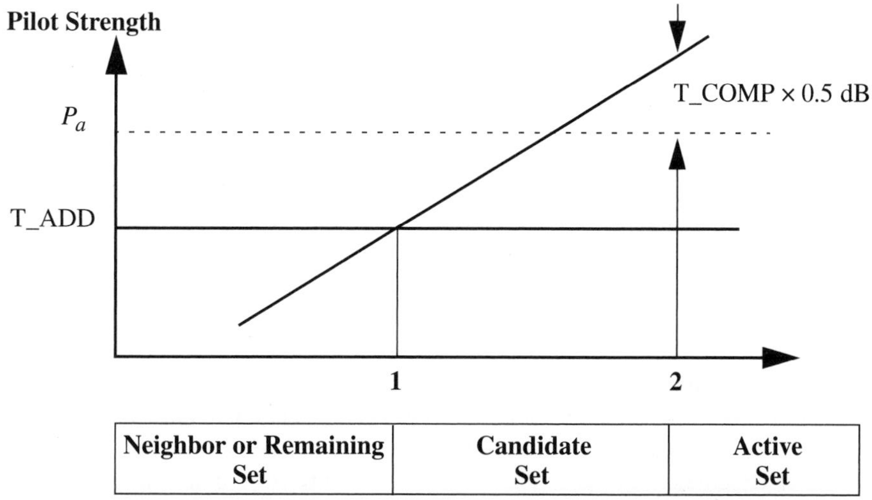

Figure 8.6 Pilot Movement from Neighbor or Remaining Set to Active Set

8.7 Handoff Procedures

8.7.1 Mobile-Assisted Soft-Handoff (MASHO) Procedures

The mobile monitors the forward pilot channel (FPICH) level received from neighboring base stations and reports to the network those FPICHs that cross a given set of thresholds [4]. Two types of thresholds are used: the first to report FPICHs with sufficient power to be used for coherent demodulation, and the second to report those FPICHs whose power has dropped to a level where it is not beneficial to use them for coherent demodulation. The margin between two thresholds provides a hysteresis to avoid a ping-pong effect brought about by variations in FPICH power. Based on this information, the network instructs the mobile to add or remove FPICHs from its active set.

The same user information modulated by the appropriate base station code is sent from multiple base stations. Coherent combining of different signals from different sectorized antennas, from different base stations, or from the same antennas but on different multiple path components is performed in the mobile by using RAKE receivers. A mobile will typically place at least one RAKE receiver finger on the signal from each base station in the active set. If the signal from the base station is temporarily weak, then the mobile can assign the finger to a stronger base station.

The signal transmitted by a mobile is processed by base stations with which the mobile is in soft handoff. The received signal from different sectors of a base station is combined in the base station on a symbol-by-symbol basis. The received signal from different base stations can be selected in the infrastructure (on a frame-by-frame basis). Soft handoff results in increased coverage range and capacity on the reverse link.

8.7.2 Dynamic Soft-Handoff Thresholds

While soft handoff improves overall system performance, it might in some situations negatively impact system capacity and network resources. On the forward link, excessive handoff reduces system capacity whereas on the reverse link, it costs more network resources (backhaul connections) [5].

Adjusting the handoff parameters at the base stations will not necessarily solve the problem. Some locations in the cell receive only weak FPICHs (requiring lower handoff thresholds) and other locations receive a few strong and dominant FPICHs (requiring higher handoff thresholds). The principle of dynamic threshold for adding FPICHs is as follows:

- The mobile detects FPICHs that cross a given static threshold, T_1. The metric for the FPICH in this case is the ratio of FPICH energy per chip to total received power (E_c/I_t).
- On crossing the static threshold, the FPICH is moved to a candidate set. It is then searched more often and tested against a second dynamic threshold, T_2.
- Comparison with T_2 determines if the FPICH is worth adding to the active set. T_2 is a function of the total energy of FPICHs demodulated coherently (in the active set).
- The condition of a FPICH for crossing T_2 is expressed as

$$10\log(P_{cj}) \geq Max\left\{ SOFT\_SLOPE \cdot 10\log\left(\sum_{i=1}^{N_A} P_{ai}\right) + ADD\_INTERCEPT, T_1 \right\} \quad (8.4)$$

where
P_{cj} = strength of the jth FPICH in the candidate set
P_{ai} = strength of the ith FPICH in the active set
N_A = number of FPICHs in the active set
SOFT_SLOPE and *ADD_INTERCEPT* = adjustable system parameters

When FPICHs in the active set are weak, adding an additional FPICH (even a weak one) will improve performance. However, when there are one or more dominant FPICHs, adding an additional weaker FPICHs above T_1 will not improve performance, but will use more network resources. The dynamic soft-handoff thresholds reduce and optimize the network resource utilization.

- After detecting an FPICH above T_2, the mobile reports it back to the network. The network then sets up the handoff resources and orders the mobile to coherently demodulate this additional FPICH. Pilot 2 is added to active set.
- When the FPICH (pilot 1) strength decreases below a dynamic threshold T_3, the handoff connection is removed. The FPICH is moved back to candidate set. The threshold T_3 is a function of the total energy of FPICHs in the active set. FPICHs not contributing sufficiently to total FPICH energy are dropped. If it decreases below a static threshold T_4, an FPICH is removed from candidate set.

- An FPICH dropping below a threshold (e.g., T_3 and T_4) is reported back to the network only after being below the threshold for a specific period. This timer allows for a fluctuating FPICH not to be prematurely reported.

Figure 8.7 shows a time representation of soft handoff and associated events when the mobile station moves away from a serving base station (FPICH 1) toward a new base station (FPICH 2). The combination of static and dynamic thresholds (versus static thresholds alone) results in a reduced soft handoff region. The major benefit of this is to limit soft handoff to areas and times when it is most beneficial.

1. When pilot 2 exceeds T_1, mobile moves it to the candidate set.
2. When pilot 2 exceeds T_2 (dynamic), mobile reports it back to the network.
3. Mobile receives an order to add pilot 2 to the active set.
4. Pilot 1 drops below T_3 (relative pilot 2).
5. Handoff timer expires on pilot 1. Mobile reports pilot strength to the network.
6. Mobile receives order to remove pilot 1.
7. Handoff timer expires after pilot 1 drops below T_4.

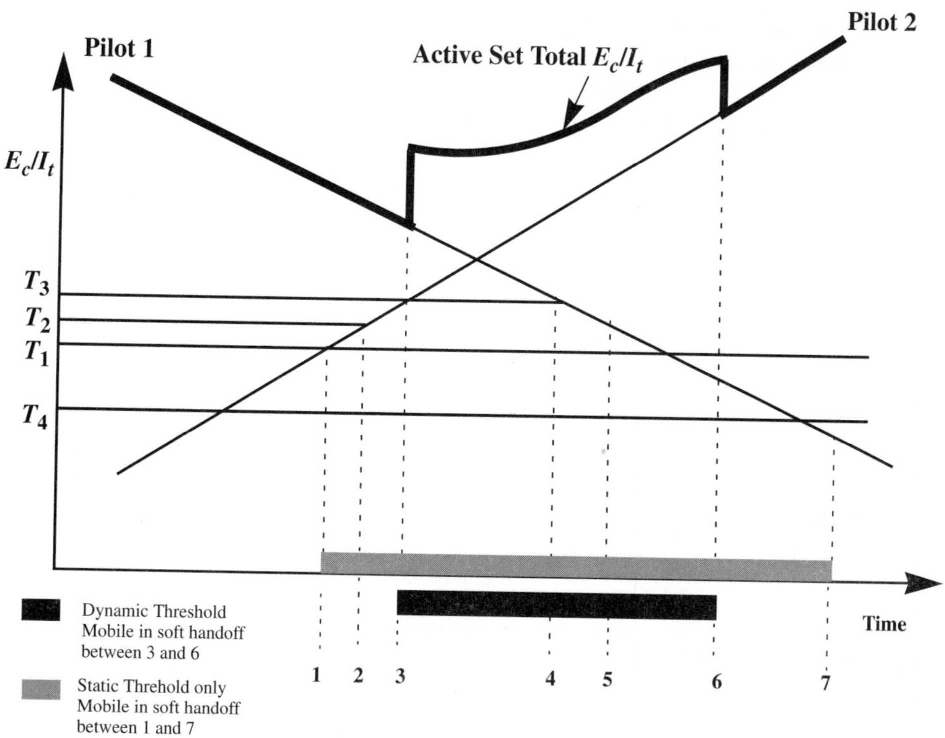

Figure 8.7 Dynamic Thresholds Handoff Procedure

8.8 Setup and End of Soft Handoff

8.8.1 Setup

One of the major benefits of a CDMA system is the ability of a mobile to communicate with more than one base station at one time during a call. This functionality allows the CDMA network to perform soft handoff. In soft handoff a controlling primary base station coordinates with other base stations as they are added or deleted during the call. This allows the base stations (up to three, total) to receive/transmit voice packets with a single mobile for a single call.

Each base station transmits the received mobile voice packets to the BSC/MSC. The BSC/MSC selects the best voice frame from one of the three base stations. This provides the public switched telephone network (PSTN) party with the best-quality voice.

Figure 8.8 shows a mobile communicating with two base stations for one call. This is called a *two-way soft handoff*. Steps of soft handoff are as follows:

- The mobile detects a pilot signal from a new cell and informs the primary base station A.
- A communications path from base station B to the original frame selector is established.
- The frame selector selects frames from both streams.
- The mobile detects that base station A's pilot is failing and requests that this path be dropped.
- The path from original base station A to the frame selector is dropped.

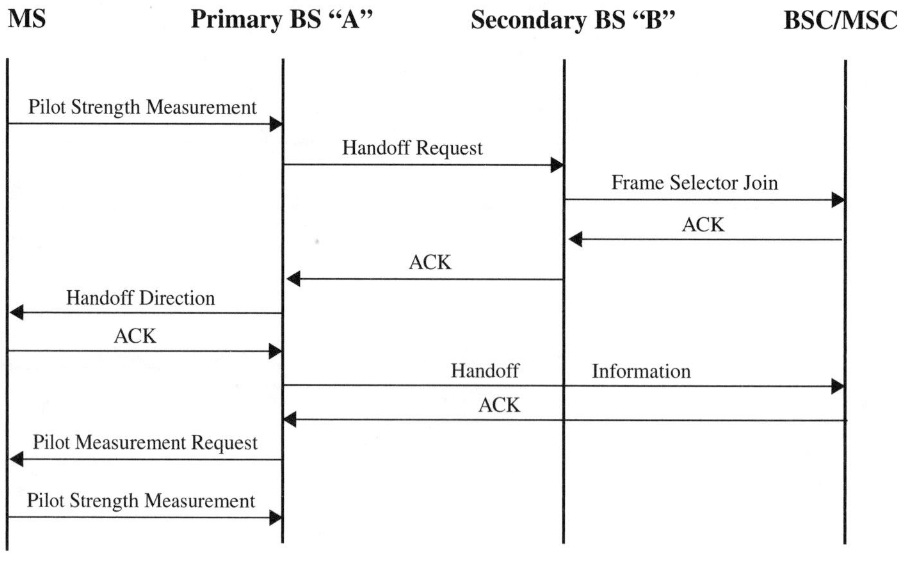

Figure 8.8 Soft Handoff Setup

Base station B provides base station A its assigned Walsh code. Base station A provides the mobile the Walsh code of B as part of the HDM. Now the mobile can listen to base station B.

Base station A gives the user's long-code mask to base station B. Now B can listen to the mobile. Both base stations A and B receive forward link power control information back from the mobile and act accordingly. The mobile receives independent puncture bits from both A and B. If directions conflict, mobile decreases power, otherwise the mobile obeys directions.

8.8.2 End of Soft Handoff

Figure 8.9 shows the process by which a mobile communicating with two base stations, A and B, to end handoff when the signal from base station A is not strong enough. When the mobile entered into soft handoff with base stations A and B, the primary base station was A. However, when the mobile drops A and starts communicating with base station B alone, B becomes the new primary base station.

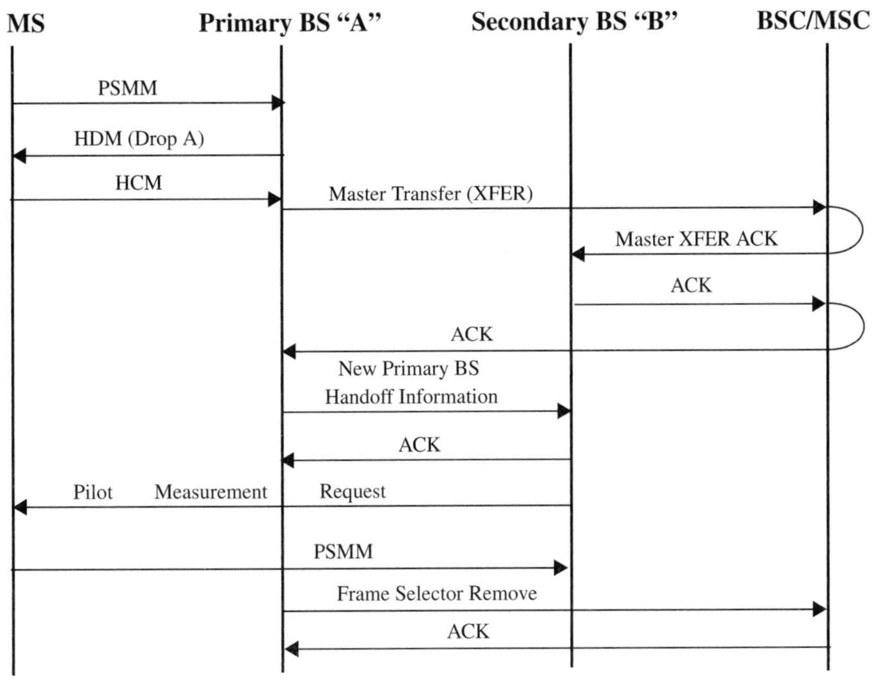

Figure 8.9 End of Soft Handoff

8.9 Pilot Set Maintenance

8.9.1 Active Set Maintenance

The active set is initialized to contain only one pilot (e.g., the pilot associated with the assigned forward traffic channel). This occurs when the mobile is first assigned a forward traffic channel. As the mobile processes HDMs, it updates the active set with the pilots listed in the HDMs.

A pilot P_c from the candidate is added to the active set when P_c exceeds a member of the active set by T_COMP. A pilot P_a from the active set is removed when P_a has dropped below T_DROP and the drop timer (T_TDROP) has expired (see Figure 8.10).

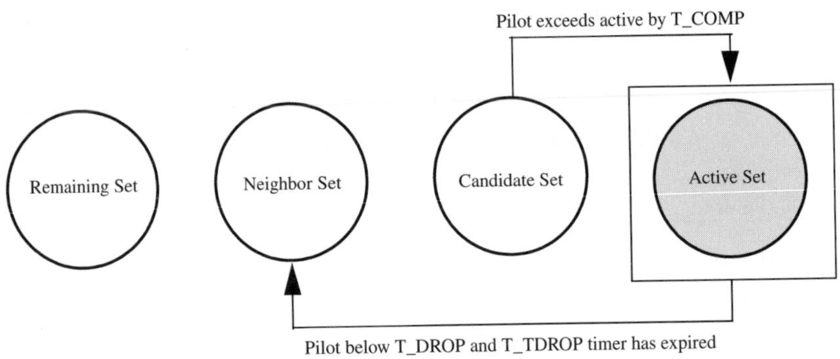

Figure 8.10 Active Set Maintenance

8.9.2 Candidate Set Maintenance

The candidate set is initialized to contain no pilot. This happens when the mobile is first assigned a forward traffic channel. A pilot P_n from the neighbor set is added to the candidate set when its strength exceeds T_ADD. Also, a pilot P_r from the remaining set is moved to the candidate set when its strength exceeds T_ADD. A pilot P_c is deleted from the candidate set when the handoff drop timer corresponding to P_c has expired. Also, when the candidate set size has been exceeded, the pilot P_c, whose handoff drop timer is close to expiring, is deleted from the candidate set (see Figure 8.11).

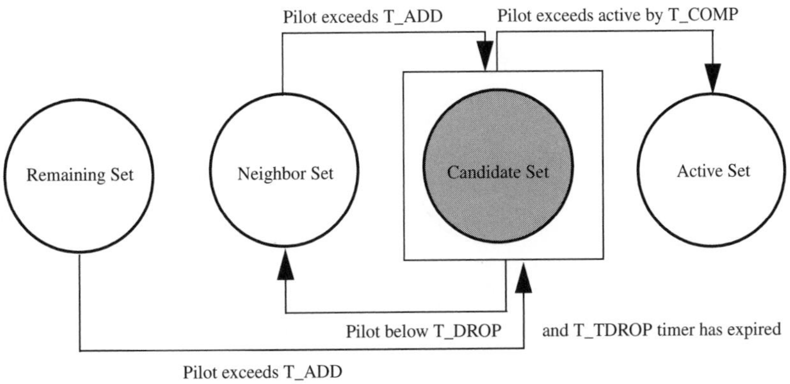

Figure 8.11 Candidate Set Maintenance

8.9.3 Neighbor Set Maintenance

The neighbor set is initialized to contain the pilots specified in the most recently received neighbor list message. This happens when the mobile is first assigned a forward traffic channel. The mobile maintains an age counter, AGE, for each pilot in the neighbor set. If a pilot moves from the active set or candidate set to the neighbor set, its counter is initialized to zero. However, if a pilot moves from the remaining set to the neighbor set, its counter is set to the maximum age value (see Figure 8.12). The mobile adds a pilot in the neighbor set under the following conditions:

- A pilot in the active set is not contained in the HDM and the corresponding handoff drop timer has expired.
- The handoff drop timer of a pilot in the candidate set has expired.
- A new pilot in the candidate set causes the candidate set size to be exceeded.
- The pilot is contained in the neighbor list message and is not already a pilot of the candidate set or neighbor set.

The mobile deletes a pilot in the neighbor set under the following conditions:

- The HDM contains a pilot from the current neighbor set.
- The strength of a pilot in the neighbor set exceeds T_ADD.
- A new pilot in the neighbor set causes the size of the neighbor set to be exceeded.
- A neighbor set pilot's AGE exceeds the maximum value of the AGE counter.

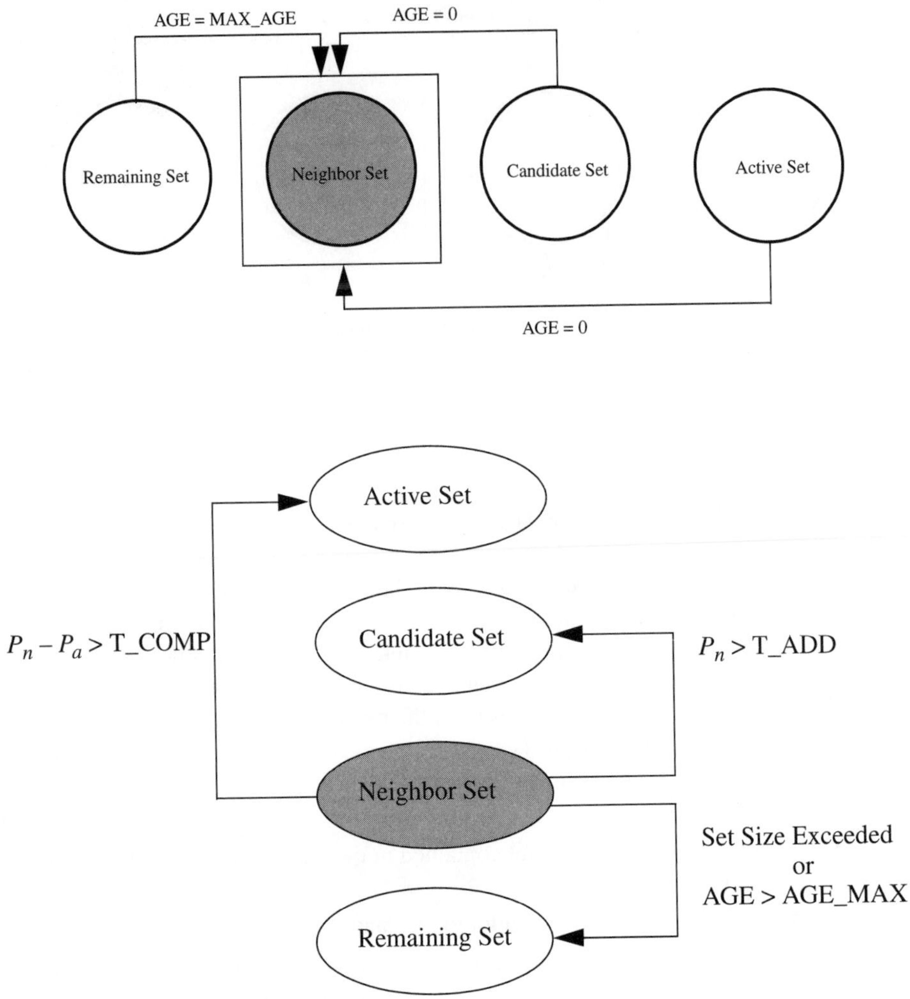

Figure 8.12 Neighbor Set Maintenance

8.10 Needs for Power Control

CDMA is an interference-limited system. Since all mobiles transmit at the same frequency, internal interference generated within the system plays a critical role in determining system capacity and voice quality. The transmit power from each mobile must be controlled to limit interference. However, the power level should be adequate for a satisfactory voice quality.

As the mobile moves around, the RF environment changes continuously owing to fast and slow fading, shadowing, external interference, and other factors. The objective of power control is to limit transmitted power on the forward and reverse links while maintaining link quality under all conditions. Due to noncoherent detection at the base station, interference on the reverse link is more critical than the forward link in IS-95. Reverse link power control is, therefore, essential for a CDMA system and is enforced by the IS-95 standard.

Power control in CDMA systems is also needed to resolve the near-far problem. To minimize the near-far problem, the goal in a CDMA system is to ensure that all mobiles achieve the same received power levels at the base station. The target value for the received power level must be the minimum level that still allows the link to meet user-defined performance objectives (bit error rate [BER], frame error rate [FER], capacity, dropped call rate, and coverage). In order to implement such a strategy, the mobiles that are closer to the base station must transmit less power than those that are far away.

Voice quality is related to FER on both the forward and reverse links. The FERs are largely related to E_b/I_t. The FER also depends on vehicle speed, local propagation conditions, and distribution of other cochannel mobiles. Since the FER is a direct measure of signal quality, the voice quality performance in a CDMA system is measured in term of FERs rather than E_b/I_t. Thus, to ensure a good signal quality, it is not sufficient to maintain a target E_b/I_t; it is also necessary to respond to specific FERs as they occur. The recommended performance bounds are as follows:

- A typical recommended range for FER: 0.2 to 3 percent (optimum power level is achieved, when FER $\leq$ 1 percent).
- A maximum length of burst error is three to four frames (optimum value of burst error $\approx$ 2 frames).

8.11 Reverse Link Power Control

The reverse link power control affects the access and reverse traffic channels. It is used for establishing the link while originating a call and reacting to large path loss fluctuations. The reverse link power control includes the open loop power control (also known as autonomous power control) and the closed loop power control. The closed loop power control involves the inner loop power control and the outer loop power control.

8.11.1 Reverse Link Open Loop Power Control

The open loop power control is based on the principle that a mobile closer to the base station needs to transmit less power as compared with a mobile that is far away from the base station or is in fade. The mobile adjusts its transmit power based on total power received (i.e., power in pilot, paging, sync, and traffic channels). This includes power received from all base stations on the forward link channels. If the received power is high, the mobile reduces its transmit power. On the other hand, if the power received is low, the mobile increases its transmit power.

In the open loop power control, the base station is not involved. The mobile determines the initial power transmitted on the access channel and traffic channel through open loop power control. A large dynamic range of 80 dB is allowed to provide an ability to guard against deep fades. The reverse open loop power control (ROPC) provides a rough estimation of transmit power based on the total received power on the forward link. The response of ROPC is rapid, because no base station transceiver (BTS) timing alignment is required. The ROPC occurs every 20 ms (i.e., 50 times per second).

The mobile acquires the CDMA system by receiving and processing the pilot, sync, and paging channels. The paging channel provides the *access parameters message* that contains the parameters to be used by the mobile when transmitting to the base station on an access channel. The access parameters are

- The access channel number
- The nominal power offset (NOM_PWR)
- The initial power offset step size
- The incremental power step size
- The number of access probes per access probe sequence
- The time-out window between access probes
- The randomization time between access probe sequences

Based on the information received on the pilot, sync, and paging channels, the mobile attempts to access the system via one of several available access channels. During the access state, the mobile has not yet been assigned a forward link traffic channel (which contains the power control bits). Since the reverse link closed loop power control is not active, the mobile initiates, on its own, any power adjustment required for a suitable operation.

The prime goal in CDMA systems is to transmit just enough power to meet the required performance objectives. If more than necessary power is transmitted, the mobile becomes a jammer to the other mobiles. Therefore, the mobile tries to get the base station's attention by first transmitting a very low power. The key rule is that the mobile transmits in inverse proportion to what it receives.

When receiving a strong pilot from the base station, the mobile transmits back a weak signal. A strong signal at the mobile implies a small propagation loss on the forward link. Assuming the same path loss on the reverse link, only a low transmit power is required from the mobile in order to compensate for the path loss.

When receiving a weak pilot from the base station, the mobile transmits back a strong signal. A weak received signal at the mobile indicates a high propagation loss on the forward link. Conversely, a high transmit power level is required from the mobile.

The mobile transmits the first access probe at a mean power level defined by

$$T_x = -R_x - K + (NOM\_PWR - 16 \times NOM\_PWR\_EXT) + (INIT\_PWR) \text{ (dBm)} \quad \textbf{(8.5)}$$

where

T_x = mean output transmit power (dBm)

R_x = mean input receive power (dBm)

NOM_PWR_EXT = nominal power for extended handoff (dB)
NOM_PWR = nominal power (–8 to 7 dB range)
INIT_PWR = initial adjustment (–16 to 15 dB range)
K = 73 for cellular; *K* = 76 for PCS

If INIT_PWR was 0, then NOM_PWR – 16 × NOM_PWR_EXT would be the correction that should provide the correct received power at the base station. NOM_PWR – 16 × NOM_PWR_EXT allows the open-loop estimation process to be adjusted for different operating environments.

The values for NOM_PWR, NOM_PWR_EXT, INIT_PWR, and the step size of a single access probe correction PWR_STEP are system parameters specified in the access parameters message. These are obtained by the mobile station prior to transmitting. If, as the result of an extended handoff direction message or a general handoff direction message, the NOM_PWR and NOM_PWR_EXT values change, the mobile uses the NOM_PWR and NOM_PWR_EXT values from the extended handoff direction message or general handoff direction message.

The total range of NOM_PWR – 16 × NOM_PWR_EXT correction is –24 to 7 dB. While operating in Band Class 0, NOM_PWR_EXT is set to 0, making the total range of correction from –8 to 7 dB. The range of INIT_PWR parameter is –16 to 15 dB, with a nominal value of 0 dB. The range of the PWR_STEP parameter is 0 to 7 dB. The accuracy of the adjustment to the mean output caused by NOM_PWR, NOM_PWR_EXT, INIT_PWR, or a single access probe correction of PWR_STEP should be ± 0.5 dB or ± 20 percent, whichever is greater.

The major flaw with this criterion is that reverse link propagation statistics are estimated based on the forward link propagation statistics. The two links are not correlated, therefore a significant error may result from this procedure. However, these errors will be corrected once the closed loop power control mechanism becomes active as the mobile seizes a forward traffic channel and begins to process power control bits.

After acknowledgment time window (T_a) expires, the mobile waits for an additional random time (RT) and increases its transmit power by a step size. The mobile tries again and the process is repeated until the mobile gets a response from the base station. However, there is a maximum number of probes per probe sequence and a maximum number of probe sequences per access attempt.

The entire process to send one message and receive an acknowledgment for the message is called an *access attempt.* Each transmission in the access attempt is referred to as an *access probe.* The mobile transmits the same message in each access probe in an access attempt. Each access probe contains an access channel preamble and an access channel capsule (see Figure 8.13). Within an access attempt, access probes are grouped into access probe sequences. Each access probe sequence consists of up to 16 access probes, all transmitted on the same access channel.

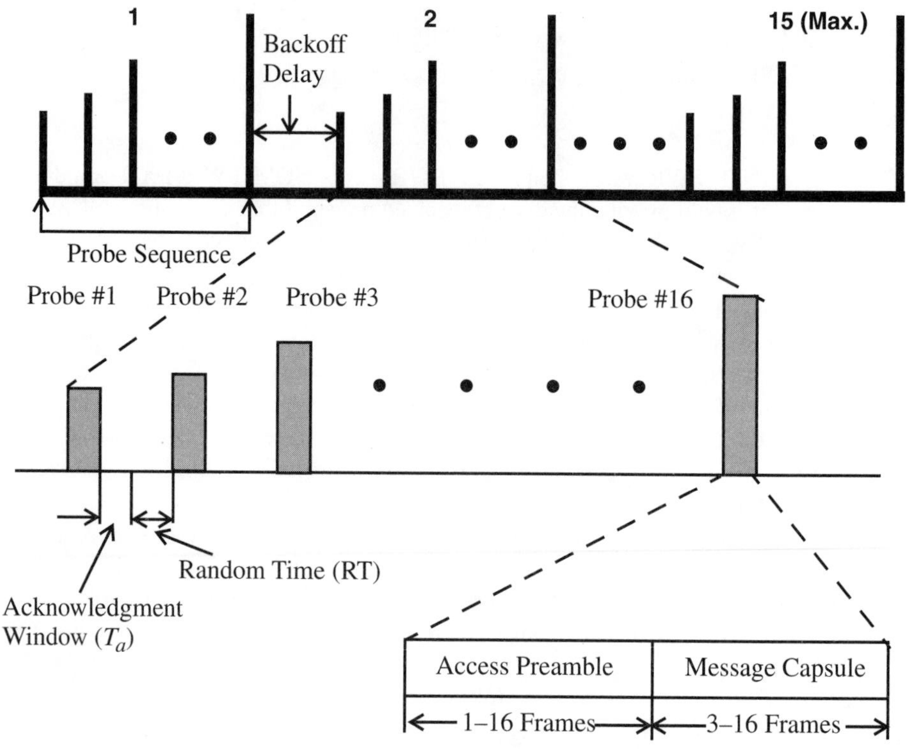

Figure 8.13 Access Attempt, Probe Sequence, and Probe in Open Loop Power Control

There are two reasons that could have prevented the mobile from getting an acknowledgment after the transmission of a probe:

1. The transmit power level was insufficient. In this case, the incremental step power strategy helps to resolve the problem.
2. There was a collision caused by the random contention of the access channel by several mobiles. In this case, the random waiting time minimizes the probability of future collisions.

The process is shown by the access probe ladder in Figure 8.14.

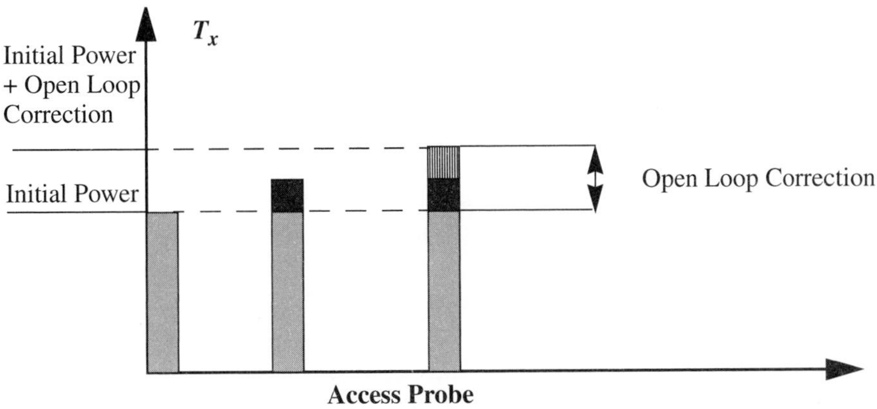

Figure 8.14 Access Probe Ladder

The transmit power is defined by

$$T_x = -R_x - K + (NOM\_PWR - 16 \times NOM\_PWR\_EXT) + (INIT\_PWR) + \text{Sum of Access Probe Corrections} \quad (8.6)$$

where access probe correction is the sum of all the appropriate incremental power steps prior to receiving an acknowledgment at the mobile.

For every access probe sequence, a backoff delay is generated pseudorandomly. Timing between access probes of an access probe sequence is also generated pseudorandomly. After transmitting each access probe, the mobile waits for T_a. If an acknowledgment is received, the access attempt ends. If no acknowledgment is received, the next access probe is transmitted after an additional random time (see Figure 8.13).

If the mobile does not receive an acknowledgment within an access attempt, the attempt is considered a failure and the mobile tries to access the system at another time. If the mobile receives an acknowledgment from the base station, it proceeds with the registration and traffic channel assignment procedures. The initial transmission on the reverse traffic channel shall be at a mean output power defined by Equation (8.6).

The mobile station supports a total combined range of initial offset parameters, closed NOM_PWR, and access probe corrections of at least ± 32 dB for mobile stations operating in Band Class 0 and ± 40 dB for mobile stations operating in Band Class 1.

The sources of error in the open loop power control are

- Assumption of reciprocity on the forward and reverse link
- Use of total received power including power from other base stations
- Slow response time ~ 30 ms to counter fast fading owing to multipath

8.11.2 Reverse Link Closed Loop Power Control

Fading sources in multipath require a much faster power control than the open loop power control. The additional power adjustments required to compensate for fading losses are handled by the reverse link closed loop power control mechanism, which has a response time of 1.25 ms for 1-dB steps and a dynamic range of 48 dB (covered in three frames). The quicker response time gives the closed loop power control mechanism the ability to override the open loop power control mechanism in practical applications. Together, two independent power control mechanisms cover a dynamic range of at least 80 dB. The closed loop power control provides correction to open loop power control. Once on the traffic channel, the mobile and base stations engage in closed loop power control.

The reverse link closed loop power control mechanism consists of two parts—the reverse inner loop power control (RILPC) and the reverse outer loop power control (ROLPC). The RILPC keeps the mobile as close to its target $(E_b/I_t)_{setpoint}$ as possible, and the ROLPC adjusts the base station target $(E_b/I_t)_{setpoint}$ for a given mobile.

To understand the operation of the closed loop power control mechanism, we review the structure of the forward traffic channel and its operation. The areas of focus are the output of the interleaver and the input to the MUX. A power control subchannel continuously transmits on the forward traffic channel. This subchannel runs at 800 power control bits per second. Therefore, a power control bit (0 or 1) is transmitted every 1.25 ms. A 0 bit indicates to the mobile to increase its mean output power level, and 1 indicates to the mobile to decrease its mean output power level.

A 20-ms frame is organized into 16 time intervals of equal duration (see Figure 8.15). These time intervals, each of 1.25 ms, are called *power control groups (PCG)*. Thus, a frame has 16 PCGs. Prior to transmission, the reverse traffic channel interleaver output data stream is gated with a time filter. The time filter allows transmission of some symbols and deletion of others. The duty cycle of the transmission gate varies with transmit data rate (i.e., the variable-rate vocoder output which, in turn, depends on the voice activity). Table 8.4 indicates the number of PCGs that are sent at different frame rates.

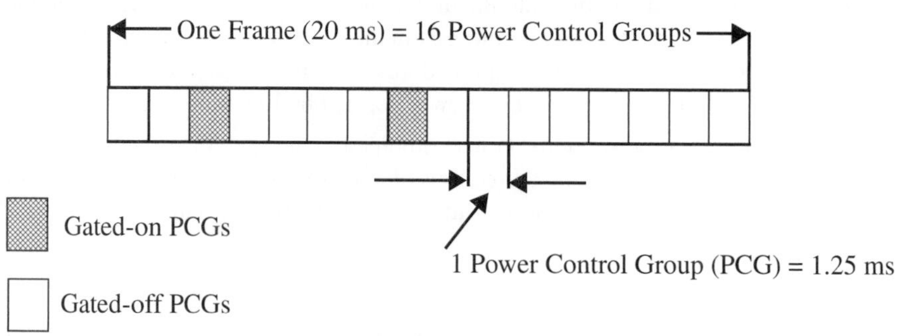

Figure 8.15 Power Control Groups in IS-95

The assignment of the gated-on and gated-off groups is determined by a data burst randomizer (DBR). At the base station, the reverse link receiver estimates the received signal strength by measuring E_b/I_t during each power group (1.25 ms):

- If the signal strength exceeds a target value, a power-down power control bit 1 is sent.
- Otherwise, a power-up control bit 0 is transmitted to the mobile via the power control subchannel on the forward link.

Table 8.4 Power Control Groups versus Frame Rate

Frame Rate	Rate (kbps)	No. of PCGs Sent
Full	9.6	16
Half	4.8	8
One-fourth (1/4)	2.4	4
One-eighth (1/8)	1.2	2

Similar to the reverse link transmission, the forward link transmissions are organized in a 20-ms frame. Each frame is subdivided into 16 PCGs. The transmission of the power control bit occurs on the forward traffic channel in the second PCG following the corresponding reverse link PCG in which the signal strength was estimated. For example, if the signal strength is estimated on PCG in #2 of a reverse link frame, then the corresponding power control bit must be sent on PCG #4 of the forward link frame (see Figure 8.16). Once the mobile receives and processes the forward link channel, it extracts the power control bits from the forward traffic channel. The power control bits then allow the mobile to fine tune its transmit power on the reverse link.

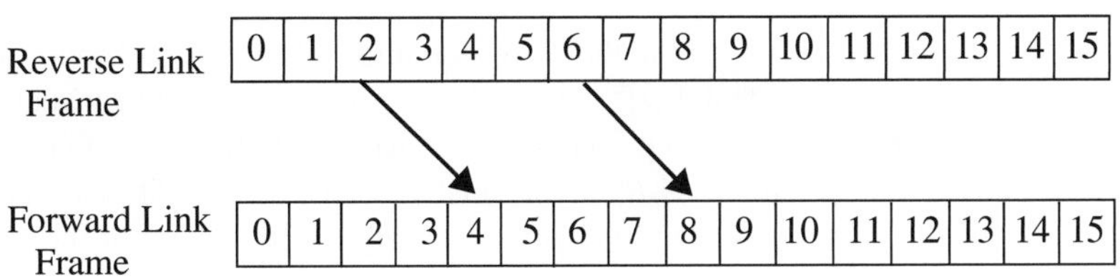

Figure 8.16 PCG Location in Reverse and Forward Link Frame

Based on the power control bit received from the base station, the mobile either increases or decreases transmit power on the reverse traffic channel as needed to approach the target value of $(E_b/I_t)_{nom}$ or set point that controls the long-term FER. Each power bit produces 1-dB change in mobile power, i.e., it attempts to bring the measured E_b/I_t value 1 dB closer to its target value. Note that it might not succeed because I_t is also always changing. Therefore, further adjustments may be required to reach the desired target. The base station, through the mobile, can only directly change E_b, not I_t, and the objective is the ratio of E_b to I_t— not any particular value for E_b or I_t.

The base station measures E_b/I_t 16 times in each 20-ms frame. If the measured E_b/I_t is greater than the current target value of E_b/I_t, the base station asks the mobile to decrease its power by 1 dB. Otherwise, the base station asks the mobile to increase its power by 1 dB (see Figure 8.17).

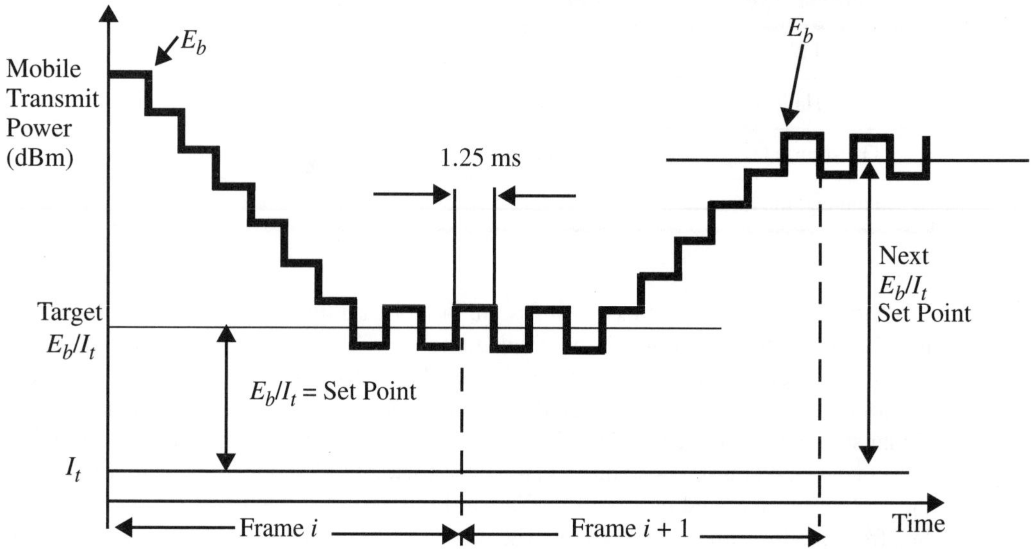

Figure 8.17 Target E_b/I_t

The relationship between E_b/I_t and the corresponding FER is nonlinear and varies with vehicle speed and RF environment. Performance deteriorates with increasing vehicle speed. Best performance corresponds to a stationary vehicle, where additive white Gaussian noise dominates. Thus, a single value of E_b/I_t is not satisfactory for all conditions. The use of a single, fixed value of E_b/I_t could reduce channel capacity by 30 percent or more by transmitting excessive, unneeded power.

The value of the variable "a" is kept very small (see Figure 8.18), so it may take 35 frames to reduce the E_b/I_t set point by 1 dB. Typically, the value of "100a" is set at about 3 dB. The set point value is reduced by "a" for each consecutive frame until a frame error occurs. The set point is then increased by a relatively large amount and the process is repeated. The set point can range from 3 dB to 10 dB. A value of $E_b/I_t \geq 5$ dB corresponds to good voice quality.

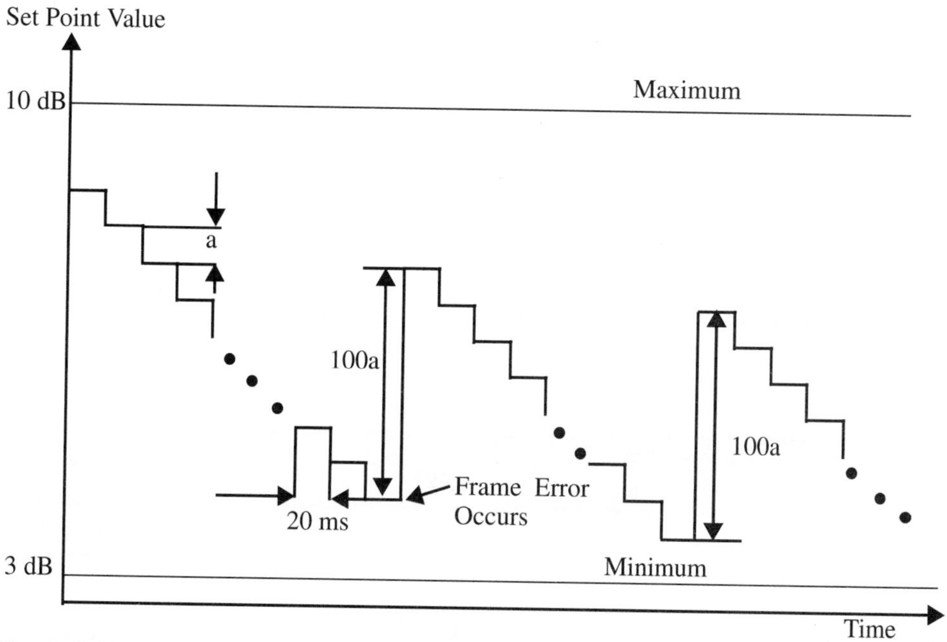

Figure 8.18 Set Point Value versus Time

Since FER is a direct measure of link quality, the system is controlled using the measured FERs rather than E_b/I_t. FER is the key parameter in controlling and ensuring a satisfactory voice quality. It is not sufficient to maintain a target E_b/I_t, but it is necessary to control FERs as they occur. The objective of the ROLPC is to balance the desired FER on the reverse link and system capacity. The system capacity can be controlled with the ROLPC parameters by increasing the acceptable FER. Change in FER can be accomplished by setting the ratio of down_frr to up_frr. The down_frr is calculated by the system by using the desired reverse FER (rfer) and up_frr as

$$\text{down\_frr} = (\text{rfer} \times \text{up\_frr}) / 2 \tag{8.7}$$

Based on simulations, the following values for up_frr are suggested:
If (0.2% $\leq$ rfer $\leq$ 0.4%), up_frr = 6000
If (0.6% $\leq$ rfer $\leq$ 1.0%), up_frr = 5000
If (1.2% $\leq$ rfer $\leq$ 2.0%), up_frr = 3000
If (2.2% $\leq$ rfer $\leq$ 3.0%), up_frr = 1000

Tables 8.5 and 8.6 lists the range and default values of different parameters for Rate Set 1 (RS1) and Rate Set 2 (RS2).

Table 8.5 ROLPC Parameters for RS1

Parameter	Range (%)	Suggested Value (%)	Description of Parameter
RFER 1	0.2–3.0	1	Target Reverse Link FER
$(E_b/I_t)_{nom\ 1}$ (dB)	3.5–8.0	6.5	Initial $(E_b/I_t)_{set\ point}$
$(E_b/I_t)_{max\ 1}$ (dB)	5.5–9.5	8.5	Maximum $(E_b/I_t)_{set\ point}$
$(E_b/I_t)_{min\ 1}$ (dB)	3.0–5.8	3.5	Minimum $(E_b/I_t)_{set\ point}$

Table 8.6 ROLPC Parameters for RS2

Parameter	Range (%)	Suggested Value (%)	Description of Parameter
RFER 2	0.2–6.0	1	Target Reverse Link FER
$(E_b/I_t)_{nom\ 2}$ (dB)	3.8–8.3	6.8	Initial $(E_b/I_t)_{set\ point}$
$(E_b/I_t)_{max\ 2}$ (dB)	5.8–9.8	8.8	Maximum $(E_b/I_t)_{set\ point}$
$(E_b/I_t)_{min\ 2}$ (dB)	3.0–5.8	3.8	Minimum $(E_b/I_t)_{set\ point}$

The inner loop power control is also responsible for detecting a mobile that fails to respond to power control, and might be causing interference to other mobiles. The base station counts the number of consecutive power decrease commands and if the count exceeds the specified threshold value, the base station will send a "*lock until power cycle*" message to the mobile. This message disables the mobile until the user turns the power off and on. Figure 8.19 gives the flow chart for the reverse link closed loop power control.

The mobile power output with both open loop and closed loop power control is given as:

$$\begin{aligned} T_x = -R_x - K &+ (NOM\_PWR - 16 \times NOM\_PWR\_EXT) \\ &+ INIT\_PWR \\ &+ \text{Sum of Access Probe Corrections} \\ &+ \text{Sum of All Closed Loop Power Control Corrections} \end{aligned} \qquad \textbf{(8.8)}$$

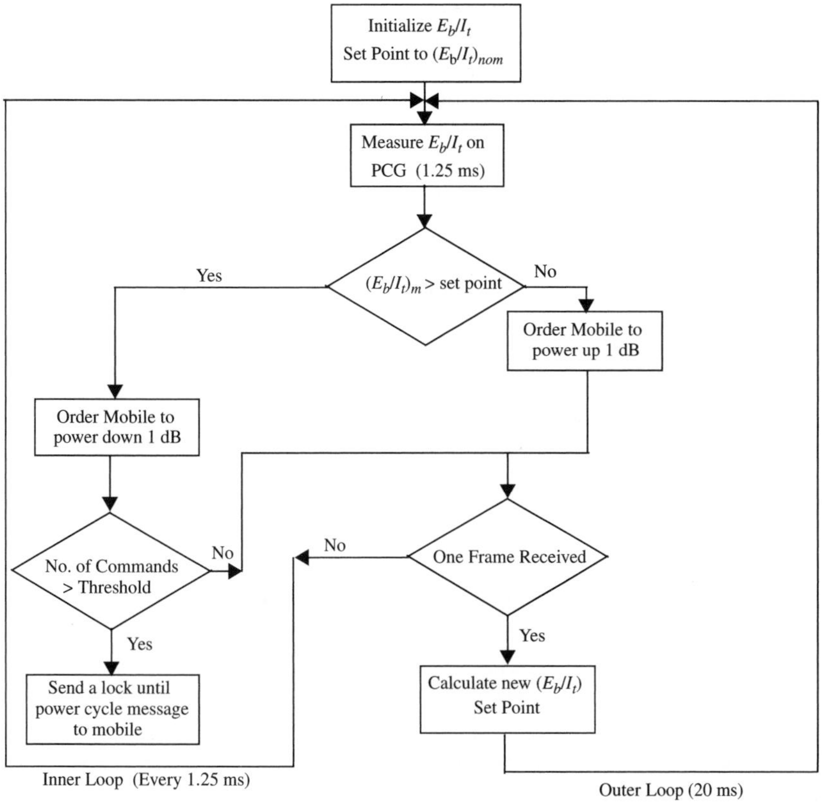

Figure 8.19 Flow Chart for Reverse Link Closed Loop Power Control

8.12 Forward Link Power Control

Forward link power control (FLPC) aims at reducing interference on the forward link. Forward link power control not only limits the in-cell interference but it is especially effective in reducing other cell/sector interference.

The FLPC attempts to set each traffic channel transmit power to the minimum required to maintain the desired FER at the mobile. The mobile continuously measures the forward traffic channel FER. It reports this measurement to the base station on a periodic basis. After receiving the measurement report, the base station takes the appropriate action to increase or decrease power on the measured logical channel. The base station also restricts the power dynamic range so that the transmitter power never exceeds a maximum value that would cause excessive interference, or falls below the minimum value required for adequate voice quality.

Since FERs are measured (not E_b/I_t as in the closed inner loop strategy), this process is a direct reflection of voice quality. However, it is a much slower process. As orthogonal Walsh codes are employed for the forward link, instead of long PN codes, cochannel interference is not an urgent issue. Therefore, slow measurements do not add much degradation to system performance. Figure 8.20 gives a flow charge for the forward link power control process.

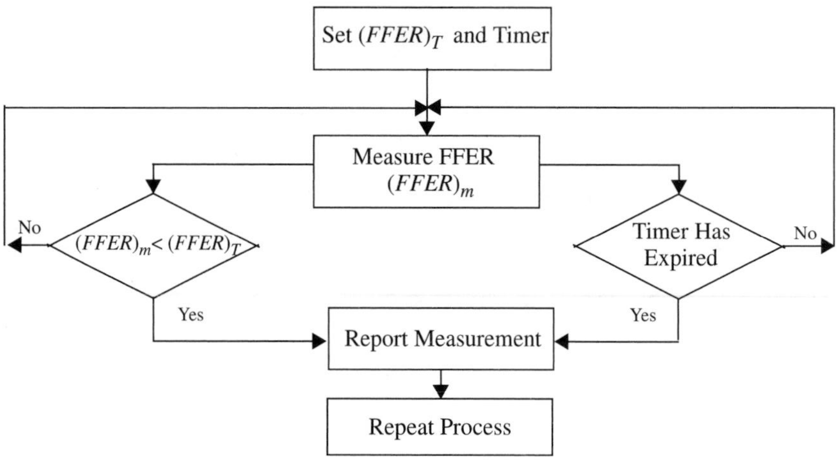

Figure 8.20 Flow Chart for Forward Link Power Control

Forward link power control is expressed in terms of parameters N, D, U, and V (see Figure 8.21), which may be adjusted to various values for the operation of an actual system.

For RS1, the power measurement report message (PMRM) contains the number of errored frames received and the total number of frames received during the interval covered by the report (frame counters then are reset for the next report interval). The FER is equal to the number of errored frames divided by the total number of frames received in the reporting interval. The followings are the steps for forward link power control for RS1 (see Figure 8.21):

Action by Mobile

- Mobile keeps track of the number of errored frame in a period of length pwr_rep_frame
- If errored frames > a specified number, the mobile send a PMRM containing
  - total number of frames in pwr_rep_frame
  - number of errored frame in pwr_rep_frame
  - FER
- If errored frame < specified number, PMRM is not sent
- After sending PMRM, the mobile waits for a period, pwr_rep_delay, before starting a new period

Action by Base Station

- On receiving the PMRM, the base station compares the reported FER as follows and adjusts traffic channel power:
  - If FER < fer_small, reduce power by *D*
  - If fer_small < FER < fer_big, increase power by *U*
  - If FER > fer_big, increase power by *V*
- If no PMRM is received:
  - Base station starts a timer fpc_step
  - When timer expires, power level is reduced by *D*
  - The timer resets after it expires or after receipt of a PMRM
- Digital gain is never set below min_gain or above max_gain
- If flpc_enable = 0, digital gain is set to nom_gain

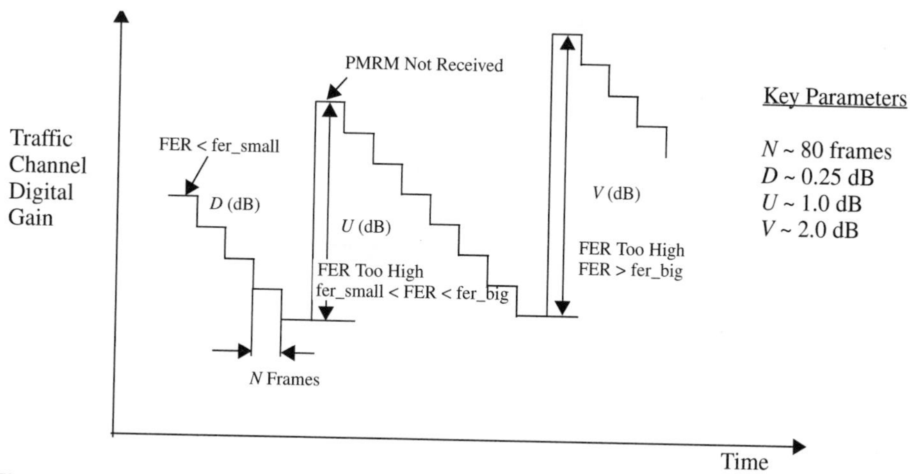

Figure 8.21 Forward Link Power Control for Rate Set 1

For RS2, one bit per reverse link frame (the *E* or erasure bit) is dedicated to inform the base station whether the last forward link frame was received without error at the mobile. This allows more rapid and precise control of forward link power than the scheme used for RS1. The followings are the steps for forward link power control for RS2 (see Figure 8.22).

Forward Link Power Control with RS2

- Uses erasure indicator bit instead of PMRM
- Much faster than RS1 implementation
  - The forward gain in RS2; forward link power control could change every two frames, thus, its response is very fast

- **Process**
  - In each frame, the mobile sends an erasure indicator bit showing whether the previous forward frame was an erasure bit
  - If an erasure is indicated by the mobile, the base station increases traffic channel digital gain by dn_adj

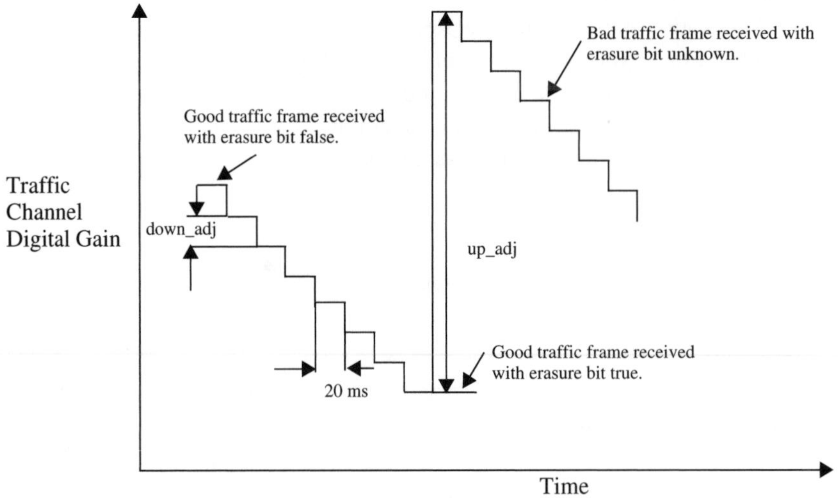

Figure 8.22 Forward Link Power Control for Rate Set 2

Tables 8.7 and 8.8 list the values of parameters for forward link power control for RS1 and RS2.

Table 8.7 Forward Link Power Control Parameters for RS1

Parameters	Range	Suggested Value	Description
FER	0.2–3%	1%	Target forward frame error rate
fer_small	0.2–5%	2%	Lower forward link FER threshold minimum PMRM FER required to increase gain by *U*
fer_big	2–10%	6%	Upper forward link FER threshold minimum PMRM FER required to increase gain by *V*
min_gain	34–50	40	Minimum traffic channel digital gain
max_gain	50–108	80	Maximum traffic channel digital gain
nom_gain	34–108	57	Nominal traffic channel digital gain
fpc_step	20–5000 ms	1600 ms	Forward power control timer value that determines when gain is decreased by *D*

Table 8.8 Forward Link Power Control Parameters for RS2

Parameter	Range	Suggested Value	Description
FER	0.2–6%	1%	Target forward frame error rate
up_adj	1–50	15	Gain increase when forward erasure is observed
dn_adj	—	N/A	Gain decrease when no forward erasure is observed
min_gain	30–50	30	Minimum traffic channel digital gain
max_gain	50–127	127	Maximum traffic channel digital gain
nom_gain	40–108	80	Nominal traffic channel digital gain

where dn_adj = (up_adj × FER) / 100

8.13 Summary

In this chapter we discussed soft handoff and power control in IS-95 CDMA. Soft handoff provides path diversity on the forward and reverse link. Diversity gains are achieved because less power is required on the forward and reverse link. This results in a reduction of the total system interference and an increase in system capacity.

Since the RF environment changes continuously owing to fast and slow fading, shadowing, external interference, and other factors, the aim of power control is to adjust the transmitted power on the forward and reverse links while maintaining link quality under all operating conditions. Power control in a CDMA system is required to resolve the near-far problem. To minimize the near-far problem, the goal in a CDMA system is to ensure that all mobile stations achieve the same received power levels at the base station.

The reverse link power control includes the open loop power control and the closed loop power control. The open loop power control is too slow to counter fast fading caused by multipath. The closed loop power control provides correction to the open loop power control. It begins after acquiring the traffic channel and is directed by the base station. The closed loop power control occurs every 1.25 ms (i.e., 800 times per second) and is much faster and more effective than the open loop power control. With the closed loop power control, power can change 16 dB per 20-ms frame.

The forward link power control includes closed loop power control. In cdma2000, the forward link power control will occur every 1.25 ms. The forward link power control is aimed at reducing interference and increasing system capacity.

8.14 Problems

1. Repeat Example 8.1 using the direct path distance of 3 km and the multipath distance of 8 km.
2. Repeat Example 8.3 with delay spread equal to 4 μs and chip rate equal to 3.84 Mcps.
3. A mobile is 1 km from the base station and operates at 880 MHz. The base station is transmitting at 47 dBm ERP on the paging channel. Assuming NOM_PWR = 3 dB, INIT_PWR = –4 dB and PWR_STEP = 3 dB, determine the mobile power level when it successfully accesses the network on the access channel, if this occurs on the second probe. Use the Hata-Okumura model for a medium-size city with base station antenna height = 30 m and mobile antenna height = 1 m.
4. Assume that the current active set consists of pilots P1, P2, P3, and P4. At a particular time, the mobile measures the signal strength of P1, P2, P3, and P4 as –95 dBm, –100 dBm, –101 dBm, and –105 dBm, respectively. Pilot P5 is contained in the candidate set. The mobile measures P5 as –102 dBm at the same time. Determine the possible values of SOFT_SLOPE and ADD_INTERCEPT that will cause the mobile to send PMRM to the base station. The mobile is TIA IS-95 B complaint. If the mobile was only IS-95 A complaint, determine the value of T_COMP that would cause the mobile to generate a power measurement report message (PMRM).

8.15 References

1. TIA/EIA/IS-95A, "*Mobile Station—Base Station Compatibility Standard for Dual-Mode Wideband Spread Spectrum Cellular Systems*," May 1995.
2. Garg, V. K., Smolik, K. F., and Wilkes, J. E., *Application of CDMA in Wireless/ Personal Communications,* Prentice Hall: Upper Saddle River, NJ, 1997.
3. TIA/EIA SP-3693, "*Mobile Station—Base Station Compatibility Standard for Dual-Mode Wideband Spread Spectrum Cellular Systems*," Nov. 18, 1997.
4. Wong, Daniel, and Lim, T. J., "Soft Handoff in CDMA Mobile Systems," *IEEE Personal Communications,* Dec. 1997 [Vol. (4)6].
5. Wang, S. W. and Wang, I., "Effects of Soft Handoff, Frequency Reuse and Non-Ideal Antenna Sectorization on CDMA System Capacity," *Proceedings of the IEEE VTC '93*, May 1993, pp. 850–854.

CHAPTER 9

Access and Paging Channel Capacity

9.1 Introduction

In IS-95, a uniform choice among a number of resources is made by using a *hash function*. The hash function allows a uniform distribution through a method that is reproducible at both the mobile station and base station.

In IS-95, the mobile station is sometimes presented with a choice of resources from which it should choose. For example, out of 32 access channels associated with a particular paging channel, which one should the mobile station use to access the network? The choice should be both uniform and random to minimize collisions and overloads on the particular channels.

In IS-95, the access channel in the reverse (up) link supports call originations, page responses, order messages, registrations, and short message services (SMS). Any excessive use of the capacity by the access channel should be limited to guarantee the capacity for the traffic channels.

In this chapter, we first present the methods used in IS-95 to generate a hash function and random number. Next, we discuss an approach to determine the access channel capacity. We also evaluate paging channel capacity in terms of various paging messages related to call processing and special services such as SMS and voice message services (VMS).

9.2 Hash Function

In IS-95, the base station (BS) or mobile station (MS) is often presented with a number of resources and must choose which one to use. For example, if there are seven active paging channels, the BS must decide which of these seven channels to use for paging a particular MS. The choice should be *uniform* and *reproducible*.

Uniform. All resources should be used uniformly to prevent collisions. For example, page messages should be evenly distributed over all paging channels and all of the slot resources within the paging channels to avoid overloading of a single channel.

Reproducible. It is advantageous for both the BS and the MS to be able to independently reproduce the choice of resources from information known to both. For example, from information known by both BS and MS, the MS should be able to reproduce the base station's choice of paging channel.

In IS-95, a uniform choice among a number of resources is made by using a hash function [1]. The hash function allows a uniform distribution through a method that is reproducible at both the MS and the BS by using (a) the number of resources and (b) a parameter (called a hash key) that is known by both the BS and MS as inputs.

Let K = the hash key parameter and N = the number of resources, with the input of a hash key K, the output of hash function $h(K)$ will be an integer value between 0 and $N-1$, thus taking N values

$$0 \le h(K) \le N-1 \qquad \textbf{(9.1)}$$

IS-95 specifies the following hash function, R as

$$R \equiv h(K) = \left\lfloor N \times \frac{(40,503 \times K) mod 2^{16}}{2^{16}} \right\rfloor \qquad \textbf{(9.2)}$$

The hash key is determined from the HASH_KEY parameter, a binary number generated from either the 32-bit binary electronic serial number (ESN) (see Figure 9.1) or the least significant 32 bits of binary encoded international mobile subscriber identity (IMSI) of the MS (see Figure 9.2).

$$\text{HASH\_KEY} = L + 216\,H \qquad \textbf{(9.3)}$$

where
L = number formed by the 16 least significant bits of the HASH_KEY
H = number formed by the 16 most significant bits of the HASH_KEY

$$K = L \oplus H \oplus DECORR \qquad \textbf{(9.4)}$$

where $DECORR$ is a modifier intended to decorate the different hash function selections made by the same MS. If L' denotes the 12 least significant bits of HASH_KEY, then Table 9.1 shows the values of parameters used in the IS-95 hash function.

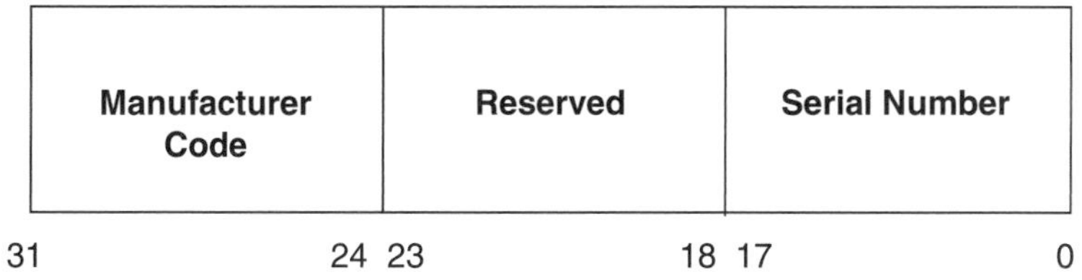

Figure 9.1 Electronic Serial Number (ESN)

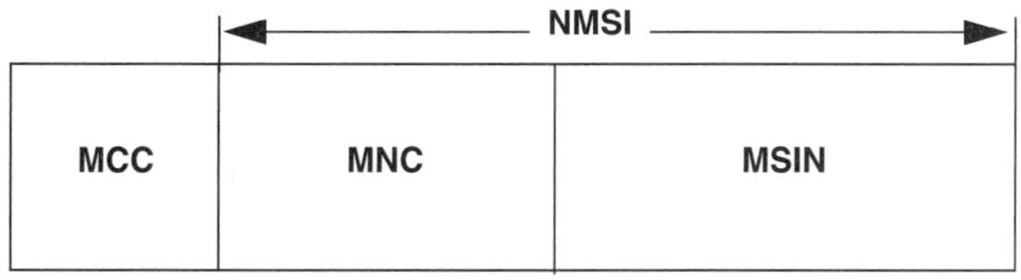

MCC: Mobile Country Code (3 digits)
MNC: Mobile Network Code
MSIN: Mobile Station Identification Number
NMSI: National Mobile Station Identification (up to 12 digits)

Figure 9.2 International Mobile Subscriber Identity (IMSI)

Table 9.1 IS-95 Hash Function Parameters

Application	*N*	*DECORR*	Return Value
CDMA channel # (central frequency)	Number of channels (up to 10) in list sent to MS	0	$R + 1$
Paging channel	Number of channels (up to 7) list sent to MS	$2 \times L'$	$R + 1$
Paging slot #	2048	$6 \times L'$	R
Access channel randomization	2^M, $M \le 512$ is in broadcast message	$14 \times L'$	R

Table 9.2 Initial Rotation of IMSI Digit Values

d_i	0	1	2	3	4	5	6	7	8	9
d'_i	9	0	1	2	3	4	5	6	7	8

This *rotation* procedure (see Table 9.2) is used to determine IMSI_S digits:

- The first three numbers (derived from the area code) are converted directly to a 10-bit binary number.
- The next three digits are converted to a 10-bit binary number.
- The first of the last four (local number) digits is converted directly to a 4-bit binary number. If the digit is a zero, it is treated as if it signified the number 10 and is converted to $(10)_2 = 1010$.
- The last three digits are converted to a 10-bit binary number in same way as the area code.

EXAMPLE 9.1

Find the paging channel assignment for the mobile that has an IMSI of 301-294-3463.

Rotation to find IMSI_S:

$301 \rightarrow 290; (290)_2 = 0\,1\,0\,0\,1\,0\,0\,0\,1\,0$
$294 \rightarrow 183; (183)_2 = 0\,0\,1\,0\,1\,1\,0\,1\,1\,1$
$(3)_2 \rightarrow 0\,0\,1\,1$
$(463) \rightarrow 352; (352)_2 = 0\,1\,0\,1\,1\,0\,0\,0\,0\,0$

IMSI_S = 0 1 0 0 1 0 0 0 1 0 0 0 1 0 1 1 0 1 1 1 0 0 1 1 0 1 0 1 1 0 0 0 0 0

HASH_KEY = 0 0 1 0 0 0 1 0 0 0 1 0 1 1 0 1 1 1 0 0 1 1 0 1 0 1 1 0 0 0 0 0

$H = 0\,0\,1\,0\,0\,0\,1\,0\,0\,0\,1\,0\,1\,1\,0\,1$
$L = 1\,1\,0\,0\,1\,1\,0\,1\,0\,1\,1\,0\,0\,0\,0\,0$
$L' = 1\,1\,0\,1\,0\,1\,1\,0\,0\,0\,0\,0$

For Paging Channel Number:

$$K = L \oplus H \oplus 2L'$$

$= 1\,1\,0\,0\,1\,1\,0\,1\,0\,1\,1\,0\,0\,0\,0\,0 \quad (L)$
$0\,0\,1\,0\,0\,0\,1\,0\,0\,0\,1\,0\,1\,1\,0\,1 \quad (H)$
$0\,0\,0\,1\,1\,0\,1\,0\,1\,1\,0\,0\,0\,0\,0\,0 \quad (2L')$

$1\,1\,1\,1\,0\,1\,0\,1\,1\,0\,0\,0\,1\,1\,0\,1; \quad K = 62{,}861$

$$R = \left\lfloor 7 \times \frac{(40503 \times 62861) mod 2^{16}}{2^{16}} \right\rfloor = \left\lfloor 7 \times \frac{51021}{65536} \right\rfloor = \lfloor 5.4496 \rfloor = 5$$

Paging channel assignment = 5 + 1 = 6

9.3 IS-95 Random Number Generator

In IS-95, the MS is sometimes presented with a choice of resources from which to choose. For example, which of the 32 access channels associated with a particular channel (distinguished by long PN offsets commanded by the mask) should the MS use when attempting an access? This choice should be uniform and random to minimize collisions and overloads on particular channels.

A good random number generator (RNG) satisfies three basic criteria:

- The generator should produce a nearly infinite sequence of numbers. This means that generator should have a nearly infinite period.
- The generator should produce a random sequence of numbers; the sequence should have no obvious pattern.
- The generator should be easily implemented on whatever processor it is to be used on. In other words, the arithmetic involved should not cause overflow or erroneous results.

$$Z_{n+1} = 16807 Z_n mod(2^{31} - 1) \qquad \textbf{(9.5)}$$

where 7^5= 16807; Z_0 = the generator seed, which is determined from information stored in the permanent memory of the mobile and information found on SYNC channel.

During mobile initiation, the mobile receives the pilot channel to obtain synchronization with the BS and obtain the SYNC channel. On the SYNC channel is found the SYNC channel message, which contains a 36-bit parameter SYS_TIME corresponding to the system time. The 32 least significant bits of SYS_TIME form a parameter RANDOM_TIME. The MS computes the generator seed by using the 32-bit ESN stored in permanent memory and the 32-bit parameter RANDON_TIME as follows:

$$Z_0 = (ESN \oplus RANDOM\_TIME) mod \text{ m} \qquad \textbf{(9.6)}$$

If Z_0 is computed to be 0, then it is replaced by $Z_0 = 1$.

EXAMPLE 9.2

A mobile station is attempting access to the base station to register. The link is poor, and the mobile must transmit three accesses before the base station receives and correctly acknowledges the access. Find the access channel number used for each of the accesses (IS-95 states that before each access, a new access channel number is randomly determined).

N = total number of access channels = 32

```
SYSTEM_TIME:  1 0 1 0 0 1 0 0 1 0 1 1 1 0 1 0 0 1 0 1 0 1 1 0 1 0 0 0 1 0 1 0 1 1 1 1
ESN ⊕ :         0 1 0 0 1 0 1 1 1 0 1 0 0 1 0 1 0 1 1 0 1 0 0 0 1 0 1 0 1 1 1 1
RANDOM_TIME:    0 1 0 0 1 0 1 1 1 0 1 0 0 1 0 1 0 1 1 0 1 0 0 0 1 0 1 0 1 1 1 1
                ------------------------------------------------------------------
                0 0 0 0 0 0 0 0 0 0 0 0 0 0 0 0 0 0 0 0 0 0 0 0 0 0 0 0 0 0 0 0
                = 0
```

Z_0 = (ESN + RANDOM_TIME) mod $m = 0 \bmod(2^{32} - 1) = 0$; $Z_0 = 1$

$Z_1 = a\, Z_0 = 16807 \times 1 \bmod 2147483647 = 16807$

The access channel for the first attempt:

$$K_1 = \left\lfloor \frac{N \times Z_1}{m} \right\rfloor = \left\lfloor \frac{32 \times 16807}{2147483647} \right\rfloor = \lfloor 0.0002504 \rfloor = 0$$

$Z_2 = a\, Z_1 \bmod m$

The access channel for the second attempt:

$$K_2 = \left\lfloor \frac{N \times Z_2}{2147483647} \right\rfloor = \left\lfloor \frac{282475249 \times 32}{2147483647} \right\rfloor = \lfloor 4.2092 \rfloor = 4$$

$Z_3 = a\, Z_2 \bmod m$

$$K_3 = \left\lfloor \frac{32 \times 1622628073}{2147483647} \right\rfloor = \lfloor 24.1794 \rfloor = 24$$

9.4 Access Channel Capacity

The access channel uses a slotted ALOHA random access protocol. Each access slot has a fixed duration and consists of multiple frames, each 20 ms in duration. The long code uniquely identifies the access channel.

The entire process of sending a message and receiving an acknowledgment for that message is called an *access attempt*. Each transmission in an access attempt is called an *access probe* [2]. The mobile station transmits the same message at each access probe in an access attempt. Each access probe contains a *preamble* and an *access channel message capsule.* The number of 20-ms frames denotes the length of the preamble 1 + PAM_SZ as well as the length of the message capsule 3 + MAX_CAP_SZ. Thus, the duration of an access probe is 4 + PAM_SZ + MAX_CAP_SZ (for PAM_SZ = 1 frame and MAX_CAP_SZ = 2 frames, the message length = 7 frames).

Within an access attempt, access probes are grouped into access probe sequences. Mobile stations send two types of messages on the access channel: a response message (i.e., the response to the base station message) or a request message. Each access attempt consists of up to MAX_REQ_SEQ (for a request access) or MAX_RSP_SEQ (for a response access) access probe sequences.

A persistence test is conducted before initiating the access probe sequence to control the rate at which a mobile station transmits requests. Assessing the appropriate range of persistence values to be assigned to mobile stations requires information about the delay found with the persistence tests. The persistence test generates a random number RP ($0 < RP < 1$) for each access channel slot. The RP is compared with a predetermined threshold P. The access probe sequence initiates if the generated random number RP is less than the predetermined threshold P. The precomputed threshold P, in general, is different, depending on the request, the access overload class, and its persistence value *psist (n)* as well as its persistence modifier.

An access probe is successful if at the base station E_b/I_t exceeds a threshold and there are no other probes received within an interval of two PN chips. A large amount of access traffic might result in unacceptable interference in the reverse link. As discussed in Chapter 8, after the first access probe fails to gain access to the network, the next access probe is sent with increased power. This results in an increase of the total power received at the base station and a reduction of the reverse link capacity. To guarantee the capacity for the traffic channels, any excessive use of the capacity by the access channel should be limited.

Within each access slot, we assume the arrival rate of the access attempt to have a Poisson distribution with a given mean. Since we are interested in the capacity of the access channel, we assume that traffic channel power is constant and the system is operating at 50 percent loading. The E_b/I_t required for the access probes to be successful is typically higher than that for the traffic channel. We assume that the E_b/I_t required for a successful access is 50 percent higher (or 1.76 dB above) than that of the traffic channel.

The access channels support call origination/termination, paging responses, orders, registrations, and short message services (SMS). Statistically, each subscriber contributes to the access traffic. Table 9.3 lists the values assumed for different access traffic types.

The number of access messages per hour, A, required to support the access attempts of the types in Table 9.3 is given as [3]

$$A = (OT + SMS + Tr + Zr + Pr) \times S \tag{9.7}$$

where S = the number of subscribers supported by a sector.

Table 9.3 Access Channel Traffic Assumptions

Access Channel Traffic Type	Value	Note
Call origination/termination (OT)	1/hr/subscriber	average number
Short message service (SMS)	m/hr/subscriber	variable
Time-based registration (Tr)	1 to 2/hr/subscriber	default value
Zone-based registration (Zr)	0.1 to 0.2/hr/subscriber	10% of time-based registration
Power on/off registration (Pr)	0	cancel out with time-based registration

If M is the access channel capacity in terms of messages per hour, then

$$M = \frac{3600}{(20 \times 10^{-3}) \times (7^*)} = 25{,}714 \text{ messages/hour} \quad \textbf{(9.8)}$$

* Note: 7 is the amount of frames per 20-ms message.

The utilization of the access channel, ρ, will be

$$\rho = \frac{A}{M} \quad \textbf{(9.9)}$$

EXAMPLE 9.3

We consider a code division multiple access (CDMA) system that uses an 8-kbps vocoder and supports 20 radio channels per CDMA carrier per sector (i.e., 13.2 Erlangs traffic at 2 percent blocking probability). The traffic is 0.02 Erlang per subscriber. Find the utilization of an access channel assuming the following traffic for the access channel:

- Call origination/termination (hr/subscriber): 1
- SMS (hr/subscriber): 3
- Time-based registration (hr/subscriber): 2
- Zone-based registration (hr/subscriber): 0.2

$$\text{Number of subscribers} = \frac{13.2}{0.02} = 660$$

$$A = (1 + 3 + 2 + 0.2) \times 660 = 4092$$

$$\rho = \frac{4092}{25714} = 12\,\%$$

9.5 Paging Channel Capacity

In a CDMA system, a paging channel carries information from base stations to mobile stations. There are three major types of call-processing–related messages.

- *Overhead message*—Contains information required for call setup (e.g., system parameter messages, channel list messages, access parameter messages, neighbor list messages, and extended system parameter messages), and is updated periodically to ensure a successful call setup.
- *Page message*—Used to page the mobile. This message is sent when a mobile switching center receives a call/service request for a mobile. Depending on the implementation, the page messages may be sent to a large area through the paging channel on all sectors.
- *Channel assignment message and order message*—Used to interact with the mobile for completing a call/service. The base station usually sends these messages only to small area (a few sectors) during the call/service setup.

As mentioned in previous chapters, the paging channel is divided into 80-ms paging slots. Each slot consists of eight half-frames, each of 10 ms in duration. Each half-frame begins with a synchronized capsule indicator (SCI) bit that is set to 1. Paging channel messages are carried in paging channel capsules that consist of the message body, an 8-bit length field to indicate the length in bits of the entire capsule, and a cyclic redundancy check (CRC) code of 30 bits.

9.5.1 Assumptions

To calculate the capacity of a paging channel, we make the following assumptions:

A.	Paging channel rate:	9,600 bps
B.	Maximum allowable utilization:	0.9
C.	Pages per user:	1.5
D.	Termination rate:	0.35
E.	Busy rate:	0.03
F.	BHCA per subscriber:	2.0
G.	No. of sectors per MSC:	200
H.	General page message length:	136 bits
I.	Overhead message length:	J + K + L + M + N
J.	System parameter message length:	264 bits
K.	Access parameter message length:	184 bits
L.	Neighbor list message length:	216 bits
M.	CDMA channel list message length:	88 bits
N.	Extended system parameter message length:	112 bits
O.	Channel assignment message length:	144 bits
P.	Order message:	102 bits

Q. Voice mail notification message length: 720 bits
R. Data burst message length: $(7X + 380)$ bits; X = no. of characters
S. Done message length: 72 bits

The following paging channel messages account for most paging channel usage:

- General page message
- Overhead message
- Order message
- Data burst message for SMS

We assume 90 percent paging channel capacity to be the maximum allowable paging channel utilization.

Next we develop equations for usage of a paging channel for the general page message, overhead messages, channel assignment and order messages, done message, voice mail service, and data burst message.

- General page message:

$$O_g = \frac{BHCA \times (D - E) \times C \times H}{3600 \times A \times B} \tag{9.10}$$

where
A, B, C, etc., refer to items listed above
$BHCA$ = busy hour call attempts

- Overhead messages:

$$O_o = \frac{I \times \dfrac{1}{[(80/1000) \times Slot\_Cycle]}}{A \times B} \tag{9.11}$$

where the overhead message will be sent within every n slots to a mobile.

- Channel assignment and order message:

$$O_{co} = \frac{[(BHCA)/G] \times [N_R \times O + P]}{3600 \times A \times B} \tag{9.12}$$

where N_R is the number of repeats for channel assignment message.

- Done message:

$$O_{ds} = \frac{\left[\frac{1}{80/1000}\right] \times S \times \left[1 \times \frac{1}{10/1000}\right]}{A \times B} \quad \textbf{(9.13)}$$

- Voice mail service (VMS):
  Two methods can be used: (1) based on call setup, procedure, and (2) based on the procedure using SMS-type transmission through paging channel.

  (1) Call setup procedure:

$$O_{v1} = \frac{[(BHCA)/F] \times No_{rate} \times 2 \times (D-E) \times \left(\frac{60}{V}\right) \times [BHCA \times E] \times H}{3600 \times A \times B} \quad \textbf{(9.14)}$$

where
V = voice mail service cycle duration
No_{rate} = number of page response rate

  (2) Procedure using SMS-type transmission through paging channel:

$$O_{v2} = \frac{BHCA \times E \times [H + Q/G]}{3600 \times A \times B} \quad \textbf{(9.15)}$$

- Data burst message:

$$O_s = \frac{\overline{M} \times (R + H \times G)}{3600 \times A \times B} \quad \textbf{(9.16)}$$

where
$\overline{M}$ = number of subscribers supported by a sector (assume ≈ 200)

EXAMPLE 9.4

Using the data listed in Section 9.5, calculate the usage of a paging channel, assuming the number of subscribers to be 100,000, BHCA per subscriber to be 2, and slot_cycle to be 8. $N_R = 4$ and $X = 7$.

$$O_g = \frac{2 \times (0.35 - 0.03) \times 1.5 \times 136 \times 100000}{3600 \times 9600 \times 0.9} = 0.42$$

$$O_0 = \frac{864 \times \left[\frac{1}{0.08 \times 8}\right]}{9600 \times 0.9} = 0.156$$

$$O_{co} = \frac{(200000/200) \times (4 \times 144 + 102)}{3600 \times 9600 \times 0.9} = 0.022$$

$$O_{ds} = \frac{(1/0.08) \times 72 + 1 \times (1/0.01)}{9600 \times 0.9} = 0.116$$

$$O_{v2} = \frac{200000 \times 0.03 \times (136 + 720/200)}{3600 \times 9600 \times 0.9} = 0.028$$

$$R = 80 \times 7 + 380 = 940 \text{ bits}$$

$$O_s = \frac{200 \times (940 + 136 \times 200)}{3600 \times 9600 \times 0.9} = 0.181$$

$$\text{Total occupancy} = 0.42 + 0.156 + 0.022 + 0.116 + 0.028 + 0.181 = 0.923 \approx 0.90$$

9.6 Summary

In this chapter we presented the IS-95 methods for generating the hash function and random number. We also discussed procedures to determine the access channel capacity and paging channel capacity in terms of various paging messages related to call processing and special services such as SMS and VMS.

9.7 References

1. Lee, J. S., and Miller, L. E., *CDMA System Engineering Handbook,* Artech House: Boston, 1998.
2. Mobile Station—Base Station Compatibility Standard for Dual-Mode Wideband Spread Spectrum Cellular System, TIA/EIA IS-95A, May 1995.
3. Kyoung, I. K., (Editor), *Handbook of CDMA System Design, Engineering, and Optimization,* Prentice Hall: Upper Saddle River, NJ, 1999.

9.8 Problems

1. Find the paging channel assignment for the mobile station having IMSI equal to 301-294-3464.
2. Solve Example 9.2 by changing the last four digits of the SYSTEM_TIME to 0 0 0 0.
3. Solve Example 9.3 using three CDMA carriers per sector.

4. Find the maximum number of subscribers per sector that can be supported based on paging channel occupancy of 90 percent. The number of sectors supported per MSC = 100, paging rate per subscriber = 1.6, termination rate = 0.3, busy rate = 0.03, BHCA per subscriber = 1.8, number of subscribers = 150,000, and number of paging channel = 1. Assume any other data that is not given.

CHAPTER 10

Reverse (Up) and Forward (Down) Link Capacity of a CDMA System

10.1 Introduction

The code division multiple access (CDMA) system is an interference-limited system in which link performance depends on the ability of the receiver to detect a signal in the presence of interference. For satisfactory performance of a CDMA link, a frame error rate (FER) is specified. Based on field trials, the required E_b/I_t values are established on the reverse link and various channels of the forward link to maintain the specified FER. The key issue in a CDMA network design is to minimize multiple access interference. Power control is critical to reduce multiaccess interference. The interference from other cells in the system must be included to determine the actual reuse factor in the CDMA system.

Trade-off betwen several system parameters is needed to achieve a balance between the system capacity, service quality, and coverage. The system capacity depends on the number of users that can be supported during the peak hours, traffic distribution, and Erlang capacity. Coverage is determined from the allowable path loss (link budget), and quality of service is based on mean opinion score (MOS), FER, E_b/I_t, and call drop rate.

In the forward direction, base station (BS) to mobile station (MS), a pilot signal is used by the mobile demodulator to provide a coherent reference that is effective even in a fading environment because the desired signal and pilot fade together. In the reverse direction (MS to BS), no pilot is used in IS-95 for power efficiency considerations, since unlike the forward direction, an independent pilot would be required for each signal. A modulation consistent with, and relatively efficient for, noncoherent detection is used for the reverse link. In wideband CDMA systems (cdma2000 and UTRA), an independent pilot would be used for each signal on the reverse link.

A CDMA system supports a different maximum number of mobiles on the forward and reverse links. The capacity of a CDMA system normally depends upon the reverse link capacity. The forward link capacity is governed by total transmitted power of the cell site and its distribution to traffic channels and other overhead channels including pilot, paging, and sync channels in IS-95. If the power amplifier cannot provide enough power to the forward traffic channels, the system capacity becomes forward link limited. Soft handoffs improve the capacity of the reverse link, however the amount and types of soft handoffs reduce the capacity of the forward link.

In this chapter, we present procedures for calculating the capacity of the reverse and forward links of a CDMA system. We establish the relationship to determine the *pole point* or *asymptotic cell capacity* that can be achieved when the received power at the base station from a mobile approaches infinity. We also relate the ratio of the received cell power to cell noise with the cell loading. We discuss a procedure to develop a *link safety margin parameter* for each of the forward link channels.

10.2 Reuse Parameters in CDMA

The own-cell interference at the BS receiver on the reverse link is the superposition of signals from other mobiles. The total interference can be modeled as band-limited white noise. Almost all the noise at the BS receiver is due to interfering mobile signals. Because of the propagation mechanism, the signal received at the BS receiver from a mobile close to the BS will be stronger than the signal received from another mobile located at the cell boundary. Therefore, the distant mobiles will be dominated by the close mobiles. To achieve a considerable capacity, all signals should arrive at the BS receiver with the same mean power, irrespective of distance. A solution to this problem is dynamic power control, which attempts to achieve a constant received mean power for each mobile (this was discussed in Chapter 8). Assuming a perfect power control for M mobiles in the cell, the own-cell interference, I_0, at the BS receiver can be given as

$$I_0 = (M-1) \cdot S \cdot v_f \tag{10.1}$$

where
S = signal power of each mobile at the BS receiver
v_f = average reverse link activity factor

In addition to the own-cell interference, signals received from other CDMA sectors and other CDMA cells act as interferers to the receiver. The interference power from other cells tends to fluctuate with variations in traffic load in the cells or sectors. We express the average other-cell interference, I_{0c}, as some fraction f of the total received own-cell power. Assuming that each interfering cell has same characteristics as the cell of interest, we can write

$$I_{0c} = f \cdot M \cdot S \cdot v_f \tag{10.2}$$

where

$$f = \frac{\text{Total other-cell received power}}{\text{Total own-cell received power}} = \text{reuse factor}$$

The total reverse link interference will be

$$I_{total} = I_0 + I_{0c} = [(1+f) \cdot M - 1] \cdot v_f \cdot S = \left[\frac{M}{\eta} - 1\right] \cdot v_f \cdot S \tag{10.3}$$

where
$\eta = 1 / (1 + f)$ = (Total own-cell power) / (Total own-cell power + other-cell power)

The η is called the CDMA *reuse efficiency*. Note that in a perfect CDMA system without any cochannel interference from the neighboring cells, the reuse efficiency would be 100 percent.

10.3 Multicell Network

CDMA multiple-cell networks use the same frequency band in all cells, as opposed to other access technologies in which the frequency used in a given cell is reused only in sufficiently distant cells to avoid cochannel interference. To compare CDMA with other multiple access schemes, capacity is determined (or measured) as the total number of users in the multiple-cell network rather than the number of users per bandwidth or per isolated cell.

The CDMA system is an interference-limited system. The CDMA link performance depends on the ability of the receiver to discern a signal in the presence of interference. For satisfactory performance of the CDMA link, an FER of about 1 percent is recommended. Field trials were conducted to establish the required E_b/I_t values for the reverse link and various channels of the forward link to maintain the recommended FER. The link budget is established to achieve the values of E_b/I_t. The required values of E_b/I_t depend on propagation environment and speed of the mobile. Based on field trials, the following values of E_b/I_t are suggested:

- Low-speed mobiles, speed $\leq$ 5 mph: 5 dB
  In this case, the duration of fades is much larger than the time between power control updates for a mobile. Thus, the effect of any fade is compensated by a quick response of the power control mechanism.
- Medium-speed mobiles, speed $\approx$ 30 mph: 7 dB
  The advantages of high or low speed are not applicable; therefore, the required E_b/I_t is somewhat higher.
- High-speed mobiles, speed $\geq$ 60 mph: 6 to 6.5 dB
  In this case, the fade duration is smaller compared with the chip length. Thus, only burst errors result on the links; they are corrected by interleaving and Viterbi decoding. Therefore, the required E_b/I_t is low.

The key issue in a CDMA network design is to minimize multiple-access interference. Power control is critical to multiaccess interference. Each cell controls the transmit power of its own mobiles. However, a serving cell is unable to control the power of mobiles in the neighboring cells. The mobiles in the neighboring cells introduce additional interference, thereby reducing the capacity of the reverse link. We include this effect by a factor f. The interference from other cells determines the actual reuse factor of the CDMA system. CDMA networks are designed to tolerate a certain amount of interference, and therefore have a capacity advantage in this regard as compared with time division mutliple access (TDMA) or frequency division multiple access (FDMA) networks.

10.4 Intercell Interference

The intercell interference factor, *f*, is difficult to evaluate because the serving cell does not have control over the power received from the mobiles in other cells. The *f* depends on the geometry of the serving cell and neighboring cells. It will be small if the serving cell radius is large, if path loss slope has a higher value, or if the standard deviation of path loss is small.

For path loss exponent $\gamma = 4$, and a standard deviation of path loss $\sigma = 8$ dB, the upper bound on *f* is 0.77. Table 10.1 lists the value of *f* with two-way and three-way soft handoff for different values of σ and $\gamma = 4$ [1].

Table 10.1 Intercell Interference Factor *f* for $\gamma = 4$

σ dB	Other Cell Interference Factor *f*	
	Two-Way SHO	Three-Way SHO
0	0.44	0.44
2	0.43	0.43
4	0.47	0.45
6	0.56	0.49
8	0.77	0.57
10	1.28	0.75
12	2.62	1.17

Cell loading is a measure of the total interference I_{total}, allowed in the system, with reference to the thermal noise, N_T (see "Cell Loading" on page 281).

$$\rho = \frac{M}{M_{max}} \approx \frac{I_{total}}{I_{total} + N_T} \tag{10.4}$$

where

M = active number of mobiles in the cell

M_{max} = maximum possible mobiles in the cell

ρ equal to 0.5 implies that the interference in the system is equal to the thermal noise level. ρ less than 0.5 implies that the system is *noise limited*, whereas ρ greater than 0.5 indicates that the system is *interference limited.* A value of ρ between 0.5 and 0.7 is typical.

10.5 Reverse Link Capacity in Single-Cell and Multicell Systems

In the absence of thermal noise, the required signal-to-noise ratio $(SNR)_{reqd}$ for a multicell system will be

$$(SNR)_{reqd} = \left(\frac{E_b}{I_t}\right)_{reqd} \cdot \frac{R}{R_c} \tag{10.5a}$$

$$\left(\frac{E_b}{I_t}\right)_{reqd} = \frac{v_f \cdot S \cdot G_p}{I_0 + I_{0c}} = \frac{v_f \cdot S \cdot G_p}{(M-1) \cdot v_f \cdot S + f \cdot M \cdot v_f \cdot S} \tag{10.5b}$$

where
G_p = processing gain = R_c/R
R = mobile transmission rate in bps
R_c = chip rate
B_w = bandwidth
I_t = total noise + interference power spectral intensity (psd)
E_b = energy per bit

$$\therefore \left(\frac{E_b}{I_t}\right)_{reqd} = \frac{G_p}{(M-1) + f \cdot M} \tag{10.6}$$

Similarly, the $(E_b/I_t)_{reqd}$ for a single-cell system will be

$$\left(\frac{E_b}{I_t}\right)_{reqd} = \frac{v_f \cdot S \cdot G_p}{(M_a - 1) \cdot v_f \cdot S} = \frac{G_p}{(M_a - 1)} \tag{10.7}$$

where
M_a = number of active users in a single-cell system

Comparing Equations (10.6) and (10.7), we get

$$\frac{M_a}{M} = 1 + f \tag{10.8}$$

From Equation (10.8), it can be noticed that the capacity of a single-cell system is higher than that of a multicell system.

10.6 Reverse Link Capacity

We consider an omnidirectional cell site serving a given set of mobiles. We divide mobiles into two groups: the mobiles that are powered up, and the mobiles that are not powered up.

The mobiles that are powered up are further divided into four subgroups:

- Active and transmitting mobiles (mobiles in conversational mode)
- Active, but not transmitting mobiles (mobiles in nonconversational mode)
- Idle and transmitting (mobiles in access mode)
- Idle and not transmitting (mobiles in nonaccess mode)

We assume that interference at the cell site by mobiles in the access mode is typically small and neglected. This may be accounted for as a source of some degradation in system quality and system capacity. We focus only on the active mobiles in our analysis. We assume there are M mobiles are transmitting at a given time in a cell. In a CDMA environment, for each mobile, there are $(M-1)$ interferers. At the cell site, the average signal power received from the ith mobile is S_{ri}. This signal power provides a bit energy equal to E_b, i.e.,

$$E_b = \frac{S_{ri}}{R} \tag{10.9}$$

where
R = mobile transmission rate in bps

The thermal noise power is $N_0 B_w$, where N_0 is the thermal noise power spectral density (psd), and B_w is the spreading bandwidth. The average interference psd at the base station is given as

$$I_0 = \frac{1}{B_w} \sum_{i=1}^{M-1} \nu_f \cdot S_{ri} \tag{10.10}$$

where
ν_f = channel activity factor

In Equation (10.10) we assume a perfect power control on the reverse link and that the signals transmitted from all the mobiles arrive at the base station with the same received power, i.e., $S_{ri} = S$ for all values of i (i.e., $1 \le i \le M-1$). The total interference and thermal noise psd will be

$$I_t = I_0 + N_0 = \frac{1}{B_w} \cdot \sum_{i=1}^{M-1} \nu_f \cdot S_{ri} + N_0 \tag{10.11}$$

Recognizing that $S_{ri} = S$, we get from Equation (10.11)

$$I_t = \frac{(M-1) \cdot \nu_f \cdot S}{B_w} + N_0 \tag{10.12}$$

The E_b/I_t will be given as

$$\frac{E_b}{I_t} = \left(\frac{B_w}{R}\right) \cdot \frac{S}{[N_0 B_w + (M-1) \cdot \nu_f \cdot S]} = G_p \cdot \frac{S}{[N_0 B_w + (M-1) \cdot \nu_f \cdot S]} \tag{10.13}$$

where
G_p = processing gain = B_w/R

Next, we express the signal strength, S, in dB as

$$S = P_m + G_m + G_b + G_{dv} + G_{sho} + L_p + M_{fade} + L_{body} + L_{pent} + L_{cable} \tag{10.14}$$

where
G_m = transmit antenna gain of the mobile (dB)
G_b = receive antenna gain of the base station (dB)
G_{dv} = base station antenna diversity (dB)
G_{sho} = soft handoff gain
L_{body} = body loss (dB)
L_{cable} = cable connection loss (dB)
L_p = path loss (dB)
L_{pent} = penetration loss through a vehicle or building (dB)
M_{fade} = log-normal shadow margin (dB)
P_m = transmit power of the mobile (dB)

Solving Equation (10.13) for M, we get

$$M = 1 + G_p \cdot \left[\frac{1}{(E_b/I_t) \cdot \nu_f}\right] - \frac{N_0 \cdot B_w}{S \cdot \nu_f} \tag{10.15}$$

and solving Equation (10.13) for S, we get

$$S = \frac{(E_b/I_t) \cdot N_0}{\dfrac{1}{R} - \dfrac{(M-1)\nu_f(E_b/I_t)}{B_w}} \tag{10.16}$$

If we include an interference factor f from the other cells, we can rewrite Equation (10.13) as

$$\frac{E_b}{I_t} = G_p \cdot \frac{S}{N_0 \cdot B_w + (M-1) \cdot \nu_f \cdot S(1+f)} \tag{10.17}$$

We also include an imperfect power control factor, η_c, and rewrite Equation (10.17) as

$$\frac{E_b}{I_t} = G_p \cdot \frac{S}{B_w \cdot N_0 + (M-1) \cdot \nu_f \cdot (S/\eta_c) \cdot (1+f)} \tag{10.18}$$

Solving Equation (10.18) for M, we get

$$M = 1 + G_p \cdot \left[\frac{\eta_c}{(E_b/I_t) \cdot \nu_f \cdot (1+f)}\right] - \frac{N_0 \cdot B_w \cdot \eta_c}{S \cdot \nu_f \cdot (1+f)} \tag{10.19}$$

Solving Equation (10.18) for S, we get

$$S = \frac{(E_b/I_t) \cdot N_0}{\frac{1}{R} - \frac{(M-1)(\nu_f)(1+f)(E_b/I_t)}{B_w \cdot \eta_c}} \tag{10.20}$$

From Equation (10.19), the maximum value of M is given as

$$M_{max} = 1 + G_p \cdot \left[\frac{\eta_c}{(E_b/I_t) \cdot (\nu_f) \cdot (1+f)}\right] \tag{10.21}$$

M_{max} is called the *pole point* or *asymptotic cell capacity* that is achieved as $S \rightarrow \infty$. For simplification we neglect 1 and rewrite Equation (10.21) as

$$M_{max} \approx G_p \cdot \left[\frac{\eta_c}{(E_b/I_t) \cdot (\nu_f) \cdot (1+f)}\right] \tag{10.22}$$

To gain a further insight into capacity dynamic, we rewrite Equation (10.20) as

$$\frac{S/\eta_c}{N_0 B_w} = \frac{1}{M_{max} \cdot \nu_f \cdot (1+f) \cdot (1-\rho)} \tag{10.23}$$

where

$\rho = M/M_{max}$ = cell loading factor

Equation (10.23) is plotted in Figure 10.1. The required SNR per mobile increases in a nonlinear fashion with ρ. The power per mobile is much larger for a heavily loaded cell than the lightly loaded cell. Practical capacity limits may be set at a point where the slope becomes too steep (such as $\rho = 0.8$). Also, for the smallest practical value of $M_{max} - M$ to be equal to 1, Equation (10.23) indicates that the maximum received power per mobile is close to the cell site noise level.

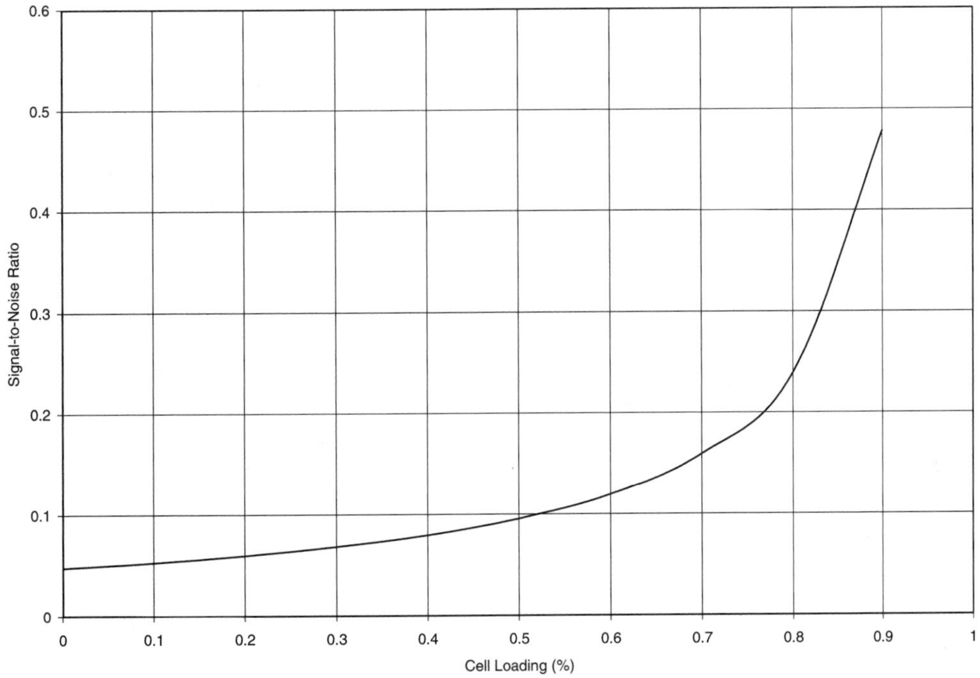

Figure 10.1 Signal-to-Noise Ratio versus Cell Loading

We further adjust Equation (10.23) to reflect the total received power (interference) equal to $P_{rec} = v_f \cdot (1+f) \cdot (S/\eta_c) \cdot M$. The ratio of P_{rec} to cell site noise can be expressed in term of the loading factor ρ. Equation (10.24) shows that the ratio of interference and noise also rises in a nonlinear fashion with ρ (see Figure 10.2).

$$\frac{I_{total}}{N_T} = \frac{P_{rec}}{N_0 B_w} = \frac{\rho}{1-\rho} \quad \textbf{(10.24)}$$

Since total power P_{total} is equal to $P_{rec} + N_0 B_w$, we can rewrite Equation (10.24) as

$$\eta = \frac{P_{total}}{N_T} = \frac{1}{1-\rho} \quad \textbf{(10.25)}$$

We define noise rise, η, as the ratio of the total received power to the thermal noise power. Equation (10.25), the load equation, predicts the amount of the noise rise over the thermal noise due to interference. The noise rise is equal to $-10\log(1-\rho)$. The interference margin in the link budget must be equal to the maximum planned noise rise. Note that when ρ approaches 1, η approaches infinity and the system reaches its pole capacity.

Two levels of simulation are performed in CDMA. The first is the *link-level* simulation, which is performed to simulate the detailed radio channel condition to subchip level so that the curves of FER versus E_b/I_t can be obtained. The second simulation is performed at the *system level* to simulate generic deployment models so that interference conditions including path loss, shadow fading, and soft handoff can be generated. System-level simulation uses link-level simulation results to compute outage for a given deployment model and traffic density. The system-level simulation is used to compute the system capacity for the given outage probability.

The required E_b/I_t value is obtained from link-level simulation and from measurements. It includes the effect of the closed loop power control and soft handoff. The effect of soft handoff is measured as the microdiversity combination gain relative to the single link E_b/I_t result.

The load equation is often used to predict average system capacity without performing system-level simulations. The load equation can be used for predicting cell capacity and planning noise rise in dimensioning process.

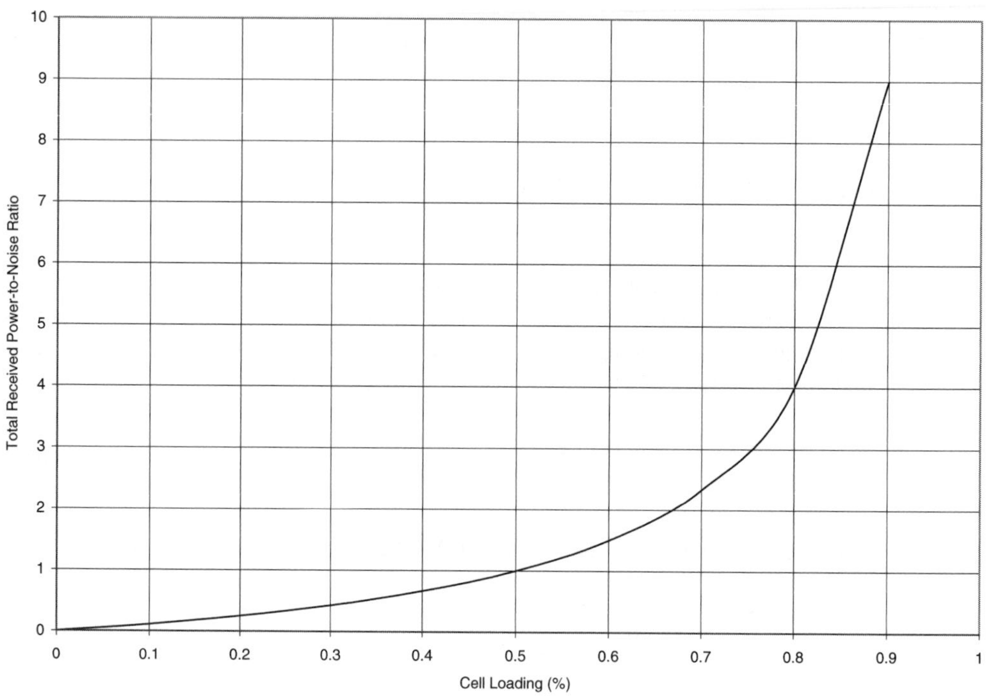

Figure 10.2 Total Received Power-to-Noise Ratio versus Cell Loading

Typical average values of different parameters for voice services in IS-95 are: $\nu_f = 0.5(0.4 \rightarrow 0.6)$, $E_b/I_t = 6 \rightarrow 7$ dB (4 to 5), $f = 0.67(0.56 \rightarrow 1.28)$ based on path loss exponent $\gamma = 4$ and standard deviation of path loss of 6 to 10 dB, and $\eta_c = 0.8(0.7 \rightarrow 0.85)$.

Using the average values, we estimate the cell capacity as

$$M_{max} = \frac{1.23 \times 10^6}{9600 \times 4} \cdot \frac{0.8}{0.5 \times (1.67)} \approx 31 \text{ (with } E_b/I_t = 6 \text{ dB)}$$

and

$$M_{max} = \frac{1.23 \times 10^6}{9600 \times 5} \cdot \frac{0.8}{0.5 \times (1.67)} \approx 25 \text{ (with } E_b/I_t = 7 \text{ dB)}$$

With a three-sector antenna, practical antenna gain of 2.55 can be achieved, and the capacity range per sector will be

$$M_{sector} = 31 \times \frac{2.55}{3} \approx 26 \text{ or } M_{sector} = 25 \times \frac{2.55}{3} \approx 21$$

The capacity range of a sector will be 21 to 26. In practice, the sector loading is often limited to 0.5 to 0.7 of the calculated pole capacities, making the average number of mobiles per sector equal to 13 to 16.

10.7 Cell Loading

The concept of cell loading is used to monitor interference levels and to select the best operating parameters for a particular cell in the system. As discussed earlier, the cell loading, ρ, is M/M_{max}. From Equation (10.19), we can write

$$\rho = \frac{M}{M_{max}} = \frac{M}{M + \dfrac{N_0 B_w}{S \cdot v_f \cdot (1+f)}} = \frac{M \cdot S \cdot v_f \cdot (1+f)}{N_0 B_w + M \cdot S \cdot v_f \cdot (1+f)} \tag{10.26}$$

$$\approx \frac{S \cdot v_f[(1+f)M - 1]}{N_0 B_w + S \cdot v_f[(1+f)M - 1]} = \frac{I_{total}}{N_T + I_{total}}, \qquad M(1+f) \gg 1 \tag{10.27}$$

or

$$\frac{I_{total}}{N_T} = \frac{\rho}{1-\rho} \tag{10.28}$$

The ρ is the convenient way to express the amount of potential capacity being used. Figure 10.2 shows a relationship between (I_{total}/N_T) and ρ.

10.8 Cell Radius

As discussed in Chapter 2, the link budget calculations (forward or reverse link) for a particular cell provide the maximum allowable path loss, $(L_p)_{max}$. Since path loss is proportional to distance, the value of maximum path loss implies a maximum distance for the link. This is the effective radius of the cell or sector in the particular direction. We use the following general relationship for the path loss in dB as a function of distance—given other system parameters such as frequency, antenna heights, cable losses, etc:

$$(L_p)_{max} = L_0 + 10\gamma \cdot \log r \qquad \textbf{(10.29)}$$

where
r = cell radius in miles (or km)
L_0 = path loss intercept, i.e., path loss in dB for reference distance equal to 1 mile (or 1 km)
γ= path loss exponent

Solving Equation (10.29) we get

$$r = 10^{\frac{(L_p)_{max} - L_0}{10\gamma}} \text{ miles (or km)} \qquad \textbf{(10.30)}$$

EXAMPLE 10.1

Calculate the required E_b/I_t using the following parameters:

B_w = 1.23 MHz
R = 9.6 kbps
P_m = 63 mW (18 dBm)
L_c = –2 dB
G_m = 0 dB
L_p = –135 dB
M_{fade} = –8 dB
G_b = 9 dB
F (noise figure) = 5 dB
T = 290 degrees K
k_b = Boltzmann's constant = 1.380662×10^{-23}
ν_f = 0.4
M = 20

Assume all other parameters to be zero.

The received signal power will be

$$S = P_m + L_{cable} + G_m + G_b + L_p + M_{fade}$$

$$S = 18 + (-2) + (0) + (9) + (-135) + (-8) = -118 \text{ dBm}$$

$$N_0 = FTk_b = 3.16228 \times 290 \times 1.380662 \times 10^{-23} = 1.266 \times 10^{-20} \text{ W} = 1.266 \times 10^{-17} \text{ mW}$$

$$\frac{E_b}{I_t} = \left(\frac{B_w}{R}\right)\frac{S}{[N_0 B_w + (M-1)\nu_f S]}$$

$$\frac{E_b}{I_t} = \frac{1.23 \times 10^6}{9.6 \times 10^3} \times \frac{10^{-11.8}}{[1.266 \times 10^{-17} \times 1.23 \times 10^6 + 19 \times 0.4 \times 10^{-11.8}]} \text{ dB}$$

$$\frac{E_b}{I_t} = \frac{128 \times 10^{-11.8}}{10^{-11}[1.5557 + 1.2045]} = \frac{128 \times 10^{-0.8}}{2.7602} = 7.345 = 8.66 \text{ dB}$$

EXAMPLE 10.2

For the IS-95 CDMA system, a chip rate of 1.2288 Mcps is specified for the data rate of 9.6 kbps (i.e., an 8-kbps vocoder). The required E_b/I_t is specified as 7.0 dB. Calculate the pole capacity. What is average number of mobiles that can be supported by a sector of the three-sector cell? Assume interference factor from the neighboring cells is $f = 0.55$, the voice activity factor $v_f = 0.5$, the power control accuracy factor is 0.80, and the gain due to sectorization is 2.55. How much reduction in sector capacity will occur with a 13.0-kbps vocoder provided all other things remain unchanged?

$$M_{max} \approx G_p \cdot \left[\frac{\eta_c}{(E_b/I_t) \cdot (\nu_f) \cdot (1 + f)}\right]$$

$$G_p = \frac{B_w}{R} = \frac{1.23 \times 10^6}{9.6 \times 10^3} = 128, \quad \frac{E_b}{I_t} = 7 \text{ dB} = 5.0$$

$$M_{max} = \frac{128 \times 0.85}{5.0 \times 0.5 \times (1 + 0.55)} = 26.42$$

$$\text{Average subscriber/sector} = \frac{26.42 \times 2.55}{3} = 22.46 \approx 22$$

With a 13.0-kbps vocoder (data rate $R = 14.4$ kbps), the processing gain will be

$$G_p = \frac{1.23 \times 10^6}{14.4 \times 10^3} = 85.4$$

The reduction in sector capacity will be

$$= \frac{128 - 85.4}{128} \times 100 = 33.28\ \%$$

EXAMPLE 10.3

A total of 36 equal-power mobiles share a frequency band through a CDMA system. Each mobile transmits information at 9.6 kbps with a DSSS BPSK-modulated signal. Calculate the minimum chip rate of the PN code needed to maintain a bit error probability of 10^{-3}. Assume that the interference factor from other cells is $f = 0.60$, voice activity factor $v_f = 0.5$, and power control accuracy factor is 0.8.

Bit error probability for BPSK $P_b = Q\left(\sqrt{\frac{2E_b}{I_t}}\right) \approx \frac{e^{-E_b/I_t}}{2\sqrt{\pi(E_b/I_t)}} = 10^{-3}$

Required $\frac{E_b}{I_t} \approx 4.8 = 6.8\ \text{dB}$

$$M = 36 = \frac{G_p}{E_b/I_t} \times \frac{1}{1+f} \times \frac{1}{v_f} \times 0.8$$

$$\frac{G_p}{4.8} \times \frac{1}{1.6} \times \frac{1}{0.5} \times 0.8 = 36$$

$$\therefore G_p = 172.8$$

Chip rate = $172.8 \times 9.6 \times 10^3 = 1.6588\ \text{Mcps}$

EXAMPLE 10.4

Calculate the Erlang capacity of a sector for the three-sector cell using the following parameters: carrier bandwidth $B_w = 1.23$ MHz; Rate Set 2 $R = 14.4$ kbps; required $E_b/I_t = 7$ dB; voice activity factor $v_f = 0.4$; interference due to other cells $f = 0.6$; three-sector antenna gain $\alpha = 2.61$; cell loading factor $\rho = 0.54$; and outage or call blocking probability $P_{out} = 2\%$.

$$G_p = \frac{B_w}{R} = \frac{1.23 \times 10^6}{14.4 \times 10^3} = 85.4$$

$$M_{max} = 1 + \frac{G_p}{(E_b/I_t)_{reqd}} \cdot \frac{1}{v_f} \cdot \frac{1}{1+f} \cdot \alpha = 1 + \frac{85.4}{5.012} \cdot \frac{1}{0.4} \cdot \frac{1}{1+0.6} \cdot 2.61 = 70.4$$

$$M_{max} \text{ per sector} = 70.4/3 = 23.48 \approx 24$$

$$M = 0.54 \times 24 \approx 13$$

From Erlang B table (see Appendix A) at 2 percent blocking probability for $M = 13$ channels, the capacity = 7.4 Erlangs.

10.9 Erlang Capacity of a Single Cell

To calculate the Erlang capacity of a single cell in a CDMA system we assume that the number of active users M can be modeled by Poisson distribution.

$$p_m = \frac{(\lambda/\mu)^M}{M!} \cdot e^{-\lambda/\mu} \tag{10.31}$$

where

λ/μ = offered average traffic load in Erlangs

λ = average arrival rate of users

$1/\mu$ = average time per call

The call service time τ per user is assumed to be exponentially distributed, so that the probability that τ exceeds T is given as

$$p_r(\tau > T) = e^{-\mu T} \qquad T > 0 \tag{10.32}$$

Using these assumptions, it has been shown in reference [1] that the blocking or outage probability p_{out} is

$$p_{out} = e^{-(\lambda v_f)/\mu} \cdot \sum_{K\lfloor \Delta_r' \rfloor}^{\infty} \left(\frac{v_f \lambda}{\mu}\right)^K \cdot \frac{1}{K!} \approx Q\left(\frac{\Delta_r' - (v_f \lambda)/\mu}{\sqrt{(v_f \lambda)/\mu}}\right) \tag{10.33}$$

where

$$\Delta'_r = \frac{G_p(1-\eta)}{(E_b/I_t)_{reqd}}$$

$\frac{1}{\eta}$ = total interference plus thermal noise power-to-thermal noise power ratio

By taking into account the interference from other cells and an imperfect power control, Equation (10.33) can be modified as

$$p_{out} \approx Q\left[\frac{\Delta_r' - v_f(\lambda/\mu)(1+f)e^{(\beta\sigma_p)^2/2}}{\sqrt{v_f \cdot (\lambda/\mu) \cdot (1+f) \cdot e^{(\beta\sigma_p)^2}}}\right] \tag{10.34}$$

where
β = (ln10)/10
σ_p = standard deviation of power control

We may invert the approximate expression for blocking probability, Equation (10.34), by solving a quadratic equation to obtain the explicit formula for normalized average user occupancy, λ/μ, in terms of Erlangs per sector [2, 3] as

$$(\lambda/\mu) \cdot v_f \cdot (1+f) = \Delta_r' \cdot F(B, \sigma_p) \tag{10.35}$$

where

$$B = \frac{[Q^{-1}(p_{out})]^2}{\Delta_r'}; \text{ and}$$

$$F(B, \sigma_p) = \frac{1}{\alpha_c} \cdot \left[1 + \frac{\alpha_c^3 B}{2}\left(1 - \sqrt{1 + \frac{4}{\alpha_c^3 B}}\right)\right] \text{ in which}$$

$$\alpha_c = e^{(\beta\sigma_p)^2/2}; \beta = (\ln 10)/10 = 0.2303$$

EXAMPLE 10.5

Find the Erlang capacity of a CDMA cell assuming the following data:

- Blocking or outage probability = 1%
- Log-normal shadowing margin = 8 dB
- Path loss exponent = 4
- Voice activity factor = 0.4
- Other cell interference factor = 0.55
- Spreading bandwidth = 1.23 MHz
- Data rate = 9.6 kbps
- $(E_b/I_t)_{reqd} = 7$ dB = 5.0
- $1/\eta = 10$
- $\sigma_p = 2$ dB = 1.5849

$$Q^{-1}(0.01) = 2.33\,; \quad \Delta_r' = \frac{G_p}{(E_b/I_t)_{reqd}} \cdot (1-\eta) = \frac{1.23 \times 10^6}{9.6 \times 10^3} \cdot \frac{(1-0.1)}{5} = 23.04$$

$$\alpha_c = e^{(0.2322 \times 1.5849)^2/2} = 1.0701$$

$$B = \frac{(2.33)^2}{23.04} = 0.2356$$

$$F(B, \sigma_p) = \frac{1}{1.0701}\left[1 + \frac{(1.0701)^3 \times 0.2356}{2}\left(1 - \sqrt{1 + \frac{4}{(1.0701)^3 \times 0.2356}}\right)\right] = 0.5494$$

$$\frac{\lambda}{\mu} = \frac{23.04 \times 0.5494}{0.4 \cdot (1 + 0.55)} = 20.42 \text{ Erlangs}$$

Number of users from Erlang B table at 1 percent blocking ~ 30.

10.10 Forward Link Capacity

An important feature of CDMA that contributes to the added capacity on the reverse link is *soft handoff* [4]. In a CDMA network, a mobile can be served by multiple cells simultaneously. However, the same feature puts an additional burden on the forward link. Since multiple cells have to provide service to the same mobile, additional resources are allocated on the forward link. The forward link performance differs vastly from that of the reverse link for the following reasons:

- Access is one-to-many instead of many-to-one
- Synchronization and coherent detections are facilitated by use of a common pilot channel
- Interference is received from a few concentrated large sources (cells) rather than many distributed small ones (mobiles)

To maximize the capacity of the forward link, it is essential to control the power of the cell so as to allocate the power to individual mobiles according to their needs. More power is provided to those mobiles that receive highest interference from the neighboring cells. Mobiles on the boundaries may be in soft handoff, in which case they are receiving signal power from two or more cells. Power control on the forward link is accomplished by measuring the mobile power received from its serving cell and the total received power. In IS-95, the information about these two power values is transmitted to the serving cell.

For the forward link a *figure of merit* is defined for various channels. The figure of merit is the difference between the received (*rec*) and specified (*sp*) E_b/I_t. The link safety margin parameter for each of the channels on the forward link in IS-95 is defined as

$$M_{pilot} = (E_c/I_t)_{rec} - (E_c/I_t)_{sp} > 0 \tag{10.36a}$$

$$M_{traffic} = (E_b/I_t)_{rec} - (E_b/I_t)_{sp} > 0 \tag{10.36b}$$

$$M_{sync} = (E_b/I_t)_{rec} - (E_b/I_t)_{sp} > 0 \tag{10.36c}$$

$$M_{paging} = (E_b/I_t)_{rec} - (E_b/I_t)_{sp} > 0 \tag{10.36d}$$

Note for the pilot channel, E_c/I_t is used instead of E_b/I_t since the pilot channel does not carry any information. Energy per chip, E_c, is used, chip rate being 1.2288 Mcps.

The forward link budget is used to confirm that quantities in Equations (10.36) are positive and there is sufficient margin for the forward link to perform efficiently. Out of M_{pilot}, $M_{traffic}$, M_{sync}, and M_{paging}, the first two are more critical. If these two are positive then the other two are also likely to be positive. For perfect link balance all margin parameters should be zero, particularly M_{pilot} and $M_{traffic}$. The suggested values for the specified E_b/I_t and E_c/I_t parameters are

- Pilot channel: $(E_c/I_t)_{sp} = -13$ dB
- Traffic channel: $(E_c/I_t)_{sp} = 7$ dB
- Sync channel: $(E_b/I_t)_{sp} = 7$ dB
- Paging channel: $(E_b/I_t)_{sp} = 7$ dB

We make the following assumptions for the CDMA forward link budget calculations:

- All mobiles are
  - at the cell edge
  - at least in two-way soft handoff

   - traveling at a medium speed
   - $(E_b/I_t) = 7$ dB for 1% FER
- Power control is working perfectly for all mobiles
- Total forward link traffic channel power is equally divided among all mobiles

The forward link capacity depends on the power that is available for the traffic channels. The power allocation to each overhead channel (i.e., P_{pilot}, P_{sync}, and P_{paging}) is determined from field tests. The suggested power allocations for the forward link channels in IS-95 are

- $P_{pilot} = 15\%$ to 20% $P_{cell\text{-}site}$
- $P_{sync} = 10\%$ of $P_{pilot} = 1.5\%$ to 2% $P_{cell\text{-}site}$
- $P_{paging} = 30\%$ to 40% of $P_{pilot} = 7\%$ $P_{cell\text{-}site}$
- $P_{traffic} = [1 - (0.2 + 0.02 + 0.07)] = 71\%$ to 76.5% $P_{cell\text{-}site}$

Note P_{paging} and $P_{traffic}$ represent the total allocated power for all the paging and traffic channels, respectively, and $P_{cell\text{-}site}$ is the total transmit power of the cell site.

$$P_{(traffic)/(mobile)} = P_{traffic}/(M_{total} \cdot \alpha_{chan}) \tag{10.37}$$

$$M_{total} = M(1 + \xi_{co}) \tag{10.38}$$

where

M = number of active mobiles per sector

ξ_{co} = channel overhead factor for extra traffic channels required for mobiles in different types of soft handoffs (see Table 10.2)

α_{chan} = channel activity factor

$$P_{(paging)/(channel)} = P_{paging}/N_p \tag{10.39}$$

where

N_p = number of paging channels

$P_{(traffic)/(mobile)}$ is a nominal value. Actual power allocated for each mobile can be up to ±4 dB, depending on the forward link power control for each mobile. On the forward link, extra traffic channels are required for the mobiles in various types of soft handoffs. The percentage of the coverage area in a handoff is a design criterion. The extra number of traffic channels in a handoff can be related to the area in the handoff. Table 10.2 provides the suggested values.

Table 10.2 Channel Overhead Factor for Various Types of Soft Handoffs

Type of Handoff	Area in Handoff (%)	ξ_{co}
Soft	25	0.25
Softer	20	0.20
Soft-soft	10	0.20
Soft-softer	10	0.20
		Total $\xi_{co} = 0.85$

10.10.1 Pilot Channel

The mobile continuously measures E_c/I_t of the pilot channel and compares it against threshold values of the handoff parameters, T_ADD and T_DROP (E_c is the energy per chip and I_t is the interference plus noise density measured on the pilot channel). The mobile reports the results of these comparisons to the serving cell. The serving cell decides whether the mobile needs handoff. The E_c/I_t of the pilot channel is needed to determine whether the mobile is within the coverage area of the particular cell. The pilot signal from a cell is transmitted at relatively higher power as compared with those of the other forward link channels (i.e., paging, sync, traffic). In order to set up a call, the mobile must successfully receive the pilot signal. The pilot channel acts as a coherent phase reference for demodulation of other channels on the forward link. Since E_c/I_t effectively determines the coverage area of a cell or sector, it is essential that the E_c/I_t be sufficiently large.

10.10.2 Traffic Channel

Let $(S_1)_m$ = power received by the mth mobile from the cell/sector providing maximum power (i.e., serving cell), and $(S_2)_m \ldots (S_Q)_m$ = power received by the mth mobile from neighboring cells. Thus,

$$(S_1)_m > (S_2)_m \ldots\ldots\ldots > (S_Q)_m \quad > 0 \tag{10.40}$$

We assume that the power received from Q cells or sectors is significant and the power from all other cells is negligible. We assume that all cell sites beyond the second ring around a serving cell contribute negligible received power, so that $Q \leq 18$. The received bit energy-to-interference plus thermal noise for the mth mobile will be [5]

$$\left(\frac{E_b}{I_t}\right)_m \geq \left(\Phi_t \cdot \frac{B_w}{R} \cdot \frac{\omega_m (S_1)_m}{\sum_{j=1}^{Q(S_j)_m} (S_j)_m + N_0 B_w} \right) \tag{10.41}$$

where
Φ_t = fraction of total cell site power assigned to traffic channels
$(1 - \Phi_t)$ = fraction of total cell power assigned to transmission of overhead channels (pilot, sync, and paging channels)
N_0 = thermal noise density
B_w = spreading bandwidth
R = data rate
G_p = processing gain = $B_w / R = R_c / R$
ω_i = fraction of total power allocated to the ith mobile
Note: the weighting factor ω_i is proportional to the total sum of other base station powers, $S_2, S_3, \ldots, S_Q$, relative to the mobile's own base station power S_1.
M = number of users in mobile's own cell or sector

$$\sum_{i=1}^{M} \omega_i \le 1 \tag{10.42}$$

From Equation (10.41) the weighting factor ω_m is given as

$$\omega_m \le \frac{(E_b/I_t)_m}{\Phi_t G_p}\left[1 + \left(\frac{\sum_{i=2}^{Q}(S_j)}{(S_1)}\right)_m + \frac{\sigma_n^2}{(S_1)_m}\right] \tag{10.43}$$

where
σ_n^2 = thermal power

Since $\Phi_t S_1$ is the maximum total power allocated to the cell/sector containing the given mobile and M is the total number of mobiles in the cell/sector, we define the relative received cell power as

$$f_m \equiv 1 + \left(\frac{\sum_{i=2}^{Q} S_j}{S_1}\right)_m \tag{10.44}$$

Next we combine Equations (10.43) and (10.44) to get

$$\sum_{i=1}^{M} f_i \le \left(\frac{G_p \Phi_t}{E_b/I_t} - \sum_{i=1}^{M} \frac{\sigma_n^2}{(S_1)_i}\right) = \Delta f \tag{10.45}$$

In general, background noise is well below the total largest received cell site signal power (the second term in Equation (10.45)) and is typically neglected relative to the first term. The capacity can be estimated from the outage or blocking probability, defined as

$$p_{out} = p_r[BER > (BER)_{sp}] \tag{10.46}$$

where

$(BER)_{sp}$ = specified bit error rate for which E_b/I_t is equal to $(E_b/I_t)_{sp}$

We compute Δ_f' for $(E_b/I_t)_{sp}$ and express the outage or blocking probability as

$$p_{out} = p_r\left[\sum_{i=1}^{M} f_i > \Delta_f'\right] \tag{10.47}$$

The distribution of $\sum_{i=1}^{M} f_i$ cannot be expressed in a closed form. The simulation results for the blocking probability for the forward link of IS-95 are shown in Figure 10.3 for $G_p = 128$, with 20 percent of the transmitted power in the cell/sector to the pilot channel, and the required $E_b/I_t = 5$ dB for the traffic channel to ensure $BER \leq 10^{-3}$ [4]. The reduction of 2 dB relative to the reverse link is justified by the coherent reception using the pilot as reference, as compared with the noncoherent detection in the reverse link. In the simulation, powers from base stations were represented as the product of the fourth order of distance and a log-normally distributed attenuation. With these parameters, the forward link can support the bit error rate of 10^{-3} for more than 99 percent of the time for 38 mobiles per sector or 114 mobiles per cell.

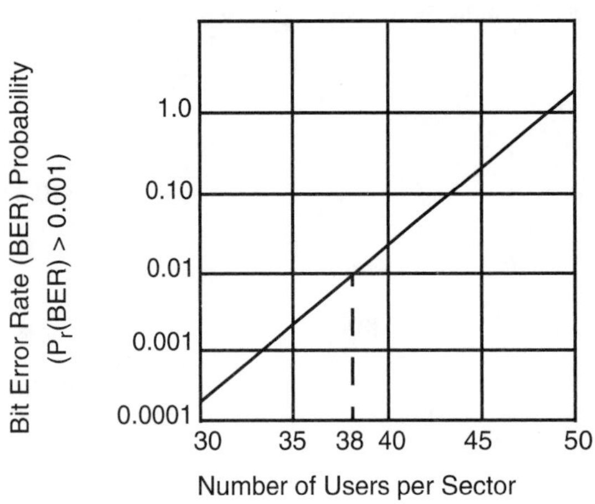

Figure 10.3 Forward Link Capacity of a Cellular CDMA System

EXAMPLE 10.6

Use the following parameters of an IS-95 CDMA system to estimate the sector capacity based on reverse and forward link performance:

- Spreading bandwidth (B_w) = 1.23 MHz
- Data Rate (R) = 9.6 kbps
- $1/\eta = 10$
- Outage or blocking probability (p_{out}) = 1%
- $(E_b/I_t)_{sp}$ for the reverse link = 7 dB = 5.0
- $(E_b/I_t)_{sp}$ for the forward link = 5 dB
- Standard deviation of power control $(\sigma_p) = 2.5$ dB = 1.7783
- Voice activity factor (v_f) = 0.5
- Interference from other cells (f) = 0.65
- Path loss exponent $(\gamma) = 4$
- Standard deviation for shadow margin = 8 dB
- Channel overhead factor for soft handoffs $(\xi_{co}) = 0.85$

Reverse Link:

$$Q^{-1}(0.01) = 2.33$$

$$\Delta_r' = \frac{1.23 \times 10^6}{9.6 \times 10^3} \times \frac{(1-0.1)}{5} = 23.04$$

$$B = (2.33)^2/23.04 = 0.2356$$

$$\alpha_c = e^{(0.2303 \times 1.7783)^2/2} = 1.8075$$

$$F(B, \sigma_p) = \frac{1}{1.0875}\left[1 + \left(\frac{(1.0875)^3 \times 0.2356}{2}\right)\left(1 - \sqrt{1 + \frac{4}{(1.0875)^3 \times 0.2356}}\right)\right] = 0.5339$$

For perfect power control: $\alpha_c = e^0 = 1$

$$F(B, \sigma_p) = 1 + \frac{0.2356}{2}\left(1 - \sqrt{1 + \frac{4}{0.2356}}\right) = 0.6183$$

Efficiency of power control: $\eta_c = \dfrac{0.5339}{0.6183} = 0.8635$

$$\frac{\lambda}{\mu} = \frac{23.04 \times 0.5339}{0.5 \times (1 + 0.65)} = 14.91 \text{ Erlangs}$$

$$\left(\frac{\lambda}{\mu}\right)_{perfect} = \frac{14.91}{0.8635} = 17.27 \text{ Erlangs}$$

From Erlang B table at 1 percent blocking with 14.91 Erlangs, ≈ 23 mobiles can be supported per sector, whereas with perfect power control we can support ≈ 27 mobiles. We lose about 15 percent of the sector capacity to imperfect power control. With a loading factor of about 70 percent, the reverse link capacity will be about 16 mobiles per sector.

Forward Link:

From Figure 10.3, the forward link can support 38 mobiles per sector with perfect power control. If we assume the same accuracy of power control as on the reverse link, the sector capacity will be reduced to

$$\text{No. of mobiles per sector} \approx 38 \times 0.8635 = 33$$

Next, we consider the effect of soft handoffs on forward link capacity. The channel overhead factor is $\xi_{co} = 0.85$. The sector capacity based on the performance of the forward link will be

$$\text{Sector capacity} = \frac{33}{1.85} \approx 18$$

In this example, the sector capacity is controlled by forward link performance. In practice, a loading factor of about 70 percent is usually suggested, which gives the sector a capacity about 13 mobiles. It should be noted that the sector capacity of the forward link is significantly affected by total percentage and distribution of soft handoffs.

10.11 CDMA Cell Size

For CDMA systems with additive white Gaussian noise (AWGN) channels, the SNR is an accurate measure of the performance [6]. The SNR, number of users in the cell/sector M, and maximum mobile transmit power establish the size of the cell on the reverse link.

In a noise (coverage) limited system, cell size rather than capacity is the main concern. The receiver sensitivity is used to calculate the size of the cell. Fading margin, propagation path loss, and minimum received signal level to achieve an acceptable performance are the key factors in establishing the cell size. The propagation path losses are determined from either statistical propagation models or from actual measurements.

In an interference (capacity) limited system, the size of the cell is determined mainly by the level of the interference from other users. On the reverse link, the maximum path loss that the cell can tolerate is determined by the carrier signal-to-interference ratio (SIR),[1] the number of simultaneous users, and the maximum power that the mobile can transmit. The maximum cell size should be such that the mobile can close the link.

In the noise-limited system, the minimum SIR can be translated to a minimum signal strength requirement on the forward link. The forward link cell boundary is defined by pilot E_c/I_t. The maximum cell size should be such that within the coverage area, the received pilot E_c/I_t should be above a predefined threshold.

10.11.1 Reverse Link Cell Size

At the cell site, the SIR per antenna can be given by

$$SIR(r) = \frac{p_m \cdot L_p(r) \cdot G'_b \cdot G'_m}{[N_0 B_w]_{cell} + [M/f_r - 1](v_f \cdot p_m) L_p(r) G'_b G'_m} \tag{10.48}$$

where
p_m = mobile's power amplifier output
G'_b = cell antenna gain including cable losses
G'_m = mobile antenna gain including cable losses
$L_p(r)$ = reverse link path loss
M = number of users in a cell/sector
v_f = voice activity factor
f_r = frequency reuse factor
$[N_0 B_w]_{cell}$ = thermal noise of the cell

The quantity $[N_0 B_w]_{cell} + [M/f_r - 1](v_f \cdot p_m) L_p(r) G'_b G'_m$ depends only on the system loading. It was shown [7] that

$$1 + \frac{[M/f_r - 1]([v_f \cdot p_m] \cdot L_p(r) G'_b G'_m)}{N_0 B_w} = \frac{1}{1 - \rho} \tag{10.49}$$

where
ρ = system loading factor

The maximum path losses that the mobile can tolerate are given by

$$L_p(r) = \frac{(SIR)[N_0 B_w]_{cell}\left(\frac{1}{1 - \rho}\right)}{p_m G'_b G'_m} \tag{10.50}$$

1. SIR includes both interference and thermal noise.

We express Equation (10.50) in dB to get

$$L_p(r) = (SIR)_{min} + [N_0B_w]_{cell} - p_m - G'_b - G'_m - 10\log(1-\rho) \tag{10.51}$$

The maximum transmission loss will be

$$T(r) = L_p(r) + G'_b + G'_m \tag{10.52}$$

10.11.2 Forward Link Cell Size

On the forward link, the parameter that determines the cell size is the pilot E_c/I_t, which is given as

$$\frac{E_c}{I_t} = \frac{\phi_p \cdot p_c \cdot L_p(r) \cdot G'_b \cdot G'_m}{[N_0B_w]_{mob} + I_{oc}(r)B_w + I_0(r)B_w} = \frac{\phi_p \cdot p_c \cdot T(r)}{[N_0B_w]_{mob} + I_0(r)B_w \cdot (1+\xi)}$$

$$= \frac{\phi_p \cdot p_c \cdot T(r)}{[N_0B_w]_{mob} + p_cT(r)(1+\xi)} \tag{10.53}$$

Solving for $T(r)$ we get

$$T(r) = \frac{(E_c/I_t)(N_0B_w)_{mob}}{p_c[\phi_p - (E_c/I_t)(1+\xi)]} \tag{10.54}$$

We express Equation (10.54) in dB to get

$$T(r) = (E_c/I_t)_{min} + (N_0B_w)_{mob} - p_c - 10\log[\phi_p - (10^{(E_c/I_t)_{min}/10})(1 + 10^{\xi/10})] \tag{10.55}$$

$$L_p(r) = (E_c/I_t)_{min} + (N_0B_w)_{mob} - p_c - 10\log[\phi_p - (10^{(E_c/I_t)_{min}/10})(1 + 10^{\xi/10})] - G'_b - G'_m \tag{10.56}$$

where
ϕ_p = portion of the cell power allocated to the pilot
p_c = cell output
G'_b = cell antenna gain including cable losses
G'_m = mobile antenna gain including cable losses
$I_{oc}(r)$ = other cell interference power spectral density
$I_o(r)$ = serving cell interference power spectral density
$\xi = I_{oc}/I_o$
$[N_0B_w]_{mob}$ = thermal noise at the mobile
$(E_c/I_t)_{min}$ = minimum required value for pilot

EXAMPLE 10.7

Calculate the transmission loss versus cell loading for the 200-mW mobile unit. Assume $(E_b/I_t)_{min} = 7$ dB, processing gain at the cell site = 21 dB, and the cell noise figure = 5 dB.

$$(SIR)_{min} + Processing\ Gain = (E_b/I_t)_{min}$$

$$(SIR)_{min} = 7 - 21 = -14 \text{ dB}$$

$$(N_0B_w)_{cell} = [3.1622 \times 290 \times 1.38066 \times 10^{-23} \times 1.2288 \times 10^6] \times 10^3 \text{ mW} = -108 \text{ dBm}$$

$$T(r) = (SIR)_{min} + (N_0B_w)_{cell} - p_m - 10\log(1-\rho) = -14 - 108 - 23 + 10\log(1-\rho)$$

$$= -145 - 10\log(1-\rho) \text{ dB}$$

Figure 10.4 shows a plot of $T(r)$ versus ρ.

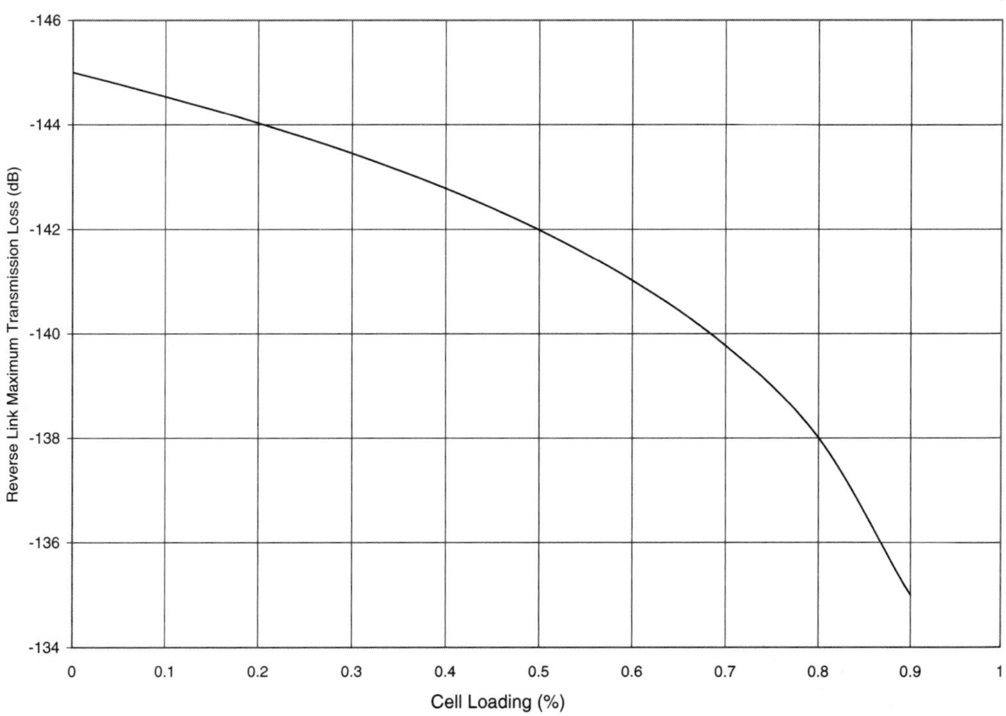

Figure 10.4 *T*(*r*) versus ρ

EXAMPLE 10.8

Plot maximum transmission loss (dB) versus percentage of power allocated to the pilot channel. Assume $(E_c/I_t)_{min} = -15$ dB, mobile noise figure = 8 dB, $I_{0c}/I_0 \sim 2.5$ dB, and cell site output = 44 dBm.

$$T(r) = (E_c/I_t)_{min} - P_c + (N_0 B_w)_{mob} - 10\log[\phi_p - (10^{(E_c/I_t)_{min}/10} \cdot (1 + 10^{I_{0c}/I_0}))]$$

$$T(r) = -15 - 44 - 105 - 10\log[\phi_p - 0.03162 \times (1 + 1.7783)]$$

$$T(r) = -164 - 10\log[\phi_p - 0.08785] \text{ dB}$$

Figure 10.5 shows a plot of $T(r)$ versus ϕ_p.

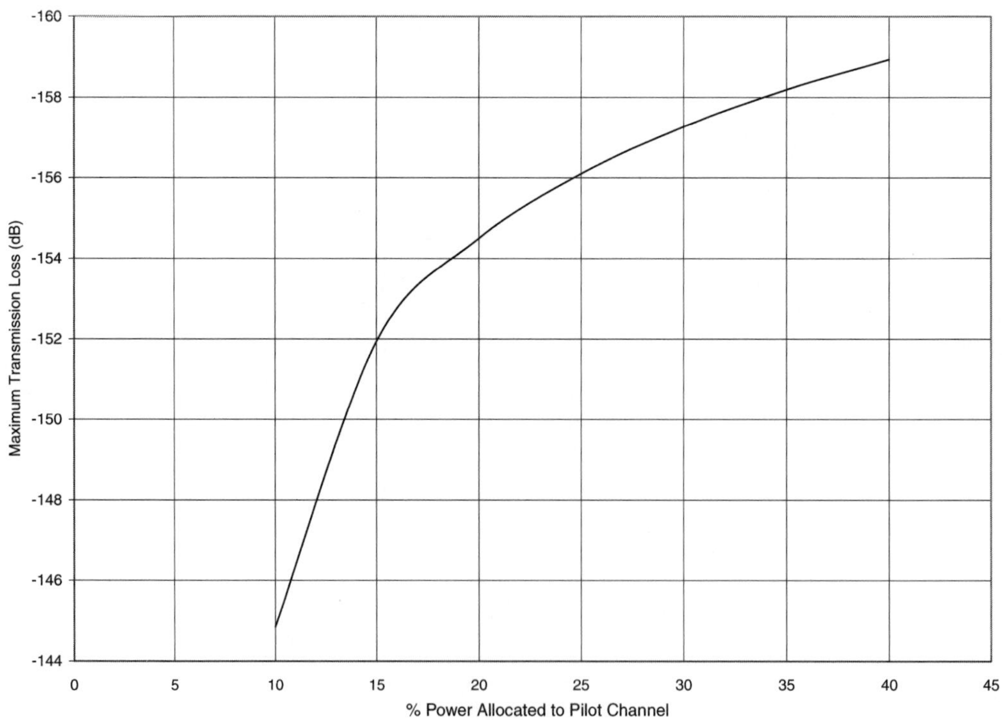

Figure 10.5 $T(r)$ versus ϕ_p

EXAMPLE 10.9

Find the cell radius corresponding to the forward link transmission using the following data:

- $p_c = 20$ W $= 43$ dBm
- $\phi_p = 0.15$
- $(E_c/I_t)_{min} = -15$ dB
- $(NF)_m = 8$ dB
- $G'_b = 6$ dBi
- $G'_m = 0.0$ dBi
- $\xi = I_{0c}/I_0 = 2.5$ dB (this reflects the situation of a mobile near the cell boundary)
- One kilometer intercept = 127
- $\gamma = 3.64$
- Bandwidth = 1.25 MHz

$$(L_p)_{max} = 43 + 10\log[0.15 - (10^{-1.5} \cdot (1 + 10^{0.25}))] - (-15) - (-105) = 151 \text{ dB}$$

$$T(r) = (L_p)_{max} + G_m' + G'_b = 151 + 0 + 6 = 157 \text{ dB}$$

$$\therefore 157 = 127 + 10 \times 3.64 \log r$$

$$\log r = 0.8242 ; \; r = 6.67 \text{ km}$$

10.12 Forward and Reverse Link Balance

A more powerful forward link results in extra interference to mobiles in other cells. On the other hand, a more powerful reverse link sacrifices capacity. It is desirable then to design the system so that two boundaries coincide. Balanced links minimize interference and eliminate associated handoff problems. The boundary of the cell on the reverse link is determined by cell loading, and the boundary on the forward link is obtained by the minimum pilot (E_c/I_t). In order to maintain two close boundaries, we need to equate the path loss on both links. We define the balance factor B_f as

$$B_f = [(SIR)_{min} - (E_c/I_t)_{min}] + [N_0F_{cell} - N_0F_{mob}] \cdot B_w + [p_c - p_m]$$
$$+ 10\log\left(\frac{\phi_p - 10^{(E_c/I_t)_{min}/10} \cdot (1 + 10^{(I_{0c}/I_0)/10})}{1-\rho}\right) \quad (10.57)$$

Based on B_f, the system designer can decide which link is the limiting factor. If $B_f < 0$, the system is forward link limited; if $B_f = 0$, links are balanced; and if $B_f > 0$, the system is reverse link limited. A more realistic rule is

$B_f < -\delta$ the system is forward link limited
$|B_f| \leq \delta$ two links are balanced
$B_f > \delta$ the system is reverse link limited

δ is the parameter that takes into account the tolerance in all factors involved in calculation of B_f. A good system design should ensure that the two links are balanced. This makes handoff transition more smooth and reduces the amount of interference.

EXAMPLE 10.10

Using the following data, calculate the power allocated to the pilot channel. What is the allocated power for the pilot channel to balance forward and reverse links?

- Maximum mobile power (p_m) = 200mW (23 dBm)
- Maximum cell site power (p_c) = 10 W (40 dBm)
- Voice activity factor (v_f) = 0.4
- Processing gain at the base station = 21 dB
- Number of users per cell (M) = 20
- Cell loading factor (ρ) = 0.5
- Cell noise figure = 5 dB
- Mobile noise figure = 8 dB
- Cell site antenna gain including cable loss (G'_b) = 6 dB
- Mobile antenna gain including cable loss (G'_m) = 0.0 dB
- I_{0c}/I_0 = 2.5 dB
- $(E_c/I_t)_{min}$ = –15 dB
- $(E_b/I_t)_{min}$ = 7 dB

$$(SIR)_{min} + processing\ gain = (E_b/I_t)_{min}$$

$$(SIR)_{min} = -21 + 7 = -14 \text{ dB}$$

Using Equation (10.51), the path loss (cell site) on the reverse link is

$$L_p(r) = -14 - 108 - 23 - 6 - 10\log(1 - 0.5) = -148 \text{ dB}$$

$$T(r) = -142 \text{ dB}$$

From Equation (10.54)

$$\phi_p = \frac{(E_c/I_t)_{min}[(N_0B_w)_m + p_c \cdot T(r)\{1 + 10^{(I_{0c}/I_0)/10}\}]}{p_c \cdot T(r)}$$

$p_c = 10^4$

$T(r) = 10^{-14.2}$

$(N_0B_w)_m = -105 \text{ dB} = 10^{-10.5}$

$(E_c/I_t)_{\min} = -15 \text{ dB} = 10^{-1.5} = 0.03163$

$$\phi_p = \frac{0.03163 \cdot [10^{-10.5} + 10^4 \cdot 10^{-14.2} \cdot (1 + 1.7783)]}{10^4 \cdot 10^{-14.2}} = 0.1078 = 10.78\%$$

For the balanced condition we use Equation (10.57) with $B_f = 0$ and solve for ϕ_p.

$$0 = [-14 - (-15)] + [-108 - (-105)] + [40 - 23] + \left[10\log\left(\frac{\phi_p - 10^{-1.5}\{1 + 1.7783\}}{0.5}\right)\right]$$

$$\therefore \phi_p = 0.1037 = 10.37\ \%$$

EXAMPLE 10.11

Consider a minicell in which the maximum transmission loss is equal to –112 dB. The other data include

- $[N_0B_w]_{mob} = -105$ dB
- $[N_0B_w]_{cell} = -108$ dB
- $[E_b/I_t]_{min} = 7$ dB
- $[E_c/I_t]_{min} = -15$ dB
- Cell loading = 50%
- $I_{0c}/I_0 = 2.5$ dB

Find the output of the mobile and cell site.

From Equation (10.51)
$-112 = -14 - 108 - p_m - 10\log(1 - 0.5)$
$p_m = -7$ dBm (= 0.2 mW)

For the minicell, the pilot strength can be determined as

$$\left(\frac{E_c}{I_t}\right)_{min} = \frac{\phi_p}{1 + I_{0c}/I_0}$$

$$\therefore \phi_p = \left(\frac{E_c}{I_t}\right)_{min}(1 + I_{0c}/I_0) = 10^{-1.5} \cdot (1 + 10^{0.25}) = 0.0879 \sim 9\%$$

From Equation (10.55)

$$T(r) = \left(\frac{E_c}{I_t}\right)_{min} - p_c + [N_0 B_w]_{mob} - 10\log[\phi_p - 10^{(E_c/I_t)_{min}/10} \cdot (1 + 10^{(I_{0c}/I_0)/10})]$$

$$\therefore -112 = -15 - p_c + (-105) - 10\log[0.09 - 10^{-1.5} \cdot (1 + 10^{0.25})]$$

$$p_c = -15 + 112 - 105 + 26.7 = 18.7 \text{ dBm } (\sim 74 \text{ mW})$$

10.13 Forward Link Budget

The forward link budget is calculated to confirm that the link safety margin parameters (Equation10.58) are positive and that there is sufficient margin for the forward link to work efficiently. M_{pilot} and $M_{traffic}$ are more critical. If both of these are positive, then M_{sync} and M_{paging} are also generally positive.

$$M_{pilot} = \left(\frac{E_c}{I_t}\right)_{rec} - \left(\frac{E_c}{I_t}\right)_{sp} \tag{10.58a}$$

$$M_{traffic} = \left(\frac{E_b}{I_t}\right)_{rec} - \left(\frac{E_b}{I_t}\right)_{sp} \tag{10.58b}$$

$$M_{sync} = \left(\frac{E_b}{I_t}\right)_{rec} - \left(\frac{E_b}{I_t}\right)_{sp} \tag{10.58c}$$

$$M_{paging} = \left(\frac{E_b}{I_t}\right)_{rec} - \left(\frac{E_b}{I_t}\right)_{sp} \tag{10.58d}$$

where *rec* means received value and *sp* means required value.

We use the following procedure to determine link safety margins:

The total cell site power P_{total} will be

$$P_{total} = 10\log[10^{0.1P_{traffic}} + 10^{0.1P_{pilot}} + 10^{0.1P_{sync}} + 10^{0.1P_{paging}}] \text{ (dBm)} \quad \textbf{(10.59)}$$

where

P_{total} = total cell site effective radiated power (ERP) (dBm)

P_{sync} = ERP of sync channel (dBm)

P_{pilot} = ERP of pilot channel (dBm)

P_{paging} = ERP of paging channel (dBm)

$P_{traffic}$ = ERP of all traffic channels (dBm)

$$P_{(traffic)/(user)} = \frac{P_{traffic}}{(M_{total} \cdot \alpha_{chan})} = P_{traffic} - 10\log\alpha_{chan} - 10\log M_{total} \text{ (dBm)} \quad \textbf{(10.60)}$$

where

$P_{traffic/user}$ = ERP of a traffic channel (dBm)

α_{chan} = channel activity factor

$M_{total} = M(1 + \eta_{co})$

η_{co} = traffic channel overhead percentage due to soft handoff

The received power at the mobile on each channel from the serving cell site will be

$$P_{r,total} = P_{total} + GL \quad \textbf{(10.61)}$$

$$P_{r,pilot} = P_{pilot} + GL \quad \textbf{(10.62)}$$

$$P_{r,(traffic)/(user)} = P_{(traffic)/(user)} + GL \quad \textbf{(10.63)}$$

$$P_{r,sync} = P_{sync} + GL \quad \textbf{(10.64)}$$

$$P_{r,paging} = P_{paging} + GL \quad \textbf{(10.65)}$$

where

$GL = G_m + L_{cable} + L_{body} + L_{pent} + M_{fade} + L_p + G_b$

in which

L_p = average propagation path loss between cell site and mobile (dB)

L_{pent} = penetration loss (dB)

L_{body} = body/orientation loss (dB)
L_{cable} = cell site feeder loss (dB)
M_{fade} = margin for lognormal shadowing (dB)
G_m = mobile antenna gain (dB)
G_b = cell site antenna gain (dB)

In-cell interference is caused by other users radiated from the same cell and is given as

$$I_{sc-ch} = 10\log(10^{0.1P_{r,total}} - 10^{0.1P_{r,ch}}) - 10\log B_w \text{ (dBm/Hz)} \quad \textbf{(10.66)}$$

where
ch is the pilot, paging, sync, or traffic/user
B_w = bandwidth

Out-of-cell interference is caused by users in other cells and is given as

$$I_{0c-ch} = I_{sc-ch} + 10\log(1/f_r - 1) \text{ dBm/Hz} \quad \textbf{(10.67)}$$

where
f_r = reuse factor

Total interference will be

$$I_{ch} = 10\log(10^{0.1I_{sc-ch}} + 10^{0.1I_{0c-ch}}) \text{ dBm/Hz} \quad \textbf{(10.68)}$$

The thermal noise density will be

$$N_0 = 10\log(290 \times 1.38 \times 10^{-23}) + N_f + 30 \text{ dBm/Hz} \quad \textbf{(10.69)}$$

where
N_f = noise figure for the mobile

The energy per bit for a channel will be:

$$E_{bch} = P_{r,ch} - 10\log R_{ch} \quad \textbf{(10.70)}$$

where
R_{ch} = data rate for the channel

The E_b/I_t for a channel can be calculated as

$$\frac{E_{bch}}{N_0 + I_{ch}} = P_{r,ch} - 10\log R_{ch} - 10\log(10^{0.1N_0} + 10^{0.1I_{ch}}) \text{ (dB)} \tag{10.71}$$

Using Equation (10.70) we can write

$$\left(\frac{E_b}{I_t}\right)_{rec,pilot} = P_{r,pilot} - 10\log B_w - 10\log(10^{0.1N_0} + 10^{0.1I_{pilot}}) \text{ (dB)} \tag{10.72}$$

$$\left(\frac{E_b}{I_t}\right)_{rec,paging} = P_{r,paging} - 10\log R_{paging} - 10\log(10^{0.1N_0} + 10^{0.1I_{paging}}) \text{ (dB)} \tag{10.73}$$

$$\left(\frac{E_b}{I_t}\right)_{rec,sync} = P_{r,sync} - 10\log R_{sync} - 10\log(10^{0.1N_0} + 10^{0.1I_{sync}}) \text{ (dB)} \tag{10.74}$$

$$\left(\frac{E_b}{I_t}\right)_{rec,traffic} = P_{r,(traffic)/(user)} - 10\log R_{traffic} - 10\log(10^{0.1N_0} + 10^{0.1I_{traffic}}) \text{ (dB)} \tag{10.75}$$

EXAMPLE 10.12

Use the following data to calculate link safety margin parameters for the forward link channels of a CDMA system:

- Pilot channel ERP (P_{pilot}) = 33.8 dBm
- Sync channel ERP (P_{sync}) = 23.8 dBm
- Paging channel ERP (P_{paging}) = 29.5 dBm
- Traffic channels ERP ($P_{traffic}$) = 41.0 dBm
- Number of users per sector on the reverse link = 13
- Channel overhead due to soft handoff = 0.85
- Path loss between cell site and mobile (L_p) = –130.2 dB
- Penetration loss (L_{pent}) = –15 dB
- Body/orientation loss (L_{body}) = –2 dB
- Fade margin (M_{fade}) = –10.3 dB
- Mobile antenna gain (G_m) = 2 dB
- Cell site antenna gain (G_b) = 13 dB
- Cable losses (L_{cable}) = –1.5 dB
- Channel activity factor (α_{chan}) = 0.42
- Bandwidth (B_w) = 1.2288 MHz

- Traffic channel rate = 9,600 bps
- Sync channel rate = 1,200 bps
- Paging channel rate = 4,800 bps
- Cell reuse factor (f_r) = 0.65

$$P_{total} = 10\log[10^{4.1} + 10^{3.38} + 10^{2.95} + 10^{2.38}] = 42.0734 \text{ dBm}$$

$$M_{total} = 13\ (1 + 0.85) = 24$$

$$P_{(traffic)/(user)} = 41.0 - 10\log(0.42) - 10\log 24 = 30.9654 \text{ dBm}$$

$$GL = 2 + 13 - 1.5 - 2.0 - 15 - 10.3 - 130.2 = -144 \text{ dB}$$

$$P_{r,\,total} = 42.0734 - 144 = -101.9266 \text{ dBm}$$

$$P_{r,\,pilot} = 33.8 - 144 = -110.2 \text{ dBm}$$

$$P_{r,\,sync} = 23.8 - 144 = -120.2 \text{ dBm}$$

$$P_{r,\,paging} = 29.5 - 144 = -114.5 \text{ dBm}$$

$$P_{r,\,(traffic)/(user)} = 30.9654 - 144 = -113.0346 \text{ dBm}$$

$$I_{sc-pilot} = 10\log(10^{-10.19266} - 10^{-11.02}) - 10\log(1.2288 \times 10^{6}) = -163.5211 \text{ dBm/Hz}$$

$$I_{0c-pilot} = -163.5211 + 10\log[1/0.65 - 1] = -166.2096 \text{ dBm/Hz}$$

$$I_{pilot} = 10\log[10^{-16.35211} + 10^{-16.62096}] = -161.65 \text{ dBm/Hz}$$

$$I_{sc-sync} = 10\log(10^{-10.19266} - 10^{-12.02}) - 60.9 = -162.8865 \text{ dBm/Hz}$$

$$I_{0c-sync} = -162.8865 - 2.6885 = -165.575 \text{ dBm/Hz}$$

$$I_{sync} = 10\log[10^{-16.28865} + 10^{-16.5575}] = -161.0157 \text{ dBm/Hz}$$

$$I_{sc-paging} = 10\log(10^{-10.19266} - 10^{-11.45}) - 60.9 = -163.0736 \text{ dBm/Hz}$$

$$I_{0c-paging} = -163.0736 - 2.6885 = -165.7621 \text{ dBm/Hz}$$

$$I_{paging} = 10\log[10^{-16.30736} + 10^{-16.57621}] = -161.2030 \text{ dBm/Hz}$$

$$I_{sc-(traffic)/(user)} = 10\log(10^{-10.19266} - 10^{-11.30346}) - 60.9 = -163.1769 \text{ dBm/Hz}$$

$$I_{0c-(traffic)/(user)} = -163.1769 - 2.6885 = -165.8654 \text{ dBm/Hz}$$

$$I_{(traffic)/(user)} = 10\log[10^{-16.31769} + 10^{-16.58654}] = -161.306 \text{ dBm/Hz}$$

$$N_0 = 10\log[290 \times 1.38 \times 10^{-23}] + 30 + 8 = -165.9772 \text{ dBm/Hz}$$

$$\left(\frac{E_c}{I_t}\right)_{rec,\,pilot} = -110.2 - 10\log 1.2288 \times 10^6 - 10\log[10^{-16.597} + 10^{-16.165}] = -10.82 \text{ dB}$$

$$\left(\frac{E_b}{I_t}\right)_{rec,\,sync} = -120.2 - 10\log(1200) - 10\log[10^{-16.597} + 10^{-16.10157}] = 8.83 \text{ dB}$$

$$\left(\frac{E_b}{I_t}\right)_{rec,\,paging} = -114.5 - 10\log(4800) - 10\log[10^{-16.597} + 10^{-16.12}] = 8.64 \text{ dB}$$

$$\left(\frac{E_b}{I_t}\right)_{rec,\,traffic} = -113.0346 - 10\log[9600] - 10\log[10^{-16.597} + 10^{-16.1306}] = 7.18 \text{ dB}$$

$$M_{pilot} = -10.82 - (-13) = 2.18$$

$$M_{traffic} = 7.18 - 7.0 = 0.18$$

$$M_{paging} = 8.64 - 7.0 = 1.64$$

$$M_{sync} = 8.83 - 7.0 = 1.83$$

Since all the link safety parameters are positive, the power allocations on forward link channels are satisfactory.

Note that link budget calculations can be easily performed using Microsoft Excel or any other spreadsheet program.

10.14 Summary

In this chapter, we developed necessary equations to calculate the reverse and forward link capacity of an IS-95 CDMA system. We found that the maximum number of mobiles that can be supported on the forward link and the reverse link of a CDMA system is different. Reverse link capacity improves with soft handoffs, however soft handoffs affect the capacity of the forward link.

In a noise-limited system, cell size rather than capacity is the main concern. The receiver sensitivity is used to calculate the size of the cell. In the interference-limited system the size of the cell is determined mainly by the level of the interference from other users.

We provided several numerical examples to illustrate the procedure and demonstrate the importance of several system parameters in capacity calculations. We concluded the chapter by presenting link safety margin parameters for the forward link channels.

10.15 Problems

1. A chip rate of 1.2288 Mcps is used for IS-95 Rate Set 2 (14.4 kbps with 13-kbps vocoder). The required E_b/I_t is 7 dB. Calculate the average number of mobiles supported by a sector of the three-sector cell and sector Erlang capacity at 2 percent blocking. Assume interference from other cells $f = 0.6$, cell loading = 0.6, voice activity factor = 0.4, power control accuracy factor = 0.90, and gain due to sectorization = 2.61.
2. Calculate the pole capacity of the IS-95 CDMA system with a chip rate of 1.2288 Mcps and Rate Set 2 (14.4 kbps). Assume interference factor due to neighboring cells $f = 0.67$, voice activity factor $v_f = 0.6$, power control accuracy factor = 0.80, and gain due to three-sector antenna = 2.55.
3. A total of 20 equal-power mobiles share a frequency band through a CDMA system. Each mobile transmits data at 16 kbps with DSSS BPSK-modulated signal. Calculate the minimum chip rate of the PN sequence needed to maintain a bit error probability of 10^{-6}. Assume the interference factor due to other cells $f = 0.6$, power control accuracy factor = 0.8, and gain due to three-sector antenna = 2.55.
4. Calculate the Erlang capacity of a sector for the three-sector cell and number of users per sector using the following parameters: carrier bandwidth = 1.23 MHz; Rate Set 1 $R = 9.6$ kbps; $(E_b/I_t)_{min} = 7$ dB; voice activity factor $v_f = 0.5$; interference due to other cells $f = 0.67$; cell loading factor $\rho = 0.5$; call blocking probability $P_{out} = 1\%$; three-sector antenna gain $\alpha = 2.55$; $1/\eta = 10$; standard deviation of power control $\sigma_p = 2$ dB.
5. Calculate total transmission loss in dB for the forward link of a CDMA system using the following data:
   - cell output power = 40 dBm
   - allocated power for pilot channel = 15% of cell output
   - mobile noise figure = 8 dB
   - $(E_c/I_t)_{min} = -13$ dB
   - $I_{oc}/I_o = 2.5$ dB
6. Calculate total transmission loss in dB for the reverse link of a CDMA system using the following data:
   - mobile output = 200 mW (23 dBM)
   - $(E_b/I_t) = 7$ dB
   - cell noise figure = 5 dB
   - cell loading $\rho = 60\%$
7. Using the data given in Problems 5 and 6, find the allocated power of the pilot channel for balancing the forward and reverse links.

8. Find the cell radius corresponding to the forward link transmission using the following data:
   - $p_c = 16$ W = 42 dBm
   - $\phi = 0.15$
   - $(E_c/I_t)_{reqd} = -15$ dB
   - $G'_b = 6$ dBi
   - $G'_m = 0.0$
   - $\zeta = 2.5$ dB
   - $B_w = 1.25$ MHz
   - $(NF)_{mobile} = 8$ dB
   - One kilometer intercept = 127 dBm
   - $\gamma = 3.64$

   Plot a graph for coverage versus ϕ.

10.16 References

1. Viterbi, A. J., *CDMA*, Addison-Wesley: New York, 1995.
2. Gilhousen, K. S., Jacobs, I. M., Padovani, R., and Weaver, L. A., "Increased Capacity Using Satellite Communications," *IEEE Transactions, Select Areas Communications*, May 1990 [Vol. JSAC-8(4), pp. 503–514].
3. Viterbi, A. M., and Viterbi, A. J., "Erlang Capacity of a Power Controlled CDMA System," *IEEE Journal of Selected Areas in Communications*, 1993 [Vol. 11(6), pp. 892–900].
4. Gilhousen, K., et al., "On the Capacity of a Cellular CDMA System," *IEEE Transactions on Vehicular Technology*, May 1991 [VT-40(2), pp. 303–312].
5. Glisic, S., and Vucctic, B., *Spread Spectrum CDMA Systems for Wireless Communications*, Artech House: Boston, 1997.
6. Borth, D. E., and Pursley, M. B., "Analysis of Direct Sequence Spread Spectrum Multiple access Communication Over Rician Fading Channels," *IEEE Transactions on Communications*, Oct. 1979 [Vol. COM-27(10), pp. 1566–1577].
7. Weber, C. L., et al., "Performance Considerations of CDMA Systems," *IEEE Transactions on Vehicular Technology*, Feb. 1981, pp. 3–9.

PART II

Evolution of 2G Systems to 3G Systems

The second part of the book focuses on the evolution of second-generation (2G) systems to third-generation (3G) systems. Within the International Telecommunications Union (ITU), the 3G systems are called International Mobile Telephony 2000 (IMT-2000). Within the IMT-2000 framework, several different air interfaces have been defined for 3G systems, based on either code division multiple access (CDMA) or time division multiple access (TDMA) technology. The 3G air interfaces are W-CDMA, cdma2000, and UWC-136 HS. The W-CDMA air interface is to be used in Europe and Asia, including Japan, Korea, and China, using the frequency bands that WARC-92 allocated for the 3G IMT-2000 system at around 2 GHz. The cdma2000 air interface is to be used in North America to evolve cdmaOne to 3G systems. The UWC-136 HS can provide 3G services with bit rates up to 384 kbps within a Global System for Mobile Communications (GSM) carrier spacing of 200 kHz. The UWC 136 HS includes advanced features that are not part of GSM; these features are intended to improve spectrum efficiency and support the new services. The cdma2000 is the multicarrier system that can be used as an upgrade solution for the existing cdmaOne operations. Chapters 11 through 15 provide details of the 3G air interfaces.

CHAPTER 11

Third-Generation Standards Activities

11.1 Introduction

The second-generation (2G) wireless systems include Global System for Mobile Communications (GSM), IS-136, IS-95 code division multiple access (CDMA), and, in Japan, Personal Digital Cellular (PDC) and Personal Handyphone System (PHS). GSM is the mobile radio standard with the highest penetration worldwide. These 2G systems are limited in maximum data rate. On the other hand, the percentage of mobile multimedia users will increase significantly after 2000. According to the Universal Mobile Telephone Services (UMTS) Forum, in 2010 about 60 percent of the traffic in Europe is likely to be created by multimedia applications [1]. A similar growth in mobile data traffic is expected worldwide, with an expected growth rate per year of 60 to 70 percent during the next five years, starting from about 3 million data users in 2000 to about 77 million data users in 2005.

More advanced services than current voice and low-data-rate services are foreseen and will bring together the three disciplines according to three basic categories:

- Computer data with Internet access, e-mail, real-time image transfer, multimedia document transfer, and mobile computing
- Telecommunications with mobility, video conferencing, GSM, and integrated services of digital networks, video telephony, and wideband data services
- Audiovisual content with video on demand, interactive video services, infotainment, electronic newspapers, teleshopping, value-added Internet services, television, and radio contribution

The GSM Association, an organization of GSM operators, is expecting a high grade of asymmetry between the demand for uplink and downlink of data transmission (e.g., Internet access), with much higher capacity needed on the downlink.

11.2 IMT-2000

The International Telecommunications Union-Radio communications (ITU-R) developed the IMT-2000 specifications. IMT-2000 is an acronym for International Mobile Telephony for the Year 2000. It was formerly called Future Public Land Mobile Telephony (FPLMTS). The ITU-R is overseeing worldwide efforts to define the third-generation (3G) wireless standards. IMT-2000 was created to facilitate the development of standards that enable a global wireless infrastructure encompassing terrestrial and satellite systems and fixed and mobile access for public and private networks. IMT-2000 has a greatly expanded range of service capabilities and covers a wide range of environments. IMT-2000 specifications aim to facilitate the introduction of new capabilities and seamless evolution from the substantial installed 2G mobile wireless base. The IMT-2000 includes a capability that enables a wireless user to communicate anywhere in the world, with complete global coverage provided by combining terrestrial and satellite stations. IMT-2000 also includes the ability to roam anywhere in the world and receive services comparable to those available in the home network. A common spectrum worldwide for IMT-2000 would facilitate global roaming and would contribute to global economies of scale for mobile stations and base stations.

IMT-2000 family members are 3G systems that are scheduled to start service in 2002, subject to market considerations. The 3G systems will provide access, by means of one or more radio links, to a wide range of telecommunications services supported by the fixed telecommunication networks, and to other services that are specific to mobile users. A range of mobile terminal types is encompassed, linking terrestrial- and satellite-based networks, and the terminals may be designed for mobile or fixed use.

The key features of IMT-2000 are

- High degree of commonality of design worldwide
- Compatibility of services within IMT-2000 and with the fixed networks
- High quality
- Small terminal for worldwide use
- Worldwide roaming capability
- Capability for multimedia applications and a wide range of services and terminals

The evolution from 2G mobile and fixed networks to 3G networks will not occur in a single leap. IMT-2000 is an important step that allows a mixture of new and emerging wireless mobile access technologies to coexist with existing wireless and fixed-access technologies. Under this specification, both the developed and developing countries of the world can deliver a wide range of voice, data, and Internet services cost effectively.

With the acceptance of the IMT-2000 family of systems concept, the development of IMT-2000 standards and specifications is now distributed over a number of international, as well as regional and national, standard forums. The ITU-R and ITU-T are addressing the overall framework of IMT-2000 radio and network interface specifications, primarily to allow interoperability between IMT-2000 family member systems. The third-generation partnership projects (3GPPs) and standard development organizations (SDOs) are working on the specifications for individual family members.

A roadmap or guide is needed by network operators and service providers who wish to implement IMT-2000 systems. The roadmap provides a guide to the key IMT-2000 standards and specifications, and gives service providers and network operators around the world direction in making critical 3G deployment decisions and planning their 3G networks.

Within the 3G systems these different service needs will be supported in a spectrum-efficient manner by a combination of frequency- and time-division duplex (FDD and TDD). The FDD mode supports wide area coverage mainly for symmetrical services, whereas TDD is suitable especially for asymmetrical services.

11.3 Technical Requirements and Radio Environments for IMT-2000

The 3G systems aim to support a wide range of bearer services from voice and low-rate to high-rate data services with up to at least 144 kbps in vehicular, 384 kbps in outdoor-to-indoor, and 2 Mbps in indoor and picocell environments. Circuit-switched and packet-switched services for symmetric and asymmetric traffic will be supported.

3G systems will be operated in all radio environments, including large metropolitan urban and suburban areas, hilly and mountainous areas, and microcell, picocell, and indoor environments. These requirements are quite well aligned in North America, Asia, Europe, and the ITU. This enables a much wider application range with 3G systems than with 2G systems. In addition, the ability for global roaming will be supported in the system design. The 3G systems will be optimized for vehicular, indoor, and fixed wireless environments.

11.4 International Standardization Activities

The international standardization activities for IMT-2000 have been mainly concentrated in the different parts of the European Telecommunication Standards Institute (ETSI) Special Mobile Group (SMG), the Research Institute of Telecommunications Transmission (RITT) in China, the Association of Radio Industries and Businesses (ARIB), and the Telecommunication Technology Committee (TTC) in Japan, the Telecommunications Technologies Association (TTA) in Korea, and the Telecommunications Industry Association (TIA) and T1P1 in the United States. The backward compatibility of the existing 2G systems (GSM, IS-136, IS-95 CDMA, and PDC) has been taken into account by the related regional standard bodies in the 3G proposals. These systems are connected to the two types of core networks, GSM Mobile Application Part (MAP) and ANSI-41, and ease their evolution to the 3G systems.

ITU communication standardization sector Task Group (TG) 8/1 called for Radio Transmission Technology (RTT) proposals (see Figure 11.1). The ETSI process started at the end of 1996 and was subdivided into phases: grouping, refinement, synthesis, and definition. In January 1998, ETSI SMG decided on the basic access scheme for wideband CDMA (W-CDMA) in paired bands (FDD) and for time division CDMA (TD-CDMA) in unpaired bands (TDD). In addition, the implementation of dual-mode terminals for FDD and TDD should be economically feasible, including the harmonization to GSM. The UMTS Terrestrial Radio Access (UTRA) concept also fits into 2 × 5 MHz spectrum allocation. Parameter harmonization of the FDD and TDD modes has been reached with respect to the implementation of terminals and achieving a worldwide standard. ETSI SMG submitted the UTRA proposal to ITU-R for IMT-2000.

China presented a TD-SCDMA proposal to ITU-R based on a synchronous TD-CDMA scheme for TDD and wireless local loop (WLL) applications.

The Japanese standardization body ARIB decided on W-CDMA. The Japanese and European W-CDMA proposals for FDD are aligned. The ARIB W-CDMA concept was submitted to ITU-R. ARIB froze the specification in the first half of 1999, so commercial services could begin in 2001.

TTA in Korea prepared two proposals for ITU-R; one is close to the W-CDMA scheme, and other is similar to the TIA cdma2000 approach.

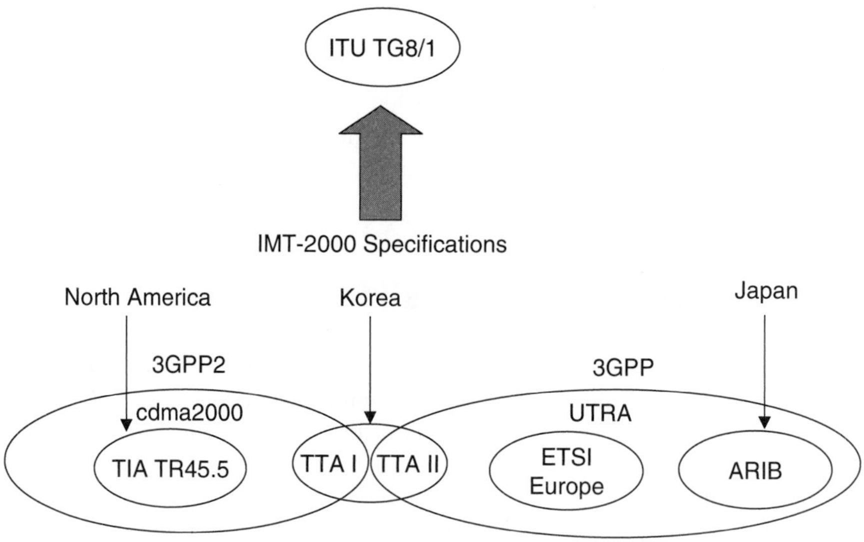

Figure 11.1 Wideband CDMA Proposals and Standards Bodies

In the United States, TIA prepared several proposals for the 3G with UWC-136 as an evolution of IS-136, cdma2000 as an evolution of IS-95, and a W-CDMA system called wireless multimedia service (WIMS). The T1P1 committee is supporting W-CDMA-NA, which corresponds to UTRA FDD. W-CDMA-NA and WIMS have been merged into wideband packet CDMA (WP-CDMA). All of these proposals were submitted to ITU-R.

Details of all proposals for IMT-2000 are available in [2]. The international consensus-building and harmonization activities between different regions and standard bodies are currently ongoing. A harmonization would lead to a quasi-worldwide standard, which would allow economic advantages for customers, network operators, and manufacturers. Therefore, two international bodies have been established:

- The Third Generation Partnership Project (3GPP) to harmonize and standardize in detail the similar ETSI, ARIB, TTC, TTA, and T1 W-CDMA and related TDD proposals.
- 3GPP2 for the cdma2000-based proposals from TIA and TTA.

In addition, major international operators have undertaken a coordination between 3GPP and 3GPP2 in the context of the ITU-R process, which will result in a globally harmonized concept.

11.5 International Frequency Allocation

Available spectrum is a prerequisite for the economic success of a new system. Therefore, the 1992 World Administration Radio Conference (WARC) specified the spectrum for the 3G mobile radio system (see Figure 11.2).

Europe and Japan basically followed these recommendations for FDD systems. In the lower band, parts of the spectrum are currently used for Digital Enhanced Cordless Telecommunications (DECT) and PHS, respectively. The Federal Communications Commission (FCC) in the United States has allocated a significant part of the WARC spectrum in the lower band to 2G personal communications services (PCS) systems. Most of the North American countries are following the FCC frequency allocation. In China, big parts of the WARC spectrum are currently allocated to WLL applications. Currently, no common spectrum is available worldwide for 3G mobile radio systems. Additional spectrum is requested from WARC 2000.

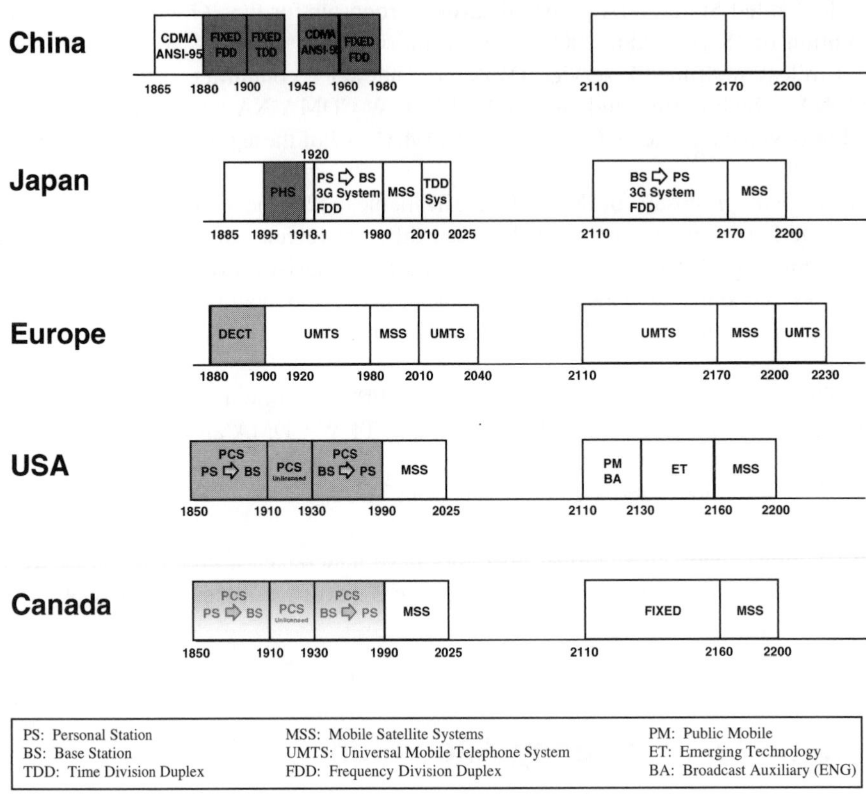

Figure 11.2 Radio Frequency Allocations (IMT-2000)

11.6 International Research Activities

11.6.1 European Research Activities

The Future Radio Wideband Multiple Access System (FRAMES) was the only Advanced Communication Technologies and Services (ACTS) project dealing with the terrestrial component of the UMTS radio interface. It is a consortium consisting of partners from manufacturers, operators, small- to medium-sized enterprises (SMEs), and research communities [3]. Its main goal was to develop a radio interface proposal that satisfies the requirements for terrestrial 3G mobile radio systems. The definition of the original FRAMES Multiple Access (FMA) scheme satisfied these requirements and took into account the big worldwide footprint of the GSM family with respect to harmonized radio parameters and the GSM core network.

Starting from the basis of the UMTS requirements and the activities and decisions in different regions of the world at the beginning of the project, the FMA scheme was completed in 1996. It combined TDMA and CDMA technologies into one harmonized platform to cope with all possible UMTS scenarios and to address possible technical solutions from a global perspective, as follows [4–6]:

- Mode 1 (FMA 1): wideband TDMA with and without spreading
- Mode 2 (FMA 2): wideband CDMA

The FRAMES project assumes that UMTS will build upon an evolved GSM platform with several hundred million subscribers at the time of launch. Working under the assumption that UMTS and GSM will operate in parallel for a long time, it is important that dual-mode terminals be easy to build. Therefore, FMA was harmonized as far as possible with the GSM network in terms of radio parameters and protocol stack.

With respect to the status of worldwide standardization activities at the time the FMA concept was defined at the end of 1996, the main goal of harmonization was to meet the different system needs of different possible access schemes in different regions and to ensure compatibility with 2G systems. Unnecessary differences from a technical point of view between TDMA- and CDMA-based schemes are avoided without sacrificing the good aspects of each technology. The harmonized approach between TDMA and CDMA modes and the GSM family minimizes terminal hardware complexity.

At the beginning of 1998 the FRAMES project adopted the ETSI consensus decision on the UTRA concept for further investigation. These studies have focused in more detail on channel coding and modulation, receiver algorithms, layer 1 aspects, and network aspects for the FDD and TDD modes to optimize the selected scheme even further. In addition, the TDD mode was developed in the FRAMES project.

The results and proposal of FRAMES were used successfully as input to the ETSI standardization and evaluation process for the UTRA concept. In addition, FRAMES partners contributed to the Japanese standardization process in ARIB, to TTA in Korea, and to the TIA in the United States. The ETSI SMG decision on the UTRA concept with the FDD mode W-CDMA is based on FMA 2, and the TDD mode TD-CDMA is based on FMA 1 with spreading. UWC-136 in the United States is based on FMA 1 without spreading for the high-speed component and on Enhanced Data Rates for GSM Evolution (EDGE).

11.6.2 Japanese Research Activities

To promote the ongoing activities at the Telecommunications Technology Council of the Ministry of Posts and Telecommunications (MPT), the IMT-2000 Study Committee (then called the FPLMTS Study Committee) was established in ARIB (then called the Research and Development Center for Radio Systems [RCR]) in 1993. Thereafter, standardization activities for 3G mobile communications systems and studies on its radio transmission technology started on a full-fledged basis. At the same time, the Telecommunication Technology Committee (TTC) of Japan established its UPT/FPLMTS working group to perform studies on network-related aspects of the next-generation system.

The IMT-2000 study committee formed the Radio Transmission Technology (RTT) Special Group in 1994 to select and evaluate the radio interface to be proposed to ITU-R from Japan. The group studied a variety of radio access candidate technologies. Initially, there were 20 proposals, including CDMA- and TDMA-based ones. Through simulation and verification experiments using test equipment, the number of candidates was narrowed to three CDMA-based proposals (Core A, Core B, and Core C) and one TDD proposal. For TDMA, the candidates were reduced to two proposals, MTDMA and BTDMA. In November 1996 the Radio Transmission Technology Special Group adopted the following conclusion: to merge the three CDMA-based proposals into one, centering on Core A, and to combine it with the TDD proposal to treat it as one candidate. Since this CDMA proposal had sufficient backing with experimental data, it was deemed feasible for implementation by 2000. Thus, the merged CDMA proposal entered into a detailed study phase, finishing its basic study activities. At the same time, the group decided to continue the studies for the two TDMA-based proposals, and judge later whether to perform detailed studies for these proposals.

With this decision, the group started its full-fledged studies on W-CDMA as the most promising RTT candidate to be submitted from Japan. When the group decided one year later not to adopt the two TDMA proposals, W-CDMA became the sole proposal to be submitted from Japan to ITU-R. The key parameters for the CDMA FDD Core A and TDD proposals, the starting point of the W-CDMA studies, are summarized in [7–9].

With the reorganization of the IMT-2000 Study Committee in 1997, several working groups were established to perform detailed studies on W-CDMA. Thereafter, the Air Interface Working Group took over the responsibility of conducting studies on W-CDMA from the RTT Special Group. In particular, air interface layer 1, which is closely related to the radio transmission technology, was studied in detail by subworking group 2 formed under the Air Interface Working Group.

At the same time, ARIB formed a coordination group under the standard subcommittee with the goal of achieving a global standard. The coordination group paid close attention to movements abroad, and carried out various activities for harmonization. The harmonization with UTRA was facilitated with proposals from some European members. The coordination group also convened regular meetings with TTA Korea. Meanwhile, harmonization with the cdma2000 concept was also discussed. For that purpose the Ad Hoc S group was formed to take care of its technical studies, and the outcome is now reflected in the W-CDMA specifications. All these study results were reflected in drafting the final Japanese proposal submitted to ITU-R in 1998.

In parallel with these standardization activities, individual companies are also involved in their own research activity to support the standardization. A 2-Mbps transmission field trial was conducted successfully with W-CDMA technology. W-CDMA system tests are now being conducted using a tentative air interface specification on ARIB W-CDMA. These system tests are conducted not only in Japan but also in Asian countries such as Singapore, Malaysia, South Korea, and China.

The 3G mobile telecommunications systems based on IMT-2000 will be introduced into service before 2002. The 3G systems will offer a plethora of telecommunications services including voice, low and high bit rate data, video to mobile users via a range of mobile terminals, and the ability to operate in both public and private environments (offices, residences, vehicles, etc.). The regional standards bodies engaged in 3G standardization activities are shown in Figure 11.3.

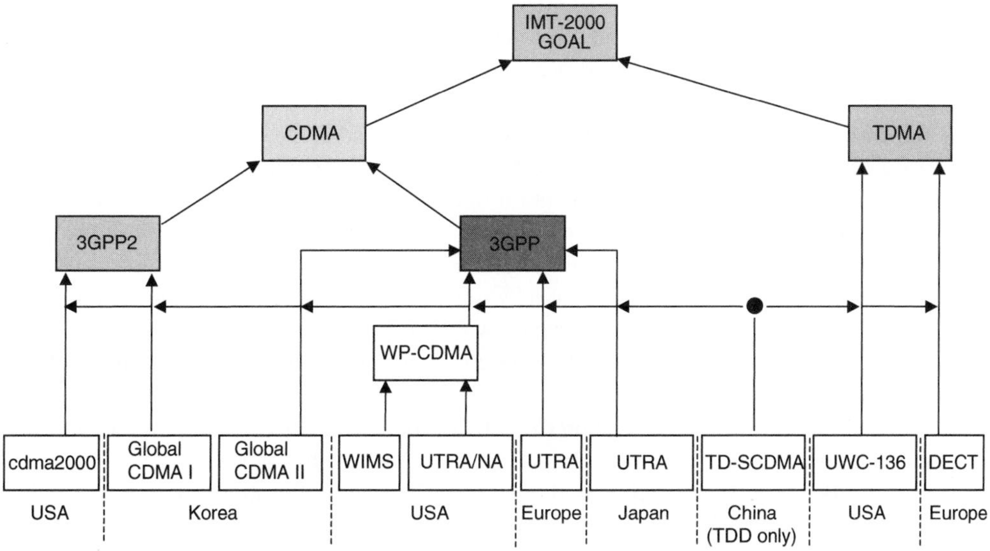

Figure 11.3 3G Standardization Activities

IMT-2000 will be an overarching federation of networks based on the "family of systems" concept. The concept was developed with the knowledge that 3G systems would evolve from the existing 2G systems, and the existing 2G systems are based on different regional standards (with ETSI GSM-MAP, ANSI-41–based, and Japan's PDC networks being the most predominant). None of the existing 2G networking protocols would be suitable as the basis for an IMT-2000 networking protocol, since no one protocol in its present form can support both GSM-MAP– and ANSI-41–based core networks.

To solve this dilemma, the family of systems concept was introduced to allow the existing operators to evolve their current networks to 3G networks and to use many existing 2G services. Since the transition from 2G to 3G services will not be instantaneous, a graceful evolution will require the continued existence of 2G systems. They will facilitate a smoother transition to 3G services and features while maintaining ubiquitous service.

The ITU, therefore, decided that it would not specify the protocol to be used within the core network; instead, it would allow regional protocols, such as GSM-MAP and ANSI-41, to evolve independently. The ITU would specify the network-to-network interface (NNI) protocol that would be used between core networks. The NNI will be used wherever end users roam between family member networks, such as between a UMTS and an ANSI-41 system.

Figure 11.3 shows the IMT-2000 radio interface convergence road map for various 3G proposals. The basic air interface and the radio access network parts of the total interfaces are being developed by Third Generation Partnership Projects (3GPP and 3GPP2).

Two different air interfaces based on CDMA access technology considered for the 3G wireless FDD systems are multicarrier (MC) mode, based on cdma2000, and direct spread (DS) mode, based on UTRA W-CDMA. The main requirement for these air interfaces is the ability to support high bit rates and new multimedia services. The goal is to support at least 384 kbps with wide area coverage and up to 2 Mbps with local area coverage

The PCS spectrum in North America currently supports 2G digital wireless systems. Accordingly, the ability to allow for a system overlay and at the same time maintain backward compatibility with the embedded base of cdmaOne (family of 2G CDMA systems) becomes a very important requirement for operators. The MC design of cdma2000 makes it possible to achieve such an overlay.

Members of the UMTS community concluded that CDMA technology is a better choice to satisfy the requirements of 3G wireless systems. Services that offer multimedia and high data rates require a wider bandwidth spectrum for CDMA. This requirement could only be satisfied by a W-CDMA system (5 MHz or more), rather than a narrowband 1.25-MHz 2G system.

The ITU World Administration Conference in 1992 (WARC-92) identified 230 MHz in the 2-GHz band for use on worldwide basis for the satellite and terrestrial components of IMT-2000. The WARC-95 revised the 2-GHz frequency allocations for mobile satellite services (MSS) to provide satellite component of IMT-2000 (see Figure 11.2).

Table 11.1 provides a summary of the current IMT-2000 3G air interface proposals to ITU and their network interfaces. Table 11.2 includes a summary of the basic parameters of ETSI UTRA and ARIB W-CDMA air interfaces [10].

Table 11.1 Current IMT-2000 Proposal

	cdma2000	ARIB/DOCOMO	UMTS	UWC-136
2G system	IS-95/cdmaOne	PDC	GSM	IS-136
3G air interface	cdma2000	W-CDMA	UTRA (W-CDMA/ TD-CDMA)	IS-136/ IS-136+/ IS-136 HS
3G network interface	Evolved ANSI-41 MAP	Evolved GSM MAP	Evolved GSM MAP	Evolved ANSI-41 MAP
Standards bodies	TI TR-45 (supported by CDG)	ARIB	ETSI	TIA TR-45 (supported by UWCC)

Table 11.2 Basic Parameters of UTRA FDD and TDD, and ARIB W-CDMA FDD and TDD

	ETSI UTRA		ARIB W-CDMA	
	FDD	**TDD**	**FDD**	**TDD**
Multiple-access method	W-CDMA	TD-CDMA	W-CDMA	TD-CDMA
Chip rate	3.84 Mcps	3.84 Mcps	3.84 (1.024/7.68/15.36) Mcps	3.84 (1.024/7.68/15.36) Mcps
Carrier spacing	5 MHz	5 MHz	5 (1.25/10/20) MHz	5 (1.25/10/20) MHz
Frame length	10 ms	10 ms	10 ms	10 ms
No. of power control groups per time slot	15	15	15	15
Time slot duration	N/A	625 μs	N/A	625 μs
Data modulation (DL/UL)	QPSK	QPSK	QPSK/BPSK	QPSK/QPSK
Spreading modulation (DL/UL)	QPSK	QPSK	QPSK/QPSK	QPSK/QPSK
Spreading factor	4-512	1, 2, 4, 8, and 16	2–512	2–512
Pulse shape	Root raised cosine $r = 0.22$	Root raised cosine $r = 0.22$	Root raised cosine $r = 0.22$	Root raised cosine $r = 0.22$

11.7 Global Partnership Projects

Two global partnership projects, 3GPP and 3GPP2, were established in late 1988 to develop regional technical standards to satisfy IMT-2000 specifications. Two different 3GPP organizations were formed with the mission of defining global wireless standards for ANSI-41 and GSM-MAP core networks.The 3GPP is developing 3G standards for the UTRA radio interface for GSM-MAP and ANSI-41 core networks. The 3GPP2 is responsible for developing standards for the cdma2000 radio interface for ANSI-41 and GSM-MAP core networks.

The formation of the partnership projects opened membership to all global operators and vendors, allowing them to participate in the definition of 3G wireless requirements. From the set of interfaces shown in Figure 11.4, a subset has been selected to define and develop specifications for the 3GPPs. These are the air interfaces and A interface (interface between radio subsystem and core network).

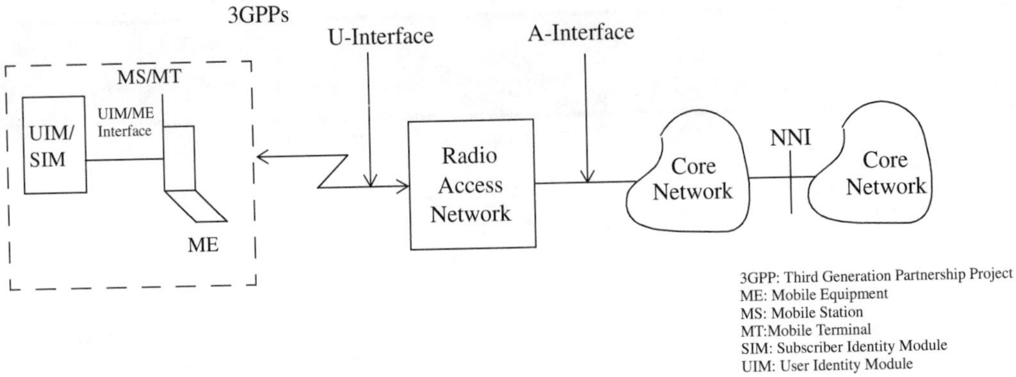

Figure 11.4 3GPPs Interfaces

11.7.1 3GPP

The 3G standards, based on an evolution of GSM-MAP and UTRA W-CDMA air interface, will be developed by the following 3GPP regional Standard Development Organization (SDO) partners.

- ARIB/TTC, CWTS, and TTA
- ETSI
- Committee T1, another ANSI-accredited SDO

Figure 11.5 shows the SDO partners of 3GPP.

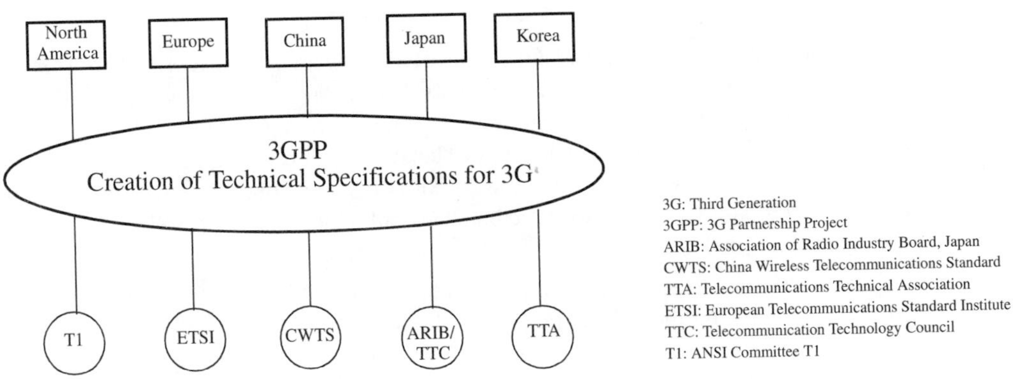

Figure 11.5 SDO Partners of 3GPP

The UTRA concept includes FDD and TDD modes to efficiently support the different UMTS service needs for symmetrical and asymmetrical services. During the evaluation of UTRA in ETSI SMG 2, investigation centered on the FDD mode. The original TD-CDMA concept (FMA 1 with spreading) was adapted for TDD, including the harmonization of parameters between FDD and TDD. Key parameters of the UTRA are given in Table 11.2.

The W-CDMA proposal of ARIB consists of FDD and TDD modes. Because of the harmonization with the W-CDMA of ETSI SMG 2, which was performed even before the ETSI decision of January 1998, the FDD mode was already quite similar to the ETSI UTRA. TDD, on the other hand, was designed around the same concept as the FDD mode, but adopted some distinctive features such as open-loop power control and transmit diversity technology. After the January 1998 ETSI SMG decision, the access scheme was renamed TD-CDMA instead of simply W-CDMA, because some TDMA features had been incorporated in order to take advantage of the technical merits of TD-CDMA (see Table 11.2). The parameters are quite well aligned between the ETSI SMG and ARIB proposals.

11.7.2 3GPP2

The 3G standards, based on ANSI-41 networks, will be created by the following SDO partners:

- The Telecommunication Technology Association (TTA) in the Republic of Korea
- The Association of Radio Industry Board (ARIB) in Japan
- The Telecommunication Technology Council (TTC) in Japan
- The Telecommunications Industry Association (TIA) in North America
- The China Wireless Telecommunication Standard (CWTS) in China

Members of the 3GPP2, shown in Figure 11.6, will develop the 3G specifications that will then be regionally standardized in the partner SDOs.

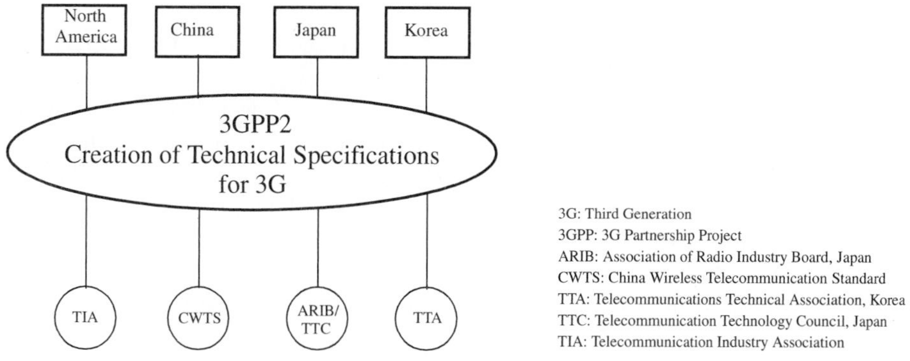

Figure 11.6 SDO Partners of 3GPP2

In the United States, different 2G mobile radio systems have been deployed, with different requirements with respect to backward compatibility. There are two fully established core networks currently being used for 2G systems in the U.S. market: ANSI-41 and GSM MAP. The ANSI-41 core network is used by AMPS, IS-136, and IS-95 air interface systems. The GSM MAP core network is used by GSM air interface-based systems. Both of these core networks will evolve into 3G, and they must interwork.

ANSI-41 specification and standardization work is being performed in TIA committee TR-45.2, and its evolution to 3G is being carried out in the technical specification groups of 3GPP2. The ANSI-41-based core network will be used by cdma2000-based radio access networks.

GSM specification work is being conducted in the ETSI SMG committees and harmonized for U.S. requirements in T1P1.5. This relationship has not changed for the 3G standardization work. GSM evolution to the 3G will be carried out in 3GPP and harmonized to U.S. requirements in T1P1. The GSM MAP-based core network will be used by UTRA-based radio access networks.

Four RTT proposals were submitted to ITU from the United States. They were W-CDMA-NA from T1P1, WIMS from TR 46.1, UWC 136 from TR 45.3, and cdma2000 from TR 45.5. Subsequently, W-CDMA-NA and WIMS converged to become WP-CDMA to strengthen W-CDMA's RTT position in the United States. As a result, WP-CDMA became the position in the United States, and it became the fifth RTT proposal submitted to ITU in January 1999. When T1 became a member of 3GPP, T1 submitted a proposal to include the packet features of WP-CDMA in 3GPP.

The GSM alliance and UWCC's joint agreement of using Enhanced Data Rates for GS Evolution (EDGE) as the common evolution path for GSM and IS-136 TDMA results in speculation that EDGE will be the 3G standard for GSM/TDMA in the United States, mainly for established operators. The logic is as follows: Both W-CDMA and EDGE are capable of delivering 384-kbps data in an outdoor/vehicular environment. W-CDMA is able to support this data rate with full coverage in the deployment area owing to its larger carrier bandwidth of 5 MHz. With a cluster of EDGE, though, it can support this data rate under good radio conditions with significantly reduced coding overhead and low co-channel interference or using a much bigger cluster size than for GSM voice or narrowband data services due to the GSM-compatible carrier bandwidth of 200 kHz. When an operator exchanges GSM voice carrier units for EDGE carrier units by keeping the frequency planning, the EDGE coverage for 384 kbps is significantly reduced from that of narrowband services.

In the United States, 384 kbps is currently seen as adequate for any known multimedia applications for wireless PCS in indoor environments, in contrast to a true IMT-2000 RTT proposal. Wireless LAN (WLAN), local multipoint distribution system (LMDS), and so on are alternatives for established operators without additional spectrum for 3G systems to support high-data-rate services in the indoor environment at multi-megabit data rates, which are deployed in new and available frequency bands.

There are three reasons to motivate the GSM operators in the United States to support both the W-CDMA effort in 3GPP and EDGE standardization in UWCC.

First, activities on 3G systems did not start in the United States prior to the ITU-R IMT-2000 RTT workshop held in Toronto, Canada, in September 1997. This was largely due to the lack of a high-data-rate market in the United States, and the ongoing deployment of PCS systems for voice and narrowband data traffic. While Japan and Europe foresaw a definite wireless data market need (e.g., for Internet access), the U.S. operators did not expect wireless applications requiring 2 Mbps data rates. Currently, U.S. operators believe that 384 kbps is adequate for the foreseeable future. The TDMA-based EDGE technology is capable of supporting 384 kbps. It should also be taken into account that most of the data applications have no hard delay requirements, which enable packet transmission. In packet-switched applications the carrier data rate is shared by different users, which results in throughput dependent on service requirements.

Second, the heavy 2G infrastructure investment has not yet depreciated. Therefore, their main concern is to meet the voice traffic demand to pay off their investment. Unlike W-CDMA, EDGE is a technology that can overlay the existing 2G mobile radio network to provide up to 384 kbps data services. This allows operators to keep on using their 2G infrastructure to provide voice and low-data-rate service. As a result, many U.S. operators see W-CDMA as a 3G technology ideal for greenfield deployment only. The same approach is true for the other IMT-2000 proposals from the United States.

Lastly, the lack of IMT-2000 frequency spectrum in the United States causes U.S. operators to consider EDGE as a logical solution to support 3G services. With new spectrum for 3G systems, the 5 MHz carrier of W-CDMA for FDD and TD-CDMA for TDD is a stumbling block to the 2 × 5 MHz license holders. These operators have to use their entire frequency block to deploy a W-CDMA carrier.

11.8 Harmonization/Consensus Building

Many companies have a strong desire to consolidate the terrestrial CDMA technologies for IMT-2000. Consolidation is also the intent of ITU's consensus-building phase. Harmonization/consensus-building activities have been proceeding since 1997 in different fora. Within ITU, consensus building is a formal part of the process and has been making progress in Working Group (WG) 5 of ITU-R Task Group 8/1. Outside the ITU, several meetings focused on harmonization within regional standards bodies such as the ARIB, ETSI, TIA, and TTA. The Operators Harmonization Group (OHG), having many operators around the world with interest in the harmonization of 3G CDMA air interfaces, was formed. 3GPP and 3GPP2 are working under the auspices of OHG to find interworking between ANSI-41 and GSM MAP systems.

Third-generation wireless technology is the primary concern of commercial operators worldwide. Ideally, these operators would like to have a single wireless standard that can be deployed globally to achieve greater economies of scale for both mobile equipment and infrastructures. With this common goal in mind, many of the world's wireless operators have formed a group to formulate a single CDMA framework for IMT-2000. Most of these operators unequivocally support this goal. They have proposed a framework—based on separate cdma2000 and W-CDMA proposals developed by 3GPP2 and 3GPP—for achieving a common global specification to provide a foundation for accelerated growth in the wireless industry. At the end of 1999 an agreement between major international operators and manufacturers was achieved on the harmonization for a Global Third-Generation (G3G) CDMA approach [11].

The technical framework was specified based on the technical merits of various options, but without considering issues that might involve intellectual property rights (IPR). The OHG formulated a technical proposal to define a superset of features for 3G wireless systems to cover most of the regional and multinational needs of various operators globally. Depending on the regional demands and competing forces, an operator can choose a subset of these features to meet region-specific requirements and economic constraints.

The OHG has defined a harmonized global specifications with a set of key radio parameters. The specifications form a basis for the harmonized G3G CDMA standard. These global specifications do not address either the UWC-136 or DECT IMT-2000 RTTs. The characteristics of these RTTs do not map in a graceful manner to the key radio parameters specified for the CDMA technology.

As a part of the harmonized G3G CDMA system, the OHG considered the following three modes of operation:

- *Multicarrier (MC) mode*—This supports $N \times 1.25$ MHz cdma2000 channels overlaid on N existing adjacent cdmaOne carriers or deployed in a clear spectrum.
- *Direct spread (DS) mode*—This has a single 5-MHz or more W-CDMA carrier.
- *Time-division duplex (TDD) mode*—This will use the same frame and slot structure as the FDD mode, in which each slot can be individually allocated to either uplink or downlink, according to operator's requirements.

The OHG decided that the harmonized CDMA standard for the MC would be based on the cdma2000 RTT proposal and the DS would be based on the UTRA W-CDMA RTT proposal. Since the key technical parameters of the harmonized TDD mode were not defined in detail (with the exception of chip rate), the RTT proposal for the TDD mode was not considered. CDMA MC and CDMA DS refer to these two standards, respectively. In addition, CDMA TDD will be used to refer to the TDD mode.

11.9 Harmonized G3G System

A harmonized G3G system based on the OHG recommendations is required to support

- High-speed data services including Internet and intranet applications
- Voice and nonvoice applications
- Global roaming
- Evolution from the embedded base of 2G systems
- ANSI-41 and GSM-MAP core networks
- Regional spectrum needs
- Minimization of mobile equipment and infrastructure cost
- Minimization of the impact of IPRs
- The free flow of IPRs
- Customer requirements on time

In addition, the OHG also recommended a set of technical parameters as guidance to define the specifications for the harmonized G3G system. These parameters include:

- **Interbase station synchronization**
  Used in the CDMA MC standard for any handoff considerations, it is based on the cdma2000 scheme. This scheme identifies different base stations by means of PN offsets. The UTRA W-CDMA is based on asynchronous scheme. It uses the CDMA DS mode, and identifies different base stations by different codes.
- **Forward link pilot structure**
  In the MC mode, it is based on the structure defined in cdma2000. The operators support the harmonized forward link pilot structure for the DS mode, which will be discussed in Chapter 12.
- **Chip rate**
  The chip rate for the CDMA MC mode is 3.6864 Mcps, whereas the chip rate for both the CDMA DS and TDD mode is 3.84 Mcps.
- **Radio frequency (RF) parameters**
  The objective for selecting the RF components for a harmonized system is to ensure greater economies of scale for the basic components used in mobile equipment and the infrastructure, which in turn would reduce the overall cost.

The OHG recommended a "pseudo-open" system architecture (see Figure 11.9) with various functional elements that would enable an operator to select a set of components and build a system to satisfy his or her requirements. This objective fosters competition among manufacturers and, at the same time allows the operators the freedom to select various system elements from different vendors. As shown in Figure 11.7, an operator can choose one or more radio access technologies that can interface with one or more backbone core networks.

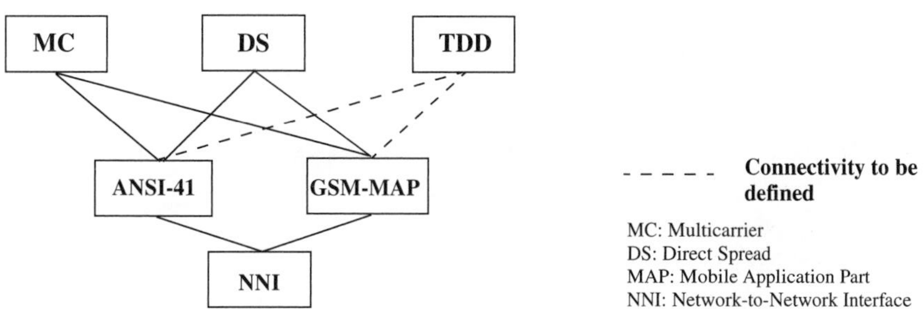

Figure 11.7 Radio Access Network and Core Network Interconnections

Based on the connections shown in Figure 11.8, an operator can select any of the following four configurations:

- Global specification for ANSI/TIA/EIA-41 network evolution to 3G global specifications for the CDMA MC mode supported by the ANSI/TIA/EIA-41 core network (see Figure 11.8(a)).
- Global specification for ANSI/TIA/EIA-41 network evolution to 3G global specifications for the CDMA DS mode supported by ANSI/TIA/EIA-41 core network (see Figure 11.8(b)).
- Global specification for GSM-MAP network evolution to 3G global specifications for the CDMA DS mode supported by GSM-MAP core network (see Figure 11.8(c)).
- Global specification for GSM-MAP network evolution to 3G global specifications for CDMA MC mode supported by the GSM-MAP core network (see Figure 11.8(d)).

A G3G operator may choose any one, or more, of the options. However, it then becomes necessary to define different protocol layers for each connectivity. Regardless of operator selection(s), the harmonized system will support functionality based on synchronized operation such as location calculation. The system should also support seamless handoff between harmonized CDMA DS and CDMA MC, including for ANSI-41 and equivalent for UMTS/GSM.

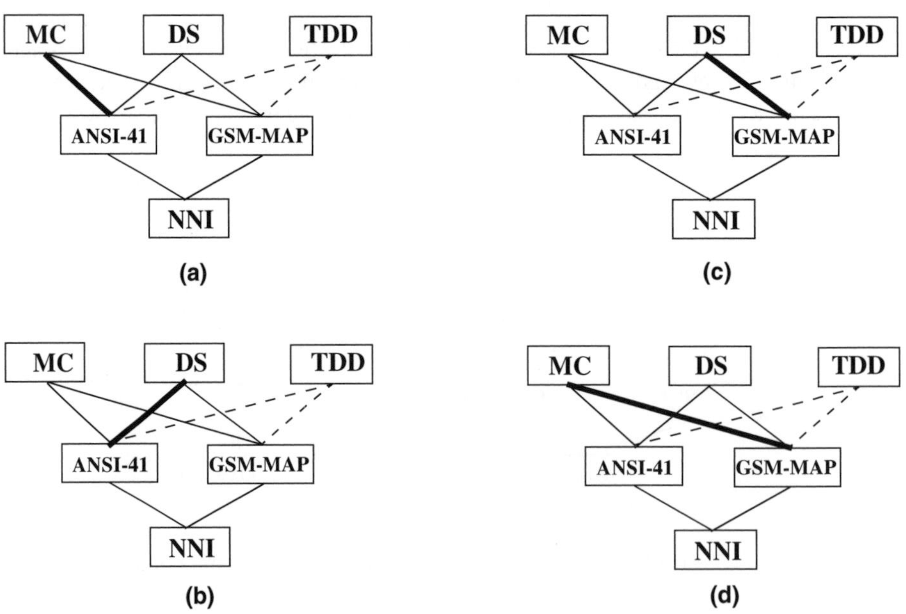

Figure 11.8 Interconnection Options

11.10 Harmonized Phased Approach

The OHG has the responsibility for all potential changes to Layer 1, Layer 2, and Layer 3 of each protocol stack for both CDMA MC and CDMA DS technologies. The conceptual stacks, with their potential hooks and extensions, are shown in Figure 11.9. Also shown in the figure is the plan of record connectivity for each radio access technology, its respective core networks, and cross-connectivity between them.

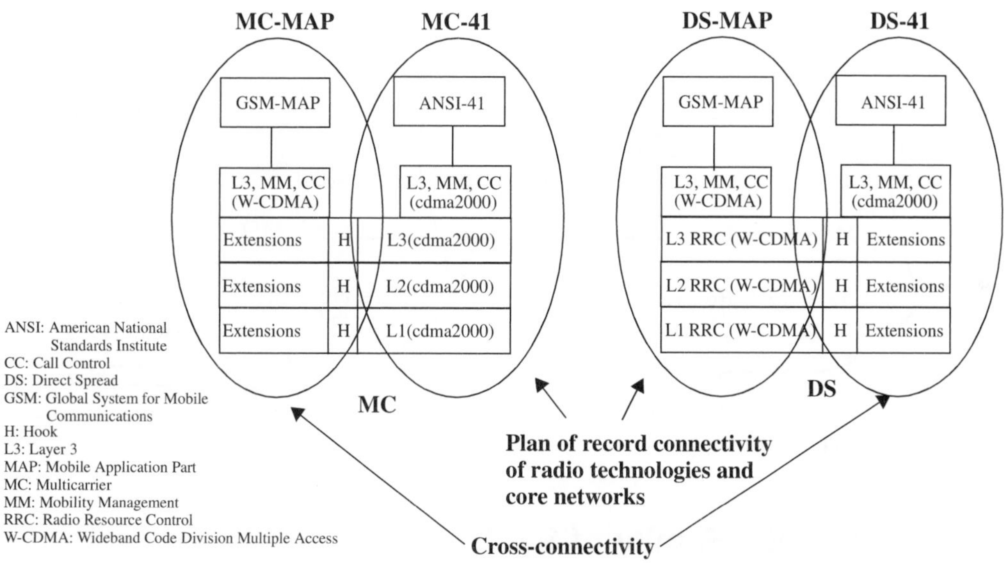

Figure 11.9 Protocol Stacks with Potential Hooks and Extensions

A *hook* is defined as any functionality that is specified for the initial release of the standard so that the extensions needed to satisfy the G3G requirements can be defined in the later release of the standard document. The hook must provide enough details to include all the functionality for MC-MAP and DS-41 connectivity. In addition, these hooks must be defined within the framework of specifications of both 3GPP and 3GPP2.

An *extension* is defined as an entity that will provide a complete specification for implementing the functionality to satisfy OHG requirements for features and cross-connectivity requirements between radio access networks and core networks. However, an extension is restricted from making major changes to the baseline protocols.

11.11 Core and Access Network

The roles of the access and core networks, as shown in Figure 11.10, are as follows:

- *Access network* contains all functions that enable a user to access services.
- *Core network* (CN) contains all switching and mobility management functions. The CN evolves independently of the access network.

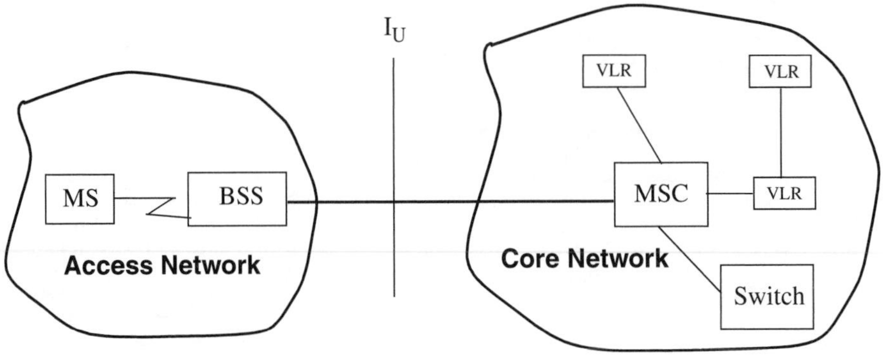

Figure 11.10 Access and Core Networks

11.12 IMT-2000 Family of Systems

Figure 11.11 identifies the functional subsystems and the associated signaling relationships (interfaces) for standardization in Capability Set-1 (CS-1).

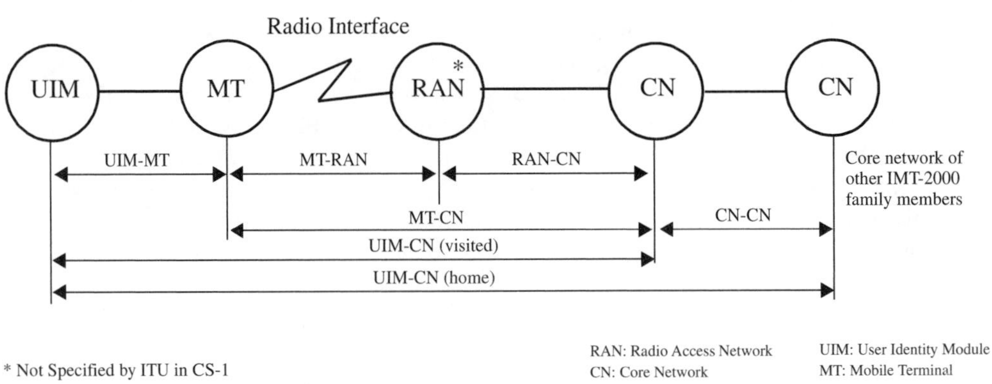

Figure 11.11 IMT-2000 Functional Subsystems

The following interfaces will be standardized by ITU-T to facilitate global roaming among IMT-2000 family members:

- UIM-MT
- MT-RAN (radio interface layers 2 and 3)
- MT-CN (radio interface layer 3)
- CN-CN (Network-to-Network Interface [NNI])

The relationship between the UIM and the CN (both visited and home) are logical interactions. In order to apply the family of systems concept to Figure 11.11, four terms used in IMT-2000 3G systems are defined (see Figure 11.12):

- Intrasubsystem
- Intersubsystem
- Intrafamily
- Interfamily

Table 11.3 lists the key terminology for the intrasubsystem, intersubsystem, intrafamily, and interfamily operations.

Table 11.3 Key Terminology

Term	Description	Responsible Bodies
Intrasubsystem	Signaling relationship contained within a specific subsystem, e.g., within CN of one family member system. An intrasubsystem signaling relationship is outside ITU-T standardization.	Family members
Intersubsystem	Signaling relationship between two subsystems, either contained in the same or different IMT-2000 family member systems, e.g., MT-RAN, etc.	Within the same family: family member Between family members: ITU-T
Intrafamily	Signaling relationship contained within the same IMT-2000 family member system.	Within the same family: family member ITU-T to provide framework for commonality
Interfamily	Signaling relationship between two subsystems contained in different IMT-2000 family member systems, e.g., CN-CN.	ITU-T (to facilitate commonality and global roaming)

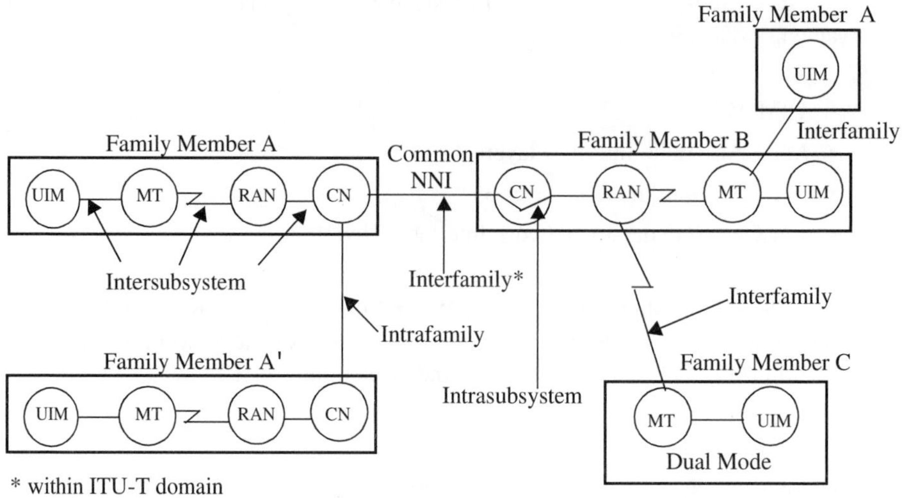

Figure 11.12 Global Roaming Possibilities

11.13 Core Network to Core Network Interface (NNI)

There is a need for common NNI in a multinetwork environment in order to derive benefit from existing fixed and mobile investments, and to support global roaming and seamless service provisioning.

Figure 11.13 shows a schematic view of interworking functions (IWFs) and a common NNI in an IMT-2000 family of systems environment. The use of a common NNI for global roaming provides a unique open interface developed by ITU-T. It provides an efficient solution for interworking between IMT-2000 core networks, since only one IWF per family member is required to interwork with all other IMT-2000 family members. It provides transparency: changes in one family member do not affect other family members. It is future-proof by easily accommodating new family members. Figure 11.14 shows how the concept indicated by Figure 11.13 is applied. It should be noted that each CN (e.g., evolved GSM MAP) may contain its own IWF for interworking with the common NNI.

The IMT-2000 family of systems concept and functional architecture are a valuable framework for planning and organizing regional SDOs' work on defining relevant standards for IMT-2000 family members and supporting ITU-T standardization activities. The interfaces and functional relationships identified in ITU-T Recommendation Q.1701 are recognized as the interfaces to be covered by ITU-T recommendations. Because of their extensive knowledge of 2G mobile systems and specific needs of individual IMT-2000 family member markets, regional SDOs are best equipped to handle intrafamily member standards matters. They also have expertise on how to evolve 2G systems toward IMT-2000 and how to interwork between 2G systems and IMT-2000. The long-term goal is to evolve toward a common ITU-T standard for IMT-2000.

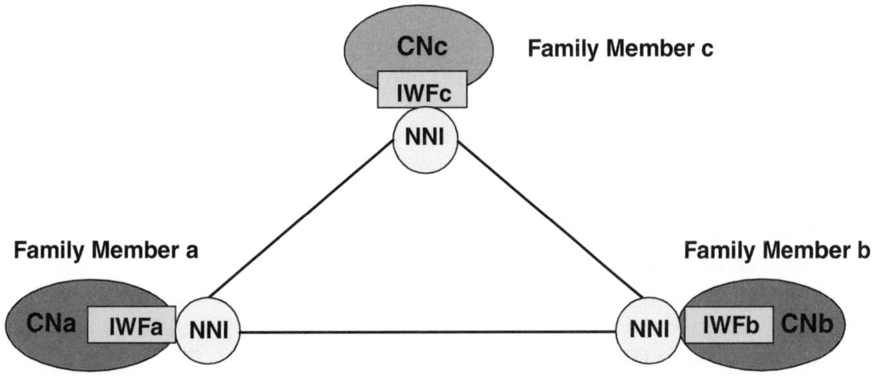

Figure 11.13 Common NNI in IMT-2000 Family Member Interconnection Model

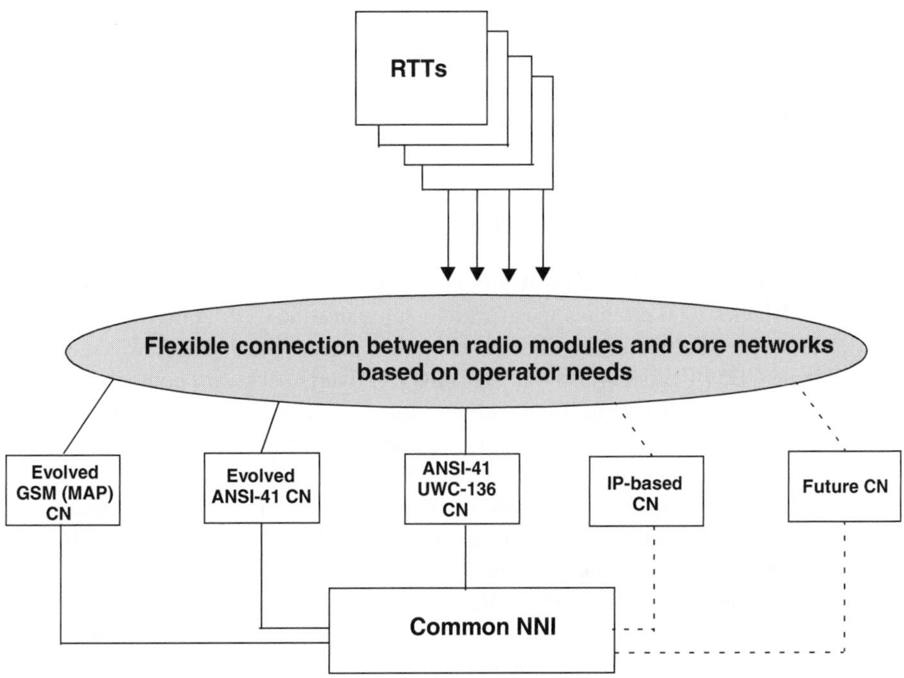

Figure 11.14 Application of IMT-2000 Family Member Concept

11.14 Evolution of 2G Systems for Higher Data Rate

GSM and IS-136 are TDMA-based systems and must evolve to CDMA using several intermediate steps, whereas evolution of cdmaOne to cdma2000 does not require these steps.

The TIA-IS-95B upgrade allows for code or channel aggregation to provide a data rate of 64–115 kbps, and also offers improvements in soft handoffs and interfrequency hard handoffs. To achieve 115 kbps rate, up to eight CDMA traffic channels each offering 14.4 kbps need to be aggregated. It is expected that an operator will initially support data rates between 28.8 kbps and 57.6 kbps on the downlink and 14.4 kbps on the uplink since mobile users generally receive more data than they send over the air.

The introduction of General Packet Radio Service (GPRS) is one of the key factors in the evolution of GSM networks to 3G capabilities. The next key step in this process is the implementation of EDGE. EDGE (often referred to as 2.5 G) will allow GSM operators to use the existing GSM radio bands to offer wireless multimedia Internet protocol (IP)-based services and applications at rate of 384 kbps with a bit rate of 48 kbps per time slot, and up to 69.2 kbps per time slot under good RF conditions.

EDGE is an evolution of GSM with high-level modulation (HLM) schemes and link adaptation features. The choice of modulation and coding scheme depends upon the data rate and environment. EDGE satisfies the following requirements of IMT-2000:

- 384 kbps capability for pedestrian (microcell) and low-speed vehicular (macrocell) environment
- 144 kbps for high-speed vehicular environment
- 2 Mbps for indoor office environment using a wideband 1.6 MHz carrier

Figure 11.15 shows the evolution path for cdmaOne, GSM, and IS-136.

The IS-2000 provides specifications for wideband CDMA that support a variety of services. The IS-2000 standard effort is being carried in two phases. Phase I is devoted to a single CDMA carrier (also known as 3G1X) that is backward compatible with cdmaOne, whereas Phase II is concerned with multiple carriers to provide wideband services.

The 3G1X of IS-2000 provides important solutions designed to enhance the system performance and capacity of the existing cdmaOne (2G) systems. These solutions can be deployed with moderate effort and complexity to existing 2G CDMA voice traffic systems. One area where important improvements are needed is power control performance, which leads to higher call capacity in the forward and in the reverse link.

The main enhancement in 3G1X is the significant increase of up to two times the Erlang capacity for voice traffic. This is due, in part, to improvements in forward and reverse power control, as defined in IS-2000.

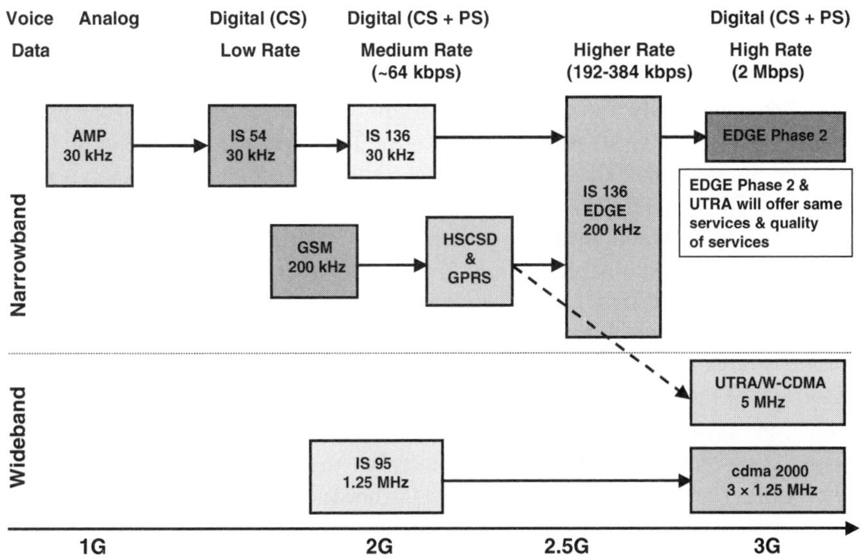

Figure 11.15 Evolution Path for 2G Systems

The introduction of a faster forward power control and pilot-assisted reverse coherent detection have relaxed the power control requirements, so that higher call capacity is achieved in the forward and reverse links. Forward link overload control provides the base station with the capability of using the transmitted power more efficiently to increase the forward link call capacity even further, while maintaining the quality of existing calls. Similarly, reverse link power overload allows the control of reverse call load in order to maintain the reverse link call quality.

3G1X voice will provide higher call capacity, with higher reliability to mobile subscribers. Since less power is required to maintain a call, the battery life of mobile is expected to increase substantially.

IS-2000 supports backward compatibility between 2G and 3G systems, such that a base station providing 3G services can serve an cdmaOne mobile, and a 3G dual-mode mobile can be served by a 2G base station. IS-2000 requires that a 3G1X base station support both cdmaOne and 3G1X services within one CDMA carrier. In addition, the 3G1X base station must provide 2G services to a 2G mobile and to a 3G1X mobile operating in 2G mode. This coexistence requirement allows the 3G1X base station to operate in the overlay mode.

IS-2000 provides new power control requirements, which impact power control at the base station.

11.15 Summary

The international consensus building process and the detailed specification of standards is facilitated by newly formed partnership projects. 3GPP deals with the proposals from Europe (UTRA), Japan (ARIB, TTC), Korea (TTA), and the United States (T1P1); 3GPP2 is working on cdma2000-based proposals. The 3GPP proposal corresponds to a combination of FDD and TDD to support all services and systems needs in a spectrum-efficient manner. For the evolution and migration from 2G systems, an evolutionary path in the core network and revolutionary approach for the radio interface are envisaged. One major evolution path to IMT-2000 is based on the GSM core network with the dominant worldwide subscriber and network base.

The OHG, in cooperation with manufacturer community, has achieved a harmonized concept for the CDMA-based proposals (G3G) by aligning radio parameters as far as possible, and a combined protocol stack to enable the development of multimode terminals and connection of this CDMA-based radio interface to evolved GSM MAP and ANSI-41 core networks. The G3G concept will support the needs of international operators and end users for terminal and global roaming. International standardization activities based on this approach are ongoing [12].

11.16 References

1. UMTS Forum, "A Regulatory Framework for UMTS," Rep. 1, June 26, 1997.
2. ETSI SMG, "Proposal for a Consensus Decision on UTRA," ETSI SMG Tdoc 032/98.
3. FRAMES, "ACTS FRAMES Workshop," *1st International Symposium on Wireless Personal Multimedia Communications*, Yokosuka, Japan, Nov. 1998, pp. 12–63.
4. DaSilva, J. S., et al., "European Third-Generation Systems," *IEEE Communications Magazine*, Oct. 1996 [Vol. 35(10), pp. 68–83].
5. Prasad, R., Konhauser, W., and Mohr, W., *Third Generation Radio Systems*, Artech House: Boston, 2000.
6. Berruto, E., et al., "Research Activities on UMTS Radio Interface, Network Architecture, and Planning," *IEEE Communications Magazine*, Feb. 1998 [Vol. 36(2), pp. 82–95].
7. Ohno, K., et al., "Wideband Coherent DS-CDMA," *Proceedings IEEE VTC '95*, July 1995, pp.779–783.
8. Adachi, F., et al., "Coherent DS-CDMA: Promising Multiple Access for Wireless Multimedia Mobile Communications," *Proceedings IEEE ISSSTA'96*, pp. 351–358.
9. Miya, K., et al., "Wideband CDMA Systems in TDD-Mode Operation for IMT-2000," *IEICE Transactions on Communications*, July 1998 [Vol. DE81 1B, No. 7, pp. 1317–1326].

10. Dahlman, E., et al., "UMTS/IMT-2000 Based Wideband CDMA," *IEEE Communications Magazine,* Sept. 1998 [Vol. 36(9), pp. 70–80].

11. OHG, "Harmonization Global 3G (G3G) Technical Framework for ITU IMT-2000 CDMA Proposal," May 1999, submitted to ITU-R, Beijing, June 1999.

12. Chaudhury, P., et al., "The 3GPP Proposal for IMT-2000," *IEEE Communications Magazine*, December 1999 [Vol. 37(12), pp. 72–81].

CHAPTER 12

Evolution of TDMA-Based 2G Systems to 3G Systems

12.1 Introduction

Third-generation (3G) wireless systems will offer access to services anywhere from a single terminal; the old boundaries between telephony, information, and entertainment services will disappear. Mobility will be built into many of the services currently considered as fixed, especially in such areas as high-speed access to the Internet, entertainment, information, and electronic commerce (e-commerce) services. The distinction between the range of services offered via wireline or wireless will become less and less clear and, as the evolution toward 3G mobile services speeds up, these distinctions will disappear within a decade.

Applications for 3G wireless networks will range from simple voice-only communications to simultaneous video, data, voice, and other multimedia applications. One of the main benefits of 3G is that it will allow a broad range of wireless services to be provided efficiently to many users.

Packet-based Internet protocol (IP) technology will be at the core of the 3G services (refer to Chapters 13 and 14 for details). Users will have continuous access to online information. E-mail messages will arrive at hand-held terminals nearly instantaneously and business users will be able to stay permanently connected to company intranets. Wireless users will be able to make video conference calls to the office and surf the Internet simultaneously, or play computer games interactively with friends in other locations. Figure 12.1 shows the bit rate requirement for various services.

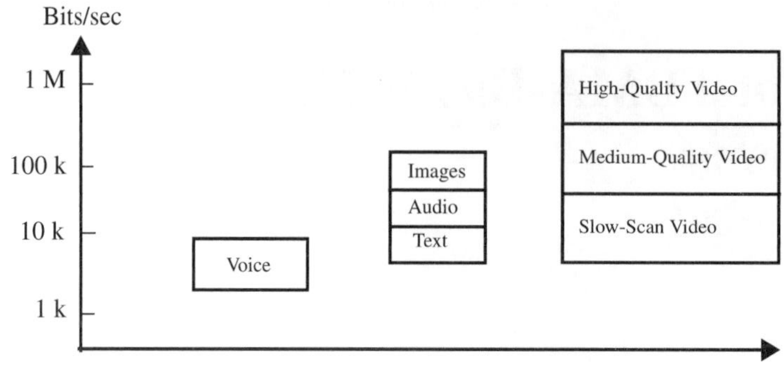

Figure 12.1 User Data Requirements

In 1997, the TIA/EIA IS-136 community, through the Universal Wireless Communications Consortium (UWCC) and Telecommunications Industry Association (TIA) TR 45.3, adopted a three-part strategy for evolving its IS-136 time division multiple access (TDMA)-based networks to 3G wireless networks in order to satisfy IMT-2000 requirements. The strategy consists of

- Enhancing the voice and data capabilities of the existing 30 kHz carrier (IS-136+)
- Adding a 200-kHz carrier (enhanced data rates for GSM evolution [EDGE]) for high-speed data (384 kbps) in high-mobility applications
- Introducing a 1.6-MHz carrier for very high-speed data (2 Mbps) in low-mobility applications (W-TDMA, FMA1 without spreading)

The highlight of this strategy was the global convergence of IS-136 TDMA with Global System for Mobile (GSM) Communications TDMA through the evolution of the 200-kHz GSM carrier for supporting high-speed data applications (384 kbps) while also improving the 30-kHz carrier for voice and mid-speed data applications.

In this chapter, we first discuss enhancements to IS-136 and then focus on general packet radio service (GPRS) for providing packet data services in IS-136 and GSM systems. We then concentrate on EDGE, which will be used for converging the IS-136 and GSM systems to offer IMT-2000-based 3G services.

12.2 IS-136+

The goal of the IS-136 voice services program is to develop a higher-quality voice service, focusing on enhancements to voice quality under fading channels, high background noise, tandeming, and music conditions. At the beginning of the IS-136+ program, it was recognized that the largest opportunity for improvement in faded channels existed in the downlink for two reasons:

1. Downlink was perceived as the limiting link in urban areas.
2. Uplink enhancements such as interference cancellation had recently been shown to provide large uplink gains.

Enhancements to IS-36 include:

- An improved channel coding (CC2) and interleaving
- An improved vocoder US1/enhanced full-rate (EFR)
- Improved modulation scheme $\pi/4$ -DQPSK and 8-PSK

The details of these enhancements are discussed below.

With the definition of a new time slot format, improved channel encoding (CC2), and interleaving options, the robust voice service mode can achieve an additional 4 dB faded channel improvement over the existing IS-641 vocoder (CC1). One of the major difference between CC1 and CC2 is that in CC2 certain fields are eliminated from the downlink (base station to mobile station) slot structure to free 18 bits for use as additional channel coding. The CC2 convolutional encoder uses a tail bit and a higher-constraint length code ($K = 7$ instead of $K = 6$ in CC1) to achieve channel coding gain over CC1. CC2 also supports a 3-slot interleaving mode for improved time diversity over the conventional 2-slot interleaving mode used in CC1.

The detailed CC2 downlink slot structure is given in Figure 12.2.

28	142	12	136	1	1	4
SYNC	Data	CDVCC	Data	F	RSVD	PRAMP

Figure 12.2 CC2 Downlink Slot Format in IS-136+

In the CC2 downlink, a 28-bit SYNC field is used by the receiver for synchronization purposes, a 142-bit and a 136-bit data field together form the total 278-bit data field, a 12-bit coded digital verification color code (CDVCC) field is used to minimize channel interference, a 1-bit fast power control (F) field is used for a faster version of uplink power control, and a 1-bit reserved (RSVD) field and 4-bit power ramp (PRAMP) field allow time for changes in downlink output power. The total number of bits in one slot is 324. The major difference between CC2 and CC1 time slot structures is that the slow associated control channel (SACCH) and coded digital control channel locator (CDL) fields are not used in CC2. The removal of the SACCH has little impact since all messages can also be sent via a fast associated control channel (FACCH) message, which replaces the voice information with signaling data. Although the FACCH replaces voice, it has been found that if the FACCH messages are sent either between talk spurts or spaced far enough in time, they are unnoticeable. Table 12.1 gives bit allocation for IS-641A algebraic code excited linear prediction (ACELP) vocoder. CC2 with $K = 7$ provides a 2-dB improvement in frame error rate (FER) over CC1 with $K = 6$ at 10 Hz Doppler shift, where K is the constraint length in channel coding.

The 1-slot format (#1) (refer to Table 12.2) has no interleaving delay, but requires 6 dB more link margin than the conventional 2-slot format to support a 1 percent Class Ia FER at 10 Hz Doppler shift. To ensure adequate voice quality, the 1-slot format is best used for indoor applications and environments in which signal-to-interference (S/I) and signal-to-noise (S/N) are relatively high (20 dB). The 3-slot format (#3) is the extra robustness mode, which, in conjunction with CC2, provides about 3.7 dB improvement in downlink performance. To minimize extra delay, 3-slot interleaving is limited to only one link at a time. This ensures that for mobile-to-mobile calls, the increase in delay over 2-slot format will be limited to 20 ms. Notice application of format #4 adds additional time diversity with 3-slot interleaving to the space diversity common in existing base stations. The additional time diversity gain is about 0.5 dB less at 1 percent FER that seen on the downlink in format #3 [1,2].

Table 12.1 Bit Allocation for IS-641-A ACELP Vocoder

Information	Number of Bits per Frame
LP filter coefficients	26
Adaptive excitation	26
Fixed or algebraic excitation	68
Gains	28
Total bits	148
Rate	7.4 kbps

Table 12.2 Interleaving Options for CC2

Format	Uplink Interleaving	Downlink Interleaving
#1	1-slot	1-slot
#2	2-slot	2-slot
#3	2-slot	3-slot
#4	3-slot	2-slot

12.2.1 US1/EFR Vocoder

The US1 vocoder is identical to the GSM enhanced full-rate (EFR) vocoder, which is also used by North American GSM1900 operators. The US1 vocoder operates at 12.2 kbps, and under high S/I and S/N offers a high-quality voice service. The US1 vocoder is identical to the IS-641 vocoder in basic structure. Both vocoders are based on ACELP, with the major difference being that the US1 vocoder employs more bits to represent the various speech parameters (see Table 12.3). Since 244 bits are generated in every 20-ms speech frame, the resulting output bit rate is 12.2 kbps.

Figures 12.3 and 12.4 show channel coding for the downlink and uplink used with US1 vocoder. The coded and interleaved bits are combined (see Figure 12.3) to form a series of 3-bit sequences or triads. For the downlink, there are 399 bits (133 triads), whereas for the uplink there are 372 bits (124 triads). For the downlink, the bits are combined such that every Class Ia bit (coded and interleaved) has one Class II bit to form the first 89 triads. The remaining 44 triads are formed from Class Ib bits.

For the uplink, the first 86 triads include two Class Ia bits and one Class II bit. The next three triads are composed of two Class Ib and one Class II bit, whereas the remaining 35 triads are formed from Class Ib bits only.

Table 12.3 Bit Allocation for US1 ACELP Vocoder

Information	Number of Bits per Frame
LP filter coefficients	38
Adaptive excitation	30
Fixed or algebraic excitation	140
Gains	36
Total bits	244
Rate	12.2 kbps

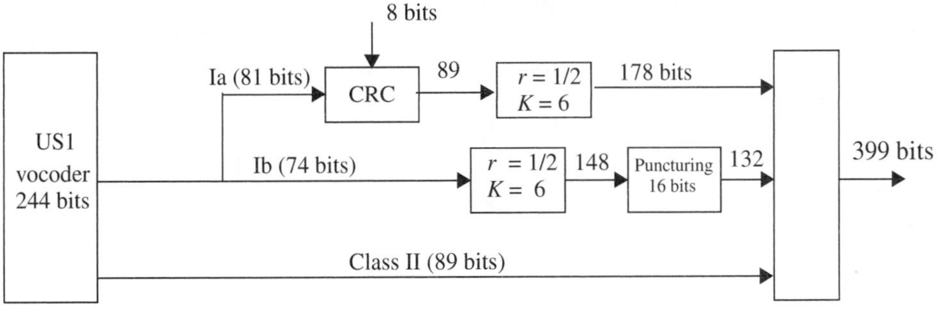

Figure 12.3 Channel Coding for Downlink with US1 Vocoder

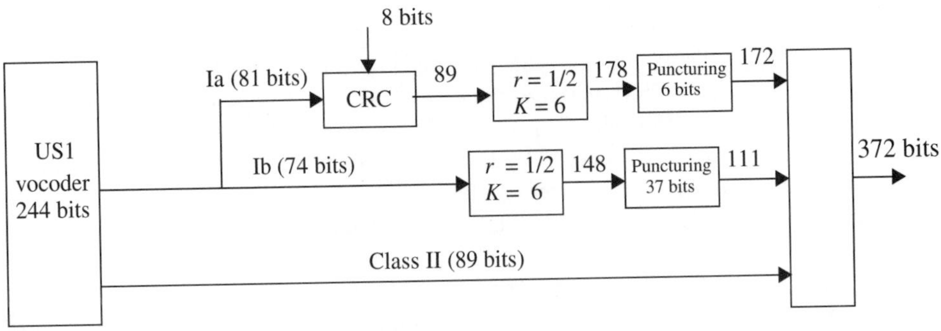

Figure 12.4 Channel Coding for Uplink with US1 Vocoder

The triads are reordered to provide additional interleaving gain [1], and then intraslot interleaving is applied. As with CC2, there are three primary interleaving options: 1-slot, 2-slot, and 3-slot. In 1-slot interleaving, there is no intraslot interleaving and current triads are simply transmitted in the current slot. In 2-slot interleaving, certain triads (from the current triad vector) are transmitted in the current slot, and remaining triads in the next slot. Thus, the current slot contains triads from the current and previous triad vector. In 3-slot interleaving, the concept is extended to include another set of triads in each transmitted slot. The current slot contains triads from the current triad vector, the previous triad vector, and the one before the previous triad vector. To minimize delay in mobile-to-mobile calls, the 3-slot option cannot be used simultaneously on both the uplink and downlink.

The selection of US1 for use in IS-136 is the initial step for the convergence of speech coding technologies between IS-136 and GSM-based systems. US1 has been deployed by both GSM1900 operators in North America and GSM900/1800 operators in Europe, Asia, and Africa.

TIA TR 45.3 committee modified the original GSM 200 kHz channel coding used with GMSK modulation by optimizing for the 30-kHz channel used in IS-136.

The goal of the European Telecommunications Standards Institute (ETSI) adaptive multirate (AMR) coder program was to develop a robust full- and half-rate solution to provide significant improvement in voice quality at low S/I and S/N. The AMR design selected by ETSI incorporates multiple submodes for use in full- or half-rate modes that are determined by the channel quality. The defined submodes, speech coder source rates, and channel coding rates for the ETSI AMR full-rate and half-rate are listed in Tables 12.4 and 12.5. The speech coder source rates common to the AMR design, US1/GSM EFR, and IS-136 full-rate speech coder (IS-641) are in parentheses. The full-rate AMR has submodes that incorporate bit-exact versions of both the 12.2-kbps US1/GSM EFR and 7.4-kbps IS-641 full-rate speech coders. Table 12.6 gives the mean opinion score (MOS) results for speech coded with background noise. Table 12.7 compares the delay in milliseconds for IS-641, GSM-EFR, and G.728 vocoders.

Table 12.4 ETSI AMR Full-Rate Submodes

Submode	Speech Coder Rate (kbps)	Channel Coding Rate (kbps)
1	12.2 (US1/GSM EFR)	10.6
2	10.2	12.6
3	7.95	14.85
4	7.4 (IS-641)	15.4
5	6.7	16.1
6	5.9	16.9
7	5.15	17.65
8	4.75	18.05

Table 12.5 ETSI AMR Half-Rate Submodes

Submode	Speech Coder Rate (kbps)	Channel Coding Rate (kbps)
1	7.95	3.45
2	7.4 (IS-641)	4.0
3	6.7	4.7
4	5.9	5.5
5	5.15	6.25
6	4.75	6.65

Table 12.6 Mean Opinion Score (MOS) for Speech Coded with and without Background Noise

Condition	Original	IS-641	G.728	GSM-EFR
Clean speech	4.34	4.09	4.23	4.26
Clean speech*		3.62	3.99	4.13
15 dB babble	3.75	3.49	3.81	3.70
15 dB babble*		3.08	3.69	3.47
20 dB car noise	3.72	3.61	3.64	3.75
20 dB car noise*		3.11	3.58	3.48
15 dB office noise	3.70	3.40	3.51	3.58
15 dB office noise*		2.75	3.55	3.31
15 dB music	3.99	3.82	3.98	3.99
15 dB music*		3.16	3.92	3.85

* With background noise

Table 12.7 Delay (ms) for Systems Based on IS-641, GSM-EFR, and G.728

Delay Cause	IS-641	G.728	GSM-EFR
Look-ahead	5	0	0
Frame size	20	20	0.625
Processing	16	16	0.5
Bitstream buffer	0	0	19.375
Transmission	26.6	6.6	6.6
Delay	67.6	42.6	27.1

12.2.2 Modulation

Differential Quadrature Phase Shift Keying (DQPSK). The modulation scheme used for the CC2 is $\pi/4$-DQPSK. In $\pi/4$-DQPSK, every two bits of information are encoded into one modulator symbol. The actual information is differentially encoded in the phase change from one symbol to the next. The most significant bit is the first bit in the input stream, and Gray code mapping is used to minimize the probability of bit error. With $\pi/4$-DQPSK modulation, the 324 input bits per slot are translated into 162 modulator output symbols. The IS-136 symbol is 24.3 kHz, and thus the raw or gross instantaneous bit rate is 48.6 kbps. The gross rate for a full-rate user is 16.2 kbps, which consists of 7.4 kbps of speech, 6.5 kbps of channel coding, and 2.3 kbps for the remaining field within the time slot.

The CC1 and CC2 provide a spectral efficiency of 1.62 b/s/Hz based on 48.6 kbps gross bit rate and an effective channel bandwidth of 30 kHz.

8-PSK. The design goal of the 8-PSK modulation was to maintain the 24.3-kHz symbol rate so that the existing transmit and receive filters (square-root raised cosine with roll-off of 0.55) could be used and the existing 30 kHz channel bandwidth is maintained. The slot length of 6.67 ms and 162 symbols per slot were also to be maintained. Thus, with 3 bits per symbol, 8-PSK modulation supports 486 bits per slot, which is consistent with the slot structures (see Figure 12.5). The instantaneous gross bit rate is 72.9 kbps, which is 50 percent more than that supported by $\pi/4$-DQPSK. For full-rate voice users using the US1 vocoder with 8-PSK modulation, the effective gross rate is 24.3 kbps. For downlink, this 24.3 kbps consists of 12.2 kbps of speech bits, 7.55 kbps of channel coding, and 4.75 kbps for the remaining field within the slot. For the uplink, the speech bit rate is the same, but the channel coding contribution is 6.0 kbps, and the remaining fields contribution is 6.1 kbps. 8-PSK modulation provides a spectral efficiency of 2.43 b/s/Hz, a 50 percent increase over $\pi/4$-DQPSK.

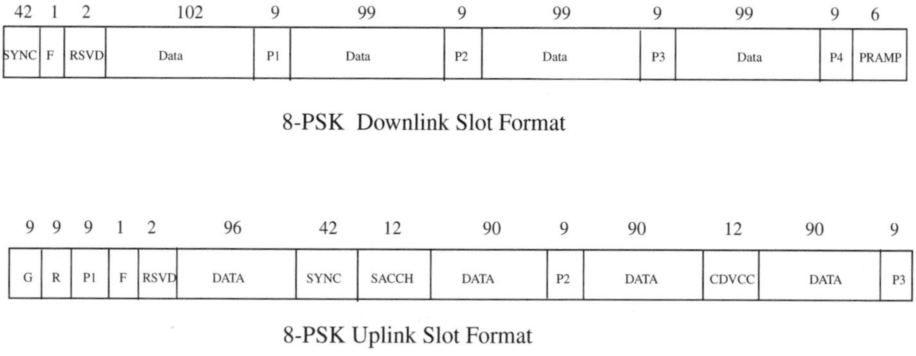

Figure 12.5 8-PSK Slot Structure

Figure 12.6 shows a pictorial representation of 8-PSK bit-to-symbol mapping.

The detailed downlink and uplink slot structures for the US1 vocoder with 8-PSK modulation are shown in Figure 12.5. One of the major difference between this slot structure and those of CC1 and CC2 slot formats is the inclusion of pilot fields, which are included to facilitate coherent detection at the receiver by developing an estimate of the channel condition.

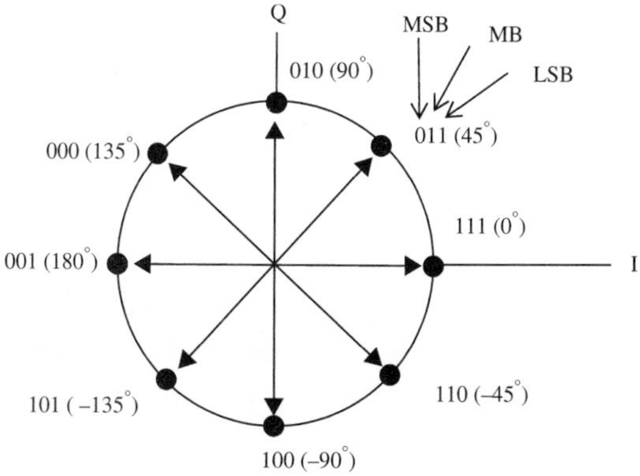

Figure 12.6 8-PSK Bit-to-Symbol Mapping

Given that the earlier versions of IS-136 allowed mobile stations to detect and utilize information (SYNC) from adjacent slots on the same RF carrier, the SYNC in 8-PSK slot format remains modulated at $\pi/4$-DQPSK. Thus, within a given time slot there are both

$\pi/4$-DQPSK and 8-PSK modulations. This produces a potential problem in decoding 8-PSK since it needs a phase reference. This could be provided by adding a reference symbol of known phase (e.g., zero phase) after the differentially encoded SYNC. This will waste 3 bits that could otherwise be used for data, and will place the importance of the phase reference for the entire slot on the reliable detection of only one symbol.

An alternate approach was used to allow channel sharing 8-PSK time slots with $\pi/4$-DQPSK slots. The SYNC remains differentially encoded from the preceding symbol in the previous time slot, and all of its values remain the same as currently defined. After differentially encoding the SYNC from the immediately proceeding symbol, a constant phase-shift equal to the absolute phase of the last symbol of the immediately proceeding time slot is added to all of the symbols following the SYNC. In this manner, the phase-shift of the receiver drives from the SYNC signal can also be removed from the data fields in order to obtain the absolute phases.

12.3 GSM Evolution for Data

From a radio access perspective, adding 3G capabilities to 2G systems mainly means supporting higher bit rates. Possible scenarios depend on spectrum availability for the network service provider. Depending on the spectrum situation, two different migration paths must be supported:

- Reframing of existing spectrum bands
- New or modified spectrum bands

Two 3G radio access schemes have been identified to support the different spectrum scenarios:

- Enhanced data rates for GSM evolution (EDGE) with high-level modulation in 200 kHz TDMA channel is based on plug-in transceiver equipment, thereby allowing the migration of existing bands in small spectrum segments.
- Universal mobile telecommunications services (UMTS) is a new radio access network based on 5-MHz wideband CDMA (W-CDMA) and optimized for efficient support of 3G services. UMTS can be used in both new and existing spectra.

From a network point of view, 3G capabilities implies the addition of packet-switched (PS) services, Internet access, and IP connectivity. With this approach, the existing mobile networks will reuse the elements of mobility support, user authentication/service handling, and circuit-switched (CS) services. PS services and IP connectivity will then be added to provide a mobile multimedia core network by evolving existing mobile network.

In Europe, GSM is moving to develop and enhance cutting-edge, customer-focused solutions to meet the challenges of the new millennium and 3G mobile services. When GSM was first designed, no one could have predicted the dramatic growth of the Internet and the rising demand for multimedia services. These developments have brought about new challenges to the world of GSM. For GSM operators, the emphasis is now rapidly changing from instigating and driving the development of technology and fundamentally enabling mobile data transmission toward improving speed, quality, simplicity, coverage, and reliability in terms of tools and services that will boost mass market application.

People increasingly demand access to information and services wherever they are and whenever they want. GSM should provide that connectivity. Internet access, Web browsing, and the whole range of mobile multimedia capability is the major driver for development of higher data speed technologies.

Current data traffic on most GSM networks is modest, less than 3 percent of total GSM traffic. But with the new initiatives coming to fruition during the course of the next two to three years, exponential growth in data traffic is forecast. Messaging-based applications may reach a penetration of up to 25 percent in developed markets by the year 2001, and 70 percent by 2003. GSM data transmission using high-speed circuit-switched data (HSCSD) and GPRS may reach a penetration of 10 percent by 2001 and 25 percent by 2003.

Today's GSM operators will have two nonexclusive options for evolving their networks to 3G wideband multimedia operation: (1) they can use GPRS and EDGE (discussed below) in the existing radio spectrum and in small amounts of the new spectrum, or (2) they can use wideband CDMA (W-CDMA) in the new 2-GHz bands, or in large amounts of the existing spectrum. Both approaches offer a high degree of investment flexibility because roll-out can proceed in line with market demand and extensive reuse of existing network equipment and radio sites.

In the new 2-GHz bands, 3G capabilities will be delivered using a new wideband radio interface that will offer much higher user data rates than are available today—384 kbps in the wide area and up to 2 Mbps in local areas. Of equal importance for such services will be the high-speed packet switching provided by GPRS and its connection to public and private IP networks.

Even without the new wideband spectrum, GSM and Digital-Advanced Mobile Phone System (D-AMPS) (IS-136) operators will be able to use existing radio bands to deliver 3G services by evolving current networks and deploying GPRS and EDGE technologies. In the early years of 3G service deployment, a large proportion of wireless traffic will still be voice-only and low-rate data. So whatever the ultimate capabilities of 3G networks, efficient and profitable ways of delivering more basic wireless services will still be needed.

The significance of EDGE for today's GSM operators is that it will increase data rates up to 384 kbps and potentially even higher in good quality radio environment, using current GSM spectrum and carrier structures more efficiently. EDGE will both complement and be an alternative to new W-CDMA coverage. EDGE will also have the effect of unifying the GSM, D-AMPS, and W-CDMA services through the use of dual-mode terminals.

12.3.1 High-Speed Circuit Switched Data (HSCSD) in GSM

HSCSD [3,4] is a feature that enables the coallocation of multiple full-rate traffic channels (TCH/F) of GSM into a HSCSD configuration. The aim of HSCSD is to provide a mixture of services with different air interface user rates by a single physical layer structure. The available capacity of a HSCSD configuration is several times the capacity of a TCH/F, leading to a significant enhancement in the air interface data transfer capability.

Ushering faster data rates into the mainstream is the new speed of 14.4 kbps per time slot and HSCSD protocols that approach wire-line access rates of up to 57.6 kbps by using multiple 14.4 kbps time slots. The increase from the current baseline 9.6 kbps to 14.4 kbps is due to a nominal reduction in the error-correction overhead of the GSM radio link protocol (RLP), allowing the use of a higher data rate. Implementation of v.4.2 bits compression could double the throughput.

For operators, migration to HSCSD brings data into the mainstream, enabled in many cases by relatively standard software upgrades to base station (BS) and mobile switching center (MSC) equipment. Flexible air interface resource allocation allows the network to dynamically assign resources related to the air interface usage according to network operator's strategy, and the end-user's request for a change in the air interface resource allocation based on data transfer needs. The provision of the asymmetric air interface connection allows simple mobile equipment (Type 1) to receive data at higher rates than would otherwise be possible with a symmetric connection.

For end-users, HSCSD enables the roll-out of mainstream high-end segment services that enable faster Web browsing, file downloads, mobile video-conference and navigation, vertical applications, telematics, and bandwidth-secure mobile LAN access. Value-added service providers (VASP) will also be able to offer guaranteed quality of service and cost-efficient mass-market applications, such as direct IP where users make circuit-switched data calls straight into a GSM network router connected to the Internet. To the end-user, the VASP or the operator is equivalent of an Internet service provider (ISP) that offers a fast secure dial-up IP service at cheaper mobile-to-mobile rates.

HSCSD is provided within the existing mobility management. Roaming is also possible. The throughput for an HSCSD connection remains constant for the duration of the call, except for interruption of transmission during handoff. The handoff is simultaneous for all time slots making up an HSCSD connection. End-users wanting to use HSCSD have to subscribe general bearer services. Supplementary services applicable to the general bearer services can be used simultaneously with HSCSD.

Firmware on most current GSM PC cards will have to be upgraded. The reduced RLP layer also means that a stronger signal strength will be necessary. Multiple time slot usage will probably only be efficiently available in off-peak times, increasing overall off-peak idle capacity usage.

12.3.2 General Packet Radio Service (GPRS) in GSM

The next phase in the high-speed road map will be the evolution of current short message services (SMS), such as smart messaging and unstructured supplementary service data (USSD), toward the new GPRS [5,6], a packet data service using TCP/IP and X.25 to offer speeds up to 115 kbps. GPRS has been standardized to optimally support a wide range of applications ranging from very frequent transmission of medium to large data volume and infrequent transmission of large data volume. Services of GPRS have been developed to reduce connection setup time and allow an optimum usage of radio resources. GPRS provides a packet data service for GSM where time slots on the air interface can be assigned to GPRS over which packet data from several mobile stations is multiplexed.

A similar evolution strategy, also adopting GPRS, has been developed for D-AMPS (IS-136). For operators planning to offer wideband multimedia services, the move to GPRS packet-based data bearer service is significant; it is a relatively small step compared with building a totally new 3G IMT-2000 network. Use of the GPRS network architecture for IS 136+ packet data service enables data subscription roaming with GSM networks around the globe that support GPRS and its evolution. The IS-136+ packet data service standard is known as GPRS-136. GPRS-136 provides the same capabilities as GSM GPRS. The user can access either X.25 or IP-based data networks.

GPRS provides a core network platform for current GSM operators not only to expand the wireless data market in preparation for the introduction of 3G services, but also a platform on which to build IMT-2000 frequencies should they acquire them.

GPRS enhances GSM data services significantly by providing end-to-end packet-switched data connections. This is particularly efficient in Internet/intranet traffic, where short bursts of intense data communications activity are interspersed with relatively long periods of inactivity. Because there is no real end-to-end connection to be established, setting up a GPRS call is almost instantaneous and users can be continuously online. Users have the additional benefit of paying for the actual data transmitted, rather than for connection time.

Because GPRS does not require any dedicated end-to-end connection, it only uses network resources and bandwidth when data is actually being transmitted. This means that a given amount of radio bandwidth can be shared efficiently and simultaneously among many users.

The implementation of GPRS has a limited impact on the GSM core network. It simply requires the addition of new packet data switching and gateway nodes, and an upgrade to existing nodes to provide a routing path for packet data between the wireless terminal and a gateway node. The gateway node provides interworking with external packet data networks for access to Internet, intranets, and databases.

A GPRS architecture for GSM is shown in Figure 12.7. GPRS will support all widely used data communications protocols, including IP, so it will be possible to connect with any data source from anywhere in the world using a GPRS mobile terminal. GPRS will support applications ranging from low-speed short messages to high-speed corporate LAN communications. However, one of the key benefits of GPRS—that it is connected through the existing GSM air interface modulation scheme—is also a limitation, restricting its potential for delivering data rates higher than 115 kbps. To build even higher rate data capabilities into GSM, a new modulation scheme is needed.

GPRS can be implemented in the existing GSM systems. It requires only minor changes in an existing GSM network. The base station subsystem (BSS) consists of base station controller (BSC) and packet control unit (PCU). The PCU supports all GPRS protocols for communication over the air interface. Its function is to set up, supervise, and disconnect packet-switched calls. PCU supports cell change, radio resource configuration, and channel assignment. The base transceiver station (BTS) is a relay station without protocol functions. It performs modulation and demodulation.

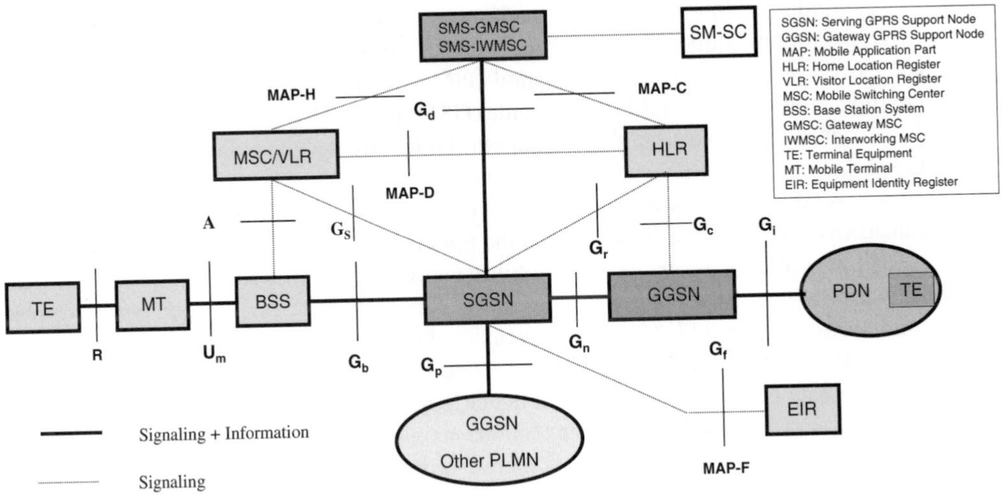

Figure 12.7 A GPRS Architecture in GSM

The GPRS standard introduces two new nodes, the serving GPRS support node (SGSN) and the gateway GPRS support node (GGSN). The home location register (HLR) is enhanced with GPRS subscriber data and routing information. Two types of services are provided by GPRS:

- Point-to-point (PTP)
- Point-to-multipoint (PTM)

Independent packet routing and transfer within the public land mobile network (PLMN) is supported by a new logical network node called the GPRS support node (GSN). The GGSN acts as a logical interface to external packet data networks. Within the GPRS networks, protocol data units (PDUs) are encapsulated at the originating GSN and decapsulated at the destination GSN. In between the GSNs, IP is used as the backbone to transfer PDUs. This whole process is referred to as *tunneling* in GPRS. The GGSN also maintains routing information used to tunnel the PDUs to the SGSN that is currently serving the mobile. All GPRS user-related data required by the SGSN to perform the routing and data transfer functionality is stored within the HLR. In GPRS, a user may have multiple data sessions in operation at one time. These sessions are called packet data protocol (PDP) contexts. The number of PDP contexts that are open for a user is limited only by the user's subscription and any operational constraints of the network. The main goal of the GPRS-136 architecture is to integrate IS-136 and GSM GPRS as much as possible with minimum changes to both technologies. In order to provide subscription roaming between GPRS-136 and GSM GPRS networks, a separate functional GSM GPRS HLR is incorporated into the architecture in addition to the IS-41 HLR.

The ETSI has specified GPRS as an overlay to the existing GSM network to provide packet data services. In order to operate a GPRS service over a GSM network, new functionality has been introduced into existing GSM network elements (NEs) and new NEs are integrated into the existing service provider GSM network.

The BSS of GSM is upgraded to support GPRS over the air interface. The BSS works with the GPRS backbone system (GBS) to provide GPRS service in a manner similar to its interaction with the switching subsystem for the circuit-switched services. The GBS manages the GPRS sessions set up between the mobile terminal and the network by providing functions such as admission control, mobility management (MM), and service management (SM). Subscriber and equipment information is shared between GPRS and the switched functions of GSM by the use of a common HLR and coordination of data between the visitor location register (VLR) and the GPRS support nodes of the GBS. The GBS is composed of two new NEs, the SGSN, and the GGSN.

The SGSN serves the mobile and performs security and access control functions. The SGSN is connected to BSS via frame-relay. The SGSN provides packet routing, mobility management, authentication, and ciphering to and from all GPRS subscribers located in the SGSN service area. A GPRS subscriber may be served by any SGSN in the network, depending on location. The traffic is routed from the SGSN to the BSC and to the mobile terminal via a BTS. At GPRS attach, the SGSN establishes a mobility management context containing information about mobility and security for the mobile. At packet data protocol (PDP) context activation, the SGSN establishes a PDP context which is used for routing purposes with the GGSN that GPRS subscriber uses. The SGSN may send in some cases location information to the MSC/VLR and receive paging requests.

The GGSN provides the gateway to external IP network, handling security and accounting functions as well as dynamic allocation of IP addresses. The GGSN contains routing information for the attached GPRS users. The routing information is used to tunnel PDUs to the mobile's current point of attachment, SGSN. The GGSN may be connected with the HLR via optional interface G_c. The GGSN is the first point of public data network (PDN) interconnection with a GSM PLMN supporting GPRS. From the external IP network's point of view, the GGSN is a host that owns all IP addresses of all subscribers served by the GPRS network.

The point-to-multipoint service center (PTM-SC) handles PTM traffic between the GPRS backbone and the HLR. The nodes will be connected by an IP backbone network. The SGSN and GGSN functions may be combined in the same physical node or separated, even residing in different mobile networks.

A special interface (G_s) is provided between MSC/VLR and SGSN to coordinate signaling for mobile terminals that can handle both circuit-switched and packet-switched data.

The HLR contains GPRS subscription data and routing information, and can be accessible from the SGSN. For the roaming mobiles, the HLR may reside in a different PLMN than the current SGSN. The HLR also maps each subscriber to one or more GGSNs.

The objective of the GPRS design is to maximize the use of existing GSM infrastructure while minimizing the changes required within GSM. The GSN contains most of the necessary capabilities to support packet transmission over GSM. The critical part in the GPRS network is the mobile-to-GSN (MS-SGSN) link, which includes the MS-BTS, BTS-

BSC, BSC-SGSN, and the SGSN-GGSN link. In particular, the U_m interface including the radio channel is the bottleneck of the GPRS network due to spectrum and channel speed/quality limitations. Since multiple traffic types of varying priorities will be supported by the GPRS network, quality of service criteria as well as resource management is required for performance evaluation.

The BSC will require new capabilities for controlling the packet channels, new hardware in the form of a PCU and new software for GPRS mobility management and paging. The BSC will also have a new traffic and signaling interface from SGSN.

The BTS will have new protocols supporting packet data for the air interface, together with new slot and channel resource allocation functions. The utilization of resources will be optimized through dynamic sharing between the two traffic types, handled by the BSC.

MS-SGSN Link. The logical link control (LLC) layer is responsible for providing a link between the mobile station (MS) and the SGSN. It governs the transport of GPRS signaling and traffic information from the MS to the SGSN. GPRS supports three service access points (SAPs) entities: the layer 3 management, subnet dependent convergence, and short message service (SMS). On the MS-BSS link, the radio link control (RLC), the media access control (MAC), and GSM RF protocols are supported (see Figure 12.8).

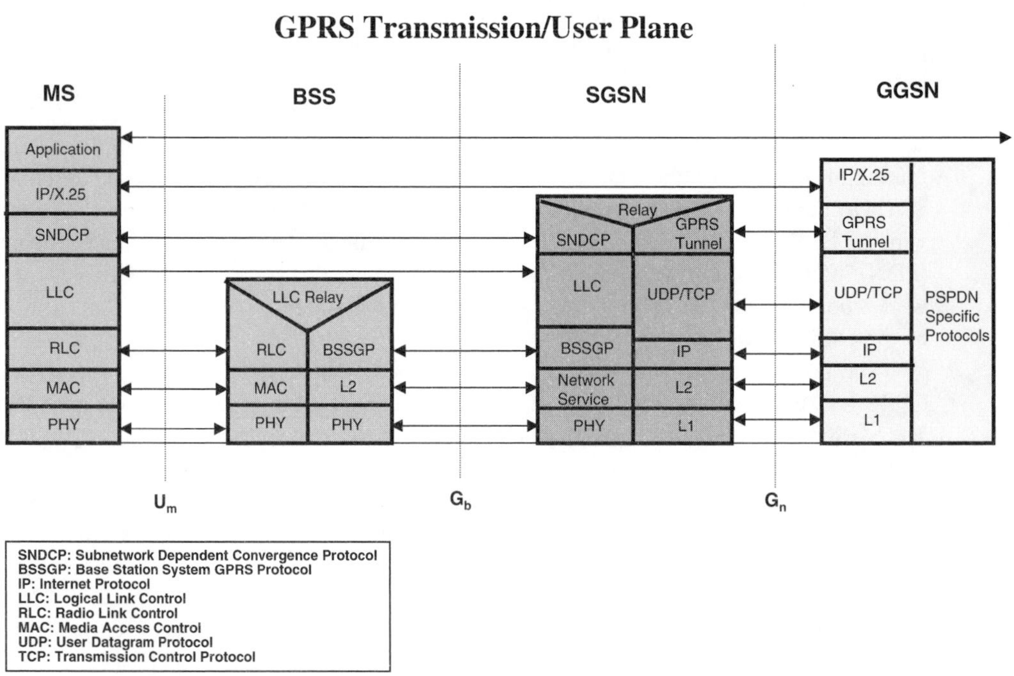

Figure 12.8 Protocol Stack in GPRS

The main drawback in implementing GPRS on an existing GSM infrastructure is that the GSM network is optimized for voice transmission (i.e., the GSM channel quality is designed for voice, which can tolerate errors at a predefined level). It is therefore expected that GPRS could have varied transmission performance in different network or coverage areas. To overcome this problem, GPRS supports multiple coding rates at the physical layer.

GPRS could share radio resources with GSM circuit-switched (CS) service. This is governed by a dynamic resource sharing based on the capacity on demand criteria. GPRS channel is allocated only if an active GPRS terminal exists in the network. Once resources are allocated to GPRS, at least one channel will serve as the *master* channel to carry all necessary signaling and control information for the operation of GPRS. All other channels will serve as *slave channels* and are only used to carry user and signaling information. If no master channel exists, all GPRS users will use the GSM common control channel (CCCH) and inform the network to allocate GPRS resources.

A physical channel dedicated to GPRS is called a packet data channel (PDCH). It is mapped into one of the physical channels allocated to GPRS. A PDCH can be used either as a packet common control channel (PCCCH) (see Figure 12.9), a packet broadcast control channel (PBCCH), or a packet traffic channel (PTCH).

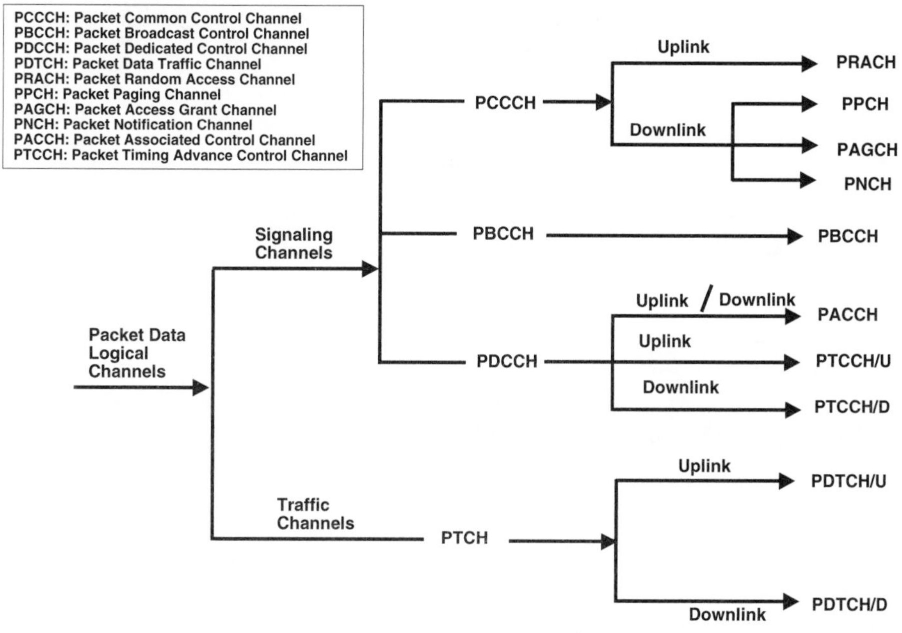

Figure 12.9 GPRS Logical Channels

The PCCCH consists of the following (see Figure 12.9):

- Packet random access channel (PRACH)—uplink
- Packet access grant channel (PAGCH)—downlink
- Packet notification channel (PNCH)— downlink
- On the other hand, the PTCH can either be:
  - Packet data traffic channel (PDTCH)
  - Packet associated control channel (PACCH)

For a given traffic characteristic, GPRS logical channels require a combination of PCCCH and PTCH. Fundamental questions such as how many PDTCHs can be supported by a single PCCCH are required in the dimensioning of GPRS.

RLC/MAC Layer. The multiframe structure of the packet data channel (PDCH) in which GPRS RLC messages are transmitted is composed of 52 TDMA frames organized into RLC blocks of four bursts resulting into 12 blocks per multiframe plus four idle frames located in the 13th, 26th, 39th, and 52nd position (see Figure 12.10).

B_0 consists of frames 1, 2, 3, and 4; B_1 consists of frames 5, 6, 7, 8, and so on. It is important that the mapping of logical channels onto the radio blocks is by means of an ordered set of blocks (B_0, B_6, B_9, B_1, B_7, B_4, B_{10}, B_2, B_8, B_5, B_{11}, B_3). The advantage of ordering the blocks is mainly to spread the locations of the control channels in each time slot reducing, the average waiting time for the users to transmit signaling packets. Secondly, it provides an interleaving of the GPRS multiframe.

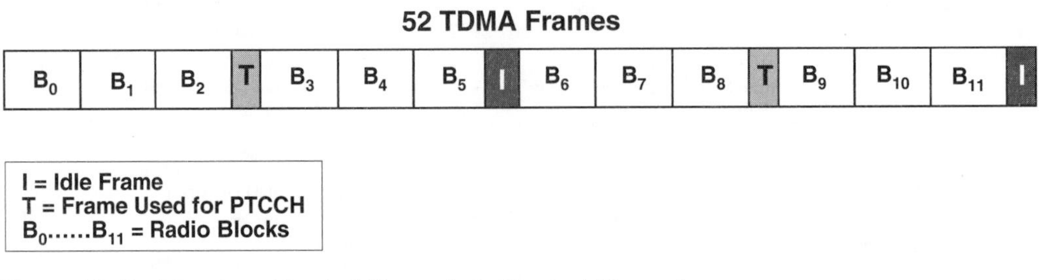

Figure 12.10 Mapping of Logical Channels to Physical Channels

GPRS uses a reservation protocol at the media access control layer. Users that have packets ready to send request a channel via the PRACHs. The random access burst consists of only one TDMA frame of duration enough to transmit an 11-bit signaling message. Only the PDCHs carrying PCCCHs contain PRACHs. The blocks used as PRACHs are indicated by an uplink state flag [uplink state flag (USF) = free] by the downlink pair channel.

Alternatively, the first K blocks following the ordered set of blocks can be assigned to PRACH permanently. The access burst is transmitted in one of the four bursts assigned as PRACH. Any packet channel request is returned by a packet immediate assignment on the PRACHs whose locations are broadcasted by PBCCH. Optionally, a packet resource request for additional channels is initiated and returned by a packet resource assignment. The persistence of random access is maintained by the traffic load and user class with a back-off algorithm for unsuccessful attempts. In the channel assignment, one or more PTCHs (time slot) will be allocated to a particular user. A user reserves a specific number of blocks on the assigned PTCH as indicated by the uplink state flag (USF). It is possible to accommodate more than one user per PTCH. User signaling is also transmitted on the same PTCH using the PAGCH, whose usage depends on the user's needs.

The performance of the media access control layer depends on the logical arrangement of the GPRS channels (e.g., allocation of random access channels, access grant channels, broadcast channels) for a given set of traffic statistics. This is determined by the amount of resources allocated for control and signaling as compared with the data traffic. A degree of flexibility is also achieved with logical channels as the traffic varies. The arrangement of logical channels is determined through the PBCCH.

LLC Layer. The LLC layer is responsible for providing a reliable link between the mobile and the SGSN. It is based on the high-level data link control (HDLC) and link access procedure on the D-channel (link access procedure on the D-channel [LAPD]) protocols. It is designed to support variable length transmission in a point-to-point or multipoint topology. It includes layer functions such as sequence control, flow control, error detection, ciphering, and recovery, as well as the provision of one or more logical link connections between two layer 3 entities. A logical link is identified by a data link control identifier (DLCI), which consists of a service access point identifier (SAPI) and terminal equipment identity (TEI) mapped on the LLC frame format. Depending on the status of the logical link, it supports an unacknowledged or an acknowledged information transfer. The former does not support error recovery mechanisms. The acknowledged information transfer supports error and flow control. This operation only applies to point-to-point operations. The LLC frame consists of an address field (1 or 5 octets), control field (2 or 6 octets), a length indicator field (2 octets maximum), information fields (1,500 octets maximum), and frame check sequence of 3 octets. Four types of control field formats are allowed; these include the supervisory functions (S format), the control functions (U), and acknowledged and unacknowledged information transfer (I and UI).

In the performance evaluation, the objective is to determine delay during the exchange of commands and responses involved in various operations supported by the LLC in relation to the transfer of an LLC PDU. The LLC commands and responses are exchanged between two layer 3 entities in conjunction with a service primitive invoke by the mobile or the SGSN.

Data Packet Routing in GPRS Network. Here, we discuss data packet routing for the mobile-originated and mobile-terminated data call scenarios [7]. In the case of mobile-originated data routing, the mobile gets an IP packet from an application and requests a channel reservation. The mobile transmits data in the reserved time slots. The packet-switched public data network (PSPDN) PDU is encapsulated into a subnetwork dependent convergence (SNDC) protocol unit that is sent via the LLC protocol over the air interface to the SGSN currently serving the mobile (see Figure 12.11).

For mobile-terminated data routing, we have two cases: routing to a home GPRS network, and routing to a visited GPRS network. In the first case, a user sends a data packet to a mobile. The packet goes through the local area network (LAN) via a router out on the GPRS context for the mobile. If the mobile is in GPRS idle state, the packet is rejected. If the mobile is in standby or active mode, the GGSN routes the packet in an encapsulated format to SGSN.

In the second case, the home GPRS network sends the data packet over the interoperator backbone network to the visiting GPRS network. The visiting GPRS network routes the packet to the appropriate SGSN (see Figure 12.11).

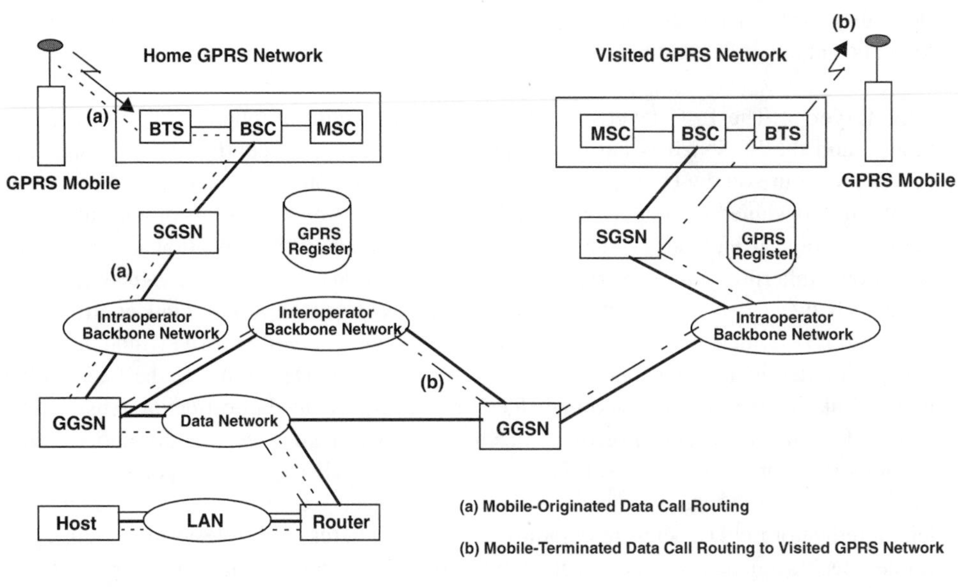

Figure 12.11 Data Call Routing in GPRS Network

Point-to-point and point-to-multipoint applications of GPRS are as follows:

- Point-to-point
  - messaging (e.g., e-mail)
  - remote access to corporate networks
  - access to the Internet
  - credit card validation (point-of-sale)
  - utility meter readings

  - road toll applications
  - automatic train control
- Point-to-multipoint
  - PTM multicast (send to all)
  - news
  - traffic information
  - weather forecasts
  - financial updates
  - PTM group call (send to some)
  - taxi fleet management
  - conferencing

GPRS will provide a service for bursty and bulky data transfer; radio resources on demand; shared use of physical radio resources; existing GSM functionality; mobile applications for the mass application market; volume-dependent charging; and integrated services, operation, and management.

12.3.3 Enhanced Data Rates for GSM Evolution (EDGE)

EDGE provides an evolutionary path that enables existing 2G systems (GSM, IS-136) to deliver 3G services in existing spectrum bands. The advantages of EDGE include fast availability, reuse of existing GSM, IS-136, and PDC infrastructure, as well as support for gradual introduction of 3G capabilities.

EDGE reuses the GSM carrier bandwidth and time slot structure. EDGE can be seen as a generic air interface for efficiently providing high bit rates, facilitating an evolution of existing 2G systems toward 3G systems.

EDGE (2.5G system) [7,8] was designed to enhance user bandwidth through GPRS. This is achieved through the use of higher-level modulation schemes. Although EDGE reuses the GSM carrier bandwidth and time slot structure, the technique is by no means restricted to GSM systems; it can be used as a generic air interface for efficient provision of higher bit rates in other TDMA systems. In the Universal Wireless Communications Consortium (UWCC), the 136 high-speed (136 HS) radio interface was proposed as a means of satisfying the requirements for an IMT-2000 RTT. EDGE was adopted by UWCC in 1998 as the outdoor component of 136 HS to provide 384-kbps data service.

The standardization effort for EDGE has two phases. In the first phase the emphasis has been placed on enhanced GPRS (EGPRS) and enhanced CSD (ECSD). The second phase is being defined with improvements for multimedia and real-time services as possible work items.

EDGE is primarily a radio interface improvement, but it can also be viewed as a system concept that allows GSM and IS-136 networks to offer a set of new services. EDGE has been designed to improve S/I by using link quality control. *Link quality control* adapts the protection of the data to the channel quality so that an optimal bit rate is achieved for all channel qualities.

The EDGE air interface is designed to facilitate higher bit rates than those currently achievable in existing 2G systems. The modulation scheme based on 8-PSK is used to increase the gross bit rate. GMSK modulation as defined in GSM is also part of the EDGE system. The symbol rate is 271 kbps for both GMSK and 8-PSK, leading to gross bit rates per time slot of 22.8 kbps and 69.2 kbps, respectively. The 8-PSK pulse shape is linearized GMSK to allow 8-PSK to fit into the GSM spectrum mask. The 8-PSK burst format is similar to GSM (see Figure 12.12).

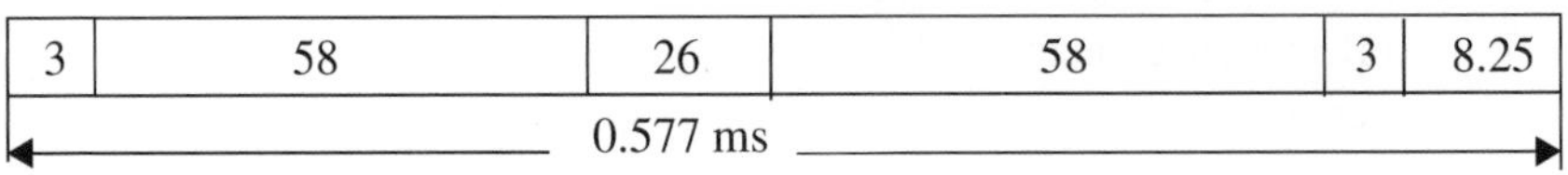

Figure 12.12 Burst Format for EDGE

In order to achieve a higher gross rate, a new modulation scheme, quaternary offset quadrature amplitude modulation (QOQAM), has been proposed for EDGE, since it can provide higher data rates and good spectral efficiency. An offset modulation scheme is proposed because it gives smaller amplitude variation than 16-QAM, which can be beneficial when using nonlinear amplifiers. EDGE will coexist with GSM in the existing frequency plan and will provide *link adaptation* (i.e., modulation and coding are adapted for channel conditions).

12.3.4 Radio Protocol Design

The radio protocol strategy in EDGE is to reuse the protocols of GSM/GPRS whenever possible, thus minimizing the need for new protocol implementation. EDGE enhances both GSM circuit-switched (HSCSD) and packet-switched (GPRS) mode operation. EDGE includes one packet-switched (PS) and one circuit-switched (CS) mode, EGPRS and ECSD, respectively.

Enhanced GPRS (EGPRS). The EDGE radio link control (RLC) protocol is somewhat different from the corresponding GPRS protocol. The main changes are related to improvements in the link quality control scheme.

A link adaptation scheme regularly estimates the link quality and subsequently selects the most appropriate modulation and coding scheme for transmission to maximize the user bit rate. The link adaptation scheme offers mechanisms for choosing the best modulation and coding alternative for the radio link. In GPRS, only the coding schemes can be changed between two consecutive link layer control (LLC) frames. In the EGPRS, even the modulation can be changed. Different coding and modulation schemes enable adjustment for the robustness of the transmission according to the environment.

Another way to handle link quality variations is *incremental redundancy*. In this scheme, information is first sent with very little coding, yielding a high bit rate if decoding is immediately successful. If decoding is not successful, additional coded bits (redundancy) are sent until decoding succeeds. The more coding that has to be sent, the lower the resulting bit rate and the higher the delay.

EGPRS will support a combined link adaptation and incremental redundancy schemes. In this case, the initial code rate of the incremental redundancy scheme is based on measurements of the link quality. Benefits of this approach are the robustness and high throughput of the incremental redundancy operation in combination with lower delays and lower memory requirements enabled by the adaptive initial code rate.

In EGPRS, the different initial code rates are obtained by puncturing a different number of bits from a common convolutional code ($r = 1/3$). The resulting coding schemes are given in Table 12.8. Incremental redundancy operation is enabled by puncturing a different set of bits each time a block is retransmitted, whereby the code rate is gradually decreased toward 1/3 for every new transmission of the block. The selection of the initial modulation and code rate is based on regular measurements of link quality.

Actual performance of modulation and the coding scheme together with channel characteristics form the basis for link adaptation. Channel characteristics are needed to estimate the effects of a switch to another modulation and coding combination; these include an estimated S/I ratio, but also time dispersion and fading characteristics (that affect the efficiency of interleaving).

In the case of GSM, EDGE with the existing GSM radio bands will offer wireless multimedia IP-based applications at the rate of 384 kbps with a bit rate of 48 kbps per time slot, and up to 69.2 kbps per time slot under good radio conditions.

EGPRS offers eight additional coding schemes. EGPRS users will have eight modulation and coding schemes available, compared with four for GPRS. Besides changes in the physical layer, modifications in the protocol structure are also needed. The lower layers of the user data plane designed for GPRS are the physical, radio link control (RLC)/media access control (MAC), and link layer control (LLC) layers. With EDGE functionality, the LLC layer will not require any modifications; however, the RLC/MAC layer has to be modified to accommodate features for efficient multiplexing and link adaptation procedures to support the essentially new physical layers in the EDGE.

Enhanced CSD (ECSD). In this case, the objective is to keep the existing GSM CS data protocols as intact as possible. In order to provide higher data rates, multislot solutions as found in ECSD are provided in EDGE. This has no impact on link or system performance.

A data frame is interleaved over 22 frames as in GSM, and three new 8-PSK channel coding schemes are defined along with the four already existing for GSM. The radio interface rate varies from 3.6 to 38.8 kbps per time slot (see Table 12.9).

Fast introduction of EGPRS/ECSD services is possible by reusing the existing transcoder rate adaptation unit (TRAU) formats and 16 kbps channel structure on the A-bis interface. Since data above 14.4 kbps cannot be rate adapted to fit into one 14.4-kbps TRAU frame, TRAU frames on several 16 kbps channels will be used to meet the increased capacity requirement. In this case, a BTS is required to handle a higher number of 16-kbps A-bis channels than time slots used on the radio interface. The benefit of using the current TRAU formats is that the introduction of new channel coding does not have any impact on the A-bis transmission, but it makes possible to hide the new coding from the TRAU unit. On the other hand, some additional complexity is introduced in the BTS owing to modified data frame handling.

Instead of reusing the current A-bis transmission formats for EDGE, new TRAU formats and rate adaptation optimized for increased capacity can be specified. The physical layer can be dimensioned statically for the maximum user rate specified for particular EDGE service or more dynamic reservation of A-bis transmission resources can be applied. The A-bis resources can even be released and reserved dynamically during the call, if the link adaptation is applied.

The channel coding schemes defined for EDGE in PS transmission are listed in Table 12.8. The schemes for EDGE in CS transmission are listed in Table 12.9.

Table 12.8 Channel Coding Scheme in EDGE (PS Transmission)

Coding Scheme	Gross Bit Rate (kbps)	Code Rate	Modulation	Radio Interface Rate per Time Slot (kbps)	Radio Interface Rate on 8 Time Slots (kbps)
CS-1	22.8	0.49	GMSK	11.2	89.6
CS-2	22.8	0.63	GMSK	14.5	116.0
CS-3	22.8	0.73	GMSK	16.7	133.6
CS-4	22.8	1.0	GMSK	22.8	182.4
PCS-1	69.2	0.329	8-PSK	22.8	182.4
PCS-2	69.2	0.496	8-PSK	34.3	274.4
PCS-3	69.2	0.596	8-PSK	41.25	330.0
PCS-4	69.2	0.746	8-PSK	51.60	412.8
PCS-5	69.2	0.829	8-PSK	57.35	458.8
PCS-6	69.2	1.000	8-PSK	69.20	553.6

Table 12.9 Channel Coding Scheme in EDGE (CS Transmission)

Channel Name	Code Rate	Modulation	Radio Interface Rate per Time Slot (kbps)
TCH/F2.4	0.16	GMSK	3.6
TCH/F4.8	0.26	GMSK	6.0
TCH/F9.6	0.53	GMSK	12.0
TCH/F14.4	0.64	GMSK	14.5
ECSD TCS-1 (NT +T)	0.42	8-PSK	29.0
ECSD TCS-2 (T)	0.46	8-PSK	32.0
ECSD TCS-3 (NT)	0.56	8-PSK	38.8

12.3.5 Services Offered by EDGE

PS Services. The GPRS architecture provides IP connectivity from mobile station to an external fixed IP network. For each service, a quality of service (QoS) profile is defined. The QoS parameters include priority, reliability, delay, and maximum and mean bit rate. A specified combination of these parameters defines a service, and different services can be selected to suit the needs of different applications.

CS Services. The current GSM standard supports both transparent and nontransparent services. Eight transparent services are defined, offering constant bit rates in the range of 9.6 to 64 kbps.

A nontransparent service uses radio link protocol (RLP) to ensure virtually error-free data delivery. For this case, there are eight services offering maximum user bit rates from 4.8 to 57.6 kbps. The actual user bit rate may vary according to channel quality and the resulting rate of transmission.

The introduction of EDGE implies no change of service definitions. The bit rates are the same, but the way services are realized in terms of channel coding is different. For example, a 57.6 kbps nontransparent service can be realized with coding scheme ECSD TCS-1 and two time slots, while the same service requires four time slots with standard GSM using coding scheme TCH/F14.4.

Thus, EDGE CS transmission makes the high-bit-rate services available with fewer time slots, which is advantageous from a terminal implementation perspective. Additionally, more users can be accepted since each user needs fewer time slots, which increases the capacity of the system.

Asymmetric Services Due to Terminal Implementation. ETSI has standardized two mobile classes: one that requires only GMSK transmission in uplink and 8-PSK in the downlink, and one that requires 8-PSK in both links. For the first class, the uplink bit rate will be limited to that of GSM/GPRS, while the EDGE bit rate is still provided in the downlink. Since most services are expected to require higher bit rates in the downlink than in the uplink, this is a way of providing attractive services with a low complexity mobile station. Similarly, the number of time slots available in uplink and downlink need not be the same. However, transparent services will be symmetrical.

12.3.6 EDGE Implementation

EDGE makes use of the existing GSM infrastructure in a highly efficient manner: radio network planning will not be greatly affected, since it will be possible to reuse many existing BTS sites. GPRS packet-switching nodes will be unaffected, because they function independently of the user bit rates. Any modifications to the switching nodes will be limited to software upgrades. There is also a smooth evolutionary path defined for terminals to ensure that EDGE-capable terminals will be small and competitively priced.

EDGE-capable channels will be equally suitable for standard GSM services, and no special EDGE, GPRS, or GSM services will be needed. From an operator viewpoint this allows seamless introduction of new EDGE services—perhaps starting with the deployment of EDGE in the service hot spots and gradually expanding coverage as demand dictates. The roll-out of EDGE-capable BSS hardware can become part of the ordinary expansion and capacity enhancement of the network. The wideband data capabilities offered by EDGE will allow a step-by-step evolution to IMT-2000, probably through a staged deployment of the new 3G air interface on the existing core GSM network. Keeping GSM as the core network for the provision of 3G wireless services has additional commercial benefits. It protects the investment of existing operators, it helps to ensure the widest possible customer base from the outset, and it fosters supplier competition through the continuous evolution of systems.

GSM operators who win licences in new 2-GHz bands will be able to introduce IMT-2000 wideband coverage in areas where early demand is likely to be greatest. Dual-mode EDGE/IMT-2000 mobile terminals will allow full roaming and handoff from one system to the other, with mapping of services between the two systems. EDGE will contribute to the commercial success of 3G system in the vital early phases by ensuring that IMT-2000 subscribers will be able to enjoy roaming and interworking globally.

Compared with establishing a total 3G system, building on an existing GSM infrastructure will be relatively fast and inexpensive. The intermediate move to GPRS and later to EDGE will make the transition to 3G easier.

While GPRS and EDGE will require new functionality in the GSM network, with new types of connections to external packet data networks, they are essentially extensions of GSM. Moving to a GSM/IMT-2000 core network will likewise be a further extension of this network.

EDGE provides GSM operators—whether or not they get a new 3G licence—with a commercially attractive solution to develop the market for wideband multimedia services. Familiar interfaces such as the Internet, volume-based charging, and a progressive increase in available user data rates will remove some of the barriers to large-scale application of wireless data services. The way forward to 3G services will be a staged evolution from today's GSM data services through GPRS and EDGE.

Increased user data rates over the radio interface will require redesign of the physical transmission methods, frame formats, and signaling protocols in different network interfaces. The extent of modification needed will depend on the user data rate requirement, i.e., whether the support of higher data is required or merely a more efficient usage of the radio time slot to support current data services is needed.

Several alternatives to cover the increased radio interface data rates on the A-bis interface for EGPRS and ECSD can be envisioned. The existing physical structure can be reused as much as possible or new transmission method optimized for EDGE can be specified.

Table 12.10 provides a comparison of GSM data services.

Table 12.10 Comparison of GSM Data Services

Service Type	Data Unit	Max. Sustained User Data Rate	Technology	Resources Used
Short message service (SMS)	Single 140 octet packet	9 bps	Simplex circuit	SDCCH or SACCH
Circuit-switched data	30 octet frames	9,600 bps	Duplex circuits	TCH
HSCSD	192 octet frames	115 kbps	Duplex circuits	1–8 TCH
GPRS	1,600 octet frames	171 kbps	Virtual circuit/ packet switching	PDCH (1–8 TCH)
EDGE		384 kbps	Virtual circuit/ packet switching	1–8 TCH

Note: SDCCH: stand-alone dedicated control channel; SACCH: slow associated control channel; TCH: traffic channel; PDCH: packet data channel (all refer to GSM logical channels).

12.4 Upgrade to UMTS (W-CDMA) in the Core GSM

A primary assumption for UMTS is that it will be based on an evolved GSM core network [7–11]. This will provide backward compatibility with GSM in terms of network protocols and interfaces (MAP, ISUP, etc.). The core network will support both GSM and UMTS/IMT-2000 services, including handover and roaming between the two (see Figure 12.13) [12]. The proposed W-CDMA-based UMTS Terrestrial Radio Access Network (UTRAN) will be connected to the GSM-UMTS core network using a new multivendor interface (I_u). The transport protocol within the new radio network and to the core network will be ATM.

There will be a clear separation between the services provided by UTRAN and the actual channels used to carry these services. All radio network functions (such as resource control) will be handled within the radio access network, and clearly separated from the service and subscription functions in the UMTS core network (UCN). The GSM-UMTS network, shown in Figure 12.14, will consist of three main parts:

- GSM-UMTS core network
- UMTS terrestrial radio access network (UTRAN)
- GSM base station subsystem (BSS)

Like the GSM-GPRS core network, the GSM-UMTS core network will have two different parts: a circuit-switched MSC and a packet-switched GRPS support node (GSN). The core network access point for GSM circuit-switched connections is the GSM MSC, and for packet-switched connection it is the SGSN.

GSM-defined services (up to and including GSM Phase 2+) will be supported in the usual GSM manner. The GSM-UMTS core network will implement supplementary services according to GSM principles (HLR-MSC/VLR). New services beyond Phase 2+ will be created using new service capabilities. These service capabilities may be seen as building blocks for application development. These include:

- Bearers defined by QoS
- Mobile station execution environment (MExE)
- Telephony value-added services (TeleVAS)
- Subscriber identity module (SIM) Toolkit
- Location services
- Open interfaces (APIs) to mobile network functions
- Downloadable application software
- Intelligent network/customized applications for mobile enhanced logic (IN/CAMEL) and service nodes

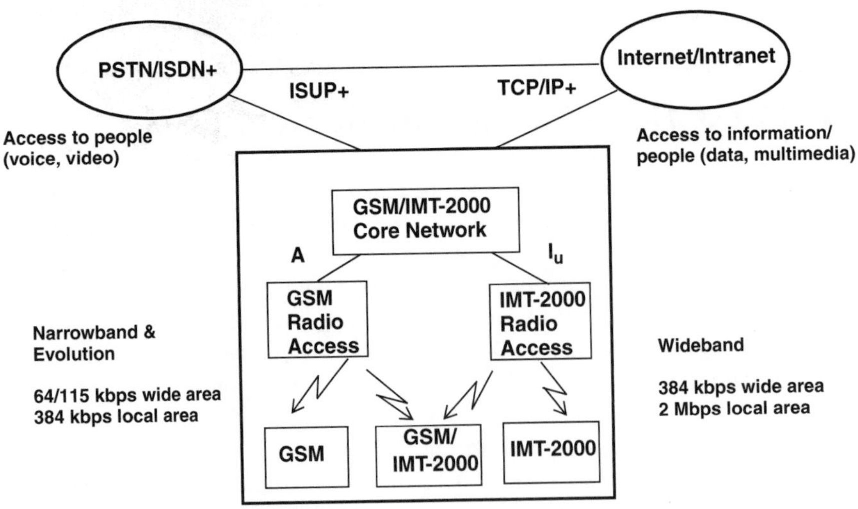

Figure 12.13 Evolution to UMTS/IMTS-2000 in a GSM Environment

In addition to new services provided by the GSM-UMTS network itself, many new services and applications will be realized using a client/server approach, with the server residing on service LANs outside the GSM-UMTS core network (see Figure 12.14). For such services, the core network will simply act as a transparent bearer. This approach is in line with current standardization activities, and will be important from a service continuity point of view. The core network will ultimately be used for the transfer of data between the end points, the client, and the server.

Intelligent network (IN) techniques are one way to provide seamless interworking across GSM-UMTS network. CAMEL already provides the basis for GSM/IN interworking. The IN infrastructure may be shared by fixed and mobile networks, and can support fixed/mobile service integration, as needed by IMT-2000. The inherent support for third-party service providers in IN means such providers could offer all or part of the integrated services. This role of IN is already apparent in services such as virtual private networks (VPN), regional subscriptions, and One Number, which are available as network-independent and customer-driven services.

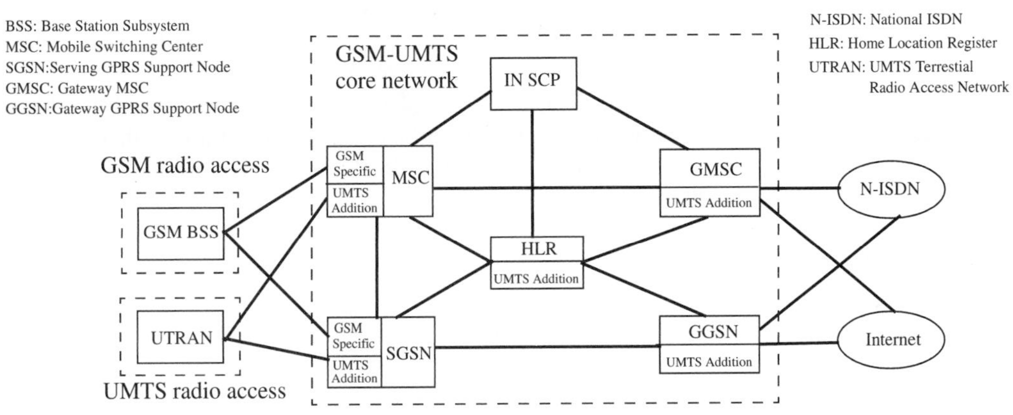

Figure 12.14 General GSM-UMTS Network Architecture

Service nodes and IN can play a complementary role. IN is suitable for subscription control and group services where high service penetration in a very wide area with frequent service invocation is more important than sophistication. Service nodes are better for providing differentiated user interfaces, e.g., personal call and messaging services that use advanced in-band processing and span several access networks.

To make the most of the new radio access network's capabilities and to cater to the large increase in data traffic volume, it is likely that ATM will be used as the transport protocol within UTRAN and toward the GSM-UMTS core network. The combination of an ATM cell-based transport network, W-CDMA's use of variable-rate speech coding with improved channel coding, and an increased volume of packet data traffic over the air interface will mean a saving of about 50 percent in transmission costs, compared with equivalent current solutions. ATM, with the newly standardized AAL2 adaptation layer, provides an efficient transport protocol, optimized for delay-sensitive speech services and packet data services. Statistical multiplexing in ATM provides maximum utilization of existing and new transmission infrastructure throughout the entire network.

In the complex multiservice, multivendor, multiprovider environment of 3G wireless services, network management will be a critical issue. The growth of packet data traffic will require new ways of charging for services and new billing systems to support them. There will continue to be a growing demand for better customer care and cost reductions in managing mobile networks, driven by the need to

- Provide sophisticated personal communications services
- Expand the customer base beyond the business user base
- Separate the service provider and network operator roles
- Provide "one-stop" billing for a range of services

New operations and management functions will be needed to support new services and network functionality. Standardization of interfaces will be critical, especially for alignment with current management interfaces in the GSM-UMTS core network. Management information will need to be part of standard traffic interfaces.

With the right service strategy and network planning, GSM operators will be able to capitalize on the wideband multimedia market through a staged evolution of their core networks with the addition of new radio access technology as it becomes available.

12.5 Summary

In this chapter, we examined the evolution of TDMA-based 2G networks to 3G networks to provide multimedia services up to 2 Mbps in the local area and up to 384 kbps in the wide area with a UMTS W-CDMA air interface. We discussed GPRS, the new packet-based data bearer service for GSM and IS-136, and building a core network capable of delivering GPRS service to meet the requirements of IMT-2000. We also discussed the use of EDGE (a 2.5-G system) to offer wireless multimedia IP-based applications of speeds up to 384 kbps. We concluded the chapter by outlining the evolutionary path of a GSM core network to UMTS to provide backward compatibility in terms of network protocol and interfaces, and to support both GSM and UMTS/IMT-2000 services with handoff and roaming between two systems.

12.6 References

1. Sollenberger, N. R., Seshadri, N., and Cox, R., "The Evolution of IS-136 TDMA for Third-Generation Wireless Services," *IEEE Personal Communications*, June 1999 [Vol. 6(3), pp. 8–18].

2. Austin, M., Buckley, A., Coursey, C., Hartman, P., Kobylinski, R., and Majmundar, M., "Service and System Enhancements for TDMA Digital Cellular Systems," *IEEE Personal Communications*, June 1999 [Vol. 6(3), pp. 20–33].
3. Digital Cellular Telecommunication System (Phase 2+), High Speed Circuit Switched Data (HSCSD)—Stage 1, Draft ETSI Document GSM 02.34 version 0.1.0, February 1995.
4. Digital Cellular Telecommunication System (Phase 2+), High Speed Circuit Switched Data (HSCSD), Service Description, Stage 2, GSM 03.34.
5. ETSI, TS 03 64 V5.10 (1997-11), Digital Cellular Telecommunications System (Phase 2+); General Packet Radio Service (GPRS). Overall Description of the GPRS Radio Interface; Stage 2 (GSM 03.64 version 5.1.0).
6. ETSI Technical Specification GSM 02.60 GPRS Service Description—Stage 1 version 5.2.1, July 1998.
7. Prasad, N. R., "GSM Evolution Towards Third Generation UMTS/IMT2000," Third ICPWC99, Feb. 1999, Jaipur, India.
8. ETSI Tdoc SMG2 95/97, EDGE Feasibility Study, Work Item 184; Improved Data Rates through Optimized Modulation," version 0.3, Dec. 1997.
9. Shanker, B., McClelland, S., "Mobilizing the Third Generation [Cellular Radio]," *Telecommunications (International Edition),* August 1997 [Vol. 31(8), pp. 27–28].
10. Garg, V. K., Halpern, S., and Smolik, K. F., "Third-Generation (3G) Mobile Communications Systems," Third ICPWC99, Feb. 1999, Jaipur, India.
11. Dahlman, E., Gudmundson, B., Nilsson, M., and Skold, J., "UMTS/IMTS-2000 Based on Wideband CDMA," *IEEE Communication Magazine*, Sept. 1998 [Vol. 36(9), pp. 48–54].
12. Ihrfors, H., "3G Wireless: What Does It Mean for GSM Core Networks?" *Mobile Communication International*, Sept. 1998, pp. 35–38.

cdma2000 System

13.1 Introduction

The cdma2000 radio transmission technology (RTT) [1,2] is a wideband, spread-spectrum radio interface that uses code division multiple access (CDMA) technology to satisfy the needs of third-generation (3G) wireless communication systems. The RTT meets all requirements specified in the International Telecommunications Union (ITU) circular letter and the corresponding documents of the International Mobile Telephony 2000 (IMT-2000) [3–6]. The service requirements are satisfied for indoor office, indoor-to-outdoor/pedestrian, and vehicular environments. The cdma2000 system will also be backward compatible with the current cdmaOne (IS-95) family of standards.

The cdma2000 system provides a wide range of implementation options to support data rates (both circuit switched and packet switched) starting from a TIA/EIA-95B-compatible rate of 9.6 kbps up to greater than 2 Mbps [7,8]. The cdma2000 system provides maximum flexibility to carriers in making engineering trade-offs between

- Channel sizes of 1, 3, 6, 9, and 12 × 1.25 MHz
- Support for advanced antenna technologies
- Cell sizes (e.g., the cdma2000 system's increased performance can be realized in terms of increased range to permit carriers to reduce the total number of cell sites)
- Higher data rates that can be supported in all channel sizes
- Support for advanced services possible or practical in other systems (e.g., high-speed circuit data, B-ISDN, or H.224/223 teleservices)

The cdma2000 system can be operated economically in a wide range of environments including

- Outdoor megacells (cell > 35-km radius)
- Outdoor macrocells (cell 1-km to 35-km radius)
- Indoor/outdoor microcells (up to 1-km radius)
- Indoor/outdoor picocells (< 50-m radius)

The cdma2000 system can be deployed in

- Indoor/outdoor environments
- Wireless local loops (WLL)
- Vehicular environments
- Mixed vehicular and indoor/outdoor environments

The cdma2000 system mobility is variable, ranging from fixed wireless to high speeds of up to 300 mph. cdma2000 provides a layered structure to support the integration of the bottom two layers of the RTT into systems that implement any network standards (e.g., ITU-T defined signaling services). It also provides backward compatibility to TIA/EIA-95B signaling and call control models. An extended cdma2000 upper layer signaling structure is capable of supporting a wide range of advanced services (e.g., multimedia) in an optimized and efficient manner.

cdma2000 supports the 3G wireless intelligent networking (WIN) services and services defined by the ITU or other international standards organizations and provides a graceful evolution from existing second-generation (2G) TIA/EIA-95B technology. It includes the following features:

- Support for overlay configurations
- Support for backward compatibility to TIA/EIA-95B signaling and network
- Support for graceful and gradual upgrade from 2G systems to 3G systems
- Sharing of common channels with an underlay TIA/EIA-95B system during transition periods

cdma2000 provides an evolutionary path by reusing existing TIA/EIA-95B standards including

- TIA/EIA-95B: mobile station and radio interface specifications
- IS-707: data services (packet, async, and fax)
- IS-127: enhanced variable-rate codec (EVRC) 8.5-kbps speech coder
- IS-733: 13-kbps speech coder
- IS 637: short message service (SMS)
- IS 638: over-the-air activation and parameter administration (supporting the configuration and service activation of mobile stations over the radio interface)
- IS-97 and IS-98 (minimum performance specifications)
- The basic TIA/EIA-95B channel structure
- Extensions to TIA/EIA-95B fundamental/supplemental channel structure, multiplex layer, and signaling to support higher-rate operation, common broadcast channels (pilot, paging, and sync)
- IS-634A: no significant changes are expected for cdma2000; the layered structure of the cdma2000 integrates smoothly with the component structure of IS-634A
- TIA/EIA-41D: no significant changes needed for the cdma2000; the layered structure of the cdma2000 offers the potential for easy integration with enhanced network services (WIN)

13.2 cdma2000 Layering Structure

13.2.1 Upper Layers

Figure 13.1 shows the layer structure of cdma2000. The *upper layers* contain three basic services:

- **Voice services**—Voice telephony services, including public switched telephone network (PSTN) access, mobile-to-mobile voice services, and Internet telephony.
- **End user data-bearing services**—Services that deliver any form of data on behalf of the mobile end user, including packet data (e.g., Internet protocol (IP) service), circuit data services (e.g., B-ISDN emulation services), and SMS. Packet data services conform to industry standard connection-oriented and connectionless packet data including IP-based protocols (e.g., transmission control protocol [TCP] and user datagram protocol [UDP]) and ISO/OSI connectionless interworking protocol (CLIP). Circuit data services emulate international standards-defined, connection-oriented services such as asynchronous (async) dial-up access, fax, V.120 rate-adapted ISDN, and B-ISDN services.
- **Signaling**—Services that control all aspects of operation of the mobile.

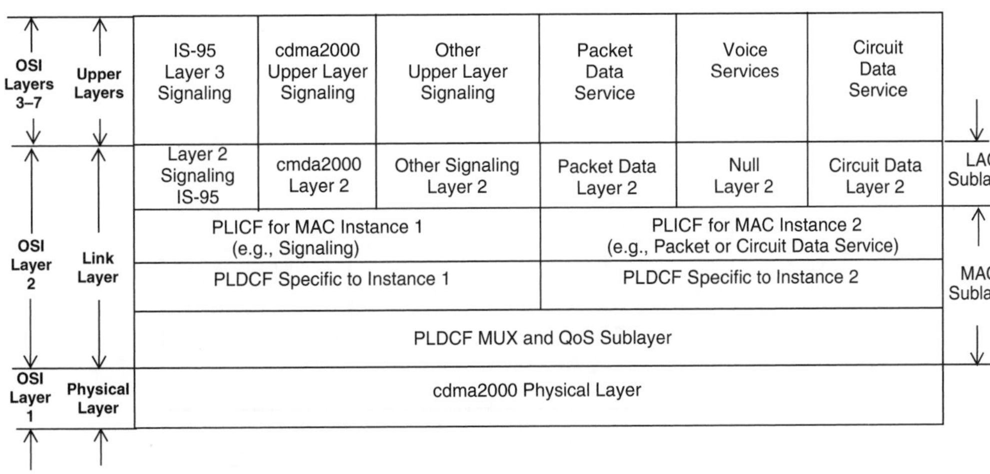

Figure 13.1 cdma2000 Layering Structure

13.2.2 Lower Layers

The *link layer* provides varying levels of reliability and quality of service (QoS) characteristics according to the needs of the specific upper layer service. It gives protocol support and control mechanisms for data transport services and performs all functions necessary to map the data transport needs of the upper layers into specific capabilities and characteristics of the physical layer. The link layer is subdivided into sublayers:

- Link access control (LAC) (see Figure 13.2)
- Media access control (MAC) (see Figure 13.3)

The LAC sublayer manages point-to-point communication channels between peer upper layer entities and provides a framework to support a wide range of different end-to-end reliable link layer protocols.

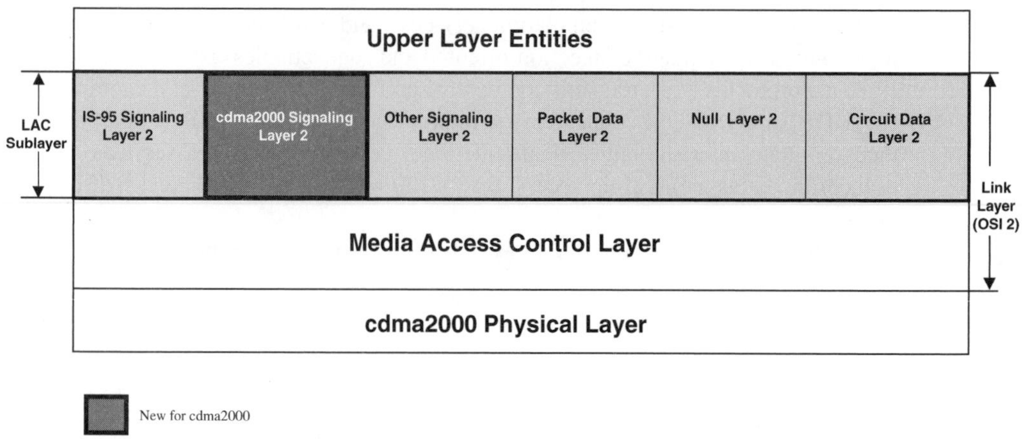

Figure 13.2 cdma2000 Link Layer

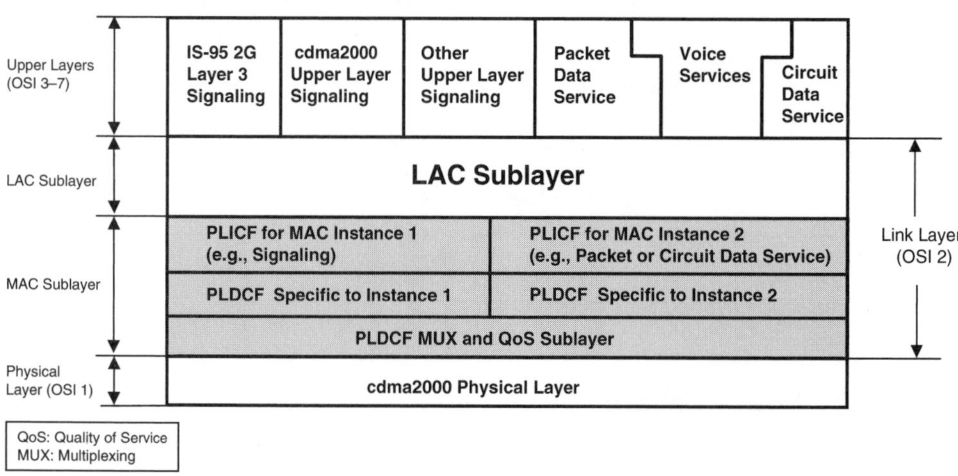

Figure 13.3 cdma2000 Media Access Control Layer

The cdma2000 system includes a flexible and efficient MAC sublayer that supports multiple instances of an advanced-state machine, one for each active packet or circuit data instance. Together with a QoS control entity, the media access control sublayer realizes the complex multimedia, multiservice capabilities of 3G wireless systems with QoS management capabilities for each active service. The media access control sublayer provides three important functions:

- **Media access control state**—Procedures for controlling the access of data services (packet and circuit) to the physical layer (including contention control between multiple services from a single user as well as between competing users).
- **Best-effort delivery**—Reasonably reliable transmission over the radio link with radio link protocol (RLP) providing a *best-effort* level of reliability.
- **Multiplexing and QoS control**—Enforcement of negotiated QoS levels by mediating conflicting requests from competing services and appropriately prioritizing access requests.

The MAC sublayer provides differing QoS to the LAC sublayer (e.g., different modes of operation). It may be constrained by backward compatibility (e.g., for IS-95B signaling layer 2) and it may have to be compatible with other link layer protocols (e.g., for compatibility with non-IS-95 air interfaces or for compatibility with future ITU-defined protocol stacks). The MAC sublayer is subdivided into

- Physical layer independent convergence function (PLICF)
- Physical layer dependent convergence function (PLDCF) is further subdivided into
  - Instance-specific PLDCF
  - PLDCF MUX and QoS sublayer

PLICF provides service to the LAC sublayer and includes all MAC operational procedures and functions that are not unique to the physical layer. Each instance of PLICF maintains service status for the corresponding service. PLICF uses services provided by PLDCF to implement actual communications activities in support of media access control sublayer service. Services used by PLICF are defined as a set of logical channels that carry different types of control or data information.The PLICF data service consists of the following states/substates (see Figures 13.4 and 13.5):

- Null state
- Initialization state
- Control hold state
  - Normal substate
  - Slotted substate
- Active state
- Suspended state
  - Virtual traffic substate
  - Slotted substate

- Dormant state
  - Dormant/idle substate
  - Dormant/burst substate

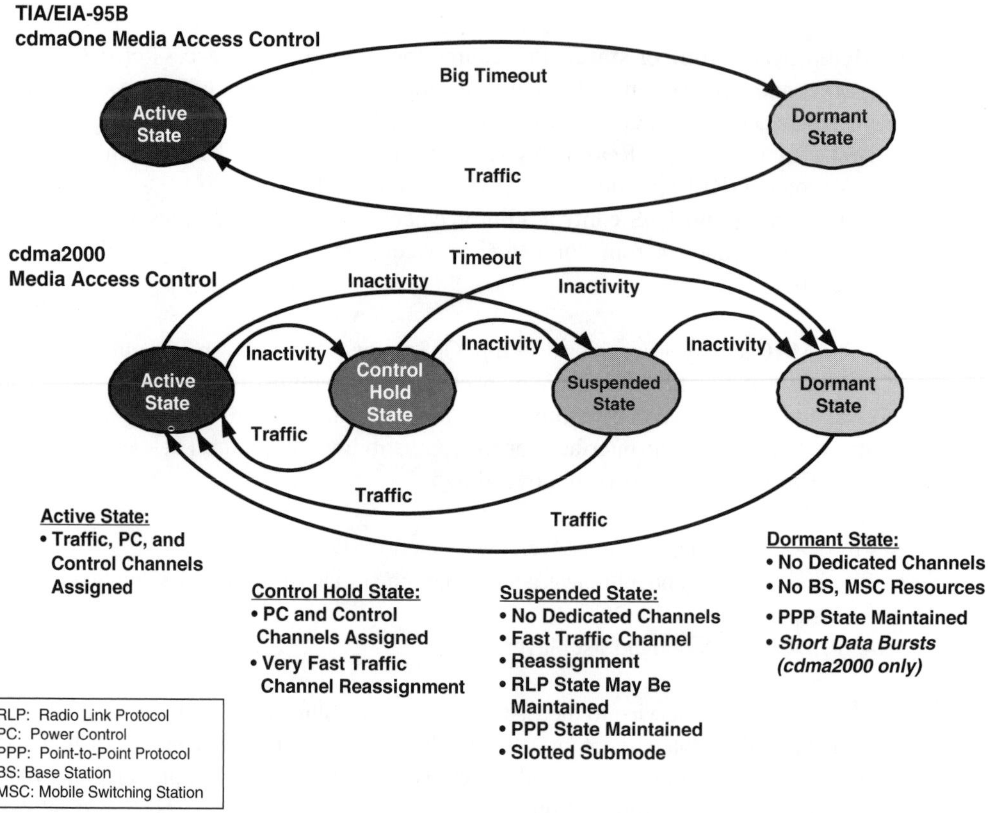

Figure 13.4 Packet Data MAC Operation States in cdma2000 and TIA IS-95B

The *null state* is considered to be the default state prior to activation of packet data service. After the packet service is invoked, a transition to *initialization state* occurs, during which an attempt is made to connect the packet service.

Traffic, power control, and control channels are assigned in the *active state*. In the *control hold state*, a dedicated control channel is maintained between the user and the base station on which any MAC command (for example, the command to begin a high-speed data burst) can be transmitted with virtually no latency. Power control is also maintained so that a high-speed burst operation can begin with no delay due to stabilization of power control.

In the *suspended hold state*, there are no dedicated channels maintained to or from the user. However, the state information for RLP is maintained, and the base station and user maintain a virtual active set that allows either one of them to know which base station can best be used (accessed by the user or paged by the base station) in the event that packet data traffic occurs for the user. This state also supports a slotted substate that permits the user's mobile device to preserve power in a highly efficient manner.

A short data burst mode is added to the cdma2000 *dormant state* to support the delivery of short messages without incurring the overhead of a transition from the dormant to the active state. Transitions between media access control states can be indicated by media access control signaling or by the expiration of timers. By properly selecting the values for the timers, the cdma2000 media access control can be adapted to a wide variety of data services and operating environments.

The states are categorized as either connected or not connected, depending on the status of data service option. The data service option is connected in the control hold state, active state, and suspended hold state. The data service option is not connected in the null state, initialization state, dormant state, or reconnect state. Figure 13.5 shows the state diagram for the PLICF data service option.

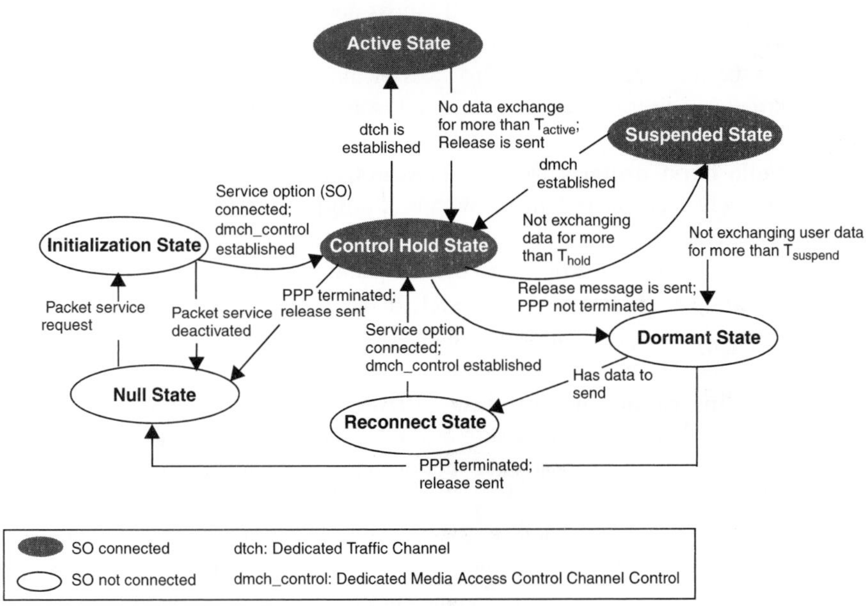

Figure 13.5 Data Services State Diagram

PLDCF performs mapping of logical channels from PLICF to logical channels supported by the specific physical layer. PLDCF performs multiplexing, demultiplexing, and consolidation of control information with bearer data from the control and traffic channels from multiple PLICF instances in the same mobile. PLDCF implements QoS capabilities, including resolution of priorities between competing PLICF instances, and maps QoS requests from PLICF instances into the appropriate physical layer service requests to deliver the desired QoS. The major functions of this sublayer are to

- Perform any required mapping of the simpler logical channels from the PLICF into the logical channels supported by physical layer.
- Perform any (optional) automatic retransmission request (ARQ) protocol functions that are tightly integrated with physical layer.
- Perform some of the physical layer-specific low-level functions of IS-95B RLP.

For cdma2000, four PLDCF specific protocols are defined:

1. **Radio link protocol (RLP)**—This protocol provides a highly efficient streaming service that makes a best effort to deliver data between peer PLICF entities. RLP provides both transparent and nontransparent modes of operation. In the nontransparent mode, RLP uses ARQ protocol to retransmit data segments that were not delivered properly by the physical layer. In the nontransparent mode, RLP can introduce some delay. In the transparent mode, RLP does not retransmit missing data segments. However, RLP maintains byte synchronization between the sender and receiver and notifies the receiver of the missing parts of the data stream. Transparent RLP does not introduce any transmission delay, and is useful for implementing voice services over RLP.
2. **Radio burst protocol (RBP)**—This protocol provides a mechanism for delivering relatively short data segments with best-effort delivery over a shared access common traffic channel (ctch). This capability is useful for delivering small amounts of data without incurring the overhead of establishing a dedicated traffic channel (dtch).
3. **Signaling radio link protocol (SRLP)**—This protocol provides best-effort streaming service for signaling information analogous to RLP, but optimized for the dedicated signaling channel (dsch).
4. **Signaling radio burst protocol (SRBP)**—This protocol provides a mechanism to deliver signaling messages with best-effort delivery analogous to RBP, but optimized for signaling information and common signaling channel (csch).

The PLDCF includes a radio link access control (RLAC) function that abstracts the RLP and RBP from the PLICF and coordinates the transmission of data (traffic or signaling) between RLP and RBP according to the current operational state of media access control (e.g., restrict the use of RBP to cases in which the PLICF is in the packet data dormant state).

The PLDCF MUX and QoS sublayer coordinates multiplexing and demultiplexing of code channels from multiple PLICF instances. It implements and enforces QoS differences between instances and maps the data streams and control information on multiple logical channels from different PLICF instances into requests for logical channels, resources, and control information from the physical layer.

13.3 cdma2000 Channels

13.3.1 Channel Naming Conventions

A logical channel is denoted by three or four lowercase acronyms followed by "ch" for *channel*. The fourth letter applies to common channels used in dormant or suspended states. Table 13.1 lists the conventions for logical channels.

Table 13.1 Logical Channel Naming Conventions

First Letter	Second Letter	Third Letter
f = forward (BS to MS) r = reverse (MS to BS)	d = dedicated c = common	t = traffic m = media access control s = signaling

A physical channel (see Table 13.2) is represented by uppercase acronyms. The first letter in the name of the channels indicates the direction of the channel, except for the paging and access channels where the direction is implicitly specified.

Table 13.2 Physical Channel Naming Conventions

Channel Name	Physical Channel
F/R-FCH	Forward/Reverse Fundamental Channel
F/R-SCCH	Forward/Reverse Supplemental Coded Channel
F/R-SCH	Forward/Reverse Supplemental Channel
F/R-DCCH	Forward/Reverse Dedicated Control Channel
F-PCH	(Forward) Paging Channel
R-ACH	Reverse Access Channel
R-EACH	(Reverse) Enhanced Access Channel
F/R-CCCH	Forward/Reverse Common Control Channel
F-DAPICH	Forward Dedicated Auxiliary Pilot Channel
F-APICH	Forward Auxiliary Pilot Channel
F/R-PICH	Forward/Reverse Pilot Channel
F-SYNC	Forward Sync Channel
F-TDPICH	(Forward) Transmit Diversity Pilot Channel
F-ATDPICH	(Forward) Auxiliary Transmit Diversity Pilot Channel
F-BCH	Forward Broadcast Channel
F-QPCH	(Forward) Quick Paging Channel
F-CPCCH	Forward Common Power Control Channel
F-CACH	Forward Common Assignment Channel

13.4 Logical Channels Used by PLICF

The following logical channels are used by PLICF:

13.4.1 Dedicated Traffic Channel (f/r-dtch)

dtch is the forward or reverse logical channel that is used to carry user data traffic. This logical channel is a point-to-point channel and is allocated for use throughout the active state of data service. It carries a data dedicated channel to a single PLICF instance.

13.4.2 Common Traffic Channel (f/r-ctch)

ctch is the forward or reverse logical channel that is used to carry short data bursts associated with the data service in the dormant/burst substate of the dormant state. This logical channel is allocated for the duration of the short burst. It shares access among many mobiles and/or PLICF instances.

13.4.3 Dedicated Media Access Control Channel (f/r-dmch_control)

dmch_control is the forward or reverse logical channel that is used to carry media access control messages. This logical channel is a point-to-point channel and is allocated throughout the active state and control hold state of data service. It carries control information dedicated to a single PLICF instance.

13.4.4 Reverse Common Media Access Control Channel (r-cmch_control)

r-cmch_control is the reverse logical channel used by a mobile while data service is in the dormant/idle substate of the dormant state or suspended state. This logical channel is used to carry media access control messages. It is shared by a group of mobiles in the sense that access to this channel is gained on a contention basis.

13.4.5 Forward Common Media Access Control Channel (f-cmch_control)

f-cmch_control is the forward logical channel used by the base station while data service is in the dormant/idle substate of the dormant state or suspended state. This logical channel is used to carry media access control messages. It is a point-to-multipoint channel.

13.4.6 Dedicated Signaling Channel (dsch)

dsch carries upper layer signaling data dedicated to a single PLICF instance.

13.4.7 Common Signaling Channel (csch)

csch carries upper layer signaling data shared access among many mobiles and/or PLICF instances.

13.5 Physical Layer

The physical layer provides coding and modulation services for a set of logical channels used by PLDCF MUX and the QoS sublayer. The physical channels (see Figure 13.6) are classified as

- **Forward/reverse dedicated physical channels (F/R-DPHCH)**—The collection of all physical channels that carry information in a dedicated, point-to-point manner between the base station and a single mobile (see Figure13.7).
- **Forward/reverse common physical channels (F/R-CPHCH)**—The collection of all physical channels that carry information in a shared access, point-to-multipoint manner between the base station and multiple mobile stations (see Figure13.8).

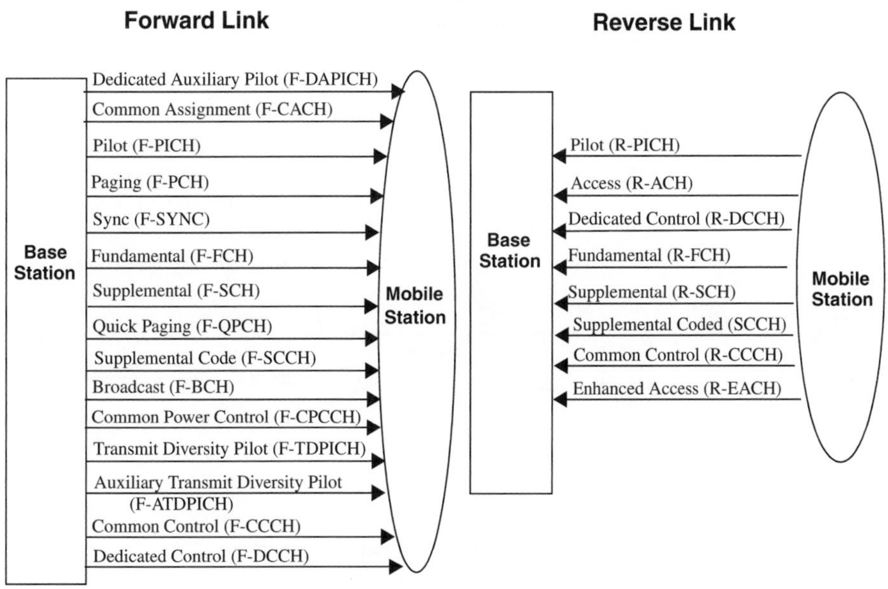

Figure 13.6 cdma2000 Physical Channels

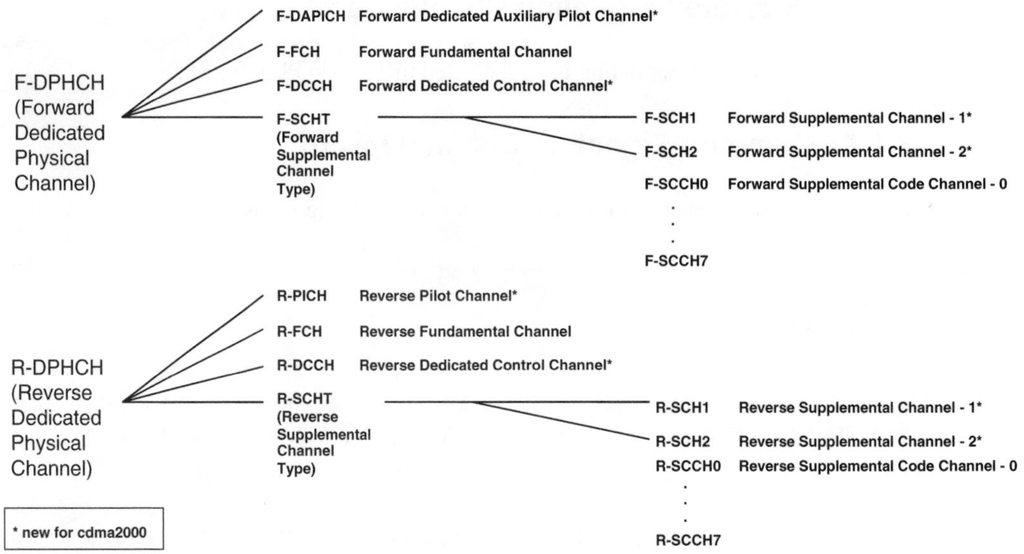

Figure 13.7 cdma2000 Overview of Dedicated Physical Channels

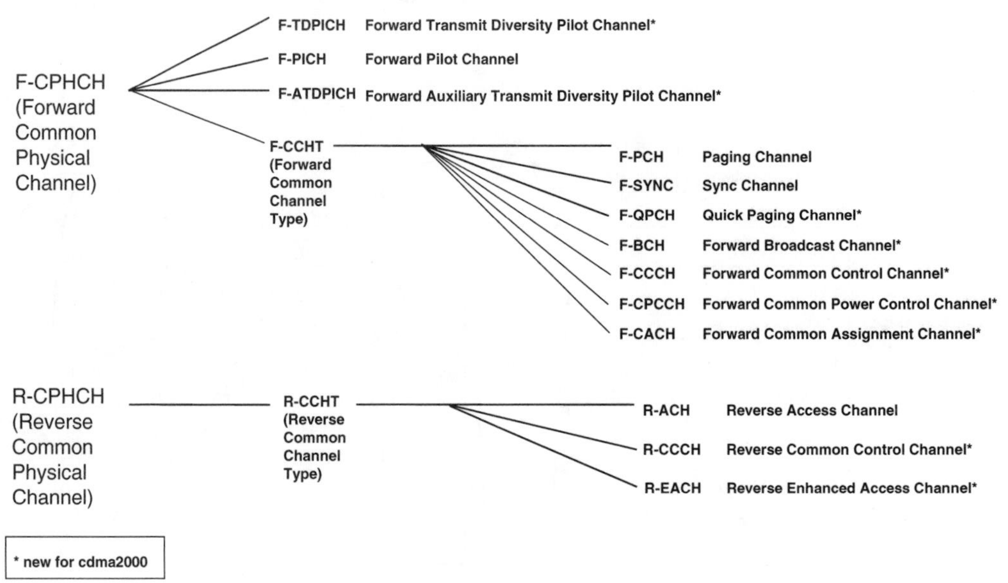

Figure 13.8 cdma2000 Overview of Common Physical Channels

13.6 Forward Link Physical Channels

Forward dedicated channels carry information between the base station and a specific mobile; common channels carry information from the base station to a set of mobiles in a point-to-multipoint manner. Table 13.3 lists these channels.

Table 13.3 Forward Link Channels

	Physical Channel	Channel Name
Forward Common Physical Channels (control and overhead channels)	Forward Pilot Channel	F-PICH
	Forward Paging Channel	F-PCH
	Forward Sync Channel	F-SYNC
	Forward Common Control Channel	F-CCCH
	(Forward) Quick Paging Channel	F-QPCH
	(Forward) Transmit Diversity Pilot Channel	F-TDPICH
	(Forward) Auxiliary Transmit Diversity Pilot Channel	F-ATDPICH
	Forward Common Power Control Channel	F-CPCCH
	Forward Common Assignment Channel	F-CACH
	Forward Broadcast Channel	F-BCH
Forward Dedicated Physical Channels	Forward Dedicated Auxiliary Pilot Channel	F-DAPICH
	Forward Dedicated Control Channel	F-DCCH
	Forward Traffic Channels • Fundamental • Supplemental • Supplemental Code	 F-FCH F-SCH F-SCCH

The cdma2000 physical layer has six types of forward dedicated channels: forward fundamental channel (F-FCH), forward dedicated control channel (F-DCCH), forward supplemental channel (F-SCH), forward supplemental code channel (F-SCCH), forward common power control channel (F-CPCCH), and forward dedicated auxiliary pilot channel (F-DAPICH).

The forward traffic channel (F-TCH) supports several distinct physical layer characteristics (spreading rate, channel modulation, coding rate, Walsh length, and a set of transmission rates). These characteristics are represented by a single parameter called the *radio configuration* (RC). This parameter uniquely identifies the physical layer characteristics of the F-TCH. Table 13.4 shows the forward RCs and the corresponding transmission rates on the F-TCH.

Table 13.4 Data Rates (kbps) of Forward Link Channel Types for SR1 and SR3

Channel Type	Data Rate (kbps), FEC, RC	
	SR1	SR3
Sync Channel (F-SYNCH)	1.2; Convolutional $r = 1/2$	1.2; Convolutional $r = 1/3$
Paging Channel (F-PCH)	9.6 or 4.8; Convolutional $r = 1/2$	N/A
Broadcast Channel (F-BCH)	19.2 (40-ms slots), 9.6 (80-ms slots), or 4.8 (160-ms slots); Convolutional $r = 1/2$	19.2 (40-ms slots), 9.6 (80-ms slots), or 4.8 (160-ms slots); Convolutional $r = 1/3$
Quick Paging Channel (QPCH)	4.8 or 2.4; None	4.8 or 2.4 (MC); None
Common Power Control Channel (F-CPCCH)	9.6; None	14.4; None
Common Assignment Channel (F-CACH)	9.6; Convolutional $r = 1/2$	9.6; Convolutional $r = 1/3$
Forward Common Control Channel (F-CCCH)	38.4 (5-, 10-, or 20-ms frames), 19.2 (10- or 20-ms frames), or 9.6 (20-ms frames); Convolutional $r = 1/4$ or 1/2	38.4 (5-, 10-, or 20-ms frames), 19.2 (10- or 20-ms frames), or 9.6 (20-ms frames); Convolutional $r = 1/3$
Forward Dedicated Control Channel (F-DCCH)	9.6 (RC 3 or RC4), 14.4 (20-ms frames), or 9.6 (5-ms frames) (RC5); Convolutional $r = 1/4$ (RC3 or RC5) $r = 1/2$ (RC4)	9.6 (RC6 or RC7), 14.4 (20-ms frames), or 9.6 (5-ms frames) (RC8 or RC9); Convolutional $r = 1/6$ (RC6) $r = 1/3$ (RC7) $r = 1/4$ (RC8, 20 ms) $r = 1/3$ (RC8, 5 ms) $r = 1/2$ (RC9, 20 ms) $r = 1/3$ (RC9, 5 ms)

Table 13.4 Data Rates (kbps) of Forward Link Channel Types for SR1 and SR3 *(cont.)*

Channel Type	Data Rate (kbps), FEC, RC	
	SR1	SR3
Forward Fundamental Channel (F-FCH)	9.6, 4.8, 2.4, or 1.2 (RC1), 14.4, 7.2, 3.6, or 1.8 (RC2), 9.6, 4.8, 2.7, or 1.5 (20-ms frames) or 9.6 (5-ms frames) (RC3 or RC4), 14.4, 7.2, 3.6, or 1.8 (20-ms frames) or 9.6 (5-ms frames) (RC5); Convolutional $r = 1/2$ (RC1, RC2, or RC4) $r = 1/4$ (RC3 or RC5)	9.6, 4.8, 2.7, or 1.5 (20-ms frames) or 9.6 (5-ms frames) (RC6 or RC7), 14.4, 7.2, 3.6, or 1.8 (20-ms frames) or 9.6 (5-ms frames) (RC8 or RC9); Convolutional $r = 1/6$ (RC6) $r = 1/3$ (RC7) $r = 1/4$ (RC8, 20 ms) $r = 1/3$ (RC8, 5 ms) $r = 1/2$ (RC9, 20 ms) $r = 1/3$ (RC9, 5 ms)
Forward Supplemental Code Channel (F-SCCH)	9.6 (RC1), 14.4 (RC2); Convolutional $r = 1/2$ (RC1 or RC2)	N/A
Forward Supplemental Channel (F-SCH)	153.6, 76.8, 38.4, 19.2, 9.6, 4.8, 2.7, or 1.5 (RC3) 307.2, 153.6, 76.8, 38.4, 19.2, 9.6, 4.8, 2.7, or 1.5 (RC4) 230.4, 115.2, 57.6, 28.8, 14.4, 7.2, 3.6, or 1.8 (RC5) Convolutional or Turbo (> 14.4) $r = 1/2$ (RC4) $r = 1/4$ (RC3 or RC5)	307.2, 153.6, 76.8, 38.4, 19.2, 9.6, 4.8, 2.7, or 1.5 (RC6), 614.4, 307.2, 153.6, 76.8, 38.4, 19.2, 9.6, 4.8, 2.7, or 1.5 (RC7), 460.8, 230.4, 115.2, 57.6, 28.8, 14.4, 7.2, 3.6, or 1.8 (RC8) 1036.8, 460.8, 230.4, 115.2, 57.6, 28.8, 14.4, 7.2, 3.6, or 1.8 (RC9) Convolutional or Turbo (> 14.4) $r = 1/3$ (RC7) $r = 1/4$ (RC8) $r = 1/2$ (RC9)

RC1 and RC2 provide backward compatibility to 3G1X with 2G, so that a cell that services 3G1X mobiles is also capable of supporting 2G mobiles. RC3, RC4, and RC5 support the 3G technology. RC3 has a more robust decoder than RC4, and requires a lower E_b/N_t that increases the forward voice capacity. However, RC4 provides twice the number of Walsh codes as RC3. For voice applications, RC3 and RC4 provide 3G technology for 8 kbps vocoders and RC5 supports 13 kbps vocoders.

The channel types on the forward link for SR1 and SR3 are listed in Table 13.5.

Table 13.5 Channel Types on the Forward Link for SR1 and SR3

Channel Type	Maximum Number	
	SR1	SR3
Forward Pilot Channel (F-PICH)	1	1
(Forward) Transmit Diversity Pilot Channel (F-TDPICH)	1	1
(Forward) Auxiliary Transmit Diversity Pilot Channel (F-ATDPICH)	Not Specified	Not Specified
(Forward) Auxiliary Pilot Channel (F-APICH)	Not Specified	Not Specified
(Forward) Sync Channel (F-SYNCH)	1	1
(Forward) Paging Channel (F-PCH)	7	None
(Forward) Broadcast Channel (F-BCH)	Not Specified	Not Specified
(Forward) Quick Paging Channel (F-QPCH)	3	3
Forward Common Power Control Channel (F-CPCCH)	7	7
Forward Common Assignment Channel (F-CACH)	7	7
Forward Common Control Channel (F-CCCH)	7	7
Forward Dedicated Control Channel (F-DCCH)	1*	1*
Forward Fundamental Channel (F-FCH)	1*	1*
Forward Supplemental Code Channel (F-SCCH) (RC1 and RC2 only)	7*	None
Forward Supplemental Channel (F-SCH) (RC3 through RC5 Only)	2*	2*

* Per forward traffic channel

The forward radio configuration of a 3G1X voice call maps to a unique reverse radio configuration. This mapping of RCs is given in Table 13.6.

Table 13.6 Mapping of the Forward and Reverse Radio Configuration for 3G1X

Forward RC	Reverse RC	Vocoder (kbps)
RC1	RC1	8
RC2	RC2	13
RC3	RC3	8
RC4	RC3	8
RC5	RC4	13

13.6.1 Forward Pilot Channel (F-PICH)

This channel is continuously broadcast throughout the cell in order to provide timing and phase information. The common pilot is an all-zeros sequence prior to Walsh spreading with Walsh 0. The F-PICH is shared by all traffic channels and is used for

- Estimating channel gain and phase
- Detecting multipath rays so that RAKE fingers are efficiently assigned to the strongest multipaths
- Cell acquisition and handoff

With a common pilot, it is possible to send the pilot signal without incurring significant overhead for each user. A system with a common pilot approach can achieve better performance than a system using a per-user pilot approach. For voice traffic, the common pilot can provide better channel estimation and lower overhead, resulting in improved receiver performance. It can also provide improved search and handoff performance.

13.6.2 Forward Sync Channel (F-SYNC)

The sync channel is used by mobiles operating within the coverage area of the base station to acquire initial time synchronization and to determine paging channel location.

13.6.3 Forward Paging Channel (F-PCH)

A cdma2000 system can have multiple paging channels per base station. A paging channel is used to send control information and paging messages from the base station to mobiles and operates at a data rate of 9.6 or 4.8 kbps (same as IS-95). The F-PCH carries overhead messages, pages, acknowledgments, channel assignments, status requests, and shared secret data (SSD) updates from the base station to the mobile.

Figures 13.9 and 13.10 show F-CPHCH (F-PICH, F-SYNC, and F-PCH) for SR1 and SR3.

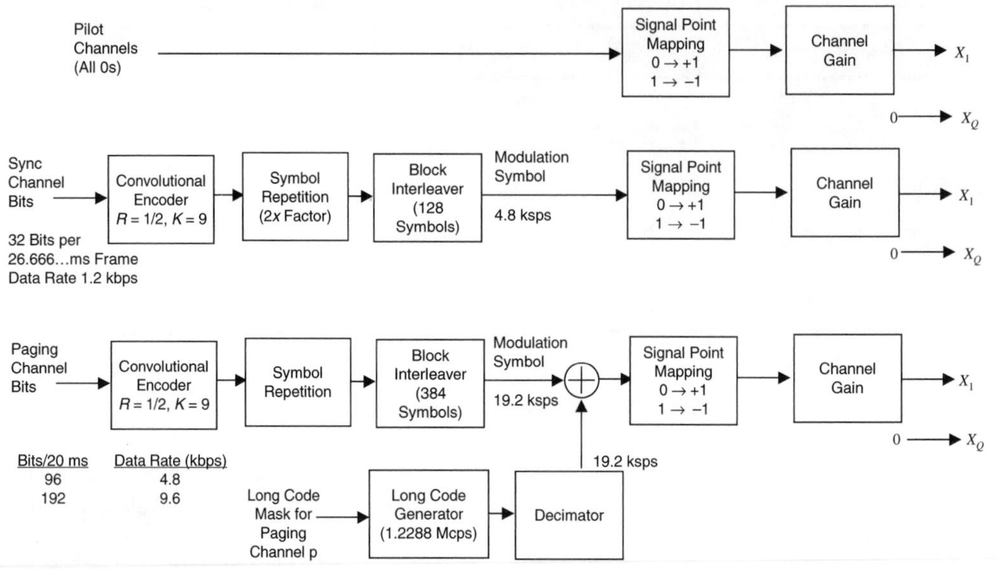

Figure 13.9 cdma2000 F-CPHCH for SR1

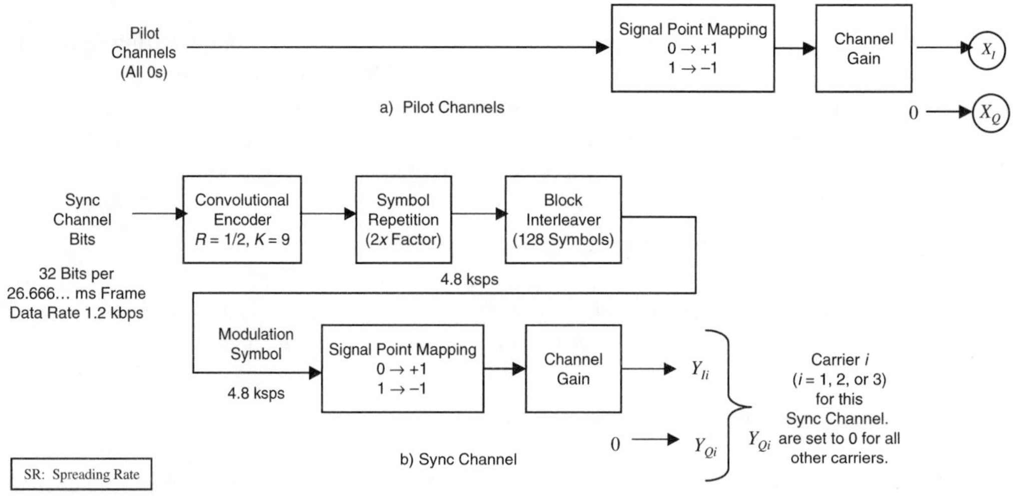

Figure 13.10 cdma2000 F-CPHCH for SR3

13.6.4 Forward Common Control Channel (F-CCCH)

The F-CCCH is a common channel used for communication of layer 3 and media access control messages from the base station to one or more mobiles. Possible frame sizes for F-CCCH are 5 ms, 10 ms, and 20 ms, depending upon the operating environment. It is identical with the F-PCH for 9.6 kbps rate (20-ms frame).

13.6.5 Forward Auxiliary Pilot Channel (F-APICH)

This channel is used with antenna beam-forming applications to generate spot beams. Spot beams can be used to increase coverage in a particular geographical area or to increase capacity toward hot spots. The F-APICH can be shared among multiple mobiles in the same spot beam.

Auxiliary pilots are code multiplexed with other forward link channels and they use orthogonal Walsh codes. Since a common pilot contains no data (all 0s), an auxiliary pilot may use a longer Walsh sequence to lessen the reduction of orthogonal Walsh codes available for traffic channels. Auxiliary pilots can also be used for orthogonal diversity transmission in the direct-spread forward link. Furthermore, if the CDMA system uses a separate antenna array to support directional or spot beams, it is necessary to provide a separate forward link pilot for channel estimation.

13.6.6 Forward Broadcast Channel (F-BCH)

This is a paging channel dedicated to carrying only the overhead messages and possible SMS broadcast messages. It removes the overhead messages from the paging channel to a separate broadcast channel. This improves the mobile initialization time and system access performance. At the same time, by reducing the number of messages on the F-PCH, the paging capacity is improved. The F-BCH has a fixed Walsh code that is communicated to the mobile on the F-SYNC.

13.6.7 (Forward) Quick Paging Channel (QPCH)

This is a new type of paging channel that is used by a base station when it needs to contact the mobile in the slotted mode. Its use reduces the time the mobile needs to be "awake," resulting in increased battery life for the mobile.

The QPCH will contain a single-bit, quick-page message to direct a slotted-mode mobile to monitor its assigned slot on the paging channel that immediately follows. The quick page message is sent up to 80 ms before the page message to alert the mobile to listen to the paging channel. The QPCH uses a different modulation, so it will appear as a different physical channel.

13.6.8 Forward Dedicated Auxiliary Pilot Channel (F-DAPICH)

An optional auxiliary pilot can be generated for a particular mobile. The F-DAPICH is used with beam-forming applications and beam-steering techniques to increase the coverage or data rate toward a particular mobile.

13.6.9 Forward Common Power Control Channel (F-CPCCH)

The F-CPCCH transmits power control bits to multiple mobiles. This is used by mobiles operating in power controlled access or reservation access mode.

13.6.10 Forward Transmit Diversity Pilot Channel (F-TDPICH)

The F-TDPICH is an unmodulated, direct-sequence spread-spectrum signal transmitted continuously by a base station to support forward link transmit diversity. The pilot channel and transmit diversity pilot channel provide phase references for coherent demodulation of forward link CDMA channels that use transmit diversity.

13.6.11 Forward Common Assignment Channel (F-CACH)

The F-CACH is used by the base station to acknowledge a mobile station accessing enhanced access channel and, in the case of reservation mode, to transmit the address of the reverse common control channel (R-CCCH) and associated common power control sub-channel.

13.6.12 Forward Auxiliary Transmit Diversity Pilot Channel (F-ATDPICH)

The F-ATDPICH is associated with the auxiliary pilot; both the auxiliary pilot channel and the auxiliary transmit diversity channel provide phase reference for coherent demodulation of those forward link channels associated with the auxiliary pilot and employ transmit diversity.

13.6.13 Forward Fundamental Channel (F-FCH)

This channel is transmitted at a variable rate as in IS-95B and consequently requires rate detection at the receiver. Each F-FCH is transmitted on a different orthogonal code channel and uses frame sizes corresponding to 20 ms and 5 ms. The 20-ms frame structure supports the data rate corresponding to RC1 and RC2, where the rates are 9.6, 4.8, 2.7, and 1.5 kbps for RC1 and 14.4, 7.2, 3.6, and 1.8 kbps for RC2. The SR1 RC1 F-FCH is shown in Figure 13.11 and SR1 RC2 F-FCH in Figure 13.12. For SR1 and RC1, a rate 1/2 convolutional encoder is used. For SR1 and RC2, a rate 1/2 or 1/3 convolutional code followed by puncturing every ninth bit effectively provides a 3/8 code rate.

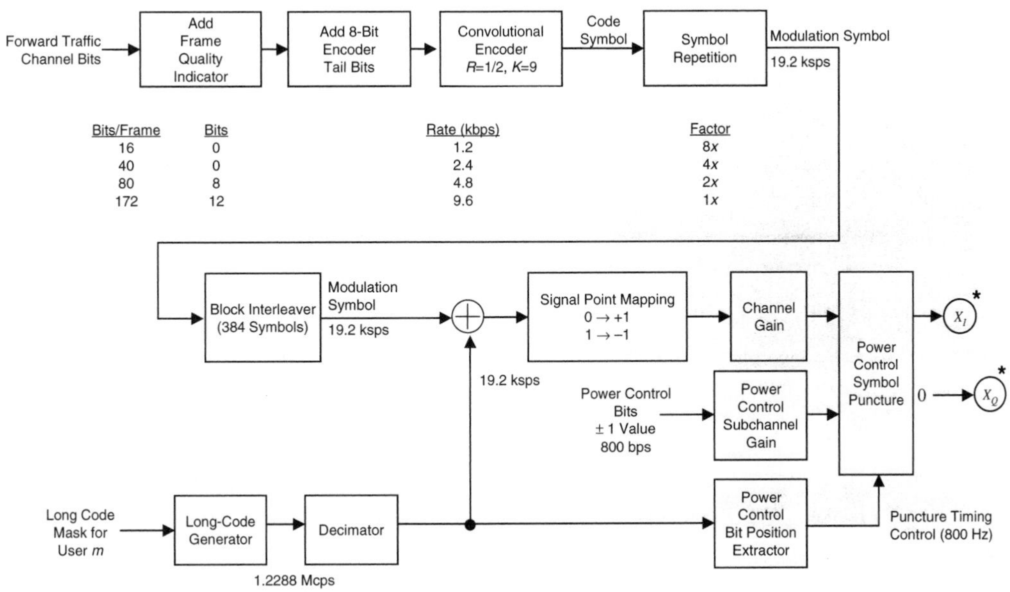

* See Figure 13.23

Figure 13.11 cdma2000 F-FCH for SR1and RC1

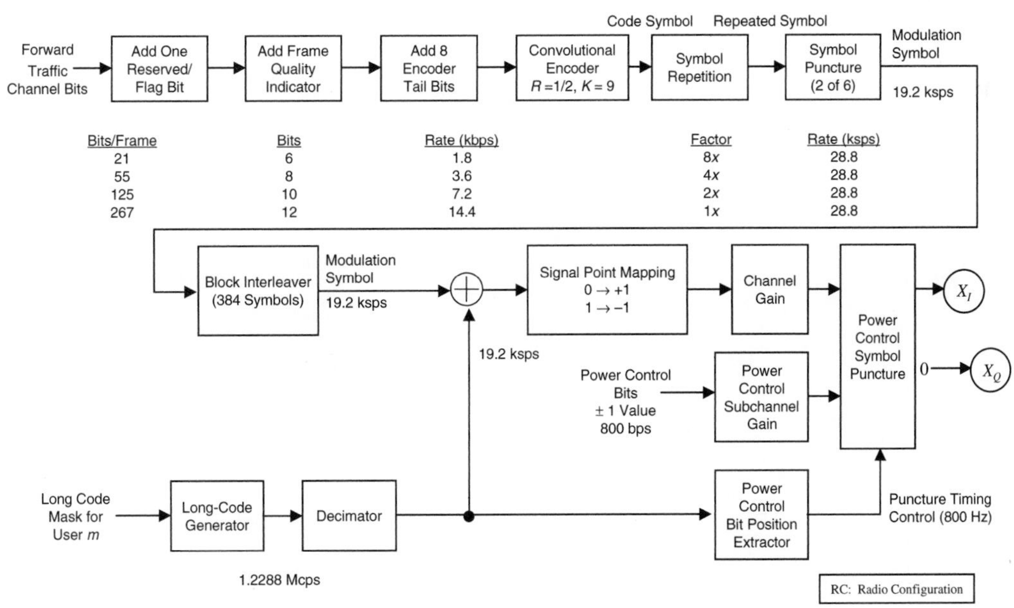

* See Figure 13.23

Figure 13.12 cdma2000 F-FCH for SR1 and RC2

For SR3, with the RC1 and RC2 F-FCHs are shown in Figures 13.13 and 13.14, respectively. For SR3 and RC1, a 1/4 code rate is used. For SR3 and RC2, code rates of 1/2 and 1/4 are supported.

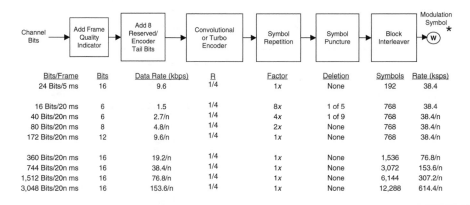

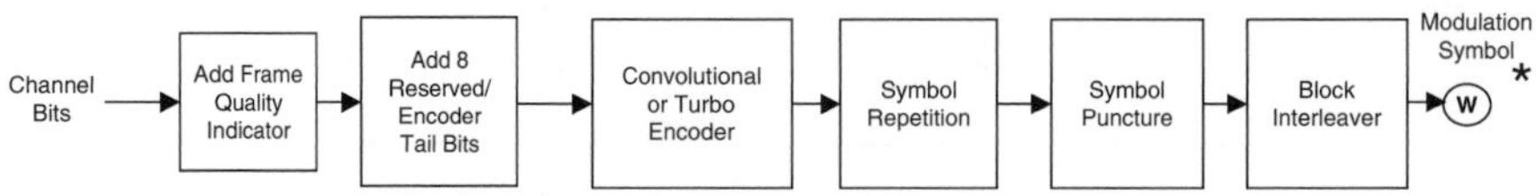

Bits/Frame	Bits	Data Rate (kbps)	R	Factor	Deletion	Symbols	Rate (ksps)
24 Bits/5 ms	16	9.6	1/4	1x	None	192	38.4
16 Bits/20 ms	6	1.5	1/4	8x	1 of 5	768	38.4
40 Bits/20n ms	6	2.7/n	1/4	4x	1 of 9	768	38.4/n
80 Bits/20n ms	8	4.8/n	1/4	2x	None	768	38.4/n
172 Bits/20n ms	12	9.6/n	1/4	1x	None	768	38.4/n
360 Bits/20n ms	16	19.2/n	1/4	1x	None	1,536	76.8/n
744 Bits/20n ms	16	38.4/n	1/4	1x	None	3,072	153.6/n
1,512 Bits/20n ms	16	76.8/n	1/4	1x	None	6,144	307.2/n
3,048 Bits/20n ms	16	153.6/n	1/4	1x	None	12,288	614.4/n

Notes:
1. The 5-ms frame is used only for F-FCHs, and only rates of 9.6 kbps or less are used for F-FCHs.
2. Turbo coding may be used for F-SCHs with rates of 19.2 kbps or more; otherwise, $K = 9$ convolutional coding is used.
3. With convolutional coding, the reserved/encoder tail bits provide an encoder tail. With turbo coding, the first two of these bits are reserved bits that are encoded, and the last six bits are replaced by an internally generated tail.
4. n is the length of frame in multiples of 20 ms. For 40 channel bits per frame, $n = 1$ or 2. For more than 40 channel bits per frame, $n = 1$, 2, or 4.

* See Figure 13.21

Figure 13.13 cdma2000 F-FCH for SR3 and RC1

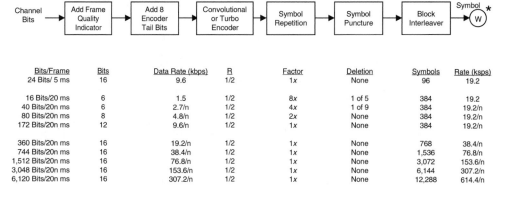

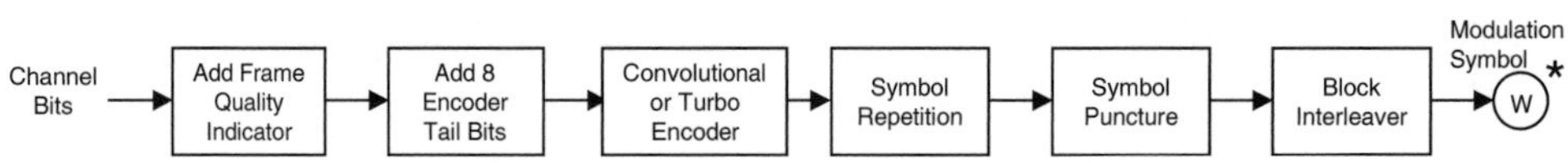

Bits/Frame	Bits	Data Rate (kbps)	R	Factor	Deletion	Symbols	Rate (ksps)
24 Bits/ 5 ms	16	9.6	1/2	1x	None	96	19.2
16 Bits/20 ms	6	1.5	1/2	8x	1 of 5	384	19.2
40 Bits/20n ms	6	2.7/n	1/2	4x	1 of 9	384	19.2/n
80 Bits/20n ms	8	4.8/n	1/2	2x	None	384	19.2/n
172 Bits/20n ms	12	9.6/n	1/2	1x	None	384	19.2/n
360 Bits/20n ms	16	19.2/n	1/2	1x	None	768	38.4/n
744 Bits/20n ms	16	38.4/n	1/2	1x	None	1,536	76.8/n
1,512 Bits/20n ms	16	76.8/n	1/2	1x	None	3,072	153.6/n
3,048 Bits/20n ms	16	153.6/n	1/2	1x	None	6,144	307.2/n
6,120 Bits/20n ms	16	307.2/n	1/2	1x	None	12,288	614.4/n

Notes:
1. The 5-ms frame is used only for F-FCHs, and only rates of 9.6 kbps or less are used for F-FCHs.
2. Turbo coding may be used for F-SCHs with rates of 19.2 kbps or more; otherwise , $K = 9$ convolutional coding is used.
3. With convolutional coding, the reserved/encoder tail bits provide an encoder tail. With turbo coding, the first two of these bits are reserved bits that are encoded, and the last six bits are replaced by an internally generated tail.
4. n is the length of frame in multiples of 20 ms. For 40 channel bits per frame, $n = 1$ or 2. For more than 40 channel bits per frame, $n = 1$, 2 or 4.

* See Figure 13.21

Figure 13.14 cdma2000 F-FCH SR3 and RC2

13.6.14 Forward Supplemental Channel (F-SCH)

The F-SCH can be operated in two distinct modes. The first mode is used for data rates not exceeding 14.4 kbps and uses blind rate detection (no scheduling or rate information provided). In the second mode, the rate information is explicitly provided to the base station. In the first mode, the variable rates provided are those derived from IS-95B RC1 and RC2. The structures for the variable rate modes are identical to the 20-ms F-FCH. In the second mode, the high-data-rate modes can have $k = 9$ convolutional coding or turbocode with $k = 4$ component encoders. For the case of convolutional codes there are 8 tail bits. For the case of turbocodes, 6 tail bits and 2 reserve bits are used.

There may be more than one F-SCH in use at a given time. The individual F-SCH target frame error rates (FERs) may be set independently with respect to the F-FCH and other F-SCHs, since optimal FER for data is different than for voice. For classes of data services that have less stringent delay requirements, the FER may also be managed by retransmission.

The F-SCH supports 20-ms frames. The F-SCH supports data rates from 9.6 to 307.2 kbps (see Figures 13.15, 13.16, 13.17, and 13.18).

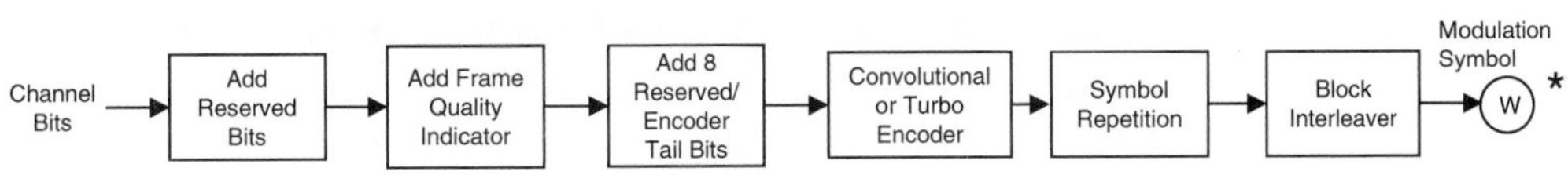

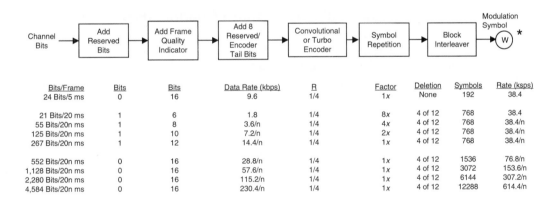

Bits/Frame	Bits	Bits	Data Rate (kbps)	R	Factor	Deletion	Symbols	Rate (ksps)
24 Bits/5 ms	0	16	9.6	1/4	1x	None	192	38.4
21 Bits/20 ms	1	6	1.8	1/4	8x	4 of 12	768	38.4
55 Bits/20n ms	1	8	3.6/n	1/4	4x	4 of 12	768	38.4/n
125 Bits/20n ms	1	10	7.2/n	1/4	2x	4 of 12	768	38.4/n
267 Bits/20n ms	1	12	14.4/n	1/4	1x	4 of 12	768	38.4/n
552 Bits/20n ms	0	16	28.8/n	1/4	1x	4 of 12	1536	76.8/n
1,128 Bits/20n ms	0	16	57.6/n	1/4	1x	4 of 12	3072	153.6/n
2,280 Bits/20n ms	0	16	115.2/n	1/4	1x	4 of 12	6144	307.2/n
4,584 Bits/20n ms	0	16	230.4/n	1/4	1x	4 of 12	12288	614.4/n

Notes:
1. The 5-ms frame is used only for F-FCHs, and only rates of 9.6 kbps or less are used for F-FCHs.
2. Turbocoding may be used for F-SCHs with rates of 19.2 kbps or more; otherwise, $K = 9$ convolutional coding is used.
3. With convolutional coding, the reserved/encoder tail bits provide an encoder tail. With turbocoding, the first two of these bits are reserved bits that are encoded, and the last six bits are replaced by an internally generated tail.
4. n is the length of frame in multiples of 20 ms. For 40 channel bits per frame, $n = 1$ or 2. For more than 40 channel bits per frame, $n = 1$, 2, or 4.

* see Figure 13.21

Figure 13.15 cdma2000 F-SCH and F-FCH for SR1 and RC5

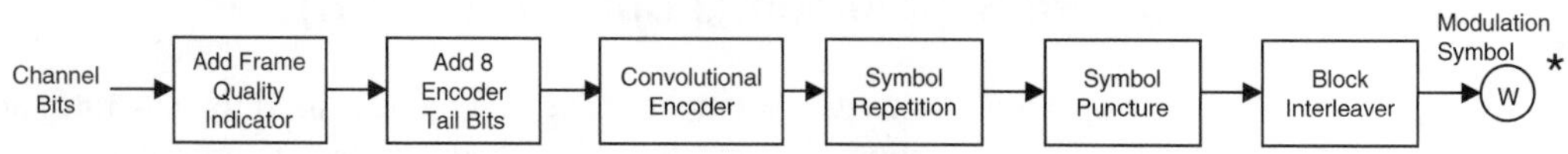

Bits/Frame	Bits	Data Rate (kbps)	R	Factor	Deletion	Symbols	Rate (ksps)
24 Bits/5 ms	16	9.6	1/6	1x	None	288	57.6
16 Bits/20 ms	6	1.5	1/6	8x	1 of 5	1,152	57.6
40 Bits/20 ms	6	2.7	1/6	4x	1 of 9	1,152	57.6
80 Bits/20 ms	8	4.8	1/6	2x	None	1,152	57.6
172 Bits/20n ms	12	9.6/n	1/6	1x	None	1,152	57.6/n
360 Bits/20n ms	16	19.2/n	1/6	1x	None	2,304	115.2/n
744 Bits/20n ms	16	38.4/n	1/6	1x	None	4,608	230.4/n
1,512 Bits/20n ms	16	76.8/n	1/6	1x	None	9.216	460.8/n
3,048 Bits/20n ms	16	153.6/n	1/6	1x	None	18,432	921.6/n
6,120 Bits/20n ms	16	307.2/n	1/6	1x	None	36,864	1,843.2/n

Notes:

1. The 5-ms frame is used only for the F-FCHs, and only rates of 9.6 kbps or less is used for F-SCHs.
2. n is the length of frame in multiples of 20 ms. For 40 channel bits per frame, $n = 1$ or 2. For more than 40 channel bits per frame, $n = 1$, 2, or 4.

* See Figure 13.21

Figure 13.16 cdma2000 F-FCH and F-SCH for SR3 and RC6

13.6.15 Forward Dedicated Control Channel (F-DCCH)

The F-DCCH supports 5-ms and 20-ms frames at a 9.6 kbps encoder input rate. Sixteen CRC bits are added to the information bits for 5-ms frames or 12 CRC bits for 20-ms frames followed by an addition of 8 tail bits, convolutional encoding, interleaving, and scrambling.

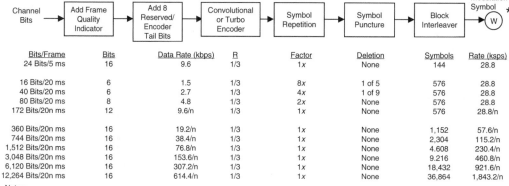

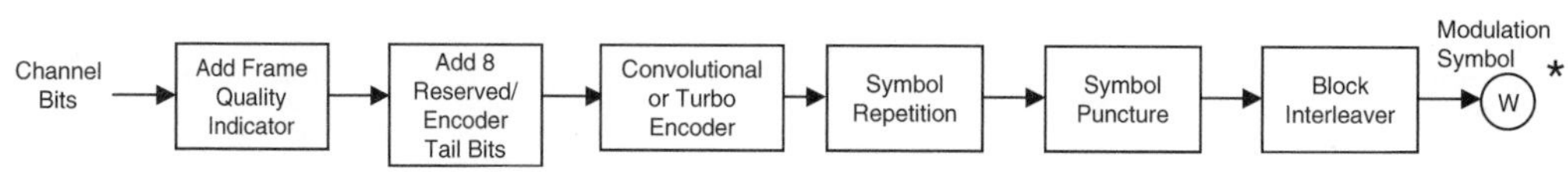

Bits/Frame	Bits	Data Rate (kbps)	R	Factor	Deletion	Symbols	Rate (ksps)
24 Bits/5 ms	16	9.6	1/3	1x	None	144	28.8
16 Bits/20 ms	6	1.5	1/3	8x	1 of 5	576	28.8
40 Bits/20 ms	6	2.7	1/3	4x	1 of 9	576	28.8
80 Bits/20 ms	8	4.8	1/3	2x	None	576	28.8
172 Bits/20n ms	12	9.6/n	1/3	1x	None	576	28.8/n
360 Bits/20n ms	16	19.2/n	1/3	1x	None	1,152	57.6/n
744 Bits/20n ms	16	38.4/n	1/3	1x	None	2,304	115.2/n
1,512 Bits/20n ms	16	76.8/n	1/3	1x	None	4.608	230.4/n
3,048 Bits/20n ms	16	153.6/n	1/3	1x	None	9.216	460.8/n
6,120 Bits/20n ms	16	307.2/n	1/3	1x	None	18,432	921.6/n
12,264 Bits/20n ms	16	614.4/n	1/3	1x	None	36,864	1,843.2/n

Notes:

1. The 5-ms frame is used only for F-FCHs, and only rates of 9.6 kbps or less are used for F-FCHs.
2. Turbocoding may be used for F-SCHs with rates of 19.2 kbps or more; otherwise, $K = 9$ convolutional coding is used.
3. With convolutional coding, the reserved/encoder tail bits provide an encoder tail. With turbocoding, the first two of these bits are reserved bits that are encoded, and the last six bits are replaced by an internally generated tail.
4. n is the length of frame in multiples of 20 ms. For 40 channel bits per frame, $n = 1$ or 2. For more than 40 channel bits per frame, $n = 1$, 2, or 4.

* See Figure 13.21

Figure 13.17 cdma2000 F-FCH and F-SCH for SR3 and RC7

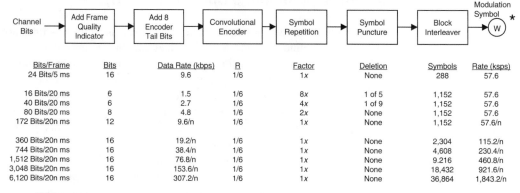

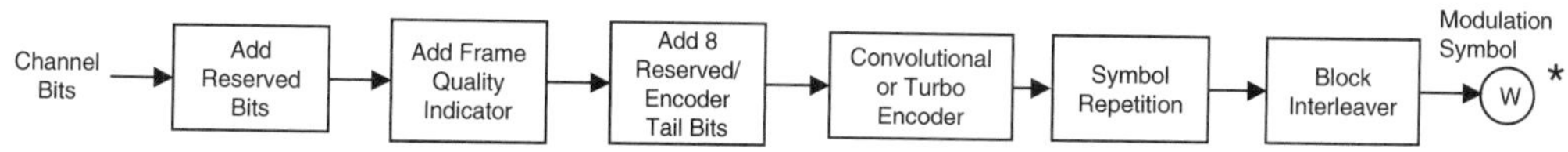

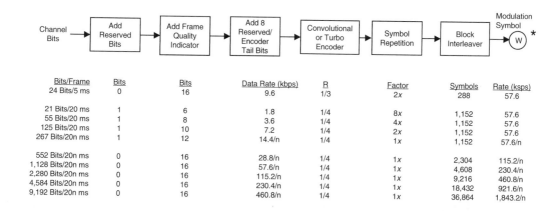

Bits/Frame	Bits	Bits	Data Rate (kbps)	R	Factor	Symbols	Rate (ksps)
24 Bits/5 ms	0	16	9.6	1/3	2x	288	57.6
21 Bits/20 ms	1	6	1.8	1/4	8x	1,152	57.6
55 Bits/20 ms	1	8	3.6	1/4	4x	1,152	57.6
125 Bits/20 ms	1	10	7.2	1/4	2x	1,152	57.6
267 Bits/20n ms	1	12	14.4/n	1/4	1x	1,152	57.6/n
552 Bits/20n ms	0	16	28.8/n	1/4	1x	2,304	115.2/n
1,128 Bits/20n ms	0	16	57.6/n	1/4	1x	4,608	230.4/n
2,280 Bits/20n ms	0	16	115.2/n	1/4	1x	9,216	460.8/n
4,584 Bits/20n ms	0	16	230.4/n	1/4	1x	18,432	921.6/n
9,192 Bits/20n ms	0	16	460.8/n	1/4	1x	36,864	1,843.2/n

Notes:
1. The 5-ms frame is only used for F-FCHs, and only rates of 9.6 kbps or less are used for F-FCHs.
2. Turbocoding may be used for F-SCHs with rates of 19.2 kbps or more; otherwise, $K = 9$ convolutional coding is used.
3. With convolutional coding, the reserved/encoder tail bits provide an encoder tail. With turbocoding, the first two of these bits are reserved bits that are encoded, and the last six bits are replaced by an internally generated tail.
4. n is the length of frame in multiples of 20 ms. For 40 channel bits per frame, $n = 1$ or 2. For more than 40 channel bits per frame, $n = 1$, 2 or 4.

* See Figure 13.21

Figure 13.18 cdma2000 F-FCH and F-SCH for SR3 and RC8

13.7 Forward Link Features

The forward link supports chip rates of N × 1.2288 Mcps (where $N = 1, 3, 6, 9, 12$). For $N = 1$, the spreading is similar to IS-95B (see Chapter 11); however quadrature phase-shift keying (QPSK) modulation and fast closed loop power control are used. For chip rates with $N > 1$, multicarrier (see Figure 13.19) option is used. The multicarrier approach demultiplexes modulation symbols onto N separate 1.25 MHz carriers ($N = 3, 6, 9, 12$). Each carrier is spread at a rate of 1.2288 Mcps.

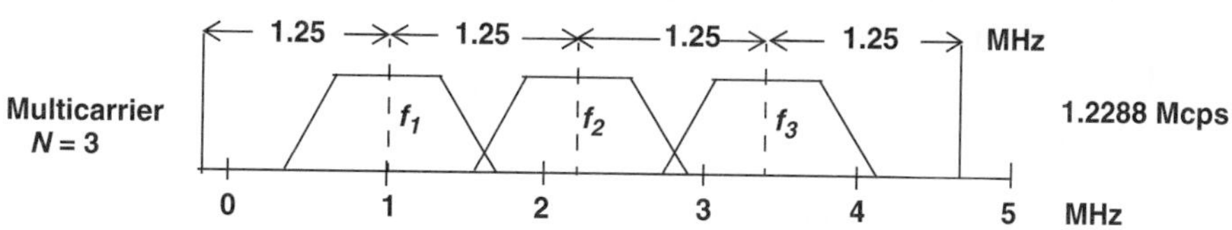

Figure 13.19 Multicarrier Approach in cdma2000

13.7.1 Transmit Diversity

Transmit diversity can reduce the required E_b/N_t (or required transmit power per channel) and thus enhance the system capacity. Transmit diversity can be implemented in the following ways:

Multicarrier Transmit Diversity. Antenna diversity can be implemented in a multicarrier forward link with no impact on the subscriber terminal, where a subset of carriers is transmitted on each antenna. The main characteristics of multicarrier approach are

- Coded information symbols are demultiplexed among multiple 1.25 MHz carriers.
- Frequency diversity is equivalent to spreading the signal over the entire bandwidth.
- Both time and frequency diversity are captured by convolutional coder/symbol repetition and interleaver.
- A RAKE receiver captures signal energy from all bands.
- Each forward link channel may be allocated an identical Walsh code on all carriers.
- Fast power control.

In 3 × 1.25 MHz multicarrier transmitter, the serial coded information symbols are divided into three parallel data streams, and each data stream is spread with a Walsh code and a long pseudorandom (PN) sequence at a rate of 1.2288 Mcps. At the output of the transmitter, there are three carriers, A, B, and C (see Figure 13.20).

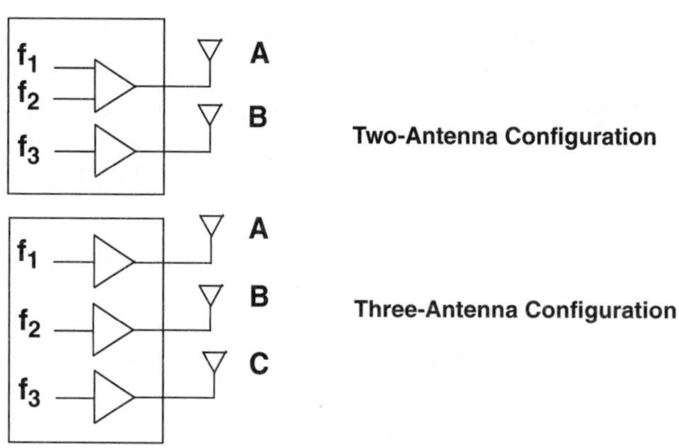

Figure 13.20 3 × 1.25 MHz Multicarrier Transmitter

After processing the serial coded information symbols with parallel carriers, the multicarrier will be transmitted by multiantenna, which is called multicarrier transmit diversity (MCTD). In the MCTD, the total carriers are divided into subsets, then each subset of the carriers is transmitted on each antenna, where frequency filtering provides near-perfect orthogonality between antennas. This provides improved frequency diversity and, hence, increases forward link capacity.

Direct-Spread Transmit Diversity. Orthogonal transmit diversity (OTD) may be used to provide transmit diversity for direct spread ($N = 1$). Coded bits are split into two data streams and are transmitted via separate antennas. A different orthogonal code is used per antenna for spreading. This maintains the orthogonality between the two output streams, and, hence, self-interference is eliminated in flat fading. Note that by splitting the coded bits into two separate streams, the effective number of spreading codes per user is the same as the case without OTD. An auxiliary pilot is introduced for the additional antenna.

13.7.2 Orthogonal Modulation

To reduce or eliminate intracell interference, each forward link physical channel is modulated by a Walsh code. To increase the number of usable Walsh codes, QPSK modulation is used before spreading. Every two information bits are mapped to a QPSK symbol. As a result, the available number of Walsh codes is increased by a factor of two relative to binary phase-shift keying (BPSK) (prespreading) symbols. Walsh code length varies to achieve different information bit rates. The forward link may be interference limited or Walsh code limited depending on the specific deployment and operating environment. When a Walsh code limit occurs, additional codes may be generated by multiplying Walsh codes by the masking functions. The codes generated in this way are called *quasiorthogonal* functions. The quasiorthogonal functions are not totally orthogonal.

13.7.3 Power Control

In IS-95, forward link power control was not considered to be as important and demanding as reverse link power control. This was supported by the argument that the base station (BS) transmits all the channels coherently in the same RF carrier, and they all fade together as the composite signal arrives at the mobile. This argument is valid as long as the thermal and background noise is negligible. For this reason, the initial IS-95A forward power control requirements were not as stringent as the ones for the reverse link. However, a particular mobile may be near a significant source of interference, or may suffer a large path loss such that the arriving composite signal is of the order of the background noise. For this reason, a faster power control is required on the forward link.

In IS-95, the forward link power control is slow and is performed at a rate not faster than 50 Hz. Its implementation depends on the rate set of traffic channel. For RC1 the mobile collects frame error rate (FER) and reports statistics to the BS, which compares the FER with a target value. The BS increases or decreases the forward link power depending on whether the FER is higher or lower than the target FER. Typically, the BS receives an FER measurement message once every 100 frames, and takes action every two seconds on average, thus providing a forward power control that runs at a rate of about 0.5 Hz for RC1.

For RC2 the mobile sends an erasure indicator bit (EIB) to the BS in every frame, indicating the quality of the frame received previously by the mobile. The BS uses the EIB to estimate the forward link FER and increases or decreases the power of the forward link traffic channels depending on the relationship of the FER to the estimated FER target value. The RC2 power control runs at a rate of 50 Hz, which represents a performance improvement over its predecessor algorithm used in the RC1.

Although there are substantial performance improvements in the forward link power control when updating the RC1 algorithm with the RC2 algorithm, the performance is still limited by the inability to cope with fast Rayleigh fading and fast-changing radio frequency (RF) conditions. The main reason for this is the slow rate at which forward link power control operates. Therefore, to improve the performance (i.e., to lower the BS average transmit power per voice channel) and to increase the forward link call capacity, a faster power control loop is required to mitigate fast Rayleigh fading.

The 3G forward link power control functions similarly to the reverse link power control in IS-95. The algorithm consists of two loops running at an effective rate of 800 Hz. This enhancement is expected to increase the user capacity of the forward link.

The forward link power control in 3G is fundamentally different from that in IS-95. Its main objective is to increase the voice call capacity in the forward link by a series of new enhancements including:

- High-speed forward power control
- Closed loop with fast time response
- Variable power step size controlled by the BS

The forward link power control operates at a high rate to track and accurately compensate for the fast Rayleigh fading on the forward link. The accurate tracking minimizes the average power transmitted by the BS to the mobile and, as a consequence, increases the forward link call capacity.

The rate of the forward link power control is increased by replacing the slow FER-based algorithm of IS-95 with a closed loop based on E_b/N_t measurements similar to the reverse link in IS-95. While the IS-95 procedure of FER statistics is a slow process with a response time of many frames, the E_b/N_t measurements are fast and easy to perform at a subframe interval. This allows the 3G forward link power control to operate at high speed.

The increase in forward link capacity is expected mostly for mobiles moving at low speed and in simplex mode (i.e., not in soft handoff) where fast Rayleigh fading can be substantial and can be mitigated effectively. For mobiles traveling at high speed, the fast power control cannot track the fast Rayleigh fading accurately, and therefore, a large increase in capacity may not be expected. In addition, the capacity increase for mobiles in soft handoff is less due to the already existing path diversity of soft handoff, which reduces Rayleigh fading.

The variable power step size has not been standardized in IS-2000 and, therefore, the BS has the flexibility to adapt the step size depending on the speed of the mobile and soft handoff status of the call. This adaptability permits minimization of the peak-to-average power ratio, and decreases the overall interference, which increases the forward link capacity.

The new fast forward power control (FFPC) algorithm on the forward link and power control for the F-FCH and F-SCH is used in cdma2000. The standards specify a fast closed loop power control at 800 Hz. Two schemes of power control for the F-FCH and F-SCH have been proposed.

- **Single channel power control**—This is based on the performance of the higher rate channel between the F-FCH and F-SCH. The gain setting for the lower rate channel is determined based on its relationship to the higher rate channel.
- **Independent power control**—In this case, gains for the F-FCH and F-SCH are determined separately. The mobile runs two separate outer loop algorithms (with different E_b/N_t targets) and sends two forward error bits to the BS.

13.7.4 Walsh Code Administration

IS-95 A/B uses fixed-length 64-chip Walsh codes. The new rate sets in cdma2000 require variable-length Walsh codes for traffic channels. In 3G1X, the Walsh codes used are from 128 chips to 2 chips in length. The F-FCH Walsh code is fixed (128 chips for RS3 and RS5, and 64 chips for RS4 and RS6), whereas the length of the Walsh codes for F-SCH decreases as the information rate increases to maintain a constant bandwidth of the modulated signal. In addition to different Walsh code lengths, the coordination of allocation of Walsh codes across the 2G and the 3G system is necessary for overlay systems.

The algorithm must ensure that Walsh codes assigned for different rate supplemental channels are always orthogonal to each other as well as to the fundamental traffic channels, paging channels, sync channel, and pilot channel. For example, if an all 0s 4-chip Walsh code (0 0 0 0) is assigned, then there are two 8-chip Walsh codes that are not to be assigned at the same time (0 0 0 0 0 0 0 0, 0 0 0 0 1 1 1 1); the remaining six 8-chip codes can be used since they are all orthogonal to it. By induction, four 16-chip, eight 32-chip, 16 64-chip, and 32 128-chip codes must also be set aside to maintain orthogonality.

The 2G and 3G Walsh code assignments must be coordinated to ensure that assigning the longer-lengths codes does not block out all of the shorter codes.

13.7.5 Modulation and Spreading

The SR1 system can be deployed in a new spectrum or as a backward-compatible upgrade anywhere an IS-95B forward link is deployed in the same RF channel. The SR1 spreading is shown in Figure 13.21.

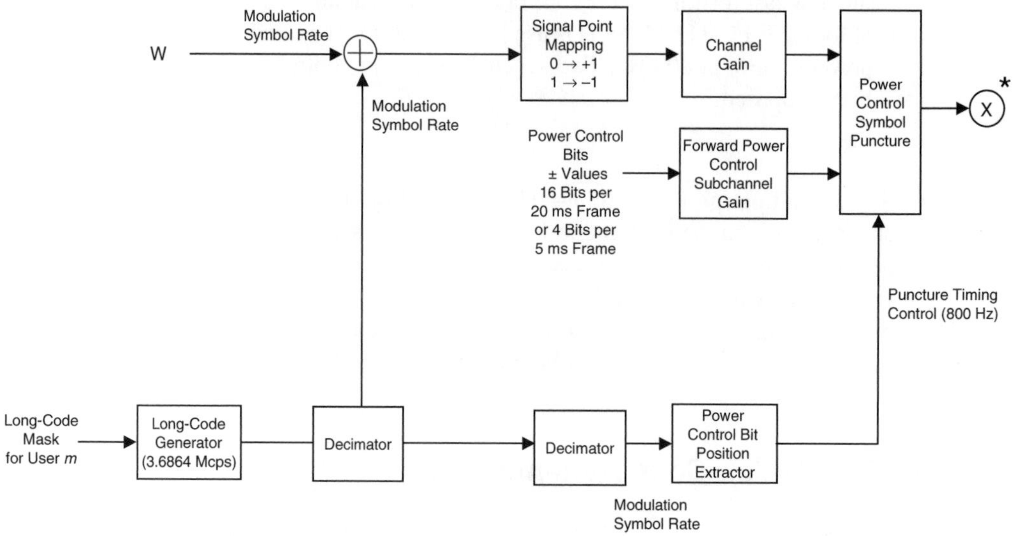

Power control symbol puncturing is on the Forward Fundamental Channels and Forward Dedicated Control Channels only.

* See Figure 13.22

Figure 13.21 SR1 PN Spreading, Baseband Filtering, and Frequency Modulation

The multicarrier system can be deployed in a new spectrum or as a backwards-compatible upgrade anywhere an IS-95B forward link is deployed in the same *N* RF channels. The new cdma2000 channels can coexist in an orthogonal manner with the code channels of the existing IS-95B system.

The overall structure of the multicarrier CDMA channel is shown in Figure 13.22. After scrambling with the long PN code corresponding to user *m*, the user data is demultiplexed into *N* carriers, where $N = 3$, 6, 9, or 12. On each carrier, the demultiplexed bits are mapped onto *I* and *Q* followed by Walsh spreading. When applicable, power control bits, for reverse closed loop power control, may be punctured onto the forward link channel at a rate of 800 Hz. The signal on each carrier is orthogonally spread by the appropriate Walsh code function in such a manner as to maintain a fixed chip rate of 1.2288 Mcps per carrier, where the Walsh code may differ on each carrier. The signal on each carrier is then complex PN spread, as shown in Figure 13.23, followed by baseband filtering and frequency modulation.

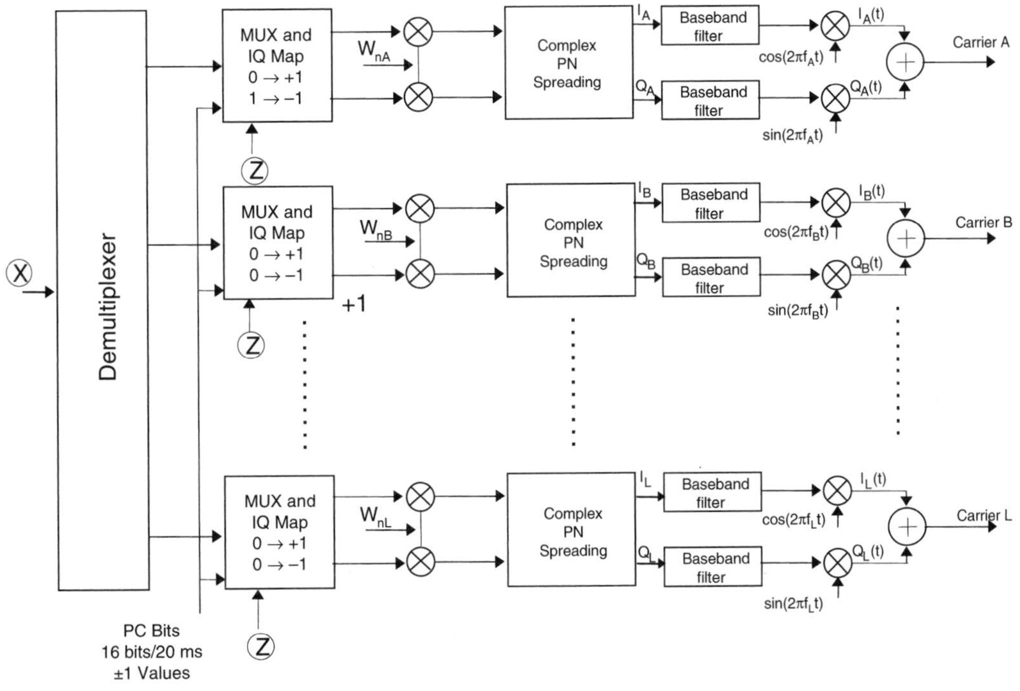

Figure 13.22 Multicarrier cdma2000 Forward Common Physical Channel Modulation and Spreading

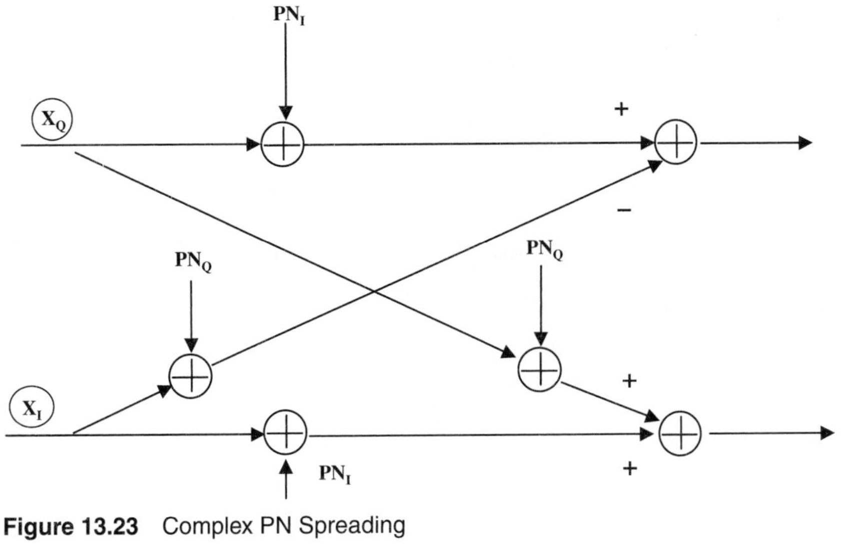

Figure 13.23 Complex PN Spreading

Figure 13.24 provides a comparison between the forward physical channels used in IS-95A and IS-95B and cdma2000.

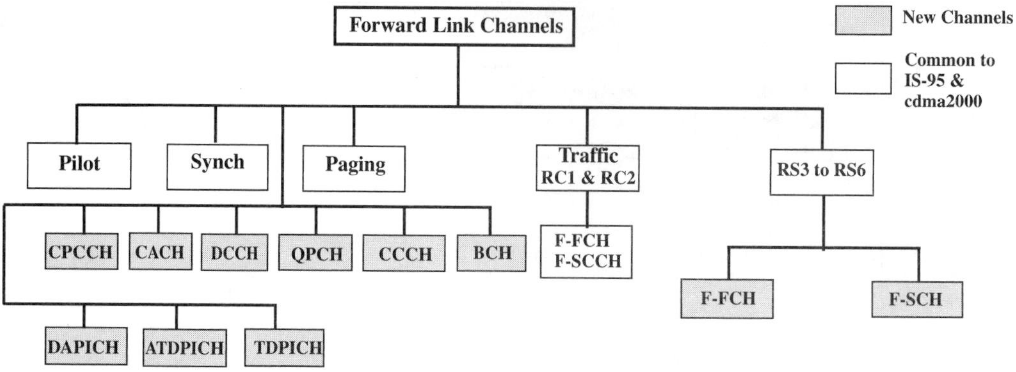

Figure 13.24 A Comparison Between Forward Physical Channels for IS-95 and cdma2000

The key characteristics of the forward link are summarized as follows:

- Channels are orthogonal and use Walsh codes. Different length Walsh codes are used to achieve the same chip rate for different information bit rates.
- QPSK modulation is used before spreading to increase the number of usable Walsh codes.
- Forward error correction (FEC) is used:
  - Convolutional codes ($k = 9$) are used for voice and data.
  - Turbocodes ($k = 4$) are used for high data rates on supplemental channels (SCHs).
- Supports nonorthogonal forward link channelization:
  - These are used when running out of orthogonal space (insufficient number of Walsh codes).
  - Quasiorthogonal functions are generated by masking existing Walsh functions.
- Synchronous forward link.
- Forward link transmit diversity.
- Fast-forward power control (closed loop) 800 times per second.
- Supplemental channel active set, subset of fundamental channel active set. The maximum data rate supported for RS3 and RS5 for supplemental channel is 153.6 kbps (raw data rate). RS4 and RS6 will be supported only for voice calls with the fundamental channel rates up to 14.4 kbps (raw data rate).
- Frame lengths:
  - 20-ms frames are used for signaling and user information.
  - 5-ms frames are used for control information.

13.8 Reverse Physical Channels

Reverse physical channels (see Figure 13.6) include dedicated channels to carry information from a single mobile to the base station and common channels to carry information from multiple mobiles to the base station. There are six radio configurations for the reverse traffic channels. A mobile station supports operation in RC1, RC3, or RC5. A mobile station may support operation in RC2, RC4, or RC6. A mobile station supporting operation in RC2 supports RC1. A mobile station supporting operation in RC4 supports RC3. A mobile station supporting operation in RC6 supports RC5. A mobile station does not use RC1 or RC2 concurrently with RC3 or RC4 on the reverse traffic channels.

Table 13.7 lists the reverse physical channels.

Table 13.7 Reverse Physical Channels

	Physical Channels	Channel Name
Reverse Common Physical Channel	Reverse Access Channel	R-ACH
	Reverse Enhanced Access Channel	R-EACH
	Reverse Common Control Channel (9.6 kbps only)	R-CCCH
Reverse Dedicated Physical Channel	Reverse Pilot Channel	R-PICH
	Reverse Dedicated Control Channel	R-DCCH
	Reverse Traffic Channel: • Fundamental • Supplemental • Supplemental Code	 R-FCH R-SCH R-SCCH

Table 13.8 provides radio configuration characteristics and data rates for reverse channels for SR1 and SR3. Table 13.9 lists channel types on the reverse link for SR1 and SR3.

Table 13.8 Radio Configuration Characteristics and Data Rates of Reverse Channels for SR1 and SR3

Channel Type		Data Rate (kbps)	
		SR1	SR3
Enhanced Access Channel (EACH)	Header	9.6	9.6
	Data	38.4 (5-, 10-, or 20-ms frames), 19.2 (10- or 20-ms frames), or 9.6 (20-ms frames)	38.4 (5-, 10-, or 20-ms frames), 19.2 (10- or 20-ms frames), or 9.6 (20-ms frames)
Access Channel (R-ACH)		4.8	N/A
Reverse Control Channel (R-CCCH)		38.4 (5-, 10-, or 20-ms frames), 19.2 (10-, or 20-ms frames), or 9.6 (20-ms frames)	38.4 (5-, 10-, or 20-ms frames), 19.2 (10-, or 20-ms frames), or 9.6 (20-ms frames)
Reversed Dedicated Control Channel (R-DCCH)	RC3	9.6	N/A
	RC4	14.4 (20-ms frames) or 9.6 (5-ms frames)	N/A
	RC5	N/A	9.6
	RC6	N/A	14.4 (20-ms frames) or 9.6 (5-ms frames)
Fundamental Channel (R-FCH)	RC1	9.6, 4.8, 2.4, or 1.2	N/A
	RC2	14.4, 7.2, 3.6, or 1.8	N/A
	RC3	9.6, 4.8, 2.7, or 1.5 (20-ms frames) or 9.6 (5-ms frames)	N/A
	RC4	14.4, 7.2, 3.6, or 1.8 (20-ms frames) or 9.6 (5-ms frames)	N/A
	RC5	N/A	9.6, 4.8, 2.7,or 1.5 (20-ms frames) or 9.6 (5-ms frames)
	RC6	N/A	14.4, 7.2, 3.6, or 1.8 (20-ms frames) or 9.6 (5-ms frames)
Supplemental Code Channel (SCCH)	RC1	9.6	N/A
	RC2	14.4	N/A

Table 13.8 Radio Configuration Characteristics and Data Rates of Reverse Channels for SR1 and SR3 *(cont.)*

Channel Type		Data Rate (kbps)	
		SR1	SR3
Supplemental Channel (SCH)	RC3	307.2, 153.6, 76.8, 38.4, 19.2, 9.6, 4.8, 2.7, or 1.5 (20-ms frames) 153.6, 76.8, 38.4, 19.2, 9.6, 4.8, 2.4, or 1.35 (40-ms frames) 76.8, 38.4, 19.2, 9.6, 4.8, 2.4, or 1.2 (80-ms frames)	N/A
	RC4	230.4, 115.2, 57.6, 28.8, 14.4, 7.2, 3.6, or 1.8	N/A
	RC5	N/A	614.4, 307.2, 153.6, 76.8, 38.4, 19.2, 9.6, 4.8, 2.7, or 1.5 (20-ms frames) 307.2, 153.6, 76.8, 38.4, 19.2, 9.6, 4.8, 2.4, 1.35 (40-ms frames) 153.6, 76.8, 38.4, 19.2, 9.6, 4.8, 2.4, or 1.2 (80-ms frames)
	RC6	N/A	1036.8, 518.4, 460.8, 259.2, 230.4, 115.2, 57.6, 28.8, 14.4, 7.2, 3.6, or 1.8

13.8.1 Reverse Pilot Channel (R-PICH)

The pilot channel is an unmodulated spread-spectrum signal used to assist the base station in detecting a mobile station transmission. The R-PICH is transmitted when the EACH, R-CCCH, or R-TCH with RC3 through RC6 is enabled. The R-PICH is also transmitted during the EACH preamble, the R-CCCH preamble, and the R-TCH preamble.

The R-PICH is used for initial acquisition, time tracking, RAKE-receiver coherent reference recovery, and power control measurements (see Figure 13.25 for R-PICH structure). The R-PICH is spread with W_0^{32}. R-PICH gating may be used only when none of the following channels is assigned: F-FCH, F-SCH, R-FCH, and R-SCH.

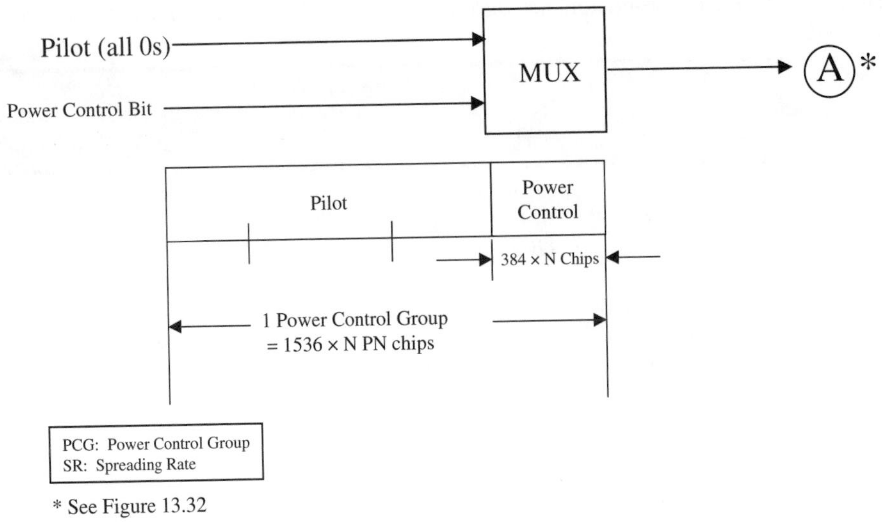

Figure 13.25 R-PICH Structure

Table 13.9 Channel Type on the Reverse Link for SR1 and SR3

Channel Type	Maximum Number	
	SR1	SR3
Reverse Pilot Channel (R-PICH)	1	1
Reverse Access Channel (R-ACH)	1	None
(Reverse) Enhanced Access Channel (EACH)	1	1
Reverse Common Control Channel (R-CCCH)	1	1
Reverse Dedicated Control Channel (R-DCCH)	1	1
Reverse Fundamental Channel (R-FCH)	1	1
Reverse Supplemental Code Channels (R-SCCH) (RC1 and RC2)	7	None
Reverse Supplemental Channels (RC3 and RC4)	2	2

13.8.2 Reverse Power Control Subchannel

The mobile station inserts a reverse power control subchannel on the R-PICH when operating on the R-TCH with RC3 through RC6. The mobile station supports both the inner power control loop and the outer power control loop for forward traffic channel power control. The subchannel provides information on the quality of the forward link at the rate of 1 bit per 1.25-ms power control group (PCG) and is used by the forward link channels to adjust their power.

The outer power control loop estimates the setpoint value based on E_b/N_t to achieve the target FER on each assigned F-TCH. These setpoints are communicated to the base station either implicitly through the inner loop or explicitly through signaling messages. The difference between setpoints helps the base station to drive the appropriate transmit power levels for F-TCHs that do not have inner loops.

The inner loop power control compares E_b/N_t of the received F-TCH with the corresponding outer power control loop setpoint to determine the value of the power control bit to be sent to the base station on the reverse power control subchannel. The mobile transmits the EIB or the quality indicator bit (QIB) on the reverse power control subchannel upon the command of the base station.

The power control symbol repetition means that the 1-bit value is constant for that repeated symbol duration. The power control bit uses the last portion of each power control group. The +1 pilot symbols and multiplexed power control symbols are all sent with the same power level. The binary power control symbols are represented with ± 1 values.

13.8.3 Reverse Access Channel (R-ACH)

The R-ACH is used by the mobile station to initiate communication with the base station and to respond to PCH messages. An ACH transmission is a coded, interleaved, and modulated spread-spectrum signal. The ACH uses a random access protocol. Access channels are uniquely identified by their long codes.

An access probe consists of an access preamble, followed by a series of access channel frames, with each carrying an SDU. The mobile station transmits information on the ACH at a fixed rate of 4.8 kbps. An ACH frame is 20 ms in duration (see Figure 13.26).

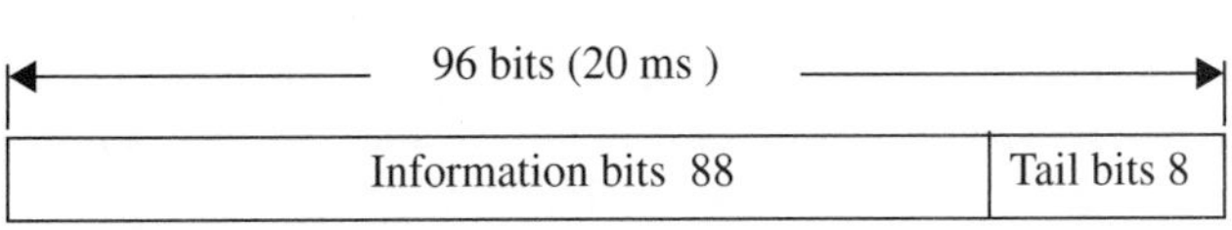

Figure 13.26 R-ACH Frame

The reverse channel may have up to 32 access channels per supported PCH. At least one ACH exists for each PCH on the corresponding forward channels. The ACH preamble consists of frames of 96 zeros that are transmitted at 4.8 kbps rate. The ACH preamble is transmitted to aid the base station in acquiring an ACH transmission.

13.8.4 (Reverse) Enhanced Access Channel (EACH)

The EACH is used by the mobile to initiate communication with the base station or to respond to a mobile-directed message. The EACH can be used in three possible modes: basic access mode, power controlled mode, or reservation access mode. Power controlled access mode and reservation access mode may operate on the same EACH. Basic access mode must operate on a separate EACH.

In the basic access mode, the mobile does not transmit the enhanced access header on the EACH. In the basic access mode, the enhanced access probe consists of an enhanced access channel preamble followed by enhanced access data.

In the power controlled access mode, the enhanced access probe consists of an enhanced access channel preamble, followed by the enhanced access header and enhanced access data.

In the reservation access mode, the enhanced access probe consists of an enhanced access channel preamble followed by an enhanced access header. Enhanced access data is sent on the R-CCCH upon receiving permission from the base station.

The EACH uses a random access protocol. Enhanced access channels are uniquely identified by their long code. Figure 13.27 shows the processing of the EACH.

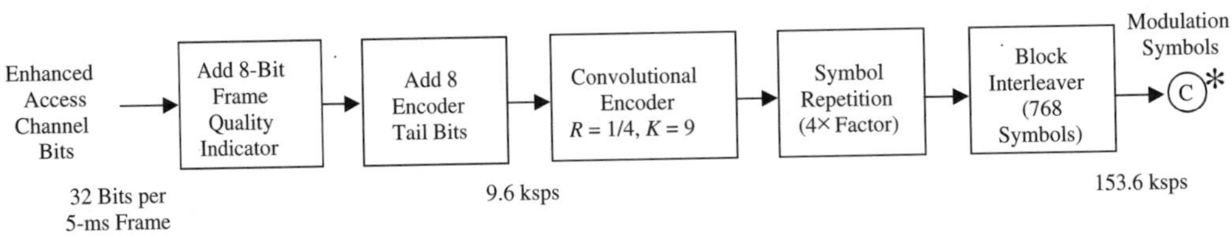

* See Figure 13.32

Figure 13.27 EACH Processing

The mobile station transmits the enhanced access header on the EACH at a fixed data rate of 9.6 kbps. The mobile station transmits the enhanced access data on the EACH at fixed data rate of 9.6, 19.2, or 38.4 kbps. The frame duration for the enhanced access header on the EACH is 5 ms. The frame duration for the enhanced access data on the EACH is 20, 10, or 5 ms. The reverse channels may contain up to 32 EACHs per supported F-CCCH. There is an F-CACH associated with every EACH operating in power controlled access mode or reservation access mode. The frame structure of the EACH is shown in Figure 13.28 and bit details are given in Table 13.10.

Information Bits	F	T

F: Frame Quality Indicator (CRC)
T: Encoder Tail bits

Figure 13.28 EACH Frame

Table 13.10 EACH Frame Structure

Frame Duration (ms)	Frame Type	Transmission Rate (kbps)	No. of Bits Per Frame			
			Information Bits	F	T	Total Bits
5	Header	9.6	32	8	8	48
20	Data	9.6	172	12	8	192
20	Data	19.2	360	16	8	384
20	Data	38.4	744	16	8	768
10	Data	19.2	172	12	8	192
10	Data	38.4	360	16	8	384
5	Data	38.4	172	12	8	192

Note: The frame quality indicator (CRC) is calculated on all bits within the frame, except the frame quality indicator itself and encoder tail bits.

13.8.5 Reverse Common Control Channel (R-CCCH)

The R-CCCH is used for the transmission of user and signaling information to the base station when R-TCHs are not in use. The R-CCCH can be used in one of two possible modes: reservation access mode and designated access mode. The R-CCCH transmission is a coded, interleaved, and modulated spread-spectrum signal. The mobile station transmits during intervals specified by the base station. The reverse common control channels are uniquely identified by their long codes. The R-CCCH processing is shown in Figure 13.29.

The mobile station transmits information on R-CCCH at variable data rates of 9.6, 19.2, and 38.4 kbps. A R-CCCH frame is 20, 10, or 5 ms in duration. The timing of the R-CCCH transmission starts on 1.25-ms increments of the system time.

The reverse cdma2000 channels may contain up to 32 R-CCCHs per supported F-CCCH and up to 32 R-CCCHs per supported F-CACH. At least one R-CCCH exists on the reverse channel for each F-CCCH on the corresponding forward channel. Each R-CCCH is associated with a single F-CCCH. Table 13.11 shows the R-CCCH bit allocations.

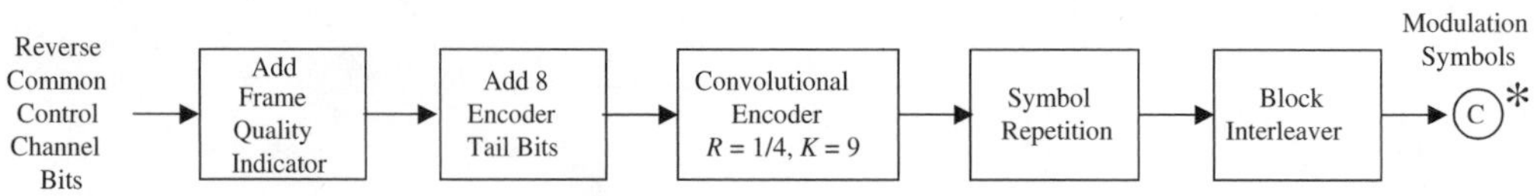

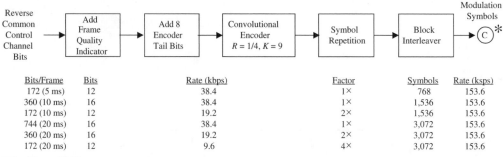

Bits/Frame	Bits	Rate (kbps)	Factor	Symbols	Rate (ksps)
172 (5 ms)	12	38.4	1×	768	153.6
360 (10 ms)	16	38.4	1×	1,536	153.6
172 (10 ms)	12	19.2	2×	1,536	153.6
744 (20 ms)	16	38.4	1×	3,072	153.6
360 (20 ms)	16	19.2	2×	3,072	153.6
172 (20 ms)	12	9.6	4×	3,072	153.6

* See Figure 13.32

Figure 13.29 R-CCCH Processing

Table 13.11 R-CCCH Frame Structure

Frame Duration (ms)	Rate (kbps)	No. of Bits Per Frame			
		Information Bits	CRC Bits	Tail Bits	Total Bits
20	9.6	172	12	8	192
20	19.2	360	16	8	384
20	38.4	744	16	8	768
10	19.2	172	12	8	192
10	38.4	360	16	8	384
10	38.4	172	12	8	192

Note: The frame quality indicator (CRC) is calculated on all bits within the frame, except the frame quality indicator itself and encoder tail bits.

13.8.6 Reverse Dedicated Control Channel (R-DCCH)

The R-DCCH is used for the transmission of user and signaling information to the base station during a call. The R-TCH may contain up to one R-DCCH. The mobile station transmits information on the R-DCCH at a fixed data rate of 9.6 or 14.4 kbps using a 20-ms frame or 9.6 kbps using a 5-ms frame. The mobile station transmits information on the R-DCCH at a data rate of 9.6 kbps for RC3 and RC5. The mobile station transmits information on the R-DCCH at a data rate of 14.4 kbps for 20-ms frames and 9.6 kbps for 5-ms frames for RC4 and RC6.

The mobile supports discontinuous transmission on the R-DCCH. The decision to enable or disable the R-DCCH is made on a frame-by-frame (i.e., 5 or 20 ms) basis. Table 13.12 provides information for R-DCCH frames for nonflexible data rates. Figure 13.30 shows the frame structure of the R-DCCH for flexible data rates. Table 13.13 lists the bits allocation of R-DCCH for flexible data rates. Figure 13.31 shows the processing of R-DCCH for RC3 and RC4.

Table 13.12 R-DCCH Frame for Nonflexible Data Rates

Frame Duration (ms)	Rate (kbps)	No. of bits per frame				
		Reserved	Information	CRC (F)	T	Total bits
20	9.6	0	172	12	8	192
20	14.4	1	267	12	8	288
5	9.6	0	24	16	8	48

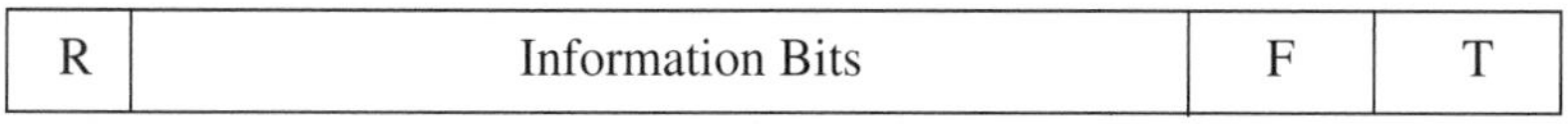

R = Radio Configuration; F = Frame quality indicator (CRC); T = Encoder tail bits

Figure 13.30 R-DCCH Frame Structure for Flexible Data Rates

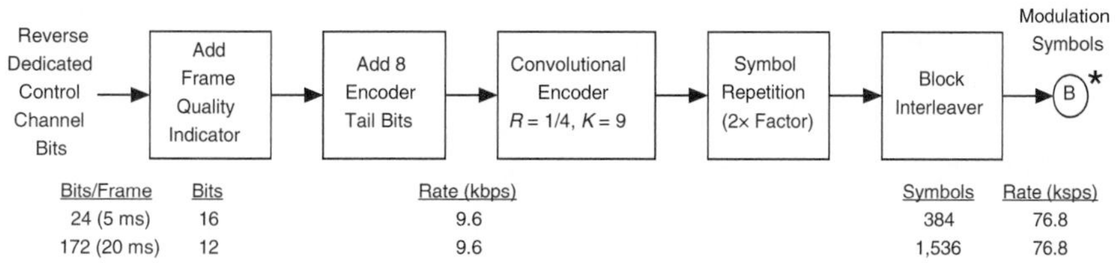

(a) Radio Configuration RC3

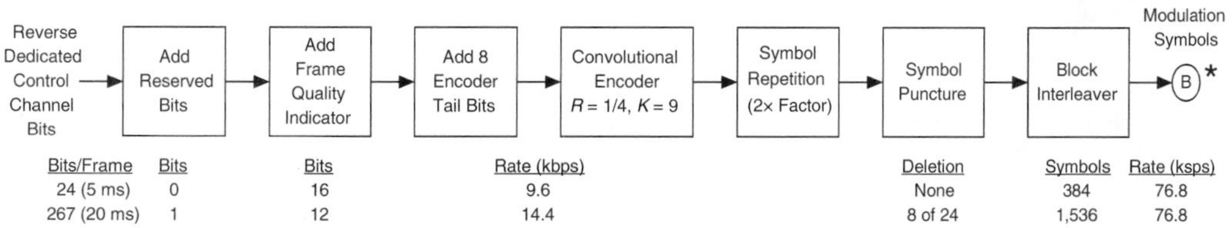

(b) Radio Configuration RC4

* See Figure 13.32

Figure 13.31 R-DCCH Processing for RC3 and RC4

Table 13.13 R-DCCH Frame for Flexible Data Rates

RC (R)	Frame Duration (ms)	Rate (kbps)	No. of Bits Per Frame				
			Reserved	Information	CRC (F)	T	Total Bits
3 and 5	20	1.05–9.6	0	1–172	12 or 16	8	21–192
4 and 6	20	1.05–14.4	0	1–268	12 or 16	8	21–288

Note: The CRC is calculated on all bits within the frame, except the frame quality indicator itself and encoder tail bits.

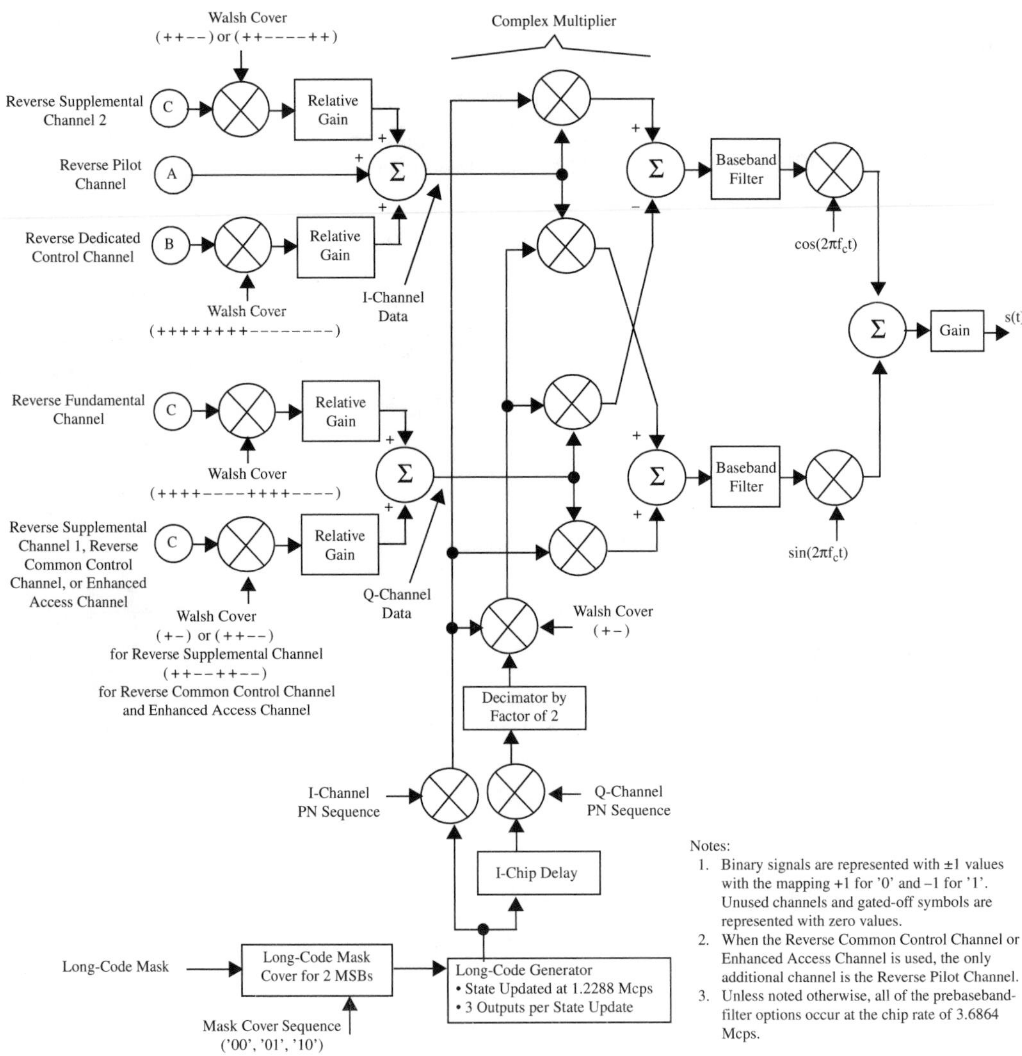

Figure 13.32 Reverse Link I and Q Mapping for SR1 and SR3

13.8.7 Reverse Fundamental Channel (R-FCH)

The R-FCH supports 5- and 20-ms frames. The 20-ms frame structures provide rates derived from the IS-95B RC1 or RC2. The 5-ms frames provide 24 information bits per frame with 16-bit CRC. Within each 20-ms frame interval, either one 20-ms R-FCH structure, up to four 5-ms R-FCH structure(s), or nothing can be transmitted. In addition, when 5-ms R-FCH structure is used, it can be on or off in each of the four 5-ms segments of a 20-ms frame interval. The R-FCH is transmitted at different rates. The rates supported for the R-FCH are different for different radio configurations (see Table 13.8). See Figure 13.34 on page 416 for R-FCH/R-SCH for RC3 and RC5, and Figure 13.35 on page 417 for RC4 and RC6.

13.8.8 Reverse Supplementary Channel (R-SCH)

The R-SCH can be operated in two distinct modes. The first mode is used for data rate not exceeding 14.4 kbps and uses blind rate detection (no scheduling or rate information). In the second mode, the rate information is explicitly known by the base station. The R-SCH is used for data calls and can operate at different prenegotiated rates (see Table 13.8). See Figure 13.34 on page 416 for R-FCH/R-SCH for RC3 and RC5, and Figure 13.35 on page 417 for RC4 and RC6.

Figure 13.33 provides a comparison between the reverse physical channels used in IS-95A, IS-95B, and cdma2000.

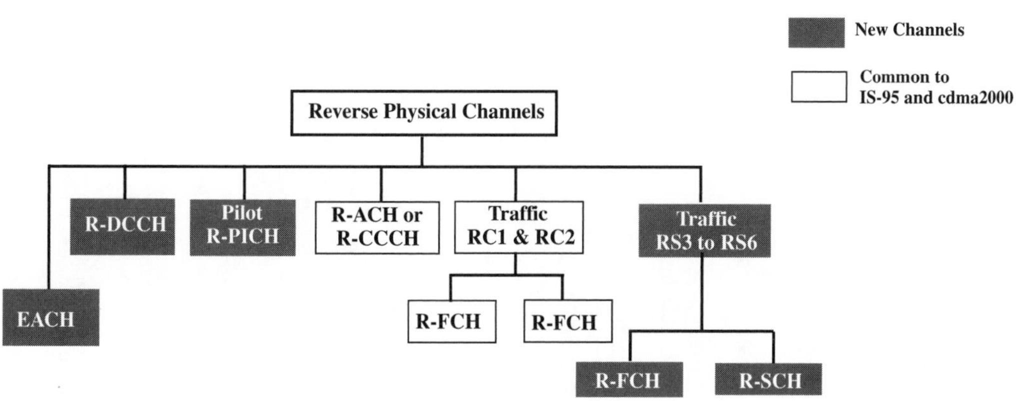

Figure 13.33 A Comparison Between Reverse Physical Channels for IS-95 and cdma2000

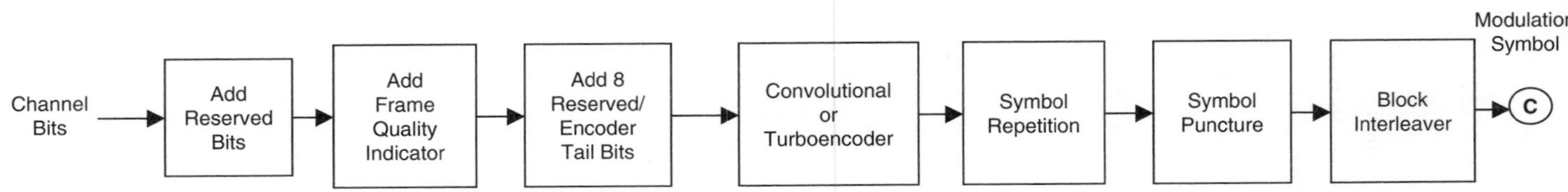

Bits/Frame	Bits	Bits	Data Rate (kbps)	R	Factor	Deletion	Symbols	Rate (ksps)
24 Bits/5 ms	0	16	9.6	1/4	2x	None	384	76.8
21 Bits/20 ms	1	6	1.8	1/4	16x	8 of 24	1,536	76.8
55 Bits/ 20n ms	1	8	3.6/n	1/4	8x	8 of 24	1,536	76.8/n
125 Bits/20n ms	1	10	7.2/n	1/4	4x	8 of 24	1,536	76.8/n
267 Bits/20n ms	1	12	14.4/n	1/4	2x	8 of 24	1,536	76.8/n
552 Bits/20n ms	0	16	28.8/n	1/4	1x	4 of 12	1,536	76.8/n
1,128 Bits/20n ms	0	16	57.6/n	1/4	1x	4 of 12	3,072	153.6/n
2,280 Bits/20n ms	0	16	115.2/n	1/4	1x	4 of 12	6,144	307.2/n
4,584 Bits/20n ms	0	16	203.4/n	1/4	1x	4 of 12	12,288	614.4/n
1 to 4,583 Bits/20n ms	0							

RC: Radio Configuration
SR: Spreading Rate
* See Figure 13.32

Notes:

1. n is the length of the frame in multiples of 20 ms. For 55 channel bits per frame, n - 1 or 2. For more than 55 channel bits per frame, n = 1, 2, or 4.
2. The 5-ms frame is only used for the reverse fundamental channel, which uses only from 21 to 268 channel bits per frame with n = 1.
3. Turbocoding may be used for the reverse supplemental channels with 552 or more channel bits per frame; otherwise, K = 9 convolutional coding is used.
4. With convolutional coding, the reserved/encoder tail bits provide an encoder tail. With turbocoding, the first two of these bits are reserved bits that are encoded and the last six bits are replaced by an internally generated tail.
5. If variable-rate reverse supplemental channel operation and/or flexible reverse link data rates are supported, the parameters are determined from the specified number of channel bits per frame, the maximum assigned number of channel bits per frame for the reverse fundamental channel or the reverse supplemental channel, and the specified frame quality indicator length.
   - When the number of channel bits per frame is 21, 55, 125, or 267 and the corresponding number of frame, quality indicator bits is 6, 8, 10, and 12, an initial reserved bit is used; otherwise, no initial reserved bits are used.
   - The frame quality indicator length is 16 for more than 288 encoder input bits per frame; 12 or 16 for more than 144 encoder input bits per frame; 10, 12, or 16 for more than 72 encoder input bits per frame; 8, 10, 12, or 16 for more than 36 encoder input bits per frame; and 6, 8, 10, 12, or 16 otherwise.
   - The code rate is 1/4. The type of encoding is convolutional if the number of encoder input bits per frame is less than 576; otherwise, it is the same as that of the maximum assigned data rate for the channel.
   - If the specified number of channel bits per frame is equal to the maximum assigned number of channel bits per frame and that number and the specified frame quality indicator length match one of the listed cases, the symbol repetition factor and symbol puncturing from that listed case are used. Otherwise, the symbol repetition factor and puncturing are calculated to achieve the same interleaver size as for the maximum assigned data rate for the channel.

Figure 13.34 R-FCH/R-SCH for RC3 and RC5

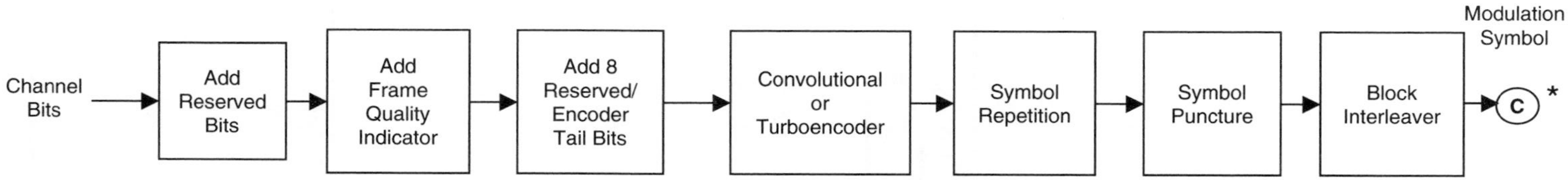

Bits/Frame	Bits	Bits	Data Rate (kbps)	R	Factor	Deletion	Symbols	Rate (ksps)
24 Bits/5 ms	0	16	9.6	1/4	2x	None	384	76.8
21 Bits/20 ms	1	6	1.8	1/4	16x	8 of 24	1,536	76.8
55 Bits/20n ms	1	8	3.6/n	1/4	8x	8 of 24	1,536	76.8/n
125 Bits/20n ms	1	10	7.2/n	1/4	4x	8 of 24	1,536	76.8/n
267 Bits/20n ms	1	12	14.4/n	1/4	2x	8 of 24	1,536	76.8/n
552 Bits/20n ms	0	16	28.8/n	1/4	1x	4 of 12	1,536	76.8/n
1,128 Bits/20n ms	0	16	57.6/n	1/4	1x	4 of 12	3,072	153.6/n
2,280 bits/20n ms	0	16	115.2/n	1/4	1x	4 of 12	6,144	307.2/n
4,584 Bits/20n ms	0	16	230.4/n	1/4	1x	4 of 12	12,288	614.4/n

SR: Spreading Rate
RC: Radio Configuration
*See Figure 13.32

Notes:

1. n is the length of the frame in multiples of 20 ms. For 55 channel bits per frame, $n = 1$ or 2. For more than 55 channel bits per frame, $n = 1$, 2, or 4.
2. The 5-ms frame is only used for the reverse fundamental channel, which uses only from 21 to 267 channel bits per frame with $n = 1$.
3. Turbocoding may be used for the reverse supplemental channels with 552 or more channel bits per frame; otherwise, $K = 9$ convolutional coding is used.
4. With convolutional coding, the reserved/encoder tail bits provide an encoder tail. With turbocoding, the first two of these bits are reserved bits that are encoded, and the last six bits are replaced by an internally generated tail.

Figure 13.35 R-FCH/R-SCH RC4 and RC6

13.8.9 FEC on Reverse Link

The reverse link uses a $k = 9$, $r = 1/4$ convolutional code for R-FCH. The distance properties of this word distance are better catered providing performance gains versus higher rate codes in fading and additive white Gaussian noise (AWGN) channel conditions. The constraint length $k = 9$, $r = 1/4$ convolutional code provides a gain of about 0.5 dB over a $k = 9$, $r = 1/2$ code even in AWGN. The R-SCH uses convolutional codes for data rates up to 14.4 kbps. Convolutional codes for higher data rates on the R-SCH are optional and the use of turbocode is preferred. A common constituent code is used for reverse link. Turbocodes of constraint length $k = 4$, $r = 1/4$, 1/3, or 1/2 are used for all R-SCH. Table 13.14 summarizes the FEC for the reverse channels.

Table 13.14 Forward Error Correction on Reverse Link Channels

Channel Type	SR1		SR3	
	FEC	r	FEC	r
Access Channel (R-ACH)	Convolution	1/3	None	None
Enhanced Access Channel (EACH)	Convolution	1/4	Convolution	1/4
Common Control Channel (R-CCCH)	Convolution	1/4	Convolution	1/4
Dedicated Control Channel (R-DCCH)	Convolution	1/4	Convolution	1/4
Fundamental Channel (R-FCH)	Convolution	1/3 (RC1) 1/2 (RC2) 1/4 (RC3 and RC4)	Convolution	1/4
Supplemental Code Channel (R-SCCH)	Convolution	1/3 (RC1) 1/2 (RC2)	None	None
Supplemental Channel (R-SCH)	Convolution or Turbo ($N \geq 360$)	1/4 (RC3, $N < 6120$); 1/2 (RC3, $N = 6120$) 1/4 (RC4)	Convolution or Turbo ($N \geq 360$)	1/4 (RC5, $N < 6120$); 1/3 (RC5,) $N \geq 6120$ 1/4(RC6, $N < 2\,0712$) 1/2(RC6, $N = 20712$)

The turboencoder encodes the data, frame quality indicator (by cyclic redundancy check [CRC]), and two reserved bits input to the turboencoder and adds an encoder output tail sequence. If the total number of data, frame quality, and reserved input bits is N_{turbo}, the turboencoder generates N_{turbo}/R encoded data output symbols followed by 6/R tail output symbols, where R is the code rate of 1/2, 1/3, or 1/4. The turbo encoder uses two systematic, recursive, convolutional encoders connected in parallel, with an interleaver (the turbo interleaver) preceding the second recursive convolutional encoder (see Figure 13.36). The two recursive convolutional codes are called the constituent codes of the turbocode. The outputs of the constituent encoders are punctured and repeated to achieve the $(N_{turbo} + 6)/R$ output symbols.

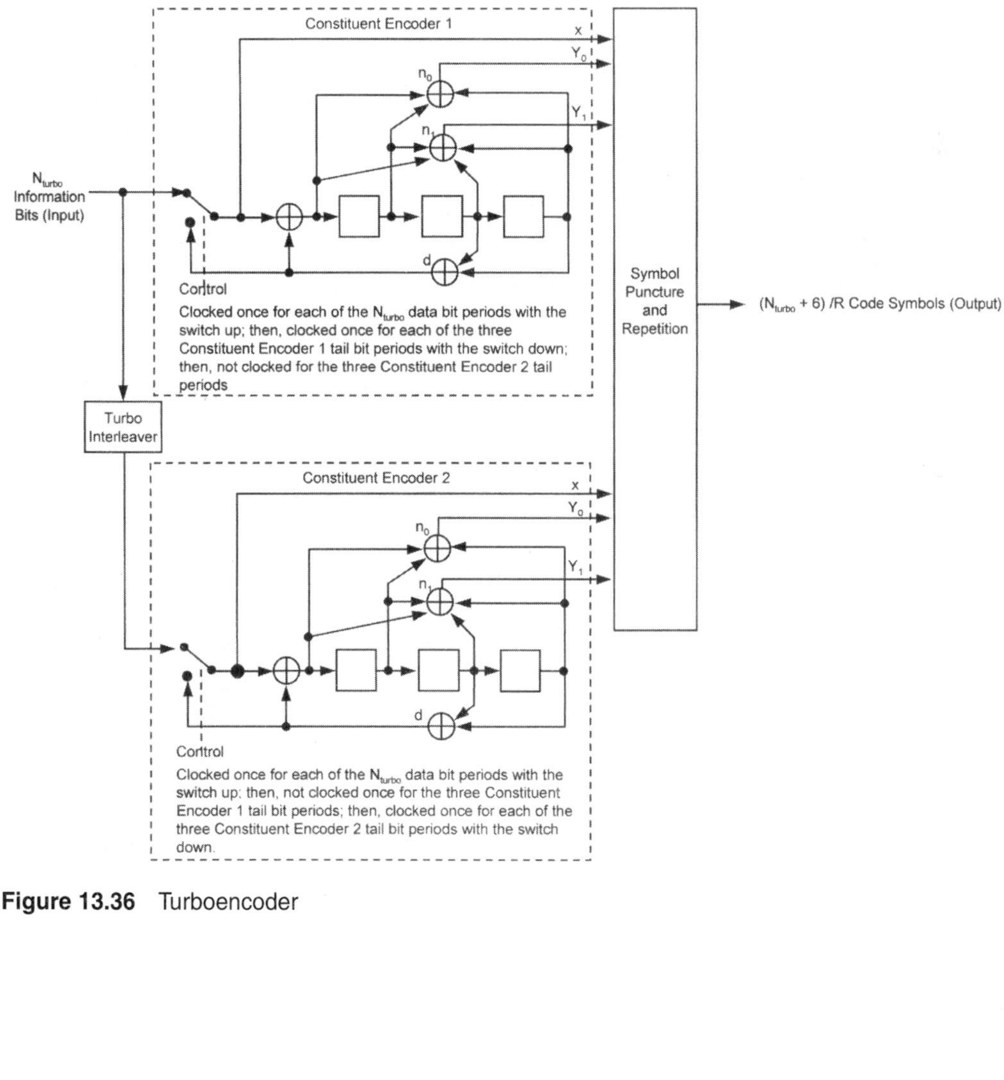

Figure 13.36 Turboencoder

13.8.10 Reverse Link Physical Layer Characteristics

Continuous Waveform. A continuous pilot and continuous data channel waveform are used for all data rates. This continuous waveform minimizes interference to biomedical devices such as hearing aids and pacemakers and permits a range increase at lower transmission rates. The continuous waveform also enables the interleaving to be performed over the entire frame rather than just the portions that are not gated off. This enables the interleaving to achieve the full benefit of the frame time diversity. The base station uses the pilot for multipath searches, tracking, and coherent demodulation as well as to measure the quality of the link for power control purposes. Separate orthogonal channels are used for the pilot and each of the data channels. Thus, the relative levels of the pilot and physical data channels can easily be adjusted without changing the frame structure or power levels of some symbols of a frame.

Orthogonal Spreading with Different-Length Walsh Sequences. The mobile uses orthogonal spreading when transmitting on the R-PICH, EACH, R-CCCH, or R-TCH with RC3 through RC6. Table 13.15 specifies the Walsh functions that are applied to the reverse link channels.

Table 13.15 Walsh Function for Reverse Link Channels

Channel Type	Walsh Function
R-PICH	W_0^{32}
EACH	W_2^{8}
R-CCCH	W_2^{8}
R-DCCH	W_8^{16}
R-FCH	W_4^{16}
R-SCH1	W_1^{2} or W_2^{4}
R-SCH2	W_2^{4} or W_6^{8}

W_m^N represents a Walsh function of length N that is serially constructed from the m-row of an $N \times N$ Hadamard matrix with the zeroth row being Walsh function 0, the first row being Walsh function 1, etc. Within Walsh function m, Walsh chips are transmitted serially from the mth row left to right. The Walsh function spreading sequence repeats with a period of $N/1.2288$ μs for SR1 and with a period of $N/3.6864$ μs for SR3.

Rate Matching. Several approaches are needed to match the data rates to Walsh spreader input rates. These include adjusting the code rate using puncturing, symbol repetition, and sequence repetition. The design approach is to first try to use a low rate code, but not to reduce the rate below $r = 1/4$ since gains of smaller rates would be small and the decoder implementation complexity would increase significantly.

Low Spectral Sidelobes. The cdma2000 system achieves low spectral sidelobes with nonideal mobile power amplifiers by splitting the physical channels into the I and Q channels and using a complex-multiply-type PN spreading approach.

Independent Data Channels. Two types of physical data channels (R-FCH and R-SCH) are used on the reverse link that can be adapted to a particular type of service. The use of R-FCH and R-SCH enables the system to be optimized for multiple simultaneous services. These channels are separately coded and interleaved and may have different transmit power levels and FER set points.

13.8.11 Reverse Power Control

The primary objective of power control in the reverse link is to resolve the near-far problem, in which mobiles that are near the BS have a better signal path than mobiles that are far away. Due to this effect, near mobiles can, in principle, raise the RF interference levels that screen out mobiles located far from the BS. This problem is resolved by controlling the reverse transmit power of each mobile, and requiring that the signal-to-noise ratio of each mobile is the same at the BS. The reverse power control must be dynamic in order to compensate for the time-dependent variations of the RF environment. In IS-95, the reverse power control has been designed to solve the following reverse link problems:

1. Choosing the initial mobile transmit power
2. Power compensation due to slow varying and log-normal shadowing effects where there is a correlation between the forward and reverse link fades
3. Measuring and maintaining a target signal-to-noise ratio at the BS for each mobile
4. Compensation of power fluctuation due to fast Rayleigh fading

The first and second items are resolved using open loop power control. The third and fourth items are resolved using a closed loop, which consists of an inner and an outer loop nested together.

In IS-95 the open loop power control is purely a mobile-controlled operation and does not involve the BS. The mobile estimates the forward path loss, and determines the transmit power based on this measurement. Since forward and reverse links operate at different carrier frequencies, the open loop power control is inadequate and too slow to compensate for fast and frequency-dependent Rayleigh fading.

The closed loop power control is used to compensate the fast Rayleigh fading. The process involves both the BS and the mobile station (MS). Once the MS is on the traffic channel, the open and closed loops work together, so that the slower open loop is able to include the faster closed loop correction.

The IS-2000 reverse power control scheme is a generalization of the IS-95 version. The system supports the same open and closed loops, but it integrates the functionality of power control of the fundamental and supplemental channels in a simple scheme that is easy to expand. The key factor in the simplification is the introduction of the reverse pilot channel (R-PICH), which is used as a reference for measuring and scaling in the open and inner loops, and then translates them to corrections that apply to F-FCH. The scaling is performed per channel and per data rate, so that rate equalization can be easily performed. This simplifies the design of the outer loop of the R-FCH, since all rates can be treated equally.

In the reverse link open loop power control, the mobile estimates the transmitted power of the reverse link channels based on the measurement of the received aggregate power. As in IS-95, the open loop functionality is located at the mobile; however, the BS must provide the value of the parameters the mobile needs to operate the open loop. This is done via signaling messages in the overhead channels and in the forward traffic channel. Open loop power control compensates for the path loss from the mobile to the base station and handles very slow fading.

IS-2000-2 [9] provides formulas for the mean output power of all the reverse channels at the mobile. The formulas for the R-ACH and R-TCHs with RC1 and RC2 are identical to the ones in IS-95. Thus, the cell support of the open loop power control for R-ACH and R-FCH with RC1 and RC2 is the same as the one used in 2G systems.

The 3G reverse closed loop power control is based on the 2G model with two main components: the inner and outer loop. Although the main functionality of the 2G and 3G closed loop power control appears to be similar, the 3G closed loop has some differences.

In 3G, the mobile transmits the R-FCH in continuous fashion for all subrates, while in 2G the reverse traffic channel is gated by the mobile depending on the subrate. In 2G, the mobile reduces the average power for subrate frames by gating off certain power control groups (PCGs) [8 for 1/2 rate, 12 for 1/4 rate, and 14 for 1/8 rate]. In this way, the power level of a full-rate frame for nongated PCG is the same as a half, quarter or one-eighth rate frame. The inner loop at the BS measures the reverse E_b/N_t of the traffic channel for every PCG. For the gated PCGs the BS sends, most probably, power-up commands to the MS. Since the MS knows which are the gated PCGs, it ignores these commands. For nongated PCGs, the BS measures the E_b/N_t of the traffic channel, which is independent of the rate of the frame.

In 3G, the reverse link transmission is continuous, and no PCG is gated. The mobile obtains a reduction in the average power of the subrate frames by repeating the PCGs more times, and reducing the power per PCG, such that the E_b received at the cell is rate independent. For this reason the inner loop cannot use the E_b/N_t per PCG measurement of the R-FCH because it is rate dependent. Since the rate of the frame is known after decoding, the 3G inner loop cannot use the R-FCH measurement, and must use the R-PICH strength measurement.

Continuous transmission provides several benefits to 3G power control. The 3G inner loop operates at an actual 800 Hz, whereas the 2G inner loop rate is lower depending on the subrate. For example, the 2G inner loop runs at 100 Hz for 1/8 rate frame. In addition, the power variance of the R-FCH in 3G is reduced. These effects improve the coherent detection, and increase the reverse link call capacity.

Since the inner loop power control uses R-PICH signal measurements and the output of the R-FCH outer loop is the full-rate E_b/N_t setpoint for the R-FCH, the output of the R-FCH must be converted to reverse pilot strength units that can be used by the inner loop.

The input to the reverse outer loop algorithm are the reverse frame errors detected by the BS in every frame. The output is the R-FCH E_b/N_t setpoint at full rate, which is converted by the BS to the R-PICH signal-to-noise ratio setpoint via offsets. The signal-to-noise ratio setpoint value is mapped to an R-PICH energy threshold using a set of numerical tables, which is compared with the R-PICH energy measurement to determine whether the reverse power control bit to be sent to the mobile on forward power control subchannel.

The key characteristics of the reverse link are summarized as follows:

- Channels are primarily code multiplexed.
- Separate channels are used for different QoS and physical layer characteristics.
- Transmission is continuous to avoid electromagnetic interference (EMI).
- Channels are orthogonalized by Walsh functions and *I*/*Q* split so that performance is equivalent to BPSK.
- Hybrid combination of QPSK and BPSK.
- Coherent reverse link with continuous pilot.
- Forward power control information is time-multiplexed with the pilot.
- By restricting alternate phase changes of the complex scrambling, power peaking is reduced and sidelobes are narrowed.
- Independent fundamental and supplemental channels with different transmit power and FER target.
- Forward error correction:
  - Convolutional codes ($k = 9$) are used for voice and data.
  - Parallel turbocodes ($k = 4$) are used for high data rates on supplemental channels.
- Fast reverse power control: 800 times per second.
- Frame lengths:
  - 20-ms frames are used for signaling and user information.
  - 5-ms frames are used for control information.

13.9 cdma2000 Media Access Control and LAC Sublayer

13.9.1 Media Access Control Sublayer

IS-200033 [10] defines two distinct planes: a *control plane* and a *data plane* (see Figure 13.37).

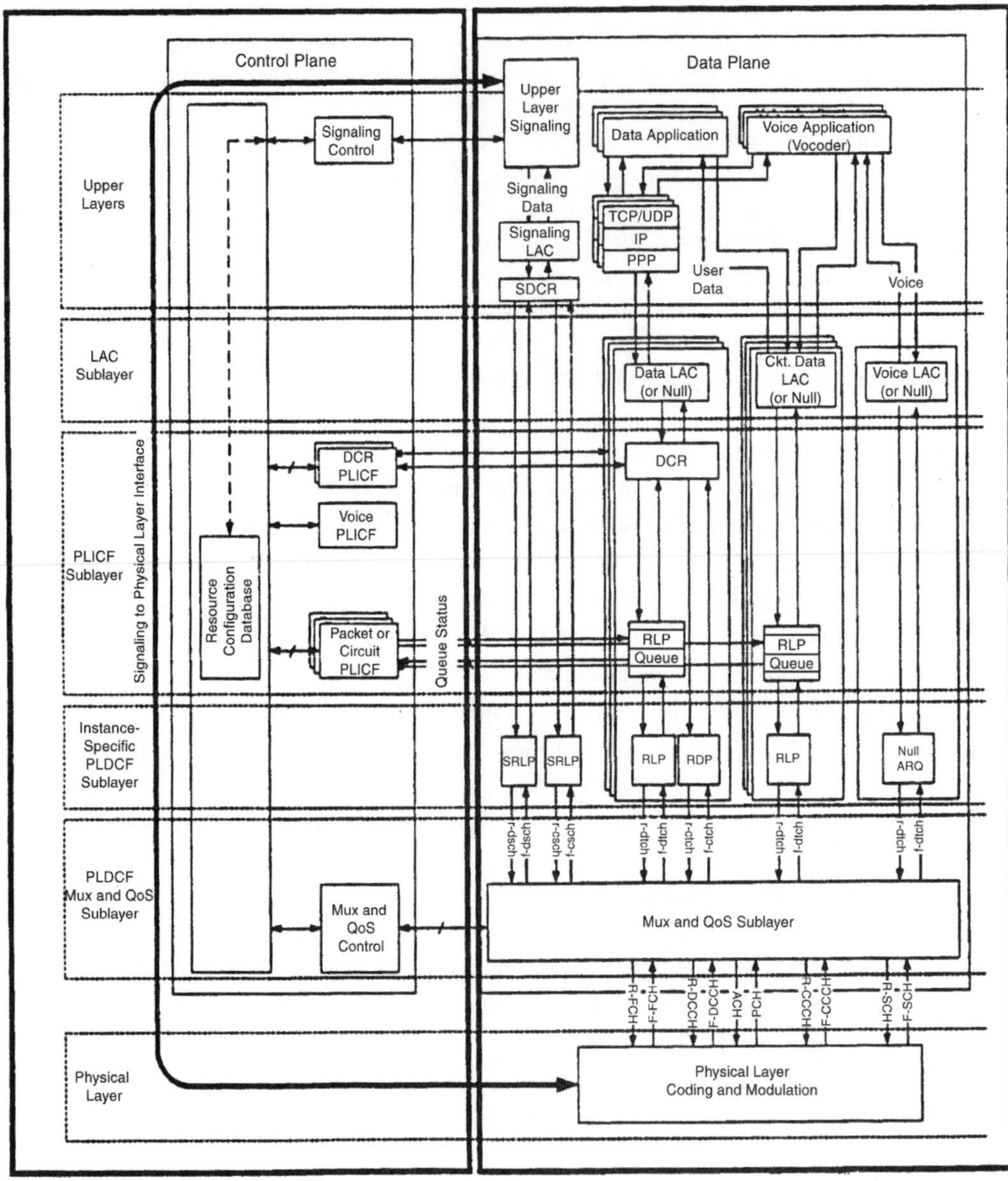

Figure 13.37 The Control Plane and the Data Plane

This approach offers a clear definition of service interfaces between all entities defined and used in cdma2000. The main functions of the control and data plane are

- **Control plane**—Responsible for all control functions including resource management, signaling control interface to resource databases, logical-to-physical channel mapping, and access control for all services. For example, the control plane entities include a state machine for packet and circuit switched data services on the media access control sublayer. The control plane is in charge of deciding how to process a service.
- **Data plane**—Responsible for the transport of all types of data including user traffic and signaling information. The data plane entities contain different protocols that are used between the mobile station and wireless access network. Protocols follow a layered model based on open system interconnect (OSI) (see Figure 13.38). All the protocol layers are located within the data plane. Service access points (SAPs) are used between different layers of the protocol stack. The data plane supports protocol layering to provide various types of services. Each sublayer within the data plane typically processes only a part of the received data corresponding to its functionality. Depending on the type of service, some sublayers may not be required. In this condition, the sublayer may be disabled by the control plane functions.

Figure 13.37 shows the control and data planes in cdma2000 and their interrelationship. The control plane has two parts: the resource management entities and the control functions for services. The resource management entities are further divided into a resource control (RC) entity and a resource configuration database (RCD) entity.

The control function entities include signaling control, data and voice service entities, and the MUX and QoS entity. Data plane entities include packet common/dedicated routing functions, related queuing functions, and protocol-specific functions.

- **Control Plane Entities**
  - **Resource control (RC)**—This entity is responsible for all interactions between service-specific control entities and the resource configuration database (RCD). These service interfaces allow a service control function to request a resource when it needs the resource or to release a resource when it no longer needs the resource due to current state. The actual allocation and de-allocation of the specific physical resources are not controlled by any control plane entity in the mobile station. The base station controls all the resource allocations and releases.
  - **Resource configuration database (RCD)**—Each mobile station has one instance of the RCD. The RCD stores all resources related information such as the logical and physical channel mapping, physical channel list, service instance information, supplemental channel, and supplemental code channels. RCD is directly controlled by the base station by means of upper layer signaling, which uses the signaling control entity to read from and write to the RCD. This architecture allows the infrastructure to control and manage the logical-to-physical channel mapping, physical channel configuration, and other data residing in the RCD.

- **Signaling control**—This entity serves as the only interface between the media access control layer and the upper layer signaling entity. The signaling control entity always interfaces with the RCD to convey the relevant upper layer signaling information received from the infrastructure. It also serves as the front end to communicate resource requests and releases back to the infrastructure.
- **Dedicated common router (DCR) PLICF**—Each instance of this entity is responsible for providing the control and status information to a corresponding DCR data plane entity instance to manage the transmission of packets on either a dedicated channel or using a common channel, depending on the state of the corresponding data service PLICF instance.
- **Voice PLICF**—This is the control entity that requests and releases resources required to provide IS-95B and cdma2000 voice services.
- **Circuit and packet data PLICF**—This entity is responsible for managing the states and interfacing with RC to request and release resources as appropriate. Even though packet and circuit switched data services have different characteristics, the same data PLICF entity definition may be used with service-specific attribute values to provide either of the two services. However, each instance of the data PLICF will handle either packet data services or circuit data services, but not both simultaneously.

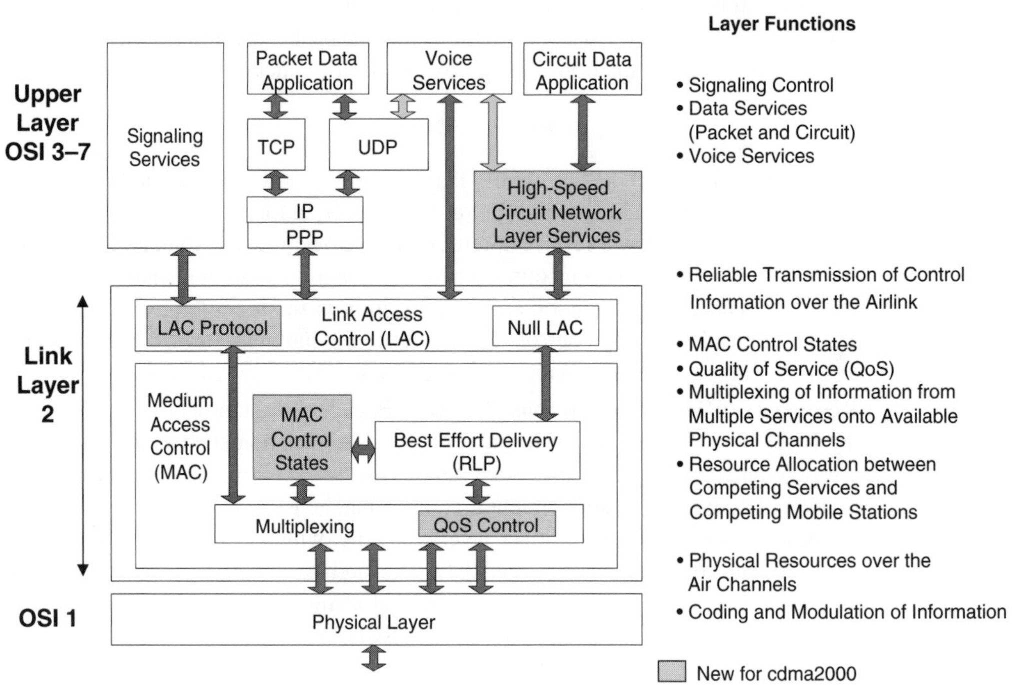

Figure 13.38 Layering Model in cdma2000

- **Data Plane Entities**
  - **Dedicated common router (DCR)**—This is responsible for routing the packets to be transmitted to either the RLP queue on dedicated channels or RBP buffers on common channels, depending on the state of the packet data service. The necessary control information is configured on the DCR by the corresponding DCR PLICF entity on the control plane. By the nature of its responsibility, DCR is not required on the receive path of packets.
  - **Radio link protocol (RLP) queue**—This entity is responsible for the buffering and handling of packets that provide all the functions to manage the data flow over a connection-oriented RLP link.
  - **Short data burst (SDB)**—The SDB provides all the necessary functions to support and manage the transport of data burst messages and traffic packets using common channels such as the access channel and the common control channel on the reverse link and paging channel and the common control channel on the forward link.

- **Resource Management Entities**
  For each mobile station, an instance of RC and an instance of RCD exists both at the mobile station and at the base station. RCD is mainly controlled by the base station. The mobile station RCD and base station RCD remain in sync at all times and synchronization is supported by upper layer signaling.
  - **Resource control (RC)**—There is one instance of the RC entity per mobile station. RC serves as the interface for all interactions between the RCD and control plane entities. The signaling control (SIG) entity has a special role to play in the resource management functions. SIG is the entity that manages the interface between the base station and the mobile station using air interface upper layer messaging. By using a single instance of RC for all accesses to RCD, RC serves as the gatekeeper for all resources, and locks related information in a centralized access scheme.
  - **Resource configuration database (RCD)**—There is one instance of RCD per mobile station. The RCD database contains a collection of all data structures related to the media access control layer that are maintained by the mobile station and the base station. RCD is arranged as a set of tables. The relationships between these tables are also maintained as part of RCD. All entries in the mobile station are controlled by the base station using upper layer signaling messages.

- **Signaling Control Entity**
  Each mobile station has one instance of the SIG entity, which serves as the interface between upper layer signaling and RCD. There are no service interfaces defined between the media access control layer and the upper layer. All the interactions between media access control layer entities and upper layers are mediated using RC and RCD entities. SIG handles the upper layer messages received and converts them into appropriate access requests for the RCD. Similarly, RC-related requests for transport over the air are made by using SIG interface primitives. SIG converts the primitive to appropriate air interface message and sends it over the air.

- **Dedicated Common Router (DCR) PLICF Entity**
  DCR PLICF is the control plane entity that deals with control and status information with data plane DCR. One DCR PLICF instance is created per packet data service instance. DCR PLICF is responsible for handling the control and status information for the DCR resulting from any changes in the RCD. DCR PLICF is responsible for determining and conveying information about the mode of transmission of data packets to DCR. DCR PLICF exists even if the corresponding packet data service does not have any resources allocated to it. In a typical scenario, DCR detects the arrival of user data and determines the volume and nature of data. DCR sends an indication to the DCR PLICF entity about the arrival of user data. This indication also recommends if the packet may be sent using a short burst or using a dedicated traffic channel. For a short data packet, DCR PLICF may recommend immediate transmission using RBP to DCR or may inform DCR to hold the data in a buffer. If the data require a dedicated channel, DCR PLICF will request that the RC entity allocate and lock the resource. It will send an indication to start the transmission on the dedicated channel once it gets an indication or confirmation of the resource allocation from the RC.

- **Voice PLICF Entity**
  Each instance of voice service requires an instance of voice PLICF. The key function of voice PLICF is that it requests the RC for the necessary voice channel resources (i.e., dedicated traffic channel). It is also responsible for any other media access control and LAC control functions to support the cdma2000 protocol stack. Mostly, for voice services, LAC could be a null function. Mobile station supports a maximum of one instance of voice PLICF at any given time.

- **Packet and Circuit Data PLICF Entity**
  Each data service instance requires an associated data PLICF instance. The defined data PLICF could be configured to support either a packet data instance or a circuit data instance. The key responsibility of the data PLICF entity is to maintain and use the RLP and RLP queue instances related to data service instance. Data PLICF is nothing but a state machine to handle the data service requirements and uses the corresponding logical channels.

13.9.2 LAC

The LAC sublayer (see Figures 13.37 and 13.38) supports the following functions on both common and dedicated channels:

- **Delivery of service data units (SDUs) across radio link**—The LAC sublayer defines and uses automatic retransmission request (ARQ) protocol to ensure reliable delivery across the radio link.

- **Layer 2 packet data units (PDUs)**—The LAC sublayer is responsible for converting SDUs from layer 3 and forming layer 2 PDUs that will be delivered to the receiving end LAC entity. On the receiving end, LAC is responsible for putting the SDUs back together from the received layer 2 PDUs.
- **Segmentation and reassembly**—The LAC sublayer is responsible for segmentation and reassembly. The LAC sublayer divides the PDUs into blocks that are compatible with size requirements imposed by the media access control sublayer. When these encapsulated PDUs are received, LAC reassembles them into the original PDU.
- **Access control**—LAC may support access control by imposing authentication on common channels at layer 2. In this regard, some IS-95B layer 3 functions have been transferred to cdma2000 layer 2. The cdma2000 LAC sublayer enforces authentication control using a global challenge procedure at call initiation.
- **Address control**—The LAC supports address control to ensure delivery of messages to appropriate mobile stations. The LAC enforces address control based on individual mobile station addressing and broadcast addressing.

The LAC supports multiple sublayers (see Figure 13.39) to provide its functions, including:

- Authentication sublayer
- ARQ sublayer
- Addressing sublayer
- Utility sublayer
- Segmentation and reassembly (SAR) sublayer

Each LAC sublayer has a well-defined function. A common feature of all these sublayers is that they add fields in the transmit direction and remove them in the receive direction. Also, each sublayer does some processing on the payload unit that they are dealing with. For example, the authentication sublayer processes the authentication-related fields in the received protocol data unit.

Although five different sublayers have been defined within the LAC, it is not necessary to have all five supported on every type of channel. For example, authentication sublayer functions are not performed on dedicated channels. Each sublayer function may be enabled or disabled using control plane functions. For example, if a service provider does not require authentication procedure in his network, he should be able to disable the authentication sublayer functions. When a sublayer is disabled, the sublayer acts as a pass-through or null sublayer and does not process the data unit at all.

IS-2000 standards do not specify the interfaces between different LAC sublayers. These interfaces may be proprietary and may be handled differently at the base station and mobile station. The implementation will also vary on these interfaces across different vendors.

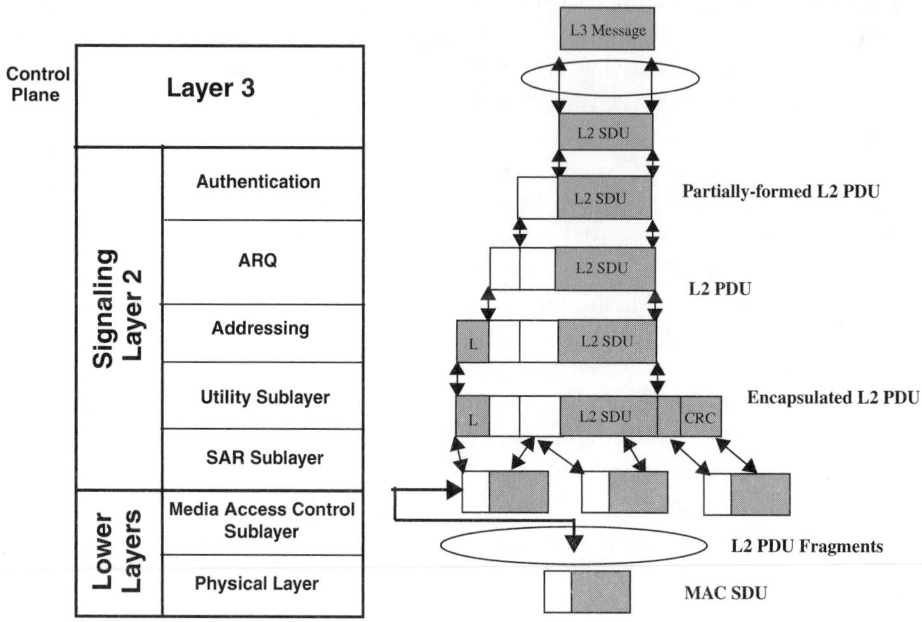

Figure 13.39 LAC Sublayers

On the infrastructure side, all these sublayers need not be implemented in a single component. The division and distribution of these sublayers and related functions are not restricted by IS-2000 LAC layer specifications.

Authentication Sublayer. This sublayer is responsible for executing authentication-related functions. Authentication may be performed either on a verify-and-pass basis or on a verify-in-parallel basis. In the verify-and-pass approach, the data unit is processed and authentication is completed on the data prior to passing it to higher layers. In the verify-in-parallel approach, the data is processed in parallel to the passing of data to the higher layers. If authentication fails, it could be followed up with the higher layers.

LAC authentication sublayer processes the AUTH_MODE, AUTHR, RANDC, and COUNT fields. The authentication sublayer is only used on the reverse link common signaling channel (r-csch).

ARQ Sublayer. This sublayer provides the functionality to support reliable exchange of upper layer SDUs. ARQ ensures reliability by using a protocol based on peer entity acknowledgment. The IS-2000 ARQ sublayer uses a selective repeat request approach to retransmit only the lost units. The receiving end is required to store the correctly received packet data units. The ARQ receiver is responsible for generating acknowledgments only when they are requested by the sender. The ARQ receiver is also responsible for detecting and discarding duplicate packets.

The ARQ sublayer supports the option of exchanging upper layers' SDUs without acknowledgment. In other words, ARQ functionality is not being used at this layer for this usage. The upper layer will indicate to the LAC layer if a given SDU needs to be sent reliably. The ARQ sublayer is used in both the forward and reverse links on the following logical channels:

- Forward and reverse common signaling channels (f/r-csch)
- Forward and reverse dedicated signaling channel (f/r-dsch)
- Forward and reverse dedicated media access control channel (f/r-dmch)

ARQ on the common signaling channels is almost identical to the access attempt procedures defined in IS-95 with slight modifications. ARQ fields used in the message are MSG_SWQ, VALID_ACK, ACK_REQ, and ACK_SEQ in both directions. In addition to these parameters, the ACK_TYPE field is used for ARQ support, but only in the reverse direction.

ARQ on dedicated channels is defined for the dedicated signaling and media access control channels on both the forward and reverse links. The fields used in the messages related to ARQ on dedicated channels include ACK_SEQ, MSG_SEQ, and ACK_REQ. ACK_TYPE is not required on dedicated channels and the VALID_ACK field is included in the ACK_SEQ field. SEQ fields are 3 bits long for signaling channels and 2 bits long for media access control channels.

The LAC layer has control functions in the control plane to manage the ARQ sublayer on the data plane. These control functions are

- *Reset:* cancels current access probes and resets the MSG_SEQ
- *Cancel:* immediately terminates the access attempt
- *Suspend:* current access probes are suspended and remain suspended until resumption is requested
- *Resume:* suspended access probe is restarted from the start of the probe
- *Restart:* suspended probe does not start from the interrupted point; instead a new access attempt starts at the first access probe sequence within the access subattempt

Addressing Sublayer. This sublayer is responsible for managing the address-related fields in the messages. The addressing sublayer adds address fields to the SDU and removes them from PDUs in transmit and receive directions, respectively. The following address types are supported:

- International mobile subscriber identifier (IMSI)
- Electronic serial number (ESN)
- Temporary mobile subscriber identifier (TMSI)
- A combination of IMSI and ESN
- A combination of IMSI_S and ESN (IMSI_S is IMSI Short)

The address fields used by the addressing sublayer are ADDRESS_TYPE, ADDR_LEN; and ADDRESS, which includes fields specific to the address type. The addressing sublayer is used only on the common signaling channels and is employed in both uplink and downlink directions.

Segmentation and Reassembly (SAR). This sublayer is responsible for segmenting the data on the transmitting side and preparing it for transmission and reassembling it to form the original PDU on the receiving side. SAR allows for efficient usage of channels and is also responsible for adding length and CRC fields. The length field is prefixed and CRC is attached at the end of the packet. Each segment includes information indicating the position of the segment. For example, on the SYNC channel, the start of message (SOM) bit indicates if it is the first segment of a sync message. When a segment is lost, it results in the loss of the entire PDU. There are no limits defined in IS-2000 on how many segments a PDU could be divided into.

Utility Sublayer. This sublayer is used on the common signaling channel, dedicated signaling channel, and dedicated media access control channel in both directions. The utility sublayer is responsible for converting/recovering the SDU_TAG of the SDU to/from the MSG_TYPE of PDU. PDUs with unknown MSG_TYPE are discarded. Encryption parameters are included whenever required.

Table 13.16 is the list of sublayers that are supported on different logical channels.

Table 13.16 Logical Channel and Sublayer Used

Logical Channel	LAC Sublayers Used
Reverse common signaling channel (*r-csch*)	Authentication, ARQ, addressing, utility, and SAR
Forward common signaling channel (*f-csch*)	Addressing, utility, and SAR
Forward and reverse signaling and media access control channels	ARQ, utility, and SAR

13.10 Data Services in cdma2000

Two types of data services are being considered—packet and high-speed circuit data services. The packet service and the media access control layer are designed to support a large number of mobile stations using packet data services. Many packet data services exhibit highly bursty traffic patterns with relatively long periods of inactivity. Due to limited air interface capacity, limited base station equipment, and constraints on mobile station power consumptions, dedicated channels for packet service users are allocated on demand and released immediately after the end of the activity period.

Releasing the dedicated channels and re-establishing them introduces latency and signaling overhead due to the renegotiation process that has to take place between the BS and the MS before user data exchange. The overhead of re-establishing the dedicated channels includes the cost of synchronization of RLP and signaling overhead associated with service negotiation to reconnect the packet service. The media access control layer avoids this latency and overhead by the BS by saving a set of state information after the initialization phase is completed.

To further reduce the overhead associated with assignment of dedicated channels, the packet service allows for exchange of short bursts of user data when no dedicated channels are present. This mode of operation may be suitable for mobile IP registration, notification services (e.g., e-mail notifications), and location tracking services where the volume of data to be exchanged is typically small.

Circuit services can be viewed as a special case of the packet services in the sense that dedicated traffic and control channels are typically assigned to the MS for extended periods of time during the circuit service sessions. This will lead to a less efficient use of the air interface capacity. However, some delay-sensitive services, such as video applications, require a dedicated channel for the duration of the call.

13.11 Mapping of Logical Channels to Physical Channels

The mapping from the logical to physical channels is not one-to-one in the sense that multiple logical channels from multiple service options may be mapped to a single physical channel. This multiplexing operation is performed at the PLDCF MUX and QoS sublayer. Figure 13.40 shows the channel structure for bearer service profiles that includes only packet and voice services.

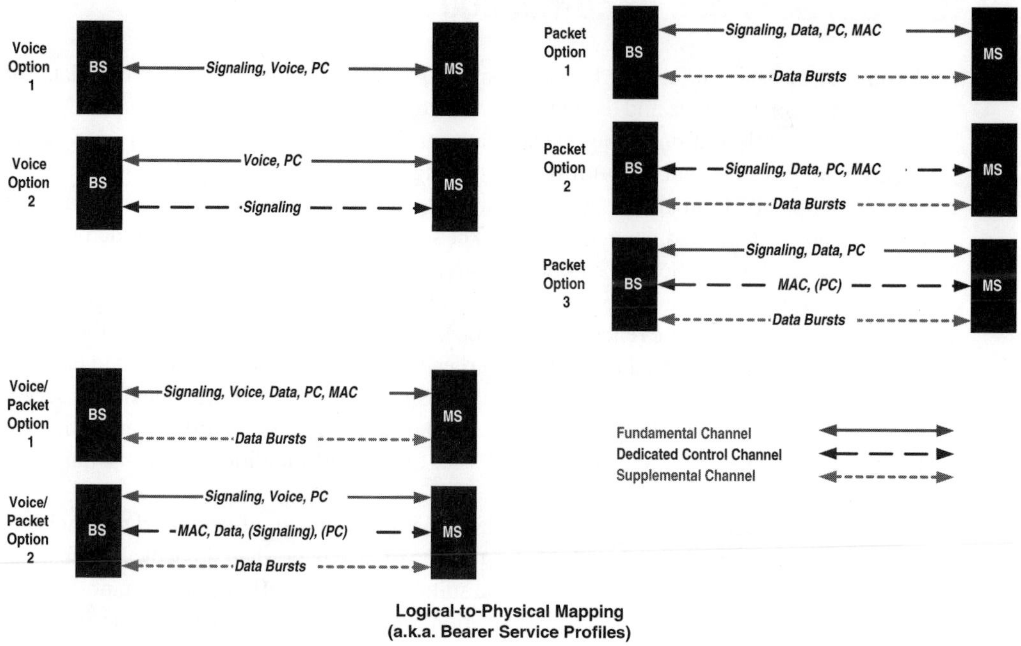

Figure 13.40 Logical-to-Physical Channel Mapping for Voice and Packet Services

13.11.1 Forward Link

Forward Link Dedicated Channel. The mapping of forward logical channels into forward physical channels is given in Table 13.17. The table shows attributes of each physical channel such as variable rate or fixed rate, physical channel frame size, and sharability of the channel.

Voice Services

In the **V1** mode, the F-SCH and F-DCCH are not used. In this mode the upper layer signaling (f-dsch), voice frames (f-dtch), and power control information are multiplexed in F-FCH (e.g., by using dim-and-burst or blank-and-burst mechanisms). Forward link continuity and outer loop power control in this mode is maintained by the F-FCH.

The **V2** mode is used to provide higher voice quality service by transmitting the upper layer signaling frames on a F-DCCH (e.g., no blank-and-burst or dim-and-burst signaling). Forward link continuity and outer loop power control in this mode is maintained by the F-FCH.

Table 13.17 Forward Dedicated Channel Structure for Different Services

Forward Link Physical Channel	Forward Link Logical Channel	Description	Services						
			V1	V2	V3	P1	P2	VP1	VP2
Fundamental Channel (F-FCH) (soft handoff)	f-dsch	Upper layer signaling messages	x		x		x	x	x
	f-dtch	RLP frames			x		x	x	
		Voice frames	x	x				x	x
	f-dmch	Media access control messages			x			x	
	pc	Power control	x	x	x		x	x	x
		Frame size	20	20	5/20		20	5/20	20
		Rates	V	V	V		V	V '	V
Supplemental Channel (F-SCH) (with or without soft handoff)	f-dtch	RLP frames Voice frames (for newly defined services)			x	x	x	x	x
		Rates			H	V/H	H	H	V/H
Dedicated control channel (F-DCCH) (with or without soft handoff)	f-dsch	Upper layer signaling messages		x		x			x
	f-dtch	RLP frames				x	x		x
	f-dmch	Media access control messages				x	x		x
	pc	Power control				x			
		Sharable		Y		N	Y		Y
		Frame size		20		5/20	5		5/20
		Rates		F		F	F		F

Key: F = Fixed (9.6 kbps or 0); V= variable (9.6, 4.8, 2.7, 1.5 kbps for RC1 and 14.4, 7.2, 3.6, 1.8 kbps for RC2); H = High Data Rates (scheduled rates) Y = Yes; N = No.

- **Packet Services**
  The packet data service, **P1** is offered on the forward direction by using F-FCH and F-SCH. The upper layer signaling message (f-dsch), MAC messages (f-dmch), and user data frames (f-dtch) are time-multiplexed in the F-FCH. In this mode, the media access control is performed in a centralized manner since media access control messages are carried on the F-FCH that is typically in soft handoff to ensure reliability of delivery for upper layer signaling messages. The 5-ms frames are used to carry short media access control messages in this mode. The F-SCH carries high-rate RLP frames containing packet data and transmission on the F-SCH is always scheduled in this mode. Lower-rate RLP frames may be carried on the F-FCH. The transmission rate of F-SCH is predefined using media access control messages. Forward link continuity and outer loop power control in this mode are maintained by the F-FCH.

  The **P2** mode of operation is an alternate basic packet data service and is similar to the **P1** mode in the sense that upper layer signaling messages (f-dsch), media access control messages (f-dmch), and user data frames (f-dtch) are time-multiplexed in one channel. However, the physical channel that is used to carry these logical channels is the F-DCCH, which may or may not be in soft handoff. Thus, the control of the media access control layer can be performed in either a distributed or centralized manner. To support the mixing of media access control signaling with RLP frames or upper layer signaling information, the F-DCCH supports dual frame size operation (5 and 20 ms). The F-SCH carries high-rate scheduled RLP frames containing packet data as well as lower-rate RLP frames. The rate for the lower-rate frames carried on F-SCH may be dynamically determined, but the transmission rate of the high-rate scheduled frames carried on the F-SCH is prespecified using media access control messages. Forward link continuity and outer loop power control are maintained by the F-DCCH and, therefore, the F-DCCH becomes unsharable in this mode.

  The **P3** mode is used for highly optimized packet data service with the potential support for distributed control of the media access control layer (i.e., the f-dmch is carried in a physical channel that can be operated with a reduced active set while upper layer signaling information is carried in a channel with a full active set). In this mode, the F-FCH is primarily used to carry high reliability, low delay upper layer messages. Power control bits are carried by F-FCH. The F-DCCH may be shared to make a more efficient use of the Walsh code resources. The F-DCCH carries the media access control signaling (f-dmch) and may not be in soft handoff. Control of the media access control layer can be performed in either a distributed or centralized manner. The F-SCH carries high-rate RLP frames containing packet data and transmission on the F-SCH is always scheduled in this mode. The transmission rate of F-SCH is prespecified using media access control messages. Lower-rate RLP frames may be carried on the F-FCH. Forward link continuity and outer loop power control in this mode are maintained by the F-FCH.

- **Concurrent Voice and Packet Services**
  The **VP1** mode offers simultaneous basic voice and packet data service by multiplexing upper layer signaling (f-dsch), media access control messages (f-dmch), voice frames (f-dtch), and potentially low-rate RLP frames (f-dtch) into the F-FCH. Control of the media access control layer is performed in a centralized manner. To support the mixing of media access control signaling with RLP frames or upper layer signaling information, the F-FCH supports dual frame size operation (5 and 20 ms). The F-SCH carries high-rate RLP frames containing packet data and transmission on the F-SCH is always scheduled in this mode. Forward link continuity and outer loop power control are maintained by the F-FCH.

  The **VP2** mode also provides simultaneous voice and packet data service. To provide a higher quality voice service in conjunction with packet data service, the media access control messages (f-dmch) and potentially upper layer signaling (f-dsch) are carried on the F-DCCH. Control of the media access control layer can be performed in either a distributed (if F-DCCH is not in soft handoff) or centralized (if DCCH is in soft handoff) manner. Power control bits are carried by the F-FCH. The F-DCCH can be shared to make a more efficient use of Walsh code resources. To support the mixing of media access control signaling with RLP frames or upper layer signaling information, the F-DCCH supports dual frame size operation (5 and 20 ms). The F-SCH carries high-rate scheduled RLP frames containing packet data as well as lower-rate RLP frames. The low-rate RLP frames may be sent on F-SCH to avoid the potential contention between voice and low-rate RLP frames on the F-FCH. Forward link continuity and outer loop power control are maintained by F-FCH.

  The reader should consult cdma2000 RTT [2] for circuit services and their combinations with voice and packet services.

Forward Link Common Channels. When neither F-DCCH nor F-FCH is allocated to the MS (e.g., in the suspended or dormant states), the upper layer signaling and media access control messages are carried to the MS using F-PCH or F-CCCH. Messages sent on these channels may be encrypted and must include mobile station ID or the packet service identifier since both F-PCH and F-CCCH are point-to-multipoint channels in the sense that there is no one-to-one mapping between the ID of these channels and mobile station ID. In addition to the control information (i.e., media access control layer or upper layer signaling), short data bursts may be carried by the F-PCH or F-CCCH. Table 13.18 shows the mapping of forward common channels to forward common physical channels.

Table 13.18 Mapping of Forward Common Logical Channels to Forward Common Physical Channels

Forward Link Physical Channels	Forward Logical Channels	Description
Common control channel (F-CCCH) or paging channel (F-PCH)	f-csch	Upper layer signaling messages
	f-ctch	RBP frames
	f-cmch	Media access control messages
Common pilot	—	Common pilot
Auxiliary pilot	—	Auxiliary pilot
Sync channel	—	Sync channel information

13.11.2 Reverse Link Channels

Reverse Link Dedicated Channels. The mapping of reverse logical channels into reverse physical channel is given in Table 13.19. The table also shows the attributes of each physical channel such as variable rate or fixed rate and physical channel frame size.

- **Voice Services**
  In **V1** mode, the R-SCH and R-DCCH are not used. In this case, the upper layer signaling (r-dsch), voice frame (r-dtch), and power control information are multiplexed in R-FCH (e.g., using dim-and-burst or blank-and-burst mechanisms).
  The **V2** mode provides a higher voice quality service by typically transmitting the upper layer signaling frames on an R-DCCH (e.g., no blank-and-burst or dim-and-burst signaling). However, if the mobile station cannot provide sufficient power to transmit on the R-DCCH, the upper layer signaling information can be transmitted on the R-FCH.

- **Packet Services**
  P1 is offered on the reverse direction by using R-FCH and R-SCH. The upper layer signaling message (r-dsch), media access control messages (r-dmch), and user data frames (r-dtch) are time-multiplexed in R-FCH. To support the mixing of media access control signaling with RLP frames or upper layer signaling information, the R-FCH supports dual frame size operation (5 and 20 ms). The R-SCH carries high-rate RLP frames containing packet data and transmission on the R-SCH is always scheduled. Lower-rate RLP frames may be carried on R-FCH.

In the **P2** mode, upper layer signaling messages (r-dsch), media access control messages (r-dmch), and user data frames (r-dtch) are time-multiplexed in R-DCCH. To support the mixing of media access control signaling with RLP frames or upper layer signaling information, the R-DCCH supports dual frame size. The R-SCH carries high-rate scheduled RLP frames containing data as well as lower-rate RLP frames.

R-FCH is primarily used to carry high reliability, low-delay upper layer messages as well power control information. The R-DCCH carries the media access control signaling (r-dmch). The R-SCH carries high-rate RLP frames containing packet data and transmission on the R-SCH is always scheduled. Lower-rate RLP frames may be carried on the R-FCH.

- **Concurrent Voice and Packet Data**
  The **VP1** mode (which offers simultaneous basic voice and packet data service) multiplexes upper layer signaling (r-dsch), media access control messages (r-dmch), voice frame (r-dtch), and, potentially, low-rate RLP frames (r-dtch) into R-FCH. To support the mixing of media access control signaling with RLP frames or upper layer signaling information, the R-FCH supports dual frame size operation (5 and 20 ms). The R-SCH carries high-rate RLP frames containing packet data and transmission on the R-SCH is always scheduled in this mode.

  In the **VP2** mode, the media access control messages (r-dmch) and potentially upper layer signaling information (r-dsch) are carried on R-DCCH (thereby, reducing potential disruption due to dim-and-burst and blank-and-burst). However, if the mobile cannot provide sufficient power to transmit on the R-DCCH, the upper signaling information can be transmitted on the R-FCH. To support the mixing of media access control signaling with RLP frames or upper layer signaling information, the R-DCCH supports dual frame size operation. The R-SCH carries high-rate scheduled RLP frames containing packet data as well as lower-rate RLP frames. The low-rate RLP frames may be sent on the R-SCH to avoid the potential contention between voice and low-rate RLP frames on the R-FCH.

Reverse Link Common Channels. When neither R-DCCH nor R-FCH is allocated to the mobile (e.g., in the suspended or dormant states), the upper layer signaling and media access control messages are conveyed to the base station using R-ACH or R-CCCH. Messages sent on these channels may be encrypted and must include the mobile ID or the packet service identifier since there is no one-to-one mapping between the identity of the R-ACH and R-CCCH channel and the mobile ID. In addition to the control information (i.e., media access control layer or upper layer signaling), short data bursts may be carried by the R-ACH or R-CCCH. Table 13.20 shows the mapping of the reverse common logical channels to reverse common physical channels.

Table 13.19 Reverse Dedicated Channel for Different Services

Reverse Link Physical Channel	Reverse Link Logical Channel	Description	Services						
			V1	V2	V3	P1	P2	VP1	VP2
Fundamental channel (R-FCH) (soft handoff)	r-dsch	Upper layer signaling messages	x	x	x		x	x	x
	r-dtch	RLP frames			x		x	x	
		Voice frames	x	x				x	x
	r-dmch	Media access control messages			x			x	
		Frame size	20	20	5/20		20	5/20	20
		Rates	V	V	V		V	V'	V
Supplemental channel (R-SCH) (with or without soft handoff)	r-dtch	RLP frames; Voice frames (for newly defined services)			x	x	x	x	x
		Rates			H	V/H	H	H	V/H
Dedicated control channel (R-DCCH) (with or without soft handoff)	r-dsch	Upper layer signaling messages		x		x			x
	r-dtch	RLP frames				x	x		x
	r-dmch	Media access control messages				x	x		x
		Frame size		20		5/20	5		5/20
		Rates		F		F	F		F

Key: F = fixed (9.6 kbps or 0); V= variable (9.6, 4.8, 2.7, 1.5 kbps for RC1, and 14.4, 7.2, 3.6, 1.8 kbps for RC2); H = high data rates (scheduled rates)

Table 13.20 Mapping of Reverse Common Logical Channels to Forward Common Physical Channels

Forward Link Physical Channels	Forward Logical Channels	Description
Common control channel (R-CCCH) or access channel (R-ACH)	r-csch	Upper layer signaling messages
	r-ctch*	RBP frames
	r-cmch	Media access control messages
Pilot	—	Pilot
	pc	Power control bits

* r-ctch is carried on the same physical channel (R-CCCH), but can be physically separated by using a different long code

13.12 Evolution of cdmaOne (IS-95) to cdma2000

Data is the word as cdmaOne operators look for a host of new network capabilities enabling them to offer new value-added services that can exploit present and future generations of technology. With the Internet and corporate intranet becoming more essential to daily business activities, the rush is on to create the wireless office that can easily tie mobile workers to the enterprise. Further, there is great potential for push technologies that deliver news and other information directly to a wireless device—this could create entirely new revenue streams for operators.

Although cdmaOne networks were not the first to offer data access, these networks are uniquely designed to accommodate data. To start with, these networks handle data and voice transmissions in much the same way. cdmaOne's inherent variable-rate transmission capability allows data rate determination to accommodate the amount of information being sent, so system resources are used only as needed. Because cdmaOne systems employ a packetized backbone for voice, packet data capabilities are already inherent in the equipment. The cdmaOne packet data transmission technology uses a TCP/IP-compliant cellular digital packet data (CDPD) protocol stack to enable seamless connectivity with enterprise networks and to expedite third-party application development.

Adding data to the cdmaOne network will allow an operator to continue using its existing radios, back-haul facilities, infrastructure, and handsets while merely implementing a software upgrade with an interworking function. Upgrade to IS-95B allows for code or channel aggregation to provide data rates of 64–115 kbps, as well as offering improvements in soft handoffs and interfrequency hard handoffs. Equipment manufacturers have already announced IS-707 packet data, circuit-switched data, and digital fax capabilities on its cdmaOne infrastructure equipment.

Mobile IP, the proposed Internet standard for mobility, is an enhancement to basic packet data services. Mobile IP lets users maintain a continuous data connection and retain a single IP address while traveling between base station controllers (BSCs) or roaming on other CDMA networks.

One of the key objectives of ITU IMT-2000 is the creation of a standard to encourage a worldwide frequency band to promote a high degree of design commonality and support high-speed data services. IMT-2000 will utilize small pocket terminals, an expanded range of operation environments, and the deployment of an open architecture that allows the graceful introduction of newly created technology. Furthermore, 3G systems promise to deliver wireless voice services with wireline quality levels, along with the speed and capacity needed to support multimedia and high-speed data applications. Location-based services, on-board navigation, emergency assistance, and other advanced services will also be supported.

The evolution of 3G will open the door of the wireless local loop (WLL) to PSTN and public data network access, while providing more convenient control of applications and network resources. It will also provide global roaming, service portability, zone-based ID and billing, and global directory access. The 3G technology is even expected to support seamless satellite interworking.

One of technical requirements for cdma2000 includes cdmaOne backward compatibility for voice services, vocoders, and signaling structure, as well as for privacy, authentication, and encryption capabilities.

Phase one of the cdma2000 effort, also known as 3G1X, employs 1.25 MHz of bandwidth and delivers a peak data rate of 144 kbps for stationary or mobile applications. Phase two of cdma2000, called 3G3X, will use 5 MHz bandwidth and is expected to deliver a peak data rate of 144 kbps for mobile and vehicular applications, and up to 2 Mbps for fixed applications. Industry insiders predict that the 3G3X phase will eventually yield up to 1 Mbps for each traffic or Walsh channel. By aggregating or bundling two channels, users can achieve the 2-Mbps peak data rate targeted for IMT-2000.

The primary difference between phase one and phase two of cdma2000 is bandwidth and resulting throughput speed, or peak data rate capability. Phase two will introduce advanced multimedia capabilities and lay the foundation for popular 3G voice services and vocoder, such as voice over IP. Since the 3G1X and 3G3X standards essentially share the same baseband radio elements, operators can take a major step toward full 3G capabilities by implementing 3G1X. cdma2000 phase two will include detailed descriptions of signal protocols, data management, and expected upscale requirements for moving from 5 MHz to 10 MHz, and 15-MHz radios in future interactions.

By migrating from the current IS-95 CDMA air interface technology to 3G1X of the cdma2000 standard, operators can achieve a twofold increase in radio capacity and ability to handle up to 144 kbps of packet data. Phase one capabilities of cdma2000 include a new physical layer for 1X and 3X 1.25 MHz channel sizes, support for multicarrier forward link 3X options, and definitions for the 1X and 3X numerology. Operators will also enjoy voice service enhancements that will produce twice the voice capacity of cdmaOne.

In the area of extended battery life, phase one will employ a quick paging channel and gated transmission of 1/8 rate to produce gains of two times the battery lives currently available. Hard handoff enhancements between 2G and 3G systems and power control enhancements will also be key factors in the improvements of voice service.

Data services will also be improved with the advent of cdma2000 phase one. Phase one will feature a media access control framework and packet data radio link protocol (RLP) definition to support packet data rates of at least 144 kbps.

Implementation of cdma2000 phase two will bring a host of new capabilities and service enhancements. Phase two will support all channel sizes (6X, 9X, and 12X) and associated numerology and a framework for advanced cdma2000 3G voice services and vocoders, including voice over IP. With phase two, true multimedia services will be available. To bring additional revenue opportunities for operators, multimedia services will be made possible through enhanced packet data media access control, full support for packet data services up to 2 Mbps, RLP support for all data rates up to 2 Mbps, and the advanced multimedia call model.

In the area of signaling and services, phase two cdma2000 will bring native 3G cdma2000 signaling structure to the link access control (LAC) and upper layer signaling structure. This structure will provide support for enhanced privacy, authentication and encryption functionality. An operator's existing architecture and network equipment can greatly affect the ease of this migration. Networks built on an open, advanced architecture with a clear upward migration pathway can attain 3G1X capabilities with a simple modular upward of the H-band operation of the radio. Networks with a less flexible architecture may be required to take the more costly steps of replacing the entire base transceiver station (BTS). To achieve the expected 144 kbps peak data rate performance, operators can make software upgrades to networks and base stations to support 3G1X data protocol.

Packet data service node (PDSN) will be required to support data connectivity to the Internet and intranet services. Many equipment vendors already offer solutions that incorporate PDSN elements, thus opening a smooth upward pathway to 3G technologies.

The recent agreement among members of the CDMA Operators Harmonization Group (OHG) proposes three optional CDMA modes and eventual development of a global standard that is compatible with both ANSI IS-41 and GSM MAP. This approach envisions the use of multimode handsets and various market-driven solutions as the surest pathway to a unified CDMA 3G standard in the next generation of wireless communications. As subscribers demand greater wireless power and convenience, the migration to 3G technology will benefit operators by supporting higher capabilities, lowering network costs, and increasing overall profitability. Figure 13.41 shows the cdmaOne evolution timeline.

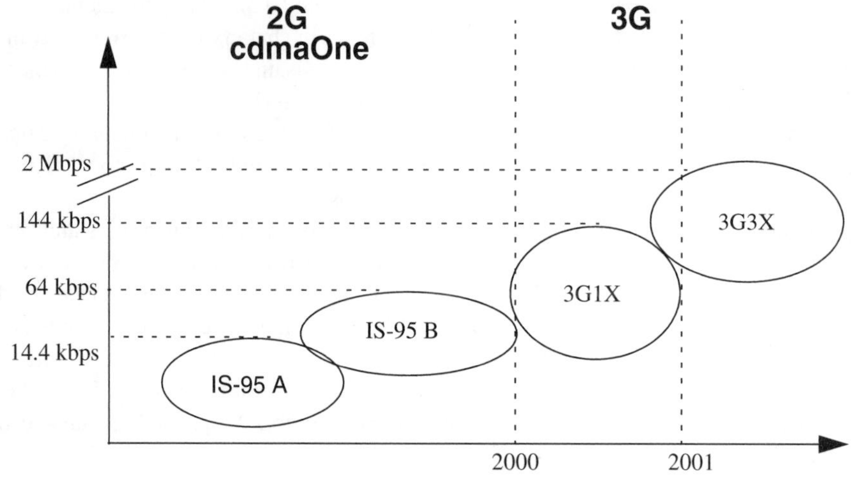

Figure 13.41 cdmaOne Evolution Timeline

cdmaOne operators will be able to upgrade to 3G without acquiring additional spectrum, a key component to minimum time to market without additional, significant investment. The design of cdma2000 will allow for deployment of the 3G enhancements while maintaining existing 2G support for cdmaOne in the spectrum an operator has today.

Both cdma2000 phase one and cdma2000 phase two can be intermingled with cdmaOne to maximize the effective use of spectrum according to the needs of an individual operator's customer base. For example, an operator that has a strong demand for a high-speed data service may choose to deploy a combination of cdma2000 phase one and cdmaOne that uses more channels for cdma2000 (see Figure 13.42). In another market, users may not be as quick to adopt high-speed data services and more channels will remain dedicated to cdmaOne services. As cdma2000 phase two capabilities become available, an operator has even more choices of ways in which to use the spectrum to support the new services (see Figures 13.43 and 13.44).

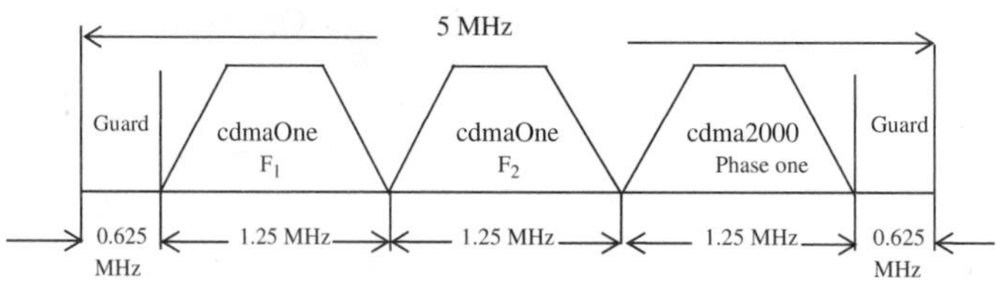

Figure 13.42 Intermixing of cdmaOne and cdma2000 Phase One (3G1X)

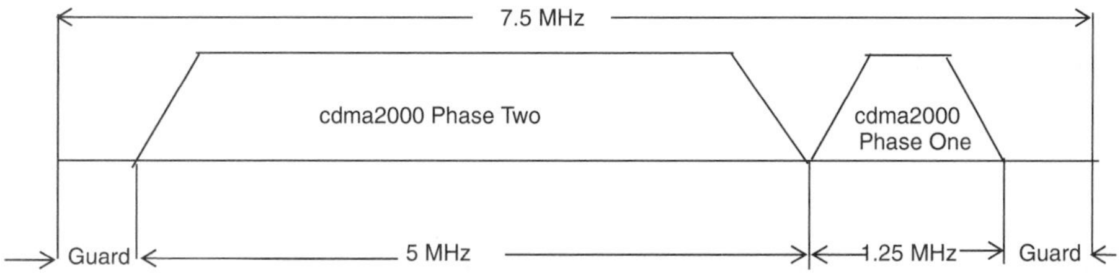

Figure 13.43 Intermixing of cdma2000 Phase One (3G1X) and cdma2000 Phase Two (3G3X)

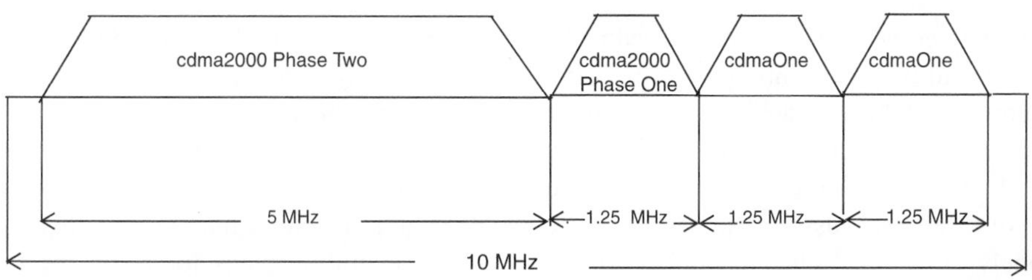

Figure 13.44 Intermixing of cdmaOne, cdma2000 Phase One (3G1X), and cdma2000 Phase Two (3G3X)

13.13 Major Technical Differences Between cdma2000 and W-CDMA

The major technical differences between the two wideband CDMA 3G proposals are listed in Table 13.21.

Table 13.21 Major Technical Differences Between cdma2000 and W-CDMA

Attribute	cdma2000	W-CDMA
Core network	ANSI-41	GSM MAP
Chip rate	3.6864 Mcps	3.84 Mcps (DOCOMO) 3.84 Mcps (UMTS)
Synchronized BS	Yes	No/but synchronized BS is optional
Frame length	20 and 5 ms	10 ms

(continued)

Table 13.21 Major Technical Differences Between cdma2000 and W-CDMA *(cont.)*

Attribute	cdma2000	W-CDMA
Multicarrier spreading option	Yes	No, direct spreading
Voice coder	Enhanced Variable Rate Codec (EVRC)	Adaptive Multi Rate (AMR)
Overhead	Low (because of shared pilot code channel)	High (because of nonshared pilot code channel)

13.14 Summary

In this chapter we first presented the cdma2000 layering structure and logical and physical channels and then concentrated on the cdma2000 physical layer, providing details of forward/reverse dedicated and common physical channel structures. We described forward and reverse link features and pointed out the improvements of cdma2000 over cdmaOne. The roles of the media access control and LAC sublayers in cdma2000 were also discussed.

We briefly discussed data services in cdma2000 and presented mapping of logical channels to physical channels. Next we presented the evolution plans for cdmaOne to cdma2000. We concluded the chapter by providing the major technical differences between cdma2000 and UTRA W-CDMA.

13.15 References

1. Rao, Y. S., and Kripalani, A., "cdma2000 Mobile Radio Access for IMT-2000," *1999 IEEE International; Conference on Personal Wireless Communications,* Feb. 1999, Jaipur, India.
2. TR 45.5 "The cdma2000 ITU-RTT Candidate Submission," TR 45-ISD/98.06.02.03, May 15, 1998.
3. Shanker, B., McClelland, S., "Mobilising the Third-Generation [Cellular Radio]," *Journal of Telecommunications (International Edition)*, August 1997.
4. Garg, V. K., Halpern, S., and Smolik, K. F., "Third Generation (3G) Mobile Communication Systems," *1999 IEEE International; Conference on Personal Wireless Communications,* Feb. 1999, Jaipur, India.
5. Dahlman, E., Gumudson, B., Nilsson, M., and Skold, J., "UMTS/IMT-2000 Based Wideband CDMA," *IEEE Communications Magazine,* Sept. 1998 [Vol. 36(9)].
6. "CDMA for Next Generation Mobile Communications Systems," *IEEE Communications Magazine,* Sept. 1998 [Vol. 36(9)].
7. Knisley, D., Quinn, L., and Ramesh, N., "cdma2000: A Third Generation Radio Transmission Technology," *Bell Labs Technical Journal,* Sept. 1998 [Vol. 3(3) J1].

8. TIA/EIA IS-2000-1 "Introduction to cdma2000 Spread Spectrum Systems," Nov. 1999.
9. TIA/EIA IS-2000-2 "Physical Layer Standard for cdma2000 Spread Spectrum Systems," Nov. 1999.
10. TIA/EIA IS-2000-3 "Medium Access Control (MAC) Standard for cdma2000 Spread Spectrum Systems," Nov. 1999.

CHAPTER 14

Third-Generation European Standards

14.1 Introduction

The International Telecommunication Union (ITU) began studies on globalization of personal communications in 1986 and identified the long-term spectrum requirements for the future third-generation (3G) mobile wireless telecommunications systems. In 1992, the ITU identified 230 MHz of spectrum in the 2-GHz band to implement the International Mobile Telecommunications-2000 (IMT-2000) [1–6] system on a worldwide basis for the satellite and terrestrial components. IMT-2000 capabilities include a wide range of voice, data, and multimedia services with the quality equivalent to or better than that of the wireline telecommunications networks in different radio environments. The aim of IMT-2000 is to provide a universal coverage enabling terminals to have seamless roaming across multiple networks. The ITU accepted the overall standardization responsibility of IMT-2000 to define radio interfaces that are applicable in different radio environments including indoor, outdoor, terrestrial, and satellite.

The 3G mobile telecommunications systems will provide worldwide access and global roaming for a wide range of services. Standards bodies in Europe, Japan, and North America are trying to achieve harmonization on the key and interrelated issues including radio interfaces; system evolution, backward compatibility, and user migration; global roaming; and phased introduction of mobile services and capabilities to support terminal mobility. The Universal Mobile Telecommunications System (UMTS) [7–11] studies were carried out by the European Telecommunications Standards Institute (ETSI) in parallel with IMT-2000 to harmonize its efforts with ITU. In Japan and North America, standardization efforts for 3G are carried out by the Association of Radio Industries and Businesses (ARIB) and the Telecommunications Industry Association (TIA) committee TR45, respectively. Two partnership projects (3GPP and 3GPP2) are involved in harmonizing 3G efforts in the Europe, Japan, and North America.

In Europe, 3G systems are intended to support a substantially wider and enhanced range of services compared with the second-generation (2G) system, the Global System for

Mobile Communications (GSM). These enhancements will include multimedia services, access to the Internet, high rate data, etc. The enhanced services will impose additional requirements on the fixed network functions to support mobility. These requirements are to be reached through an evolutionary path to capitalize on the investments for 2G systems in Europe, Japan, and North America.

In North America, the 3G wireless telecommunication system, cdma2000, [12–15] has been proposed to the ITU to meet most of the IMT-2000 requirements in the indoor office, indoor-to-outdoor pedestrian, and vehicular environments. In addition, the cdma2000 satisfies the requirements for 3G evolution of the 2G TIA/EIA 95 family of standards (cdmaOne). The cdma2000 details are given in Chapter 13.

In Japan, evolution of the GSM platform is planned for the IMT-2000 (3G) core network due to its flexibility and widespread use around the world. Smooth migration from GSM to IMT-2000 is possible. The service area of 3G systems will be overlaid with the existing 2G public digital cellular (PDC) systems. The 3G systems will connect and interwork with 2G systems through interworking function (IWF). IMT-2000-PDC dual-mode and IMT-2000 single-mode terminals will be deployed.

14.2 Third-Generation European Systems

Within the European research program, Advanced Communication Technologies and Services (ACTS), the project Future Radio Wideband Multiple Access System (FRAMES) was initiated with the objective of defining the radio interface for UMTS. During the first year of the project, a comprehensive evaluation of different multiple-access technologies was carried out.

As a result, FRAMES multiple access (FMA) was selected. FMA consists of two modes:

- FMA1: wideband TDMA with and without spreading
- FMA2: direct-sequence wideband CDMA

FRAMES was mainly targeted to contribute to the UMTS standardization process. Within ETSI Special Mobile Group 2 (SMG 2), the following five concepts subgroups were set up in the UMTS radio interface selection process:

1. Wideband code division multiple access CDMA (W-CDMA)
2. Orthogonal frequency division multiple access (OFDMA)
3. Wideband time division multiple access (W-TDMA)
4. TDMA/CDMA
5. Opportunity-driven multiple access (ODMA)

Two options of FMA1 wideband TDMA with and without spreading were contributed to the SMG 2, W-TDMA, and TDMA/CDMA concept subgroups, respectively. FMA2 wideband CDMA was contributed to the W-CDMA concept subgroup and also as an input to the ARIB for the Japanese 3G W-CDMA standardization. In the United States, the wideband TDMA concept based on FMA1 without spreading was proposed for standardization as an IS-136 high-speed (HS) data solution for outdoor and indoor applications.

In January 1998, a consensus agreement on the UMTS radio interface was reached in ETSI SMG. The proposed solutions are based on the W-CDMA and TD/CDMA concepts. The W-CDMA solution is proposed for frequency-division duplex (FDD) operation in paired frequency bands, and TD/CDMA is proposed for time-division duplex (TDD) mode in the unpaired frequency band.

ETSI UMTS Terrestrial Radio Access (UTRA)/FDD proposal based on W-CDMA has key parameters similar to those of ARIB W-CDMA [5]. As far as the TDD mode is concerned, the ARIB W-CDMA proposal has the same key parameters as the FDD mode including chip rate, frame length, and modulation/demodulation schemes; however, it is not aligned with the ETSI TD/CDMA proposal. In the 3G Partnership Project (3GPP) the efforts for harmonization are going on to achieve commonality between W-CDMA/TDD and UTRA/TDD. Although most of the key parameters in the UTRA W-CDMA/FDD and ARIB W-CDMA/FDD are similar, some differences do exist. Table 14.1 compares these two systems.

Table 14.1 Comparison of ARIB W-CDMA and ETSI UTRA (FDD Mode)

Attribute	ARIB (W-CDMA)	ETSI UTRA
Multiple access	DS-CDMA	DS-CDMA
Bandwidth	5 MHz (1.25/10/20)	5 MHz (10/20)
Chip rate	3.84 Mcps (1.024 Mcps)	3.84 Mcps
Inter BS timing	Asynchronous (sync. optional)	Asynchronous (sync. optional)
Cell search scheme	3-step code acquisition based on nonscrambled symbols	3-step code acquisition based on nonscrambled symbols
Frame length	10 ms	10 ms
Handover	SHO	SHO
Data modulation	QPSK	QPSK
Spreading modulation	QPSK	QPSK
Downlink pilot structure	Common pilot channel	Common pilot channel
Detection	Pilot symbol based coherent	Pilot symbol based coherent
Data modulation	BPSK	BPSK
Detection	I/Q Multiplexed	I/Q Multiplexed

Table 14.1 Comparison of ARIB W-CDMA and ETSI UTRA (FDD Mode) *(cont.)*

Attribute	ARIB (W-CDMA)	ETSI UTRA
Uplink pilot structure	Pilot symbol based coherent	Pilot symbol based coherent
Power control	Open loop (initial, RACH), closed loop (1,500 times/sec DCH SIR based)	Open loop (initial, RACH), closed loop (1,500 times/sec DCH SIR based)
Channel coding	Convolutional codes, turbocodes	Convolutional codes, RS codes, turbocodes
Interleaving periods	10/20/40/80 ms	10/20/40/80 ms

SHO: Soft Handover
QPSK: Quadrature Phase Shift Keying
CH: Channel
DCH: Dedicated channel
RACH: Random Access Channel
BPSK: Binary Phase Shift Keying
SIR: Signal-to-Interference Ratio

Table 14.2 provides the main parameters of FMA1 and FMA2 and compares them with 2G GSM and D-AMPS (IS-136) systems. In the design of FMA much attention was given on the harmonization of FMA1 and FMA2. The most important aspect of the harmonization is the backward compatibility with 2G systems, in particular with the widely used GSM. In addition, FMA1 and FMA2 are also harmonized with each other.

14.3 FMA1

In FMA1, [6] users are separated orthogonally into time slots, and within each time slot spreading can provide additional separation. In FDD mode, all time slots of a frame are assigned to either uplink or downlink. In TDD mode, time slots are dynamically divided between uplink and downlink (i.e., the position of the switching point can be varied). For TDD mode, the frame duration is 4.615 ms (same as GSM) and can consist of 1/64, 1/16, and 1/8 slots fitting together in the frame. In TDD mode, the minimum length of uplink and downlink parts is 1/8 of frame duration (577 μs). The nonspread bursts of FMA1 are assigned to the 1/64 and 1/16 slots, whereas the spread bursts are assigned to the 1/8 slots. The 1/8 and 1/64 slots can be used for every service from low-rate speech and data to high-rate data services. The 1/16 slot is used for medium- to high-rate data services. A basic physical channel is one time slot in the case of FMA1 without spreading, and one time slot and one spreading code in the case of FMA1 with spreading. In FMA1, user bit rates from a few kbps up to 2 Mbps are achieved by allocating different numbers of basic channels (i.e., different numbers of time slots and/or spreading codes) to a user.

Table 14.2 The Main Parameters of 2G and 3G Systems

Main Multiple Access Parameters and Physical Layer	2G		3G		
	GSM	DAMPS	FMA1 without spreading	FMA1 with spreading	FMA2
Multiple access	TDMA	TDMA	W-TDMA	TD/CDMA	W-CDMA
Duplexing	FDD	FDD	FDD and TDD		FDD
Channel spacing	200 kHz	30 kHz	1.6 MHz		4.4–5 MHz with 200 kHz multiple
Carrier chip/bit rate	271 Kbps	48.6 Kbps	N/A	2.167 Mcps	3.84 Mcps
Time slot structure	8 slots (FR) or 16 slots (HR)/Frame	6 slots/frame	16 or 64 slots/TDMA frame	8 slots/TDMA frame	N/A
Spreading	N/A	N/A	N/A	Orthogonal, 16 chips/symbol	spreading factor 4–512, short codes for DL and UL (with MUD); and long code optional for UL
Frame length	4.615 ms	6.667ms	4.615 ms		10 ms
Multirate concept	Multislot	Multislot	Multislot	Multislot & Multicode	Variable rate spreading
FEC code	Convolutional	Convolutional	Convolutional Codes, $R = 1/4$ to 1 puncturing/repetition, turbocodes		Convolutional codes; UL, $R = 1/2$; DL, $R = 1/3$ puncturing/repetition, turbocodes
Data modulation	GMSK	$\pi/4$-DQPSK	BOQAM/QOQAM	QPSK/ 16-QAM	QPSK
Spreading modulation	N/A	N/A	N/A	Linearized GMSK	Dual channel/balanced QPSK or complex four-phase spreading (UL)
Pulse shaping	Root raised cosine, roll-off = 0.35	Root raised cosine, roll-off = 0.35	Root raised cosine, roll-off = 0.35	(Linearized GMSK)	Root raised cosine, roll-off = 0.22
Detection	coherent	coherent	Coherent, based on training sequence		coherent detection UL; reference bit based; DL-pilot/reference bit based
Other diversity means	Frequency hopping	N/A	Frequency/time hopping per frame or slot		Macrodiversity
Power control	N/A	N/A	Slow power control, 50 dB dynamic range		UL: open loop (80 dB dynamic range) and fast closed loop, DL fast closed loop (2 dB dynamic range); 1500 Hz

Synchronization and handover measurements are based on either a slotted discontinuous wideband broadcast control channel (BCCH) or a continuous narrowband (200 kHz) BCCH. For FMA1 with spreading, a continuous wave pilot signal with high spreading ratio is also considered for BCCH. For slotted wideband BCCH, special bursts are defined including frequency correction burst, synchronization burst, and access burst. The frequency correction burst is of 72 μs and contains 171 fixed symbols. Two types of synchronization bursts are defined: one of 72 μs duration and another of 288 μs duration. Two types of access bursts are defined: one with a 72 μs duration and another with 288 μs duration. The guard periods of the shorter and longer access bursts allow reception of initial random access messages for maximum cell radii of 5 km and 36 km, respectively.

FMA1 without spreading uses binary offset quadrature amplitude modulation (BOQAM) and quaternary offset QAM (QOQAM) modulation. BOQAM allows near constant envelope signals. The channel spacing of 1.6 MHz and spectral properties of modulated FMA1 carrier allows flexible frequency planning for FMA1 carriers within a UMTS frequency band as well as coexistence with GSM carriers within GSM900/1800 frequency bands. FMA1 uses the hybrid automatic repeat request (ARQ) protocol for error correction.

Both real-time (RT) (circuit-switched) and non-real-time (NRT) (packet-switched) services can be provided by FMA1. The RT services are protected by forward error correction (FEC) codes. Advanced coding schemes such as turbocodes and rate-compatible punctured convolution (RCPC) codes will be used to meet the quality of service (QoS) requirements. Also, concatenated codes (e.g., Reed-Solomon codes concatenated with convolutional Viterbi [RSV] codes) can be used to achieve very low bit error rates (BERs). For RT services, frequency and time hopping together with interleaving are used to improve frequency diversity, and to average out interference from other users.

14.4 FMA2 (W-CDMA)

The objectives of W-CDMA are

- Support of high-speed data (> 384 kbps with wide area coverage and up to 2 Mbps for indoor/local outdoor coverage)
- High service flexibility with support of multiple parallel variable-rate services on each connection
- Efficient packet access
- High initial capacity and coverage with built-in support for future capacity/coverage—enhancing technologies, such as smart antennas and advanced receiver structures (multiuser detection [MUD])
- Support for interfrequency handover for operation with hierarchical cell structures
- Easy implementation of dual-mode UMTS/GSM terminals as well as handover between UMTS and GSM

To use the capabilities of W-CDMA efficiently, a radio interface protocol has been designed to fully incorporate the W-CDMA physical layer. The protocol stack for FMA1 and FMA2 are harmonized as far as possible in different layers (see Figure 14.1). In the FMA concept, the goal is to reuse as much of the mode-specific protocols as possible when designing the other mode.

Although there are basic differences in radio link control (RLC) and media access control (MAC) layers between FMA1 and FMA2, it is expected that protocol structure and handling of information exchange can be further harmonized. The logical link control (LLC) provides the same functionality within the protocol stack, and it is assumed to be made independent apart from some differences for internal parameters. The radio network layer (RNL) is different for the two modes. The radio resource control (RRC) [16] provides some fundamental differences between handover for FMA2 and mobile assisted hard handover for FMA1, while most functions common to both FMA modes are provided with the radio bearer control (RBC) function. Both RBC and RRC could be presented as a single RRC protocol as well.

In UMTS Terrestrial Radio Access Network (UTRAN), overall protocol structure is divided into layers corresponding to the open system interconnect (OSI) model [17]. These include the following:

- Physical layer (layer 1 [L1])
- Media access control sublayer (lower part of layer 2 [L2])
- Radio link control sublayer (upper part of layer 2)
- Radio network layer (layer 3 [L3])

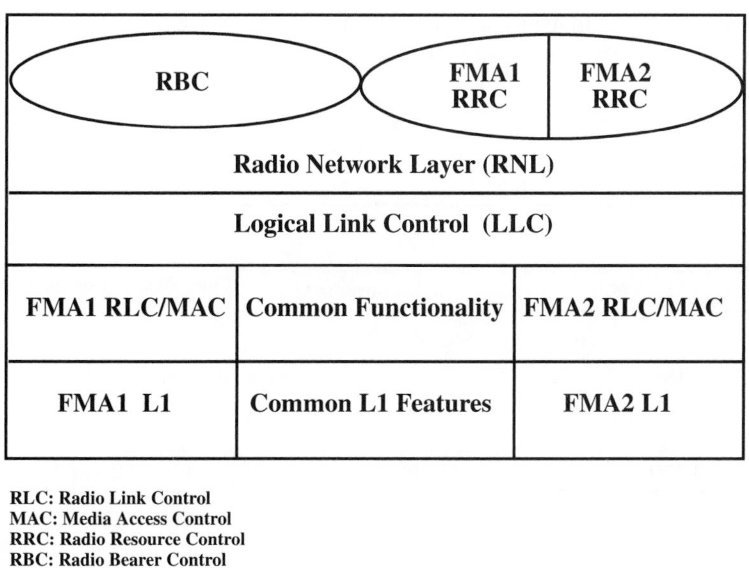

Figure 14.1 Layers of FMA1 and FMA2

The physical layer offers information transfer service over its W-CDMA radio medium for the upper layers. The MAC sublayer provides data transfer services for RLC and reallocates radio resources [18,19]. On request, MAC may also provide to the higher layer information about traffic volume and quality indication. Automatic retransmission request (ARQ) functionality is realized in the RLC sublayer. The retransmission protocol ensures that the optimum utilization of the available radio resources is achieved without incurring excessively long delays. The RRC provides general broadcast and notification services to all mobiles. It also provides services for establishing, maintaining, and releasing of a user equipment (UE)/UTRAN connection and transfer of messages using this connection. The RRC functionality also includes the establishment, maintenance, and release of a radio access bearer and RRC mobility control (soft/hard handover procedures). Other radio resource control functions performed by RRC are arbitration of radio resource allocation between the cells, control of requested QoS, and outer loop power control (seeking target of the closed loop power control).

For data flow in W-CDMA (see Figure 14.2), the network layer packet data units (N-PDUs) are first segmented into smaller packets and transformed into link access control (LAC) PDUs. The LAC overhead (~3 octets) typically consists of at least a service access points identifier and sequence number for higher level ARQ and other fields. The LAC PDUs are segmented into smaller packets, RLC-PDUs, corresponding to physical layer transport blocks. Each RLC-PDU contains sequence number used for the low-level fast ARQ. A cyclic redundancy check (CRC) for error detection is calculated and appended to each RLC-PDU by physical layer.

The data flow of W-CDMA system is similar to the data flow of general packet radio service (GPRS). One important difference is that in the GPRS system, a RLC-PDU always consists of four bursts, while the code rate may vary. In W-CDMA, all RLC-PDUs have the same size regardless of transmission rate. This means that since the transmission rate may change every 10 ms, the number of RLC-PDUs transferred each 10 ms varies.

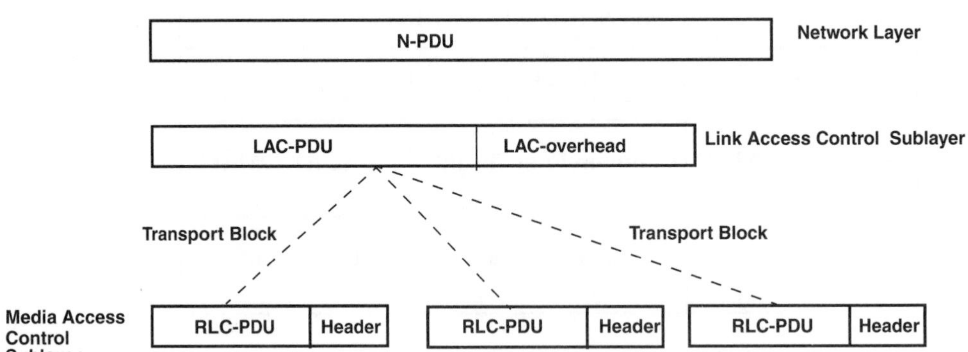

Figure 14.2 Data Flow in W-CDMA

14.5 Physical Layer

Access to the physical layer services is achieved through the use of *transport channels* via the MAC sublayer [18]. The *transport channels* in W-CDMA are as follows:

- *Dedicated channel (DCH)*—a channel dedicated to one UE used in uplink or downlink.
- *Fast uplink signaling channel (FAUSCH)*—an uplink channel used to allocate dedicated channels in conjunction with a FACH.
- *Opportunity-driven dedicated channel (ODCH)*—a channel dedicated to one UE used in relay link.
- *ODMA random access channel (ORACH)*—a contention-based channel used in relay-link.
- *Broadcast channel (BCH)*—a downlink channel used to broadcast system information into an entire cell.
- *Synchronization channel (SCH)*—a downlink channel used to broadcast synchronization information into an entire cell in TDD mode.
- *Random access channel (RACH)*—a contention-based uplink channel used for transmission of a relatively small amount of data, e.g., for initial access or non-real-time dedicated control or traffic data. RACH uses open-loop power control.
- *Forward access channel (FACH)*—a common downlink channel without closed-loop power control used to transmit a relatively small amount of data.
- *Downlink shared channel (DSCH)*—a downlink channel shared by several UEs carrying dedicated control or traffic data.
- *DSCH control channel*—a downlink channel associated with a DSCH used for signaling of DSCH resource allocation.
- *Uplink shared channel (USCH)*—an uplink channel shared by several UEs carrying dedicated control or traffic data; used in TDD mode only.
- *Common packet channel (CPCH)*—a contention-based channel used to transmit bursty data traffic, only used in FDD mode in the uplink direction. The CPCH is shared by the UEs in a cell; therefore, it is a common resource. The CPCH is fast power controlled.
- *Paging channel (PCH)*—a downlink channel used to broadcast control information into an entire cell to allow efficient UE sleep mode procedures.

The characteristics of a transport channel are defined by its transport format (or format set), which specifies the physical layer processing to be applied to the transport channel, such as coding and interleaving, and any service-specific rate matching as needed. The physical layer operates according to layer 1 (L1) radio frame timing. A transport block is defined as the data accepted by the physical layer to be encoded. The transport block timing is tied exactly to the L1 frame timing, e.g., every transmission block is generated precisely every 10 ms, or multiple of 10 ms. Transport channels are always unidirectional and either common (i.e., shared among several users) or dedicated (i.e., used for a specific user).

A mobile can set up multiple transport channels simultaneously, each having its own transport characteristics (e.g., offering different error correction capability). Each transport channel can be used for information stream transfer of one radio bearer or for layer 2 (L2) and higher layer signaling messages. The multiplexing of the transport channel onto the same or different physical channel is carried out by L1. In addition, the transport format combination indication (TFCI) field uniquely identifies the transport format used by each transport channel within the current radio frame.

Physical channels are defined in the physical layer (see Figure 14.3). Two duplex modes FDD and TDD, are used. In the FDD mode, a physical channel is defined by the code, frequency, and in the uplink (mobile station [MS] to base station [BS]) by the relative phase (I/Q). In the TDD mode, the physical channel is characterized by time slot. The RRC controls the physical layer.

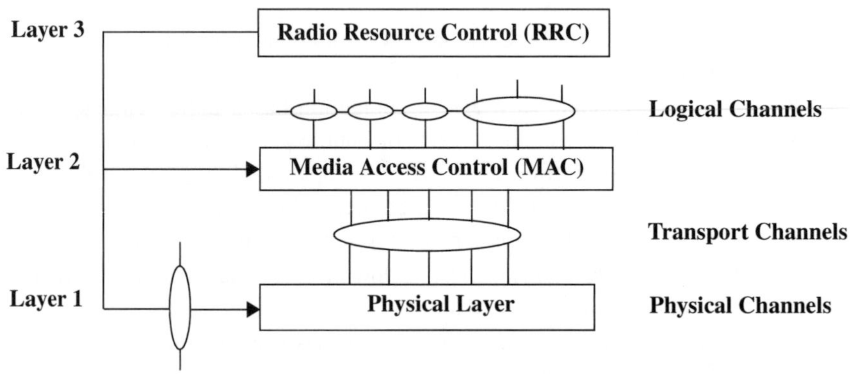

Figure 14.3 Radio Interface Protocol Architecture Around Physical Layer

The physical layer provides data transport services to higher OSI layers. In order to provide data transport services, the following functions are performed by the physical layer:

- Macrodiversity distribution/combining and soft handover execution
- Error detection on transport channels and indication to higher layers
- Forward error correction encoding/decoding of transport channels
- Multiplexing of transport channels and demultiplexing of coded composite transport channels
- Rate matching (data multiplexing on DCH)
- Mapping of coded composite transport channels on physical channels
- Power weighting and combining of physical channels
- Modulation and spreading/demodulation and despreading of physical channels
- Frequency and time (chip, bit, slot, frame) synchronization
- Measurements of frame error rate (FER), signal-to-interference ratio (SIR), interference power, etc., and indication of measurements to higher layers

- Closed-loop power control
- Radio frequency (RF) processing

The *physical channels* include *dedicated physical channels* and *common physical channels*. The common physical channels do not have soft handover, but some of them can have fast power control.

The dedicated physical channel (DPCH) is mapped onto two physical channels: the dedicated physical data channel (DPDCH) and dedicated physical control channel (DPCCH). The DPDCH carries higher layer information including user data, and the DPCCH carries the necessary physical layer control information. These two dedicated channels are used to support efficiently the variable bit rate in the physical layer. The bit rate of the DPCCH is constant, whereas the bit rate of the DPDCH can vary from frame to frame.

The following are the physical channels:

- Common pilot channel (CPICH)
- Primary common control physical channel (primary CCPCH)
- Secondary common control physical channel (secondary CCPCH)
- Physical random access channel (PRACH)
- Physical common packet channel (PCPCH)
- Dedicated physical data channel (DPDCH)
- Dedicated physical control channel (DPCCH)
- Physical downlink shared channel (PDSCH)
- Page indication channel (PICH)
- Acquisition indication channel (AICH)
- Synchronization channel (SCH)

14.5.1 Uplink Dedicated Physical Channels

Dedicated Physical Data Channel (DPDCH). DPDCH is used to transmit data generated at L2 and above (i.e., dedicated control and traffic channels). There may be zero, one, or several DPDCHs on each L1 connection. The DPDCH frame carries 10×2^k ($k = 0 \ldots,$ 6) bits, corresponding to a spreading factor of $256/2^k$ with 3.84 Mcps. Multiple parallel variable-rate services can be multiplexed within each DPDCH frame. The overall DPDCH bit rate is variable on a frame-by-frame basis. UE provides data information (transport format combination information [TFCI]) to detect with a orthogonal variable spreading factor (OVSF) on the DPDCH. In most of the cases, only one DPDCH is allocated per connection, with multiple services sharing the same physical channel. However, multiple DPDCHs can also be allocated to avoid too low a spreading factor at high data rates. The required instantaneous received power of the DPDCH may also vary on a frame-by-frame basis (see Figure 14.4). The frame structure of DPDCH is shown in Figure 14.5. Each frame of length 10 ms is divided into 15 slots, each slot with $T = 2{,}560$ chips, corresponding to a power control period of 0.6667 ms. A superframe consists of 72 consecutive frames, i.e., the superframe duration is 720 ms. The DPDCH spreading factor may range from 256 down to 4.

Dedicated Physical Control Channel (DPCCH). DPCCH is used to transmit control information generated at L1. The control information consists of known pilot bits to support channel estimation for coherent detection, transmit power control (TPC) commands, feedback information, and an optional transport format combination indicator (TFCI). The TFCI informs the receiver about the instantaneous parameters of different transport channels multiplexed on the uplink DPDCH, and corresponds to the data transmitted in the same frame. The UTRAN determines if a TFCI should be transmitted, thus making it mandatory for all UEs to support the use of TFCI in the uplink. Fast power control is based on measurements of the received DPCCH connections with a variable-rate DPDCH. Although the DPCCH frame structure is fixed for a given connection, it may vary between different environments. As an example, in an indoor environment, with semistationary terminals, the pilot power and rate of power control commands may be significantly reduced, resulting in a reduced overhead and increased capacity. There is one and only one uplink DPCCH on each L1 connection. (See Figure 14.5 for DPCCH frame structure.) The field order of N_{pilot}, N_{TPC}, N_{FBI}, N_{TFCI}, and total number of bits/slot are fixed, though the number of bits per field may vary during a connection. The N_{FBI} bits are used to support techniques requiring feedback between the UE and UTRAN access point (i.e., cell transceiver), including feedback indicator (FBI) mode transmit diversity and site selection diversity (SSDT). The *S* field is used for SSDT signaling and *D* field is used for FB mode transmit diversity signaling (see Figure 14.6). Each of the *S* and *D* fields can be of length 0, 1, or 2, with a total FBI field size equal to 0, 1, or 2 bits. Simultaneous use of SSDT power control and FB mode transmit diversity requires that both the *S* and *D* fields be of length 1. There are two types of uplink dedicated physical channels: those that include TFCI (e.g., for several simultaneous services) and those that do not include TFCI (e.g., for fixed-rate services). Multicode operation is possible for the uplink dedicated physical channels. When multimode transmission is used, several parallel DPDCHs are transmitted using different channelization codes. Table 14.3 provides uplink DPDCH data rates. Table 14.4 includes details of DPCCH fields for SF of 256 and slot format 0 to 5.

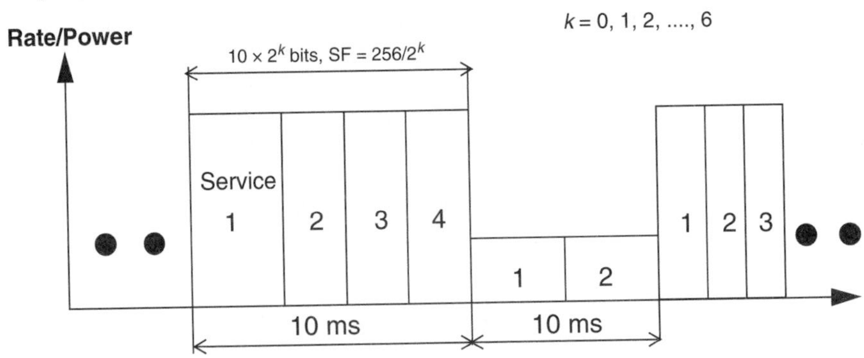

Figure 14.4 Uplink DPDCH Frame

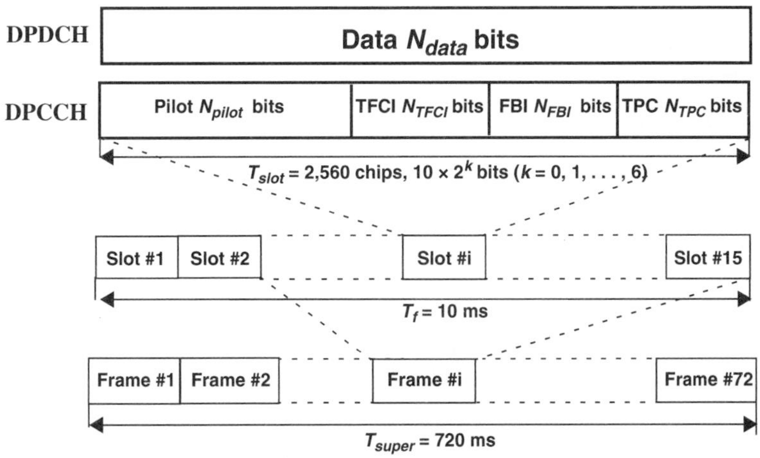

Figure 14.5 Uplink DPCCH Frame Structure

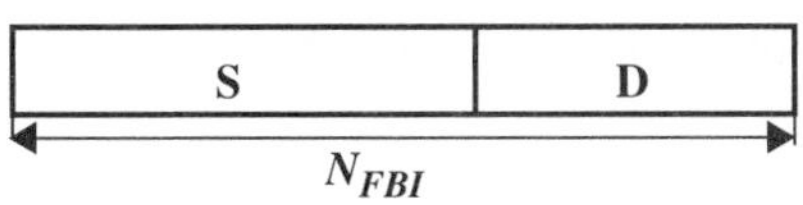

Figure 14.6 Details of FBI Field

Table 14.3 Uplink DPDCH Data Rates

Slot Format	SF	Bits/Slot	Data/Slot	DPDCH Rate (kbps)	Max. user data rate (kbps) with r =1/2
0	256	10	10	15	7.5
1	128	20	20	30	15
2	64	40	40	60	30
3	32	80	80	120	60
4	16	160	160	240	120
5	8	320	320	480	240
6	4	640	640	960	480
	4 with 6 parallel codes	3,840	3,840	5,760	2,304

Table 14.4 Uplink DPCCH Fields for SF 256

Slot Format	Bits/Slot	N_{pilot}	N_{TPC}	N_{TFCI}	N_{FBI}
0	10	6	2	2	0
1	10	8	2	0	0
2	10	5	2	2	1
3	10	7	2	0	1
4	10	6	2	0	2
5	10	5	1	2	2

14.5.2 Uplink Common Physical Channels

Physical Random Access Channel (PRACH). PRACH is used to carry the RACH. The random access transmission is based on a slotted ALOHA scheme where a random access burst can be transmitted in different access slots. There are 15 access slots per two frames and they are spaced 5120 chips apart. (see Figures 14.7 and 14.8.) Before a random request can be carried out, the UE must acquire the following information from the BCH of the target cell:

- The cell-specific spreading codes available for the preamble and message parts
- The signatures and access slots available in the cell
- The spreading factors (SFs) allowed for the message part
- The primary CCPCH transmit power level
- The uplink interference level at the base station

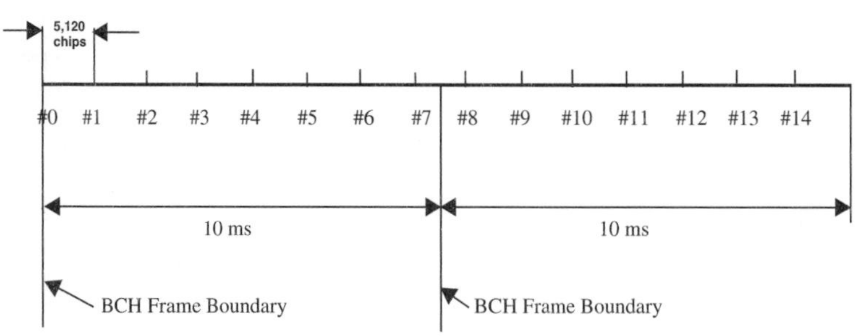

Figure 14.7 RACH Access Slot Numbers and Their Spacing

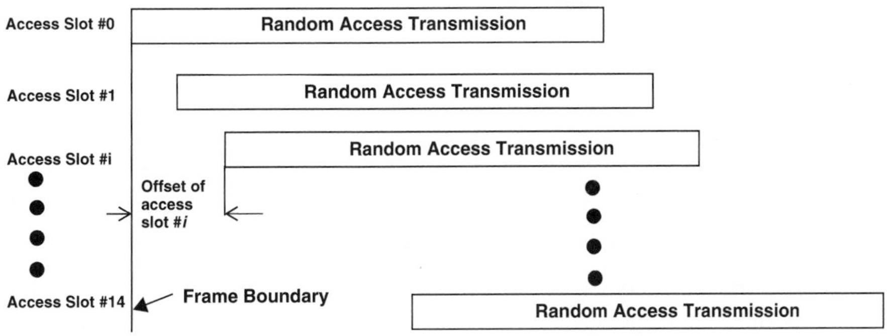

Figure 14.8 PRACH Allocated for RACH Access Slots

The random access transmission consists of one or several preambles that are 4,096 chips in length and a message 10 ms in length (see Figure 14.9). The 10-ms message is divided into 15 slots, each of length $T = 2{,}560$ chips. Each slot contains two parts, a data part that carries L2 information and a control part that carries L1 control information. The data and control parts are transmitted in parallel. The data part has 10×2^k bits ($k = 0, 1, 2, 3$). This corresponds to a spreading factor of 256, 128, 64, and 32, respectively, for the message part.

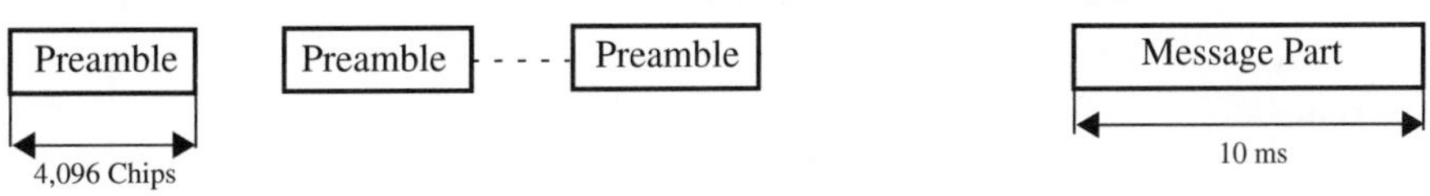

Figure 14.9 Structure of the Random Access Transmission

The control part has 8 known pilot bits to support channel estimation for coherent detection and 2 TFCI bits. This corresponds to an SF of 256 for the message control part. The total number of TFCI bits in the random access message is $15 \times 2 = 30$. The TFCI value corresponds to a certain transport format of the current random access message (see Figure 14.10).

Upon receiving the random access burst, the base station responds with an access grant message on the FACH. In case the random access request is for a dedicated channel (circuit-switched or packet) and the request is granted, the access grant message includes a pointer to the dedicated physical channel(s) to use. As soon as the mobile has moved to the dedicated channel, closed loop power control is activated. The following steps are carried out during a random access burst:

1. The mobile selects the spreading codes to be used for preamble and message parts. The mobile also selects the SF (i.e., channel bit rate) for the message part.
2. The mobile randomly selects the signature and access slot to be used for the random access burst.
3. The mobile estimates the downlink path loss and calculates the required uplink transmit power to be used for the random access burst.
4. The mobile transmits the random access burst.
5. The mobile waits for an acknowledgment on a corresponding downlink FACH. If no acknowledgment is received within a predefined time-out period, the random access procedure of step 2 is repeated.

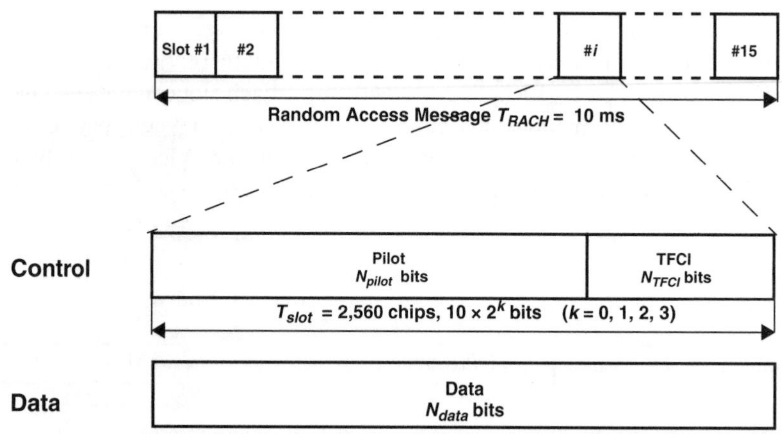

Figure 14.10 Structure of Random Access Message Part

Physical Common Packet Channel (PCPCH). PCPCH is used to carry the CPCH. The structure of the CPCH random access transmission is shown in Figure 14.11. The CPCH random access transmission contains one or several access preambles that are 4,096 chips in length, one collision-detection preamble (CD-P) that is 4,096 chips in length, a 10-ms DPCCH power-control preamble (PC-P), and a message of variable length $N \times 10$ ms. Each message consists of N_Max_frames 10-ms frames. Each 10-ms frame is divided into 15 slots, each of length $T = 2{,}560$ chips. Each slot contains two parts, a data part to carry L2 information and a control part to carry L1 control information. The data and control parts are transmitted in parallel. The data part consists of 10×2^k bits ($k = 2, 3, 4, 5, 6$), corresponding to SFs of 64, 32, 16, 8, and 4, respectively. The SF for the message control part is 256.

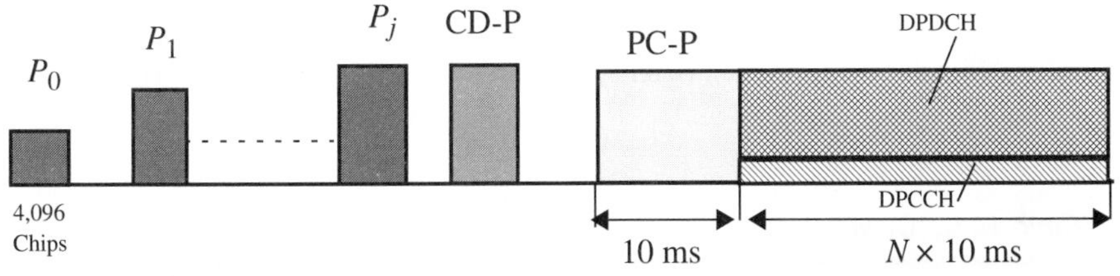

Figure 14.11 Structure of CPCH Random Access Transmission

14.5.3 Downlink Dedicated Physical Channels

There is only one type of downlink DPCH. The DPCH carries dedicated data generated at L2 and above, i.e., the dedicated transport channel (DTCH). Information is transmitted in time-multiplex with control data generated at L1 (i.e., known pilot bits, TPC commands, and an optional TFCI). The downlink DPCH can be seen as a time multiplex of a downlink DPDCH and a downlink DPCCH. The UTRAN determines if a TFCI should be transmitted, thus making it is mandatory for all UEs to support the use of TFCI in the downlink.

Figure 14.12 shows the frame structure of the downlink DPCH. Each frame of 10 ms duration is divided into 15 slots, and each slot has 2,560 chips, corresponding to one power group of 0.667 ms. A superframe contains 72 consecutive frames. The parameter k determines the total number of bits per downlink DPCH slot. It is related to spreading factor as $SF = 512/2^k$ ($k = 0, 1, \ldots, 7$). The SF may range from 512 down to 4.

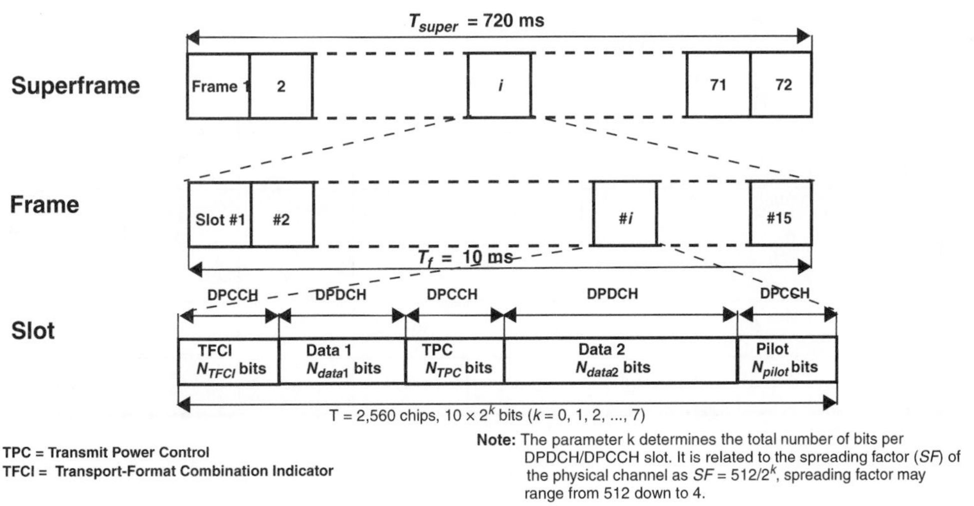

Figure 14.12 Frame Structure of Downlink DPCH

The overhead due to the DPCCH transmission is negotiated at the connection setup and is renegotiated during communication to match the particular propagation conditions. Tables 14.5 and 14.6 provide details of downlink DPDCH and DPCCH.

Table 14.5 Downlink Dedicated DPDCH and DPCCH

Slot Format	SF	Bits/ Slot	DPDCH		DPCCH		
			Data1	Data2	N_{TFCI}	N_{TPC}	N_{Pilot}
0	512	10	2	2	0	2	4
1	512	10	0	2	2	2	4
2	256	20	2	14	0	2	2
3	256	20	0	14	2	2	2
4	256	20	2	12	0	2	4
5	256	20	0	12	2	2	4
6	256	20	2	8	0	2	8
7	256	20	0	8	2	2	8
8	128	40	6	28	0	2	4
9	128	40	4	28	2	2	4
10	128	40	6	24	0	2	4
11	128	40	4	24	2	2	8
12	64	80	4	56	8	4	8
13	32	160	20	120	8	4	8
14	16	320	48	240	8	8	16
15	8	640	112	496	8	8	16
16	4	1,280	240	1,008	8	8	16

Table 14.6 Downlink DPDCH Symbol and Bit Rates

Spreading Factor (SF)	Channel Symbol Rate (ksps)	Channel Bit Rate (kbps)	DPDCH Channel Bit Rate Range (kbps)	Max. User Data Rate with $r = 1/2$ Coding (approx) (kbps)
512	7.5	15	3–6	1–3
256	15	30	12–24	6–12
128	30	60	42–51	20–24
64	60	120	90	45
32	120	240	210	105
16	240	480	432	215
8	480	960	912	456
4	960	1,920	1,872	936
4 with 3 parallel codes	2,880	5,760	5,616	2,300

14.5.4 Downlink Common Control Physical Channels

Primary Common Control Physical Channel (P-CCPCH). P-CCPCH is a fixed-rate (30 kbps, $SF = 256$) downlink physical channel used to carry the BCH. The frame structure (Figure 14.13) differs from the downlink DPCH in that no TPC commands, no TFCI, and no pilot bits are transmitted. The P-CCPCH is not transmitted during the first 256 chips of each slot. Instead, primary and secondary SCH are transmitted during this period.

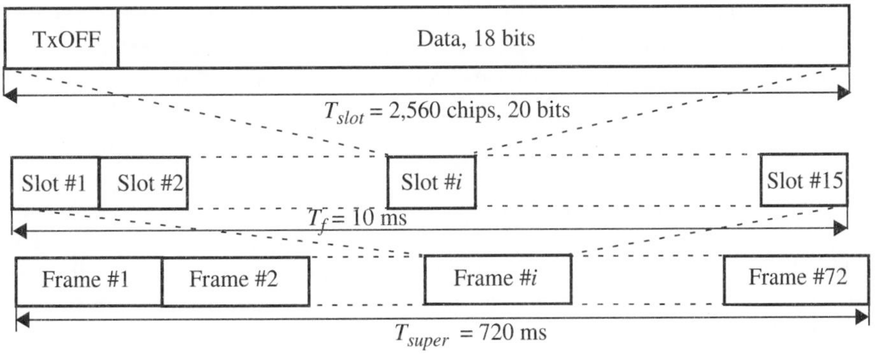

Figure 14.13 Frame Structure of Primary Common Control Physical Channel (P-CCPCH)

Secondary Common Control Physical Channel (S-CCPCH). S-CCPCH is used to carry the FACH and PCH. The FACH and PCH are time multiplexed on a frame-by-frame basis within the superframe structure. The frames allocated to FACH and PCH, respectively, are broadcast on the BCH. There are two types of S-CCPCH: one that includes TFCI, and another that does not. It is the UTRAN that decides if a TFCI should be transmitted, thus making it mandatory for all UEs to support the use of TFCI. The main difference between a CCPCH and a downlink dedicated physical channel is that a CCPCH is not closed-loop power controlled. The main difference between the primary and secondary CCPCH is that the primary CCPCH has a fixed predefined rate, whereas the secondary CCPCH can support variable rate with the help of the TFCI field included. A primary CCPCH is continuously transmitted over the entire cell while the secondary CCPCH is only transmitted when there is data available and may be transmitted in a narrow beam in the same way as a dedicated physical channel (see Figure 14.14).

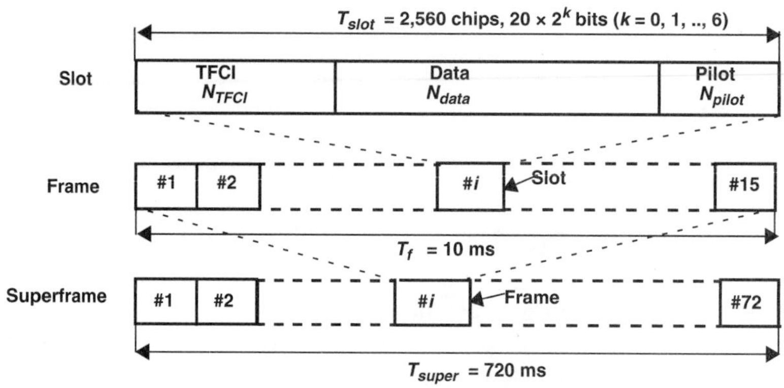

Figure 14.14 Frame Structure of Secondary CCPCH

Synchronization Channel (SCH). The SCH is a downlink channel used for cell search. The SCH consists of two subchannels: primary and secondary SCH. The primary SCH consists of a modulated Gold code 256 chips in length, the *primary synchronization code,* transmitted once every slot. The primary synchronization code is the same for every base station in the system, and is transmitted time-aligned with the period where the primary CCPCH is not transmitted. The secondary SCH consists of repeatedly transmitting 15-sequence modulated orthogonal Gold code 256 chips in length, the *secondary synchronization code,* transmitted in parallel with the primary synchronization channel. The secondary synchronization code is chosen from a set of 15 different codes $\{c_1, c_2, c_3, \ldots, c_{15}\}$, 256 chips in length. The sequence on the secondary SCH indicates to which of the 64 different codes the cell's downlink scrambling code belongs. Sixty-four sequences are used to encode the 64 different code groups, each containing eight scrambling codes (see Figure 14.15).

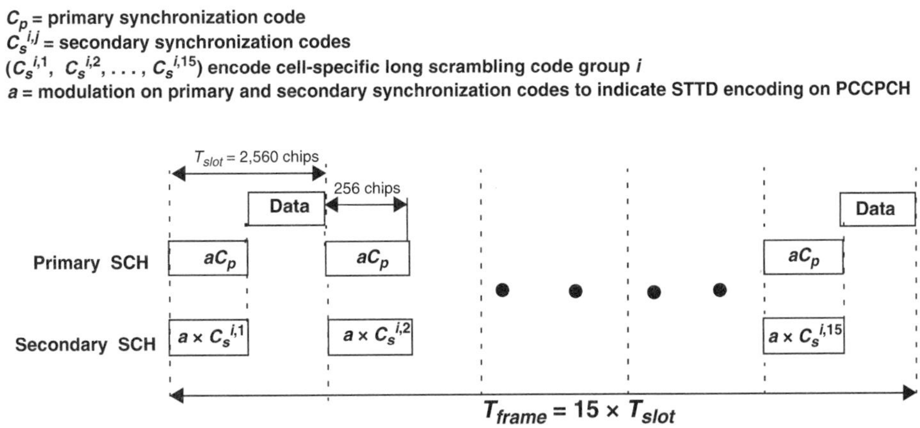

Figure 14.15 Structure of Synchronization Channel (SCH)

Physical Downlink Shared Channel (PDSCH). PDSCH is used to carry the downlink shared channel (DSCH). It is shared by users based on code multiplexing and is always associated with DPCH. The PDSCH does not carry any pilot symbols. The frame structure of the DSCH, when associated with a DCH is shown in Figure 14.16. The DSCH transmission with associated DCH is a special case of multicode transmission. The channels do not necessarily have the same SF and for DSCH the SF may vary from frame to frame. The relevant L1 control information is transmitted on DCH; the PDSCH does not contain DPCCH information. For DSCH, the allowed SFs vary from 256 to 4. DSCH may consist of multiple parallel codes as well as those negotiated at a higher layer prior to starting data transmission. In such a case the parallel codes are operated with frame synchronization between each other.

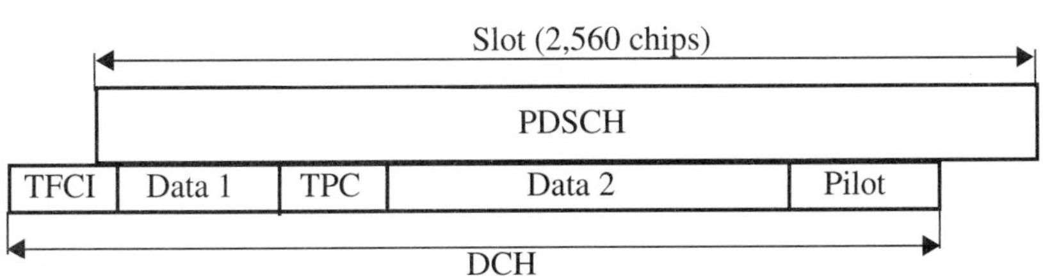

Figure 14.16 Frame Structure for DSCH When Associated with a DCH

Acquisition Indication Channel (AICH). AICH is used to carry acquisition indicators (AIs). Acquisition indicator AI_i corresponds to signature i on the PRACH or P-CCPCH. Note that for P-CCPCH, the AICH is in response to either an access preamble or a CD preamble. AP-AICH corresponds to the access preamble AICH and CD-AICH corresponds to CD preamble AICH. The AP-AICH and CD-AICH use different channelization codes. Figure 14.17 shows the frame structure of AICH. The two AICH frames, 20 ms in total duration, consist of 15 access slots (AS), each 20 symbols (5120 chips) in length. Each AS consists of two parts, an AI part and an empty part. The empty part of the AS consists of four zeros. The phase reference for the AICH is the CPICH.

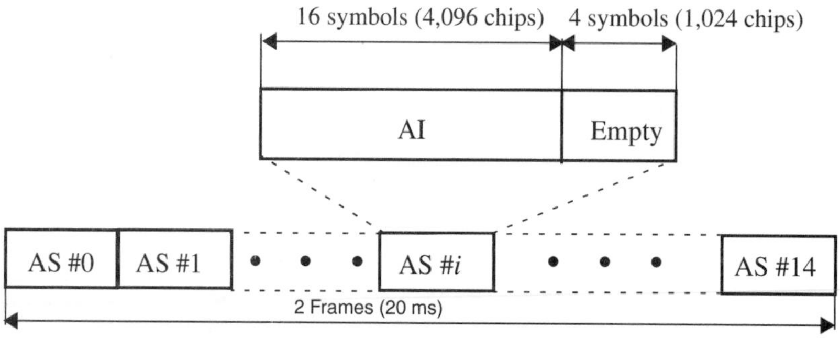

Figure 14.17 Structure of Acquisition Indicator Channel (AICH)

Page Indication Channel (PICH). PICH is a fixed-rate ($SF = 256$) channel used to carry the page indicator (PI). The PICH is always associated with an S-CCPCH, to which a PCH transport channel is mapped. The PICH frame 10 ms in duration consists of 300 bits. Of these, 288 bits are used to carry page indicators. The remaining 12 are not used. If a paging indicator in a certain frame is set to 1, it is an indication that UEs associated with this page indicator should read the corresponding frame of the associated S-CCPCH.

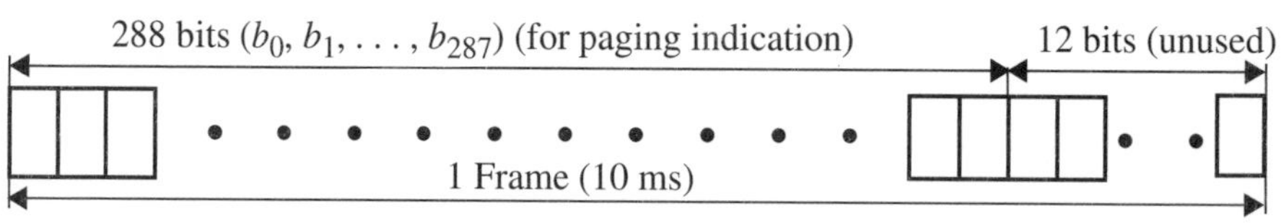

Figure 14.18 Structure of Page Indicator Channel (PICH)

Common Pilot Channel (CPICH). CPICH is an unmodulated code channel that is scrambled with the cell-specific primary scrambling code. The CPICH is used for channel estimation at the UE for the dedicated channel and provides the channel estimation reference for the common channels when they are not associated with the dedicated channels or not involved in adaptive antenna techniques. There are two types of common pilot channels: primary and secondary. The primary CPICH is always under the primary scrambling code with a fixed channelization code allocation. There is only one primary CPICH per cell or sector. The primary CPICH is used to take measurements for handover and cell selection/reselection. By adjusting the primary CPICH power level, the cell load can be adjusted between different cells. A reduction in the primary CPICH power causes part of the UEs to hand over to other cells, whereas an increase in the primary CPICH power initiates more UEs to hand over to the cell as well as makes their initial access to the network in that cell. The secondary CPICH may have channelization code of length 256 and may be under a secondary scrambling code. Typical usage of the secondary CPICH would be the operation of narrow antenna beams intended for service provision at specific "hot spots" or areas with high traffic intensity. The CPICH does not carry any higher level information, nor is there any transport channel mapping to it. The CPICH uses the spreading factor of 256. It may be sent from two antennas in the case of transmission diversity methods used in the base station. In this case, the transmission from the two antennas are separated by a simple modulation pattern on the CPICH transmitted from the diversity antenna, called diversity CPICH. The diversity pilot is used with open-loop and closed-loop transmission diversity schemes.

14.6 Logical Channels

A set of *logical channel* types is defined for different kinds of data transfer services as offered by MAC [18]. A general classification of logical channels can be organized into two groups: *control channels* (for transfer of control information) and *traffic channels* (for transfer of user information). The control channels are

- *Broadcast control channel* (BCCH)—a downlink channel for broadcasting system control information; can be divided into two types: BCCH-constant (BCCH-C) and BCCH-variable (BCCH-V). BCCH-C transmits mostly layer 3 information elements that do not change, expect for change of system information. BCCH-V transmits layer 3 information elements that change frequently and that a UE has to receive in a short time, e.g., downlink power level, uplink interference level, etc.
- *Paging control channel* (PCCH)—a downlink channel that transfers paging information.
- *Common control channel* (CCCH)—a bidirectional channel for transmitting control information between UTRAN and UEs.
- *Dedicated control channel* (DCCH)—a point-to-point bidirectional channel to transmit dedicated control information between a UE and the UTRAN.
- *Synchronization control channel* (SCCH)—a downlink channel to broadcast synchronization information; used in TDD mode only.

- *ODMA common control channel* (OCCCH)—a bidirectional channel to transmit control information between UEs.
- *ODMA dedicated control channel* (ODCCH)—a point-to-point bidirectional channel to transmit control information between UEs; established through RRC connection setup procedure.

The traffic channels are

- *Dedicated traffic channel* (DTCH)—a point-to-point channel dedicated to a single UE to transfer user information. A DTCH can exist in both uplink and downlink.
- *ODMA dedicated traffic channel* (ODTCH)—a point-to-point channel dedicated to one UE to transfer user information between UEs. An ODTCH exists in relay link.
- *Common traffic channel* (CTCH)—a point-to-multipoint unidirectional channel to transfer dedicated user information for all or a group of specified UEs.

14.7 Mapping Between Logical and Transport Channels

The following mapping between logical channels and transport channels exists [20].

- BCCH is connected to BCH
- PCCH is connected to PCH
- CCCH is connected to RACH and FACH
- SCCH is connected to SCH (TDD mode only)
- DTCH can be connected to either RACH and FACH, to RACH and DSCH, to DCH (FDD only), or to USCH (TDD only)
- CTCH can be connected to DSCH, FACH, or BCH
- DCCH can be connected to either RACH and FACH, to RACH and DSCH, to a DCH, to a CPCH (FDD only), to FAUSCH, to a CPCH (TDD only), or to a USCH (TDD only)

Figure 14.19 shows the mapping between logical channels and transport channels. It should be noted that RRC allocates radio resources on a "slow" basis. It decides and assigns the transport format for the service bearer possible in a service life cycle to meet individual users' QoS requirements. The MAC sublayer controls radio resource on a "fast" basis, in the sense that, given the transport format combination set assigned by RRC, MAC selects the appropriate transport format within an assigned transport format set for each active transport channel depending on source rate and total interference threshold level.

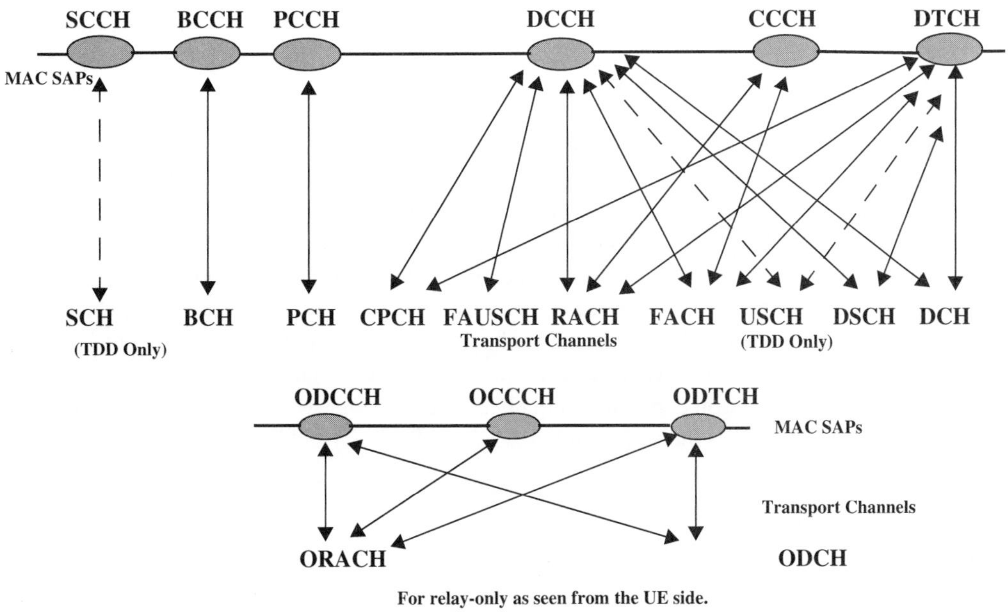

Figure 14.19 Mapping Between Logical Channels and Transport Channels

14.8 Mapping of Transport Channels onto Physical Channels

The mapping of transport channels onto physical channels is shown in Figure 14.20.

Transport Channels	Physical Channel
	Common Pilot Channel (CPICH)
BCH	Primary Common Control Physical Channel (Primary CCPCH)
FACH PCH	Secondary Common Control Physical Channel (Secondary CCPCH)
RACH	Physical Random Access Channel (PRACH)
CPCH	Physical Common Packet Channel (PCPCH)
DCH	Dedicated Physical Data Channel (DPDCH) Dedicated Physical Control Channel (DPCCH)
DSCH	Physical Downlink Shared Channel (PDSCH) Page Indication Channel (PICH) Synchronization Channel (SCH) Acquisition Indication Channel (AICH)

Figure 14.20 Mapping of Transport Channels to Physical Channels

14.9 Channelization Codes

14.9.1 Downlink Channelization Codes

The channelization codes are orthogonal variable spreading factor (OVSF) codes that preserve the orthogonality between a user's different physical channels [21]. The OVSF codes are defined using the code tree of Figure 14.21. Each level in the code tree defines channelization code of length SF. All codes within the code tree cannot be used simultaneously within one cell. A code can be used in a cell if and only if no other code on the path from specific code to the root of the tree or in the subtree below the specific code is used in the same cell. This means that the number of available channelization codes is not fixed but depends on the rate and spreading factor of each physical channel.

The channelization codes are uniquely defined as $C_{SF,k}$, where SF is the spreading factor of the code and k is the code number $0 \le k \le SF - 1$. The channelization codes are generated as follows:

$$C_{1,0} = 1$$

$$\begin{bmatrix} C_{2,0} \\ C_{2,1} \end{bmatrix} = \begin{bmatrix} C_{1,0} & C_{1,0} \\ C_{1,0} & -C_{1,0} \end{bmatrix} = \begin{bmatrix} 1 & 1 \\ 1 & -1 \end{bmatrix}$$

$$\begin{bmatrix} C_{2^{(n+1)},0} \\ C_{2^{(n+1)},1} \\ C_{2^{(n+1)},2} \\ \bullet \\ \bullet \\ C_{2^{(n+1)},2^{(n+1)}-2} \\ C_{2^{(n+1)},2^{(n+1)}-1} \end{bmatrix} = \begin{bmatrix} C_{2^n,0} & C_{2^n,0} \\ C_{2^n,0} & -C_{2^n,0} \\ C_{2^n,1} & C_{2^n,1} \\ \bullet & \bullet \\ \bullet & \bullet \\ C_{2^n,2^n-1} & C_{2^n,2^n-1} \\ C_{2^n,2^n-1} & -C_{2^n,2^n-1} \end{bmatrix} \tag{14.1}$$

The leftmost value in each channelization code word refers to the chip transmitted first in time. For the DPCCH and DPDCHs, the following conditions are applied:

- The DPCCH is always spread by code $C_{256,0}$, i.e., $C_{ch,0} = C_{256,0}$.
- When only one DPDCH is to be transmitted, DPDCH_1 is spread by $C_{ch,1} = C_{SF,k}$ where SF is the spreading factor of DPDCH, and $k = SF/4$.

- When more than one DPDCH is to be transmitted, all DPDCHs have SFs equal to 4. DPDCH is spread by the code $C_{ch,n} = C_{4,k}$ when $k = 1$ if $n \in (1, 2)$, $k = 3$ if $n \in (3, 4)$, and $k = 2$ if $n \in (5, 6)$.

For multicode transmission, each additional DPDCH may be transmitted on either I or the Q branch. For each branch, each additional DPDCH should be assigned its own channelization code. DPDCHs on different branches may share a common channelization code.

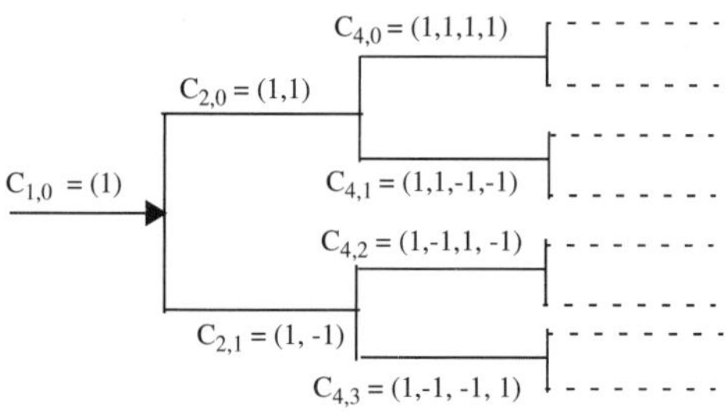

Figure 14.21 Channelization Code Tree

14.9.2 Uplink Channelization Codes

For the uplink, the channelization codes are the same type of OVSF codes. Each connection is allocated at least one uplink channelization code to be used for the DPDCH. Further uplink channelization codes may be allocated for a DPDCH. Additional uplink channelization codes may be assigned if more than one DPDCH is required

For the uplink, the restrictions on the allocation of channelization codes are valid only within one mobile station. As different mobiles use different uplink scrambling codes, the uplink channelization codes may be allocated without coordination between different connections. The uplink channelization codes are, therefore, always allocated in a predetermined order. The mobile and network only need to agree on the number of uplink channelization codes. The exact codes to be used are then implicitly given.

The channelization code for the primary CPICH is fixed to $C_{256,0}$, and the channelization code for the primary CCPCH is fixed to $C_{256,1}$.

The channelization code for the BCCH is a predefined code that is the same for all cells within the system. The channelization code(s) used for the secondary CCPCH is broadcast on the BCCH. The channelization codes for the downlink DPCH are decided by UTRAN. The mobile is informed about what downlink channelization codes to receive in the downlink access grant message that is a base station's response to an uplink random access request. The set of channelization codes may be changed during the duration of a connection, typically as a result of change in service or intercell handover. A change of downlink channelization codes is negotiated over the DCCH.

14.10 Scrambling Codes

14.10.1 Uplink Scrambling Codes

There are 2^{24} *uplink scrambling codes*. Either short or long scrambling codes can be used on the uplink. The short scrambling codes are used with MUD only. The uplink scrambling code is decided by the network. The mobile is informed in the downlink access grant message about what scrambling code to use. The scrambling code may, in rare cases, be changed during the duration of a connection. A change of uplink scrambling code is negotiated over the DCCH. The uplink scrambling generator (short or long) is initialized by a 25-bit value. One bit indicates selection of short or long codes (short = 1, long = 0). Twenty-four bits are used to generate scrambling codes (see Figure 14.22).

Figure 14.22 Initialization Code for Uplink Scrambling Code Generator

The short and long scrambling codes are generated as follows:

$$C_{scramb} = C_1(w_0 + jC'_2 w_1) \tag{14.2}$$

where w_0 and w_1 are chip rate sequences defined as repetitions of
$w_0 = \{1 \quad 1\}$
$w_1 = \{1 \quad -1\}$
C_1 is a real chip rate code and C_2' is a decimated version of the real chip rate code C_2.
With a decimated factor of 2, C_2' is given as:
$C_2'(2k) = C_2'(2k+1) = C_2(2k), \quad k = 0, 1, 2, \ldots$

The code C_2, used in generating the quadrature component of the complex spreading code, is a 16,777,232 chip shifted version of the code C_1, used in generating the in-phase component. The uplink scrambling code word has a period of one radio frame.

14.10.2 Downlink Scrambling Codes

For the *downlink scrambling codes*, a total of $2^{18} - 1 = 262{,}143$ scrambling codes can be generated. However, not all the scrambling codes are used. The scrambling codes are divided into 512 sets, each consisting of a primary scrambling code and 15 secondary scrambling codes.

The primary scrambling codes consist of scrambling codes $n = 16 \times i$, where $i = 0, 1, \ldots, 511$. The ith set of secondary scrambling codes consists of scrambling codes $16 \times i + k$, where $k = 1, 2, \ldots, 15$. There is one-to-one mapping between each primary scrambling code and 15 secondary scrambling codes in a set such that the ith primary scrambling code corresponds to the ith set of scrambling codes. Thus, scrambling codes $m = 0, 1 \ldots, 8{,}191$ are used.

The set of primary scrambling codes is further divided into 64 scrambling code groups, each consisting of eight primary scrambling codes. The jth scrambling code group consists of primary scrambling codes $16 \times 8 \times j + 16 \times k$, where $j = 0, 1, \ldots, 63$ and $k = 0, 1, \ldots, 7$.

Each cell is allocated one and only one primary scrambling code. The primary CCPCH is always transmitted using the primary scrambling code. The other downlink physical channels can be transmitted with either the primary scrambling code or a secondary scrambling code from the set associated with the primary scrambling code of the cell. The mixture of primary and secondary scrambling codes for one coded composite transport channel (CC-Tr-CH) is allowed.

The scrambling code sequences are formed by combining two real sequences into a complex sequence. Each of the two real sequences are constructed as the positionwise modulo-2 sum of 38,400 chip segments of two binary m-sequences generated by two generator polynomials of degree 18. The resulting sequence constitutes segments of a set of Gold sequences [21]. The scrambling codes are repeated for every 10-ms radio frame. Table 14.7 summarizes the channelization and scrambling codes.

Table 14.7 Channelization and Scrambling Codes

	Channelization Code	Scrambling Code
Use	**Uplink:** Separation of DPDCH and DPCCH from same UE **Downlink:** Separation of DL connections to different users within one cell	**Uplink:** Separation of UEs **Downlink:** Separation of cells
Length	**Uplink:** 4–256 **Downlink:** 4–512	**Uplink:** 10 ms 38,400 chips (long) or 66.7 ms 256 chips (short) (used with MUD) **Downlink:** 10 ms 38,400 chips
Number of codes	Number of codes under one scrambling code = spreading factor	**Uplink:** several million **Downlink:** 512
Code family	Orthogonal variable spreading factor (OVSF)	**Long codes:** 10 ms **Gold codes** **Short codes:** extended S(2) code family
Spreading	Yes, increases transmission bandwidth	No, does not affect transmission bandwidth

14.10.3 Random Access Codes

For the scrambling code of the *preamble part*, the code generating method is the same as for the real part of long codes on dedicated channels. Only the first 4,096 chips of the code are used for preamble spreading with the chip rate of 3.84 Mcps. The long code C_1 for the in-phase component is used directly on both in-phase and quadrature branches without offset between branches. The preamble scrambling code is defined as the positionwise modulo-2 sum of 4,096 chips segments of two binary *m*-sequences generated by means of two generator polynomials of degree 25.

The preamble part consists of 256 repetitions of a length 16 signature, $<P_0, P_1, P_2, \ldots, P_{15}>$. Before scrambling, the preamble is therefore

$$P_0, P_1, P_2, \ldots, P_{15}, P_0, P_1, P_2, \ldots, P_{15}, P_0, P_1, P_2, \ldots, P_{15} \tag{14.3}$$

The signature, s, is from the set of 16 Hadamard codes of length 16. The preamble signature, s, $1 \le s \le 16$, points to one of the 16 nodes in the code tree that corresponds to channelization codes of length 16. The subtree below the specified node is used for spreading of the *message part*. The *control part* is spread with channelization code $C_{ch,c}$ of the spreading factor 256 in the lowest branch of the subtree, i.e.,$C_{ch,c} = C_{256,m}$ where $m = 16(s - 1) + 15$. The spread control part is mapped to the Q-branch, similar to the DPCCH for the dedicated channel.

The data part uses any of the channelization codes from spreading factor 32 to 256 in the uppermost branch of the subtree. The data part is spread by channelization code $C_{ch,d}$ where $C_{ch,d} = C_{SF,m}$, and SF is the spreading factor used for the data part and $m = SF \times (s - 1) / 16$.

The message part is also subject to scrambling with a 10-ms complex code. The scrambling is cell-specific and has one-to-one correspondence to the scrambling code used for the preamble part.

14.11 Spreading/Modulation

14.11.1 Uplink Dedicated Physical Channels

Figure 14.23 shows the uplink spreading of DPCCH and DPDCH. The binary DPCCH and DPDCH are represented by real-valued sequences, i.e., the binary value 0 is mapped to the real value +1, whereas the binary value 1 is mapped to the real value –1. The DPCCH is spread to the chip rate by the channelization code $C_{ch,0}$, whereas the nth DPDCH$_n$ is spread to the chip rate by the channelization code $C_{ch,n}$. One DPCCH and up to six parallel DPDCHs can be transmitted simultaneously, i.e., $0 \le n \le 6$. After channelization, the real-valued spread signals are weighted by gain factor β_c for DPCCH and β_d for all DPDCHs. After the weighting, the stream of real-valued chips on the I and Q branches are then summed and treated as a complex-valued stream of chips. The complex-valued signal is scrambled by complex-valued scrambling code C. After pulse-shaping, QPSK modulation is performed (see Figure 14.24).

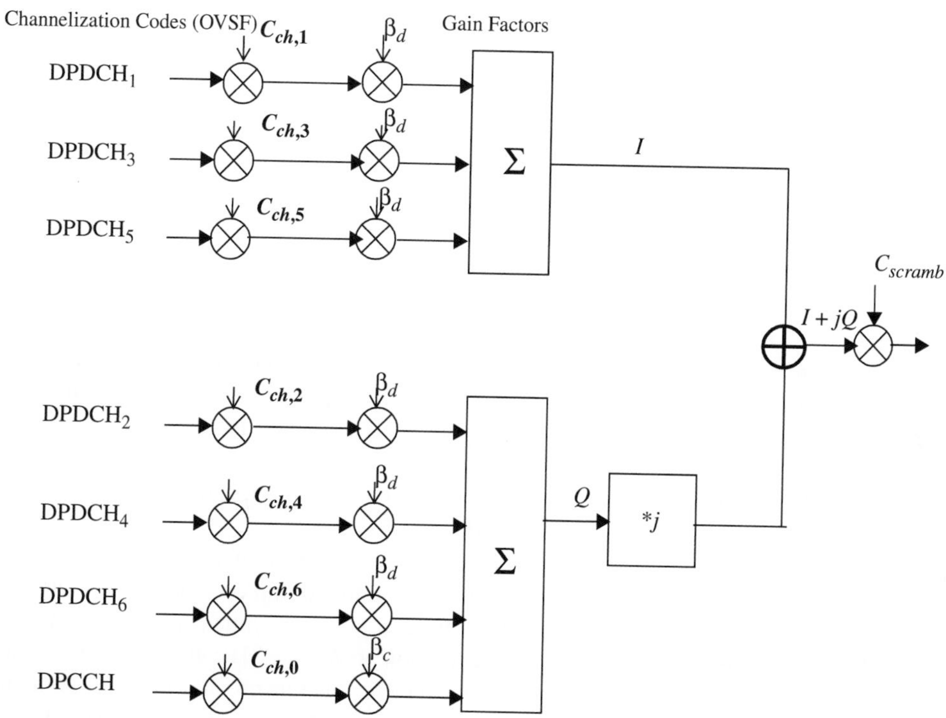

Figure 14.23 Spreading for Uplink DPCCH and DPDCH

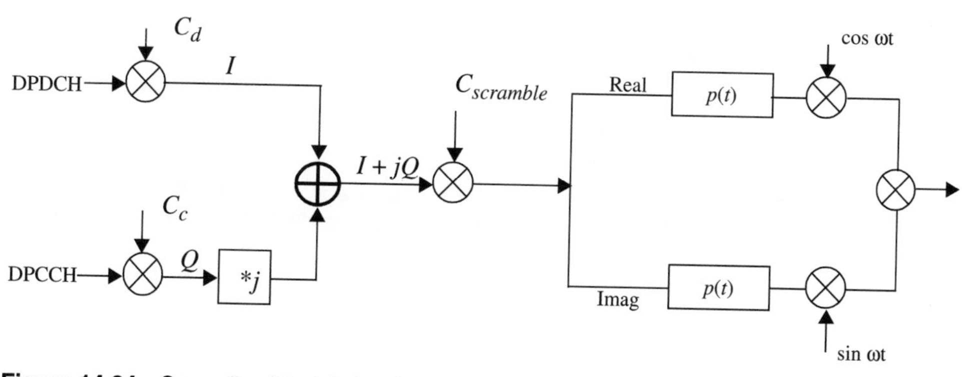

Figure 14.24 Spreading/Modulation for Uplink DPCCH and DPDCH

14.11.2 Physical Random Access Channel

The spreading and modulation of the message part of the physical random access channel (PRACH) is basically the same as for the uplink dedicated physical channels, where the uplink DPDCH and the uplink DPCCH are replaced by the data part and control part respectively. The scrambling code for the message part is selected based on the preamble code.

14.11.3 Downlink Dedicated Physical Channel

Figures 14.25 and 14.26 show the spreading for the downlink DPCH. Data modulation is QPSK where each pair of two bits are serial-to-parallel converted and mapped to the I and Q branch, respectively. The I and Q branches are then spread to the chip rate with same channelization code C_{ch} (real spreading) and then scrambled by the scrambling code $C_{scramble}$ (complex scrambling).

Spreading/modulation of CPICH, secondary CCPCH, PICH, and AICH is done in an identical manner as for the downlink DPCH.

Spreading/modulation of the primary CCPCH is performed in the same way as for the downlink DPCH, except that the primary CCPCH is time-multiplexed after spreading. Primary SCH and secondary SCH are code-multiplexed and transmitted simultaneously during the first 256 chips of each slot. The transmission power of SCH can be adjusted by gain factor $G_{P\text{-}SCH}$ and $G_{S\text{-}SCH}$, respectively, independent of transmission power of primary CCPCH. The SCH is nonorthogonal to the other downlink physical channels (see Figure 14.27).

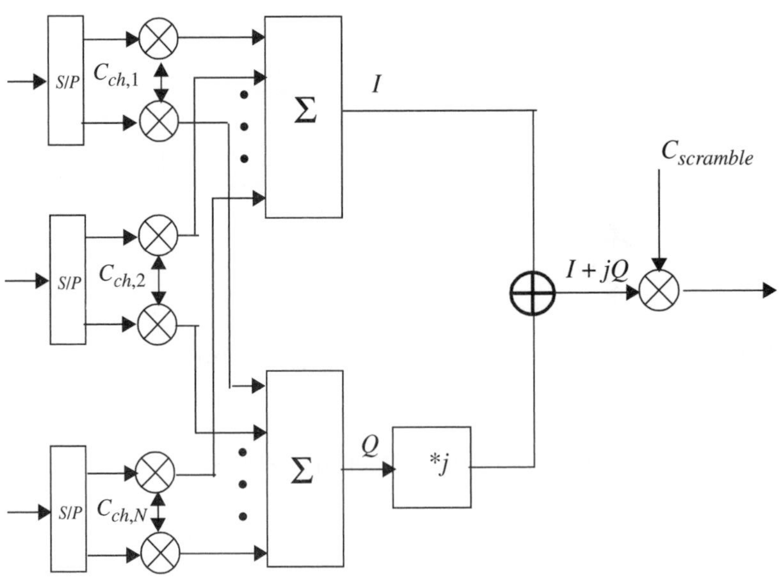

Figure 14.25 Spreading for Downlink DPCH

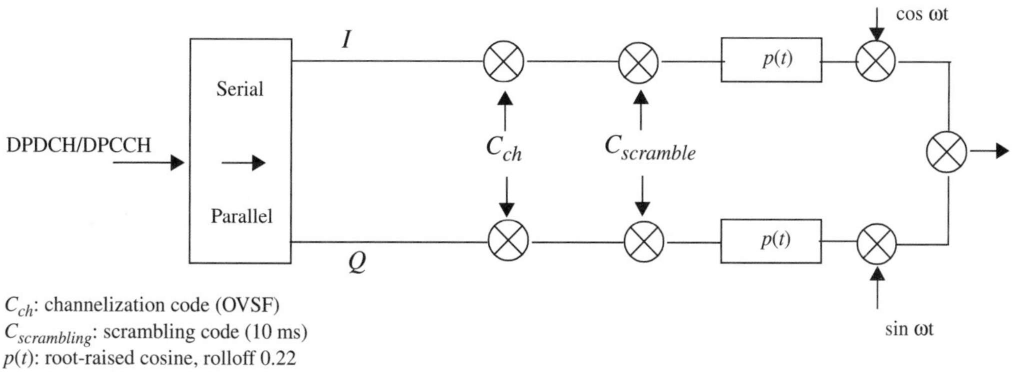

Figure 14.26 Spreading/Modulation for Downlink Dedicated Physical Channel (DPCH)

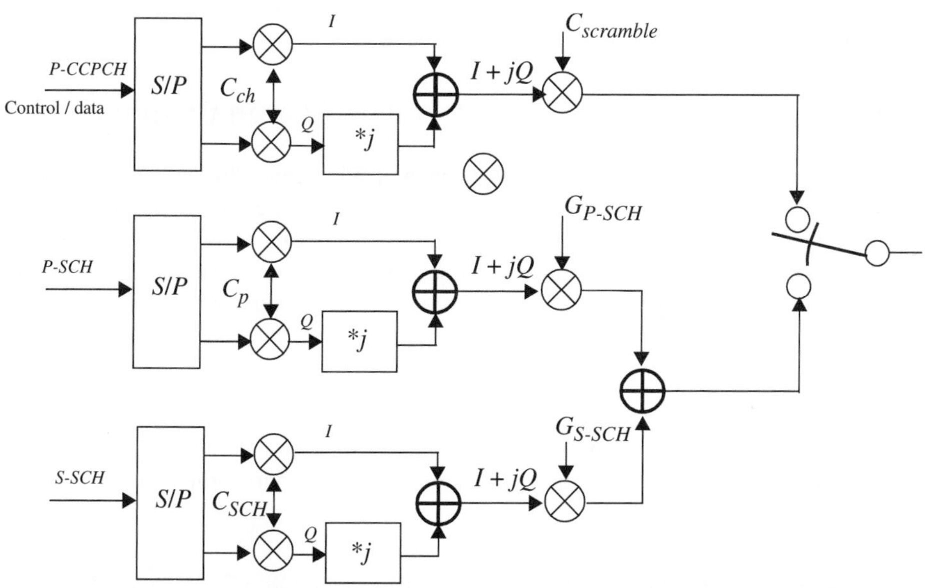

Figure 14.27 Spreading and Modulation for SCH and P-CCPCH

14.12 MAC and RLC Protocols

14.12.1 MAC Multiplexing

In addition to the design of the physical layer, there has been significant effort devoted to the design of the higher layers of UTRA. MAC functions include control of transport format, priority handling between voice/data services of a mobile, priority handling among all mobile common channel message scheduling, and mobile identification. The RLC and MAC sublayers are responsible for efficiently transferring data of real-time and non-real-time services. The transfer of non-real-time data includes the possibility of low-level automatic repeat request (ARQ), offering higher protocol layers reliable data transfer. The service multiplexing in the physical layer means that different services from a mobile may possibly use one channelization code that is handled by the physical layer (A UMTS/W-CDMA feature that differs from cdma2000). UMTS MAC controls the service multiplexing but does not carry out the multiplexing of data streams originating from different services. For example, it allows several upper-layer services (RLC instances) to be mapped onto the same transport channel. The MAC layer provides data transfer services on logical channels.

The MAC protocol is part of the link layer. It resolves contention between users accessing the same physical resource and manages the packet access procedure. Since the 3G systems offer a multitude of services to customers at widely varying QoS requirements, the MAC offers capabilities to manage the access demands of different users and different service classes. This is performed using reservation and priority schemes. Services with delay constraints can use a reservation scheme to reserve capacity to guarantee the QoS. Priority schemes can be used to schedule the requests from different services.

14.12.2 MAC Multiplexing for Downlink Shared Channel

In UMTS, there appears to be an orthogonal channelization code shortage problem on the downlink (there is no such problem on the uplink; as scrambling code is mobile-specific, each mobile manages its orthogonal channelization code tree). The problem will become worse if each packet data user is allocated one permanent channelization code. DSCH, which allows multiple mobiles to share the same code, is used to resolve this problem. The benefits of DSCH are

- Minimize the impact of L1 configuration
- Simplify orthogonal spreading code allocation
- Utilize the existing TFCI definition by maintaining a transport format combination set for all the users using DSCH

Figure 14.28 shows a DSCH time-multiplexing scheme [18]. The users on the DSCH transport channel are multiplexed at the MAC layer according to the selected transport format. The multiplexed users (e.g., user 1, user 2, user 3, and so on) have a common transport format combination set that defines the valid set of transport formats for the DSCH transport channel. With every addition and subtraction of users on the DSCH transport channel, the DSCH combination set is updated. The user-multiplexed transport block set (e.g., in the first 10-ms transmission interval there are two user 1 transport blocks U1_TB, one user 2 transport block U2_TB, and two user 3 transport blocks U3_TB) is delivered to the common coding unit in the physical channel, which is configured by the information contained in the transport format selected by the TFCI. A similar process takes place in all other 10-ms transmission intervals. At the mobile MAC sublayer, the transport block set from each user is demultiplexed accordingly. From the knowledge of the number of transport blocks and transport block size available from the transport format, the transport block set belonging to the mobile can be extracted.

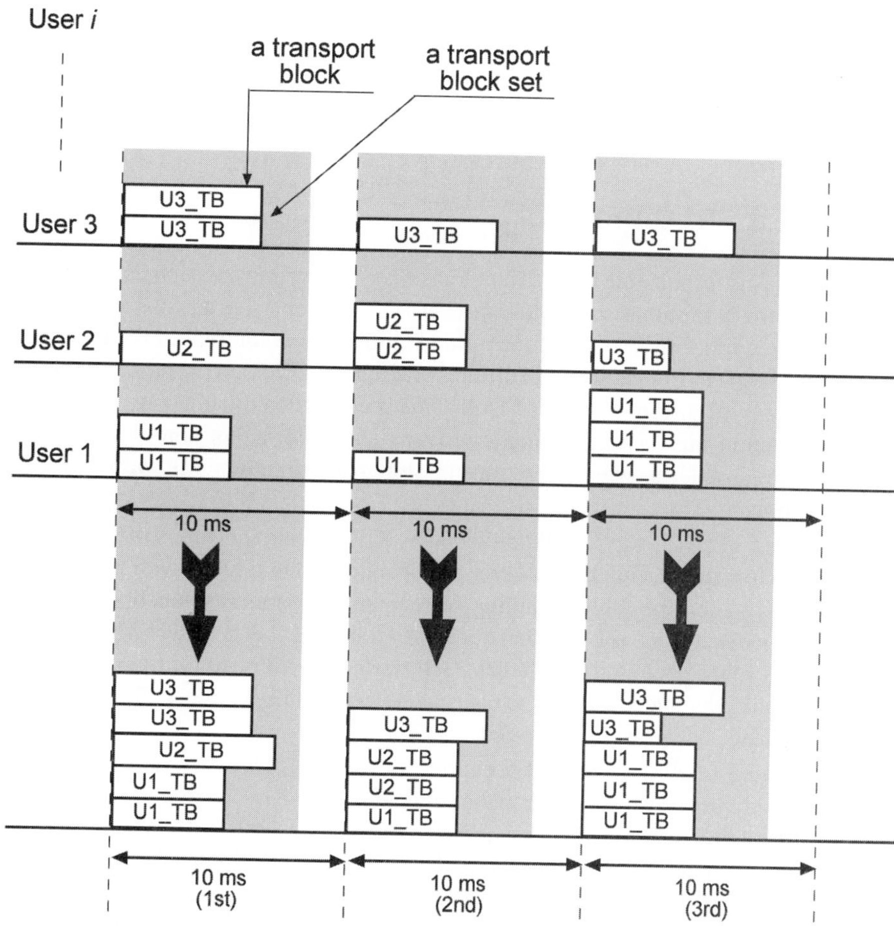

Figure 14.28 DSCH Time Multiplexing

The advantages of the DSCH are clear. First, all mobiles using the transport channel share the same channelization code. Therefore, the downlink channelization code shortage problem is relieved. Second, it simplifies channelization code tree management in the downlink (for UMTS, the code tree management is very intensive since it requires management of the code set every 10 ms for each user). Third, multicast services are supported. Since all users on the DSCH use the same code, multicast service is well supported. Fourth, it is easy to prioritize user data through the selection of transport format by the MAC sublayer. Finally, power control can be supported on the DSCH.

W-CDMA is based on 5 MHz with a basic chip rate of 3.84 Mcps. The frame duration is 10 ms, allowing for low-delay speech and fast control messages. Each radio frame is divided into 15 time slots 0.667 ms in length, corresponding to one power-control period. On the downlink, L2 dedicated data is time-multiplexed with L1 control information within each time slot. The L1 control information contains known pilot bits for uplink closed-loop power control, and a TFCI. The number of bits per downlink slot is not fixed but may vary from 10–1,280 corresponding to 3.2–2,048 kbps data rate.

14.12.3 RLC Protocol

RLC [19] is responsible for efficient transmission or retransmission under variable bit rate. Therefore, a minimal segmentation overhead, a simple retransmission protocol, and an optimized transmission or retransmission unit size are required for the RLC design in different radio environments (e.g., different fading scenarios). RLC protocol is configured by L3 to provide different levels of QoS. This is controlled by adjusting the maximum number of retransmissions according to service delay criteria. For non-real-time bearers, RLC provides low-level selective retransmission ARQ functionality with CRC-based error detection.

Three types of services are provided by RLC to higher layers. *Transparent mode* offers service for transmitting higher-layer protocol data units (PDUs) without adding any protocol information (possibly including segmentation/reassembly functionality). *Unacknowledged mode* offers service for transmitting higher layer PDUs without guaranteeing delivery to the peer entity. *Acknowledged mode* offers service for transmitting higher layer PDUs and guarantees delivery to the peer entity. For this service, both in-sequence and out-of-sequence delivery are supported.

Figure 14.29 shows the RLC entities for UTRAN. Three types of service access points (SAPs) are used corresponding to the services provided by the RLC: *transparent mode SAP*, *unacknowledged mode SAP*, and *acknowledged mode SAP*. RLC control SAP may be used by RRC to request status report (e.g., buffer status). Transparent mode entity controls the data flow for BCCH, PCCH, and DTCH logical channels. The entity includes segmentation/reassembly function and transmitter/receiver buffer. Both unacknowledged mode and acknowledged mode entity include segmentation/reassembly function, RLC header addition/removal function, and transmitter/receiver buffer. In addition, acknowledged mode entity includes retransmission buffer. The retransmission buffer receives acknowledgment from the receiving side. The acknowledgment is used to indicate retransmission of RLC PDUs and to delete a PDU from a retransmission buffer. A number of RLC protocols were proposed for UMTS. A modified service-specific connection-oriented protocol (SSCOP) is used [19].

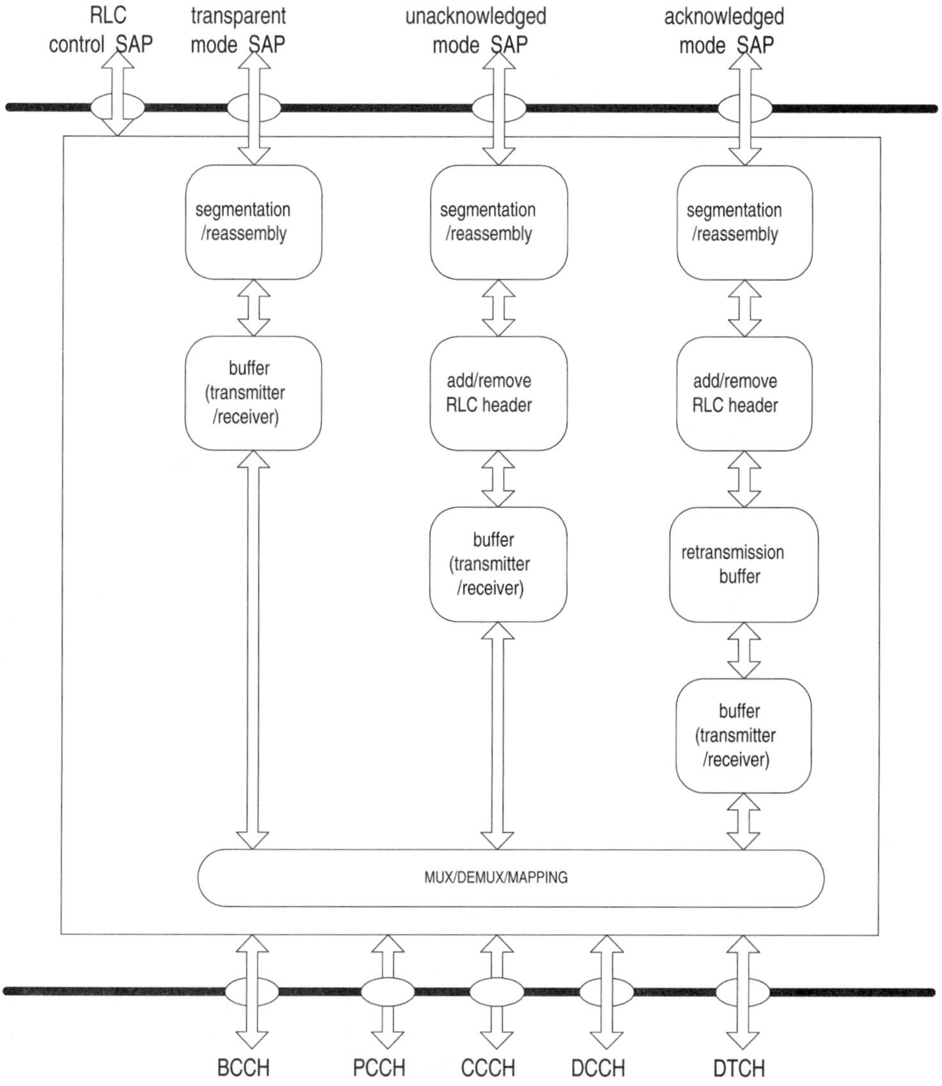

Figure 14.29 RLC Protocol Model

14.13 Transport Channels

Parallel transport channels (TrCh-1 and TrCh-M) are separately channel-coded and interleaved (see Figure 14.30). The code transport channels are then time-multiplexed into a coded composite transport channel (CC-Tr-Ch). Interframe (10 ms) interleaving is carried out after transport channel multiplexing. Different coding and interleaving schemes (see Figure 14.31) can be applied to a transport channel depending on the specific requirements in terms of error rates, delay, and so forth. These include the following:

- A rate of 1/3 convolutional coding is typically used for low-delay services such as voice with moderate error rate requirements (BER ~ 10^{-3}).
- A concatenation of rate 1/3 convolutional coding and Reed-Solomon coding plus interleaving can be used for high-quality service (BER ~ 10^{-6}).
- Turbocodes are used for high-rate quality services.

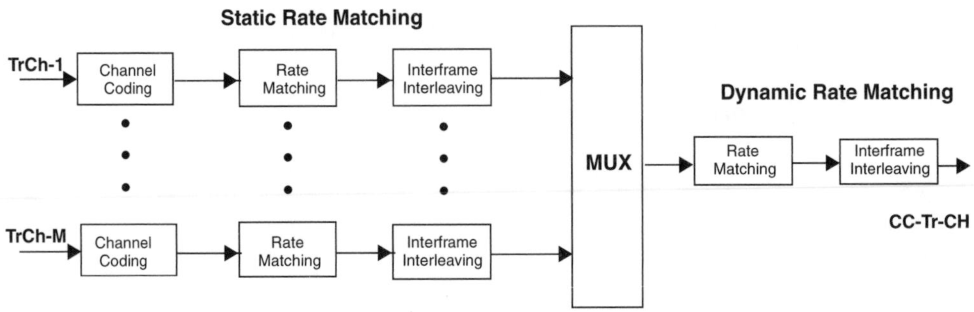

Figure 14.30 Parallel Transport Channels

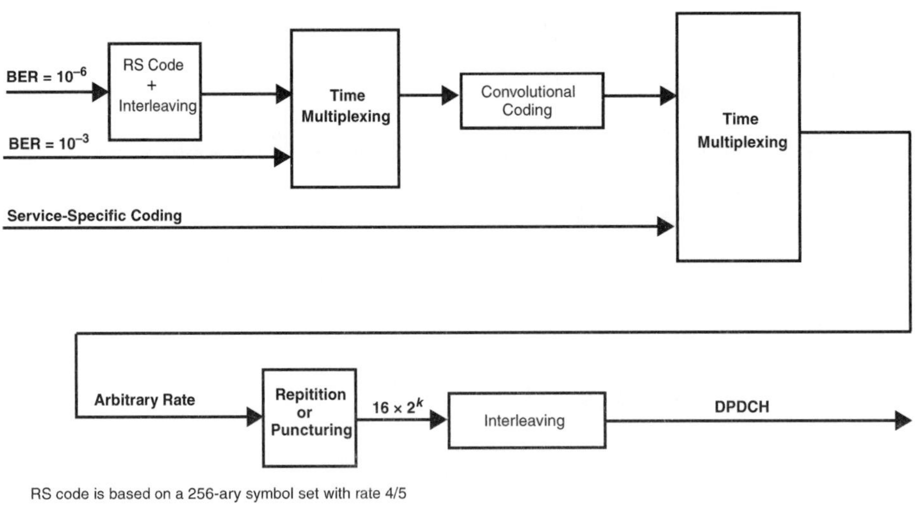

Figure 14.31 Transport Channel Coding and Multiplexing

14.14 Rate Matching

Rate matching is applied to match the bit rate of the CC-Tr-Ch to one of the limited set of bit rates of the uplink or downlink physical channel. Two different rate matching steps are used: *static rate matching* and *dynamic rate matching*. Static rate matching is carried out with the addition, removal, or redefinition of a transport channel (i.e., on a very slow basis). Static rate matching is applied after channel coding and uses code puncturing to adjust the channel coding rate of each transport channel so that maximum bit rate of the CC-Tr-CH is matched to the bit rate of the physical channel. Static rate matching is applied on both uplink and downlink. On the downlink, the static rate matching is used to, if possible, reduce the CC-Tr-CH rate to the closest lower physical channel rate (next higher spreading factor), thus avoiding the overallocation of orthogonal codes on the downlink and reducing the risk of a code-limited downlink capacity. Static rate matching is distributed between parallel transport channels in such a way that the transport channel fulfill their quality requirements at approximately the same channel SIR; that is, static rate matching also performs SIR matching.

Dynamic rate matching is carried out once every 10-ms radio frame (i.e., very rapidly). Dynamic rate matching is applied after transport channel multiplexing and uses symbol repetition so that the instantaneous bit rate of the CC-Tr-CH is exactly matched to the bit rate of the physical channel. Dynamic rate matching is only applied to the uplink. On the downlink, discontinuous transmission within each slot is used when the instantaneous rate of the CC-Tr-CH does not exactly match the bit rate of the physical channel.

It should be noted that although transport channel coding and multiplexing are carried out by the physical layer, the process is fully controlled by the radio resource controller in terms of, for example, choosing the appropriate coding scheme, interleaving parameters, and rate matching parameters.

14.15 Uplink and Downlink Multiplexing

The services in the uplink direction are multiplexed dynamically to achieve a continuous data stream with the exception of zero rate [22]. After receiving a transport block from higher layers, a CRC is attached for error checking of the transport block at the receiving end (see Figure 14.32). The CRC length has four different values: 8, 12, 16, and 24 bits. After adding CRC bits, the transport blocks are either concatenated together or segmented to different coding blocks. This depends on whether the transport block fits the available code block size as defined for the channel coding method. If the transport block with CRC attached does not fit into the maximum available code block, it is divided into several code blocks.

Channel coding is performed for forward error correcting on the coding blocks after concatenation or segmentation. For some services or bit classes, no channel coding is used. For example, the adaptive multiple rate (AMR) coding class C bits are sent without channel coding. In such a case there is no limitation on the coding block size.

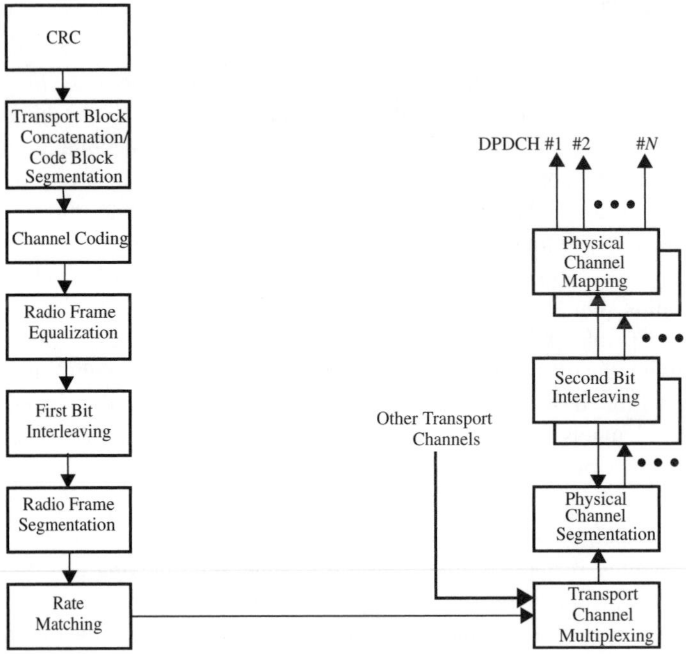

Figure 14.32 Uplink Multiplexing and Channel Coding

Radio frame equalization is used to ensure that data can be divided into equal-sized blocks when transmitting over more than a single 10-ms radio frame. This is achieved by padding the necessary number of zero bits so the data can be placed in equal-sized blocks per frame.

The first bit interleaving or interframe interleaving is used when the delay budget allows more than 10 ms of interleaving. Rate matching is used to match the number of bits to be transmitted to the number available on a single frame. This is achieved either by puncturing or by repetition. Repetition is preferred in the uplink direction. The uplink rate matching is a dynamic operation that may vary on a frame-by-frame basis.

The different transport channels are multiplexed together. Each transport channel provides data in 10-ms blocks. In case more than one physical channel (spreading code) is used, physical channel segmentation is used. This operation divides the data evenly on the available spreading codes. The use of serial multiplexing also means that with multicode transmission, the lower rates can be implemented by sending fewer codes with the full rate.

The second interleaving is applied separately for each physical channel in case more than a single code channel is used. The output bits of the second interleaver are mapped on the physical channels.

The multiplexing operation in the downlink (see Figure 14.33) is similar to that in the uplink but there are some functions that are performed differently. As in the uplink, the bit interleaving is implemented in two parts, covering both intraframe and interframe interleaving. The differences are in the order in which rate matching and segmentation functions are performed. The discontinuous transmission (DTX) indication insertion point establishes whether fixed or flexible bit positions are used. The DTX indication bits are not transmitted over the air. They are inserted to inform the transmitter at which bit positions the transmission should be turned off. They are not needed in the uplink, where the rate matching is done in a more dynamic fashion, always filling the frame when there is something to transmit on the DPDCH. The use of fixed positions means that for a given transport channel, the same symbols are always used. With flexible positions the channel bits unused by one service may be utilized by another service. The concept of fixed versus flexible positions in the downlink is shown in Figure 14.34.

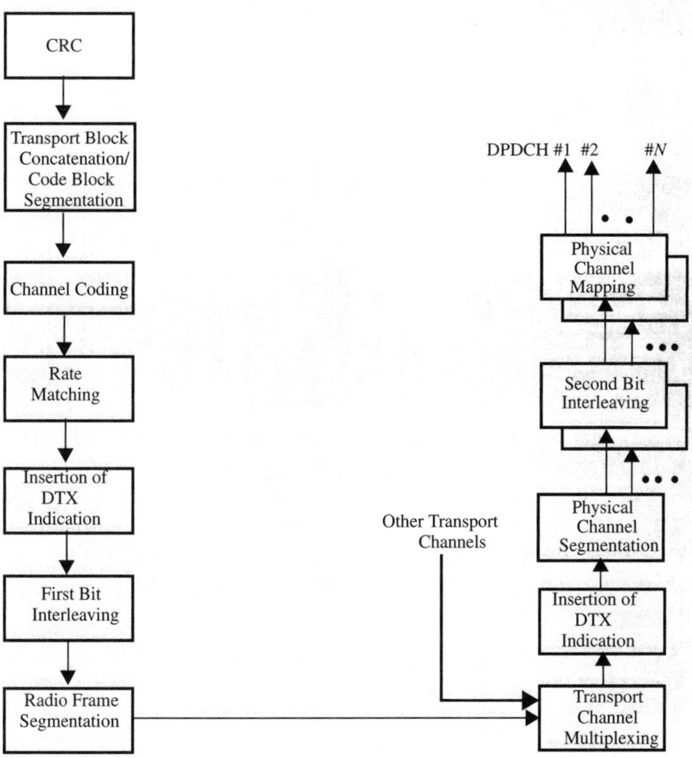

Figure 14.33 Downlink Multiplexing and Channel Coding

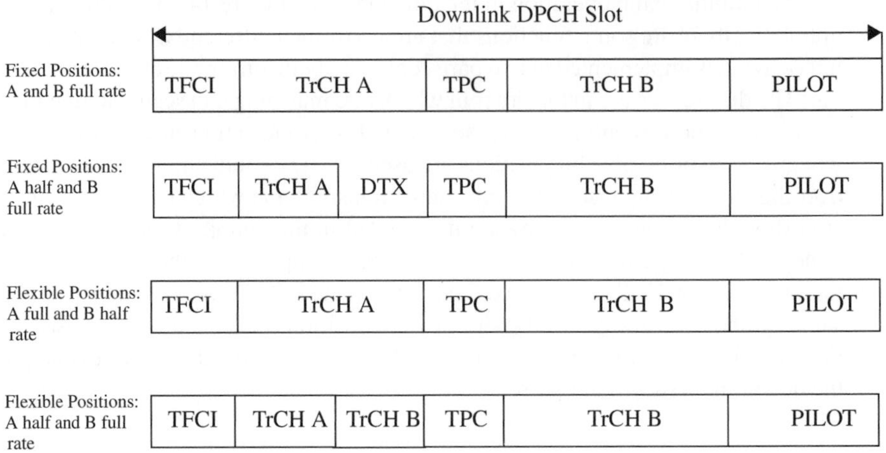

Figure 14.34 Flexible and Fixed Transport Channel Slot Positions in the Downlink

14.16 Frame Controller Header

The frame controller header (FCH) (see Figure 14.35) indicates the transmission rate for the current rate on the DPDCH. The coding for the 6-bit FCH is mapped to bidirectional Walsh functions of length 2 that represent the 64 different values for FCH. The FCH data is interleaved and multiplexed over the entire DPCCH frame. For multiple variable rate services, the FCH indicates the rate or number of bits per frame for each service. The actual code rate for FCH depends on the number of different combinations of transmission rates to be supported with services. In the case when the supported number of different service combinations is less than 64, a subset of the maximum of 64 code word is used. The supported number of different data rates or their combinations is defined with higher layer protocol during initial service negotiation at connection setup and may be adjusted with service negotiation when new service is added or the properties of the current service are changed during connection.

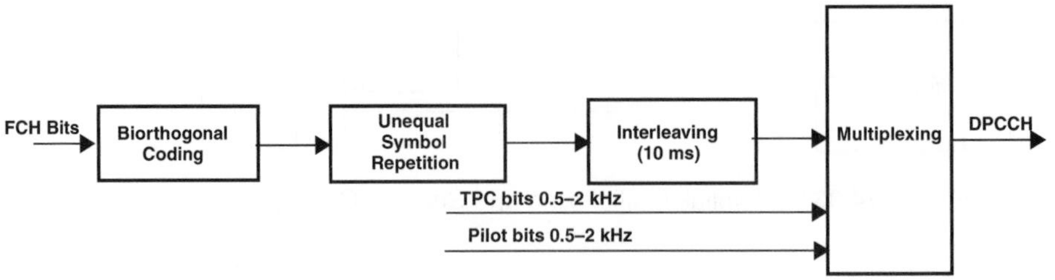

Figure 14.35 FCH Coding and Multiplexing with Other L1 Control Data on the DPCCH

14.17 Power Control

The uplink closed loop power control adjusts the mobile transmit power to keep received uplink SIR at a given SIR target. The base station estimates the received DPCCH power after RAKE combining of the connection to be power controlled. Simultaneously, the base station estimates the total uplink received interference in the current frequency band. The base station generates TPC commands according the following rule:

$$SIR_{est} > SIR_{target,UL} \rightarrow \text{TPC command = "down"}$$
$$SIR_{est} < SIR_{target,UL} \rightarrow \text{TPC command = "up"}$$

Upon receiving a TPC command, the mobile adjusts the transmit power of both the DPCCH and DPDCH in the given direction with step of Δ_{TPC} dB. The step size Δ_{TPC} is a parameter that may differ between different cells. In the case of soft handover, the mobile should adjust the power with the largest step in the "down" direction ordered by a TPC command received from each base station in the active set.

The outer loop power control adjusts the SIR target used by the closed loop power control. The SIR target is independently adjusted for each connection based on the estimated quality of the connection. In addition, the power offset between the uplink DPCCH and DPDCH may be adjusted. The quality estimation differs for different service combinations. Typically, a combination of estimated BER and FER is used.

The open loop power control is used to adjust the transmit power of the physical random access channel (RACH). Before the transmission of a random-access frame, the mobile measures the received power of the downlink primary CCPCH over a sufficiently long period of time to remove any effect of the nonreciprocal multipath fading. From power estimate and knowledge of primary CCPCH transmit power (broadcast on BCH), the downlink path-loss including shadow fading is determined. From this path loss estimate and the knowledge of the uplink interference level and required SIR, the transmit power of the RACH is established. The uplink interference level as well as the required SIR are broadcasted on the BCH.

The downlink closed loop power control adjusts the base station transmit power to keep the received downlink SIR at a given SIR target. The mobile estimates the received DPCCH power after RAKE combining of the connection to be power controlled. The mobile also estimates the total downlink received interference in the frequency band. The mobile generates a TPC command according to the following rule:

$$SIR_{est} > SIR_{target,DL} \rightarrow \text{TPC Command = "down"}$$
$$SIR_{est} < SIR_{target,DL} \rightarrow \text{TPC Command = "up"}$$

Upon receiving of a TPC command, the base station adjusts the transmit power in the given direction Δ_{TPC} dB. The step size Δ_{TPC} is a parameter that may differ between different cells.

The outer loop adjusts the SIR target used by the closed loop power control. The SIR target is independently adjusted for each connection based on the estimated quality of the connection. In addition, the power offset between the downlink DPCCH and DPDCH may be adjusted. The quality estimate differs for different service combinations. Typically, a combination of estimated BER and FER is used.

14.18 UTRAN Procedures

14.18.1 Initial Cell Search Procedure

Initial cell search [23] is carried out in three steps. During the first step (see Figure 14.36) the mobile uses the primary SCH to acquire slot synchronization to the strongest base station. This is done with a single matched filter matched to primary synchronization code (PSC) C_p, which is the same for all base stations. The output of the matched filter will have peaks for each ray of each base station within the range of the mobile. Detecting the position of the strongest peak gives the timing of the strongest base station modulated slot length.

During the second step, the mobile uses the secondary SCH to find frame synchronization and identify the code group of the base station found in the first step. This is done by correlating the received signal at the positions of secondary synchronization code (SSC) with all possible 15 SSCs. The correlation with 15 different SSCs gives 15 different demodulated sequences. To achieve frame synchronization, the 15 demodulated sequences are correlated with the 15 different cyclic shifts of secondary SCH modulation sequence, giving a total 225 correlation values. By identifying the code/shift pair that gives the maximum correlation value, the code group as well as frame synchronization is achieved.

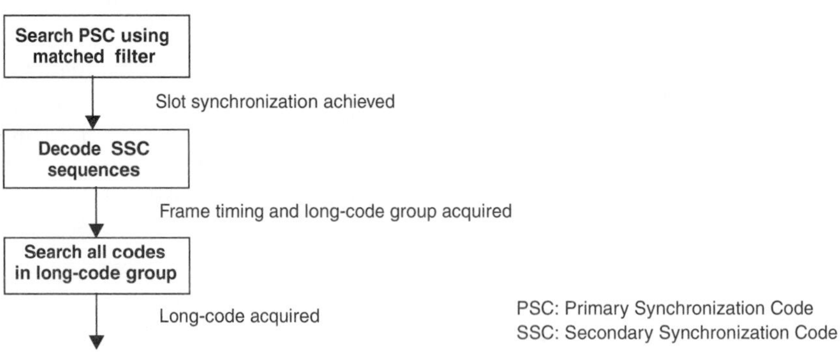

Figure 14.36 Initial Cell Search

During the third step, the mobile finds the exact scrambling code used by the determined base station. The scrambling code is identified through symbol-by-symbol correlation over the primary CCPCH, with all scrambling codes within the code group identified in step 2. After the scrambling code has been identified, the primary CCPCH can be detected, superframe synchronization can be acquired, and system- and cell-specific BCCH information can be obtained. Both primary and secondary synchronization code are unmodulated orthogonal Gold codes 256 chips in length.

14.18.2 Paging Channel Procedure

The paging channel (PCH) is operated with PICH to provide UEs with efficient sleep mode operation. The paging indicators use a channelization code 256 chips in length. The paging indicators occur once per slot on the corresponding physical channel, PICH. A UE, once registered to a network, is allocated a paging group. For the paging group there are paging indicators (PIs) that appear periodically on PICH when there are paging messages for any of the UEs belonging to that paging group. Once the PI has been detected, the UE decodes the next PCH frame transmitted on the secondary CCPCH to find out if there was a paging message intended for it. For detection of the PICH, the UE needs to obtain the phase reference from the CPICH, as with the AICH. The PICH needs to be heard by all UEs in the cell and therefore is sent at higher power without power control.

14.18.3 Acquisition Indication Channel Procedure

Along with the RACH, the AICH is used by the base station to indicate the reception of the random access channel signature sequence. The AICH uses the identical signature sequence as RACH on one of the downlink channelization codes of the base station to which RACH belongs. After the base station has detected the preamble with a random access attempt, the same signature sequence used on the preamble is sent back on AICH. The AICH uses a SF of 256 and 16 symbols as signature sequence. For detection of AICH, the UE needs the phase reference with respect to the CPICH. Since the AICH needs to be heard by all UEs, typically it is sent at high power without power control. The AICH is not visible to higher layers, but it is controlled directly by the physical layer in the base station.

14.18.4 Random Access Channel Procedure

The random access channel (RACH) is typically used for signaling purposes, to register the UE with the network after power-on, to perform a location update after moving from one location to another, or to initiate a call.

The UE decodes the BCH to find out the available RACH subchannels and their scrambling codes and sequence signatures. The UE randomly selects one of the RACH sub-channels from the group its access class allows it to use. The sequence signature is also selected randomly from the available signatures. The UE measures downlink power level and sets the initial power of the RACH with the proper margin according to the open loop power control accuracy. A 1-ms RACH preamble is sent with the selected sequence signature. The UE decodes the AICH to find out whether the base station has detected the preamble. In case no AICH is detected within a certain time, the UE increases the preamble transmission power by a step given by the base station, in multiples of 1 dB. The preamble is retransmitted in the next available access slot. When an AICH transmission is detected from the base station, the UE transmits the 10-ms message part of the RACH transmission.

14.18.5 Interfrequency Handover

Interfrequency handover typically occurs in the following situations:

- Handover between cells to which different numbers of carriers have been allocated, e.g., due to different capacity requirements (hot-spot scenarios)
- Handover between cells of different overlapping orthogonal cell layers using different carrier frequencies
- Handover between different operators/systems using different carrier frequencies, including handover to GSM

A key requirement for support of seamless interfrequency handover is the possibility for the mobile to carry out cell search on a carrier frequency different from the current one, without affecting the ordinary data flow. W-CDMA supports interfrequency cell search using either *dual receiver approach* or *slotted downlink transmission* approach. For a mobile with receiver diversity, one of the receiver branches is temporarily assigned to carry out reception on a different carrier. This is referred to as dual receiver approach.

With slotted downlink transmission, the information normally transmitted during a 10-ms frame is compressed in time, either by code puncturing or by reducing the SF by a factor of 2. In this way, a time period of up to 5 ms is created, during which the mobile receiver is idle and can be used for intrafrequency measurements.

14.18.6 Transmit Diversity Procedure

The downlink DPCH can use either the open loop or closed loop transmit diversity procedure to improve link performance. This procedure is not required from the network side but it is mandatory in the UE. In the open loop transmit diversity procedure, the coded information is sent from two antennas. The method in the 3GPP specification is called the space time block coding based transmit diversity (STTD).

In the case of the closed loop transmit diversity procedure, the base station uses two antennas to transmit user information. The use of two antennas is based on feedback from the UE, transmitted in the uplink DPCCH. The closed loop transmit diversity has two modes of operation. In mode 1, the UE feedback commands control the phase adjustments to maximize the power received by the UE. The base station maintains the phase with antenna 1 and adjusts the phase of antenna 2 based on the sliding averaging over two consecutive feedback commands. With this method, four different phase settings are applied to antenna 2.

In mode 2 operation, the amplitude is adjusted in addition to the phase adjustment. The same signaling rate is used, but the command is spread over four bits in four uplink DPCCH slots, with a single bit for amplitude and three bits for phase adjustment. This method gives eight different phase and two different amplitude combinations, giving a total of 16 combinations for transmission from the base station. The amplitude values are defined to be 0.2 and 0.8 and the phase values for the antenna phase offsets are evenly distributed from –135 to 180 degrees. In this mode, the last three slots of the frame contain only phase information; amplitude information is taken from the previous four slots. This allows the command period to work with 15 slots as with mode 1. The average at the frame boundary is slightly modified by averaging the commands from slot the 13 and slot 0 in order to avoid discontinuity in the adjustment process.

14.19 Packet Mode Operation

See Figure 14.37 for a demonstration of packet mode operation.

- **Common Channel Packet Transmission**—An uplink packet is appended directly to a random access burst. Common channel packet transmission is typically used for short infrequent packets, where link maintenance (power control and pilot symbols to preserve power control and synchronization) needed for a dedicated channel would lead to unacceptable overhead. Also, delay associated with a transfer to a dedicated channel is avoided. Common channel packet transmission is limited to short packets that use only a limited amount of a capacity.
- **Dedicated Channel Packet Transmission**—An initial random access request is used to set up a dedicated channel for the packet transmission. On this dedicated channel, closed loop power control is performed. The dedicated channel can be either set up for the transmission of a single packet or for the transmission of a sequence of packets. Single packet transmission is used for transmission of large infrequent packets. The initial random access request includes the amount of the data to be transmitted. The network may respond to access request in two different ways:
  - *Short acknowledgment*—A scheduling message is sent to the mobile at the time when the actual packet transmission can begin. The scheduling message includes the transfer format, e.g., bit rate to be used.
  - *Immediate scheduling message*—This allows for immediate transmission or indicates at what time in the near future the mobile can start its transmission.
- **Multipacket Transmission**—For multipacket transmission on a dedicated channel, an initial random access request is used to set up a dedicated packet channel. On this channel, a short packet can be transmitted without scheduling, similar to the common channel packet transmission. Large packets may require that an access request is first sent by the mobile on the dedicated channel. The network responds to this request in the same way as for the single packet case.

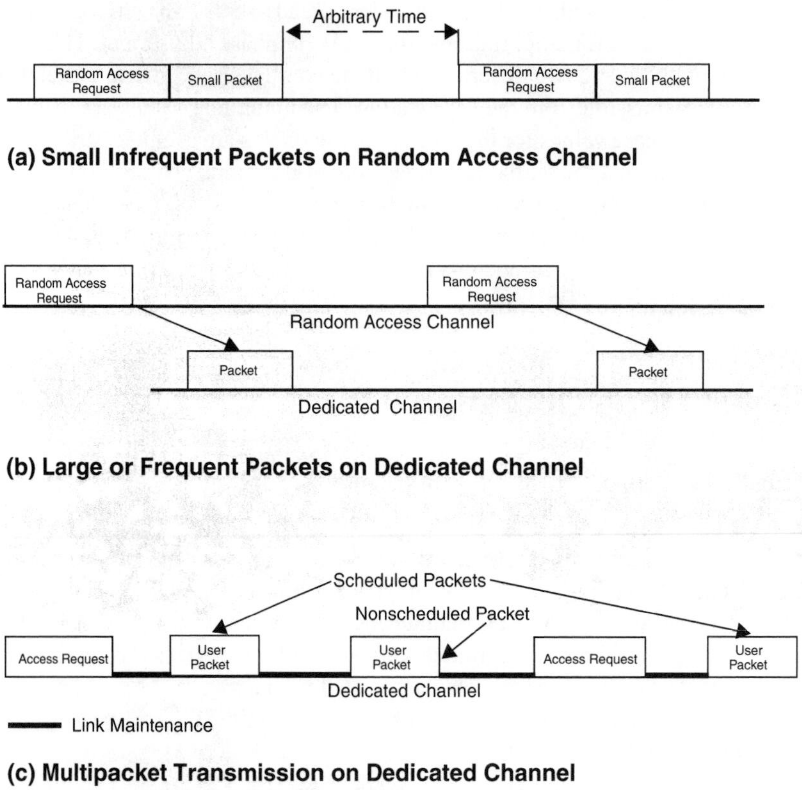

Figure 14.37 Packet Mode Operation

14.20 UMTS Network Reference Architecture

A UMTS system can be divided into a set of domains and the reference points that interconnect them. Figure 14.38 shows these domains and reference points.

A simplified mapping of functional entities to the domain model is shown in Figure 14.39. Note that this is a reference model and does not represent any physical architecture. The l_u is split functionally into two logical interfaces: the l_{ups} connecting the packet switched domain to the access network, and the l_{ucs} connecting the circuit switched domain to the access network. The standards do not dictate that these are physically separate, but the user plane for each is different and control plane may be different. The l_{ur} logically connects radio network controllers (RNCs) but could be physically realized by a direct connection between RNCs or via the core network.

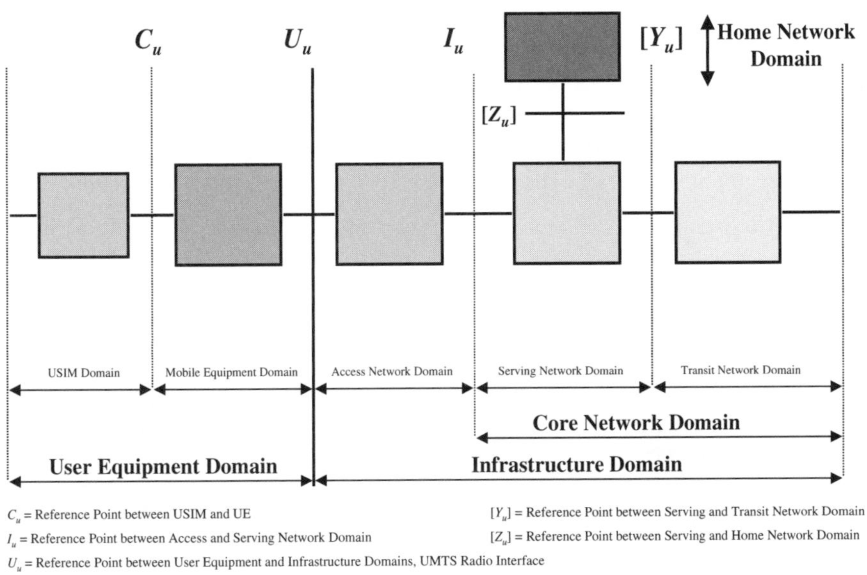

Figure 14.38 UMTS Domains and Reference Points

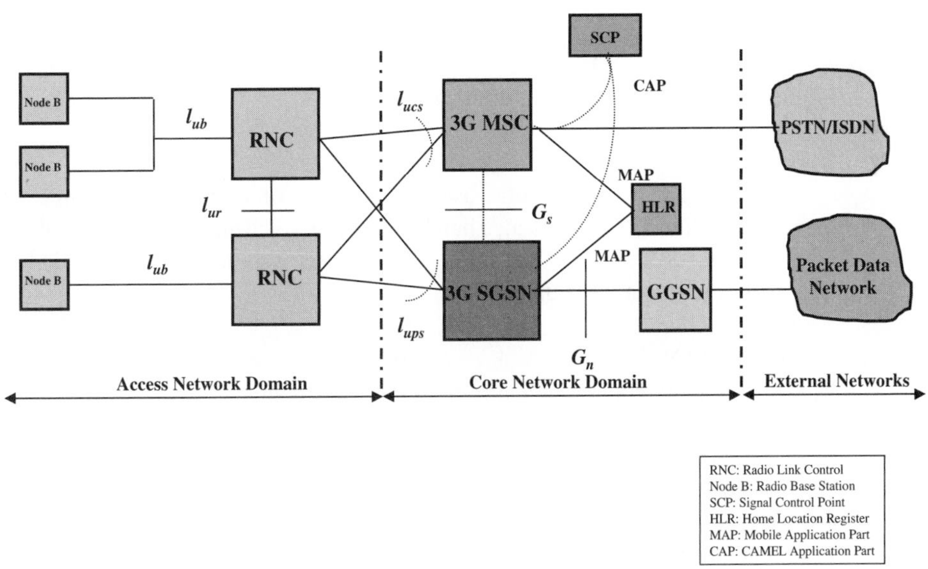

Figure 14.39 Simplified UMTS Network Reference Model

14.21 UMTS Terrestrial Radio Access Network Overview

The UMTS Terrestrial Radio Access Network (UTRAN) consists of a set of radio network subsystems (RNSs) (see Figure 14.40). The RNS has two main elements: Node B and radio network controller (RNC). The RNS is responsible for the radio resources and transmission/reception in a set of cells. A cell (sector) is one coverage area served by a broadcast channel.

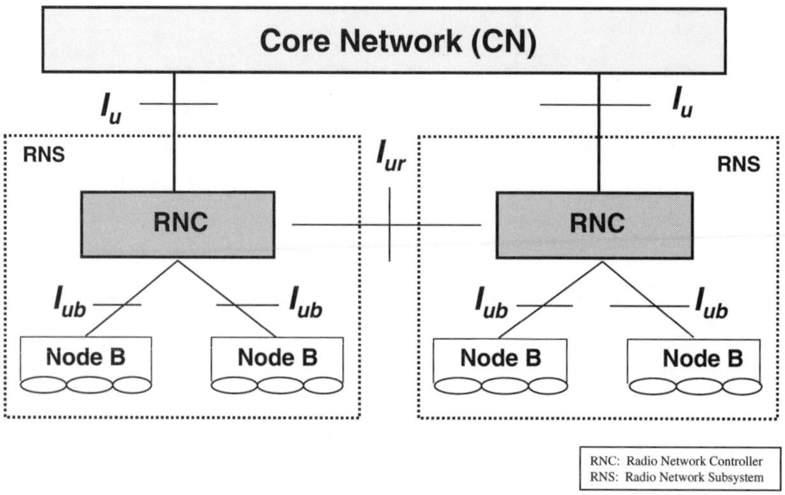

Figure 14.40 UTRAN Logical Architecture

An *RNC* is responsible for the use and allocation of all the radio resources of the RNS to which it belongs. The RNC also handles the user voice and packet data traffic, performing the actions on the user data streams that are necessary to access the radio bearers. The responsibilities of an RNC include:

- Intra-UTRAN handover
- Macrodiversity combining/splitting of I_{ub} data streams
- Frame synchronization
- Radio resource management
- Outer loop power control
- I_u interface user plane setup
- Serving RNS (SRNS) relocation
- Radio resource allocation (allocation of codes, etc.)
- Frame selection/distribution function necessary for soft handover (functions of UMTS radio interface physical layer)

- UMTS radio link control (RLC) sublayers function execution
- Termination of media access control (MAC), RLC, and radio resource control (RRC) protocols for transport channels; i.e., DCH, DSCH, RACH, FACH
- I_{ub}'s user plane protocols termination

A *Node B* is responsible for radio transmission and reception in one or more cells to/from the user equipment (UE). The logical architecture for Node B is shown in Figure 14.41.

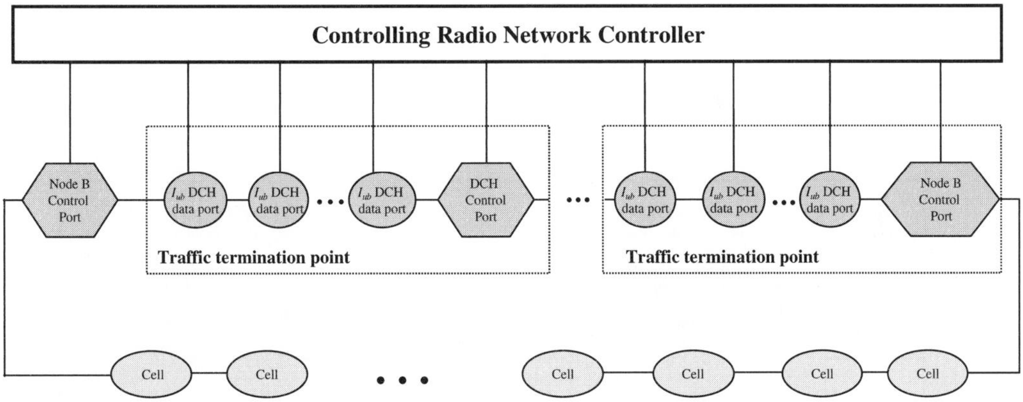

Figure 14.41 Node B Logical Architecture

The following are the responsibilities of Node B:

- Termination of I_{ub} interface from RNC
- Termination of MAC protocol for transport channels RACH, FACH
- Termination of MAC, RLC, and RRC protocols for transport channels BCH, PCH
- Radio environment survey (BER estimate, receiving signal strength, etc.)
- Inner loop power control
- Open loop power control
- Radio channel coding/decoding
- Macrodiversity combining/splitting of data streams from its cells (sectors)
- Termination of U_u interface from UE
- Error detection on transport channels and indication to higher layers
- FEC encoding/decoding and interleaving/deinterleaving of transport channels
- Multiplexing of transport channels and demultiplexing of coded composite transport channels
- Power weighting and combining of physical channels

- Modulation and spreading/demodulation and despreading of physical channels
- Frequency and time (chip, bit, slot, frame) synchronization
- RF processing

14.21.1 UTRAN Logical Interfaces

In UTRAN, protocol structure is designed so that layers and planes are logically independent of each other and, if required, parts of protocol structure can be changed in the future without affecting other parts [24–26].

The protocol structure contains two main layers: the radio network layer (RNL) and the transport network layer (TNL). In the RNL, all UTRAN-related functions are visible, whereas the TNL deals with transport technology selected to be used for UTRAN but without any UTRAN-specific changes. A general protocol model for UTRAN interfaces is shown in Figure 14.42.

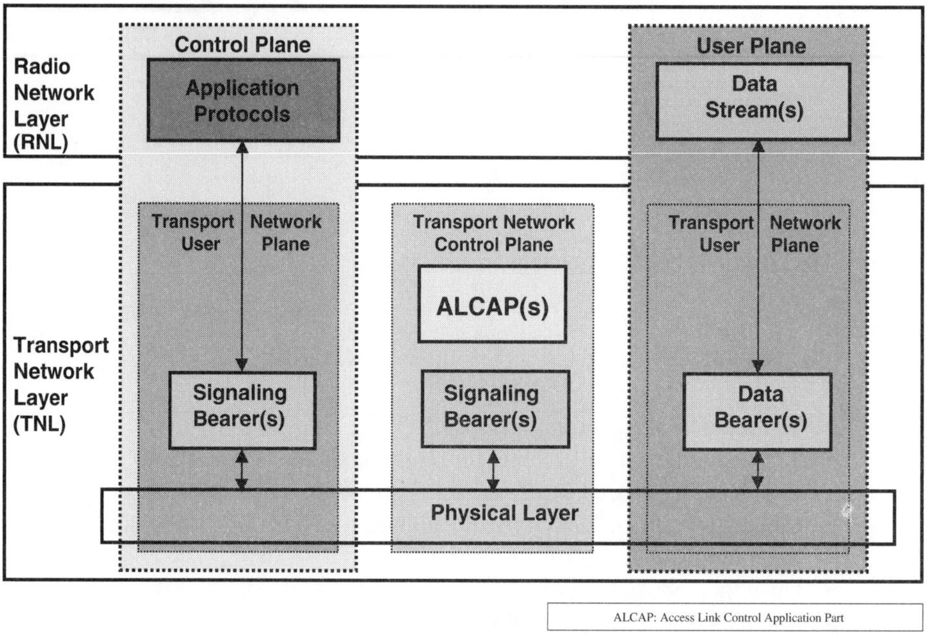

Figure 14.42 General Protocol Model for UTRAN Interfaces

The control plane is used for all UMTS-specific control signaling. It includes the application protocol (i.e., radio access network application part [RANAP] in I_u, radio network subsystem application part [RNSAP] in I_{ur}, and Node B application part [NBAP] in I_{ub}). The application protocol is used for setting up bearers to the UE. In the three-plane structure, the bearer parameters in the application protocol are not directly related to the user plane technology, but rather they are general bearer parameters.

User information is carried by the user plane. The user plane includes data stream(s), and data bearer(s) for data stream(s). Each data stream is characterized by one or more frame protocols specified for that interface.

The transport network control plane carries all control signaling within the transport layer. It does not include radio network layer information. It contains access link control application part (ALCAP) required to set up the transport bearers (data bearer) for the user plane. It also includes the signaling bearer needed for the ALCAP. The transport plane lies between the control plane and the user plane. The addition of the transport plane in UTRAN allows the application protocol in the radio network control plane (RNCP) to be totally independent of the technology selected for data bearer in the user plane.

With the transport network control plane (TNCP), the transport bearers for data bearers in the user plane are set up in the following way. First, there is a signaling transaction by application protocol in the control plane that initiates setup of the data bearer by the ALCAP protocol specific for the user plane technology. The independence of the control plane and user plane assumes that an ALCAP signaling occurs. The ALCAP may not be used for all types of data bearers. If there is no ALCAP signaling transaction, the transport network control plane is not required. This situation occurs when preconfigured data bearers are used. Also, the ALCAP protocols in the transport network control plane are not used to set up the signaling bearer for the application protocol or the ALCAP during real-time operation.

I_u Interface. The UMTS I_u interface is the open logical interface that interconnects one UTRAN to the UMTS core network (UCN). On the UTRAN side, I_u interface is terminated at the RNC. At the UCN side, it is terminated at U-MSC. The I_u interface consists of three different protocol planes—the RNCP, the TNCP, and the user plane (UP).

The RNCP performs the following functionality:

- It carries information for the general control of UTRAN radio network operations.
- It carries information for control of UTRAN in the context of each specific call.
- It carries user call control (CC) and mobility management (MM) signaling messages.

The control plane serves two service domains in the core network: the packet-switched (PS) domain and circuit-switched (CS) domain. The CS domain supports circuit-switched services. Some examples of CS services are voice and fax. The CS domain can also provide intelligent services such as voice mail and free phone. The CS domain connects to PSTN and ISDN. The CS domain is expected to evolve from the existing 2G GSM public land mobile network (PLMN).

The PS domain deals with PS services. Some examples of PS services are Internet access and multimedia services. Since Internet connectivity is provided, all services currently available on the Internet, such as search engines and e-mail, are available to mobile users. The PS domain connects to IP networks. The PS domain is expected to evolve from the GPRS PLMN.

The I_u circuit switched (CS) and packet switched (PS) protocol architectures are shown in Figures 14.43 and 14.44.

The control plane protocol stack consists of RANAP on top of signaling system 7 (SS7) protocols. The protocol layers are the signaling connection control part (SCCP), the message transfer part (MTP3-B), and signaling asynchronous transfer mode (ATM) adaptation layer for network-to-network interface (SAAL-NNI). The SAAL-NNI is divided into the service-specific coordination function (SSCF), the service-specific connection-oriented protocol (SSCOP), and ATM adaptation layer 5 (AAL5) layers. The SSCF and SSCOP layers are specifically designed for signaling transport in ATM networks, and take care of signaling connection management functions. AAL5 is used for segmenting the data to ATM cells.

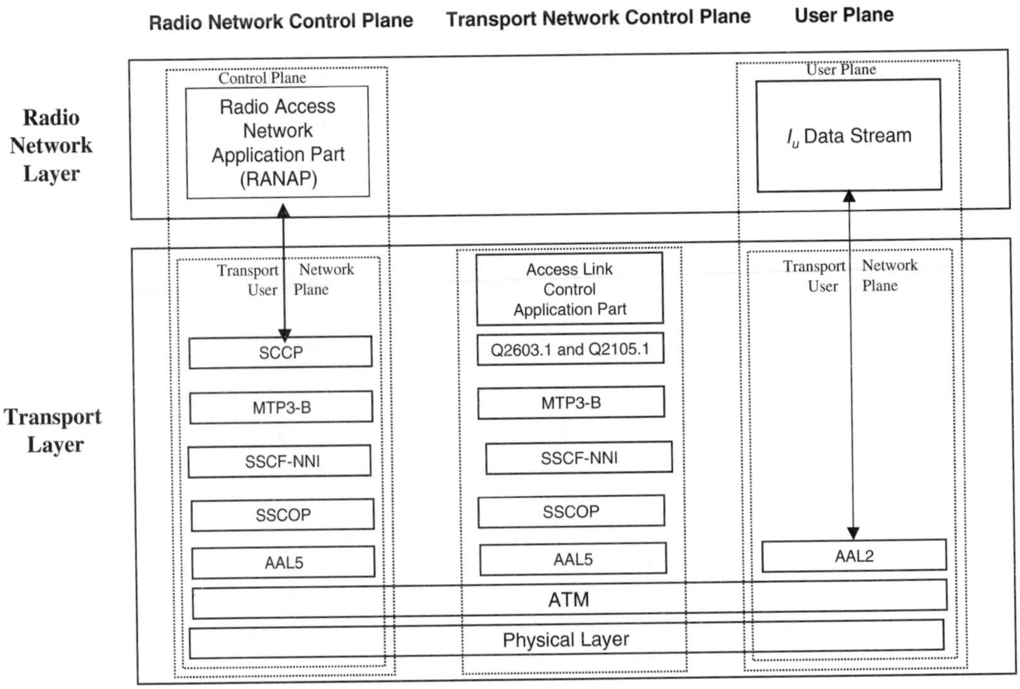

Figure 14.43 CS Protocol Architecture on I_u Interface

As an alternative, an IP-based signaling bearer is specified for the I_u PS control plane. The IP-based signaling bearer consists of SS7 MTP3–user adaptation layer (M3UA), simple control transmission protocol (SCTP), IP, and AAL5. The SCTP layer is specifically designed for signaling transport on the Internet.

TNCP carries information for the control of transport network used within UCN.

UP carries user voice and packet data information. The ATM adaption layer 2 (AAL2) is used for the following services: narrowband speech (e.g., EFR, AMR); unrestricted digital information service (up to 64 kbps, i.e., ISDN B channel); any low to average bit rate CS service (e.g., modem service to/from PSTN/ISDN). AAL5 is used for the following services: non-real-time PS data service (i.e., best effort packet access) and real-time PS data.

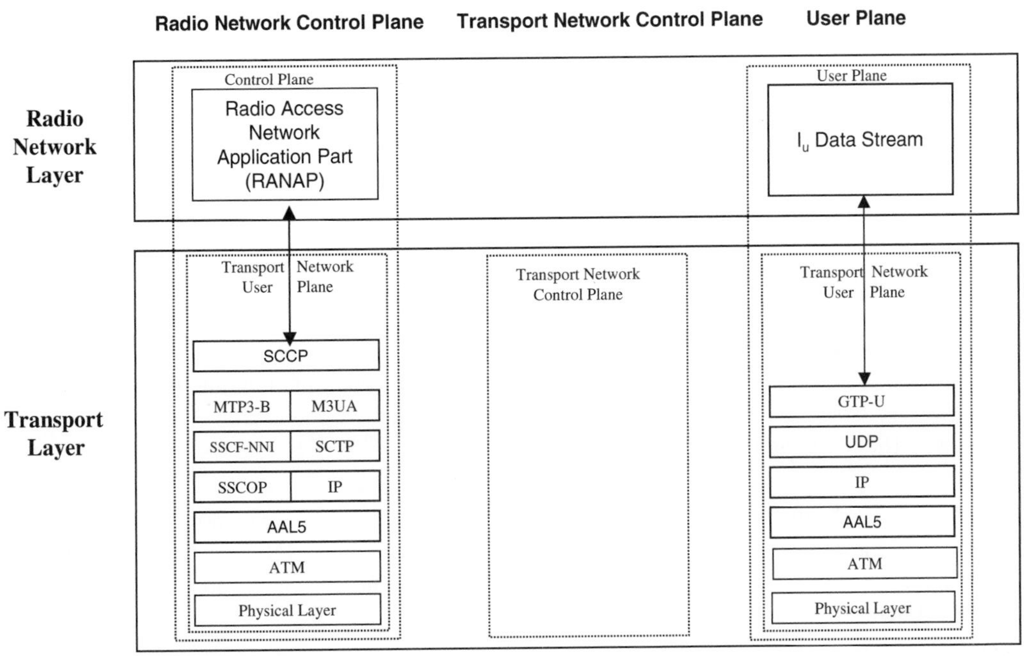

Figure 14.44 PS Protocol Architecture on I_u Interface

I_{ur} Interface. The connection between two RNCs (serving RNC [SRNC] and drift RNC [DRNC]) is the I_{ur} interface. It is used in soft handover scenarios when different macro diversity streams of one communication are supported by Node Bs that belong to different RNCs. Communication between one RNC and one Node B of two different RNCs are realized through the I_{ur} interface. Three different protocol planes are defined for it:

- Radio network control plane (RNCP)
- Transport network control plane (TNCP)
- User plane (UP)

The I_{ur} interface is used to carry

- Information for control of radio resources in the context of a specific service request of one mobile on RNCP
- Information for control of a transport network used within UTRAN on TNCP
- User voice and packet data information on UP

The protocols used on this interface are

- Radio access network application part (RANAP)
- DCH frame protocol (DCHFP)
- RACH frame protocol (RACHFP)
- FACH frame protocol (FACHFP)
- Access link control application part (ALCAP)
- Q.aal2
- Signaling connection control part (SCCP)
- Message transfer part 3-B (MTP3-B)
- Signaling ATM adaptation layer for network to network interface (SAAL-NNI). SAAL-NNI is further divided into service-specific coordination function for network to network interface (SSCF-NNI), service-specific connection oriented protocol (SSCOP), and AAL5.

The bearer is AAL2. The protocol structure of the l_{ur} interface is shown in Figure 14.45.

Initially, this interface was designed to support the inter-RNC soft handover, but more features were added during the development of the standard. The I_{ur} provides the following functions:

1. Basic inter-RNC mobility support
   - support of SRNC relocation
   - support of inter-RNC cell and UTRAN registration area update
   - support of inter-RNC packet paging
   - reporting of protocol errors
2. Dedicated channel traffic support
   - establishment, modification, and release of dedicated channel in the DRNC due to hard and soft handover in the dedicated channel state
   - setup and release of dedicated transport connections across the I_{ur} interface
   - transfer of DCH transport blocks between SRNC and DRNC
   - management of radio links in the DRNS via dedicated measurement report procedures and power setting procedures.
3. Common channel traffic support
   - setup and release of the transport connection across the I_{ur} for common channel data streams
   - splitting of MAC layer between the SRNC (MAC-d) and DRNC (MAC-c and MAC-sh); the scheduling for downlink data transmission is performed in the DRNC
   - flow control between the MAC-d and MAC-c/MAC-sh
4. Global resource management support
   - transfer of cell measurements between two RNCs
   - transfer of Node B timing between two RNCs

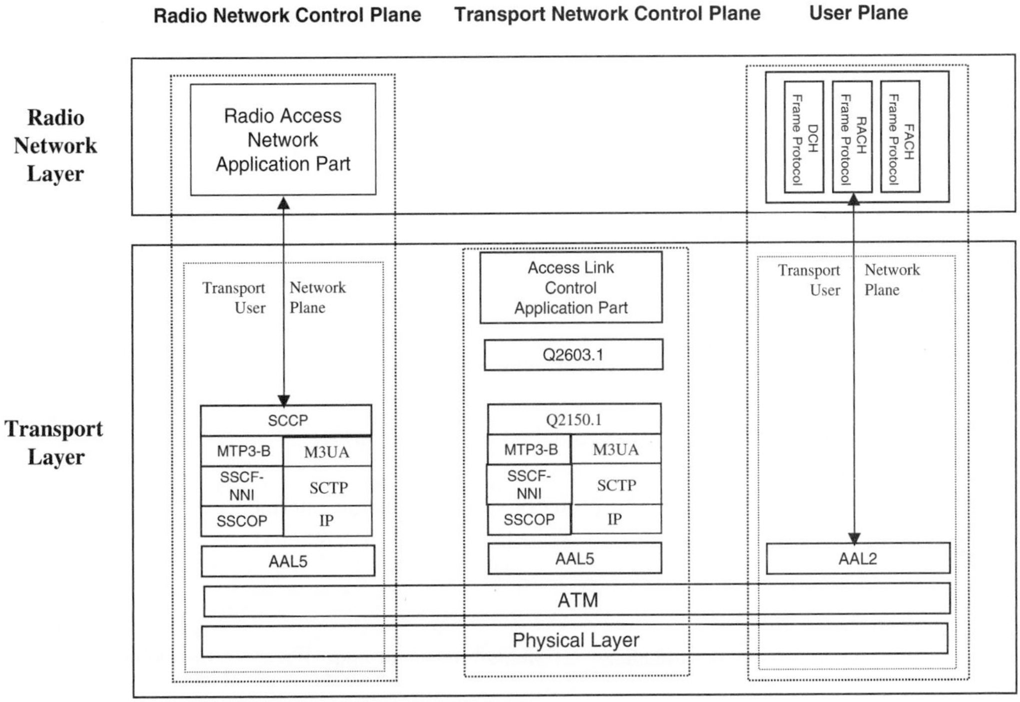

Figure 14.45 Protocol Structure of I_{ur} Interface

I_{ub} Interface. The connection between the RNC and Node B is the I_{ub} interface. There is one I_{ub} interface for each Node B. The I_{ub} interface is used for all of the communications between a Node B and the RNC of the same RNS. Three different protocol planes are defined for it:

- Radio network control plane (RNCP)
- Transport network control plane (TNCP)
- User plane (UP)

The I_{ub} interface is used to carry

- Information for the general control of Node B for radio network operation on RNCP
- Information for the control of radio resources in the context of a specific service request of one mobile on RNCP
- Information for the control of transport network used within UTRAN on TCNP
- User CC and MM signaling message on RNCP
- User voice and packet data information on UP

The protocols used on this interface include

- Node B application part protocol (NBAP)
- DCH frame protocol (DCHFP)
- RACH frame protocol (RACHFP)
- FACH frame protocol (FACHFP)
- Access link control application part (ALCAP)
- Q.aal2
- SCCP or TCP/IP
- MTP3-B
- SAAL-UNI (SSCF-NNI, SSCOP, and AAL5)

When using multiple low-speed links in the I_{ub} interface, Node B supports inverse multiplexing for ATM (IMA).

The bearer is AAL2. The protocol structure for the interface I_{ub} is shown in Figure 14.46.

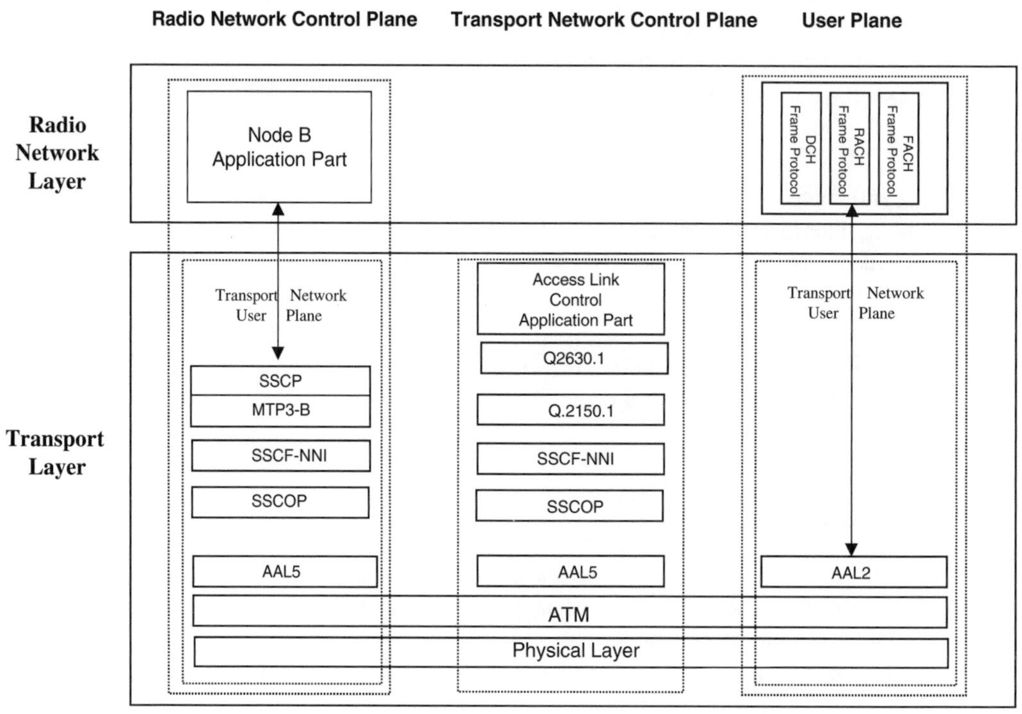

Figure 14.46 Protocol Structure of I_{ub} Interface

U_u Interface. The UMTS U_u interface is the radio interface between a Node B and one of its UEs. The U_u is the interface through which UE accesses the fixed part of the system.

14.21.2 Distribution of UTRAN Functions

Located in the RNC:

- Radio resource control (L3 Function)
- Radio link control (RLC)
- Macrodiversity combining
- Active cell set modification
- Assign transport format combination set (centralized data base function)
- Multiplexing/demultiplexing of higher layer PDUs into/from transport block delivered to/from the physical layer on shared dedicated transport channels (used for soft handover)
- L1 function: macrodiversity distribution/combining (centralized multipoint termination)
- Selection of the appropriate transport format for each transport channel depending upon the instantaneous source rate—collocate with RRC
- Priority handling between data flows of one user

Located in Node B:

- Scheduling of broadcast, paging, and notification messages; location in Node B—to reduce data repetition over l_{ub} and reduce RNC CPU load and memory space
- Collision resolution on RACH; location in Node B—to reduce nonconstructive traffic over l_{ub} interface and reduce round trip delay
- Multiplexing/demultiplexing of higher layer PDUs to/from transport blocks delivered to/from the physical layer on common transport channels

14.22 UMTS Core Network Architecture

Figure 14.47 shows the UMTS core network (UCN) in relation to all other entities within the UMTS network and all of the interfaces to the associated networks.

The UCN consists of a CS entity for providing voice and CS data services and a PS entity for providing packet-based services. The logical architecture offers a clear separation between the CS domain and PS domain. The CS domain contains the functional entities: mobile switching center (MSC) and gateway MSC (GMSC). The PS domain comprises the functional entities: serving GPRS support node (SGSN), gateway GPRS support node (GGSN), domain name server (DNS), dynamic host configuration protocol (DHCP) server, packet charging gateway, and firewalls. The core network can be split into the following different functional areas:

- Functional entities needed to support PS services (e.g., 3G-SGSN, 3G-GGSN)
- Functional entities needed to support CS services (e.g., 3G-MSC/VLR)
- Functional entities common to both types of services (e.g., 3G-HLR)

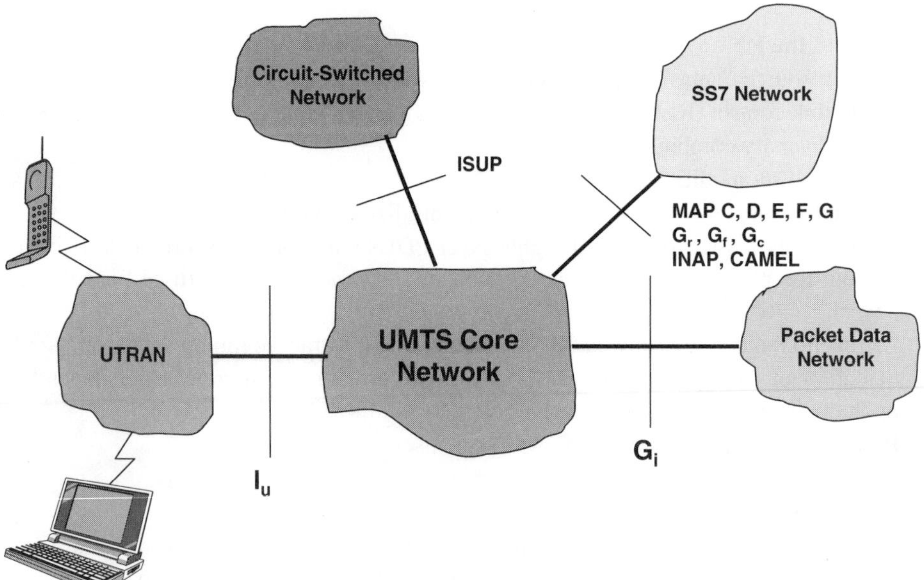

Figure 14.47 UMTS Architecture

Other areas that can be considered part of core network include

- Network management systems (billing and provisioning, service management, element management, etc.)
- Intelligent network (IN) system (SCP, SSP, etc.)
- ATM/SDH/IP switch/transport infrastructure

Figure 14.48 shows all the entities that connect to the core network—UTRAN, PSTN, Internet, and the logical connections between terminal equipment (MS, UE) and the PSTN/Internet.

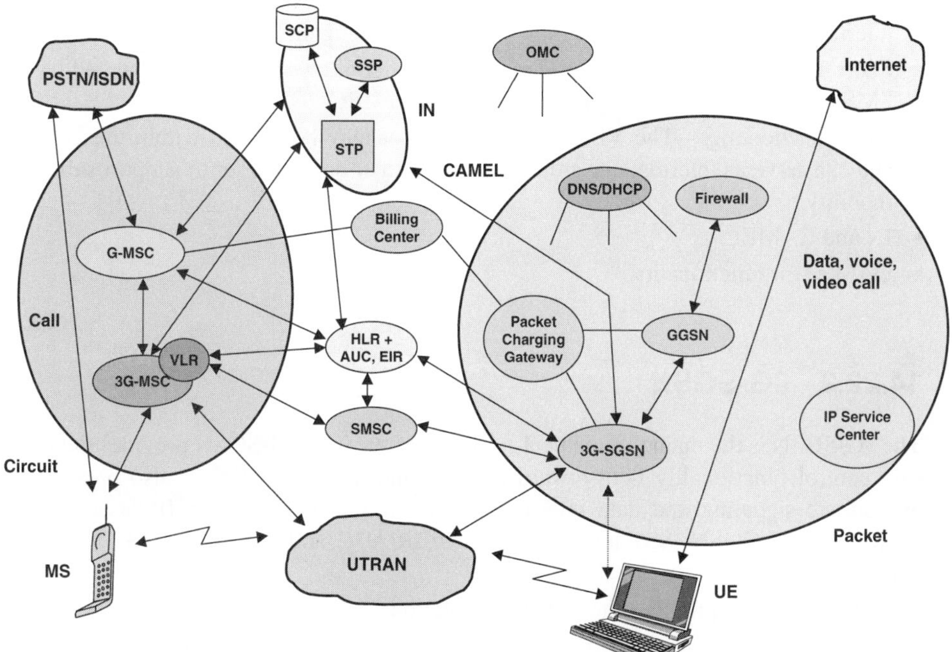

Figure 14.48 Logical Architecture of the UMTS Core Network

14.22.1 3G-MSC

The 3G-MSC is the main CN element to provide CS services. The 3G-MSC also provides the necessary control and corresponding signaling interfaces including SS7, MAP, ISUP, etc. The 3G-MSC provides the interconnection to the external networks like PSTN and ISDN. The 3G-MSC provides the following functions:

- Mobility management—Handles attach, authentication, and updates to the HLR, SRNS relocation, and intersystems handover.
- Call management—Handles call setup messages from/to the UE.
- Supplementary services—Handles call-related supplementary service such as call waiting.
- CS data services—The interworking function (IWF) provides rate adaptation and message translation for circuit-mode data services, such as fax.
- Vocoding.
- SS7, MAP, and RANAP interfaces—The 3G-MSC is able to complete originating or terminating calls in the network in interaction with other entities of a mobile network, e.g., HLR and AUC. It also controls/communicates with RNC using RANAP, which may use the services of SS7.

- ATM/AAL2 connection to UTRAN for transportation of user plane traffic across the I_u interface. Higher CS data rates may be supported using different adaptation layer.
- Short message services (SMS)—This functionality allows the user to send and receive SMS data to and from the SMS-GMSC/SMS-IWMSC.
- VLR functionality—The VLR is a database that may be located within the 3G-MSC and can serve as intermediate storage for subscriber data in order to support subscriber mobility.
- IN and CAMEL.
- OAM agent functionality.

14.22.2 3G-SGSN

The 3G-SGSN is the main CN element for PS services. The 3G-SGSN provides the necessary control functionality both toward the UE, and the 3G-GGSN. It also provides the appropriate signaling and data interfaces including connection to an IP-based network toward the 3G-GGSN, SS7 toward the HLR/EIR/AUC and TCP/IP, or SS7 toward the UTRAN.

The 3G-SGSN provides the following functions:

- Session management—Handles session setup messages to and from the UE and the GGSN, and operates admission control and QoS mechanisms.
- I_u, G_n, MAP interfaces—The 3G-SGSN is able to complete originating or terminating sessions in the network by interaction with other entities of a mobile network, e.g., GGSN, HLR, or AUC. It also controls/communicates with UTRAN using RANAP.
- ATM/AAL5 physical connection to the UTRAN for transportation of user data plane traffic across the I_u interface using GPRS tunneling protocol (GTP).
- Connection across the G_n interface toward the GGSN for transportation of user plane traffic using GTP. Note that no physical transport layer is defined for this interface.
- SMS—This functionality allows the user to send and receive SMS data to and from the SMS-GMSC/SMS-IWMSC.
- Mobility management—Handles attach, authentication, and updates to the HLR and SRNS relocation, and intersystem handover.
- Subscriber database functionality—This database (similar to the VLR) is located within the 3G-SGSN and serves as intermediate storage for subscriber data to support subscriber mobility.
- Charging—The SGSN collects charging information related to radio network usage by the user.
- OAM agent functionality.

14.22.3 3G-GGSN

The GGSN provides interworking with the external PS network. It is connected with SGSN via an IP-based network. The GGSN may optionally support an SS7 interface with the HLR to handle mobile terminated packet sessions.

The 3G-GGSN provides the following functions:

- Maintain information location at SGSN level (macromobility).
- Gateway between UMTS packet network and external data networks (e.g., IP, X.25).
- Gateway-specific access methods to intranets (e.g., PPP termination).
- User data screening/security—This can include subscription-based, user-controlled, or network-controlled screening.
- User-level address allocation—The GGSN may have to allocate (depending on subscription) a dynamic address to the UE upon packet data protocol (PDP) context activation. This functionality may be carried out by use of the DHCP function.
- Charging—The GGSN collects charging information related to external data network usage by the user.
- OAM functionality.

14.22.4 SMS-GMSC/SMS-IWMSC

The overall requirement for these two nodes is to handle the SMS from point to point. The functionality required can be split into two parts. The SMS-GMSC is an MSC capable of receiving a terminated short message from an service center, interrogating an HLR for routing information and SMS information, and delivering the short message to the SGSN of the recipient UE. The SMS-GMSC provides the following functions:

- Reception of short message packet data unit (PDU)
- Interrogation of HLR for routing information
- Forwarding of the short message PDU to the MSC or SGSN using the routing information

The SMS-IWMSC is an MSC capable of receiving an originating short message from within the PLMN and submitting it to the recipient service center. The SMS-IWMSC provides the following functions:

- Reception of the short message PDU from either the 3G-SGSN or 3G-MSC
- Establishing a link with the addressed service center
- Transferring the short message PDU to the service center

Note: The service center is a function that is responsible for relaying, storing, and forwarding of a short message. The service center is not a part of UCN, although the MSC and the service center may be integrated.

14.22.5 Firewall

This entity is used to protect the service providers' backbone data networks from attack by external packet data networks. The security of a backbone data network can be ensured by applying packet filtering mechanisms based on access control lists (ACLs) or any other methods deemed suitable.

14.22.6 DNS/DHCP

The DNS server is used, as in any IP network, to translate host names into IP addresses, i.e., logical names are handled instead of raw IP addresses. Also, the DNS server is used to translate the access point name (APN) into the GGSN IP address. It may optionally be used to allow the UE to use logical names instead of physical IP addresses.

A dynamic host configuration protocol server is used to manage the allocation of IP configuration information, by automatically assigning IP addresses to systems configured to use DHCP.

14.23 Adaptive Multirate (AMR) Codec for UMTS

The AMR codec has eight source rates: 4.75, 5.15, 5.90, 6.70 (PDC-EFR), 7.40 (IS-641), 7.95 (VSELP), 10.2, and 12.2 kbps (GSM-EFR). The AMR codec rates are controlled by radio access network and do not depend on the speech activity as in cdma2000. During high cell loading, such as during busy hours, the AMR codec uses lower bit rates to offer higher capacity while providing slightly lower speech quality. Also, if a mobile is running out of the cell coverage area and using maximum transmission power, a lower AMR bit rate can be used to extend the cell coverage area. With the AMR speech codec, it is possible to achieve a trade-off among capacity, coverage, and speech quality as per a service provider's requirements. The AMR speech codec is capable of switching bit rate every 20-ms frame upon command.

The AMR codec operates on speech frames of 20 ms corresponding to 160 samples at the sampling rate of 8,000 samples per second. The AMR uses algebraic code excited linear prediction (ACELP) coding. Every 160 speech samples, the speech signal is analyzed to determine the parameters of the CELP model (i.e., LP filter coefficients, adaptive and fixed codebooks' indices and gains). The speech parameter bits delivered by the speech encoder are rearranged according to their subjective importance before sending them to the network. The rearranged bits are further classified into class Ia, Ib, and II bits based on their sensitivity to errors. The most sensitive and the strongest channel coding is applied to class Ia bits. The AMR codec uses the following functions to provide discontinuous transmission (DTX) activity:

- Voice activity detector (VAD) on the transmission (Tx) side
- Background acoustic noise evaluation on the Tx side to transmit characteristics parameters to the receiving (Rx) side

- Comfort noise information transmission to the Rx side through regular use of the silence descriptor (STD) frame
- Comfort noise generation on the Rx side during periods when no normal speech frames are received

DTX prolongs terminal battery life and reduces the average required bit rate, providing a lower interference and an increase in system capacity. The AMR also deals with error concealment. The purpose of frame substitution is to conceal the effect of the lost speech frames. The purpose of muting the output, in the case of several lost frames, is to indicate the breakdown of the channel to the user and to avoid generating possibly annoying sounds as a result of the frame substitution procedure. The AMR speech codec can tolerate about a 1 percent frame error rate (FER) (about a bit error rate of 10^{-4}) of class Ia bits without any deterioration of speech quality. For class Ib and class II bits, a higher FER is allowed.

14.24 UMTS Bearer Service

Network services are end-to-end, i.e., from terminal equipment (TE) to another piece of TE. An end-to-end service has a certain QoS provided to the user by the network. To realize a certain network QoS, a bearer service with well-defined characteristics and functionality must be set up from the source to the destination of the service. UMTS bearer service layered architecture is shown in Figure 14.49. Each bearer service on layer N offers its service by using the services provided by (N – 1) layers below.

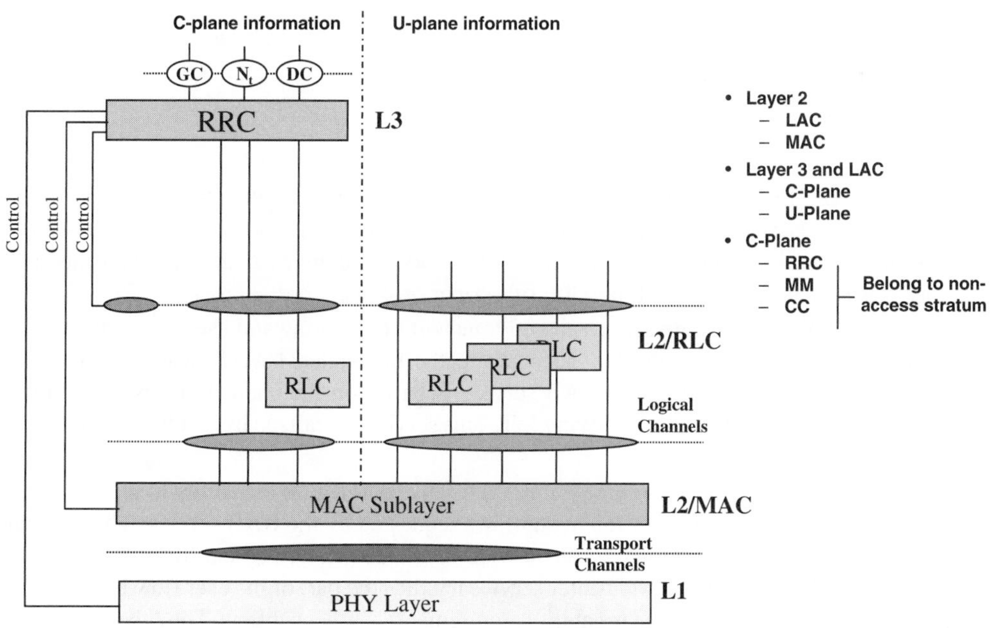

Figure 14.49 UMTS Bearer Service Layered Architecture

The traffic passes through different services of the network on its way from one piece of TE to another (see Figure 14.50). The end-to-end service used by the TE is realized by using a combination of TE/MT local bearer service, a UMTS bearer service, and an external bearer service. We focus only on the bearer service that provides the UMTS QoS.

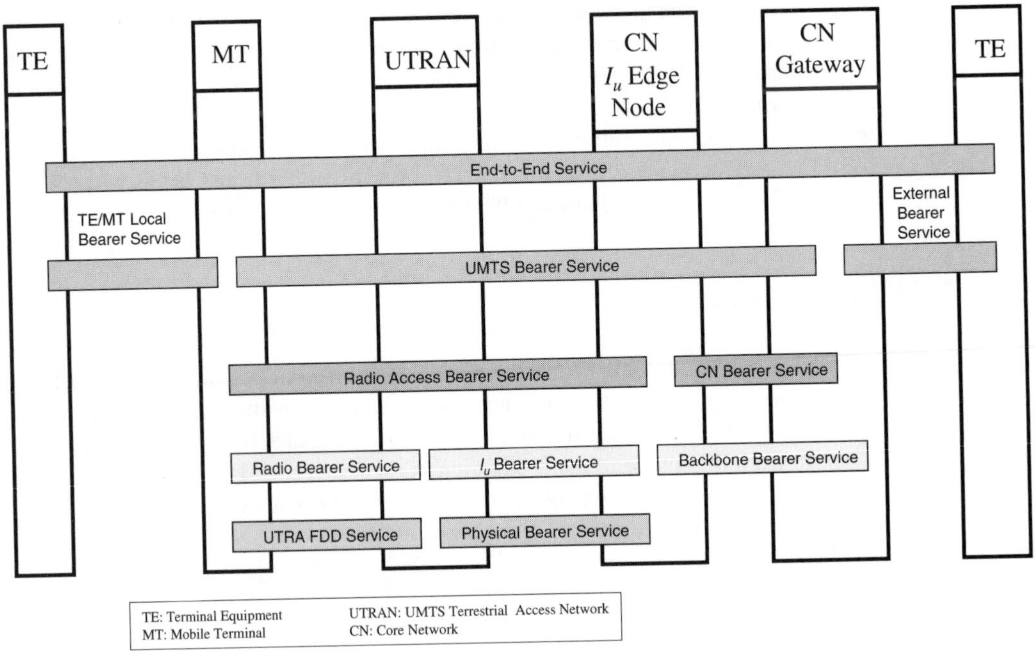

Figure 14.50 End-to-End UMTS Bearer Service

The UMTS bearer service has two parts: the *radio access bearer (RAB)* service and the *core network bearer (CNB)* service. Both services are aimed at optimizing the UMTS bearer service over the respective wireless network topology by taking into consideration aspects such as mobility and mobility subscriber profiles.

The RAB service provides a secured transport of signaling and user data between MT and CN I_u edge node with QoS adequate to the negotiated UMTS bearer service with default QoS for signaling. The RAB service is based on the characteristics of the radio interface and maintained for a moving MT. This service is realized by a radio bearer service and an I_u bearer service.

To support unequal error protection, UTRAN and MT have the ability to segment/reassemble the user flows into different subflows requested by the RAB service. The segmentation/reassembly is provided by the service data unit (SDU) payload format signaled at RAB establishment. The radio bearer service handles the part of the user flow belonging to one subflow, according to the reliability requirements for that subflow. The I_u bearer service for packet traffic provides different bearer services for a variety of QoS.

The CNB service of the CN connects the CN I_u edge node with CN gateway to the external network. The role of this service is to effectively control and use the backbone network to provide the contracted UMTS bearer service. The UMTS packet CN supports different backbone bearer services for different QoS.

The CN bearer service uses a generic network service. The backbone network service covers the layer 1/layer 2 functionality and is selected according to service provider's choice to satisfy the QoS requirements of the CN bearer service.

14.25 QoS Management

UMTS allows a user/application to negotiate bearer parameters that are most appropriate to carry the information. During an active session, it is possible to change bearer properties using a bearer renegotiation procedure. Bearer negotiation is initiated by an application, whereas renegotiation may be initiated either by the application or by the network. An application-initiated renegotiation is similar to a negotiation procedure that occurs in the bearer establishment phase. The application requests a bearer depending on its needs, and the network checks the available resources and the user's service class subscription to respond. The user either accepts or rejects the offer. The bearer class, bearer parameters, and parameter values are directly related to an application as well as to the networks that lie between the sender and the receiver.

14.25.1 Functions for UMTS Bearer Service in the Control Plane

Control plane QoS management functions support the establishment and modification of a UMTS bearer service by signaling/negotiating with the UMTS external services and by the establishment or modification of all UMTS internal services with the required characteristics. The control plane QoS management functions include [27]:

- Service manager coordinates the functions of the control plane for establishing, modifying, and maintaining the service. It provides all user plane QoS management functions with the relevant attributes.
- Translation function performs conversion between UMTS bearer service attributes and QoS parameters of external networks service control protocol.
- Admission/capability control maintains information about all available resources of a network entity and about all resources allocated to UMTS bearer services. It determines for each UMTS bearer service request whether the required resources can be provided by this entity. It reserves the resources if allocated to the UMTS bearer service.
- Subscription control checks the administrative rights of the UMTS bearer service user to use the requested service with the specified QoS attributes.

14.25.2 Functions for UMTS Bearer Service in the User Plane

User plane QoS management functions maintain the signaling and user data traffic within certain limits, as defined by specific QoS attributes. UMTS bearer services with different QoS attributes values are supported by the QoS management functions. These functions ensure the provision of the QoS negotiated for a UMTS bearer service and include

- Mapping function, which provides each data unit with the specific marking required to receive the intended QoS at the transfer by a bearer service.
- Classification function, which assigns data units to the established service of an MT according to the related QoS attributes if the MT has multiple UMTS bearer services established. The appropriate UMTS bearer service is derived from the data unit header or from traffic characteristics of the data.
- Resource manager, which distributes the available resources among all services sharing the same resources. The resource manager distributes the resources according to the required QoS.
- Traffic conditioner, which provides conformance between the negotiated QoS for a service and the data unit traffic. Policing or traffic shaping performs traffic conditioning. The policing function compares the data unit traffic with the related QoS attributes. The traffic shaper forms the data unit traffic according to the QoS of the service.

14.26 Quality of Service in UMTS

14.26.1 QoS Classes

The QoS in a UMTS network must take into account the restrictions and limitations of the air interface. In CDMA systems, all users use the same bandwidth at the same time; therefore, users interfere with one another. Because of the propagation mechanism, the signal received by the base station from a user close to the base station will be stronger than the signal received from another user located at the cell boundary. Hence, the close users will dominate the distant users. This situation becomes much worse if the close user happens to be a high-speed data user. To achieve a considerable capacity and QoS, all signals should arrive at the base station with the same mean power irrespective of distance. A solution to this problem is a precise power control, which attempts to achieve a constant mean power for each user. The UTRA/W-CDMA air interface uses power control on both uplink and downlink at 1,500 Hz.

In defining the UMTS QoS classes, the restrictions and limitations of the air interface have to be taken into account. The QoS mechanisms for the wireless network must be robust and capable to provide reasonable QoS resolution. UMTS defines four different QoS classes: the *conversational class, streaming class, interactive class,* and *background class.* The main distinguishing factor among these classes is how delay sensitive the traffic is. The conversational class is meant for traffic, which is very delay sensitive, whereas the background class is the most delay insensitive class.

The conversational and streaming classes are mostly used to carry real-time traffic flows. The main difference between them is based on how delay sensitive the traffic is. The conversational real-time services, such as video telephony or speech, are the most delay sensitive applications and require data streams to be carried in conversational class.

The interactive class and background class are mainly used for traditional applications such as Web browsing (WWW), e-mail, Telnet, data exchange via file transfer protocol (FTP), and news. Due to less stringent delay requirements as compared with the conversational and streaming classes, both classes provide better error rate by means of channel coding and retransmission. The main difference between the interactive and background classes is that the interactive class is used mainly for interactive applications (e.g., e-mail or interactive Web browsing), and the background class is meant for background traffic (e.g., background of e-mails, or background of downloading). Separating interactive and background applications ensures better response for the interactive applications. Traffic in the interactive class has higher priority in scheduling than the background class traffic, while the background applications use transmission resources only when interactive applications do not need them.

14.26.2 QoS Attributes

The defined UMTS bearer attributes ranges and radio access bearer attributes ranges and their relationships for each bearer class are summarized in Tables 14.8 and 14.9.

Table 14.8 UMTS Bearer Attributes for Each Bearer Class

Traffic Class	Conversational Class	Streaming Class	Interactive Class	Background Class
Maximum bit rate (kbps)	< 2,000	< 2,000	< 2,000 – overhead	< 2,000 – overhead
Delivery order	Yes/No	Yes/No	Yes/No	Yes/No
Maximum SDU size (octets)	< 1,500	< 1,500	< 1,500	< 1,500
SDU format information	(1)	(1)	(1)	(1)
SDU error ratio	10^{-2}, 10^{-3}, 10^{-4}, 10^{-5}	10^{-2}, 10^{-3}, 10^{-4}, 10^{-5}	10^{-3}, 10^{-4}, 10^{-5}	10^{-3}, 10^{-4} 10^{-5}
Residual bit error ratio	5×10^{-2}, 10^{-2}, 10^{-3}, 10^{-4}	10^{-2}, 10^{-3}, 10^{-4}, 10^{-5}	4×10^{-3}, 10^{-5}, 6×10^{-8}	4×10^{-3}, 10^{-5}, 6×10^{-8}
Delivery of erroneous SDUs	Yes/No	Yes/No	Yes/No	Yes/No
Transfer delay (ms)	100, maximum	500, maximum		
Guaranteed bit rate (kbps)	< 2,000	< 2,000		
Traffic handling priority			TBD	

Note: (1) Definition of possible values of exact SDU sizes for which UTRAN can support transparent RLC protocol mode.

Table 14.9 Radio Access Bearer Attributes for Each Bearer Class

Traffic Class	Conversational Class	Streaming Class	Interactive Class	Background Class
Maximum bit rate (kbps)	< 2,000	< 2,000	< 2,000 – overhead	< 2,000 – overhead
Delivery order	Yes/No	Yes/No	Yes/No	Yes/No
Maximum SDU size (octets)	< 1,500	< 1,500	< 1,500	< 1,500
SDU format information	(1)	(1)	(1)	(1)
SDU error ratio	10^{-2}, 10^{-3}, 10^{-4}, 10^{-5}	10^{-2}, 10^{-3}, 10^{-4}, 10^{-5}	10^{-3}, 10^{-4}, 10^{-5}	10^{-3}, 10^{-4}, 10^{-5}
Residual bit error ratio	5×10^{-2}, 10^{-2}, 10^{-3}, 10^{-4}	5×10^{-2}, 10^{-2}, 10^{-3}, 10^{-4}, 10^{-5}	4×10^{-3}, 10^{-5}, 6×10^{-8}	4×10^{-3}, 10^{-5}, 6×10^{-8}
Delivery of erroneous SDUs	Yes/No	Yes/No	Yes/No	Yes/No
Transfer delay (ms)	80, maximum	500, maximum		
Guaranteed bit rate (kbps)	< 2,000	< 2,000		
Traffic handling priority			TBD	

Note: (1) Definition of possible values of exact SDU sizes for which UTRAN can support transparent RLC protocol mode.

14.27 Summary

In this chapter, we discussed the architecture of the UMTS including its domains and interfaces. We outlined the channel structure in the UTRAN. The channel structure in the UTRAN is a three-tier system that includes logical, transport, and physical channels. The mapping of the logical to transport and the transport to physical channels was presented. A detailed description of the control, transport, and user plane in the I_u, I_{ur}, I_{ub} interfaces was provided. The role of each network entity in the UMTS was presented. The chapter was concluded with a discussion of the UMTS bearer service and UMTS QoS.

14.28 References

1. Special Issue on IMT-2000: Standard Efforts of the ITU, *IEEE Personal Communications*, July 1997 [Vol. 4].

2. Mohr, W., "ACTS FRAMES Project Towards IMT-2000/UMTS," *1999 IEEE International Conference on Personal Wireless Communications*, Feb. 1999, Jaipur, India.
3. Dahlman, E., Gumumdson, B., Nilsson, M., and Skold, J., "UMTS/IMTS-2000 Based Wideband CDMA," *IEEE Communications Magazine*, Sept. 1998 [Vol. 36(9)].
4. Prasad, N. R., "GSM Evolution Towards Third Generation UMTS/IMT-2000," *1999 IEEE International Conference on Personal Wireless Communications*, Feb. 1999, Jaipur, India.
5. Adachi, F., Sawahashi, M., and Suda, H., "Wideband DS-CDMA for Next Generation Mobile Communications Systems," *IEEE Communications Magazine*, Sept. 1998 [Vol. 36(9)].
6. Klein, A. et al., "FRAMES Multiple Access Mode 1—Wideband TDMA With and Without Spreading," *Proceedings of the IEEE International Conference on Personal Indoor and Mobile Radio Communications,* PIMRC '97, Helsinki, Finland, Sept. 1997, pp. 37–41.
7. Holma, H., and Toskala, A., *WCDMA for UMTS*, John Wiley and Sons: New York, 2000.
8. Universal Mobile Telecommunications System (UMTS), Requirements for the UMTS Terrestrial Radio Access System (UTRA), ETSI Technical Report, UMTS 21.01 version 3.0.1, Nov. 1997.
9. Universal Mobile Telecommunications System (UMTS), Selection Procedure for the Choice of Radio Transmission Technologies of the UMTS, ETSI Technical Report, UMTS 30.03 version 3.1.0, Nov. 1997.
10. Universal Mobile Telecommunications System (UMTS), Requirements for the UMTS Terrestrial Radio Access System (UTRA) Concept Evaluation, ETSI Technical Report, UMTS 30.06 version 3.0.0, Dec. 1997.
11. Ojanpera, T., and Prasad, R., "An Overview of Third-Generation Wireless Personal Communications: European Perspective," *IEEE Personal Communications*, Dec. 1998 [Vol. 5(6), pp. 59–65].
12. Rao, Y.S., and Kripalani, A., "cdma2000 Mobile Radio Access for IMT-2000," *1999 IEEE International Conference on Personal Wireless Communications,* Feb. 1999, Jaipur, India.
13. TR 45.5 "The cdma2000 ITU-RTT Candidate Submission," TR45-ISD/98.06.02.03, May 1998.
14. Knisley, D., Quinn, L., and Ramesh, N., "cdma2000: A Third Generation Radio Transmission Technology," *Bell Labs Technical Journal,* Sept. 1998 [Vol. 3(3), J1].
15. TIA/EIA IS-2000-1 "Introduction to cdma2000 Spread Spectrum Systems."
16. 3G TS 25.331 RRC Protocol Specification.
17. 3GPPTechnical Specification 25.401 UTRAN Overall Description.
18. 3G TS 25.321 MAC Protocol Specification.
19. 3G TS 25.322 RLC Protocol Specification.
20. 3GPPTechnical Specification 25.211, Physical Channels and Mapping of Transport Channels onto Physical Channels (FDD).
21. 3GPP Technical Specification 25.213, Spreading and Modulation (FDD).

22. 3GPP Technical Specification 25.212, Multiplexing and Channel Coding (FDD).
23. 3GPP Technical Specification 25.214, Physical Layer Procedures (FDD).
24. 3GPPTechnical Specification 25.410 UTRAN I_u Interface: General Aspects and Principles.
25. 3GPPTechnical Specification 25.420 UTRAN I_{ur} Interface: General Aspects and Principles.
26. 3GPPTechnical Specification 25.430 UTRAN I_{ub} Interface: General Aspects and Principles.
27. 3GPP Technical Specification 23.107, QoS Concept and Architecture.

CHAPTER 15

Wireless Data in CDMA

15.1 Introduction

Wireless data offerings are now evolving to suit the consumer market for the simple reason that the Internet is becoming an everyday tool and users are demanding data mobility. Currently, wireless data represents about 10 percent of all air time. While success has been concentrated in vertical markets such as public safety, health care, and transportation, the horizontal market (i.e., consumers) for wireless data is growing. Some 17 million people are expected to use wireless e-mail by 2002. The Internet has changed user expectations of what data access means. The ability to retrieve information via Internet has been "an amplifier of demand" for wireless data applications.

More than 75 percent of Internet users are also wireless users, and a mobile subscriber is four times more likely than a nonsubscriber to use the Internet. Such keen interest in both industries is prompting user demand for converged services. With more than half a billion Internet users expected by 2005, the potential market for Internet-related wireless data services is large.

Operators of cdmaOne networks are well positioned to capitalize on these trends, because cdmaOne is ready-made for data. The technology converts voice into packetized data for delivery, and cdmaOne handsets already integrate modems. Also, bandwidth on demand is inherent in CDMA and space is not wasted during idle periods. The cdmaOne community is moving ahead for a data-centric future. The group is advancing five sets of data-enabled service offerings:

1. Voice service, including high-quality voice coders and command features such as voice-dialing
2. Wireless information such as e-mail, news, weather, sports, location-based offerings, basic Internet access, and other services that do not include images or video
3. Services that include images and video, such as mobile fax and image transfers

4. Network-based agent services that might include media translation services, such as computer-generated voice translations of e-mail when a user is not able to read e-mail text, or contact management applications
5. Multimedia, which is a combination of the other four categories but offered as an on-demand service, with the different modes being accessible simultaneously as needed

Two critical factors for ongoing success of wireless data are cost containment and sustained revenue growth. cdmaOne is uniquely positioned to satisfy both aims, with the technology's capacity attributes helping to keep down costs and the evolutionary higher-speed data path creating new opportunities for sales of value-added services.

Currently, cdmaOne service providers are operating under the IS-95A standard; this allows circuit-switched data service up to 14.4 kbps using network software changes. Service providers and vendors are working within this standard to enhance the data and fax capabilities of cdmaOne. An enhancement, IS-707A [1], supports analog faxing. In some wireless local loop applications, handsets are being designed with analog-to-digital converters to allow interaction with standard Group 3 (G3) fax technology.

The next step on the cdmaOne evolutionary path takes a service provider to IS-95B [2], which delivers packet data at a sustained bit rate of 64 kbps. IS-95B requires software and hardware changes to mobiles as well as software infrastructure changes.

cdma2000 technology is such that second-generation (2G) and third-generation (3G) standards and services can exist in the same spectrum, allowing a cdmaOne operator ultimate flexibility in high-speed data deployment. The first phase of cdma2000 is the cdmaOne 3G solution. Phase one of cdma2000 Basic or 1XRTT, also called 3GIX [3], promises packet data at a sustained bit rate of 144 kbps with software and hardware changes to handsets and infrastructure. 3G1X is a preliminary step to 3G3X, envisioned to provide data rates up to 2 Mbps.

cdmaOne service providers have a number of other data enhancement options from which to choose. Quick Net Connect (QNC), a fast wireless Internet access technology developed by Qualcomm, 3Com, and Unwired Planet, and implemented using the circuit-data standard, is already available in the cdmaOne community. Designed for mobile-originated calls, QNC bypasses the public-switched telephone network (PSTN) to link directly with packet data networks, thus saving on modem training time at call setup and eliminating the need for modem pools.

Another technology, Qualcomm's High Data Rate (HDR), offers fast data over a dedicated 1.25-MHz channel that is not shared with voice traffic. HDR requires access by HDR-compatible handsets.

Mobile internet protocol (IP) and IP enhancement are being specified by the Internet Engineering Task Force (IETF) to be tied into 1X RTT implementations. A primary benefit of mobile IP is that it maintains data through handoffs by letting access devices roam while maintaining the same IP address. This is essential because packets of Internet information are delivered to a specific IP address. Without mobile IP, a mobile host gets a new IP address at each point of network attachment and data delivery cannot continue.

While cdmaOne network standards are moving quickly up the evolutionary path, terminal vendors are trying to develop complementary end-user devices. One example is Qualcomm's pdQ smart phone, which combines a cdmaOne handset with a Palm Pilot type organizer. Because pdQ is based on the popular Palm Computing platform, more than 1,000 productivity applications are immediately available to the device's user, including those enabling data synchronization between the pdQ and a personal computer. The pdQ itself has three new Qualcomm-created applications for short message service (SMS) alert management, e-mail, and Web browsing. Since the evolutionary path of cdmaOne to cdma2000 allows both groups of standards to coexist on the same network in the same spectrum, service providers should be able to make incremental implementations of new technology as needed. Such flexibility can help new and incumbent cdmaOne service providers future-proof their network as the data market gradually fulfills its long-held promise.

Though cdmaOne began to offer SMS and asynchronous data later than the Global System for Mobile Communications (GSM) network did, the two technologies are on the same timeline when it comes to high-speed data and packet data development. Although the cdmaOne network was not the first to offer data access, such networks are uniquely designed to accommodate data. To start with, the networks handle data and voice transmission in much the same way. Further, cdmaOne's inherent variable-rate transmission capability allows data rate determination to accommodate the information being sent, so system resources are used only as needed. Since cdmaOne systems employ a packetized backbone for voice, packet data capabilities are already inherent to the equipment.

In this chapter, we first introduce basic principles of data networking and introduce the transmission control protocol/Internet protocol (TCP/IP) suite. We then discuss the data capabilities of the cdmaOne system. We focus on IS-95 rev B, which allows for code or channel aggregation to provide data rates of 64–115 kbps. The network reference models for packet data services and protocol options specified in TIA EIA/IS-657 [4] are also included. The packet mode data service features are then discussed. We also focus on 3G1X data services of the 3G system designed to support increased data rates. The 3G1X system will allow service providers the ability to offer packet mode data services for demanding applications that require higher speeds, such as World Wide Web (WWW), Internet access, e-mail, etc.

15.2 Data Communication Services

The end-to-end communication services are classified as either *synchronous* (synch) or *asynchronous* (asynch). A synch communication service delivers a bit stream with a fixed delay and a given bit error rate. Voice communication is an example of the synch communication service. The synch delivery of a 64 kbps voice bit stream can be implemented by dividing the bit stream into packets that are received with random delays and are stored in a buffer to hold the bits until they are delivered. This implementation of a synch transmission service is called *packetized-voice*. In packetized-voice, a buffer is used to absorb the random fluctuations in the packet transmission delays. Another implementation of the synch transmission of the bit stream is to use a dedicated coaxial cable that propagates the bits one after the other, all with the same delay.

In an asynch communication service, the bit stream to be transferred is divided into packets. The packets are received by the destination with varying delays, and a fraction of them may not be received correctly at the destination. An asynch communication service is evaluated by its quality of service (QoS). The QoS deals with parameters such as the packet error rate, the delay, the throughput, the reliability, and the security of the communication. There are two classes of asynch communication services: *connection-oriented* and *connectionless.* A connection-oriented communication service delivers the packet in sequence, i.e., in correct order and with confirmed delivery. Depending on the QoS requirements, the delivery may be guaranteed to be error free. Thus, connection-oriented service looks from end to end like a dedicated link, which may be noiseless or noisy. A connectionless communication service delivers the packets individually. The packets can be delivered out of order, some may contain errors, and others may be lost. Some connectionless services provide an acknowledgment (ACK) of the correctly delivered packets. Thus, connectionless services are similar to mail service provided by the post office: letters may be delivered out of order, and normal mail delivery neither guarantees nor acknowledges the delivery. Another class of communication service is also used in some applications. It is called *expedited data*, and corresponds to a potentially faster delivery of packets, usually by making them jump to the head of the queues of packets that are waiting to be transmitted.

Communication services are implemented by transporting bits over the network. One essential objective of bit transport is the connectivity where one network user should be able to exchange information with many other users. It should be possible to *route* the bits of one user to any one of a large number of other users. The property to vary the path followed by the bits is called *switching.* Three basic methods used for switching bits in communication networks are

- Circuit switching
- Virtual circuit packet switching
- Datagram packet switching

In circuit switching, the switch connects transmission paths to establish a circuit between transmitter and receiver. Circuit switching is quite suitable for continuous data transmission services.

A packet-switched network uses another scheme. The nodes of the network, *packet-switching nodes*, play a role similar to that of switches in circuit-switched networks. Packet-switched networks can use two different methods for selecting the path followed by packets: *virtual circuit* (VC) and *datagram.* In the VC transport, the different packets that are part of the same information transfer are sent along the same path. The packets follow one another as if they were using a dedicated circuit even though they may be interleaved with other packet streams. Some implementations of VC perform an error control on each link between successive nodes. Thus, not only are the packets delivered in sequence from node to node along the path, but they are also transmitted without errors. This is implemented by each node checking the correctness of the packets it receives and asking the previous node to retransmit any incorrect packets. The VC packet switching does not need a buffer at the destination.

Since multiple VCs may exist between the source–destination pair, routing cannot be done on the basis of source–destination address only; data packets must also carry a VC identification. Routing is done on the basis of explicit route number and destination address. An explicit routing table at each node associates an appropriate outgoing transmission group with the destination address and explicit route number. By changing the explicit route number for a given destination, a new path will be followed. This introduces alternative route capability. If a link or node along the path becomes inoperative, any session using that path can be re-established on an explicit route by bypassing the failed element. Explicit routes can also be assigned on the basis of type of traffic, type of physical media along the path (satellite or terrestrial, for example), or other criteria. Routes could also be listed on the basis of cost, the smallest cost route being assigned first, then the next smallest cost route, and so forth.

In datagram packet switching, the bits are grouped as packets. Each packet is labeled with the address of its destination. The packets are routed independently of one another and arrive at their destination out of sequence. The datagram packet switching requires buffers at the source and destination. In datagram packet-switching networks, each network node keeps a complete (global) topological database that is updated regularly as topological changes occur. Generally, the routing philosophy of datagram networks is to route packets (datagram) along paths of minimum time delay.

15.3 OSI Upper Layers

The role of the open system interconnect (OSI) upper layers (transport, session, presentation, and application) can be summarized as follows:

- The *transport layer* segments the messages into packets of acceptable sizes and performs the reassembly at the destination. It may multiplex many low-rate transmissions onto one virtual circuit or divide a high-rate transmission into parallel virtual circuits. The transport layer controls transmission errors and requests retransmissions of packets corrupted by transmission errors. In addition, the flow may be controlled by some mechanism to prevent one host from sending data faster than the destination host can receive.
- The *session layer* sets up the call and takes care of the user authentication and billing. The session layer supervises the synchronization (packet numbering) and the recovery in case of failures. The session layer closes the session at the end of transmission.
- The *presentation layer* asks the session layer to set up a call. It specifies the destination's name and type of transmission (e.g., datagram, high priority). The presentation layer translates between the local syntax used by the application process and transfer syntax. It also performs the required encryption and data compression.
- The *application layer* provides information transfer services for user application programs. The user interacts with the application layer through a user interface. The application layer is composed of specific application service elements (SASEs) that use the services of common application service elements (CASEs). A CASE establishes an association between SASEs and may include an association control service element (ACSE), a remote operation service element (ROSE), and a commitment concurrency and recovery (CCR) element.

15.4 Transmission Control Protocol

The transmission control protocol (TCP) is the connection-oriented transport layer protocol designed to operate on the top of the datagram network layer IP. The two widely used protocols are known under the collective name TCP/IP. TCP provides a reliable end-to-end byte stream transport. The segmentation and reassembly of the messages are handled by IP, not by TCP.

TCP uses the *selective repeat protocol* (SRP), with positive acknowledgments and time-out. Each byte sent is numbered and must be acknowledged. A number of bytes can be sent in the same packet, and the acknowledgment (ACK) then indicates the sequence number of the next byte expected by the receiver. ACK carrying sequence number "m" provides acknowledgment for all packets up to, and including, the packet with sequence number "$m - 1$". The receiver sends duplicate ACKs for any lost packets.

The TCP header is at least 20 bytes (16 bits each) in length, and contains 16 error detection bits for the data and the header. The error detection bits are calculated by summing the 1's complements of the groups of 16 bits that make up the data and the header, and by taking the 1's complement of that sum. The amount of data that can be sent before being acknowledged is limited by the window size (W_{max}), which can be adjusted either by the sender or the receiver to control the flow based on the available buffers and the congestion. Initial sequence numbers are negotiated by means of a three-way handshake at the outset of connection. Connections are released by means of a three-way handshake.

TCP transmitter (Tx) uses an adaptive window-based transmit strategy. Tx does not allow more than W_{max} unacknowledged packets outstanding at any given time. With the congestion window lower edge at time t equal to $X(t)$, packets up to $X(t) - 1$ have been transmitted and acknowledged. Tx can send starting from $X(t)$. $X(t)$ has a nondecreasing sample path. With congestion window width at time t equal to $W(t)$, this is the amount of packets Tx is allowed to send starting $X(t)$. $W(t)$ can increase or decrease (because of window adaptation), but can never exceed W_{max}.Transitions in $X(t)$ and $W(t)$ are triggered by receipt of ACK. Receiver (Rx) of an ACK increases $X(t)$ by an amount equal to the amount of data acknowledged. Change in $W(t)$, however, depends on the version of TCP and congestion control process.

Tx starts a timer each time a new packet is sent. If the timer reaches a round trip time-out (RTO) value before the packet is acknowledged, a time-out occurs. Retransmission is initiated on time-out. RTO value is derived from a round trip timer (RTT) estimation procedure. RTO is sent only in multiples of a timer granularity.

Window adaptation procedure is as follows:

1. Slow start phase:
   If $W < W_{th}$, $W \leftarrow W + 1$ for each ACK received, W_{th} is the slow-start threshold
2. Congestion avoidance phase:
   If, $W \leftarrow W + 1/W$ for each ACK received
3. Upon time-out:
   $W^+ \leftarrow 1$ and $W^+_{th} \leftarrow W/2$

ISO defined five classes (0 to 4) of connection-oriented transport services (ISO 8073). We briefly describe class 4, which transmits packets with error recovery and in the correct order. This protocol is known as Transport Protocol Class 4 (TP4) and is designed for unreliable networks. The basic steps in a TP4 connection follow:

- **Connection establishment**—This is performed by means of a three-way handshake to agree on connection parameters, such as a credit value that specifies how many packets can be sent until the next credit arrives, the connection number, the transport source and destination access points, and a maximum time-out before ACK.
- **Data transfer**—The data packets are numbered sequentially. This allows resequencing. ACKs may be done for blocks of packets. There is a provision for expedited data transport in which the data packets are sent and acknowledged one at a time. Expedited packets jump to the head of the queues. Flow is controlled by windows or by credits.
- **Clear connection**—Connections are released by an expedited packet indicating the connection termination. The buffers are then flushed of the data packets corresponding to that connection.

In practice, TCP has been tuned for traditional networks consisting of wired links and stationary hosts. TCP assumes that congestion in the network is the primary cause for packet losses and unusual delay. TCP performs well over wired networks by adapting to end-to-end delays and congestion losses. TCP reacts to packet losses by dropping its transmission (congestion) window size before retransmitting packets, initiating congestion control, or by using avoidance mechanisms. These measures result in a reduction in the load on the intermediate links, thereby controlling the congestion in the network.

In a wireless network, packet losses can occur because of handoff or fading and can be random. When TCP responds to packet losses by invoking congestion control or avoidance algorithms, a degraded end-to-end performance in wireless networks results. The wireless environment violates many of the assumptions made by TCP. Many approaches have been used to improve end-to-end TCP performance over wireless links. They can be classified into three categories: (1) end-to-end TCP protocols, where loss recovery is performed by the sender, such as explicit loss notification (ELN) option; (2) link-layer protocols that provide local reliability using techniques such as FEC and retransmission of lost packets in response to automatic repeat request (ARQ) messages; and (3) split TCP connection protocol that breaks the end-to-end TCP connection into two parts at the base station, one between the sender and the base station and the other between the base station and the receiver.

In the first case, a link layer ARQ mechanism is used to improve error rate seen by TCP. IS-95 CDMA data stack uses this approach. In the second case, network layer software is modified at the base station to monitor every passing packet in either direction. Cache packets are used at the base station and local retransmissions are performed across wireless links. With split TCP, TCP may get ACK even before the packet is successfully delivered to the receiver. It also involves software overhead.

15.5 User Datagram Protocol

The connectionless transport service user datagram protocol (UDP) uses the IP network layer. Together, these two protocols are known as UDP/IP. UDP uses 16-bit port numbers. The UDP header contains 16 error detection bits which are set to zero when they are not used. UDP adds multiplexing capabilities and also the possibility of error detection to IP.

15.6 Network Layer on the Internet

The Internet protocol (IP) provides the basis for the interconnections of the Internet. IP is a datagram protocol. The packets contain an IP header. The basic header, without options, is shown in Figure 15.1.

The *version* field allows new versions of IP to be installed while the network is operational. The *Internet header length* (IHL) field indicates the header length on that packet. The *type of service* field indicates the QoS desired (e.g., low delay, high throughput, high reliability). *Identification, flag,* and *fragment offset* allow reassembly of a fragmented datagram. The *time to live* field indicates how long this packet can remain in the network. It is reduced at each hop, and the packet is discarded when this field reaches zero. The *protocol* field indicates what higher level protocol is contained in the data portion of the IP packet. The *header checksum* field is a checksum of the bytes in the IP header only. High level protocols must be concerned with error checking of their data. The *source* and *destination* Internet addresses indicate the sending host and the intended recipient host for this datagram.

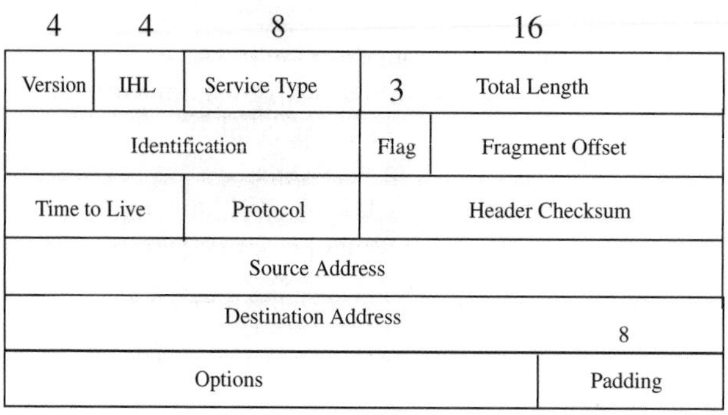

Figure 15.1 IP Header

A symbolic address, or name, of the form *user@domain* can be used instead of an Internet address. It is translated into an Internet address by directory tables that are organized along the same hierarchy as the addresses. Typically, the *domain* is of the form *machine.institution.type.country*. The type is *edu* for educational institutions, *com* for companies, *gov* for governmental agencies, *org* for nonprofit organizations, or *mil* for military. The *country* field is omitted for the United States and is a two-letter country code for the other countries (e.g., *fr* for France). For instance, the author's address is *vgarg@uic.edu.*

With best-effort delivery service (optional QoS), IP packets may be lost, corrupted, delivered out of order, or duplicated. The upper layer entities should anticipate and recover on an end-to-end basis.

Standard methods of sending IP over any point-to-point (PPP) link include dial-up modems (asynch framing), leased lines (bit synchronous framing), and ISDN, IS-99 CDMA (octet-synchronous framing). The link control protocol (LCP) runs during initial link establishment and negotiates link-level parameters (e.g., maximum frame size, etc.). The IP control protocol (IPCP) establishes the IP address of a client (the PPP server allocates a temporary address or the client notifies the server of its fixed address) and negotiates the use of TCP/IP header compression.

A special protocol, the Internet control message protocol (ICMP) takes care of packet delivery problems that arise when links or stations are down. The protocol operates by sending ICMP messages to report delivery errors, such as unreachable destinations, overly high delivery rates, and changes in interconnection. These messages are then used by the Internet software.

Mobile IP is based on the concept of a "home agent," which tracks a mobile host's location and is affiliated with a static IP address on the home network and a foreign agent that supports mobility on a foreign network by providing routing to a visiting mobile host. Networks supporting mobile IP will have to create foreign agents to deliver packets of information to the mobile host. Mobile IP is fundamental to the goal of providing the successful model for wireless data, wherein the traditional wired connection to the corporate intranet is made wireless.

15.7 Internet Reference Model

The Internet reference model is shown in Figure 15.2. It includes:

- The *application layer*, which covers OSI application and presentation layers. Some of the application layer protocols are HyperText Transfer Protocol (HTTP), File Transfer Protocol (FTP), Simple Mail Transfer Protocol (SMTP), Post Office Protocol (POP), etc.
- The *end-to-end layer*, which includes OSI transport and session layers. The end-to-end protocols are TCP and UDP.
- The *Internet layer*, which corresponds to the upper part of OSI network layer. The protocol is IP.
- The *subnet layers*, which include the lower part of the OSI network layer, link layer, and physical layer.

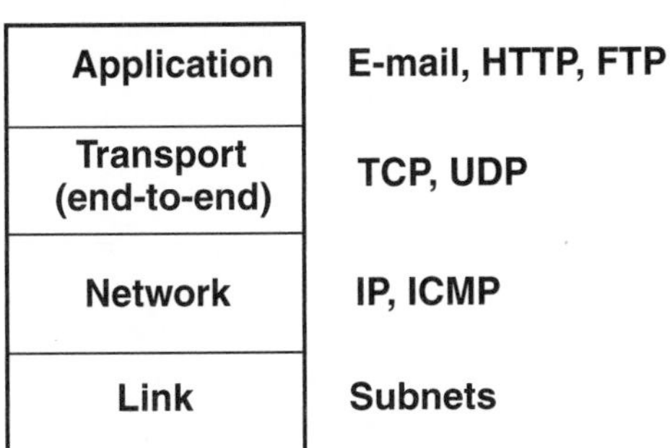

Figure 15.2 Internet Reference Model

15.8 TCP/IP Suite

The TCP/IP suite (Figure 15.3) occupies the middle five layers of the 7-layer OSI model (see Figure 15.4). The TCP/IP layering scheme combines several of the OSI layers. From an implementation standpoint, the TCP/IP stack encapsulates the network layer (OSI layer 3) and transport layer (OSI layer 4). The physical layer, the data-link layer (OSI layers 1 and 2, respectively), and the application layer (OSI layer 7) at the top can be considered non-TCP/IP–specific. TCP/IP can be adapted to many different physical media types.

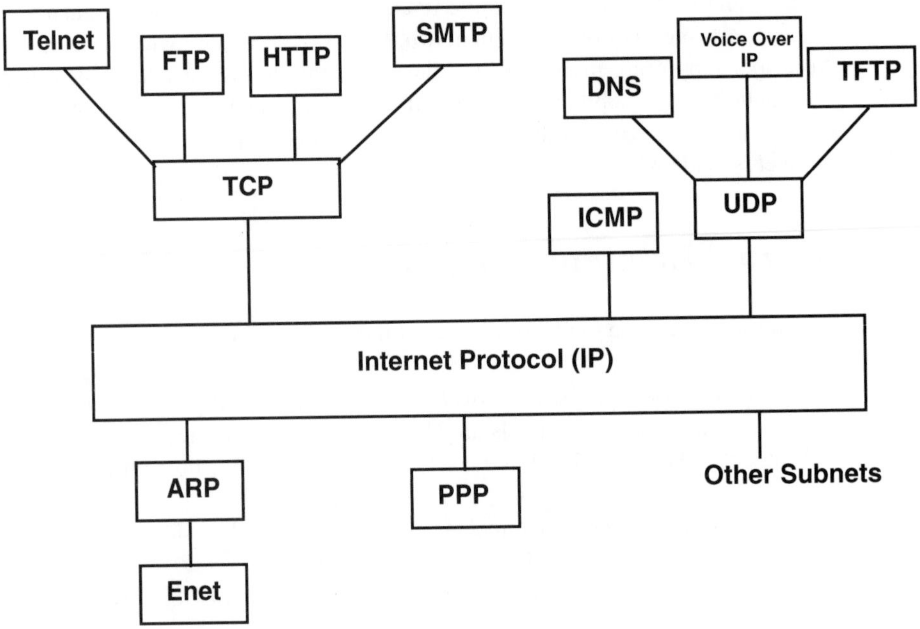

TCP: Transmission Control Protocol
TFTP: Trivial File Transfer Protocol
ARP: Address Resolution Protocol
FTP: File Transfer Protocol
HTTP: HyperText Transfer Protocol
SMTP: Simple Mail Transfer Protocol
ICMP: Internet Control Message Protocol
UDP: User Datagram Protocol
PPP: Point-to-Point Protocol
Enet: Ethernet

Figure 15.3 TCP/IP Protocol Suite

TCP/IP serves as a conduit to and from devices, enabling the sharing, monitoring, or controlling of those devices. A TCP/IP stack can have a tremendous effect on a device's memory resources and CPU utilization. Interactions with other parts of the system may be highly undesirable and unpredictable. Problems in TCP/IP stacks can render a system inoperable. The TCP/IP stack should become a fairly integrated part of a product.

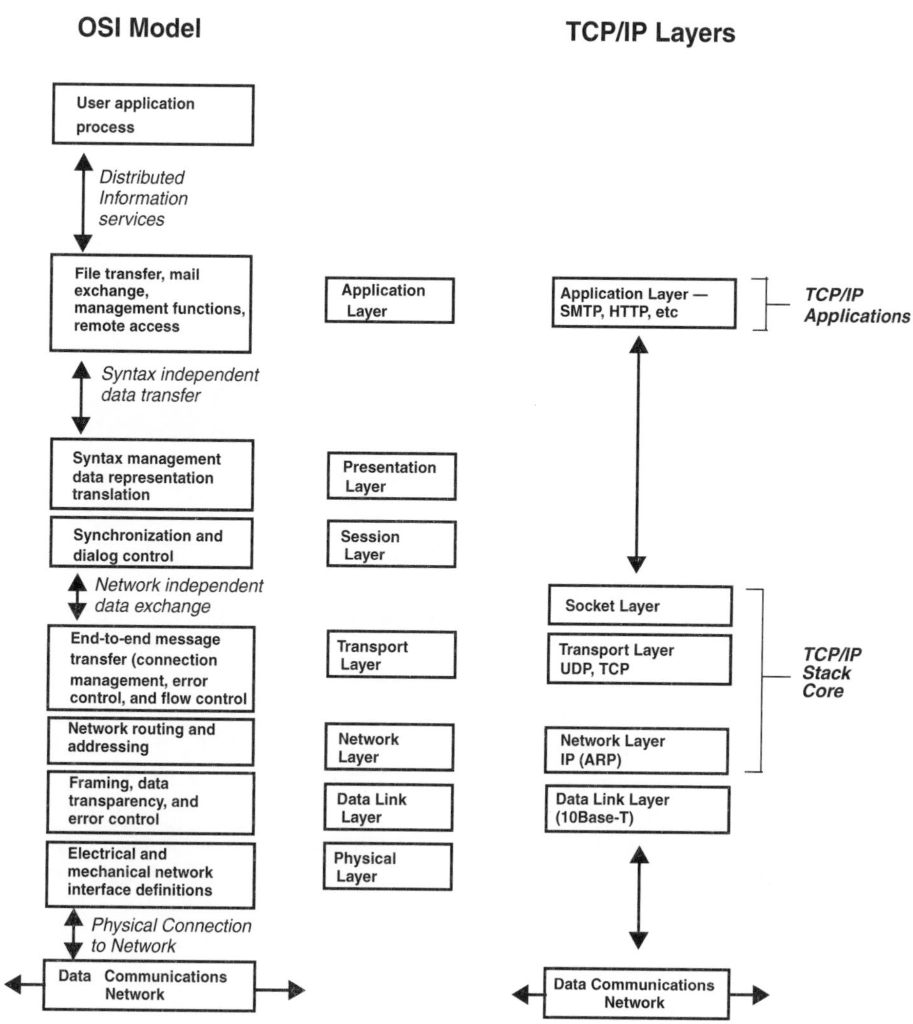

Figure 15.4 A Comparison of the OSI Model and TCP/IP Protocol Layers

15.9 cdmaOne Data Rate

The data services in IS-95 include circuit-switched async data, circuit-switched digital fax, packet data, and analog fax. The current data standards are as follows:

- **IS-658**
  - Defines the interface between the MSC and an external data interworking function (IWF)
  - Applies to circuit-switched async data, digital fax, and packet data services
- **Medium data rates (MDR)**
  - Included as part of IS-95B standardization
- **IS-707** [5–11]
  - Rate Set 1 Service Options
    - SO 4100: Async data
    - SO 4101: Digital fax
    - SO 4103: IP/Mobile IP
    - SO 4104: CDPD
  - Rate Set 2 Service Options
    - SO 12: Async data
    - SO 13: Digital fax
    - SO 15: IP/Mobile IP
    - SO 16: CDPD

The cdmaOne packet data technology uses a TCP/IP-compliant cellular digital packet data (CDPD) protocol stack to provide seamless connectivity with enterprise networks and expedite third-party application development. Adding data to cdmaOne networks allows an operator to continue using its existing radios, backhaul facilities, infrastructure, and handsets while merely implementing a software upgrade with an IWF.

IS-95B allows for code or channel aggregation to provide data rates of 64 to 115 kbps, as well as offering improvements in soft handoffs and interfrequency hard handoffs. To achieve a 115-kbps rate, up to eight CDMA traffic channels each offering 14.4 kbps are needed to be aggregated.

Mobile IP allows users to maintain a continuous data connection and retain a single IP address while traveling between base station controllers (BSCs) or roaming in other CDMA networks.

Packet data transmission is supported on TIA/EIA IS-95B traffic channels (TCHs) using primary or secondary traffic. For packet data transmission, the radio link protocol (RLP) Type 2 specified in IS-707.A.8 [10] is used. IS-707A specifies a packet bearer service for communication between terminal equipment (TE) and a packet IWF via base station/mobile switching center (BS/MSC). It provides procedures to apply to multiple packet data services (e.g., CDPD and mobile IP).

Packet data service options provide a means of establishing and maintaining TCHs for packet data service. Service options 22, 23, 24, and 25 are used to request packet data service through an IWF supporting an Internet standard PPP interface to network layer protocols [12]. Service options 26, 27, 28, and 29 are used to request packet data service through an IWF supporting CDPD data service over a PPP interface.

Packet data services can be of two types: (1) Type 1 packet data service provides connections based on CDPD protocol stack and (2) Type 2 packet data service provides connections based on an Internet and ISO standard protocol stack.

TIA/EIA IS-707 defines two CDMA packet mode data services for Rate Set I (9.6 kbps): service option 4103, Internet standard protocol data services, and service option 4104, CDPD data services. Rate Set II (14.4 kbps) packet mode data service options are also defined in TIA/EIA IS-707 [13]: service option 15 for Internet standard protocols data services and service option 16 for CDPD data services.

A CDMA-CDPD network interconnects with public packet data network (PPDN) via an IWF supporting CDMA-CDPD data services (see Figure 15.5). The IWF provides network transport services as well as mobility management. Other administrative functions such as authentication and accounting are provided, but may or may not physically reside on the IWF.

Mobility management allows subscribers to maintain data connectivity over a wide geographic area, even while traveling outside the service area of the CDMA-CDPD service provider. If the mobile subscriber moves into a geographic area serviced by a different CDMA-CDPD service provider, roaming agreements between the two service providers must be in place to continue data connectivity.

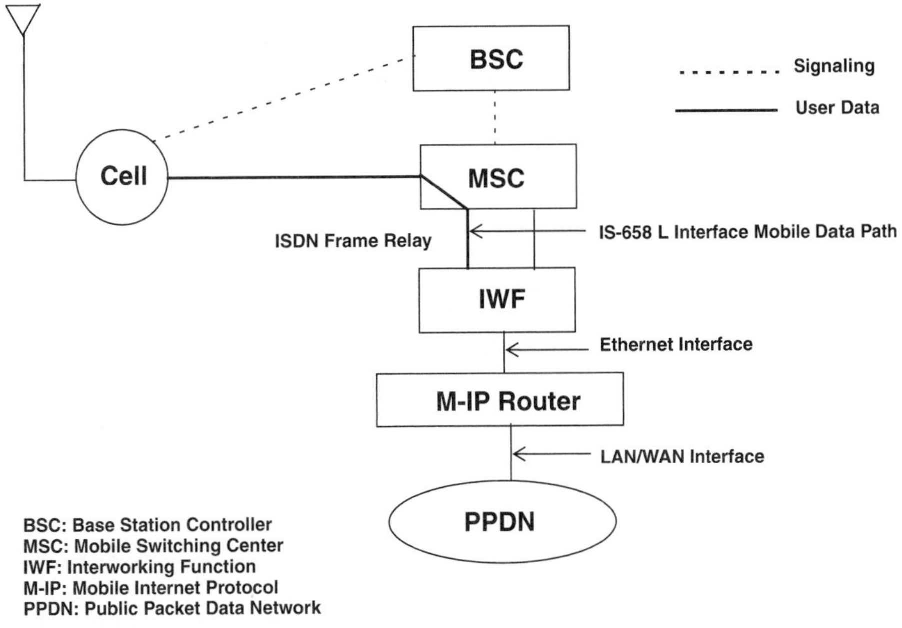

Figure 15.5 CDMA Data Service System Architecture

Before a mobile user's data application starts to run, the mobile must make a request for a packet data service and send a CDPD registration to the IWF. The mobile station deregisters from the CDPD network when the last data application is terminated. While a data application is running, a link layer connection is established for the mobile to send and receive packets to and from the CDPD network. In order to transmit data, a packet data call is established for the mobile by allocating necessary resources to connect the mobile and IWF. When there is no data to send for a specified period of time, the network resources are released until new data arrives. Without a packet data call, the link layer connection remains dormant. The link layer connection is reactivated when user data arrives. When a mobile moves, the link layer connection with the serving IWF is dropped and a new link layer connection is established with the new IWF. User packets may be lost during the transition. However, the lost packets are detected and retransmitted as needed by upper layer protocols without disruption to user applications. The interface between the IWF and MSC is specified in TIA/EIA IS-658.

15.10 Network Reference Model

The network reference model for packet data services and protocol options specified in IS-657 is shown in Figure 15.6.

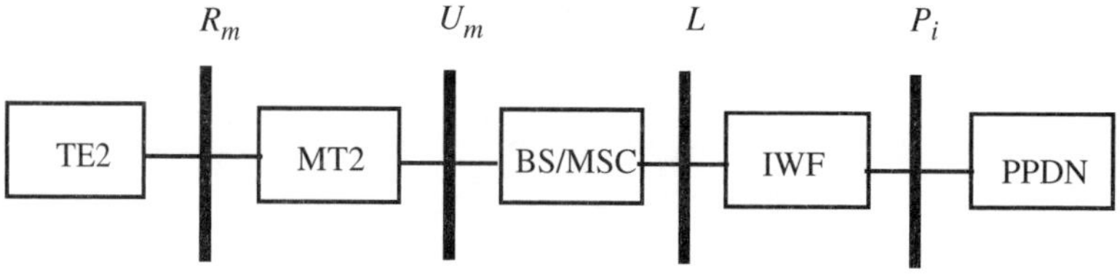

Figure 15.6 Reference Model

15.10.1 Network Element

The reference model elements are as follows:

- **Terminal equipment 2 (TE2)**—A TE2 is a data terminal device that has a non-ISDN user network interface.
- **Mobile termination 0 (MT0)**—An MT0 is a self-contained mobile termination that does not support an external interface.
- **Mobile termination 2 (MT2)**—An MT2 provides a non-ISDN (R_m) user interface.
- **Base station (BS)**—A base station represents the equipment on the land side of the U_m interface; this includes radio processing and management and protocol processing and management.

- **Mobile switching center (MSC)**—The MSC represents the functions provided by the cellular switch, including circuit-switched call management, mobile location, and mobile management.
- **Interworking function (IWF)**—An IWF provides functions needed for terminal equipment connected to a mobile termination to interface with other networks such as PSTN or CDPD networks. A CDPD mobile data intermediate system (MD-IS) is an example of an IWF.
- **Public packet data network (PPDN)**—A public packet data network (such as the Internet) provides a transport mechanism for packet data between processing elements capable of using such service.

15.10.2 Network Reference Points

The reference points are

- **Reference point (R_m)**—A physical interface connecting a TE2 to an MT2.
- **Reference point (U_m)**—A physical interface connecting an MT0 or MT2 to a BS/MSC. This is an air interface.
- **Reference point (L)**—A physical interface connecting a BS/MSC to an IWF.
- **Reference point (P_i)**—A physical interface connecting IWF to a PPDN.

15.11 Protocol Options

TIA/EIA IS-657 defines the requirements for communication protocols on the links between a mobile and IWF, including requirements for R_m, U_m, and L interfaces. The relay layer provides lower-layer communication and packet framing between the entities of the packet data service reference model. Over the R_m interface between the TE2 and the MT2, the relay layer is a simple EIA/TIA 232E interface. Over the U_m interface, the relay layer is a combination of RLP (defined in TIA/EIA IS-99 [2]) and TIA/EIA IS-95A [14] protocols. On the L interface, the relay layer uses the protocols defined in TIA/EIA IS-687. The two options for protocol stacks are discussed below.

- **Relay Layer (R_m) Interface Protocol Option**
  The relay layer R_m interface protocol option supports TE2 applications in which the TE2 is responsible for all aspects of packet data service mobility management and network address management (e.g., IP control protocol [IPCP] and CDPD registration and authentication protocols). The link layer is implemented using PPP. When using the relay layer R_m interface protocol option, the link layer connection is between the TE2 and the IWF. The network layer includes protocols such as IP and connectionless network protocol (CLNP), and packet data network registration and authentication protocols such as mobile network registration protocol (MNRP) (see Figure 15.7).

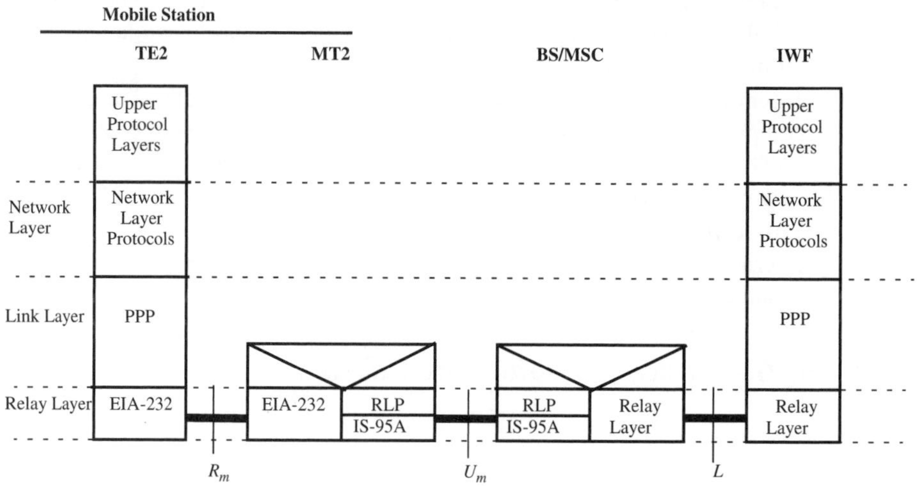

Figure 15.7 Relay Layer (R_m) Interface Protocol Option

- **Network Layer (R_m) Interface Option**
  The network layer R_m interface protocol option supports TE2 applications in which MT2 is responsible for all aspects of packet mobility management and network address management (e.g., IPCP and CDPD registration and authentication protocols). In this option, there are independent link layer connections between the TE2 and the MT2, and between the TE2 and IWF. The IWF link layer (between the MT2 and IWF) is implemented using Internet PPP. The R_m link layer (between MT2 and TE2) can be implemented using the Internet PPP protocol to support the IP network layer protocol. For this option, the network layer also provides independent service between the TE2 and MT2, and between MT2 and IWF. The TE2 includes routing protocols, and operates as if locally connected to a network routing server. The MT2 includes both routing and packet data network registration and authentication protocols (see Figure 15.8).

15.11.1 Radio Link Protocol

The radio link protocol (RLP) employs the link layer retransmission approach to improve TCP performance. Each TCP packet is segmented into several 20-ms radio link frames (192 bits at 9.6 kbps transmission rate). The RLP improves the frame error rate seen by the TCP layer. The RLP is a negative acknowledgment (NAK)-based selective repeat ARQ scheme and performs partial link recovery through a limited number of retransmissions, *n,* in the case of frame error. The RLP frame size is 20 bytes. About 30 RLP frames per TCP segment are used.

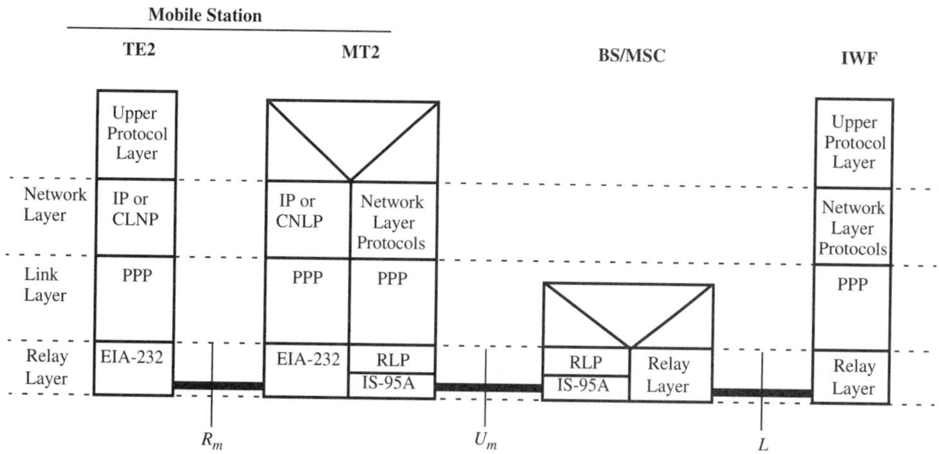

Figure 15.8 Network Layer *(R)* Interface Protocol Option

15.11.2 Applicable Mobile Type

The CDMA-CDPD feature applies to mobiles that comply with IS-657 and support service negotiation and configuration. The mobiles must support the optional RLP data frame encryption as specified in IS-657.

Two mobile configurations are defined in IS-657: relay layer (R_m) protocol option and network layer (R_m) protocol option. From the system's perspective, both mobile configurations are supported equally and transparently. However, the network layer option has an advantage over the relay layer option; since its mobility management protocols reside with the mobile unit (MT2), only the standard Internet protocol stacks run on a laptop (TE2).

15.12 Packet Data Protocol States

The IWF and the mobile use a link layer connection to transmit and receive data packets. The IWF link layer connection is opened when a packet data service option is first connected. Once an IWF link layer connection is opened, bandwidth (in form of traffic channel assignment) is allocated to the connection on an as-needed basis. The IWF link layer connection can exist in any of the following states:

- **Closed**—The IWF link layer connection is closed when the IWF has no link layer connection state information for the mobile. In this state, the mobile does not provide packet data service.

- **Open**—The IWF layer connection is open when the IWF has link layer connection state information for the mobile. The open state has two substates:
  1. **Active**—An open IWF link layer connection is active when there is an *L* interface virtual circuit for the mobile and the mobile is on a traffic channel with packet data service option connected.
  2. **Dormant**—An open IWF link connection is dormant when there is no *L* interface virtual circuit for mobile and the mobile is not on a traffic channel with packet data service option connected.

The BS/MSC and IWF maintain the state of link layer connection. The mobile maintains the state of PPP link control protocol (LCP) and manages the IWF link layer connection using LCP opening and closing procedures.

With the IWF link layer connection in dormant state, if either the mobile or BS/MSC has data to send, it is not necessary to reopen the link layer connection or to reinitialize any upper layer protocols, provided the packet data service type has not changed since the link layer last entered the dormant state. The mobile and BS/MSC can freely mix packet data service requests using any supported rate set within a service type.

15.12.1 Mobile Station Packet Data Service States

Mobile station packet data service states are [15]

- **Inactive state**—The mobile does not provide packet data service.
- **Active state**—The mobile provides packet data service.

15.12.2 Mobile Station Packet Data Service Call Control Functions

The mobile performs the packet data service call control function using the following states (see Figure 15.9):

- **Null state**—The packet data service call control function is in this state when packet data service has been activated.
- **Initialization/idle state**—In this state the mobile attempts to establish a traffic channel for the purpose of initiating packet data service.
- **Initialize/traffic state**—In this state the mobile communicates with BS/MSC on a traffic channel, and attempts to connect a packet data service option for the purpose of initiating packet data service.
- **Connected state**—In this state a packet data service option is connected. The mobile can transfer packet data.
- **Dormant/idle state**—In this state the mobile is not on a traffic channel. The mobile cannot transfer packet data.

- **Dormant/traffic state**—In this state the mobile is communicating with BS/MSC on a traffic channel, but packet data service option has been disconnected. The mobile can not transfer packet data.
- **Reconnect/idle state**—In this state the mobile attempts to establish a traffic channel.
- **Reconnect/traffic state**—In this state the mobile communicates with BS/MSC on a traffic channel and attempts to connect a packet data service option.

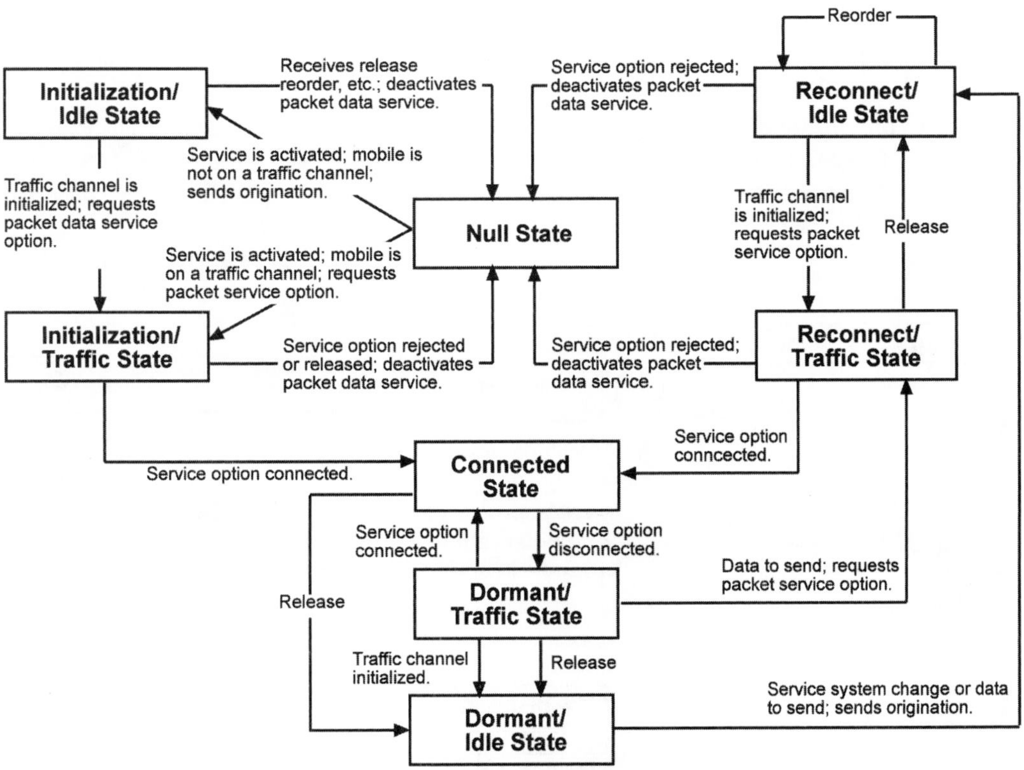

Figure 15.9 Mobile Station Packet Data Service Call Control Functions

15.12.3 BS/MSC Packet Data Service States

Packet data service processing in the BS/MSC consists of the following two states:

- **Inactive state**—In this state, the BS/MSC does not provide packet data service to the mobile.
- **Active state**—In this state, the BS/MSC provides packet data service to the mobile.

15.12.4 BS/MSC Packet Data Service Call Control Functions

The BS/MSC performs the packet data service call control function, which consists of the following five states (see Figure 15.10):

- **Null state**—In this state, BS/MSC has no connection of a packet data service option to the mobile.
- **Paging state**—In this state, the IWF has requested that the BS/MSC connect a packet data service option to the mobile for delivery of packet data and the BS/MSC pages the mobile.
- **Initialization/idle state**—In this state, the BS/MSC is awaiting initialization of a traffic channel with the mobile.
- **Initialization/traffic state**—In this state, the mobile is on a traffic channel. The BS/MSC awaits connection of a packet data service option.
- **Connected state**—In this state, a packet data service option has been connected. Packet data is exchanged with the mobile.

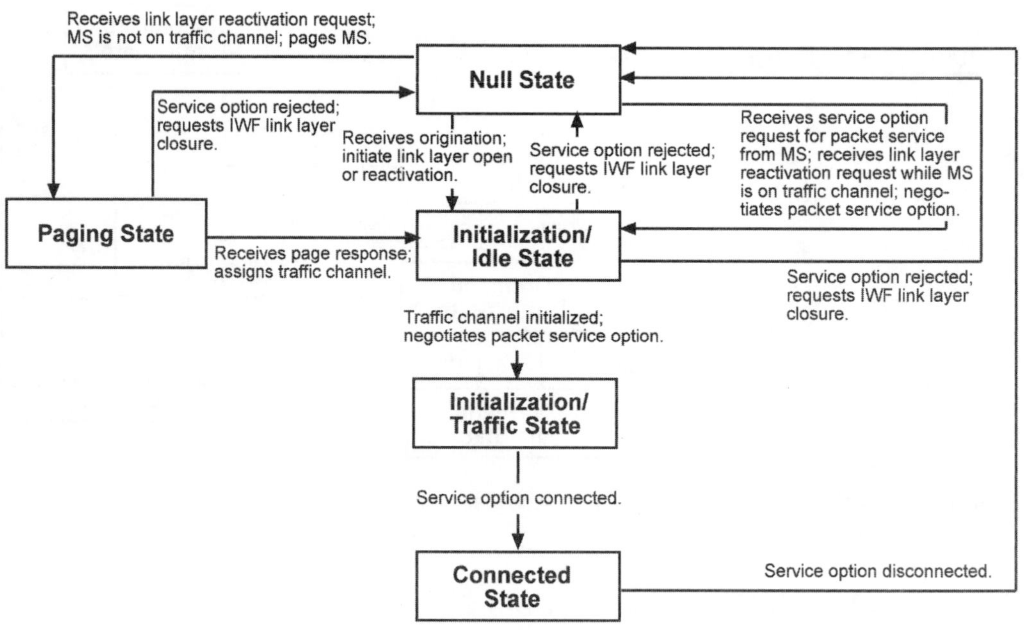

Figure 15.10 Packet Data Service Call Control States in BS/MSC

15.13 Packet Mode Data Service Features

- **Dual 9.6 and 14.4 kbps speed**—The CDMA packet data services support both 9.6 and 14.4 kbps data rates.
- **Bandwidth management**—Dormant mode is supported to ensure that bandwidth is not wasted if an endpoint enters an inactive state.
- **Mobility**—Mobility support for packet mode data services allows the user's data application to continue correct operation during movement of the mobile. Subscriber mobility is supported for intra-MSC and inter-MSC. Inter-MSC support requires that a unique system identification/network identification (SID/NID) be assigned for each MSC. This is because a new IWF will provide the packet data service when a mobile moves into a new MSC serving area, so an IWF link layer connection transfer must take place to maintain the user's packet data connectivity. According to IS-657, the mobile is required to reopen the link layer connection when it detects a change in the SID or NID of the serving system. Thus, inter-MSC packet data mobility is easily supported in a network where each MSC has a unique SID/NID. Intersystem mobility is supported if both the CDMA and CDPD have roaming agreements and the systems use shared secret data (SSD).
- **Fraud containment**—The CDPD authentication is performed as a part of CDPD registration procedures when a mobile requests the CDPD service. The mobile user is validated with the home system based on a triplet formed by the network entity identifier (NEI), the authentication sequence number (ASN), and the authentication random number (ARN) assigned to the mobile. The per-NEI authentication procedure applies to both intra- and intersystem service requests. The user data sent over the air interface are encrypted.
- **Accounting/billing interface**—CDPD accounting records consist of the data packet count, data octet count, control packet count, control octet count, discarded packet count, and packet data service connection time. The packet data service connection time starts when the mobile first registers with the CDPD network and ends when the mobile deregisters from the CDPD network. The accounting information is collected at the serving IWF and distributed to the home IWF in real time. The CDMA network will continue to collect the accounting records in the same way as the voice and circuit-mode data calls. It collects only air link connection time, not the data volume. Additional information may be added to the automatic message accounting (AMA) records for packet data calls to keep track of active and dormant links.
- **CDMA data encryption**—CDMA data encryption is performed on mobile radio link protocol (RLP) data frames carried on the digital traffic channel for IS-657 or IS-99 conforming mobiles according to "Common Cryptographic Algorithms, Revision A.1," as defined in TIA TR45.0A [2]. The data encryption feature increases security of user data transmitted over the air interface. It is mandatory for the CDMA-CDPD services (i.e., service options 7 and 15). The feature requires activation of the CDMA authentication feature with global authentication because the DataKey and *L* table used in the encryption process is calculated for each data service connection only after a successful CDMA authentication. CDMA authentication is necessary to ensure that the authentication center or VLR has the same SSD and therefore, the same DataKey and *L* table as mobile.

- **Voice or circuit-mode data service while packet data is dormant**—The feature allows subscribers with a multifunction packet data with voice and/or circuit-data capable mobile devices to use voice or circuit-data services while the packet data is in dormant mode. The mobile maintains the packet data service call control state information while handling the voice call and restarting the packet data protocol stack when the dormant link layer connection is reactivated. It is assumed that the mobile will be either manually or automatically set up in the proper mode for the requested service. The system is not involved in setting up the service mode on the mobile. The dormant link layer connection is supported with some limitations during a voice call. The limitation is that a request to reactivate the dormant link layer connection will be rejected if there is a voice call. When a packet data call is rejected, the IWF will gracefully discard the data received. This may eventually cause the user data application to be terminated if the voice call continues longer than is allowed by the user application. When a link layer connection enters a dormant state, subscriber call features that were deactivated for packet data calls will be reactivated. Tones and announcements will be played for the voice call. When a packet data call is established, subscriber call features that are not supported for packet data calls will be deactivated, and tones and announcements will be blocked.
- **Simultaneous voice and data support**—This feature allows voice and packet data traffic to be carried simultaneously within a single traffic channel. Subscribers must have a mobile device that is capable of simultaneous voice and data service. Normal voice calls are allowed and carried as primary traffic at any time, whether the mobile is connected to a packet data call or not. If a new voice call is requested while a packet data call is connected, the packet data call will be dropped to start the voice call. The packet data call will then be added using service negotiation procedures to provide a service configuration containing voice as primary traffic and packet data as secondary traffic. The transition of packet data from primary to secondary traffic may cause loss of user data, but this will be detected and the data retransmitted as necessary by the upper layer protocols without disruption to user application. Thus, during a voice call, a link layer connection may be opened and a packet data call is established, and data is carried as secondary traffic without disrupting the voice call. Both 9.6 and 14.4 kbps data rates are supported for this feature.
- **Service options 4103 and 15**—Service option 4103 provides the standard IP packet mode data services instead of the CDPD data service. Service option 4103 is for data rate of 9.6 kbps, whereas service option 15 is for the same service at 14.4 kbps. Service options 4103 and 15 may be provided with or without mobility support. Mobility management for service options 4103 and 15 will be based on mobile IP protocol. Service options 4103 and 15 without mobility are suitable for users who require only fixed services (i.e., users not moving while data applications are running).

15.14 3G1X Data Services

The cdma2000 RTT [3] allows support of increased data rates to satisfy the demand for higher-speed data services. The *CDMA 3G One Carrier Wireless System* is referred to as the "3G 1.25-MHz" system or the 3G1X system. The 3G1X system is intended to provide 3G technology while protecting the investments already made in the 2G systems. The key features of the 3G1X system are

- Significant increase in the voice capacity of the current IS-95B systems
- High-speed data capabilities up to 144 kbps on the reverse and forward links
- Backward compatibility with IS-95B signaling

Service providers are also looking for ways to add new services to the existing wireless network, so they can differentiate themselves and generate additional revenue. The large growth in Internet use is also driving service providers' interest in providing cost-effective wireless Internet access services. The 3G1X system will give service providers the ability to offer packet mode data services for demanding applications that require higher speeds, such as WWW/Internet access, Telnet, FTP, e-mail, universal mailbox, and remote data access services, all using the existing IS-95A and IS-95B networks.

IS-2000 includes the complete specifications for 3G1X system. Spreading rate (SR) is a new parameter introduced in IS-2000. SR is the PN chip rate of the system defined as the multiple of 1.2288 Mcps. Two different SRs are covered by IS-2000.

- SR1: 1.2288 Mcps-based system that uses a direct-spread single carrier. SR1 is used for IS-95A and IS-95B and will be used by the 3G1X system.
- SR3: 3.6864 Mcps-based system that uses three 1.2288-Mcps carriers on the forward channel and a 3.6864-Mcps direct-spread carrier on the reverse channel. SR3 will be used with 3G3X (or 3G 5-MHz) system.

In the 3G1X system, traffic channel data rates are grouped into radio configurations RC1 and RC2. These are the rate sets supported by IS-95A IS-95B and used with 8-kbps and 13-kbps vocoders, respectively. The 8-kbps vocoders (basic variable-rate codec [VRC] and enhanced VRC [EVRC]) and 13-kbps vocoders will continue to be used for 3G1X but new rate sets are defined to take into account the different SRs and modulations. RC3 and RC5 are derived from RC1 and are used with 8-kbps vocoders, and RC4 and RC6 are derived from RC2 and used with 13-kbps vocoders. These new rate sets use the same spreading rate, SR1 as RC1 and RC2, but have different modulation and allow for higher-rate supplemental channels. RC3 and RC4 use convolutional coding, and RC5 and RC6 allow for convolutional or turbocoding.

The high-level architecture for 3G1X high-speed packet data (HSPD) service is very similar to that of the CDMA low-speed packet data (LSPD) service (see Figure 15.5 for the major network entities for HSPD service). The IWF acts as an application processor (AP) to the MSC.

The 3G1X HSPD service differs from CDMA LSPD in that multiple channels may be used for one traffic channel. For the case of HSPD, two basic changes must be accommodated. First, a traffic channel for a single service could consist of up to two channels. The HSPD service implements a minimum of one channel, the fundamental channel (FCH), per traffic channel. Second, when higher data bandwidth is needed, a supplementary channel (SCH) either in the forward or reverse direction is added on-demand for the duration of a data burst. A wide range of data rates (effective data rates from 8 kbps to 144 kbps) is supported over each SCH in the 3G1X HSPD.

The IWF supports forward and reverse bursts. The MSC supports the detection and initiation of forward burst, as well as the correlation of SCH with FCH for forward and reverse bursts. The MSC also supports the origination of calls with the new service class. The cell site supports power control across SCH and FCH, determination and assignment of burst resources, and the unidirectional nature of SCH.

The 3G1X HSPD service provides a subscriber the ability to transmit and receive bursts of data with effective data rates of up to 144 kbps over a packet data network via 3G1X CDMA air interface. The HSPD service enables mobile users with laptop computers or other data devices conforming to the IS-2000 and IS-707B standards to access various data applications such the Internet, intranets, remote databases, e-mail, and file transfer at higher speeds.

The 3G1X HSPD service is built on the CDMA LSPD, discussed earlier in the chapter. The operation of FCH and packet data call setup and tear-down procedures in HSPD are almost the same as LSPD service when there is no data burst in progress. To end users, the most visible advantage of HSPD service over LSPD service is speed. Both services support dormant mode and full mobility. Although simultaneous voice and data are not supported in either 3G1X HSPD or LSPD service, voice or circuit-mode data service can be provided while packet data is in dormant mode. The CDMA packet data standard specifies two types of service with mobility

- CDPD emulation
- Mobile IP

In 3G1X air interface, a burst mode capability is defined to allow better interference management and capacity utilization. An active high-speed packet data mobile always has a traffic channel with a fundamental code. The channel is called the FCH. An active HSPD call with need for higher bandwidth in either the forward or reverse direction could be allocated an additional channel for the duration of a data burst; this is performed on the order of seconds. The additional channel allocated during this state is called the SCH. One SCH is assigned per data service. An SCH with data rate of 19.2 kbps or higher is equivalent to multiple voice calls in air interface capacity.

The data rate and duration of the burst (i.e., SCH) is dynamically determined by infrastructure and is dependent on load, interference, and resource availability conditions. Static allocation of multiple codes to a small number of users can result in inefficient use of air interface capacity. Dynamic infrastructure-controlled burst allocation makes it possible to share bandwidth efficiently among several high-speed packet data mobiles. The burst allocation scheme is designed to maximize utilization of channel bandwidth and system resources.

The SCH does not offer a guaranteed bit rate. The data rate offered by the FCH with a raw data rate of 9.6 kbps is always guaranteed to the HSPD user. During call admission, an HSPD call has the same priority as a voice call. The FCH gets the same treatment as a voice call. The allocation of data burst (i.e., the allocation of SCH) has lower priority than call admission. Voice calls are never terminated to accommodate an HSPD call. Also, the impact on the quality of voice calls due to HSPD calls is minimized. To achieve this

- Burst admission takes into consideration capacity that may be needed for incoming voice calls or handoffs.
- The burst duration is kept short (a few seconds) in order to facilitate fairness of resource usage among multiple contending data users and to ensure that high-speed data users do not adversely impact voice service.
- SCHs will be released at the end of the burst or may be released prematurely before the assigned burst duration is over if power, interference, or traffic conditions change during the burst. The load threshold for premature burst termination will be typically set at a value lower than the threshold for call admission.

The HSPD service and the present CDMA data service can be physically constructed in the same manner in the MSC with differences in the type of protocol handler (PH) for a particular service (provided information processing is done in the MSC). The new PH for HSPD services will terminate the RLP protocol from the mobile. The PH for HSPD services will correlate multiple frame selectors associated with each RLP termination where multiple channels are assigned to a specific traffic channel for a data burst.

For the forward direction, the burst allocation is triggered when data gets backlogged in the system. For the reverse direction, data builds up at the mobile which, in turn, sends an SCH request message to the system, triggering the burst allocation procedure.

The service could be asymmetric, i.e., high-speed packet data mobile, at any given time, may be assigned different bandwidths on the forward and reverse link. This maximizes use of bandwidth in both directions by meeting the bandwidth demand of end users in each direction.

15.14.1 3G1X HSPD Service in Forward Direction

The system supports origination and page response messages with the service option designated for 3G1X mobile IP service. It also negotiates with mobiles the maximum data rates for a connection, taking into account mobile capability, subscriber, and system-provisioned parameters. The system supports forward and reverse data on the FCH in the same way as with LSPD service. The FCH is allocated when an HSPD call is set up. The FCH stays up for the entire duration of the active call. The system allows 3G1X HSPD calls with a need for higher bandwidth in the forward direction by allocating an SCH with the allowed data rates for the duration of the data burst. The data rates supported on forward SCH (F-SCH) are 9.6, 19.2, 38.4, 76.8, and 153.6 kbps.

To conserve air link capacity and other system resources, the system supports a dormant mode by maintaining a link layer connection record and a packet data inactivity timer while the call is active for each mobile IP service user. Mobility support for 3G1X mobile IP service allows users' data applications to continue correct operation while moving. There are two levels of mobility involved: link layer mobility (supported by CDMA wireless network) and network layer mobility (supported by the mobile IP protocol). The CDMA network maintains the link layer connection records and handles IWF transfer to maintain user packet data connectivity while mobiles roam across IWF serving areas. Soft handoffs and semi-soft handoffs are supported for 3G1X HSPD calls. Hard handoffs for 3G1X HSPD calls will be rejected. User data lost to premature call terminations will be retransmitted as necessary by the upper layer protocols. Neither 2G to 3G nor 3G to 2G handoff is supported for packet data calls. When a mobile roams from 3G to 2G cell or from 2G to 3G cell, the handoff request will be rejected, resulting in the mobile losing its connectivity to the mobile IP packet data network. HSPD calls with mobile IP service on 3G air interface cannot be handed off to LSPD calls with CDPD service on 2G air interface.

15.14.2 Mobile IP Network Architecture

The mobile IP protocols provide mobility to Internet access. Two methods have been proposed: (1) Internet access configuration using a mobile IP tunnel, and (2) Internet access configuration using voluntary L2TP tunnels and mobile IP.

Internet Access Configuration Using a Mobile IP Tunnel. Figure 15.11 provides an overview. Mobility is provided through the mobile IP protocol. Forward traffic is routed between the home and foreign agents using a mobile IP tunnel. Reverse traffic is routed directly from the foreign agent to the remote server.

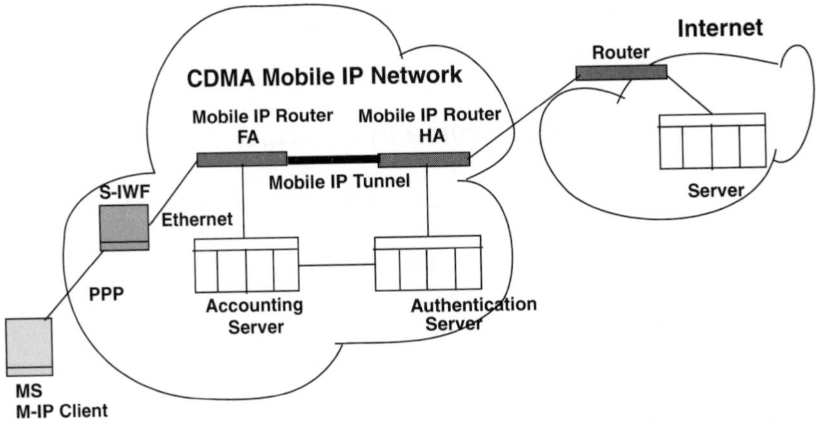

Figure 15.11 Internet Access Network Architecture

- **Functional Allocation**
  This configuration supports mobile IP as follows:
  - A mobile IP-capable router on the visited network provides the foreign agent (FA) function.
  - A mobile IP-capable router on the home network provides the home agent (HA) function.
  - An external authentication server may be connected to the HA router to maintain subscriber information and perform mobile IP authentication.
  - An external accounting server may be connected to interface with the FA routers to collect and store accounting records. A single external server may perform both the authentication and accounting functions.
  - The IWF in the serving system (S-IWF) and FA router reside on the same LAN and are interconnected by Ethernet.
  - The S-IWF terminates PPP protocol and relays IP datagrams received from the mobile to the designated mobile-IP FA router on the local network.
  - The HA router is fixed, and its IP address is conveyed to the FA by the mobile in the mobile IP registration message.

- **Protocol Stack**
  Figure 15.12 shows the protocol stacks provided by each of the network elements in a mobile IP implementation.

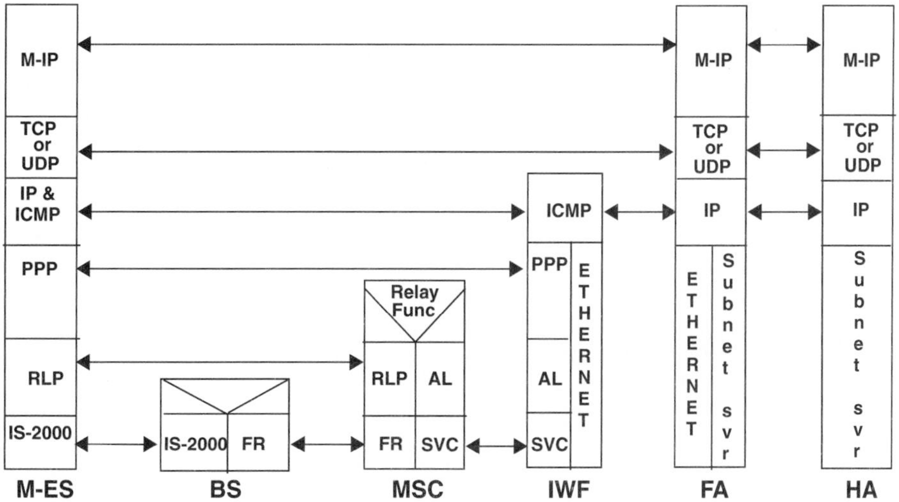

FR = Frame relay
SVC = Switched virtual circuit
FA = Foreign agent
HA = Home agent

Figure 15.12 Internet Access Protocol Stack—Mobile IP Implementation

- **Security**
  - Three levels of authentication and authorization validation are provided: IS-41 service authorization validation (mandatory), IS-41 authentication (optional), and mobile HA authentication for mobile IP registration/reply messages (mandatory).
  - Data privacy over the air link can optionally be provided by using RLP encryption (ORYX) between the mobile and packet-switching unit (PSU). RLP encryption requires the IS-41 authentication feature to be activated.
  - Identification inserted in the mobile IP registration request/reply provides antireply protection for registration messages.

- **Packet Routing**
  - Datagrams sent by the mobile are delivered to the S-IWF for relaying to the FA. The FA routes datagrams from the mobile to the destination.
  - The HA encapsulates datagrams destined for the mobile in another IP header (i.e., IP-in-IP encapsulation) and forwards them to the FA.
  - The FA decapsulates the datagrams from the HA over the mobile IP tunnel and relays the original datagrams destined for the mobile to the S-IWF for delivery to the mobile.
- **Accounting**
  - FA routers or IWFs collect accounting records and send them to designated accounting servers for storage.

Intranet Access Configuration using Voluntary L2TP Tunnels and Mobile IP. Figure 15.13 shows the intranet access network architecture using voluntary L2TP tunnels and mobile IP. Mobility is provided by mobile IP protocols. However, a voluntary tunnel is established by the client between itself and an L2TP server. All forward and reverse traffic is sent through the L2TP tunnel. Voluntary tunneling is client-initiated tunneling in which the client encapsulates and encrypts the data to be transmitted right from his laptop. Mobile users initiate voluntary tunneling by invoking L2TP client software on the laptop, which directly interacts with L2TP server software on the L2TP network server (LNS) over the mobile IP connection.

- **Protocol Stack**
  Figure 15.14 shows the protocol stacks provided by each network element in a mobile IP implementation using a voluntary L2TP tunnel.

- **Addressing**
  - Private network assigns address to terminal by LNS using PPP contained within voluntary tunnel.
  - Wireless session protocol (WSP) assigns IP address to user for link between mobile and wireless network (i.e., the address used for mobile IP tunneling).

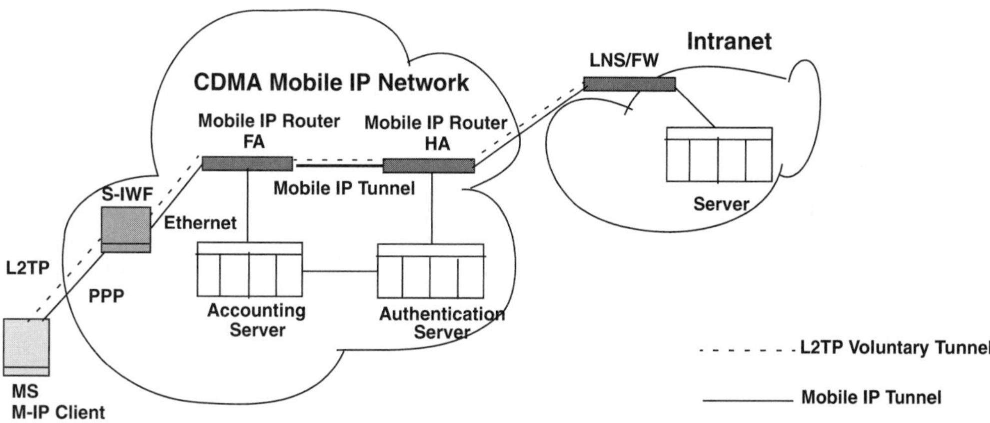

Figure 15.13 Intranet Access Network Architecture

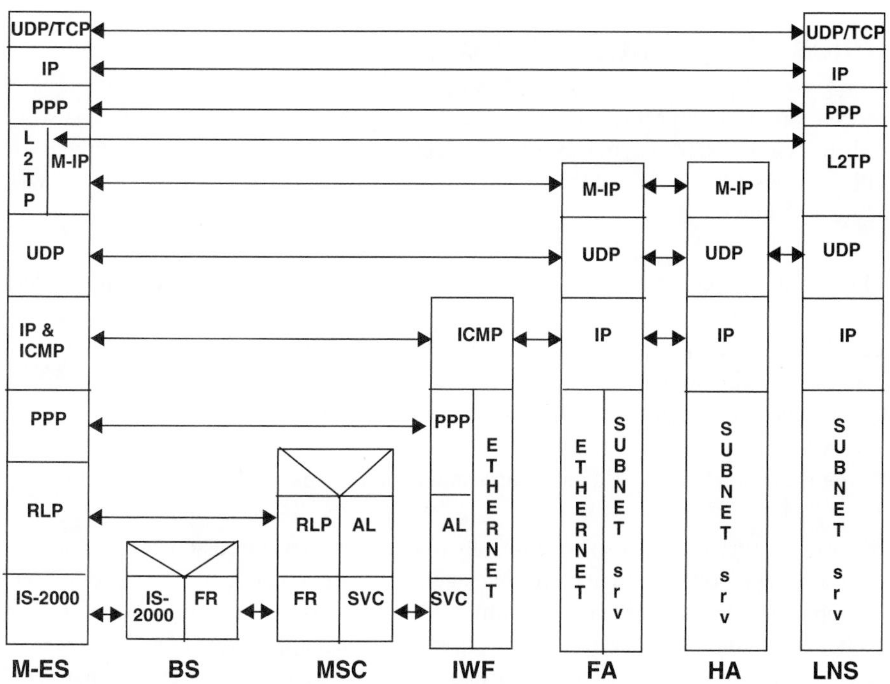

Figure 15.14 Intranet Access Protocol Stack with Voluntary L2TP Tunnels and Mobile IP

- **Security and Firewall Traversal**
  - Two levels of authentication are provided: IS-41 authentication (optional) and mobile HA authentication during mobile IP registration (mandatory).
  - Identification inserted in the mobile-IP registration request/reply provides antireply protection.
  - LNS authenticates terminals using PPP authentication integrity such as challenge handshake authentication protocol (CHAP).
  - Alternatively, end-to-end authentication, integrity, and confidentially can be provided if the terminal and LNS support IP.

- **Protocol Overhead**
  Overhead due to voluntary tunnel is on the order of 7 to 10 percent for e-mail/Web/FTP applications. This is attributable to the inclusion of the L2TP protocol layer.

- **Accounting**
  FA routers or IWFs collect accounting records and send them to designated accounting servers for storage.

15.15 Wireless Application Protocol

The wireless application protocol (WAP) specifies an application framework and network protocols for wireless devices such as mobile telephones, pagers, and personal digital assistants (PDAs). WAP resulted from the work of the WAP Forum. The objectives of the WAP Forum are as follows:

- To bring Internet content and advanced data services to digital cellular phones and other wireless terminals
- To create a global wireless protocol specification that will work across different wireless network technologies
- To enable the creation of content and applications that scale across a very wide range of bearer networks
- To embrace and extend existing standards and technologies wherever appropriate

Some of the services that can benefit from WAP technology are customer care and provisioning, message notification and call management, e-mail, telephony value-added services, mapping and locator services, e-commerce transactions and banking services, online address books and directory services, and corporate intranet applications.

The "always-on" design of the GPRS packet-switched data service means that WAP terminals using GPRS bearer services can remain logged on to a WAP server to transfer data such as e-mail. Users will be billed for the amount of data transferred, not for the time spent in transferring data or waiting for it.

Currently, GSM phone primarily delivers information via "alert type" short message service (SMS) messages, which typically consist of stock prices, sport results, weather reports, diary reminders, news headlines, bank balances, telemetry data, and e-mail notification. WAP microbrowsers will allow interactivity with the information using online browsing capabilities. Most of the GSM phones released during 2000 are WAP-compliant and by the first quarter of 2001 were also able to use GPRS and circuit-switched data facilities. In the WAP e-commerce area, there will be advanced forms of access to online payments, purchase of tickets and goods, and even WAP-based gambling—all in a secure encrypted manner. A plethora of WAP servers developed jointly by telecoms and traditional IT vendors is appearing, providing "mobile LAN" access for employees to corporate intranets, internal e-mail, diary synchronization, screen text headers for voice mail messages, unified messaging, receipt and updating of customer job information, and access to company phone and database directories.

The WAP programming model is similar to the WWW programming model, and it offers many benefits to the application developer community. These include a familiar programming model, a proven architecture, and the ability to leverage existing tools (e.g., Web servers, Extensible Markup Language [XML] tools etc.). WAP defines a set of standard components that enable communication between mobile terminals and network servers, including

- Standard naming model—WWW-standard URLs are used to identify WAP content on origin servers.
- Content typing—All WAP content is given a specific type consistent with WWW typing, thus allowing WAP users to correctly process the content based on its type.
- Standard communication protocols—WAP communication protocols enable the communication of browser requests from the mobile station to the network Web server.
- Proxy technology—The WAP uses proxy technology to communicate between the wireless domain and the Internet. The WAP proxy typically consists of the following:
  - Protocol gateway—Translates requests from the WAP protocol stack to a WWW protocol stack.
  - Content encoders and decoders—Translate WAP content into compact encoded formats to reduce the size of data over the network.

This information enables a mobile user to browse a wide variety of WAP content and applications, and ensures that an application author can build services and applications that run on a wide base of mobile terminals. More details about WAP are provided in Chapter 17.

15.16 Summary

In this chapter we presented the data capabilities of the cdmaOne system, which allows for code or channel aggregation to achieve data rates of 64–115 kbps. We discussed the network reference model for packet mode data services and outlined the packet mode data service features in IS-95B. We then presented 3G1X data services of the 3G system, designed to support increased data rates. The 3G1X system will allow service providers to offer packet data services for demanding applications that require higher speeds, such as WWW/Internet access, e-mail, or remote data access. We discussed the mobile IP protocols for providing mobility to Internet access. Two proposed methods—(1) Internet access configuration using a mobile IP tunnel, and (2) Internet access configuration using voluntary L2TP tunnels and mobile IP—were presented. A brief introduction of WAP was also provided.

15.17 References

1. TIA/EIA/IS-707-A.7, "Data Service Options for Wideband Spread Spectrum Systems: Analog FAX Service," July 1998.
2. TIA/EIA-95B, "Mobile-Station-Base Station Compatibility Standard for Dual-Mode Spread Spectrum System," August 1998.
3. TIA/EIA TR45-ISD/98.06.02.03, "The cdma2000 ITU-R RTT Candidate Submission," July 1998.
4. TIA/EIA/IS-657, "Packet Data Service Options for Wideband Spread Spectrum Systems," 1996 .
5. TIA/EIA/IS-707-A.1, "Data Service Options for Spread Spectrum Systems: Introduction and Service Guide," July 1998.
6. TIA/EIA/IS-707-A.2, "Data Service Options for Wideband Spread Spectrum Systems: Radio Link Protocol," July 1998.
7. TIA/EIA/IS-707-A.3, "Data Service Options for Wideband Spread Spectrum Systems: At Command Processing and R_m Interface," July 1998.
8. TIA/EIA/IS-707-A.4, "Data Service Options for Wideband Spread Spectrum Systems: Aysnc Data and FAX Services," July 1998.
9. TIA/EIA/IS-707-A.5, "Data Service Options for Wideband Spread Spectrum Systems: Packet Data Services," July 1998.
10. TIA/EIA/IS-707-A.8, "Data Service Options for Wideband Spread Spectrum Systems: Radio Link Protocol Type 2," July 1998.
11. TIA/EIA/IS-707-A.9, "Data Service Options for Wideband Spread Spectrum Systems: High Speed Packet Data Services," July 1998.
12. TIA/EIA/IS-99, "Data Service Option Standard for Wideband Spread Spectrum Digital Cellular System," 1995.
13. TIA/EIA/IS-658, "Data Services Networking Function Interface for Wideband Spread Spectrum Systems," 1996.
14. TIA/EIA/IS-95-A, "Mobile Station-Base Station Compatibility Standard for Dual-Mode Wideband Spread Spectrum Cellular System," May 1995.
15. TIA/EIA/IS-634-A, "MSC-BS Interface for Public Mobile and Personal Communication Systems," 1998.

PART III

Wireless Networks

In the third part of this book, several important topics in wireless networking are discussed. These include wireless local loop (WLL), wireless application protocol (WAP), Bluetooth, and wireless local area network (WLAN). WLL technology is considered because radio systems can be rapidly developed, easily extended, and are distance insensitive. WLL eliminates the wires, poles, and ducts essential for a wireline network. In areas isolated from land, the WLL network can be installed rapidly. An additional advantage of the WLL system is that it reduces the cost of operation and maintenance since there is no copper wire to maintain. WAP has become an industry standard for providing data to wireless handheld devices. Bluetooth enables users to connect a wide range of computing and telecommunication devices without using any additional or proprietary cables. A WLAN is a flexible data communications system implemented as an extension of, or as an alternative for, a wired local area network (LAN).

Wireless networks require large numbers of translations and other parameters to define their physical and logical configurations. In wireless systems, data changes are frequent for several reasons: system growth, handoffs, changes in patterns of subscriber density, quasi-dynamic frequency channel assignments, trunk assignments, and ongoing fine-tuning of the system to improve quality of service. Because of these reasons, an efficient network management scheme is essential for the health of wireless networks. Chapter 20 is devoted to the discussion of network management functionalities.

Two chapters (Chapters 18 and 19) in this part present wireless network planning and radio frequency (RF) optimization to highlight the steps that are important for code division multiple access (CDMA)-based networks.

Wireless Local Loop

16.1 Introduction

Rural and isolated areas, as well as regions subject to natural disasters such as earthquakes, floods, and hurricanes, introduce greater challenges to providing telephone services. A wireless local loop (WLL) system provides a better solution to alleviate these difficulties. WLL provides two-way communication services to near-stationary users within a small service area. It refers to the use of radio and wireless technologies in the local loop instead of copper wire technologies to provide access to the public switched telephone network (PSTN). WLL technology is attractive because radio systems can be rapidly developed, easily extended, and are distance insensitive. WLL eliminates the wires, poles, and ducts essential for a wireline network. In areas isolated from land, the WLL network can be installed rapidly. An additional advantage of the WLL system is that it reduces the cost of operation and maintenance since there is no copper wire to maintain.

In many cases a WLL system is economically preferable to a wired network, and as such is a direct substitute for the wireline local loop [1]. In addition, wireless technology offers a platform on which to build additional mobility-related services that may generate additional revenue to service providers.

There are many applications using WLL radio technology in urban, suburban, and high-density areas. Generally, wireline equipment is overengineered to fulfill future traffic demand and minimize frequent costly deployment. A wireline system takes about three to five years to fully deploy. The practical fill-rate for a wireline network is less than 70 percent. In contrast, a wireless network can be deployed within two to six months with a practical fill-rate of greater than 90 percent. The basic attributes of a WLL system are the following:

- Coverage increase. The path loss is close to free-space loss.
- Capacity increase. The required signal-to-interference ratio for a WLL $(S/I)_{wll}$ is much lower than the required $(S/I)_{cell}$ for a cellular system.
- High-gain directional antennas can be used. Thus, the interference decreases and frequency reuse distance is reduced (i.e., capacity gain).

Technology alternatives for WLL can be classified in four general categories: point-to-point systems, cellular systems modified for WLL applications, satellite-based systems, and systems designed especially for WLL applications. There are a number of systems designed especially for WLL applications; some are based on standards, whereas others are proprietary solutions. Three low-powered systems based on time division multiple access (TDMA) air interface standards especially designed for WLL applications are the European Digital Enhanced Cordless Telecommunications (DECT), the Japanese Personal Handy-phone System (PHS), and the Personal Access Communication System (PACS) (refer to [2,3] for a comparison of these low-power WLL systems).

In this chapter, we focus on the WLL system based on code division multiple access (CDMA) technology. We present user requirements and services that a WLL system provides. We then provide details of the TR-45.1 architectural reference model for a WLL. We also present details of air interface and system architecture of WLL systems based on the W-CDMA technology.

16.2 User Requirements for a WLL System

To be feasible, a WLL system must provide at least the same functionality as a wired network, if not more, with the same or better quality of service. The following features characterize the end-user requirements of residential and small business customers:

- Voice service
- Voiceband data service (access to Internet via modem)
- Fax service
- Data service (e.g., Internet access)
- ISDN service (basic rate access [BRA] $2B + D = 144$ kbps)

A service provider trying to meet end-user requirements demands a greater flexibility from the WLL system. The services that can be offered by a WLL system include:

- **Voice service:** 64-kbps pulse code modulation (PCM), 32-kbps adaptive differential PCM (ADPCM) or 16-kbps low-delay codebook excited prediction (LD-CELP)
- **Voiceband data service:** ~56 kbps (fax, modem)
- **Data service:** ~115 kbps (Internet service, PC communication)
- **ISDN service:** 144 kbps ($2B + D$)

It may be noted that there are many other user requirements that might be suitable for a WLL system. An example would be a high-speed link (up to 100 Mbps) between the end user and service provider's network. This type of service is more suitable for large business customers and future residential services directed toward voice, data, and video integration.

16.3 WLL Systems

WLL system can be deployed in two ways: over a personal communication services (PCS) system or over a local multipoint distribution services (LMDS) system.

- **PCS system architecture**—The architecture of a PCS system has three main components: base transceiver station (BTS), base station controller (BSC), and mobile switching center (MSC) (see Figure 16.1).

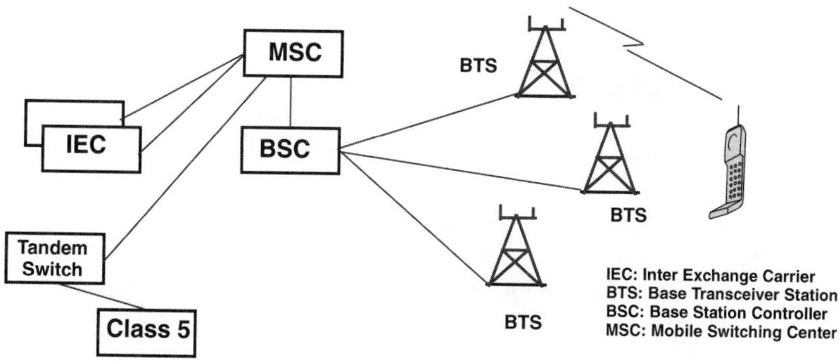

Figure 16.1 PCS System Architecture

- **LMDS architecture**—An overall architecture of an LMDS system is shown in Figure 16.2. The digital connection unit (DCU) takes the digital signals from the telecommunications switching unit (TSU) and video provisioning unit (VPU), combines them in a common broadband digital signal, assigns the signal a time division multiplexing (TDM) form for transmission, and places it in the appropriate radio frequency (RF) format for transmission to the base station unit (BSU). The input from the BSU is passed through the digital connection unit (DCU), then it is fed to the TSU if it is data or voice, or to the VPU if it is a video control signal. The BSU transmits in the LMDS band to the network interface unit (NIU) located at the end user premises. The signal from the BSU is TDM and the NIU returns as TDMA.

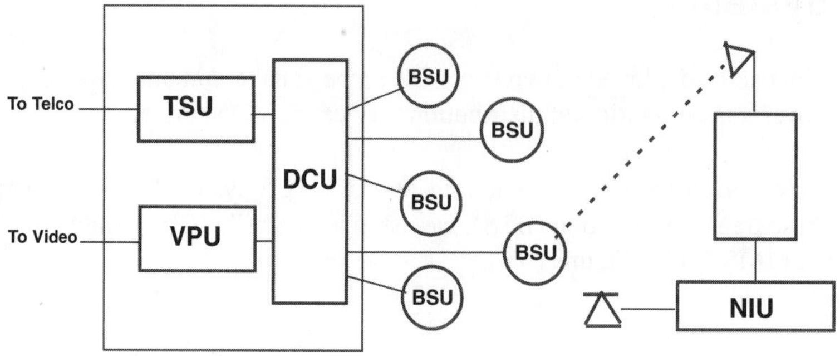

Figure 16.2 LMDS System Architecture

16.4 WLL Architecture

A simplified TR-45.1 architectural reference model for WLL is shown in Figure 16.3. The wireless access network unit (WANU) consists of the BTS or radio ports (RPs), the radio port controller unit (RPCU), access manager, and home location register (HLR), as required. The interface between the WANU and the switch is called A_{wll} and can be IS-634, IS-653, GR-303, V5.1, ISDN BRA, and so on.

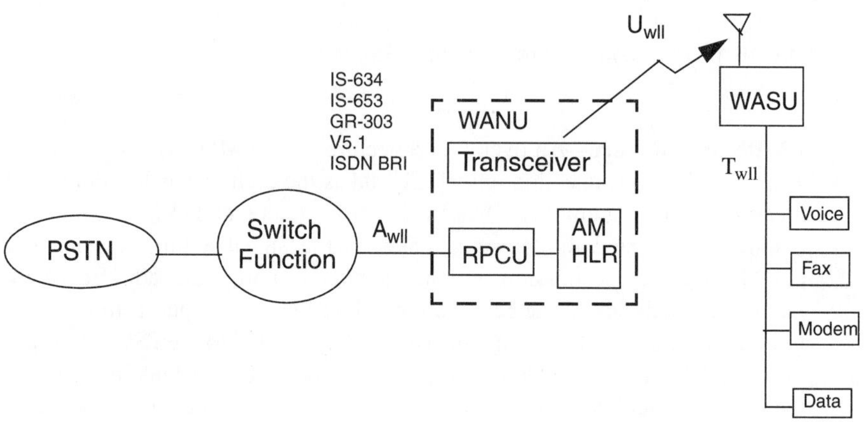

Figure 16.3 TR-45 Wireless Local Loop Reference Model

The air interface between the WANU and the user side is called *U*. The WANU should provide for the authentication and privacy of the air interface, radio resource management, limited mobility management, and over-the-air registration of subscriber units (SUs). It may also be required to provide operation, administration, maintenance, and planning

(OAM&P); as well as routing, billing and switching functions as necessary or appropriate. The WANU also provides protocol conversion and transcoding of voice and data. To support voiceband data and Group 3 fax, an interworking function (IWF) may also be required.

The TR-45.1 reference model combines the BTS and radio port control unit (RPCU) or BSC within a single functional network element, i.e., the WANU. The transmission backhaul between the base station (BS) and switching system can be leased lines, cables, or microwave. Compared with the leased lines and cable approaches, microwave has the advantage of speed and flexibility in deployment, and right of way when crossing third-party property is not an issue. However, when deploying microwave, consideration must be given to frequency availability, tower height restrictions, and the limitations of antenna size.

The wireless access subscriber unit (WASU) provides an air interface toward the network and a traditional interface T_{wll} to the subscriber. This interface includes protocol conversion and transcoding, authentication functions, local power, OAM&P, dual-tone multifrequency (DTMF) dial tone, and RJ-11 functions. A modem function may also be required to support voiceband data so that analog signals such as data and fax can be transported over the air digitally and reconstructed by IWF in the network. The T_{wll} interface can be RJ-11 or RJ-45. An O_{wll} interface is defined to provide the OAM&P interface to the WLL system.

The switch can be a digital switch with or without advanced intelligent network (AIN) capability, an ISDN switch, or an MSC. In general, the WLL features and services can be those supported by the switch, as shown in Figure 16.3. The interface between the radio system network elements and non-AIN digital switch can be via a GR303 interface with robbed bit signaling allowing WLL to interface with any end office digital switch. The WLL services are transparent to the switch, and the subscriber data bases reside in the switch. All switch-based services can be supported. The switch sees a logical line appearance, that is, a line per subscriber with no concentration at the A_{wll} interface in the TR.45 reference model. The radio system supports both fixed and mobile subscribers. However, mobility is limited to the areas served by a single WLL controller; wide area roaming is not supported. The access manager/home location register (AM/HLR) handles authentication and privacy. The maximum data rate supported on the interface is 56 kbps; clear channel data, at 64 kbps, cannot be supported. Mobility management signaling between RPCU and AM/HLR may use any available transport (e.g., TCP/IP, X.25, etc.). Alternatively, the services and features can be supported by HLR or an AIN service control point (SCP). The architecture shown in Figure 16.4 takes advantage of the AIN capabilities of the switch. The subscriber data bases reside in the HLR. Features supported by the system are not the same as the features supported by the switch. Wide area roaming and mobility can be supported. The WLL controller uses the existing call control and mobility management protocols. Mobility management protocol between the access manager (AM) and the VLR may use ISDN non-call–associated signaling (NCAS), signaling system 7 (SS7) TCAP, or other transport options. Mobility management protocol between the WLL controller and the AM may use any available transport option.

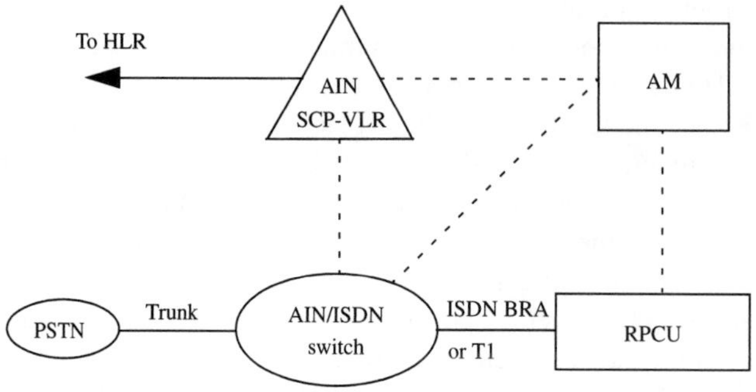

Figure 16.4 WLL System with Features Derived from HLR

In order for the air interface to pass voiceband data and Group 3 fax, a network data IWF is usually required. Figure 16.5 provides a general view of the network architecture for supporting interworking of these wireless and wireline data services. One approach that has been described in detail is to use X.25 on ISDN B- or D-channels for the intermediate network and to assume that the remote network is the PSTN. The radio access system refers to the radio devices and the (nonswitched) wireline backhaul necessary to connect the radio ports or base stations to their controllers. The controllers are connected via an intermediate network to the data IWFs. The IWF is required to convert the digital data on the air interface. This may include rate adaptation and the termination of a specialized error and flow control radio link protocol. The IWF must also be on a network from which the desired data application (represented as a host) is reachable. The wireline network on which the IWF and host reside is called the remote network.

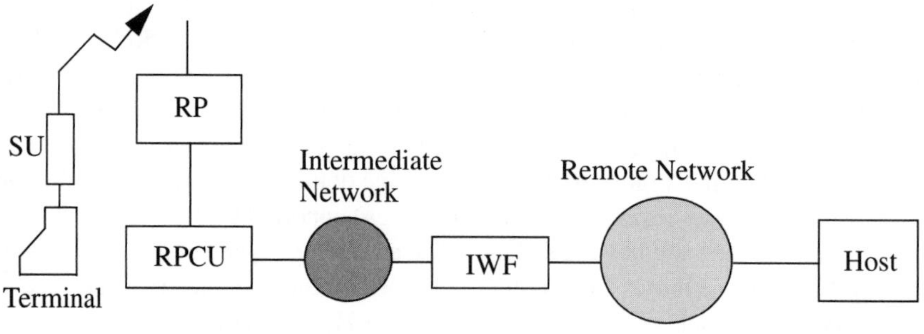

Figure 16.5 Generalized Network Architecture for Wireless-to-Wireline Data Interworking

16.5 Capacity of a CDMA WLL

Since in a CDMA system every cell uses the same RF channel, the reuse factor N is one. The number of users, M, in a cell is a function of S/I [4]. We consider a scenario shown in Figure 16.6, with the mobile unit at position A, the worst case. The serving cell and two close-in interfering cells I_1 and I_2 are at the same distance R from the mobile. The S/I at position A can be expressed as:

$$\frac{S}{I} = \frac{S}{I_s + I_a} = \frac{E_b/N_0}{G_p} \tag{16.1}$$

where

I_s = self-interference

I_a = adjacent interference

E_b = energy per bit

N_0 = (interference + noise) density

$G_p = B_w/R$ = processing gain

Assuming that the dominant adjacent interferers are I_1 and I_2

$$\left[\frac{S}{I}\right]_{wll} = \frac{P_t R^{-\gamma}}{I_s + I_a} = \frac{P_t R^{-\gamma}}{(M_{wll} - 1)P_t R^{-\gamma} + 2M_{wll} P_t R^{-\gamma}} \approx \frac{1}{3M_{wll} - 1} \tag{16.2}$$

where

$I_s = (M_{wll} - 1)\, P_t\, R^{-\gamma}$

$I_a = I_1 + I_2 + \Delta = 2\, M_{wll}\, P_t\, R^{-\gamma} + \Delta \sim 2\, M_{wll}\, P_t\, R^{-\gamma}$

P_t = signal power

R = distance from transmitter

γ = path loss exponent

From Equation (16.2), the total number of traffic channels in a WLL cell is

$$M_{wll} = \frac{1}{3}\left[\frac{1}{(S/I)wll} + 1\right] \tag{16.3}$$

Similarly, we can write that the total traffic channels in a cellular cell is

$$M_{cell} = \frac{1}{3}\left[\frac{1}{(S/I)cell} + 1\right] \tag{16.4}$$

Since $(S/I)_{wll} > (S/I)_{cell}$, therefore $M_{wll} > M_{cell}$.

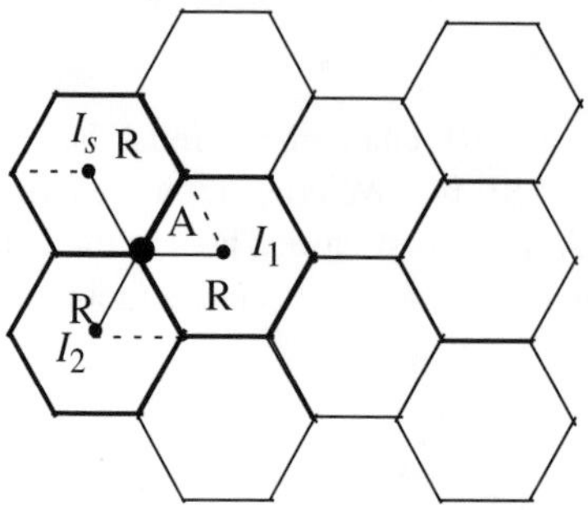

Figure 16.6 CDMA System with Its Interference

Substituting Equation (16.1) into Equation (16.3), and introducing power control efficiency and voice activity factor, we get

$$M_{wll} = \frac{\lambda\alpha}{\nu_f}\left[\frac{1}{3}\left\{\frac{1}{\frac{(E_b/N_0)_{wll}}{G_p}} + 1\right\}\right] \tag{16.5}$$

where
λ = three-sector antenna gain (~2.55)
α = power control efficiency (~0.9)
ν_f = voice activity factor (~0.5)

EXAMPLE 16.1

Compare the capacity of a CDMA WLL cell with that of a CDMA cellular cell, assuming the following data:

Bandwidth (B) = 1.23 MHz
Information rate = 9.6 kbps
No. of sectors = 3
$(E_b/N_0)_{wll}$ = 5 dB
$(E_b/N_0)_{cell}$ = 7 dB
Voice activity = 0.5
Power control efficiency = 0.9

$$M_{wll} = \frac{2.55 \times 0.9}{0.5}\left[\frac{1}{3}\left\{\frac{128}{3.16} + 1\right\}\right] = 63.5$$

and

$$M_{cell} = \frac{2.55 \times 0.9}{0.5}\left[\frac{1}{3}\left\{\frac{128}{5} + 1\right\}\right] = 40.7$$

$$\frac{M_{wll}}{M_{cell}} = \frac{63.5}{40.7} = 1.56$$

The CDMA system can be used in high-capacity applications because the frequency reuse factor N is one. In deploying a CDMA WLL, power control is used to reduce near-far interference. The power control is set at the beginning of the service, according to all the fixed-to-fixed links, and leaves them untouched until the environment changes. No frequency planning is required. Increasing the processing gain and reducing the required $(E_b/N_0)_{wll}$ increase system capacity. A high-gain directional antenna (smart antenna) can also be used to further reduce the adjacent interference.

16.6 W-CDMA WLL

16.6.1 WLL Based on Korean WLL Specifications

The air interface, *Wireless Local Loop Specifications* [5], is applied to a wireless connection between a terminal and base station. The air interface specifies a CDMA-based protocol operating in the 2.3-GHz to 2.4-GHz band. The standard supports telephone, facsimile, and ISDN services through WLL for PSTN users. The standard specifies wireless packet transmission mode to support high-speed data services efficiently through the WLL. The standard specifies a 5-MHz or 10-MHz bandwidth. The WLL system described in this chapter uses 10-MHz bandwidth.

The standard specifies requirements for physical layer (layer 1), the data link layer (layer 2), and the network layer (layer 3) of the WLL connection [6]. The terminals are the existing fixed PSTN and ISDN terminals and the terminal interface module connects the existing fixed terminals with the wireline network through a radio interface. The base station provides the function of network radio interface termination and the function of wired interface with the switching system. The switching system provides a local switching or toll switching function and is connected to a fixed communication network such as PSTN, ISDN, etc.

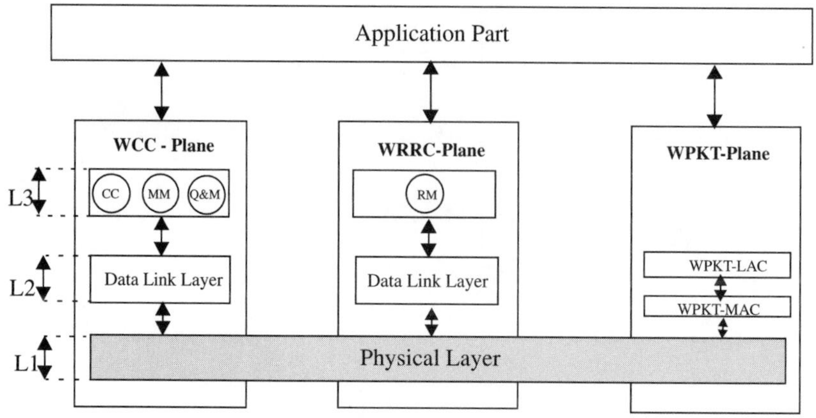

Figure 16.7 WLL Air Interface Protocol Architecture

Wireless protocol consists of

- **WLL communication control (WCC) plane**—Responsible for the bearer connection control (CC) and call control and mobility management (MM) function
- **WLL radio resource control (WRRC) plane**—Responsible for radio resource management (RM)
- **WLL packet control (WPKT) plane**—Supports radio packet data transmission

Each signaling plane has a general hierarchical structure pursuant to the open system interconnect (OSI reference model.

Physical Layer. The physical layer (L1) includes the functions of modulation/demodulation, channel coding, and decoding. These functions of the physical layer apply to both the WRRC and WCC planes.

The physical layer consists of a forward link (from the BS to the terminal) and a reverse link (from the terminal to the BS). The forward link has pilot, sync, paging, forward traffic, and forward power control and signaling channels. Also, the forward packet traffic channel can be included in forward channel types according to user request. The reverse link includes an access and a reverse traffic channel. In addition, a packet access channel and a reverse packet traffic channel can be included according to user request. Reverse link includes single signaling mode in which one terminal is used by several users.

Data Link Layer. The data link layer (L2) is used to deliver the information from the network layer with a high degree of reliability. Main functions are connectionless operation, data retransmission, point-to-point data delivery between originating and terminating points, error control, and sequence control. Basic function of this layer is applied equally to WRRC and WCC planes. The WRRC plane supports only connectionless mode, which does not set up data links for the sync channel (SYCH), access channel (ACH), or paging channel (PCH). Logical channels used in data link layer of the WRRC plane are SYCH, ACH, PCH, and signaling channel (SGCH). Separate functions for connectionless access mode are provided for wireless packet data transmission. The data link layer in the WCC plane supports only connection-oriented mode, which sets up the data link for SGCH. The logical channels used are SGCH and traffic channel (TCH).

Network Layer. The network layer of the WRRC plane performs the control procedure from power-on of the terminal until assignment of the subscriber-dedicated radio channel resource. It supports bearer mode change procedure between telephone and fax. For this purpose, it provides synchronization information control, system information control, random access control, and radio resource management, and then performs interworking functions with the WCC plane. If several fixed terminals are connected to the interface device, the signaling is done through ACH and PCH to the fixed terminal, which generates the first call. The second or later calls generated while the first call is connected can use the dedicated SGCH already assigned to the first call.

The network layer of the WCC plane performs signaling function (WRRC plane RM) for bearer connection control and the signaling function (WCC plane CC) for subscriber registration, authentication, encryption, attachment, and detachment. It has interworking function with the WRRC plane. Call control for the existing PSTN and ISDN terminals performed after bearer connection is included in the connection control (CC) part of the communication control signaling plane. CC has bypass mode that directly delivers information from the terminal to the base station without signaling an information-processing procedure in the wireless section. The specification does not specify call control separately and abides by the current subscriber-signaling scheme as much as possible. The application part in the WLL air interface protocol is defined according to the implementation environment of the service provider. It may include the following functions with respect to the viewpoint of wireless connection:

- Overall state management of function entities (CC/MM/RM) included in WLL communication control signaling plane and WLL radio resource control plane
- Signaling information interchange and protocol conversion between WCC plane and WRRC plane
- Supervision on type/state/operation of terminal device connected to the interface device

System Architecture. The WLL system includes a WLL gateway switching system that connects the radio system (RS) to PSTN [7]. The radio port controller (RPC) provides concentration and control functions to a number of radio ports (RPs) (e.g., base stations). radio interface units (RIUs) are the fixed units attached to the residential or commercial buildings. The radio port operation and maintenance (RPOM) unit is responsible for maintaining and managing the radio network elements. The IWF unit is used as a gateway to data services such as the Internet and the public-switched packet data network (PSPDN). The WLL system consists of various components that can be configured according to market needs.

The RIU can connect up to 32 lines. Subscribers can access PSTN, ISDN, Internet, or other data networks via regular phone, a fax terminal, a modem, an ISDN phone, or a personal computer (PC), any of which may be connected to the RIU. The RIU can be configured to support 1, 2, 4, or 32 subscriber lines and can provide various services requested by the subscribers.

The RP can be deployed indoors or outdoors, depending on the needs. The RP can be equipped with a sectorized antenna, which is more suitable for urban areas. The RPC interacts with the IWF and the switch. The interface between the switch and RPC can be configured as E1/V5.2 interfaces, two-wire analog subscriber lines, or multiple ISDN (2B + D) lines. The interface between the switch and public network can be configured as PSTN or ISDN trunks.

The RPC interworks with the RPOM for network management, the IWF for Internet service and PC communication, and the WLL gateway. The RPC includes several interface units including network interface units (NIUs), line interface units (LIUs), and Ethernet interface units (EIUs). These provide the necessary functions to interface with the gateway switch and IWF, the RPs, and the RPOM, respectively. The RPC also includes a multiprocessor main control unit (MCU) with an active standby processor to handle call processing and other management functions. Other primary units included in the RPC are the transcoder and echo canceller unit (TREU) and the time switch unit.

Due to W-CDMA, the RP provides high-speed data service and compensates for noisy radio environments. Since the antenna provides diversity by using one transmit (Tx) antenna and two receive (Rx) antennas, the RP can provide high-quality voice and data transmission. The RP can have either omnidirectional or sectorized antennas and can be installed indoors or outdoors. The physical interface with the RPC can be based on an E1 interface or an HDSL interface.

The architecture of RP includes RPC interface units (RCIUs) to interface with RPC and radio frequency units (RFUs) to provide the necessary functions to receive and transmit RF signals over the air. The MCU provides primary processing functions for the RP. Other units in the RP are the analog channel unit (ACU), which provides conversion between the analog and digital signals, and the baseband transmitter unit (BTU) and baseband receiver unit (BRU), which are used for channel coding/decoding, interleaving/deinterleaving, and spreading/despreading spectrum.

The RIU is connected to the RP via a W-CDMA (2.3 GHz) interface. On the subscriber side, the RIU is connected to PSTN phones, ISDN phones, facsimile terminals, modems, and PCs to provide voice service, voiceband data service, Internet service, PC communication, and ISDN service. The RIU can be configured to support 1, 2, 4, or 32 lines, and the distance between the RP and RIU can be up to 8 km. The voice coder in the RIU can be 64-kbps PCM, 32-kbps ADPCM, or 16-kbps LD-CELP. The RIU includes the radio frequency unit (RFU), baseband unit (BBU), analog channel unit (ACU), and the main control unit (MCU). The RIU has a number of interface units to interface with the user equipment: a serial interface unit (SIU) providing RS-232C functions, basic rate access (BRA) providing ISDN functions, and a subscriber loop interface circuit (SLIC) providing telephony interface functions.

Table 16.1 provides key characteristics of this WLL system.

Table 16.1 Key Characteristics of WLL System Based on Korean Specifications

RPC	
Subscriber line capacity	Max. 20,000 (@ 0.1 Erlang)
No. of RPs supported	Max. 32
No. of E1 supported	Max. 64
RP	
Channel capacity	80
RF transmit power	Max. 20 W
RIU	
1 line	POTS or data service
2 lines	2 POTS or 1 POTS +1 data
4 lines	4 POTS or 3 POTS + 1 data or 3 POTS + ISDN
32 lines	32 POTS or 24 POTS + 4 data + 4 ISDN
PROM	
Subscriber line capacity	Max. 500,000 subscribers
No. of RPCs supported	Max 16 RPCs
IWF	
Data service line capacity	Max. 8,100 ports
Switching System	
Subscriber line capacity	Max. 250,000
Trunk capacity	Max. 60,000

16.6.2 Interworking with the Data World

The IWF unit of the WLL system provides a gateway to external data networks including the Internet, X.25, and frame relay networks. The IWF provides necessary functions to perform data and protocol conversion and intersystem interface matching when a WLL subscriber tries to search for database information of the PSPDN, or to get onto the Internet. Figure 16.8 shows the architecture of the WLL system providing support for public database and Internet access.

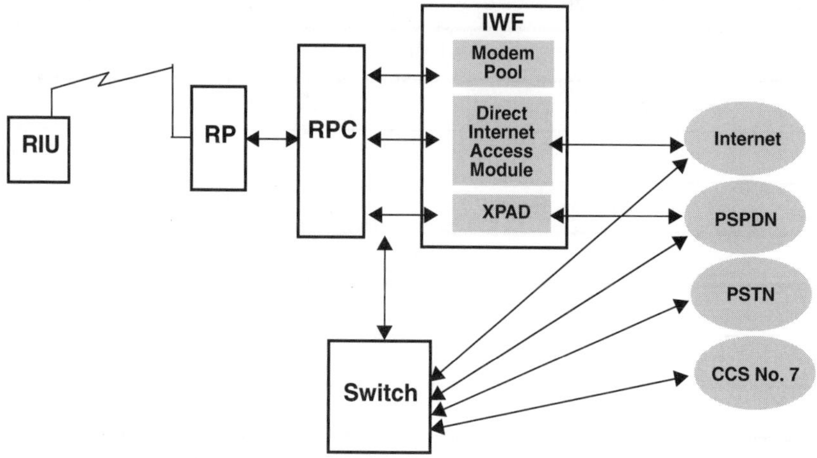

Figure 16.8 WLL Architecture Highlighting IWF Interfaces

The IWF is composed of three main modules:

- **Modem pool module**—Enables data transmission between a subscriber's PC and the PSTN/PSPDN. The subscriber's PC equipped with a modem is connected to the RJ-11 port on the RIU. The modem pool module supports popular public modem protocols such as V.42, V.42bis, V.24, and MNP2-5. Other functions provided are error correcting and compressing, and 64-kbps PCM modulation and demodulation.
- **Direct Internet access module**—Directly connects the subscriber's PC to the Internet. The subscriber's PC is connected to the RS-232C port on the RIU. The direct Internet access module provides the following functions: protocol conversion from point-to-point protocol (PPP) to transmission control protocol/Internet protocol (TCP/IP) and TCP/IP to PPP; terminal server; and router.
- **XPAD module**—Used when the PC subscriber wants to connect to a PSPDN network with an RS-232C port. This module provides packet framing/deframing, packet network (X.25) interface, and E1 interface functions.

The RPC plays a critical role in processing data service calls. The RPC and the IWF assign the resources (i.e., modem pool or direct Internet access module) according to the required service options. Figure 16.9 shows a typical connection for the WLL packet mode data call to provide Internet or X.25 service to WLL users.

The PC is assumed to have a full Internet stack including TCP/IP, PPP, e-mail, Telnet, FTP, etc. The RIU supports PPP on the RS-232 interface. In addition, the RIU interprets the information carried in PPP frames and interacts with the air interface protocol to set up the traffic channel or to transfer the information on the air interface. The packet mode data calls have no impact on the RP. The RPC supports the air interface-specific protocol stack on the interface between the RP and itself. For the interface between the IWF and the RPC, the RPC supports the link access protocol for D (LAPD).

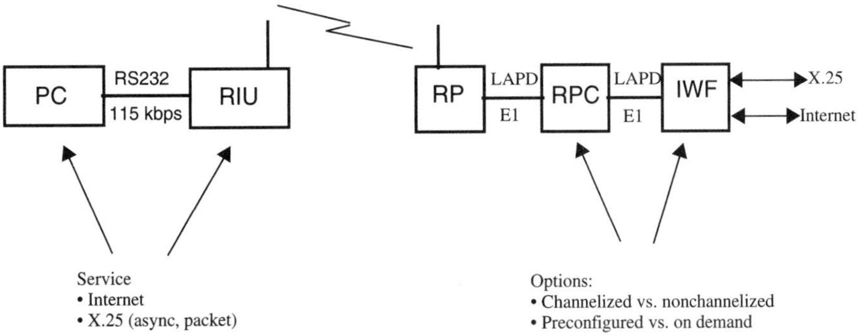

Figure 16.9 Reference Configuration for IWF-Related Communications

16.7 Airloop WLL System

The Lucent airloop is a W-CDMA-based system developed for a wide range of customers. It operates mainly in the 3.4-GHz band using 5-MHz carriers, each supporting 115 16-kbps channels [8]. To support 32-kbps ADPCM, two channels are used simultaneously. The spreading code is 4.096 Mcps; thus, for a 16-kbps data rate, a spreading factor of 256 is used. The high-level network architecture is shown in Figure 16.10.

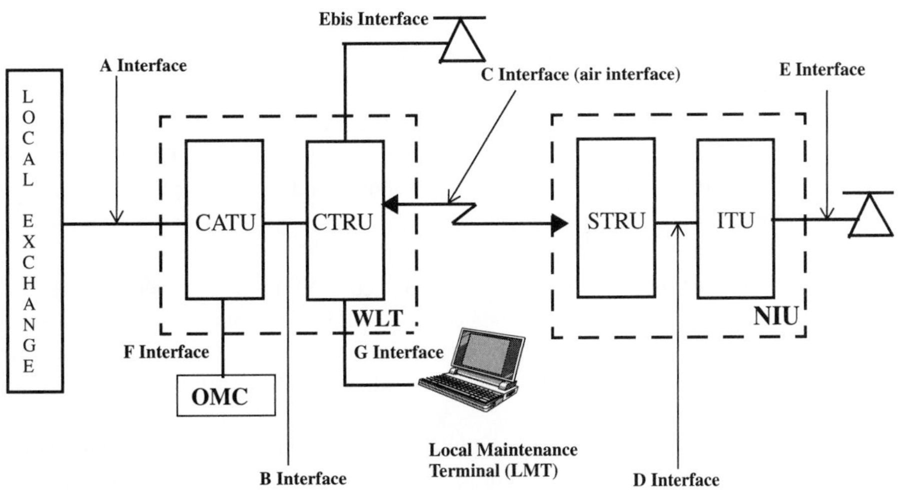

Figure 16.10 Airloop Architecture

The airloop system is composed of two elements: the wireless line transceiver (WLT) and the network interface unit (NIU). The WLT is the central network equipment that provides the interface to the local exchange. The NIU is located on the subscriber's premises and provides the interface to the subscriber's equipment. An air interface provides the radio link between the WLT and NIU.

The system uses a network of radio base station (RBS) to provide coverage of the intended service area. The main functional blocks are

- **Central office**—Contains digital switching and network routing facilities required to connect the radio network to ISDN and the Internet.
- **Central access and transcoding unit (CATU)**—Co-located with the switch and performs two functions:
  - It controls the allocation of radio resources and ensures that allocation is appropriate to the service being provided; for example, 64-kbps digital data, 32-kbps speech, ISDN, etc.
  - For speech services, it provides transcoding between various speech-coding rates and the switched 64-kbps PCM.
- **Central transceiver unit (CTRU)**—Located at the RBS. A single CTRU provides a single CDMA radio channel. It performs the following functions:
  - Provides the CDMA air interface.
  - Transfers ISDN and POTS signaling information transparently between the air interface and the CATU.

  The connection between the CATU and CTRU is provided by a 2-Mbps transmission capability and is normally made using direct connection by point-to-point microwave links.
- **Network interface unit (NIU)**—Connects the subscribers to a radio network. The NIU consists of two functional blocks, the intelligent telephone socket (ITS) and the subscriber transceiver unit (STRU). The ITS provides the point of connection to the subscriber's terminal equipment; for example, PBX, telephone, or LAN. The ITS is available in dual-line or multiline configuration. The STRU is located on the outside of the subscriber's building and consists of an integrated antenna and radio transceiver. The STRU provides the interface between the ITS and the CDMA air interface. The STRU is connected to the ITS by a standard four-wire telephone or data networking cable.

The number of subscriber connections to each NIU is determined by the type of service being provided by the connection. The basic NIU connection provides a single ISDN (2B + D) connection, effectively giving two unrestricted 64-kbps channels. The same unit also can be configured as either two or eight individual POTS lines using ADPCM and LD-CELP speech coding, respectively.

The modulation technique used first takes a 16-kbps channel and adds error-correction coding to achieve 32 kbps. It then uses Walsh spreading with a spreading factor of 128 to reach a transmitted data rate of 4 Mbps. Finally, it multiplies that by one of a set of 16 PN code sequences, also at 4 Mbps, which does not change the output data rate but provides for interference from adjacent cells. During the design of the network, each cell must have a PN code sequence number assigned to it such that neighboring cells do not have the same number. Table 16.2 provides key parameters of the airloop system.

Table 16.2 Key Parameters of Airloop System

Services	Performance
Telephony: Yes, good quality	Range (radius): 4 km
ISDN: Yes	Cells per 100 km: 2
Fax: Yes	Capacity/cell, 2 × 1 MHz: 5.75
Data: Up to 128 kbps	
Videophone: No	
Supplementary services: Good	
Multiple lines: Yes, 2-line or 8-line units	

The airloop system supports various protocols for communicating with the local exchange, i.e., call-associated signaling (CAS), Q.931, V5.1, and V5.2. Owing to signaling constraints, CAS and Q.931 can support 30 subscribers per E1 link to the local exchange, for a maximum of 480. V5.1 can support a minimum of 28 subscribers per E1 link to the local exchange and a maximum of 30, depending on the allocation of control channels. V5.2 is a concentrating protocol.

16.7.1 Interfaces

- **A interface** is an E1 interface. Sixteen E1 links are supported on the A interface, i.e., there may be up to 16 E1 links between the CATU and the local exchange.
- **B interface** provides the E1 link to the CTRU. The B interface carries logical channel information, CTRU signaling, subscriber signaling, and subscriber traffic.
- **C interface** is the air interface between the CTRU and the STRU. This is orthogonal CDMA.
- **D interface** is a physical interface that connects the STRU and ITU. This carries serial data and power.
- **E interface** is the physical connection between the ITU and the subscriber equipment.
- **Ebis interface** provides connection for wireless subscribers directly from the CTRU.
- **F interface** provides communication between WLT and OMC for the purposes of network management, control, and alarm reporting. The interface uses simple network management protocol (SNMP) driven by a physical Ethernet connection.
- **G interface** provides the connection for the LMT into the CTRU central processor card serial b port. It utilizes the RS232 protocol.

16.7.2 Frequency Bands and Channel Arrangements

The system is designed to use a 3.6- to 4-GHz band (see Figure 16.11), but the frequency plan is centered on 3.8 GHz. The spectrum requires band sharing with bands specified by CCIR (Comité Consultatif International des Radio Communications) Rec. 635, CCIR Report 933 and 934 for fixed links. There is a 1.1-GHz guard band between two 100-MHz bands, one for the uplink and one for the downlink. These are

- Downlink band: 3.85–3.95 GHz
- Uplink band: 3.64–3.74 GHz

j channels allow 10 full-duplex frequencies (see Figure 16.11).

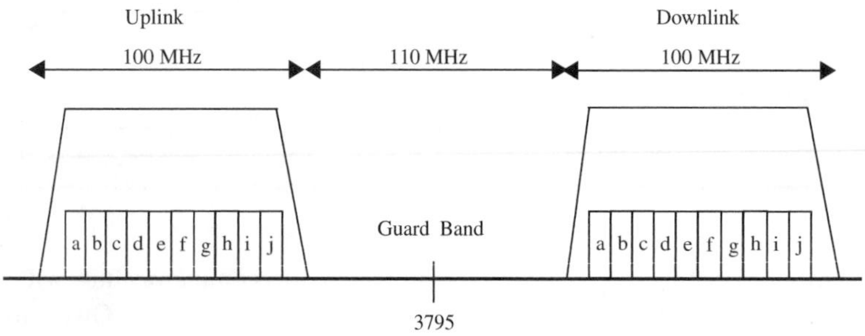

Figure 16.11 4-GHz Frequency Plan

The channel center frequencies are given in Table 16.3.

Table 16.3 Channel Frequencies

Channel	Uplink (MHz)	Downlink (MHz)
a	3645	3855
b	3655	3865
c	3665	3875
d	3675	3885
e	3685	3895
f	3695	3905
g	3705	3915
h	3715	3925
i	3725	3935
j	3735	3945

16.7.3 Capacity

Capacity considerations are fundamental to CDMA planning and operation [9,10]. For the purpose of this discussion, *capacity* will be defined simply as the number of users that can be simultaneously supported. Forward and reverse link capacity is addressed separately.

Both the forward and reverse links of the airloop system employ orthogonal spreading for each user, i.e., codes that when exactly time-aligned allow the demodulation of any single code such that no interference is experienced due to any of the other codes. In the downlink of a CDMA system, codes are naturally time-aligned when transmitted from the base station, thus preserving this feature of the codes.

In the reverse link conventional mobile CDMA systems, the subscriber mobility is such that the spreading codes transmitted by each user arrive with random time-alignment at the base station; thus, each appears as random interference to all other user signals. The level of interference is proportional to the number of simultaneous users. So, eventually, the addition of more users causes interference to rise to unacceptable levels. This defines the reverse link capacity.

The airloop system differs significantly from such systems since the static nature of the users allows the use of time control of the users' transmitted signals to align all users' spreading codes as received at the base station. This process allows the reverse link to approach orthogonality by suppressing the interference experienced due to other simultaneous users to a fraction of that in conventional CDMA systems. This substantially increases the capacity of the reverse link.

Many factors influencing CDMA capacity give rise to a desirable flexibility in system operation. Since capacity is dependent upon interference levels, a cell's capacity is inherently dynamic; i.e., a cell can naturally absorb more users if neighboring cells are lightly loaded. Finally, capacity limits are soft rather than hard because system capacity can be increased by lowering voice quality requirements. In this procedure, more users are supported at the expense of slightly degrading the call quality of all users.

Upper limits on forward link capacity are fundamentally determined by restrictions on cell site radiated power. The forward link signal is composed of message traffic for subscribers, a sector-specific signal (pilot) used by all subscriber equipment, and miscellaneous signals (e.g., sync, paging). Total power is allocated among these functions. Further users cannot be supported when the sum of allocations required exceeds the available transmit power.

Required allocations are governed by the need for a minimum signal-to-interference ratio at each subscriber's equipment. The power allocated to other users within the cell, as well as the received power from neighbor cell sites, contributes to this interference. Interference from same-cell users is partly mitigated by the use of orthogonal codes, which allow the receiver to suppress these signals; however, multipath effects limit the extent to which this interference can be screened out.

Forward link power distributions are further restricted by the requirement that a generous fraction of power must be allocated to the sector pilot. The sector pilot is important because it is used by all subscriber equipment in site acquisition. Capacity limits are therefore reached when the remaining power, distributed among all users, is not enough to meet subscriber equipment signal-to-interference requirements.

16.7.4 Radio Range

For a system at full capacity (i.e., where all communication channels are in use), the antenna is mounted below roof height. With ISDN circuit quality (i.e., bit error rate [BER] is less than 10^{-6}), the typical radio ranges for 95 percent coverage are given in Table 16.4.

Table 16.4 NIU Radio Range

Environment	Range
Urban	2.5 km
Suburban	4 km
Rural	6 km

16.7.5 Power Control

Capacity can be maximized by minimizing the total level of system interference, i.e., by controlling all CDMA signals to be at the lowest level necessary to meet signal-to-interference requirements. Power control ensures that each signal meets minimum requirements for communication while not causing undue levels of interference to other signals.

Control is accomplished via a closed-loop algorithm on the forward link and open- and closed-loop algorithms on the reverse link. The open-loop mechanisms are based on measurements of parameters known to influence the desired output, whereas the closed-loop mechanisms are based on direct measurements of the output itself.

The objective of reverse link control is to ensure the minimum necessary signal-to-interference ratio at the cell site for each piece of subscriber equipment. In the open-loop path, the subscriber equipment makes power adjustments based on its estimate for path loss from cell site to the subscriber equipment. This estimate is based on the subscriber equipment's measurement of received cell site pilot power. These adjustments compensate for path loss variations that are correlated between the forward and reverse links. In the closed-loop path, the cell site compensates for uncorrelated path loss variations (e.g., multipath fading) and additional sources of interference by measuring the received signal-to-interference from the subscriber equipment and transmitting appropriate power adjustment commands. The final value of the subscriber equipment transmit power is jointly determined by these two control paths.

The objective of the forward link control is to ensure the minimum necessary signal-to-interference ratio at each mobile from the cell site. In the closed-loop mechanism, the subscriber equipment requests forward link power adjustments based on its received signal.

16.8 Summary

In this chapter, we presented WLL systems based on wideband CDMA technology. The unique features and benefits of CDMA make it a good technology choice for fixed wireless telephone systems. W-CDMA technology used by the WLL system has been selected to provide primarily ISDN-like services and data rates to subscribers while providing a smooth transition into third-generation wireless technology. The following two air interfaces used in the WLL systems were discussed:

1. Korean national standard, which specifies a CDMA-based protocol operating in the 2.3–2.4 GHz band. This standard allows two types of bandwidth, 5 MHz and 10 MHz.
2. Lucent airloop technology based on W-CDMA air interface, which operates in the 3.4 GHz band using 5-MHz wide channels, each supporting 115 16-kbps channels.

Wireless technology in the local loop can make telephone service rapidly available to millions of people in those countries where there is a limited infrastructure. It also has obvious cost savings benefits in countries with established telephone infrastructures. WLL is an economical alternative to the copper-based wireline local loop. There is a large set of technologies available, ranging from proprietary systems to systems based on standards. Operators should carefully consider the implications of different alternatives, especially the spectrum and standards issues involved, when deciding which technology choice is right for their particular needs.

16.9 References

1. Calhoun, C., *Wireless Access and the Local Telephone Network,* Artech House: Boston, 1992.
2. Yu, C., et al., "Low-Tier Wireless Loop Radio Systems," *IEEE Communications Magazine,* May 1997.
3. Tuttleby, W., ed., *Cordless Telecommunications World Wide,* Springer-Verlag: Berlin, 1996.
4. Lee, W. C. Y., "Spectrum and Technology of a Wireless Local Loop System," *IEEE Personal Communications*, Feb. 1998 [Vol. 5(1)].
5. TTA, "Wideband CDMA Air Interface Compatibility Interim Standard for 2.3 GHz Band WLL System," TTA KO-06.0015/0016, Dec. 1997.
6. Ulema, M., and Yoon, Y.-K., "A CDMA Based WLL System Protocol," *Proceedings of the IEEE Wireless Communications Networking Conference*, Sept. 1999.
7. Yoon, Y. K., and Ulema, M., "A Wireless Local Loop System Based on Wideband CDMA Technology," *IEEE Communications Magazine,* Oct. 1999 [Vol. 37(10)].
8. Webb, W., *Introduction to Wireless Local Loop*, Artech House: Boston, 1998.
9. Lee, W. C. Y., "Applying CDMA to the WLL," *Cellular Business*, Oct. 1995.
10. Margiotta, D., "Wireless Fixed Loop Access," *Telephony*, Aug. 2, 1993.

CHAPTER 17

Wireless Application Protocol, Bluetooth, and Wireless Local Area Network

17.1 Introduction

In this chapter, we discuss three important topics in wireless communications and wireless computing: the wireless application protocol (WAP), Bluetooth, and the wireless local area network (WLAN). WAP is becoming an industry standard for providing data to wireless handheld devices. Bluetooth enables users to connect a wide range of computing and telecommunication devices, without any additional or proprietary cables. A WLAN is a flexible data communications system implemented as an extension of or as an alternative for a wired local area network (LAN). In the first part of the chapter, the goal of WAP, the WAP programming model, and WAP architecture are presented. Bluetooth, including protocol stack, security, and usage model, is discussed. The chapter is concluded by presenting WLAN topology, technologies, and standards.

17.2 Wireless Application Protocol

WAP [1, 2, 3–9, 10] has become the de facto global industry standard for providing data to wireless handheld mobile devices. WAP takes a client-server approach and incorporates a relatively simple microbrowser into the mobile phone, requiring only limited resources on the mobile phones. WAP puts the intelligence in the WAP gateways while adding just a microbrowser to the mobile phones themselves. Microbrowser-based services and applications reside temporarily on servers, not permanently in phones. WAP is aimed at turning mass-market phones into *network-based smart phones*. The philosophy behind WAP's approach is to use as few resources as possible on the handheld device and compensate for the constraints of the device by enriching the functionality of the network.

WAP specifies a thin client microbrowser using a new standard called wireless markup language (WML) that is optimized for wireless handheld mobile devices. WML is a stripped-down version of HyperText Markup Language (HTML).

WAP specifies a proxy server that acts as a gateway between the wireless network and the wireline Internet, providing protocol translation and optimizing data transfer for the wireless handset. WAP also specifies a computer-telephony integration application programming interface (API), called wireless telephony application interface (WTAI), between data and voice. This enables applications to take full advantage of the fact that this wireless mobile device is most often a phone and a mobile user's constant companion. Onboard memory on a WAP phone can be used for off-line content, enhanced address books, bookmarks, and text input methods.

The importance of WAP can be found in the fact that it provides an evolutionary path for application developers and network operators to offer their services on different network types, bearers, and terminal capabilities. The design of the WAP standard separates the application elements from the bearer being used. This helps in the migration of some applications from short message service (SMS) or circuit-switched data to general packet radio service (GPRS), for example.

17.3 Goals of WAP

The goals of the WAP are

- Independence from wireless network standards.
- Interoperability: Terminals from different manufacturers must be able to communicate with services in the mobile network.
- Overcoming the boundaries of wireless networks: Low bandwidth, high latency, less connection stability.
- Overcoming the boundaries of wireless devices: Small display, limited input facilities, limited memory and CPU, limited battery power.
- Increased efficiency: Provide quality of service (QoS) suitable to the behavior and characteristics of the mobile world.
- Increased reliability: Provide a consistent and predictable platform for deploying services.
- Security: Enable services to be extended over potentially unprotected mobile networks while preserving the integrity of data.
- Scalability: Applications scale across transport options and device types.
- Extensibility: Adaptability over time to new networks and transport.

The WAP is envisioned as a comprehensive and scalable protocol designed for use with

- Any mobile device, from low-end devices such as those with a one-line display, to high-end devices such as smart phones.
- Any existing or planned wireless service such as the SMS, circuit-switched data, unstructured supplementary services data (USSD), and GPRS.

- Any mobile network standard such as code division multiple access (CDMA), global system of mobile communications (GSM), or universal mobile telephone system (UMTS). WAP has been designed to work with all cellular standards and is supported by major worldwide wireless leaders such as AT&T wireless and NTT DoCoMo.
- Multiple input terminals such as keypads, keyboards, touch-screens, etc.

17.4 WAP Programming Model

Before presenting the WAP programming model, we briefly discuss the World Wide Web (WWW) model that is the basis for the WAP model.

17.4.1 WWW Model

The Internet WWW architecture provides a flexible and powerful programming model. Applications and content are presented in standard data formats, and are browsed by applications known as Web browsers. The Web browser is a network application, i.e., it sends requests for named data objects to a network server and the network server responds with data encoded using the standard formats.

The WWW standards specify several mechanisms necessary to build a general-purpose application environment. These include

- **Standard naming model**—All servers and content on the WWW are named with an Internet-standard *uniform resource locator* (URL).
- **Content typing**—All content on the WWW is given a specific type, thereby allowing Web browsers to correctly process the content based on its type.
- **Standard content formats**—All Web browsers support a set of standard content formats. These include HTML, the JavaScript scripting language (ECMAScript, JavaScript), and a large number of other formats.
- **Standard protocols**—Standard networking protocols allow any Web browser to communicate with any Web server. The most commonly used protocol on the WWW is the HyperText Transport Protocol (HTTP). This infrastructure allows users to easily reach a large number of third-party applications and content services. It also allows application developers to easily create applications and content services for a large community of clients. The WWW protocols define three classes of servers:
  - **Origin server**—The server on which a given resource (content) resides or is to be created.
  - **Proxy**—An intermediary program that acts as both a server and a client for the purpose of making requests on behalf of other clients. The proxy typically resides between clients and servers that have no means of direct communication, e.g., across a firewall. Requests are either serviced by a proxy program or passed on with possible translation to other servers. A proxy must implement both the client and server requirements of the WWW specifications.

- **Gateway**—A server that acts as an intermediary for some other server. Unlike a proxy, a gateway receives requests as if it were the origin server for the requested resource. The requesting client may not be aware that it is communicating with a gateway.

17.4.2 WAP Model

The WAP programming model (see Figure 17.1) is similar to the WWW programming model. This provides several benefits to the application developer community, including a familiar programming model, a proven architecture, and the ability to leverage existing tools (e.g., Web servers, Extensible Markup Language [XML] tools, etc.). Optimization and extensions have been made in order to match the characteristics of the wireless environment. Wherever possible, existing standards have been adopted or have been used as the starting point for WAP technology.

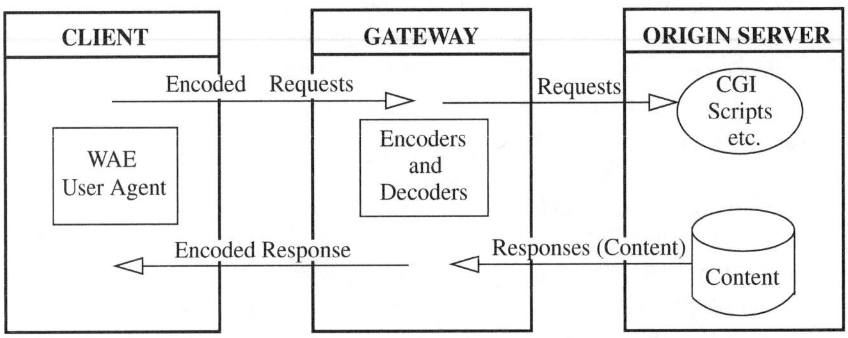

Figure 17.1 WAP Programming Model

WAP content and applications are specified in a set of well-known content formats based on those on the WWW. Content is transported using a set of standard communication protocols based on WWW communication protocols. A microbrowser in the wireless terminal coordinates the user interface and is analogous to a standard Web browser. WAP defines a set of standard components that enable communication between mobile terminals and network servers, including the following:

- **Standard naming model**—WWW-standard URLs are used to identify WAP content on origin servers. WWW-standard uniform resource identifiers (URIs) are used to identify local resources in a device, e.g., call control functions.

- **Content typing**—All WAP content is given a specific type consistent with WWW typing. This allows WAP user agents to correctly process the content based on its type.
- **Standard content formats**—WAP content formats are based on WWW technology and include display markup, calendar information, electronic business card objects, images, and scripting language.
- **Standard communication protocols**—WAP communication protocols enable the communication of browser requests from the mobile terminal to network Web server. The WAP content types and protocols have been optimized for mass market, handheld wireless devices. WAP utilizes proxy technology to connect between the wireless domain and the WWW.

The WAP proxy is typically comprises the following functionality:

- **Protocol gateway**—The protocol gateway translates requests from the WAP protocol stack to the WWW protocol stack (HTTP and TCP/IP).
- **Content encoders and decoders**—The content encoders translate WAP content into compact encoded formats to reduce the size of data over the network. This infrastructure ensures that mobile terminal users can browse a wide variety of WAP content and applications, and that the application author is able to build content services and applications that run on a large base of mobile devices. The WAP proxy allows content and applications to be hosted on standard WWW servers and to be developed using proven WWW technologies such as common gateway interface (CGI) scripting. While the nominal use of WAP will include a Web server, WAP proxy, and WAP client, the WAP architecture can easily support other configurations. It is possible to create an origin server that includes the WAP proxy functionality. Such a server might be used to facilitate end-to-end security solutions or applications that require better access control or a guarantee of responsiveness.

17.5 WAP Architecture

WAP architecture provides a scalable and extensible environment for application development of mobile communication devices. This is achieved using a layered design of the entire protocol stack (see Figure 17.2). Each of the layers of the architecture is accessible by the layers above, as well as by other services and applications. The WAP layered architecture enables other services and applications to utilize the features of the WAP stack through a set of well-defined interfaces. External applications may access the session, transaction, security, and transport layers directly. The layered design and functions of the WAP protocol stack resemble the layers of the open system interconnect (OSI) model. The following sections provide a description of the various elements of the WAP protocol stack.

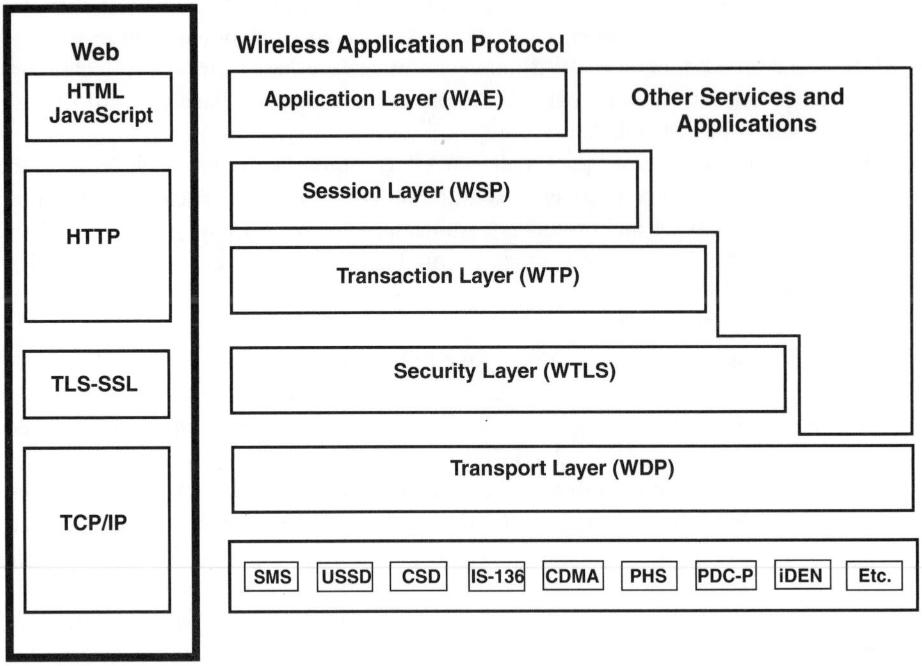

Figure 17.2 WAP Architecture

17.5.1 Wireless Application Environment (WAE)

The uppermost layer in the WAP protocol stack, WAE is a general-purpose application environment based on a combination of WWW and mobile telephony technologies. The primary objective of the WAE is to establish an interoperable environment that will allow operators and service providers to build applications and services that can reach a wide variety of different wireless platforms in an efficient and useful manner. Various components of WAE are

- **Addressing model**—WAP uses the same addressing model as the one used on the Internet, i.e., URL. A URL uniquely identifies a resource on a server that can be retrieved using standard protocols. WAP also uses URI. A URI is used for addressing resources that are not necessarily accessed using known protocols. An example of using a URI is local access to a wireless device's telephony functions.
- **Wireless Markup Language**—WML is WAP's analogy to HTML and is based on the XML. It is WAP's answer to problems such as creating services that fit on small handheld devices, low-bandwidth wireless bearer, etc. WML uses a deck/card metaphor to specify a service. A card is typically a unit of interaction with the user, i.e., either presentation of information or request for information from the user. A collection of cards is called a deck, which usually constitutes a service. This approach ensures that a suit-

able amount of information is displayed to the user at any given time, thereby allowing interpage navigation to be avoided to the maximum possible extent. Key features of WML include variables, text formatting features, support for images, support for soft buttons, navigation control, control of browser history, support for event handling (e.g., telephony services), and different types of user interactions (e.g., selection lists and input fields). One of the key advantages of WML is that it can be binary encoded by the WAP gateway/proxy in order to save bandwidth in the wireless domain.

- **WMLScript**—WMLScript is used for enhancing services written in WML in that it adds intelligence to the services. WMLScript can be used for validation of user input. Since WML does not provide any mechanisms for achieving this, a round trip to the server would be needed in order to determine whether user input is valid if scripting was not available. Access to local functions in a wireless device is another area where WMLScript is used; for example, access to telephony-related functions. WMLScript libraries contain functions that extend the basic WMLScript functionality. This provides a means for future expansion of functions without having to change the core of WMLScript. Just as with WML, WMLScript can be binary encoded by the WAP gateway/proxy in order to minimize the amount of data sent over the air.

17.5.2 Wireless Telephony Application

The wireless telephone application (WTA) environment provides a means to creating telephony services using WAP. WTA utilizes a user agent, which is based on the WML user agent but extends its functionality such that it meets the special requirements of telephony services. This functionality includes

- **Wireless telephony application interface**—WTAI is an interface toward a set of telephony-related functions in a mobile phone that can be invoked from WML and/or WMLScript. These functions include call management, handling of text messages, and phone book control.
- **Repository**—Occasionally it is not feasible to retrieve content from a server, so the repository makes it possible to store WTA services persistently in the device in order to enable access to them without accessing the network.
- **Event handling**—Typical events in a mobile network are incoming calls, call disconnect, and call answering. The event handling within WTA enables WTA services stored in the repository to be started in response to such events.
- **WTA service indication**—This is a content type that allows a user to be notified about the events of different kinds of services (e.g., new voicemails) and be given the possibility to start the appropriate service to handle the event. In the most basic form, WTA service indication makes it possible to send a URL and a message to a wireless device.

WTAI enables access to functions that are not suitable for allowing common access to them, e.g., setting up calls and manipulating the phone book without user acknowledgment. The WTA framework relies on a dedicated WTA user-agent to carry out these functions.

Only trusted content providers should be able to make content available to the WTA user agent. Thus, it must be possible to distinguish between servers that are allowed to supply the user agent with services containing these functions and those that are not. To accomplish this, the WTA user agent retrieves its services from the WTA domain, which, in contrast to the Internet, is controlled by the network operator. WTA services and other services are separated from each other using WTA access control based on port numbers.

17.5.3 Wireless Session Protocol

The wireless session protocol (WSP) provides a means for organized exchange of content between cooperating client/server applications. Its features are as follows:

- It establishes a reliable session from client to server and releases the session in an orderly manner.
- It agrees on a common level of protocol functionality using capability negotiation.
- It exchanges content between client and server using compact encoding.
- It suspends and resumes the session.
- It provides HTTP 1.1 functionality.
- It exchanges client and server session headers.
- It interrupts transactions in process.
- It pushes content from server to client in a unsynchronized manner.
- It negotiates support for multiple, simultaneous asynchronous transactions.

The core of the WSP design is a binary form of HTTP. Consequently, all methods defined by HTTP 1.1 are supported. In addition, capability negotiation can be used to agree on a set of extended request methods, so that full compatibility to HTTP applications can be retained. HTTP content headers are used to define content type, character set encoding, language, etc. in an extensible manner. However, compact binary encoding is defined for the well-known headers to reduce protocol overhead.

The life cycle of a WSP session is not tied to the underlying transport protocol. A session can be suspended while the session is idle to free up network resources or save battery power. A lightweight session re-establishment protocol allows the session to be resumed without the overload of full-blown session establishment. A session may be resumed over a different bearer network.

WSP allows extended capabilities to be negotiated between peers. This allows for both high-performance, feature-full implementation as well as simple, basic, and small implementations. WSP provides an optimal mechanism for attaching header information to the acknowledgment of a transaction. It also optionally supports asynchronous requests, so that a client can submit multiple requests to the server simultaneously. This improves utilization of air time and latency as the result of each request being sent to the client when the client becomes available.

17.5.4 Wireless Transaction Protocol

The wireless transaction protocol (WTP) does not have security mechanisms. WTP has been defined as a lightweight transaction-oriented protocol that is suitable for implementation in "thin" clients and operates efficiently over wireless datagram networks. Reliability is improved through the use of unique transaction identifiers, acknowledgments, duplicate removal, and retransmissions. There is an optional user-to-user reliability function in which a WTP user can confirm every received message. The last acknowledgment of the transaction, which may contain out-of-band information related to the transaction, is also optional. WTP has no explicit connection setup or tear-down phases, as this would impose excessive overhead on the communication link. This improves efficiency over connection-oriented services. The protocol provides mechanisms to minimize the number of transactions being replayed as a result of duplicate packets.

WTP is designed for services oriented toward transactions, such as browsing. The basic unit of interchange is an entire message rather than a stream of bytes. Concatenation may be used, where applicable, to convey multiple packet data units (PDUs) in one service data unit (SDU) of the datagram transport. WTP allows asynchronous transactions. There are three classes of transaction service:

- Class 0: Unreliable "send" with no result message. No retransmission if the sent message is lost.
- Class 1: Reliable "send" with no result message. The recipient acknowledges the sent message; otherwise the message is resent.
- Class 2: Reliable "send" with exactly one reliable result message. A data request is sent and a result is received which finally is acknowledged by the initiating part.

Note that with reliable "send," both success and failure is reported.

17.5.5 Wireless Transport Layer Security

The purpose of wireless transport layer security (WTLS) is to provide transport layer security between a WAP client and the WAP gateway/proxy. WTLS is a security protocol based on the industry standard transport layer security (TLS) protocol with new features such as datagram support, optimized handshake, and dynamic key refreshing. The WTLS layer is modular, and whether it is used depends on the required security level of the given application, or characteristics of the underlying network. In addition, WTLS provides an interface for managing secure connections. The primary goal of WTLS is to provide the following features between two communicating applications:

- **Data integrity**—WTLS contains facilities to ensure that data sent between the terminal and an application server are unchanged and uncorrupted.
- **Privacy**—WTLS contains facilities to ensure that data transmitted between the terminal and an application server are private and cannot be understood by any intermediate parties who may have intercepted the data stream.

- **Authentication**—WTLS contains facilities to establish the authenticity of the terminal and application server.
- **Denial-of-service protection**—WTLS contains facilities for detecting and rejecting data that is replayed or not successfully verified. WTLS makes many typical denial-of-service attacks harder to accomplish and protects the upper protocol layers.

WTLS protocol is optimized for low-bandwidth bearer networks with relatively long latency. These features make it possible to certify that the sent data has not been manipulated by a third party, that privacy is guaranteed, that the author of a message can be identified, and that both parties cannot falsely deny having sent their messages. WTLS is optional and can be used with both connectionless and the connection mode WAP stack configuration.

17.5.6 Wireless Datagram Protocol

The wireless datagram protocol (WDP) offers a consistent service to the upper layer protocols of WAP and communicates transparently over one of the available bearer services. The services offered by WDP include application addressing by port numbers, optional segmentation and reassembly, and optional error detection.

WDP supports several simultaneous communication instances from a higher layer over a single underlying WDP bearer service. The port number identifies the higher layer entity above the WDP. Reusing the elements of the underlying bearers and supporting multiple bearers, WDP can be optimized for efficient operation within the limited resources of a mobile device.

The WDP adaptation layer is the layer of the WDP protocol that maps the WDP protocol functions directly onto a specific bearer. The adaptation layer is different for each bearer and deals with the specific capabilities and characteristics of that particular bearer service. At the gateway, the adaptation layer terminates and passes the WDP packets on to a WAP proxy/server via a tunneling protocol.

If WAP is used over a bearer user data protocol (UDP), the WDP layer is not needed. On other bearers, such as GSM SMS, the datagram functionality is provided by WDP. This means that WAP uses a datagram service, which hides the characteristics of different bearers and provides port number functionality.

Processing errors can occur when WDP datagrams are sent from one WDP provider to another. For example, a wireless data gateway is unable to send the datagram to the WAP gateway, or the receiver does not have enough buffer space to receive large messages. The wireless control message protocol (WCMP) provides an efficient error-handling mechanism for WDP.

17.5.7 Optimal WAP Bearers

The WAP protocols are designed to operate over a variety of different services, including short message service, circuit switched data, and packet data. The bearers offer differing levels of quality of service with respect to throughput, error rate, and delays. The WAP protocols are designed to compensate for or tolerate these varying levels of service.

- **Short message service**—Given its limited length of 160 characters per short message, the overhead of the WAP protocol that would be required to be transmitted in an SMS message would mean that several SMS messages may have to be sent for even the simplest of transactions.
- **Circuit switched service (CSD)**—Most of the trial-based services use CSD as the underlying bearer. CSD lacks immediacy. A dial-up connection taking about 10 seconds is required to connect the WAP client to the WAP gateway, and this is the best-case scenario when there is a complete end-to-end digital call.
- **Unstructured supplementary services data**—USSD is a means of transmitting information or instructions over a GSM network. In USSD, a session is established and the radio connection stays open until the user, application, or time-out releases it. USSD text messages can be up to 182 characters in length. USSD is preferable over the structured services due to the following reasons:
  - Turnaround response times for interactive applications are shorter for USSD.
  - Users need not access any particular phone menu to access services with USSD; instead they can enter commands directly from the initial mobile phone screen.
  - Works exactly the same way when users are roaming.

  USSD can be an ideal bearer of WAP on GSM networks.
- **General packet radio service**—GPRS is a new bearer that is immediate, relatively fast, and supports virtual connectivity, thereby allowing relevant information to be sent from the network even as it is being generated. There are two efficient means of delivering proactively by sending (pushing) content to a mobile phone: by SMS, which is of course one of the WAP bearers, or by the user maintaining a more or less permanent GPRS session with the content server. WAP incorporates two different connection modes: WSP connection mode or WSP connection protocol. This is similar to the two GPRS point-to-point services: connection oriented and connectionless. For the interactive menu-based information exchanges that WAP anticipates, GPRS and WAP can be ideal bearers for each other.

17.6 Bluetooth

Bluetooth [11–17] enables users to connect a wide range of computing and telecommunication devices easily and simply, without having to buy additional or proprietary cables. The cable solution is complicated since it might require a cable specific to the device that is being connected. The infrared solution eliminates the cable, but requires a clear line of sight. The Bluetooth standard was developed to solve all these problems .

The Bluetooth system operates in the 2.4-GHz industrial scientific medicine (ISM) band. In a vast majority of countries around the world, the range of this frequency band is 2.4–2.4835 GHz. The ISM band is open to any radio system such as cordless phones, garage door openers, and microwaves. It is susceptible to strong interference.

Bluetooth radio technology provides a universal bridge to existing data networks, a peripheral interface, and a mechanism to form small private ad hoc groupings of connected devices away from fixed network infrastructures. Bluetooth radio uses a fast acknowledgment and frequency hopping scheme to make the link robust. Bluetooth typically hops faster and uses shorter packets. Short packets and fast hopping limit the impact of interference from other radio systems that use the same frequency band. The use of a forward error correction (FEC) scheme limits the impact of random noise on long-distance links. Table 17.1 lists the parameters of Bluetooth air interface.

Table 17.1 Bluetooth Air Interface Details

Feature	Values	Notes
Frequency range: • USA, Europe, and most other countries • Spain • France	 2.4–2.4835 GHz 2.445–2.475 GHz 2.4465–2.4835 GHz	 79 RF channels 23 RF channels 23 RF channels
Bandwidth of each RF channel	1 MHz	
Gross data rate	1 Mbps (initial) 2 Mbps (latter)	
One-to-one connection allowable maximum data rate	721 kbps	3 voice channels
Frequency hopping rate	1,600 hops/second	
Range	10–100 meters	

Transmitter equipment is classified into three power classes (see Table 17.2). A power control is required for power class 1 equipment. The power control is used for limiting the transmitted power over 0 dBm. Power control capability under 0 dBm is optional and could be used for optimizing the power consumption and overall interference level. The power steps form a monotonic sequence, with a maximum step size of 8 dB and a minimum step size of 2 dB. Class 1 equipment with a maximum transmit power of 20 dBm must be able to control its transmit power down to 4 dBm or less. Equipment with power control capability optimizes the output power in a link.

The actual sensitivity level is defined as the input level for which a raw bit error rate (BER) of 0.1 percent is met. The requirement for a Bluetooth receiver is an actual sensitivity level of –70 dBm or better. The receiver must achieve –70 dBm sensitivity level with any Bluetooth transmitter compliant to the transmitter specification given in Table 17.2.

Table 17.2 Transmitter Characteristics

Power Class	Maximum output power (P_{max})	Nominal Output Power	Minimum Output Power*	Power Control
1	100 mW (20dBm)	N/A	1 mW (0 dBm)	P_{min} < 4 dBm to P_{max} Optional: P_{min}† to P_{max}
2	2.5 mW (4 dBm)	1 mW (0 dBm)	0.25 mW (–6dBm)	Optional: P_{min}† to P_{max}
3	1 mW (0 dBm)	N/A	N/A	Optional: P_{min}† to P_{max}

Note:
* Minimum output power at maximum power setting.
† The lower power limit $P_{min} < -30$ dBm is suggested but it is not mandatory, and may be chosen according to application needs.

17.7 Definitions of the Terms used in Bluetooth

- **Piconet**—A collection of devices connected via Bluetooth technology in an ad hoc fashion. A piconet starts with two connected devices, such as a PC and cellular phone, and may grow to eight connected devices. All Bluetooth devices are peer units and have identical implementations. However, when establishing a piconet, one unit will act as a master for synchronization purposes, and the other(s) as slave(s) for the duration of the piconet connection.
- **Scatternet**—Two or more independent and nonsynchronized piconets that communicate with each other. A slave as well as a master unit in one piconet can establish this connection by becoming a slave in the other piconet.
- **Master unit**—The device in the piconet whose clock and hopping sequence are used to synchronize all other devices in the piconet.
- **Slave units**—All devices in a piconet that are not the master (up to seven active units for each master).
- **Media access control address**—A 3-bit media access control address used to distinguish between units participating in the piconet.
- **Parked units**—Devices in a piconet that are synchronized but do not have a media access control address.
- **Sniff and hold mode**—Devices that are synchronized to a piconet and that have temporarily entered power-saving mode, in which device activity is reduced.

17.8 Bluetooth Protocol Stack

The Bluetooth protocol stack allows devices to locate, connect, and exchange data with each other and to execute interoperable, interactive applications against each other. Bluetooth protocol stack can be divided into three groups: transport protocol group, middleware protocol group, and application group (see Figure 17.3).

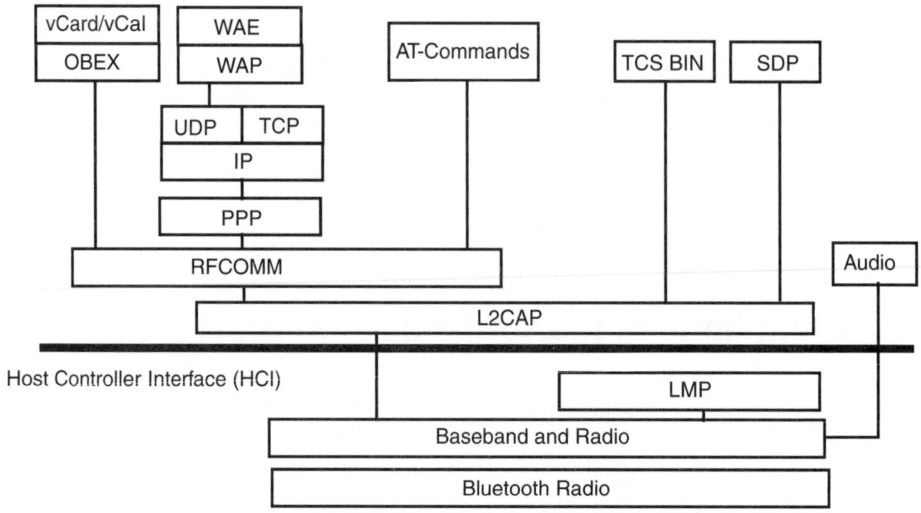

Figure 17.3 Bluetooth Protocol Stack

17.8.1 Transport Protocol Group

The protocols in this group are designed to allow Bluetooth devices to locate and connect to each other. These protocols carry audio and data traffic between devices and support both asynchronous transmissions and synchronous communications for telephony-grade voice communication. Audio traffic is treated with high priority in Bluetooth. Audio traffic bypasses all protocol layers and goes directly to the baseband layer, which then transmits it in small packets directly over Bluetooth's air interface.

The protocols in this group are also responsible for managing the physical and logical links between the devices so that the layers above and applications can pass data through the connections. The protocols in this group are radio, baseband, link manager, logical link, and host controller interface (HCI).

- **Logical link control and adaptation protocol (L2CAP) layer**—All data traffic is routed through the L2CAP layer. This layer shields the higher layers from the details of the lower layers. The higher layers need not be aware of the frequency hops occurring at the radio and baseband levels. It is also responsible for segmenting larger packets from higher layers into smaller packets, which are easier to handle by the lower layers. The L2CAP layer in two peer devices facilitates the maintenance of the desired grade of service. The L2CAP layer is responsible for admission control based on the requested level of service and coordinating with the lower layers to maintain this level of service.

- **Link manager protocol (LMP) layer**—The link manager layers in communicating devices are responsible for negotiating the properties of the Bluetooth air interface between them. These properties may be anything from bandwidth allocation to support services of a particular type to periodic bandwidth reservation for audio traffic. This layer is responsible for supervising device pairing. Device pairing is creation of a trust relationship between the devices by generating and storing an authentication key for future device authentication. This is an important step in establishing a communication between two devices. If this fails, the communication link might get severed. The link managers are also responsible for power control and request adjustments in power levels.
- **Baseband and radio layers**—The baseband layer is responsible for the process of searching for other devices and establishing a connection with them. It is also responsible for assigning the master and slave roles. This layer also controls the Bluetooth unit's synchronization and transmission frequency hopping sequence. This layer also manages the links between the devices and is responsible for determining the packet types supported for synchronous and asynchronous traffic.
- **Host controller interface layer**—The HCI allows higher layers of the stack, including applications, to access the baseband, link manager, etc., through a single standard interface. Through HCI commands, the module may enter certain modes of operation. Higher layers are informed of certain events through the HCI. HCI is not a required part of the specification. It has been developed to serve the purpose of interoperability between host devices and Bluetooth modules. Bluetooth product implementation need not be HCI compliant to support a fully compliant Bluetooth air interface.

17.8.2 Middleware Protocol Group

This group is composed of the protocols needed for existing applications to operate over Bluetooth links. The protocols in this group can be third-party and industry-standard protocols and protocols developed specifically by the Special Interest Group (SIG) for Bluetooth Wireless Communication. The protocols in this group can include TCP, IP, PPP, etc. A serial port emulator protocol known as RFCOMM enables applications that normally would interface with a serial port to operate with Bluetooth links. A packet-based telephony control protocol for advanced telephony operations is also present. This group has a service discovery protocol (SDP) that lets devices discover each other's services.

- **RFCOMM layer**—Serial ports are the most common communication interface in use today. These serial ports invariability involve the use of cable. Bluetooth's prime aim is to eliminate cables, and support serial communication without cables. RFCOMM provides a virtual serial port to applications. The advantage provided by this layer is that it is easy for applications designed for cabled serial ports to migrate to Bluetooth. The applications can use RFCOMM much like a serial port to accomplish scenarios such as dial-up networking. RFCOMM is an important part of the protocol stack because of the function it performs.

- **Service discovery protocol layer**—In Bluetooth wireless communications, any two devices can start communicating on the spur of the moment. Once a connection is established, there is a need for the devices to find and understand the services the other devices has to offer. This is taken care of in this layer. The SDP is a standard method for Bluetooth devices to discover and learn about the services offered by the other device(s). Service discovery is important in providing value to the end user.
- **Infrared data association (IrDA) interoperability protocols**—The Bluetooth SIG has adopted some IrDA protocols to ensure interoperability between applications. IrDA and Bluetooth share some important attributes. The infrared object exchange protocol (IrOBEX) is designed to enable units supporting infrared communication to exchange a wide variety of data and comments.
- **Infrared object exchange protocol**—IrOBEX (often abbreviated as OBEX) is a session protocol developed by the Infrared Data Association to exchange objects in a simple and spontaneous manner. OBEX provides the same basic functionality as HTTP but in a much lighter fashion. It uses a client/server model and is independent of the transport mechanism and transport API, provided it realizes a reliable transport base. In addition, the OBEX protocol defines a folder-listing object, which is used to browse the contents of folders on remote device(s).
- **Networking layers**—Bluetooth wireless communication uses a peer-to-peer network topology rather than a LAN-type topology. Dial-up networking uses the attention (AT) command layer. In most cases, the network that is being accessed is an IP network. Once a dial-up connection to an IP network is established, then standard protocols like TCP, UDP, and HTTP can be used. A device can also connect to an IP network using a network access point. The Internet PPP is used to connect to the access point. The specification does not define a profile that uses the TCP/IP directly over Bluetooth links.
- **Telephony control specification (TCS) layer and audio**—This layer is designed to support telephony functions, which include call control and group management. These are associated with setting up voice calls. Once a call is established, a Bluetooth audio channel can carry the call's voice content. TCS can also be used to set up data calls. The TCS protocols are compatible with International Telecommunications Union (ITU) specifications. The Bluetooth SIG also considered a second protocol called TCS-AT, which is a modem control protocol. AT commands over RFCOMM are used for some applications. Audio traffic is treated separately in Bluetooth. Audio traffic is isochronous, meaning that it has a time element associated with it. Audio traffic is routed directly to the baseband. Special packets called synchronous connection-oriented packets are used for audio traffic. Bluetooth audio communication takes place at a rate of 64 kbps using one of the two data encoding schemes—8-bit logarithmic pulse code modulation or continuous variable slope delta modulation.

17.8.3 Application Group

This group consists of actual applications that make use of Bluetooth links and refers to the software that exists above the protocol stack. The software uses the protocol stack to provide some function to the user of the Bluetooth devices. The most interesting applications are those that instantiate the Bluetooth profiles. The Bluetooth SIG does not define any application protocols nor does it specify any API. Bluetooth profiles are developed to establish a base point for use by the protocol stack to accomplish a given usage case.

17.9 Bluetooth Link Types

The Bluetooth baseband technology supports two link types: synchronous connection-oriented (SCO) type, which are used primarily for voice, and asynchronous connectionless (ACL) type, which are used primarily for packet data.

Different master/slave pairs of the same piconet can use different link types and the link type may change arbitrarily during a session. Each link type supports up to 16 different packet types. Four of these are control packets and are common for both SCO and ACL links. Both link types use a time division duplex (TDD) scheme for full-duplex transmission.

The SCO link is symmetric and typically supports time-bounded voice traffic. SCO packets are transmitted over reserved intervals. Once the connection is established, both master and slave units may send SCO packets without being polled. The SCO link type supports circuit-switched, point-to-point connections, and is used often for voice traffic. The data rate for SCO links is 64 kbps.

The ACL link is packet-oriented and supports both symmetric and asymmetric traffic. The master unit controls the link bandwidth and decides how much piconet bandwidth is given to each slave, and the symmetry of the traffic. Slaves must be polled before they can transmit data. The ACL link also supports broadcast messages from the master to all slaves in the piconet. Multislot packets can be used in ACL and they can reach maximum data rates of 721 kbps in one direction and 57.6 kbps in the other direction if no error correction is used.

Data packets are protected by an automatic retransmission query (ARQ) scheme. Thus, when a packet arrives, a check is performed on it. If there is an error detected, the receiving unit indicates this in the return packet. In this way, retransmission is done only for the faulty packets. Retransmission is not feasible for voice, so a better error protection is used.

17.10 Bluetooth Security

Bluetooth security features support authentication and encryption. These features are based on a secret link key that is shared by a pair of devices. A pairing procedure is used when two devices communicate for the first time to generate this key. There are three security modes to a device:

- **Nonsecure**—A device will not initiate any security procedure.
- **Service level enforced security**—A device does not initiate security procedures before channel establishment at L2CAP level. This mode allows different and flexible access policies for applications, especially when running applications with different security requirements in parallel.
- **Link level enforced security**—A device initiates security procedures before the link setup at the LMP is completed.

Figure 17.4 shows the Bluetooth security architecture.

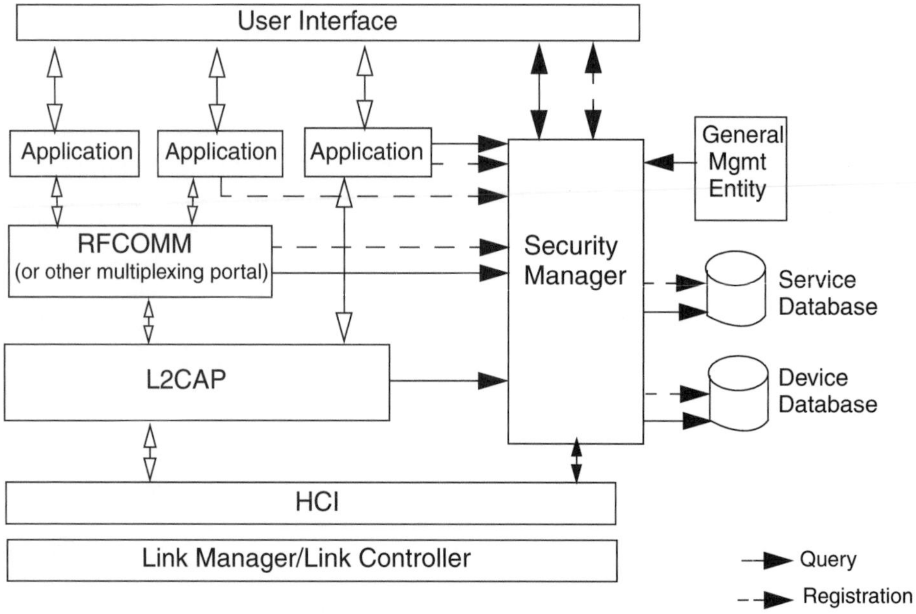

Figure 17.4 Bluetooth Security Architecture

17.10.1 Security Levels

There are two kinds of security levels:

- **Authentication**—*Authentication* basically verifies who is at the other end of the link. In Bluetooth, this is achieved by the authentication procedure based on the stored link key or by pairing. To meet different requirements on availability of services without user intervention, authentication is performed after determining what the security level of the requested service is. Thus, authentication cannot be performed when the ACL link is established. The authentication is performed when a connection request to a service is submitted. The following procedures are performed (see Figure 17.5):

1. Connect request to L2CAP
2. L2CAP requests access from the security manager
3. Security manager lookup in service database
4. Security manager lookup in device database
5. If necessary, security manager enforces authentication and encryption
6. Security manager grants access
7. L2CAP continues to set up connection

Authentication can be performed in both directions; client authenticates server and vice versa.

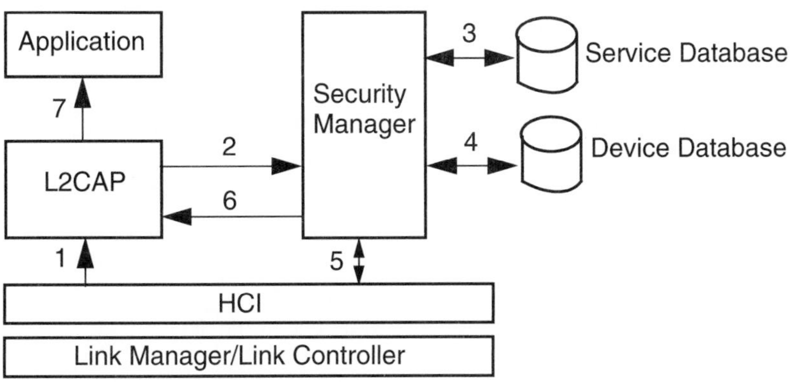

Figure 17.5 Authentication Procedures

- **Authorization**—When one device is allowed to access the other, the concept of trust comes into existence. Trusted devices are allowed access to services. Likewise, untrusted devices might require authorization based on user interaction before access to services is granted. There are two kinds of device trust levels:
  - **Trusted device**—Device with fixed relationship (paired) that is trusted and unrestricted access to all services.
  - **Untrusted device**—The device has been previously authenticated and a link key is stored, but the device is not marked as trusted in the device database. An unknown device is also an untrusted device. No security information is available for this device.

For services, the requirement for authorization, authentication, and encryption are set independently (although some restrictions apply). The access requirements allow the definition of three security levels:

- Services that require authorization and authentication—automatic access is only granted to trusted devices. Other devices need a manual authorization.
- Services that require authentication only—authorization is not necessary.
- Services open to all devices—authentication is not required, no access approval required before service access is granted.

A default security level is defined to serve the needs of legacy applications. This default policy will be used unless other settings are found in a security database related to a service.

Bluetooth security has its limitations:

- Only a device is authenticated and not its user.
- There is no mechanism to preset authorization per service. However, a more flexible security policy can be implemented with the present architecture without the need to change the Bluetooth protocol stack.
- It is not possible to enforce unidirectional traffic.

17.11 Network Connection Establishment in Bluetooth

Before any connection in a piconet is created, all devices are in STANDBY mode. In this mode, an unconnected unit periodically listens for messages every 1.28 seconds. Each time a device wakes up, it listens on a set of 32 hop frequencies defined for that unit. The number of hop frequencies varies in different geographic regions.

The connection procedure is initiated by any one of the devices, which then becomes the master. A connection is made by a PAGE message if the address is already known, or by an INQUIRY message followed by a subsequent PAGE message if the address is unknown (see Figure 17.6).

In the initial PAGE state, the master unit sends a train of 16 identical PAGE messages on 16 different hop frequencies defined for the device to be paged (i.e., the slave unit). If no response is received, the master transmits a train on the remaining 16 hop frequencies in the wake-up sequence. The maximum delay before the master reaches the slave is twice the wake-up period (2.56 seconds) while the average delay is half the wake-up period (0.64 seconds).

The INQUIRY message is typically used for finding Bluetooth devices, including public printers, fax machines, and similar devices with unknown addresses. The INQUIRY message is similar to the PAGE message, but may require one additional train period to collect all responses.

A power-saving mode can be used for connected units in a piconet if no data is to be transmitted. The master unit can put slave units into HOLD mode, where only an internal timer is running. Slave units can also demand to be put into HOLD mode. Data transfer restarts instantly when units transition out of HOLD mode. The HOLD is used when connecting several piconets or managing a low-power device such as a temperature sensor.

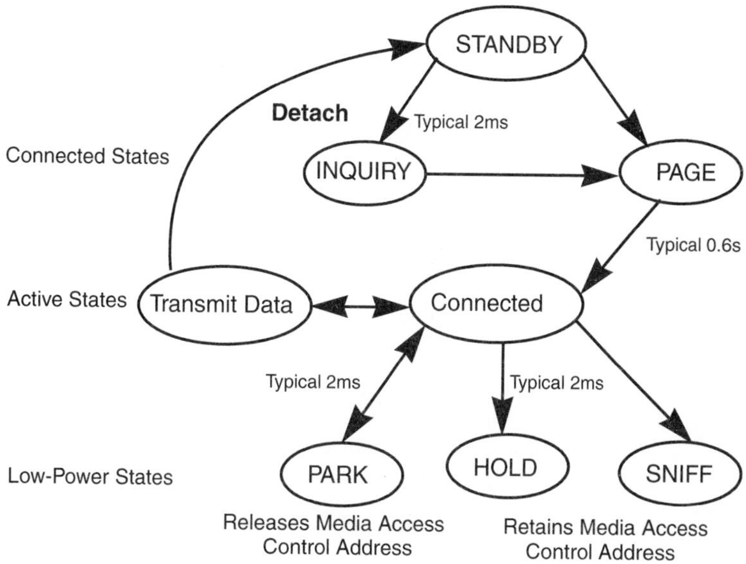

Figure 17.6 Device States in Bluetooth

Two more low-power modes are also available, the SNIFF mode and the PARK mode. In the SNIFF mode, a slave device listens to the piconet at reduced rate, thus reducing its duty cycle. The SNIFF interval is programmable and depends on the application. In the PARK mode, a device is still synchronized to the piconet but does not participate in the traffic. Parked devices have given up their media access control address and occasionally listen to the traffic of the master to resynchronize and check on broadcast messages.

If we list the low-power modes in increasing order of power efficiency, then the SNIFF mode has the highest duty cycle, followed by the HOLD mode with a lower duty cycle, and the PARK mode with the lowest duty cycle.

17.12 Error Correction in Bluetooth

Three error correction schemes are defined for the Bluetooth baseband controller:

- 1/3-rate forward error correction (FEC) code
- 2/3-rate forward error correction code
- Automatic retransmission query (ARQ) scheme for data

The purpose of the FEC scheme on the data payload is to reduce the number of retransmissions. However, in a reasonably error-free environment, FEC creates unnecessary overhead that reduces the throughput. Therefore, the packet definitions have been kept flexible as to whether or not to use FEC in the payload. The packet header is always protected by a 1/3-rate FEC. It contains link information and should survive bit errors. An unnumbered ARQ scheme is applied in which data transmitted in one slot is directly acknowledged by the recipient in the next slot. For a data transmission to be acknowledged, both the header error check and the cyclic redundancy check (CRC) must be satisfied, otherwise a negative acknowledgment is returned.

17.13 Network Topology in Bluetooth

Bluetooth devices can create both point-to-point and point-to-multipoint connections. A connection with two or several (maximum eight) devices is a piconet, where all devices follow the same frequency hop scheme. To avoid interference between devices, one of the devices automatically becomes the master of the piconet. In each slot, a packet can be exchanged between the master and one of the slaves. Packets have a fixed format (see Figure 17.7).

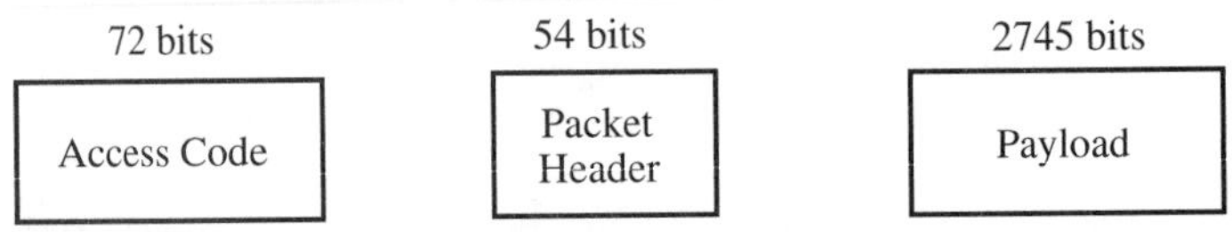

Figure 17.7 Packet Format in Bluetooth

Each packet begins with a 72-bit access code that is derived from the master identity and is unique for the channel. Every packet exchanged on the channel is preceded by this access code. Recipients on the piconet compare incoming signals with access codes. If the two do not match, the received packet is not considered valid on the channel and the rest of its contents are ignored. Besides packet identification, the access code is also used for synchronization and compensating for offset. The access code is robust and resistant to interference.

If two or more piconets communicate with each other, they are then called a *scatternet.*

17.14 Bluetooth Usage Models

In this section, we discuss some of the Bluetooth usage models. Each usage model has one or more profiles.

- **Three-in-one phone**—The three-in-one phone usage model describes how a telephone handset may connect to three different service providers. The telephone may act as a cordless telephone connecting to the public switched telephone network (PSTN) at home, charged at a fixed line charge. The telephone can also connect directly to other telephones acting as a walkie-talkie or handset extension. Finally, the telephone may act as a cellular phone connecting to the cellular infrastructure.

- **File transfer**—The file transfer usage model offers the capability to transfer data objects from one Bluetooth device to another. Files, entire folders, directories, and streaming media formats are supported in this model. The model also offers the possibility of browsing the contents of folders on a remote device.
- **Synchronization**—The synchronization usage model provides the means for automatic synchronization between, for instance, a desktop PC, a portable PC, a PDA, and a notebook. The synchronization requires business card, calendar, and task information to be transferred and processed by computers, cellular phones, and PDAs utilizing a common protocol and format.
- **Internet bridge**—The Internet bridge usage model describes how a mobile phone or cordless modem provides a PC with dial-up networking capabilities without the need for a physical connection to the PC. This networking scenario requires a two-piece protocol stack, one for AT commands to control the mobile phone and another stack to transfer payload data.
- **Ultimate handset**—The ultimate handset usage model defines how a Bluetooth-equipped wireless handset can be connected to act as a remote unit's audio input and output interface. This unit is probably a mobile phone or a PC for audio input and output.

17.15 WAP and Bluetooth

Bluetooth can be used with WAP like any other wireless network. Bluetooth wireless networks can be used to transport data from a WAP client to WAP server. The WAP client can make use of Bluetooth's service discovery protocol to find the WAP server/gateway. This is very useful when the WAP device is a mobile phone and when it comes into range of a WAP server, it can use Bluetooth's SDP to discover the gateway. The Bluetooth SDP must be able to provide some details about the WAP server to the WAP client.

The other feature that can be supported is the reverse of the above. The WAP server can periodically check for the availability of WAP-enabled clients in its range. It can use Bluetooth's SDP to do this. If there are any clients, the server can push any data to the client. The client, of course, is not required to accept the data pushed to it.

17.16 Applications of WAP

The first and foremost application of WAP is accessing the Internet from mobile devices. This is already in use in many mobile phones. This application is gaining popularity by the day and many Web sites already have a WAP version of their site. For example, an application is sending sale offers to mobile customers through WAP. The user's phone will be able to receive any sale prices and offers from the Web site of a store.

Games can be played from mobile devices over the wireless devices. This application has been implemented in certain countries and is under development in many others. This is an application that is predicted to gain high popularity.

Applications to access timesheets and filing expense claims via mobile handsets are currently being developed. When implemented, these applications will be a breakthrough in the business world.

Applications to locate WAP customers geographically are also being developed. Applications to help users who are lost or stranded by guiding them using their locations are also being considered.

WAP also provides short messaging, e-mail, and weather and traffic alerts based on the geographic location of the customer. These applications are available in some countries but soon will be provided in all countries.

One of the biggest applications of WAP under consideration is banking from mobile devices. These applications will be very popular if they are implemented in a secured manner.

The mobile industry appears to be moving forward, putting aside the issues of network and air interface standards and instead concentrating on laying the foundations for service development, regarded by many as the key driver to multimedia on the move and third-generation mobile systems. From that point of view, WAP and Bluetooth will play fundamental roles in the near future.

17.17 Wireless Local Area Network

A WLAN [18–26] is a flexible data communications system implemented as an extension of or as an alternative for a wired LAN. Using radio frequency (RF) technology, WLANs transmit and receive data over the air, minimizing the need for wired connections. Thus, WLANs combine data connectivity with user mobility.

The importance of WLAN technology, however, goes far beyond just the absence of wires. The advent of WLAN opens up a whole new definition of what a network infrastructure can be. No longer does an infrastructure need to be solid and fixed, difficult to move, and expensive to change. Instead, it can move with the user and change as fast as the organization changes.

Recently, manufacturers have deployed WLANs for process and control applications. Retail applications have expanded to include wireless point of sale (WPOS). The healthcare and education industries are also fast growing markets for WLANs. WLANs provide high-speed, reliable data communications in a building or campus environment as well as coverage in rural areas. WLANs are simple to install and do not incur monthly user fees or data transmission charges.

In a WLAN the connection between the client and the user is accomplished by the use of wireless medium such RF or infrared (IR) communications instead of a cable. This allows a remote user to stay connected to the network while mobile or not physically attached to the network. The wireless connection is usually accomplished by the user having a handheld terminal or laptop that has a RF interface card installed inside the terminal or through the PC card slot of the laptop. The client connection from the wired LAN to the user is made through an access point (AP) that can support multiple users simultaneously. The AP can reside at any node on the wired network and acts as a gateway for wireless users' data to be routed onto the wired network (see Figure 17.8).

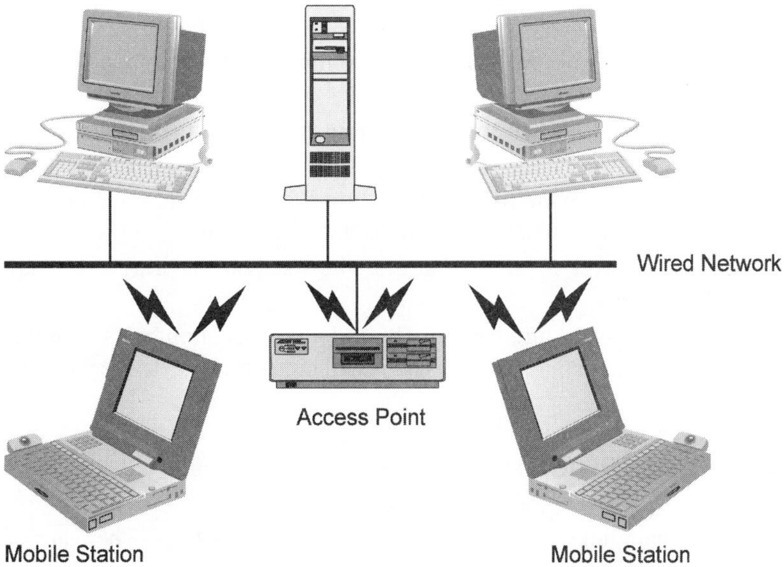

Figure 17.8 WLAN Application

The range of a WLAN depends on the actual usage and environment of the system and can vary from 100 feet inside a solid walled building to several thousand feet outdoors, with a direct line of sight. Much like cellular phone systems, the WLAN is capable of roaming from the AP and reconnecting to the network through other APs residing at other points in the network. This can allow the wired LAN to be extended to cover a much larger area than the existing coverage by the use of multiple APs.

An important feature of the WLAN is that it can be used independently of a wired network. It can be used anywhere as a stand-alone network to link multiple computers together without having to build or extend a wired network. WLANs are billed on the basis of installed equipment cost; once in place, there are no charges for the use of the network. The network communications take place in a part of the radio spectrum that is designed as license free. In this band, 2.4–2.5 GHz, users can operate without a license as long as they use equipment of the type approved for use in the license free band. The 2.4-GHz band has been designated as license free by the ITU and is available as such in most countries of the world.

A WLAN is capable of operating at speeds in the range of 1–2 Mbps depending on the actual system; both of these speeds are supported by the standard for WLAN networks defined by the Institute of Electrical and Electronic Engineers (IEEE). The fastest WLANs use a 802.11b high-rate standard to move data through air at a maximum speed of 11 Mbps. The IEEE established the 802.11b standard for wireless networks and the Wireless Ethernet Compatibility Alliance (WECA) to assure that WLAN products are interoperable from manufacturer to manufacturer. Any LAN application, network operating system, or protocol, including TCP/IP, will run on a 802.11b-compliant WLAN as easily as they run over Ethernet.

The following are some advantages of deploying a WLAN:

- Mobility improves productivity with real-time access to information, regardless of worker location, for faster and more efficient decision-making.
- Cost-effective network setup for hard-to-wire locations such as older buildings and solid wall structures.
- Reduced cost of ownership, particularly in dynamic environments requiring frequent modification due to minimal wiring and installation costs per device and per user.

17.18 WLAN Equipment

There are three main links that form the basis of the wireless LAN. These are:

- **LAN adapter**—Wireless adapters are made in the same basic form as their wired counterparts: PCMCIA, Cardbus, PCI, and USB. They also serve the same function, enabling end users to access the network. In a wired LAN, adapters provide interface between the network operating system and the wire. In a WLAN, they provide the interface between the network operating system and an antenna to create a transparent connection to the network.
- **Access point**—The AP is the wireless equivalent of a LAN hub. It receives, buffers, and transmits data between the WLAN and the wired network, supporting a group of wireless user devices. Typically, an AP is connected with the backbone through a standard Ethernet cable, and communicates with wireless devices by means of an antenna. The AP or antenna connected to it is generally mounted on a high wall or on the ceiling. Like cells in a cellular phone network, multiple APs can support handoff from one AP to another as the user moves from area to area. APs have range from 20 to 500 meters. A single AP can support between 15 and 250 users, depending on technology, configuration, and use. It is relatively easy to scale WLANs by adding more APs to reduce network congestion and enlarge the coverage area. Large networks that require multiple APs create overlapping cells for constant connectivity to the network. A wireless AP can monitor movement of clients across its domain and permit or deny specific traffic or clients from communicating through it.

- **Outdoor LAN bridges**—Outdoor LAN bridges are used to connect LANs in different buildings. When the cost of buying a fiber optic cable between buildings is considered, particularly if there are barriers such as highways or bodies of water in the way, a WLAN can be an economical alternative. An outdoor bridge can provide a less expensive alternative to recurring leased line charges. WLAN bridge products support fairly high data rates and ranges of several miles with the use of line-of-sight directional antennas. Some APs can also be used as a bridge between buildings of relatively close proximity.

17.19 WLAN Topologies

WLANs can be built with either of three topologies:

- Peer-to-peer topology
- AP-based topology
- Point-to-multipoint bridge topology

In a peer-to-peer topology, client devices within a cell communicate directly to each other as shown in Figure 17.9.

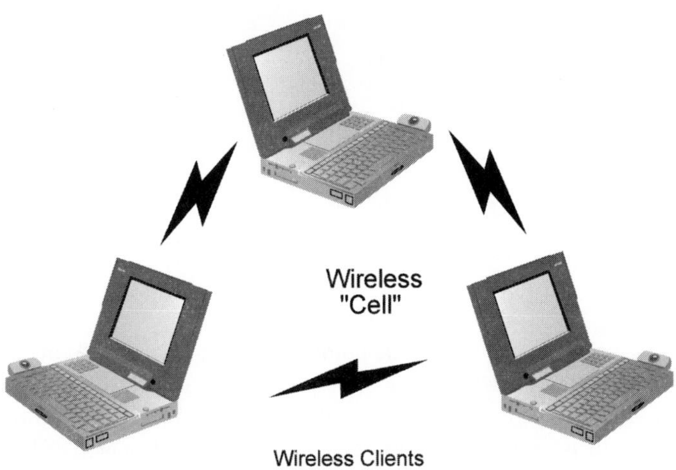

Figure 17.9 Peer-to-Peer Topology

An AP-based technology uses access points to bridge traffic onto a wired (Ethernet or token ring) or a wireless backbone, as shown in Figure 17.10. The AP enables a wireless client device to communicate with any other wired or wireless device on the network. The AP topology is more commonly used, demonstrating that WLANs do not replace wired LANs; instead, they extend connectivity to mobile devices.

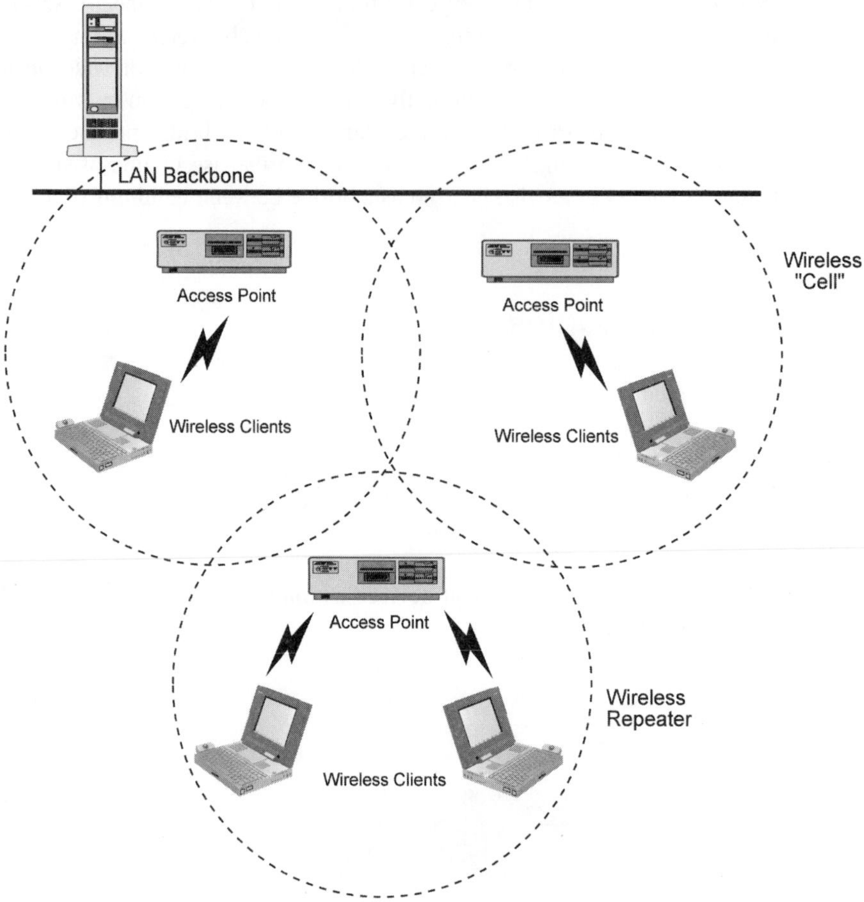

Figure 17.10 AP-Based Topology

Another wireless network topology is the point-to-multipoint bridge. Wireless bridges connect a LAN in one building to a LAN in another building even if the buildings are miles apart. These conditions receive a clear line of sight between buildings. The line-of-sight range varies based on the type of wireless bridge and antenna used as well as the environmental conditions.

17.20 WLAN Technologies

The technologies available for use in WLANs are infrared, UHF (narrowband), and spread-spectrum implementation. Each implementation comes with its own set of advantages and limitations.

Infrared Technology. Infrared is an invisible band of radiation that exists at the lower end of the visible electromagnetic spectrum. This type of transmission is most effective when a clear line-of-sight exists between the transmitter and the receiver.

Two types of infrared WLAN solutions are available: diffused-beam and direct-beam (or line-of-sight). Currently, direct-beam WLANs offer a faster data rate than the diffused-beam networks, but diffused-beam networks are more directional since they use reflected rays to transmit/receive a data signal, thus achieving lower data rates in the 1–2 Mbps range. Infrared is a short-range technology. When used indoors, it can be limited by solid objects such as doors, walls, merchandise, etc. In addition, the lighting environment can affect signal quality. For example, loss of communications might occur if there is a large amount of sunlight or background light in an environment. Fluorescent lights also might contain large amounts of infrared. This problem may be solved by using a high signal power and an optimal bandwidth filter, which reduces the infrared signals coming from outside sources. In an outdoor environment, snow, ice, and fog may affect the operation of an infrared-based system. Table 17.3 gives considerations for choosing infrared technology.

Table 17.3 Considerations for Choosing Infrared Technology

Advantages	No government regulations controlling use
	Immunity to electromagnetic (EM) and RF interference
Disadvantages	Generally a short-range technology (30–50 ft radius under ideal conditions)
	Signals cannot penetrate solid objects
	Signal affected by light, snow, ice, fog
	Dirt can interfere with infrared

UHF Narrowband. UHF wireless data communication systems have been available since the early 1980s. These systems normally transmit in the 430–470 MHz frequency range, with rare systems using segments of the 800-MHz range. The lower portion of this band (430–450 MHz) is referred to as the unprotected (unlicensed) band, and 450–470 MHz is referred to as the protected (licensed) band. In the unprotected band, RF licenses are not granted for specific frequencies and anyone is allowed to use any frequency, giving customers some assurance that they will have complete use of that frequency.

Because independent narrowband RF systems cannot coexist on the same frequency, government agencies allocate specific radio frequencies to users through RF site licenses. A limited amount of unlicensed spectrum is also available in some countries. In order to have many frequencies that can be allocated to users, the bandwidth given to a specific user is very small.

The term *narrowband* is used to describe this technology because the RF signal is sent in a very narrow bandwidth, typically 12.5 kHz or 25 kHz. Power levels range 1 to 2 watts for narrowband RF data systems. This narrow bandwidth combined with high power results in larger transmission distances than are available from 900-MHz or 2.4-GHz spread-spectrum systems, which have lower power levels and wider bandwidths. Table 17.4 lists the advantages and disadvantages of UHF technology.

Table 17.4 Considerations for Choosing UHF Technology

Advantages	Longest range
	Low cost solution for large sites with low to medium data throughput requirements
Disadvantages	Large radio and antennas increase wireless client size
	RF site license required for protected bands
	No multivendor interoperability
	Low throughput and interference potential

Many modern UHF systems use synthesized radio technology, which refers to the way channel frequencies are generated in the radio. The crystal-controlled products in legacy UHF products require factory installation of unique crystals for each possible channel frequency. Synthesized technology uses a single, standard crystal frequency and drives the required channel frequency by dividing the crystal frequency down to a small value, then multiplying it up to the desired channel frequency. The division and multiplication factors are unique for each desired channel frequency, and are programmed into digital memory in the radio at the time of manufacturing.

Synthesized UHF-based solutions provide the ability to install equipment without the complexity of hardware crystals. Common equipment can be purchased and the specific UHF frequency used for each device can be tuned based upon specific location requirements. Additionally, synthesized UHF radios do not exhibit the frequency drift problem experienced in crystal-controlled UHF radios, a feature that eliminates tuning problems after installations have been running for a period of time.

Modern UHF systems allow APs to be individually configured for operation on one of the several preprogrammed frequencies. Terminals are programmed with a list of all frequencies used in the installed APs, allowing them to change frequencies when roaming. To increase throughput, APs may be installed with overlapping coverage but using different frequencies.

Spread-Spectrum Technology. Most WLAN systems use spread-spectrum technology, a wideband radio frequency technique that uses the entire allotted spectrum in a shared fashion as opposed to dividing it into discrete private pieces as with narrowband. Spread spectrum spreads the transmission power over the entire usable spectrum. This is obviously a less efficient use of the bandwidth than the narrowband approach. However, spread spectrum is designed to trade off bandwidth efficiency for reliability, integrity, and security as per its military origins. The bandwidth trade-off produces a signal that is easier to detect, provided that the receiver knows the parameters of the spread-spectrum signal being broadcast. If the receiver is not tuned to the right frequency, a spread-spectrum signal looks like background noise.

By operating across a broad range of radio frequencies, a spread-spectrum device could communicate clearly despite interference from other devices using the same spectrum in the same physical location. In addition to its relative immunity to interference, spread spectrum makes eavesdropping and jamming inherently difficult.

In commercial applications, spread-spectrum techniques currently offer data rates up to 2 Mbps. Because the U.S. Federal Communication Commission (FCC) does not require site licensing for the bands used by spread-spectrum systems, this technology has become the standard for high-speed RF data transmission. Two modulation schemes are commonly used to encode spread-spectrum signals, direct-sequence spread-spectrum (DSSS) and frequency hopping spread-spectrum (FHSS).

17.21 High-Rate WLAN Standard

The 802.11b standard focuses on the bottom two layers of the open system interconnect (OSI) model, the physical layer and data link layer (see Figure 17.11). The 802.11b standard allows for two types of transmissions: FHSS and DSSS. Both types of spread spectrum use more bandwidth than a typical narrowband transmission, but enable a strong signal that is easier for the receiver to detect than the narrowband signal.

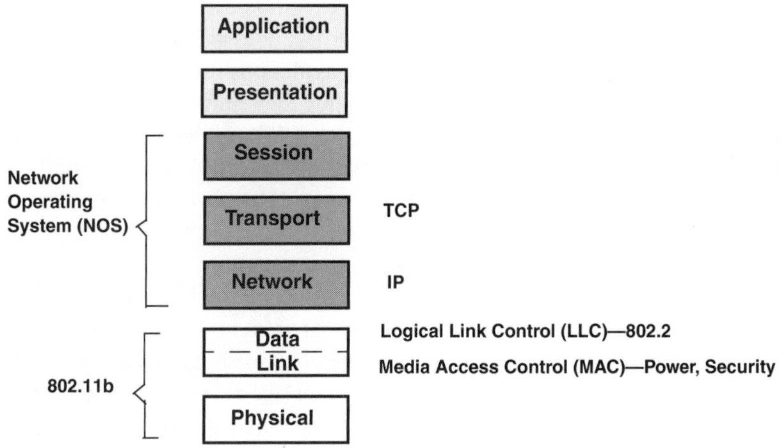

Figure 17.11 802.11b and OSI Model

17.21.1 802.11b Operating Modes

802.11b defines two types of equipment: a wireless station which is usually a PC equipped with wireless network interface card (NIC) and an AP. The 802.11b standard defines two modes: *infrastructure mode* and *ad hoc mode*. In the infrastructure mode, the wireless network consists of at least one access point connected to the wired network infrastructure and a set of wireless end stations. This configuration is called a basic service set. An extended service set (ESS) is a set of two or more basic service sets forming a single subnetwork. Since most corporate WLANs require access to the wired LAN for services (file servers, printers, Internet links), they operate in infrastructure mode.

Ad hoc mode (also called peer-to-peer mode or an independent basic service set) is simply a set of 802.11b wireless stations that communicate directly with one another without using an access point or any connection to a wired network. This mode is useful for quickly and easily setting up a wireless network anywhere that a wireless infrastructure does not exist or is not required for services, such as a hotel room, convention center, or airport where access to wired network is barred.

17.21.2 802.11b Physical Layer

The three physical layers originally defined in 802.11b included two spread-spectrum radio techniques and a diffuse infrared specification. The *radio-based* standards operate within the 2.4-GHz unlicensed ISM band. The original 802.11b wireless standard defines data rates of 1 Mbps and 2 Mbps via radio waves using FHSS or DSSS. Note that FHSS and DSSS are fundamentally different mechanisms and will not interoperate with one another. In the FHSS, the 2.4-GHz band is divided into 75 1-MHz subchannels. The sender and receiver agree on a hopping pattern, and data is sent over a sequence of the subchannels. Each conversation within 802.11b network occurs over a different hopping pattern, and patterns are designed to minimize the chance of two senders using the same subchannel simultaneously.

FHSS techniques allow for a relatively simple radio design, but are limited to speeds of no higher than 2 Mbps. This limitation is driven primarily by FCC regulations that restrict subchannel bandwidth to 1 MHz. These regulations force FHSS systems to spread their usage across the entire 2.4-GHz band, meaning they must hop often, which leads to a high amount of hopping overhead. In contrast, the DSSS techniques divide the 2.4-GHz band into 14 22-MHz channels. Adjacent channels overlap one another partially, with three of the 14 channels being completely nonoverlapping. Data is sent across one of these 22 MHz channels. To compensate for noise on a given channel, *spreading* (chipping) is used. Each bit of user data is converted into a series of bit patterns called *chips*. The strength of the DSSS system is that when the ratio between the original signal bandwidth and the spread signal bandwidth is large, the system offers a great immunity to interference.

The infrared physical layer operates in the near-visible light range of 850–950 nanometers. Diffuse transmission is used so the transmitter and receiver do not have to point to each other. The transmission distance is limited to the range of 10 to 20 meters, and the signal is contained by walls and windows. This feature has the advantage of isolating the transmission systems in different rooms. The system cannot operate outdoors. The transmission system uses pulse-position modulation (PPM), in which the binary data is mapped into symbols that consist of group of slots.

17.21.3 802.11b Data Link Layer

The data link layer within 802.11b consists of two sublayers: logical link control (LLC) and media access control (MAC). 802.11b uses the same 802.2 LLC and 48-bit addressing as other 802 LANs, allowing for simple bridging from wireless to IEEE wired networks, but the MAC is unique to WLANs. The 802.11b MAC is similar in concept to 802.3 in that it is designed to support multiple users on a shared medium by having the sender sense the medium before accessing it. For 802.3 Ethernet LANs, the carrier sense multiple access with collision detection (CSMA/CD) protocol regulates how Ethernet stations establish access to the network and how they detect and handle collisions that occur when two or more devices try to simultaneously communicate over the LAN. In an 802.11b WLAN, collision detection is not possible due to the near-far problem. To detect a collision, a station must be able to transmit and listen at the same time, but in radio systems the transmission drowns out the ability of the station to hear a collision.

To account for this difference, 802.11b uses a slightly modified protocol known as carrier sense multiple access with collision avoidance (CSMA/CA) or distributed coordination function (DCF). CSMA/CA attempts to avoid collisions by using explicit packet acknowledgment (ACK), which means an ACK packet is sent by the receiving station to confirm that the data packet arrived intact.

CSMA/CA works as follows: A station wishing to transmit senses the air, and, if no activity is detected, the station waits an additional, randomly selected period of time and then transmits if the medium is still free. If the packet is received intact, the receiving station issues an ACK frame that, once successfully received by the sender, completes the process. If the ACK frame is not detected by the sending station, either because the original data packet or the ACK was not received intact, a collision is assumed to have occurred and the data packet is retransmitted after waiting another random amount of time. CSMA/CA provides a way to share access over the air. This explicit ACK mechanism also handles interference and other radio-related problems very effectively. However, it does add some overhead to 802.11b that 802.3 does not have, so that an 802.11b LAN will always have slower performance than equivalent Ethernet LAN.

Another MAC layer problem specific to wireless is the *hidden node* issue, in which two stations on opposite sides of an access point can both hear activity from an AP, but not from one another, usually because of distance or an obstruction. To solve this problem, 802.11b specifies an optional request to send/clear to send (RTS/CTS) protocol at the MAC layer. When this feature is in use, a sending station transmits an RTS and waits for the AP to reply with a CTS. Since all stations in the network can hear the AP, the CTS causes them to delay any intended transmissions, allowing the sending station to transmit and receive a packet acknowledgment without any chance of collision. Since RTS/CTS adds additional overhead to the network by temporarily reserving the medium, it is typically used only on the largest-sized packets, for which transmission would be expensive from a bandwidth standpoint.

The 802.11b MAC layer provides for two other robustness features: *CRC checksum* and *packet fragmentation*. Each packet has a CRC checksum calculated and attached to ensure that the data was not corrupted in transmit. This is different from Ethernet, where higher-level protocols such as TCP handle error checking. Packet fragmentation allows large packets to be segmented into smaller units when sent over the air, which is useful in very congested environments or when interference is a factor, since large packets have a better chance of being corrupted. This technique reduces the need for retransmission in many cases and improves overall wireless network performance. The MAC layer is responsible for reassembling fragments received, rendering the process transparent to higher-level protocols.

17.21.4 Association, Cellular Architecture, and Roaming

The 802.11b MAC layer is responsible for how a client associates with an AP. When an 802.11b client enters the range of one or more APs, it chooses an AP to associate with (also known as joining a basic service set), based on signal strength and observed packet error rates. Once accepted by the AP, the client tunes to the radio channel to which the AP is set. Periodically it surveys all 802.11b channels in order to access whether a different AP would provide it with better performance characteristics. If it determines that this is the case, it reassociates with the new AP, tuning to the radio channel to which that AP is set. Reassociation usually occurs because the wireless station has physically moved away from the original AP, causing the signal to be weakened. In other cases, reassociating occurs because of changes in radio characteristics in the building or high network traffic on the original AP. In the latter case, this function is known as load balancing since its primary function is to distribute the total WLAN load most efficiently across the available wireless infrastructure.

The process of dynamically associating and reassociating with APs allows network managers to set up WLANs with very broad coverage by creating a series of overlapping 802.11b cells throughout a building or across a campus. To be successful, the information technology (IT) manager ideally will employ channel reuse, taking care to set up each access point on an 802.11b DSSS channel that does not overlap with a channel used by neighboring AP (see Figure 17.12).

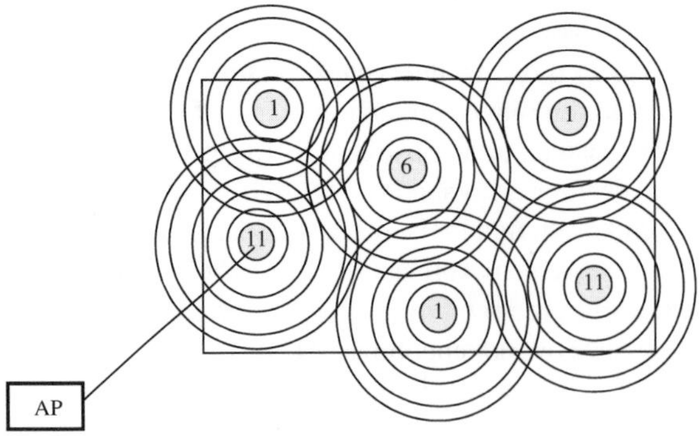

Figure 17.12 DSSS Channel Without Overlap with a Channel Used by Neighboring AP

As noted above, while there are 14 partially overlapping channels specified in 802.11b DSSS, there are only three channels that do no overlap at all; these are the best to use for multicell coverage. If two APs are in range of one another and are set to the same or partially overlapping channels, they may cause some interference for one another, thus lowering the total available bandwidth in the area of overlap.

17.21.5 Power Management

Power management is necessary to minimize power requirements for battery powered portable mobile units. The standard supports two power-utilization modes, *continuous aware mode* and *power save polling mode*. In the former, the radio is always on and draws power, whereas in the latter, the radio is dozing with the AP queuing any data for it.

A power saver mode or sleep is defined when station is not transmitting in order to save battery power. However, critical data transmissions cannot be missed. Therefore APs are required to have buffers to queue messages. Sleeping stations are required to periodically wake up and retrieve messages from the AP. Power management is more difficult for peer-to-peer inter basic service set configurations without central AP. In this case, all clients in the inter basic service set must be awakened when periodic beacon is sent. Clients randomly handle the task of sending out the beacon. An announcement traffic information message (ATIM) window commences. During this period, any station can go to sleep if there is no announced activity for it during this short period.

17.21.6 Security

802.11b provides for media access control and encryption mechanisms, which are known as *wired equivalent privacy* (WEP), with the objective of providing WLANs with security equivalent to their wired counterparts. For the access control, extended service set identification (ESSID), also known as a WLAN service area ID, is programmed into each AP and is required in order for a wireless client to associate with an AP. In addition, there is provision for a table of MAC addresses called an *access control list* to be included in the AP, restricting access to clients whose MAC addresses are not on the list.

For data encryption, the standard provides for optional encryption using a 40-bit shared-key RC4 PRNG algorithm from RSA data security. All data sent and received while the end station and AP are associated can be encrypted using this key. In addition, when encryption is in use, the access point will issue an encrypted challenge packet to any client attempting to associate with it. The client must use its key to encrypt the correct response in order to authenticate itself and gain network access.

Beyond layer 2, 802.11b WLANs support the same security standards supported by other 802 LANs for access control (such as network operating system logins) and encryption (such as IPSec or application-level encryption). These higher-level technologies can be used to create end-to-end secure networks encompassing both wired LAN and WLAN components, with the wireless piece of the network gaining unique additional security from the 802.11b feature set.

17.21.7 Extensions to 802.11b

In September 1999, IEEE ratified the 802.11b high-rate amendment to the standard, which added two higher speeds (5.5 and 11 Mbps) to 802.11b. The key contribution of the 802.11b addition to the WLAN standard was to standardize the physical layer support to two new speeds, 5.5 Mbps and 11 Mbps. To accomplish this, DSSS had to be selected as the sole physical layer technique for the standard since, as noted above, frequency hopping cannot support the higher speeds without violating current FCC regulations. The implication is that 802.11b systems will interoperate with 1 Mbps and 2 Mbps 802.11b DSSS systems, but will not work with 1 Mbps and 2 Mbps FHSS systems.

The original 802.11b DSSS standard specifies an 11-bit chipping called a *Barker sequence* to encode all data sent over the air. Each 11-chip sequence represents a single data bit (1 or 0), and is converted to a waveform, called a symbol, that can be sent over the air. These symbols are transmitted at a 1 million symbols per second (Msps) rate using binary phase-shift keying (BPSK). In the case of 2 Mbps, a more sophisticated implementation based on quadrature phase-shift keying (QPSK) is used. This doubles the data rate available in BPSK, via improved efficiency in the use of the radio bandwidth.

To increase the data rate in the 802.11b standard, advanced coding techniques are employed. Rather than the two 11-bit Barker sequences, 802.11b specifies complementary code keying (CCK), which consists of a set of 64 8-bit code words. As a set, these code words have unique mathematical properties that allow them to be correctly distinguished from one another by a receiver, even in the presence of substantial noise and multipath interference. The 5.5-Mbps rate uses CCK to encode 4 bits per carrier, while the 11-Mbps rate encodes 8 bits per carrier. Both speeds use QPSK modulation and signal at 1.375 Msps. This is how the higher data rates are obtained. Table 17.5 lists the specifications.

To support very noisy environments as well as extended range, 802.11b WLANs use dynamic rate shifting, allowing data rates to be automatically adjusted to compensate for the changing nature of the radio channel. Ideally, users connect at full 11-Mbps rate. However, when devices move beyond the optimal range for 11-Mbps operation, or if substantial interference is present, 802.11b devices will transmit at lower speeds, falling back to 5.5, 2, and 1 Mbps. Likewise, if a device moves back within the range of a higher-speed transmission, the connection will automatically speed up again. Rate shifting is a physical layer mechanism transparent to the user and upper layers of the protocol stack.

Table 17.5 802.11b Data Rate Specifications

Data Rate	Code Length	Modulation	Symbol Rate	Bits/Symbol
1 Mbps	11 (Barker sequence)	BPSK	1 Msps	1
2 Mbps	11 (Barker sequence)	QPSK	1 Msps	2
5.5 Mbps	8 (CCK)	QPSK	1.375 Msps	4
11 Mbps	8(CCK)	QPSK	1.375 Msps	8

17.22 Other WLAN Standards

HyperLAN Family of Standards. The HyperLAN committee in European Telecommunications Standards Institute (ETSI), referred to as Radio and Equipment Systems (RES) 10, worked with the European Conference of Postal and Telecommunications Administration (CEPT, a committee of PTT and other administration representatives) to identify its target spectrum. The CEPT identified the 5.15 to 5.25-GHz band (this allocation allows three channels), with an optional expansion to 5.30 GHz (extension to five channels). Any country in the CEPT area (which covers all Europe, as well as other countries that implement CEPT recommendations) may decide to implement this recommendation. Most of the CEPT countries permit HyperLAN systems to use this 5-GHz band. In America, the FCC roughly followed the European model. The unlicensed National Information Infrastructure band (U-NII) covers approximately 300 MHz in three different bands between 5.1 GHz and 5.8 GHz. The regulators in Japan are likely to align with the 5-GHz band too. In addition, a second band from 17.1 to 17.3 GHz was identified by CEPT, but so far no systems have been defined to use this band.

HyperLAN/1 is aligned with the IEEE 802 family of standards and is very much like a modern wireless Ethernet. HyperLAN/1, a standard completed and ratified in 1996, defines the operation of lower portion of the OSI reference model, namely the data link layer and physical layer. The data link layer is further divided into two portions: the channel access control (CAC) sublayer and media access and control (MAC) sublayer. The CAC sublayer defines how a given channel access attempt will be made depending on whether the channel is busy or idle and at what priority level attempt will be made. The HyperLAN media access control layer defines the various protocols which provide the HyperLAN/1 features of power conservation, security, and multihop routing (i.e., support for forwarding), as well as the data transfer service to the upper layers of protocols. HyperLAN/1 uses the same modulation technology that is used in GSM, Gaussian minimum shift keying (GMSK). It has an over air data rate of 23.5 Mbps and maximum user data rate (per channel) of over 18 Mbps. The range in typical indoor environments is 35–50 meters.

Open Air, Shared Wireless Access Protocol (SWAP) and Bluetooth. The SWAP open air industry standard, unveiled in 1996, is 2.4-GHz frequency-hopping spread-spectrum architecture that operates at a data rate of 1.6 Mbps per channel, with 15 independent channels (hopping patterns) available. This multichannel architecture enables multiple independent WLANs to operate in the same physical space, significantly increasing aggregate network bandwidth. The specification is designed to be transparent to users on a network, working with standard network operating environments including Microsoft Windows NT, UNIX, Novell NetWare, Banyan Vines, IBM LAN Server, and protocols including IPX, IP, NetBEUI, and DECnet.

The Home RF Working Group published and ratified the SWAP specification in late 1998 for wireless voice and data communication in homes and small offices. SWAP is an industry-driven standard. The Home RF Working Group was founded by a core team of companies including Compaq Computer Corporation, Ericsson Enterprise Networks, Hewlett-Packard, IBM, Intel, Microsoft, Motorola, Philips Consumer Communications, and Symbionics.

In May 1998, the Bluetooth SIG announced their intention to publish an open wireless specification for very low-cost, short-range voice and data communications. Bluetooth is also an industry-driven standard and involves Ericsson, IBM, Intel, Nokia, Motorola, Lucent, Toshiba, and many others. Table 17.6 provides a summary of the key standards of WLAN.

Table 17.6 Key Standards of WLAN

Standard	Scope
Wireless Industry Forum, Open Air	2.4-GHz frequency-hopping spread-spectrum
IEEE 802.11b	Defines WLAN interoperability among multivendor products, infrared, 2.4-GHz frequency-hopping, and 2.4 GHz DSSS
Home Radio Frequency Working Group	SWAP for wireless networking within a home
Bluetooth Consortium	Short-range radio links using 2.4 GHz FHSS

17.23 Summary

In this chapter we discussed three important topics that are bridging mobile computing and mobile communications. WAP specifies a thin-client microbrowser using a new standard called Wireless Markup Language (WML) that is optimized for wireless handheld mobile devices. WML is a stripped down version of HyperText Markup Language (HTML). Bluetooth technology allows for replacing many proprietary cables that connect one device to another with one universal short-range radio link.

Bluetooth radio technology provides a universal bridge to existing data networks, a peripheral interface, and a mechanism to form small private ad hoc groupings of connected devices away from fixed network infrastructures. The Bluetooth technology has a number of advantages, including minimal hardware dimensions, low cost of components, and low power consumption. These advantages make it possible to introduce Bluetooth in many types of devices at a low cost. The 720-kbps data capability provided by Bluetooth can be used for cable replacement and several other applications such as LAN.

IEEE 802.11b WLANs are already commonly used in several large vertical markets. The IEEE 802.11b was the first standard to make WLANs usable in the general workplace by providing robust and reliable 11-Mbps data performance, five times faster than the original standard. The new standard will also give WLAN customers the freedom to choose flexible, interoperable solutions from multiple vendors, since it has been endorsed by most major networking and PC vendors. Broad manufacturer acceptance and certifiable interoperability means users can expect to see affordable, high-speed wireless solutions proliferate throughout the large enterprise, small business, and home markets. This global WLAN standard opens exciting new opportunities to expand the potential of network computing.

17.24 References

1. Wireless Application Protocol Forum, "Wireless Application Protocol (WAP)," *Bluetooth Special Interest Group,* version 1.0, 1998.
2. Wireless Application Protocol Forum, "WAP Conformance," Draft version 27, May 1998.
3. http://www.wapforum.org/.
4. WAP Forum, "Wireless Application Protocol White Paper," available at http://www.wapforum.org/what/WAP_white_pages.pdf.
5. Laurence, L. M. E., "Introduction to WAP Architecture," available at http://www.fit.qut.edu.au//DataComms/itn540/gallery/a100/lee/INTROD~1.HTM.
6. Buckingham, S., "Introduction to WAP," available at http://www.gsmworld.com/technology/yes2wap.html.
7. Kanjilal, J., WAP: Internet over Wireless Networks," available at http://www.infocommworld.com//99sep/cover02.htm.
8. Shah, R., "Wireless Application Protocol Set to Take Over," available at http://www.sunworld.com/sunworldonline/swol-01-2000/swol-01-connectivity.htm.
9. Wireless Markup Language (WML), available at http://www.cellular.co.za/wml.htm.
10. Heijden, M. V. D., and Taylor, M., *Understanding WAP, Wireless Applications, Devices and Services*, Artech House: Boston, 2000.
11. Sand, K., Tik-111.550 Seminar on Multimedia: "Bluetooth," March 4, 1999.
12. Simpson, W., ed., "The Point-to-Point Protocol (PPP)," *STD 50, RFC1661, Day-dreamer,* July 1994.
13. Simpson, W., ed., "PPP in HDLC Framing," *STD51, RFC1662, Day-dreamer,* July 1994.
14. Moran, Paul, ed., "Bluetooth LAN Access Profile using PPP," *Bluetooth Special Interest Group*, version 1.0, 1999.
15. Bluetooth Specification Release 1.0 section F:4, "Interoperability Requirements for Bluetooth as WAP Bearer."
16. http://www.bluetooth.com/.
17. Muller, N. J., *Bluetooth Demystified*, McGraw Hill: New York, 2000.
18. Pahlavan, K., "Trends in Local Wireless Networks," *IEEE Communications Magazine*, March 1995, pp. 88–95.
19. Bantz, D. F., "Wireless LAN Design Alternative," *IEEE Network,* March/April 1994, pp. 43–53.
20. Jain, R., "Wireless Local Area Network Recent Developments," *Wireless Seminar Series Electrical and Computer Engineering Department,* Ohio State University, Feb. 19, 1998.
21. Intel Corporation, "IEEE 802.11b High Rate Wireless LAN," available at http://www.intel.com/network/white.paper/wireless lan/.
22. Breeze Wireless Communications, Inc., "Network Security in a Wireless LAN," available at http://www.breezecom.com/pdfs/security.pdf.

23. Intermec Technologies Corporation, "Guide to Wireless Technologies," available at http://www.intermec.com/datactr/wlan_wp.pdf.

24. Proxim, Inc., "What is a Wireless LAN?," available at http://www.proxim.com/wireless/whiteppr/whatwlan.shtml.

25. NDC Communications, Inc., "Wireless LAN Systems—Technology and Specifications," available at http://networking.ittoolbox.com/peer/.

26. IEEE, "Wireless Medium Access Control (MAC) and Physical Layer (PHY) Specifications," P802.11D6.2, July 1998.

CHAPTER 18

Planning of a CDMA System

18.1 Introduction

A code division multiple access (CDMA) system represents a major departure from conventional analog systems such as the advanced mobile phone system (AMPS), and digital time division multiple access (TDMA) systems such as the Global System for Mobile Communications (GSM) and IS-54. The CDMA system has a number of unique features that make its infrastructure design significantly different from previous standards. In this chapter, we present an overview for planning a CDMA system for a personal communication services (PCS) (1,900-MHz) system, though the approach for an 800-MHz cellular system is quite similar. We first discuss steps that should be used in the planning of a CDMA network. In the latter part of the chapter, we focus on pseudorandom (PN) offset planning for IS-95/cdma2000 systems and discuss the major related issues. The chapter also presents necessary steps used in the planning of a wideband CDMA (W-CDMA) system and examines the role of multiuser detection (MUD) technique.

18.2 Planning of a CDMA Network

A systematic approach that should be used in planning of a CDMA network is given below.

18.2.1 Classification of Environment

The system coverage area is divided into land usage categories. Four categories are often used: dense metropolitan, urban, suburban, and rural [1]. These categories are intended to define propagation characteristics and/or the population density of subscribers, which defines the system capacity required to serve the area. One location may belong to two different categorizations. As an example, a beach area may be classified as suburban from a propagation standpoint, with few tall buildings, but dense metropolitan from a population density (system capacity) standpoint, particularly in summer. We describe a typical set of the categories as shown in Table 18.1.

Table 18.1 Typical Set of Categories for the Four Environments

Category	Description
Dense metropolitan	Dense business districts with skyscrapers (10–20 stories and more) along with high-rise apartment buildings. Population density over 20,000 subscribers per square mile.
Urban	Urban residential and office areas. Typical buildings are 5–10 stories, such as hotels, hospitals, etc. Population density 7,500 to 20,000 subscribers per square mile.
Suburban	Mix of residential and business communities. Structures include houses (1–2 stories, 50 m apart), and shops and offices (2–5 stories). Population density 500 to 7,500 subscribers per square mile.
Rural	Open farm land, large open spaces, and sparsely populated residential areas. Typical structures are 1–2 stories, such as houses, barns, etc. Population density under 500 subscribers per square mile.

18.2.2 Signal Strength

Signal strength in a given area is a function of effective radiated power (ERP) and path loss (see Chapter 2). Signal strength is defined as the power that would be received by copolarized dipole antenna at a standard height above ground into a 50-Ohm load. Path loss is determined by propagation characteristics, distance, and obstructions. Propagation characteristics are defined in terms of 1 mile (or 1 km) intercept and path-loss slope (refer to Chapter 2). Actual terrain propagation models, as discussed in Chapter 2, should be used. The sensitivity of a subscriber's unit is the required signal strength when the subscriber is standing in the open area. If the users are likely to be within enclosed areas, a penetration loss must be added to the propagation path loss (see Chapter 2). In rural areas, most users are in open areas, in vehicles, or in small buildings where less than 10–12 dB attenuation allowance may be adequate, whereas 15–20 dB or more may be necessary to penetrate deeply into large office and industrial buildings in a dense metropolitan urban area [2].

Both the coverage effectiveness and deployment cost of the system depend upon the accuracy of path-loss slope, 1 mile (or 1 km) intercept, and penetration allowance selected for each land usage category. If too low an intercept, too steep a slope, or too high a penetration allowance is chosen, more cell sites will be required to cover a given area, and system costs will rise significantly. Conversely, if too high an intercept, too slight a slope, or too little penetration allowance is used, the signal strength will be inadequate and will result in customers' dissatisfaction. The cost of fixing an existing, under designed system might far exceed that of deliberately being somewhat conservative initially.

Because of the high cost of an under- or overdesigned system, necessary steps must be taken to assure the accuracy of initial propagation predictions. After land usage is categorized, propagation testing should be undertaken for each land usage category in the immediate vicinity where the system is to be deployed. The cost of propagation testing is often a cheap insurance compared with the cost of a major design flaw.

In some special cases, even the use of different propagation parameters for each land usage category will be inadequate. Some cell sites might be in unusual locations, and might require individualized propagation testing prior to final selection and construction. To identify these exceptional cases, the designer's experience and power of observation are important.

Just as propagation is not always predicted accurately by land usage categorization, system capacity cannot be planned solely on the basis of any one source of data. Census population density data is often a good predictor, but major business districts have little or no one in residence. Many downtown areas appear rural in terms of census population density. Major highways, interstates, intersections, and shopping malls are not represented in census data either, but they are all major sources of system traffic. Multiple data sources, including common sense, should be used in estimating traffic.

18.2.3 Signal Quality

In designing a CDMA system, it is the quality of signal, not just the signal strength, that defines an acceptable call. An IS-95 CDMA system should be designed for an E_b/I_T in the 5–7 dB range (see Chapter 6), usually with the goal of achieving a 1 percent or less frame error rate (FER) to produce toll-quality speech at the vocoder output. Because of features in signal interleaving, coding, and signal spreading, the required signal quality for CDMA is driven by the speed at which a subscriber moves. Rayleigh fading frequency (see Chapter 2) driven by subscriber speed, can increase the static E_b/I_T requirement used in link budget (see Chapter 2). Thus, the selection of nominal E_b/I_T has a major impact on the deployed system cost, as it affects both system capacity per radio frequency (RF) carrier and cell count.

18.2.4 Nominal Cell Configurations

For each land usage category, a nominal cell configuration should be designed. A 300-ft tower with physically large, high-gain antennas is often cost-effective in rural areas. Towers of intermediate height (150–200 ft) are acceptable in suburban areas, and zoning concerns may make 100-ft towers (or rooftop) with physically small, low-gain antennas the standard for urban and dense metropolitan urban areas.

18.2.5 Subscriber Unit

The ERP of the subscriber unit and receiver performance constrain the link budget in both forward and reverse directions. Handheld CDMA PCS subscriber units generally have 200 mW RF output into a dipole antenna. They have about 8 dB receiver noise figure, and a three-finger RAKE receiver, giving them ability to capture and coherently combine signals from up to three cell sites in soft handoff. Body/orientation loss is typically 2–3 dB. Typical height above ground for propagation modeling is about 1.5 m, roughly that of a subscriber seated in the car.

18.2.6 Cell Overhead Channels

In the IS-95 system, CDMA cell sites transmit *pilot, sync,* and *paging* signals that serve a function similar to control channels in an AMPS or TDMA system. These signals provide the subscriber unit with synchronization and allow it to measure signal quality, which is important in managing soft handoffs. Since these channels are co-channel with the traffic channels, they must be considered in self-interference calculations.

Typical power allocations for synchronous systems such as IS-95 or cdma2000, on a forward link operating at the design loading as a percentage of the total transmitted power are as follows:

- Pilot channel: 15–20%
- Sync channel: 2%
- Paging channel(s): 6–7%
- Traffic channels: 71–77%

When few or no traffic channels are active, total transmitted carrier power should be reduced and the percentage of allocated power for traffic channels should fall, but the ratio between pilot, sync, and paging signals should be maintained. Figure 18.1 shows the forward link transmit path for an IS-95 CDMA system. Note that digital gain affects individual channel and baseband combiner and radio (BCR) affects all channels. If during optimization, the power in each channel is changed by adjusting the BCR attenuation, then all channel powers are changed automatically by the same ratio. However, if the digital gain of any one of the channels is reduced or increased, the digital gain of others must be adjusted in the same ratio to preserve the power percentages in relation to the pilot channel and the maximum power.

The base station output power calibration is performed to make sure that when the particular values of BCR and digital gains are set, the required output power is seen at the antenna (connector J4). BCR potentiometer is adjusted to change the output power during calibration. Once the transmit path has been calibrated, the power can be changed electronically by setting the values of BCR and digital gains (dg).When the BCR attenuation value is reduced, each traffic channel is allocated more power. Likewise, when the BCR attenuation is increased, each traffic channel is allocated less power. If the maximum power is not changed then the effect of BCR attenuation change is as follows:

- BCR reduced → Capacity reduced
- BCR increased → Capacity increased

The maximum output power is used for forward link overload control to protect against overdrive of the transmit amplifier. The maximum power and pilot power must be changed by the same ratio to maintain E_c/I_t.

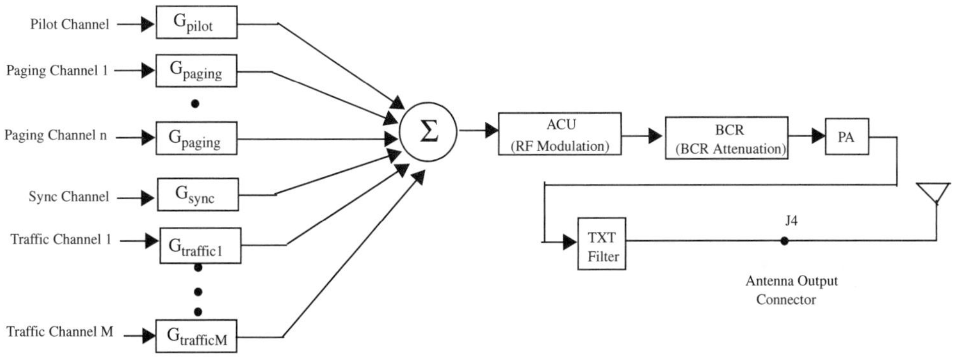

Figure 18.1 Forward Link Transmit Path

EXAMPLE 18.1

Using the following data, determine power allocations for the pilot, paging, sync, and traffic channels of an IS-95 PCS minicell with maximum transmit power of 8 W. Assume pilot gain = 108 digital gain unit (dgu), paging gain = 64 dgu, sync gain = 34 dgu, and traffic gain = 72 dgu. The forward BCR attenuation, *bcr*, is 8 dB and scale factor, *scale*, is given as 0.0006491. What is the capacity of the forward link with 40% overhead due to soft handoff? Assume the voice activity factor = 0.45 and upper power threshold is 85%.

$$\text{Power} = scale \times 10^{-(bcr)/10} \times [\upsilon_{pl} G^2_{pilot} + \upsilon_{pg} G^2_{page} + \upsilon_{sy} G^2_{sync} + \upsilon_{tr} G^2_{traffic,\,m}]$$

where

G_{pilot} and υ_{pl} = gain and channel activity of pilot channel
G_{page} and υ_{pg} = gain and channel activity of paging channel
G_{sync} and υ_{sy} = gain and channel activity of sync channel
$G_{traffic}$ and υ_{tr} = gain and channel activity of traffic channel

Channel activity factor (1 for pilot, sync, and paging channels and 0.45 for the traffic channels):

$$\therefore 8 = 0.0006491 \times 10^{-0.8} \times [108^2 + 64^2 + 34^2 + X]$$

$$\therefore X = 60,843 \text{ dgu}^2$$

The power threshold on the forward link is provided to limit the number of users to protect the amplifier from overdrive. An upper power threshold value of 85% of the available power for traffic channels is given.

Power available for traffic channels = $0.85 \times 60{,}843 = 51,716.6$ dgu^2

Since each call uses 72 dgu with voice activity factor of 0.45 then,

$$\text{Number of legs} = \frac{51,716.6}{(72)^2 \times 0.45} = 22.17 \approx 22$$

Capacity of forward link with 40% overhead due to soft handoff = $22/1.40 \approx 16$

$$P_{pilot} = 0.0006491 \times 10^{-0.8} \times 108^2 = 1.2 \text{ W (15\% of total power)}$$

$$P_{page} = 0.006491 \times 10^{-0.8} \times 64^2 = 0.42 \text{ W (5.3\% of total power)}$$

$$P_{sync} = 0.0006491 \times 10^{-0.8} \times 34^2 = 0.12 \text{ W (1.5\% of total power)}$$

$$P_{traffic} = 100 - (15 + 5.3 + 1.5) = 78.2\% \text{ of total power}$$

18.2.7 Cell Loading and Cell Placement

The cell loading should be limited to 50 to 60 percent of the pole capacity. In designing for CDMA systems, it is important to place cells in such a manner as to allow soft handoff—up to three cells providing RF coverage in handoff regions—but to avoid covering an area with more than three cells or sectors.

18.2.8 Noise Floor

The noise floor is easily manageable in relatively flat areas, but hills or ridges that see signal from too many cells might have no useful forward channel coverage. Mobiles operating on elevated spots or ridge lines might illuminate an entire basin or valley on the reverse channel, raising the noise floor over a large area and affecting many cells. Therefore, careful attention to site selection, downtilt of antennas, and use of minimum ERP are important in managing the noise floor.

18.2.9 Coverage Reliability

Very deep fades in excess of 30–40 dB occur due to Rayleigh fading. It is unrealistic and uneconomical to design a CDMA system for these conditions. Most system designs allow for log-normal fading with an 8-dB standard deviation (see Chapter 2). With a 12-dB standard deviation to include building penetration effects, an 8-dB fade margin is adequate for 90 percent area coverage of the cell and 75 percent perimeter coverage of the cell boundary. Cells should generally be designed with 10 to 15 percent overlapping area coverage to account for *"cell breathing."* An 8-dB fade margin often provides better than 90 percent area coverage.

18.2.10 Phase Approach

Given the link budget, nominal cell configuration, and local propagation parameters, it is possible to calculate the nominal coverage radius of a cell, i.e., the nominal maximum radius at which the link budget is satisfied. While this radius will not be accurate for many cells because of terrain configurations, it provides a rough idea of the area a cell can cover.

Next, we identify the areas where the most subscribers are to be served with the fewest cells. If the service goal is the wireless local loop, this may mean bedroom communities and industrial areas. If the goal is to compete with conventional cellular, then major roads, city centers, and industrial areas are likely targets.

After assigning coverage priority, we decide to build in phases (usually by year) and draw each phase of system coverage on a map. The designer uses this map as a blueprint, first designing a system to cover the first phase region, and then adding cells to meet the coverage and capacity goals of each successive phase. This process lends itself better to different cell placement than that of a single, massive phase. The final cell count may be higher. The advantage is that coverage gaps are avoided in each individual phase.

18.2.11 Nominal Design

Once nominal cell configurations are assigned for each land usage type and basic propagation information is obtained and nominal cell coverage radii are calculated for each nominal cell configuration, it then becomes possible to develop a nominal design.

After the coverage area is overlaid with land usage information, cells are placed for deployment in phase 1, beginning with dense metropolitan cells, then urban, suburban, and rural. Cells are placed as simple circles of the calculated nominal radius on an essentially hexagonal grid with 10–15 percent overlap. At this stage, cells are placed without any consideration for minor terrain characteristics. After the first phase is deployed, cells are placed for the remaining deployment phases.

At this point, a rough idea of cell count for each deployment phase can be achieved for budgeting purposes. A software propagation model can be run to get an idea of cell layout. The resulting nominal design is the "straw man" from which the preliminary design is developed.

18.2.12 Preliminary Design

Working with an RF modeling tool, the designer returns to phase 1 cells. The designer carefully adjusts the placement of each cell after evaluating the terrain characteristics that affect the coverage of available existing sites (towers, buildings, municipal water tanks, and so on). The designer makes an effort to ensure that high-traffic areas such as interstate highways, shopping areas, and so on have full coverage in addition to meeting overall coverage and capacity goals. As cells are placed, successive runs of propagation models are made and cell configurations are refined. Tower heights may deviate from nominal values, cells may be added or deleted as necessary, and cell spacings are adjusted so that a portion of the coverage area is carried in some form of soft handoff. In a CDMA system, roughly 30–40 percent of calls are assumed to be carried in soft handoff.

After phase 1 cells are placed, the designer proceeds to the successive phases, adding further deployment phases as required. This results in the preliminary design. At this point, site search area maps are developed and actual sites are sought. The preliminary design gradually leads to actual implementation as sites are found, evaluated, bought, and final unique site configurations are designed.

The optimization of a CDMA system has many more degrees of freedom than an AMPS or TDMA system does. Some CDMA parameter adjustments allow manipulation of cell coverage areas to cover flaws in the initial design without physically moving cells. On the other hand, peak traffic could create coverage gaps in city centers at busy hours if the system is improperly configured. Education, training, and a full understanding of the properties of CDMA systems are vital to the successful design and operation of CDMA networks. CDMA offers much in terms of capacity and performance, and in turn demands much of those who design and operate it.

18.3 CDMA Uplink and Downlink Load Factors

18.3.1 Uplink Load Factor

We define the $(E_b/I_t)_i$ for the ith user as [3]

$$\left(\frac{E_b}{I_t}\right)_i = \frac{R_c}{R_i \cdot v_i} \cdot \frac{S_i}{I_{total} - S_i} \qquad (18.1)$$

where

R_c = chip rate

S_i = received signal power from ith user

v_i = channel activity factor of ith user

R_i = bit rate of the ith user

I_{total} = total received power including thermal noise power at the base station

Solving Equation (18.1) for S_i, we get

$$S_i = \frac{1}{1 + \dfrac{R_c}{(E_b/I_t)_i \cdot R_i \cdot v_i}} \cdot I_{total} \qquad (18.2)$$

Let $S_i = \phi_i \cdot I_{total}$, then

$$\phi_i = \frac{1}{1 + \dfrac{R_c}{(E_b/I_t)_i \cdot R_i \cdot v_i}} \qquad (18.3)$$

where ϕ_i is the load factor of the ith connection.

The total received interference, excluding the thermal noise N_T, can be expressed as the sum of the received powers from all M users in the same cell.

$$I_{total} - N_T = \sum_{i=1}^{M} S_i = \sum_{i=1}^{M} \phi_i \cdot I_{total} \qquad \textbf{(18.4)}$$

We define the *noise-rise* as the ratio of the total received power to the thermal noise power.

$$Noise\text{-}Rise = \frac{I_{total}}{N_T} \qquad \textbf{(18.5)}$$

Using Equation (18.4), we can obtain

$$Noise\text{-}Rise = \frac{I_{total}}{N_T} = \frac{1}{1 - \sum_{i=1}^{M} \phi_i} = \frac{1}{1 - \eta_{UL}} \qquad \textbf{(18.6)}$$

where we define the uplink load factor η_{UL} as

$$\eta_{UL} = \sum_{i=1}^{M} \phi_i$$

When η_{UL} approaches 1, the corresponding noise-rise approaches infinity and the system reaches its pole capacity. Additionally, the interference from other cells must be included in the load factor. We define β as the interference factor due to other cells.

$$\beta = \frac{other\ cell\ interference}{own\ cell\ interference}$$

The uplink load factor can then be written as

$$\eta_{UL} = (1+\beta) \cdot \sum_{i=1}^{M} \phi_i = (1+\beta) \cdot \sum_{i=1}^{M} \frac{1}{1 + \dfrac{R_c}{(E_b/I_t)_i \cdot R_i \cdot v_i}} \qquad \textbf{(18.7)}$$

The load equation predicts the amount of noise-rise over thermal-noise due to interference. The noise-rise is equal to $-10\log(1-\eta_{UL})$. The interference margin in the link budget must be equal to the maximum planned noise-rise.

The required E_b/I_t can be derived from link level simulations and from measurements. It includes the effect of the closed-loop power control and soft handoff. The effect of soft handoff is measured as macrodiversity combining gain relative to the single link result. The other cell-to-own (serving) cell interference ratio β is a function of cell environment or cell isolation (e.g., macro/micro, urban/suburban) and antenna pattern (e.g., omni, three-sector or six-sector).

The load equation is generally used to make semianalytical predictions of the average capacity of a CDMA cell without going into system-level capacity simulations. The load equation can be used for the purpose of predicting cell capacity and planning noise-rise for dimensioning purposes.

For a classic all-voice-service network, where M users in the cell have a low bit rate R, we can note that

$$\frac{R_c}{(E_b/I_t) \cdot R \cdot v} \gg 1$$

and the uplink load equation can be approximated and simplified to

$$\eta_{UL} = \frac{(E_b/I_t)}{(R_c/R)} \cdot M \cdot v \cdot (1 + \beta) \quad \textbf{(18.8)}$$

The parameters used in uplink load equation are further explained in Table 18.2 for the W-CDMA UMTS system [4–7].

Table 18.2 Parameters for Load Factor Calculation of Uplink in a UMTS System

Parameter	Definitions	Recommended Value
M	Number of users per cell	
v_i	Activity factor of the *i*th user at physical layer	**0.67** for speech assuming 50 percent voice activity and DPCCH overhead during DTX **1.0** for data
E_b/I_t	Signal energy per bit divided by (noise + interference) spectral density required to meet a predefined QoS (e.g., BER)	Dependent on service, bit rate, multipath fading channel, receive antenna diversity, mobile speed. The recommended values for UMTS are: • **Speech: 5 dB** • **144-kbps circuit-switched data: 1.5 dB** • **384-kbps packet-switched data: 1.0 dB**
R_c	W-CDMA chip rate	**3.84 Mcps**
R_i	Bit rate of *i*th user	Dependent on service
β	Other cell-to-own cell interference ratio seen by the base station receiver	Macrocell with omnidirectional antennas: **0.55** Microcell with sector antennas: **0.60–0.87**

EXAMPLE 18.2

Assuming a spreading rate of 3.84 Mcps and $\beta = 0.60$, develop a curve showing the relationship between uplink noise-rise versus uplink data throughput.

We first calculate the pole throughput per cell from Equation (18.7) and use $\eta_{UL} = 0.4988$ at a noise-rise of 3.0 dB. [See Equation (18.6).]

$$0.4988 = (1+0.6)\cdot\frac{1}{1+\dfrac{R_c}{Sum}} = (1+0.6)\cdot\frac{1}{1+\dfrac{3840}{Sum}}$$

$$Sum = 1740 \text{ kbps}$$

This is the pole throughput. A relationship between noise-rise versus total data throughput is given in Table 18.3 and shown in Figure 18.2.

Table 18.3 Uplink Noise-Rise versus Uplink Data Throughput

Noise-Rise (dB)	η_{UL}	Uplink Throughput (kbps)
1	0.2057	358
2	0.369	642
3	0.4988	868
4	0.6019	1047
5	0.6838	1190
6	0.7488	1303
7	0.8055	1402
8	0.8415	1464
9	0.8741	1521
10	0.90	1566

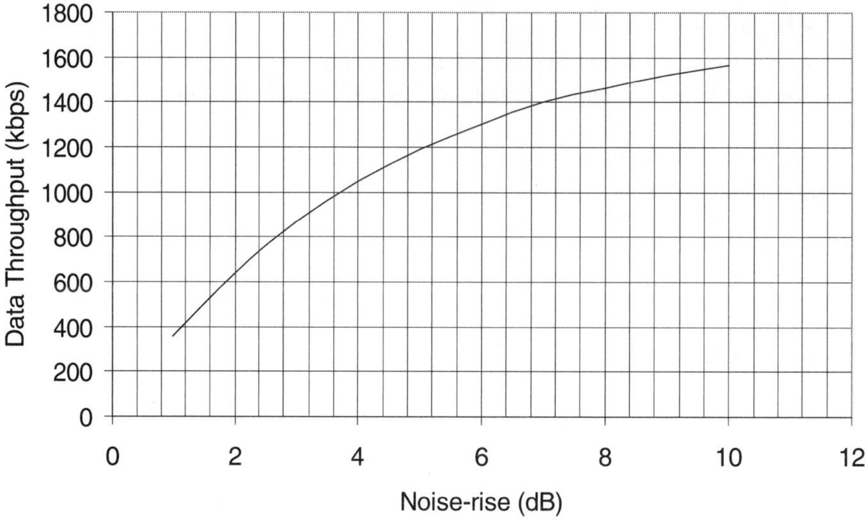

Figure 18.2 Noise-Rise versus Total Data Throughput

18.3.2 Downlink Load Factor

The downlink load factor η_{DL} is defined similarly to the uplink load factor. However, the parameters are slightly different. The downlink load factor η_{DL} shows very similar behavior to the uplink load factor η_{UL} in the sense that when approaching unity, the system reaches its pole capacity and noise rise over thermal goes to infinity.

$$\eta_{DL} = \sum_{i=1}^{M} v_i \cdot \frac{(E_b/I_t)_i}{R_c/R_i} \cdot [(1-\psi_i)+\beta_i] \quad \textbf{(18.9)}$$

where

$-10\log(1-\eta_{DL})$ is the noise-rise over thermal noise due to multiple access interference

ψ_i = orthogonality factor in the downlink. W-CDMA uses orthogonal codes in the downlink to separate mobile users. Without any multipath propagation, the orthogonality remains when the base station signal is received by the mobile. When there is delay spread in the radio channel, the mobile sees part of the base station signal as multiple access interference. The orthogonality factor of 1 corresponds to perfectly orthogonal users. Typical value lies between 0.4 and 0.9 in multipath channels.

β_i = other cell-to-own cell interference

For downlink dimensioning, it is important to estimate the total transmit power of the base station. This should be based on *average* transmission power for the user, not on the *maximum* transmission power at the cell edge.

The minimum required transmission power for each user is obtained by average attenuation between the base station transmitter and mobile receiver, that is $\bar{L}$, and the mobile sensitivity, in the absence of multiaccess interference (intra-or intercell). The effect of noise-rise due to interference is added to this minimum power. The total represents the transmission power required for a user at an *average* location in the cell. The total base station transmission power can be expressed as

$$BS\_T_xP = \frac{N_{rf} \cdot R_c \cdot \bar{L} \cdot \sum_{i=1}^{M} v_i \cdot \frac{(E_b/I_t)_i}{(R_c/R_i)}}{(1 - \bar{\eta}_{DL})} \qquad \textbf{(18.10)}$$

where

N_{rf} = noise spectral density of the mobile receiver front end

$$N_{rf} = kT + N_f = (-174 + N_f) \text{ dBm/Hz (assuming T = 290° K)}$$

in which

k = Boltzmann constant = 1.381×10^{-23} J/K

T = temperature in Kelvin

N_f = mobile station noise figure with typical values of 5–9 dB

The load factor in Equation (18.9) can be approximated by using the average values of ψ and β in the cell, i.e.,

$$\bar{\eta}_{DL} = \sum_{i=1}^{M} v_i \cdot \frac{(E_b/I_t)_i}{(R_c/R_i)} \cdot [(1 - \bar{\psi}) + \bar{\beta}] \qquad \textbf{(18.11)}$$

Note that both uplink and downlink load factors affect the coverage but the effects are not the same. The parameters used in downlink load equation are further explained in Table 18.4 for the W-CDMA UMTS system.

In a CDMA system, the forward link capacity is limited by available forward link power. The forward link overload control assures us that when the forward link thresholds are reached, no more users are added to the system. On the other hand, the reverse link capacity is limited by tolerable amount of interference. The reverse link overload control ensures that when the reverse link interference levels reach an intolerable level, no more users are added to the system.

Table 18.4 Parameters for Load Factor Calculation of Downlink in a UMTS System

Parameters	Definitions	Recommended Values for Dimensioning
M	Number of connections per cell = Number of users per cell × (1+ Soft handoff overhead)	
v_i	Activity factor of *i*th user at physical layer	**0.67** for speech assuming 50 percent voice activity and DPCCH overhead during DTX **1.0** for data
E_b/I_t	Signal energy per bit divided by (noise + interference) spectral density required to meet a predefined QoS (e.g., BER)	Dependent on service, bit rate, multipath fading channel, transmit antenna diversity, mobile speed, etc. • **Speech: 5 dB** • **144-kbps circuit-switched data: 1.5 dB** • **384-kbps packet-switched data: 1.0 dB**
R_c	W-CDMA chip rate	**3.84 Mcps for UMTS**
R_i	Bit rate of the *i*th user	Dependent on service
ψ_i	Orthogonality of channel of the *i*th user	Dependent on the multipath propagation **1: fully orthogonal 1-path channel** **0: no orthogonality**
$\bar{\psi}$	Average orthogonality factor in the cell	**0.6 to 0.9**
β_i	Ratio of the other cell-to-own cell base station power, received by the *i*th user	Each user experiences different β_i, depending on its location in the cell and lognormal fading
$\bar{\beta}$	Average ratio of other cell-to-own cell base station power received by user	Macrocell with omnidirectional antennas: **0.55** Macrocell with sector antennas: **0.60–0.87**

18.4 Multiuser Detection

The interference caused by the presence of other users in the cell is called *multiple access interference* (MAI). Conventional signal detectors detect only a single user's signal. When there are multiple users in the same environment, the conventional detectors treat other users' signals as noise or interference. MAI affects system capacity and system performance. When there are more users, the MAI is high. The system performance is also affected by the near-far problem. Mitigation of the MAI is possible by

- Good cross-correlation code waveform design
- Open-loop power control for mobiles and closed-loop power control for the base station
- Forward error correction codes
- Sectorized or adaptive antennas that focus reception over a narrow desired angle range

A new detection technique aimed at detecting all signals from all users jointly and simultaneously has been developed. These detectors are called *multiuser detectors* (MUD). MUDs reduce and eliminate MAI. In this detection technique, code and timing information of multiple users are jointly used to better detect each individual user. The information of amplitude and phase of the signals may also be taken into consideration. The MUDs are used in conjunction with conventional receivers. MUDs can be classified into two categories based on the type of transformations applied to the signal outputs from the correlators in the detection phase: (1) linear multiuser detectors, and (2) subtractive interference cancellation detectors.

In the *linear multiuser detectors*, a linear mapping (transformation) is applied to the soft outputs of the conventional detector to produce a new set of outputs that provide better performance than the conventional detectors. The main types of detectors in this category are decorrelating detectors, minimum mean squared error (MMSE) detectors, and polynomial expansion (PE) detectors.

In the *subtractive interference cancellation detectors*, estimates of interference are generated and subtracted out. The main types of detectors in this category are successive interference cancellation (SIC) detectors, parallel interference cancellation (PIC) detectors, and zero forcing decision feedback (ZF-DF) detectors.

The advantages of MUD are many and they outweigh their limitations.

- **Advantages**
  - Improved system capacity.
  - Efficient power utilization—The reduction in interference on the uplink can be interpreted either as increased coverage region with the same power or reduced transmitting power for the mobile.
  - Reduced need for precision in power control—Due to reduced MAI and near-far effect, the need for all users to arrive at the receiver at the same power is reduced.
  - Improved utilization of the spectrum in the uplink—Bandwidth saved in the uplink can be used to improve downlink capacity or can be used to support higher data rates.

- **Limitations**
  - Boundary on capacity improvement of the cellular environment—The existence of MAI in the other cells places a boundary on the capacity improvement. The intercell interference still needs to be addressed. Inclusion of signals from other cells in the MUD algorithm can reduce this boundary, although it cannot eliminate it.

- Complexity of the design on the downlink—Implementing the MUD on the downlink is complex. Primary issues of cost and size at the mobile prevent designers from implementing MUD at the mobile.

The use of multiuser detection techniques has been suggested in the W-CDMA UMTS system. Multiuser detection (MUD) and interference cancellation (IC) techniques improve system performance by canceling the intracell interference. The actual capacity improvement depends on the efficiency of the MUD algorithm, the radio environment, and the system load. In addition to capacity improvement, MUD and IC help to alleviate the near-far problem. Multiuser detection scheme processes the signal from the correlators jointly to eliminate the unwanted multiple-access interference from the desired signal.

Since with MUD the interference from a single cell (i.e., intracell interference) is canceled, the intercell (outer cells) interference limits the achievable capacity. If the intercell interference is β times the intracell interference, then the upper limit for capacity increase will be $(1 + \beta) / \beta$. With a propagation path loss exponent of 4, intercell interference is 67 percent of intracell interference; therefore, ideally, MUD should improve the capacity of the system by a factor of $1.67 / 0.67 = 2.49$ compared with a system without MUD. However, in practice, the MUD efficiency is not 100 percent. It depends on the detection scheme, channel estimation, delay estimation, and power control error.

Since MUD efficiency varies in different radio environments, the capacity improvement attainable by MUD is not fixed. The impact of MUD on coverage introduces a new variable to the network planning process, since MUD efficiency needs to be taken into account in the coverage design. The efficiency of MUD is estimated from the load that can be supported with a specific E_b/I_t value with a multiuser receiver. In the analysis, we denote the number of users with a RAKE receiver by M_{RAKE} and those with a MUD receiver by M_{MUD}. The efficiency of the MUD receiver, η, at a given E_b/I_t is [3]

$$M_{RAKE} = (1 - \eta) M_{MUD} \tag{18.12}$$

The capacity of the network with a MUD receiver in a base transceiver station (BTS) is defined as

$$\frac{E_b}{I_t} = \frac{S \cdot G_p \cdot \alpha_c}{(1-\eta) I_{intra} + I_{inter} + N_0} \tag{18.13}$$

but

$$\mu = \frac{I_{intra}}{I_{intra} + I_{inter}}$$

$$\therefore I_{inter} = \frac{(1-\mu)}{\mu} I_{intra} \tag{18.14}$$

Substituting Equation (18.14) into Equation (18.13) and neglecting the effect of thermal noise, we get

$$\frac{E_b}{I_t} = \frac{SG_p\alpha_c}{(1-\eta)(M-1)S + \left(\frac{1-\mu}{\mu}\right)MS} \quad \textbf{(18.15)}$$

where
S = received signal power
G_p = processing gain = R_c/R
M = number of users associated with the BTS
R_c = chip rate
R = information rate
η = efficiency of the MUD
α_c = power control efficiency

Solving Equation (18.15) for M, we get

$$M = \frac{\mu[\alpha_c G_p(E_b/I_t)^{-1} + (1-\eta)]}{(1-\mu\eta)} \quad \textbf{(18.16)}$$

The η with value 0 represents the capacity of the conventional RAKE receiver-based system. We use the following data and develop capacity as function of MUD efficiency (see Table 18.5).

- Chip rate = 3.84 Mcps, bandwidth 5 MHz
- Information rate = 12.2 kbps
- $\mu = 0.67$
- $\eta = 0, 0.2, 0.4, 0.5$
- E_b/I_t =5 dB
- $\alpha_c = 0.85$

Table 18.5 Capacity Comparison with Multiuser Detection (MUD)

MUD Efficiency (η)	RAKE Receiver	MUD		
		0.2	0.4	0.5
No. of users per cell	57.8	66.12	78.04	85.81
No. of users/MHz/cell	11.48	13.22	15.61	17.16
Capacity improvement due to MUD with respect to RAKE	0	15.16	35.98	49.48

In an interference-limited CDMA system, the achievable range depends on cell loading. If the number of users increases, the cell range decreases. In the following section, we calculate the decrease in cell range when the network load increases and analyze the effect of a base station MUD receiver in increasing the range in a loaded network. Base station MUD receiver is shown to lower the mobile station transmission power in a loaded network, thus making uplink range and average mobile transmission power less sensitive to network load.

In an unloaded network, the uplink limits the achievable range and coverage, as the maximum transmission power of the mobile station is lower compared with the maximum transmission power of the base station in the downlink. In a loaded network, the downlink may limit the range if there is more load and thus more interference in the downlink than the uplink.

The received signal-to-interference ratio at the base station is given as

$$\frac{E_b}{I_t} = \frac{E_{b,\,loaded}}{I_{intra} + I_{inter} + N_0} \tag{18.17}$$

where E_b is the received energy per bit, I_{intra} is the intracell interference from own cell mobiles, I_{inter} is the interference from the mobiles not connected to this particular base station, and N_0 is the thermal noise.

In case of an unloaded network $I_{intra} = 0$, $I_{inter} = 0$, and the required E_b/N_0 for range calculations is equal to E_b/I_t. In the loaded network, the fraction of own-cell interference from total interference is defined as

$$\bar{\mu} = \frac{I_{intra} + S}{I_{intra} + S + I_{inter}}$$

where $S = E_b/G_p$ is the received signal power from one user and G_p is the processing gain. $\bar{\mu}$ depends upon propagation environment. The higher the path-loss attenuation factor, the higher the $\bar{\mu}$. I_{inter} can be expressed in term of I_{intra} as

$$I_{inter} = I_{intra}\left(\frac{1}{\bar{\mu}} - 1\right) + \frac{E_{b,\,loaded}}{G_p}\left(\frac{1}{\bar{\mu}} - 1\right) \tag{18.18}$$

but

$$I_{intra} = (M-1)\frac{E_{b,\,loaded}}{G_p}$$

$$\therefore (I_{inter} + I_{intra}) = \left(\frac{M}{\bar{\mu}} - 1\right)\frac{E_{b,\,loaded}}{G_p} \tag{18.19}$$

$$\frac{E_{b,\,loaded}}{I_t} = \frac{E_{b,\,loaded}}{\left(\frac{M}{\bar{\mu}} - 1\right)\frac{E_{b,\,loaded}}{G_p} + N_0} = \left(\frac{E_b}{N_0}\right)_{unloaded} \tag{18.20}$$

Solving for the required E_b/N_0 in the loaded case gives

$$\left(\frac{E_b}{N_0}\right)_{loaded} = \frac{1}{\left(\frac{E_b}{N_0}\right)^{-1}_{unloaded} - \left(\frac{M}{\bar{\mu}} - 1\right)\frac{1}{G_p}} \tag{18.21}$$

The effect of the MUD receiver can be taken into account by using the efficiency of MUD η as a measure of performance of MUD receiver. With MUD receiver, the intracell interference $I_{intra,MUD}$ can be written as

$$I_{intra,MUD} = (1-\eta)I_{intra} = (1-\eta)(M-1)\frac{E_b}{G_p} \tag{18.22}$$

and

$$I_{inter} = \left(\frac{1}{\bar{\mu}} - 1\right)(1-\eta)(M-1)\frac{E_b}{G_p} + \frac{E_b}{G_p}\left(\frac{1}{\bar{\mu}} - 1\right) \tag{18.23}$$

The total interference will be

$$I_{inter} + I_{intra,MUD} = \frac{E_{b,loaded}}{G_p}\left[\frac{M(1-\eta)+\eta}{\bar{\mu}} - 1\right] \tag{18.24}$$

The required E_b/N_0 in the loaded network with MUD receiver becomes

$$\left(\frac{E_b}{N_0}\right)_{loaded,MUD} = \frac{1}{\left(\frac{E_b}{N_0}\right)^{-1}_{unloaded} - \left[\frac{M(1-\eta)+\eta}{\bar{\mu}} - 1\right]\frac{1}{G_p}} \tag{18.25}$$

The transmit power from a mobile is given as

$$S_{TX,MS} = \frac{E_b}{N_0} + R_b + N_f + kT - G_{HO} - G_{MS} - G_{BS} \tag{18.26}$$

In Equation (18.26) all the terms are the same except for E_b/N_0, regardless of the base station receiver algorithm. $S_{TX,MS}$ is determined only from the E_b/N_0 requirement. The decrease in the required transmission power with MUD receiver is thus given as

$$\frac{S_{TX,MS}}{S_{TX,MS,MUD}} = \frac{\left(\frac{E_b}{N_0}\right)_{loaded}}{\left(\frac{E_b}{N_0}\right)_{loaded,MUD}} = \frac{\left(\frac{E_b}{N_0}\right)^{-1}_{unloaded} - \left[\frac{M(1-\eta)+\eta}{\bar{\mu}} - 1\right]\frac{1}{G_p}}{\left(\frac{E_b}{N_0}\right)^{-1}_{unloaded} - \left[\frac{M}{\bar{\mu}} - 1\right]\frac{1}{G_p}} \tag{18.27}$$

EXAMPLE 18.3

The maximum number of users in a cell of a WCDMA system at 12.2-kbps voice service is 140. The required E_b/N_0 in the unloaded condition is 5 dB. Assume $\bar{\mu} = 0.67$ and develop a curve to show range as a function of uplink load. Use the data given in Table 18.6. Also, develop a curve to show range as a function of uplink load for the MUD receiver at 40 percent efficiency.

We first calculate $(E_b/N_0)_{loaded}$ and $(E_b/N_0)_{loaded,MUD}$ as a function of uplink load using Equation (18.21) and Equation (18.25) (see Table 18.6).

Table 18.6 E_b/N_0 with Respect to Uplink Cell Loading

Uplink Load (M)	$(E_b/N_0)_{loaded}$ (dB)	$(E_b/N_0)_{loaded,MUD}$ (dB)
0	5.00	5.00
14 (10%)	5.97	5.56
28 (20%)	7.29	6.24
42 (30%)	9.20	7.03
56 (40%)	12.69	8.01

Next we calculate the allowable path loss with and without MUD and determine the cell range (see Table 18.7). We use the information in Table 18.8 to calculate allowable loss and cell range.

Table 18.7 Cell Loading versus Cell Range with and without MUD

Cell Load	Without MUD		With MUD	
	Allowable Path Loss (dB)	Cell Range (km)	Allowable Path Loss (dB)	Cell Range (km)
0	136.70	2.122	136.70	2.122
10%	135.73	1.997	136.14	2.049
20%	134.41	1.838	135.46	1.963
30%	132.50	1.631	134.67	1.868
40%	128.75	1.310	133.69	1.757

The cell range with MUD and without MUD is given in Figure 18.3. Note that cell range with MUD is higher.

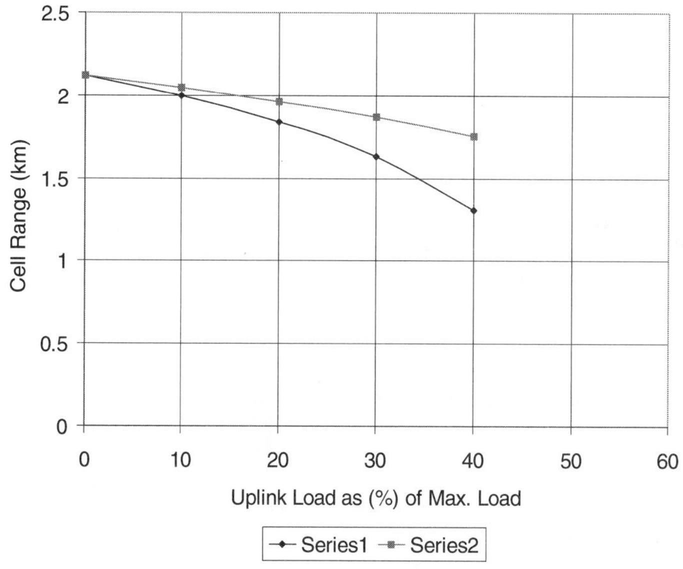

Figure 18.3 Cell Range with and without MUD

18.5 Radio Link Budgets and Coverage Efficiency of CDMA

There are several CDMA-specific parameters in the link budget that are not included in a TDMA-based radio-access system. The important parameters are as follows:

- **Interference Margin**
  The interference margin is required in the W-CDMA link budget because the loading of the cell (the load factor) affects the coverage. The more loading allowed in the system, the larger the interference margin needed in the uplink, and the smaller is the coverage area. For *coverage-limited* systems, a smaller interference margin is suggested, whereas for *capacity-limited* systems, a larger interference margin should be used. In the coverage-limited systems the cell radius is limited by maximum allowable path loss in the link budget, and the maximum air interface capacity of the base station is not used. Typical values for interference margin in the coverage-limited systems are 1 to 3 dB, corresponding to 20 to 50 percent cell loading.
- **Fast Fading Margin**
  Some margin is required in the mobile unit transmission power to maintain adequate closed-loop fast power control. This applies particularly to slow-moving mobiles where fast power control is able to effectively compensate the fast fading. Typical values for the fast fading margin are 2 to 5 dB for slow-moving mobiles.
- **Soft Handoff Gain**
  Soft handoff provides an additional macrodiversity gain against fast fading by reducing the required E_b/I_t relative to a single radio link, due to the effect of macrodiversity combining. The total soft handoff gain may be between 2.0 and 3.0 dB.

We provide three examples of link budgets for typical UMTS services. These are 12.2-kbps voice services using an adaptive multirate (AMR) speech codec, 144-kbps circuit-switched data, and 384-kbps packet-switched data in an urban macrocellular environment at a planned uplink noise rise of 3 dB (refer to Tables 18.8, 18.9, and 18.10).

Table 18.8 Reference Link Budget with AMR Speech Codec 12.2-kbps Voice Service 120 km/h in Car Users with Soft Handoff

Transmitter (Mobile Station)			
a	Maximum mobile Tx power (dBm)	21.0	
b	Mobile antenna gain (dBi)	0.0	
c	Body/orientation loss (dB)	3.0	
d	Equivalent isotopic radiated power [ERIP] (dBm)	18.0	d = a + b – c
Receiver (Base Station [BS])			
e	Thermal noise density (dBm/Hz)	–174.0	
f	BS receiver noise figure (dB)	5.0	
g	Receiver noise density (dBm/Hz)	–169.0	g = e + f
h	Receiver noise power (dBm)	–103.2	h = g + 10log(3840000)
i	Interference margin (50% loading) (dB)	3.0	
j	Receiver interference power (dBm)	–103.2	$j = 10\log[10^{(h+i)/10} - 10^{(h/10)}]$
k	Total effective noise + interference (dBm)	–100.2	$k = 10\log[10^{(h/10)} + 10^{(j/10)}]$
l	Processing gain (dB)	25.0	l = 10log[3840/12.2]
m	Required E_b/I_t	5.0	
n	Receiver sensitivity (dBm)	–120.2	n = m – l + k
o	BS antenna gain (dBi)	18.0	
p	BS cable/connection losses (dB)	2.0	
q	Fast fading margin (dB)	0.0	
r	Max. path loss (dB)	154.2	r = d – n + o – p – q
	Coverage Probability	90	
	Standard deviation for log-normal fading (dB)	8	
s	Log-normal fading margin (dB)	10.5	
t	Soft handoff gain (dB)	3.0	
u	Penetration loss in car (dB)	10	
v	Allowable path loss for cell range (dB)	136.7	v = r – s + t – u
w	One-mile path-loss intercept (dB)	132.2	
x	Path-loss exponent	3.67	
y	Cell range (mile)	1.326	$y = 10^{(v - w)/10 \times x}$
z	Cell range (km)	2.122	$z = 1.6 \times y$

Table 18.9 Reference Link Budget of 144-kbps Circuit-Switched Data Service (3 km/h, Indoor User)

Transmitter (Mobile Station)			
a	Maximum mobile Tx power (dBm)	24.0	
b	Mobile antenna gain (dBi)	2.0	
c	Body/orientation loss (dB)	0.0	
d	Equivalent isotopic radiated power [ERIP] (dBm)	26.0	d = a + b – c
Receiver (Base Station [BS])			
e	Thermal noise density (dBm/Hz)	–174.0	
f	BS receiver noise figure (dB)	5.0	
g	Receiver noise density (dBm/Hz)	–169.0	g = e + f
h	Receiver noise power (dBm)	–103.2	h = g + 10log(3840000)
i	Interference margin (50% loading) (dB)	3.0	
j	Receiver interference power (dBm)	–103.2	$j = 10\log[10^{(h+i)/10} - 10^{(h/10)}]$
k	Total effective noise + interference (dBm)	–100.2	$k = 10\log[10^{(h/10)} + 10^{(j/10)}]$
l	Processing gain (dB)	14.3	l = 10log[3840/144]
m	Required E_b/I_t	1.5	
n	Receiver sensitivity (dBm)	–113.0	n = m – l +k
o	BS antenna gain (dBi)	18.0	
p	BS cable/connection losses (dB)	2.0	
q	Fast fading margin (dB)	4.0	
r	Max. path loss (dB)	151.0	r = d – n + o – p– q
	Coverage probability	90	
	Standard deviation for log-normal fading (dB)	8	
s	Log-normal fading margin (dB)	10.5	
t	Soft handoff gain (dB)	2.0	
u	Indoor penetration loss (dB)	16	
v	Allowable path loss for cell range (dB)	126.5	v = r – s + t – u
w	One mile path loss intercept (dB)	129.4	
x	Path-loss exponent	3.52	
y	Cell range (mile)	1.2089	$y = 10^{(v - w)/10 \times x}$
z	Cell range (km)	1.9342	z = 1.6 × y

Table 18.10 Reference Link Budget of 384-kbps Packet-Switched Data Service (3 km/h, Outdoor User)

Transmitter (Mobile Station)			
a	Maximum mobile Tx power (dBm)	24.0	
b	Mobile antenna gain (dBi)	2.0	
c	Body/orientation loss (dB)	0.0	
d	Equivalent isotopic radiated power [ERIP] (dBm)	26.0	d = a + b – c
Receiver (Base Station [BS])			
e	Thermal noise density (dBm/Hz)	–174.0	
f	BS receiver noise figure (dB)	5.0	
g	Receiver noise density (dBm/Hz)	–169.0	g = e + f
h	Receiver noise power (dBm)	–103.2	h = g + 10log(3840000)
i	Interference margin (50% loading) (dB)	3.0	
j	Receiver interference power (dBm)	–103.2	$j = 10\log[10^{(h+i)/10} - 10^{(h/10)}]$
k	Total effective noise + interference (dBm)	–100.2	$k = 10\log[10^{(h/10)} + 10^{(j/10)}]$
l	Processing gain (dB)	10.0	l = 10log[3840/384]
m	Required E_b/I_t	1.0	
n	Receiver sensitivity (dBm)	–109.2	n = m – l + k
o	BS antenna gain (dBi)	18.0	
p	BS cable/connection losses (dB)	2.0	
q	Fast fading margin (dB)	4.0	
r	Max. path loss (dB)	147.2	r = d – n + o – p – q
	Coverage probability	90	
	Standard deviation for log-normal fading (dB)	8	
s	Log-normal fading margin (dB)	10.5	
t	Soft handoff gain (dB)	0.0	
u	Penetration loss in car (dB)	0.0	
v	Allowable path loss for cell range (dB)	136.7	v = r – s + t – u
w	One mile path loss intercept (dB)	134.2	
x	Path-loss exponent	3.87	
y	Cell range (mile)	1.1604	$y = 10^{(v - w)/10 \times x}$
z	Cell range (km)	1.8566	$z = 1.6 \times y$

18.6 Cell Coverage for Cellular and PCS CDMA

The uplink E_b/I_t depends on many factors. The received signal strength depends on the mobile power, path loss, and the gains of the mobile station and base station antenna systems. The interference level depends on the base station receiver noise figure, the interference from other users, and the user speech activity, i.e., the channel activity factor (see Figure 18.4).

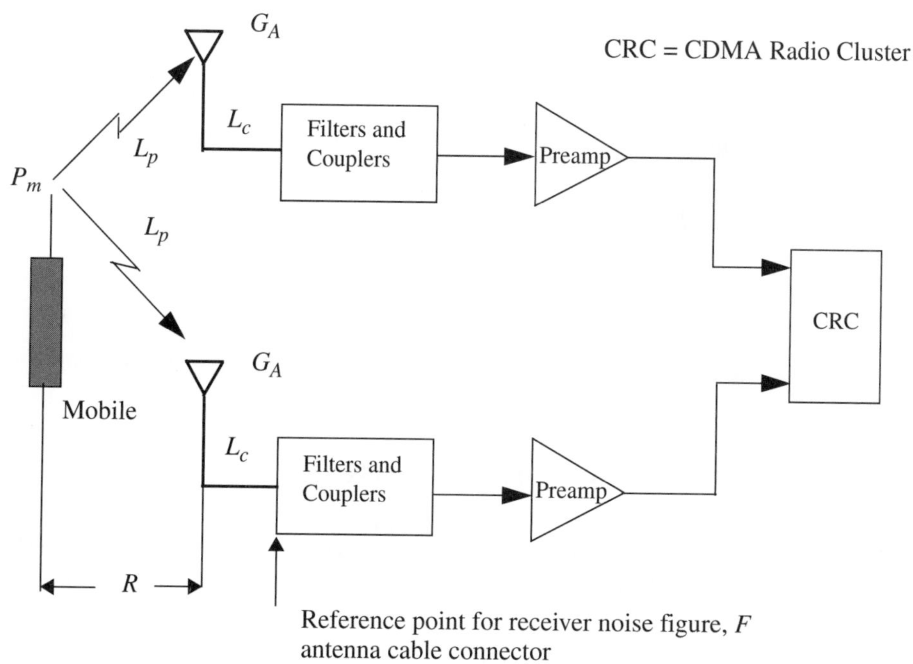

Figure 18.4 Model of CDMA Uplink

EXAMPLE 18.4

Refer to Figure 18.4 and use the following data to determine the maximum range for a handheld mobile with IS-95 Rate Set 1 (9.6 kbps):

- Effective radiated power of mobile (P_m): 200 mW (23 dBm)
- Base station antenna gain (G_A): 15 dBi
- Base station receiver antenna cable loss (L_c): 2.5 dB
- Body loss (L_{body}): 2 dB

- Receiver noise figure (F_{dB}): 5.5 dB
- Information rate (Rate Set 1) (R_b): 9,600 bps
- Required E_b/I_t $(E_b/I_t)_{reqd}$: 7.0 dB
- Fading margin for 90% confidence (M_{fade}): 10.2 dB

Base station noise floor:

$$N_0 = 10\log(kT) + F_{dB}$$
$$= 10\log(1.38 \times 10^{-23} \times 290) + 30 + 5.5 = -168.5 \text{ dBm/Hz}$$

$$(E_b)_{min} = N_0 + \left(\frac{E_b}{I_t}\right)_{reqd} = -168.5 + 7.0 = -161.5 \text{ dBm/Hz}$$

$$S_{min} = (E_b)_{min} + 10\log R_b = -161.5 + 10\log 9600 = -121.7 \text{ dBm}$$

$$S_{min} = P_m - L_p + G_A - L_c$$

$$\therefore L_p = P_m - S_{min} + G_A - L_c = 23 - (-121.7) + 15 - 2.5 = 157.2 \text{ dB}$$

To provide a margin for shadow fading and body loss, the maximum allowable path loss (L_{pa}) is

$$L_{pa} = L_p - M_{fade} - L_{body} = 157.2 - 10.2 - 2.0 = 145.0 \text{ dB}$$

From the allowable path loss, we can determine the maximum radius for a CDMA PCS handheld mobile in a suburban outdoor environment using the following PCS propagation path-loss model (see Figure 18.5).

$$L_{pa} = 118 + 38.4\log R$$

$$\therefore 145 = 118 + 38.4\log R$$

or $R = 5.0$ mile

Assuming a penetration loss for a typical building equal to 20 dB, the allowable path loss will be 145.0 – 20.0 = 125 dB.

$$\therefore 125 = 118 + 38.4\log R$$

or $R = 1.5$ mile

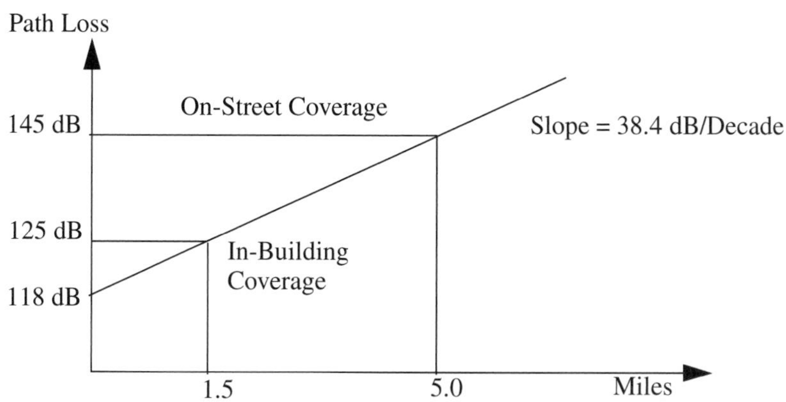

Figure 18.5 PCS Propagation Path Loss Model

The corresponding cellular propagation path-loss model in an outdoor suburban environment is

$$L_{pa} = 109 + 38.4\log R$$

$$\therefore 145 = 109 + 38.4\log R$$

or $R = 8.66$ miles

Similarly, in a building environment,

$$125 = 109 + 38.4\log R$$

or $R = 2.6$ miles

The results are summarized in Table 18.11.

Table 18.11 Maximum Range for Handheld Mobile (Rate Set 1)*

Max. Allowable Path Loss	Environment	PCS (1900 MHz)	Cellular (850 MHz)
145 dB	Outdoor suburban	5.0 miles	8.66 miles
125 dB	Indoor suburban	1.5 miles	2.60 miles

* Figures assume no co-channel interference, and maximum ranges are subject to shrinkage caused by traffic loading.

EXAMPLE 18.5

Using the following data, calculate the maximum allowable path loss in a typical suburban environment of an IS-95 CDMA system. What is the maximum cell radius?

BS Parameters

- Thermal noise density (N_0): –174 dBm/Hz
- Information rate (9,600 bps) (R_b): 39.8 dB
- Noise figure (F_{dB}): 8 dB
- Required E_b/I_t $(E_b/I_t)_{reqd}$: 7 dB
- BS received antenna gain (G_A): 15 dB
- BS cable and connectors loss (L_c): 2.5 dB
- Log-normal fade margin (M_{fade}): 8 dB
- Soft-handoff gain (S_{ho}): 4 dB
- System interface margin (I_{sm}): 3 dB (nominal for 50% cell loading)
- Building penetration loss (L_{pent}): 10 dB
- Cell transmit ERP (3.55 W) ($P_{traffic}$): 35.5 dBm
- Number of traffic channels (50% loading per sector, three-sector case: 13 traffic channel elements [TCEs]) ($N_{traffic}$): 11.1 dB
- Percent of total power for traffic channels – 75% ($p_{traffic}$): 1.3 dB

Subscriber Unit Parameters

- Mobile transmit power (P_m): 23 dBm
- Mobile antenna gain (G_m): 0 dB
- Body loss (L_{body}): 3 dB

Propagation Parameters

- One mile intercept: –75 dBm
- Slope: 38.4 dB/decade
- Tower height: 200 feet

Note: Reference tower height = 150 ft and reference cell transmit ERP = 100 W

SOLUTION:

Subscriber ERP = $P_m - L_{body} = 23 - 3 = 20$ dBm

BS sensitivity = Thermal noise density + Information rate + Noise figure
+ Required E_b/I_t + Cable and connector loss + Log-normal fade margin
+ System interference margin + Building penetration – Antenna gain
– Soft handoff gain

$$= N_0 + R_b + F_{dB} + (E_b/I_t)_{reqd} + L_c + M_{fade} + I_{sm} + L_{pent} - G_A - S_{ho}$$

$$= -174 + 39.8 + 8 + 7 + 2.5 + 8 + 3 + 10 - 15 - 4 = -114.7 \text{ dB}$$

Maximum allowable path loss = 20 – (–114.7) = 134.7 dB

ERP per traffic channel element (TCE) = Cell ERP per traffic channel
– No. of traffic channels
– Percentage of total power
– Cable/connector loss
+ Antenna gain
$= P_{traffic} - N_{traffic} - p_{traffic} - \mathrm{L_c} + \mathrm{G_A}$

ERP per TCE = 35.5 – 11.1 – 1.3 – 2.5 + 15 = 35.4 dBm

Signal strength at cell boundary = ERP per TCE – Maximum allowable path loss
= 35.4 – 134.7 = –99.3 dBm

For a tower height of 200 feet, the distance (R) from the cell will be

$$-99.3 = -75 - 38.4\log(R) + 10\log(3.55/100) + 15\log(200/150)$$

$$\log R = 0.3040$$

$$\therefore R = 2.014 \text{ miles}$$

EXAMPLE 18.6

You are responsible for designing a CDMA cellular network in an area with a population of 2 million residents. Each user averages 1 call per hour with an average call duration of 1.73 minutes during a busy hour. Based on market survey, employment data, and other statistics, marketing has indicated that they can sell service to enough people so that the busy-hour penetration rate is 2 percent of the population. For a grade of service (GoS) of 1 percent, determine the required number of cells in the system and number of channels per cell to serve the estimated population. Assume that the busy-hour traffic is uniformly distributed in the service area.

Total population = 2 million
Busy-hour penetration rate = 2%
Number of busy-hour users = $0.02 \times 2 \times 10^6 = 40{,}000$

$$\text{Traffic per user} = \frac{1 \times 1.73 \times 60}{3600} = 0.0288 \text{ Erlangs}$$

$$\text{Total traffic} = 0.0288 \times 40{,}000 = 1153.3 \text{ Erlangs}$$

Assuming the RF coverage layout indicates that the whole service area can be covered by 200 omnicells, then the traffic per cell will be

$$\text{Traffic per cell} = \frac{1153.3}{200} = 5.767 \text{ Erlangs}$$

For GoS = 0.01 and 5.767 Erlangs, the number of channel elements per cell = 12

In this example, we assume that the traffic is uniformly distributed. In practical situations the load is unevenly distributed in the service area. This means that some areas would need more cells or more channels per cell to service the demand. A traffic demand map, which provides the demand in each geographical region, is used to accurately plan the network to service the uneven distribution of traffic.

EXAMPLE 18.7

An IS-95 CDMA system is engineered for the busy-hour traffic. RF design must provide both coverage and capacity in the service area. Cell breathing in CDMA can cause capacity and coverage problems. Initial engineering procedure deals with the following:

- Determination of soft capacity limit per sector
- Determination of demand (busy hour call attempts [BHCA], Erlangs)
- Design of system with sufficient number of cells, sectors, and carriers to meet the demand
- Engineering of cell site
- Engineering of links between cell site and MSC
- Engineering of MSC

We will demonstrate the procedure to engineer a mini-cell cellular system using the following data:

Subscriber Statistics
- BHCA:100,000
- Average call holding time per call: 140 seconds

Network Design Criteria
- Blocking probability target = 2%
- 13-kbps vocoder with 14.4-kbps data rate
- Network is designed at 54% loading
- Required $E_b/I_t = 7$ dB

RF Design Information
- RF coverage design has been developed using 60 three-sector cells such that the traffic load carried by each cell is the same
- Traffic is assumed to be uniformly distributed in each cell
- An average of 40% users are assumed to be in soft handoff

Assume $v_f = 0.4$, $\beta = 0.6$, and $\lambda = 0.85$.

Equipment Availability

- DS_0 links between cell site and MSC are 64 kbps

$$M_{max} = \frac{(1.2288 \times 10^6)/(14.4 \times 10^3)}{0.4 \times 5 \times (1 + 0.6)} \times 0.85 + 1 = 23.67 \approx 24$$

$$M = \rho \cdot M_{max} = 0.54 \times 24 = 12.96 \approx 13$$

Erlangs supported per sector at 2% blocking = 7.4 Erlangs

$$\text{Total traffic load} = \frac{100,000 \times 140}{3600} = 3888.9 \text{ Erlangs}$$

$$\text{Load per sector} = \frac{3888.9}{3 \times 60} = 21.6 \text{ Erlangs}$$

Air interface limit per carrier = 7.4 Erlangs

$$\text{Number of carriers required} = \frac{21.6}{7.4} = 2.92 \approx 3$$

Traffic channel elements per sector from Erlang B table = 30

Channel elements per sector for users in soft handoff = 12 (40%)

1 channel element per sector per carrier for pilot channel (3 per sector)

1 channel element per sector per carrier for sync, paging, and access (3 per sector)

Total channel elements = 30 + 12 + 3 + 3 = 48

Channel Elements Pooling in a Sectored Cell

The channel element pooling in a sectored cell allows sharing of channel elements between sectors. With 2% blocking, the number of channel elements required in a cell to support 64.8 Erlangs = 76.

Add 40% channel elements for handoff = $1.4 \times 76 \approx 106$

Assuming every carrier uses its own paging and pilot channel, then
Channel elements for overhead channel per carrier = $2 \times 3 = 6$
Total number of channel elements for overhead channels = $3 \times 6 = 18$

$$\text{Total channel elements required per carrier} = 106 + 18 = \frac{124}{3} \approx 41$$

A saving of seven channel elements per carrier.

18.7 PN Offset Planning for CDMA

In IS-95 on the forward link, physical channels (i.e., pilot, paging, sync, or traffic channels) are separated from each other by using different Walsh codes. In addition, a physical channel is also multiplied by a short PN sequence (2^{15} chips). In fact, every physical channel on the forward link is multiplied by the same PN sequence that is assigned to that particular base station or sector. A mobile differentiates the pilots by their PN offset. A proper PN offset planning is necessary to avoid an alias of PN sequences in a CDMA network.

In PN offset planning, the first parameter to be determined is PILOT_INC. A large PILOT_INC will increase the adjacent PN offset separation, but will reduce the number of valid PN offsets, which will reduce the reuse distance. A low setting will do just the opposite. The lower bound on PILOT_INC is

$$PILOT\_INC \times 64 > max\{W_R, W_N\} \quad \textbf{(18.28)}$$

where
W_R, W_N = search window sizes for remaining and neighboring set

If Equation (18.28) is violated, the two adjacent search windows will overlap and the measured value of the PN offset of a pilot found in the overlapping region will have an ambiguity because it belongs to both search windows.

The length of short PN sequence is 2^{15}, or 32,768 chips. A shift of 1 chip corresponds to a propagation distance of 244 m [$(1/1.2288 \times 10^6$ sec) $\times$ (3×10^8) m/sec ~ 244 m]. In order to provide enough isolation between PN sequences, IS-95 specifies a minimum separation of 64 chips. Each usable PN sequence is defined by a PN offset. There are a total of 512 usable PN sequences (32,768/64 = 512). The separation can be increased further by using the PILOT_INC parameter specified in IS-95. If PILOT_INC is 1, then the minimum separation is 1 × 64 = 64 chips. If the PILOT_INC is 2, the minimum separation will be 2 × 64 = 128 chips. Thus, the total number of usable PN sequences, $(PN)_{Total}$ is given by

$$(PN)_{Total} = \frac{32768}{(PILOT\_INC) \times 64} \quad \textbf{(18.29)}$$

To distinguish a pilot from the remote cell site with a multipath component of the serving pilot, enough separation between the adjacent PN offsets must be provided to avoid *adjacent PN offset* alias. In the cases where PN offset values are reused, the reused distance must be large enough to avoid *Co-PN Offset alias*. A third type of PN alias is called the *handoff alias*. We consider two sectors, A and B, with the same PN offset. Sector A is close to the mobile and is included in the neighbor list. Sector B is not included in the neighbor list; however, due to different antenna orientations or terrain conditions, the signal from sector B is much stronger than the signal arriving from sector A. If the pilot of sector B falls into the mobile's neighbor set search window, W_N, the mobile will confuse the pilot B as the pilot A and will handoff to sector A while despreading the signal from sector B. This will result in a strong forward link interference.

For a network where most of the cells are of the same size, the PN offset planning is relatively simple. The short PN code provides a sufficient number of distinct PN offsets to accommodate a large PN reuse pattern while maintaining reasonably large adjacent PN separations. In a typical network, the cells in rural and highway areas are much larger compared with the cells in urban areas. For such networks the PN offset planning becomes much more cumbersome. The *Co-PN offset alias* is the major concern for the small cell cluster, whereas the *adjacent PN offset alias* is more likely to occur in the large cells. If the small cell clusters and the large cell clusters are made to use the same reuse pattern, then it may be possible that a sufficient reuse distance cannot be maintained between the small and large cell clusters.

A CDMA signal that is not inside the *active set search window* (surcharging) cannot be despreaded by the fingers of a RAKE receiver. It will only contribute to the background noise. On the other hand, a CDMA signal in the search window despreaded by the fingers of the RAKE receiver will receive a 19.3 dB processing gain with the 13.3-kbps vocoder. If a remote CDMA signal does not belong to the serving cell site but falls into the SRCH_WIN_A, it then becomes one of the three strongest components. The mobile will treat this remote signal as one of the multipath components of the serving cell signal and will perform coherent combining on the two unrelated signals. This will result in strong interference. This type of PN offset alias must be avoided by the proper PN offset planning.

18.7.1 Adjacent PN Offset Alias

The adjacent PN-offset alias occurs because of large differences in propagation delay. We consider two pilots with adjacent PN offsets. When they reach the mobile, the pilot with an earlier phase propagates an extra distance that is large enough so that its phase shifts behind. If this pilot falls into the search window of the pilot with a later phase, the mobile will confuse these two pilots with two multipath components of the same pilot.

Let us consider two base stations, i and j, that are using adjacent PN sequences (i.e., PN sequences are separated by I = PILOT_INC × 64 chips). The mobile m is located at the boundary of the base station j and is served by that base station. The distance between the mobile m and the base station j is such that the propagation delay is Y chips. The distance between the mobile m and the base station i is such that propagation delay is X chips (see Figure 18.6).

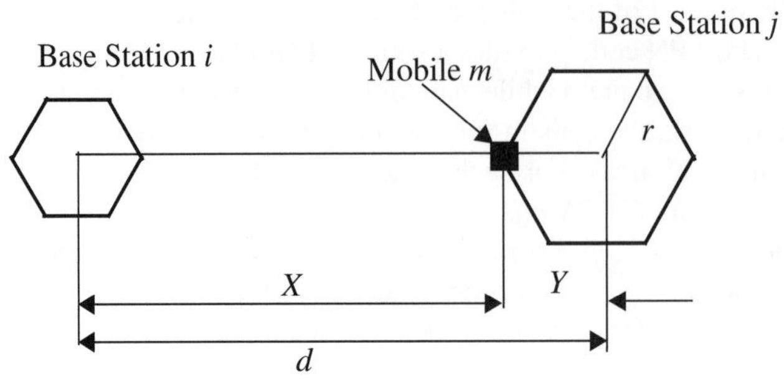

Figure 18.6 Two Base Stations and a Mobile

Next, we refer to Figure 18.7 that shows the PN sequences in time domain. PN sequence *i*, $(PN)_i$, and PN sequence *j*, $(PN)_j$, are not aligned in time when they are transmitted from their respective base stations. There is a shift of *I* chips between them. The $(PN)_i$ experiences a delay of *X* chips before it is received by the mobile *m*, whereas $(PN)_j$ is subjected to a delay of *Y* chips before it is captured by the mobile. Since the mobile is served by base station *j*, SRCH_WIN_A is centered on the received $(PN)_j$ (assuming this pilot component is the earliest arriving multipath). If the received $(PN)_i$ falls in SRCH_WIN_A of the mobile, the signal will be interpreted by the mobile as a multipath component of $(PN)_j$. Since the alias $(PN)_i$ arrives before $(PN)_j$, the mobile would move its SRCH_WIN_A to center on $(PN)_i$ and attempt to demodulate and combine both pilot signals. The result is interference, and usually a dropped call. The mobile mistakenly thinks that received $(PN)_i$ is an earlier arriving multipath component of $(PN)_j$.

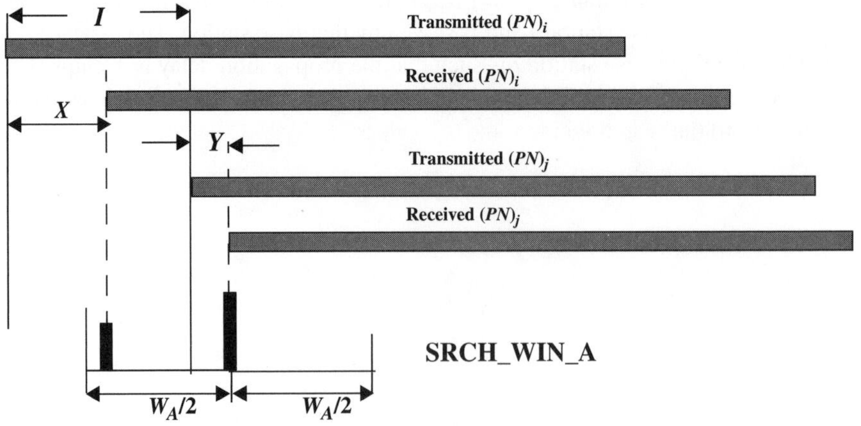

Figure 18.7 PN Sequence in Time Domain

In order to avoid PN offset alias, we should have

$$X < I + Y - W_A/2 \tag{18.30}$$

$$X + Y < I + 2Y - W_A/2 \tag{18.31}$$

$$D < I + 2R - W_A/2 \tag{18.32}$$

$$d < 244(I - W_A/2 + 2R) \tag{18.33}$$

or

$$d < 244I - 122W_A + 2r \tag{18.34}$$

where
D = separation (chips)
d = separation (m)
r = radius of base station j (serving base station) (m)
W_A = size of SRCH_WIN_A (chips)

Equation (18.34) represents the condition for no alias between two base stations using adjacent PN sequences separated by I.

Another way to avoid alias between the two base stations is to have the remote signal from base station i be at least 21 dB weaker than the serving base station j.

$$d > r \cdot [10^{21/(10\gamma)} - 1] \tag{18.35}$$

where
γ = propagation path loss exponent

EXAMPLE 18.8

Consider a CDMA network with a maximum cell radius of 15 km. The path-loss exponent γ is 3.2 and the search window W_A is 28 chips. Find PILOT_INC and the number of valid PN offsets.

$$d \geq R \cdot \left[10^{\frac{21}{10 \times 3.2}} - 1 \right] = 15 \times 3.5316 \approx 53 \text{ km}, D = \frac{53 \times 1000}{244} \approx 217 \text{ chips}$$

$$D = PILOT\_INC \times 64 - \frac{W_A}{2}$$

$$PILOT\_INC = \frac{D + W_A/2}{64} = \frac{217 + 14}{64} \approx 4$$

$$\text{PN offsets} = \frac{32768}{4 \times 64} = 128$$

These will yield a total of 42 cells with distinct PN offsets. For uniform networks, we may deploy 42 cells/cluster, or 37 cells/cluster leaving some reserved PN offsets for future use. It is desirable to reserve a certain number of PN offsets so that a new cell can be inserted into the existing cluster without disturbing the others.

18.7.2 Co-PN Offset Alias

Refer to Figure 18.8, in which cell A and cell B use the same PN offset. The Co-PN offset alias will occur if pilots from two different cells fall into the same SRCH_WIN_A of the mobile and both become one of the three strongest signal components. If one pilot propagates a longer distance than the other, that the difference in propagation delay causes the remote pilot to "fall out" of the mobile's SRCH_WIN_A, the mobile search finger will never find the remote pilot and Co-PN offset alias will be avoided. We assume that the distance between cell A and cell B as D. The mobile is located between the two cells, and its distance from the serving cell A is d_h. The distance to the remote cell B will be $D - d_h$. To ensure that the remote pilot falls out of the active search window, we need

$$(D - d_h) - d_h \geq W_A/2 \tag{18.36}$$

Since $d_h \leq R$

$$D \geq 2R + \frac{W_A}{2} \tag{18.37}$$

or

$$d \geq 122 W_A + 2r \tag{18.38}$$

where
d = separation (m)
r = coverage radius of serving base station (m)
W_A = size of SRCH_WIN_A (chips)

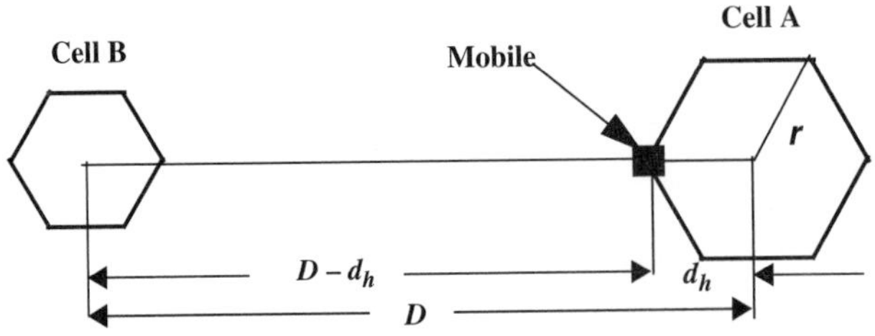

Figure 18.8 PN Sequences with the Same Offset

When cell A and cell B have different radii, *r*, in Equation (18.38) should be the maximum cell radius. Assuming the number of cells in a PN reuse cluster is *K*, then the reuse distance will be

$$d_c \approx r\sqrt{3K} \tag{18.39}$$

Equation (18.39) is proportional to cell radius *r*. Since the search window W_A is independent of *r*, the required Co-PN offset distance, d_c, in Equation (18.39) does not decrease proportionally to *r*. Thus, the small cells are much more vulnerable to Co-PN offset alias since they tend to have a small reuse distance.

Separation in time domain is not the only way to avoid PN offset alias. We can also use the received pilot strength to separate two pilots having the same PN offset. If the path loss between cell B and the mobile is sufficiently large, then $(PN)_B$ would undergo high attenuation before reaching the mobile. Then, even if $(PN)_B$ falls within the search window, $(PN)_B$ would have very low pilot strength and the mobile would not be able to demodulate it.

18.7.3 PN Offset Allocation Schemes

For a nonuniform network, it is often not desirable to force large cell clusters and small cell clusters into the same reuse pattern. Typically, the shape of the small cell clusters and large cell clusters are different; small cell clusters usually cover an area (two-dimensional), but the large cells often cover highways (one-dimensional). It is recommended to divide the total available PN offset into two disjointed sets: one for small cells, and the other for large cells. In this way, the small cells and large cells no longer share the PNs from the same set, so the reuse distance from large cells to small cells is no longer a concern. Also, the small cell clusters can have a different reuse pattern from the large cell clusters.

The number of cells per reuse cluster for small cells must be much higher than that for large cells, since small cells are more vulnerable to Co-PN alias. Depending on the actual cell configuration, $K_S = 27$ to 32 cells per reuse cluster for small cells is recommended [8]. For the large cells, the number of cells per cluster K_L can be much smaller, particularly if large cells are used only to cover the highways. Usually $K_L = 7$ to 12 is sufficient [8]. The resulting reuse distance must satisfy Equation (18.38) and $K_L + K_S \le 42$.

The adjacent sectors are least vulnerable to adjacent PN alias. A large path difference is a necessary condition for adjacent pilot alias and it is least likely for two pilots from adjacent sectors to produce a large path difference. Since the two sectors are facing different directions, in order for two adjacent pilot starting from the same point (BTS tower) and ending at the same point (mobile), one pilot must go through at least one reflection and become much weaker. Also, the pilot with earlier phase has to travel an extra distance of about 59 km [$\Delta D = (4 \times 64 - 14) \times 244 \approx 59km$], which is highly unlikely. Therefore, one should allocate adjacent PN offsets to adjacent sectors and allocate nonadjacent PNs to remote sectors. It is more likely for large cells with high antennas to cause interference to small cells, since signals from high antennas can propagate much farther. Therefore, it is best to allocate the PNs with later phases to large cells and those with earlier phases to small cells. This further reduces the likelihood of alias because the propagation delay will tend to shift the phase behind. So a pilot from a large cell with later phase, after propagation delay, will appear with an even later phase and will be even less likely get alias with pilot from small cells with early phases. The small cell pilots, with lower antennas, cannot propagate very far. The only expectation is that because the short code is periodic with period = 512 × 64 chips, a PN offset of 512 × 64 is the same as a PN offset of 0. Thus, it is not desirable to use a PN offset of 0. The first PN offset value should start at 4[× 64 chips], and the last one at 508[× 64 chips] [8].

The general expression for allocating PILOT_PN to each sector is

$$\{\alpha, \beta, \gamma\} - Sector\text{'s offset} = (PN_0)_j + 12k - \{8, 4, 0\} \quad \textbf{(18.40)}$$

where
$(PN_0)_j$ = first PN offset value from the sets
$k = 1, 2, \ldots, K_j$
K_j = number of cells in reuse cluster

Equation (18.40) is applicable for both small and large cell clusters. For small cell cluster, the subscript $j = S$ and for large cell cluster $j = L$.

18.8 Search Windows

In cdmaOne and cdma2000, the mobile uses the following three search windows to track the received pilot signals:

- SRCH_WIN_A: search window size for the active and candidate sets
- SRCH_WIN_N: search window size for the neighbor set
- SRCH_WIN_R: search window size for remaining set

18.8.1 SRCH_WIN_A

SRCH_WIN_A is the search window that the mobile uses to track the active and candidate set pilots. This window should be set according to the propagation environment. This window should capture all usable multipath signal components of a base station, and, at the same time, the window should be small enough to maximize searcher performance.

The search window should be large enough to accommodate the maximum expected arrival time difference between the pilot's usable multipath components (i.e., the pilot's maximum delay spread) [9].

$$T_{d,A} > \frac{2 \times (\tau_d)_{max}}{T_{chip}} \quad \textbf{(18.41)}$$

where
$T_{d,A}$ = required delay budget (chips)
$(\tau_d)_{max}$ = maximum delay spread in seconds
T_{chip} = chip time (813.8 nanoseconds)

EXAMPLE 18.9

Consider the propagation environment of a cdma2000 network, where the signal with a direct path travels 3 km to the mobile, and the multipath component travels 5 km before reaching the mobile. What should be the size of SRCH_WIN_A?

Direct path travels a distance of $= \frac{3000}{244} = 12.3$ chips

Multipath travels a distance of $= \frac{5000}{244} = 20.5$ chips

The difference in distance travelled between the two paths = 20.5 – 12.3 = 8.2 chips

The window size $\geq 2 \times 8.2 = 16.4$ chips

Use window size = 17 chips

EXAMPLE 18.10

Consider cells A and B of a cdma2000 network, separated by a distance of 20 km. The mobile travels from cell A to cell B. The RF engineer wishes to contain the soft handoff region between points X and Y located at distances 8 and 12 km from cell A (see Figure 18.9). What should be the search window size?

At point X the mobile is 8,000/244 = 32.8 chips from cell B
At point X the mobile is 12,000/244 = 49.2 chips from cell A

Path difference = 49.2 – 32.8 = 16.4 chips

At point Y the mobile is 12,000/244 = 49.2 chips from cell B
At point Y the mobile is 8,000/244 = 32.8 chips from cell A

Path difference = 49.2 – 32.8 = 16.4 chips

The SRCH_WIN_A $> 2 \times 16.4 > 32.8$ chips (use window size = 33 chips)

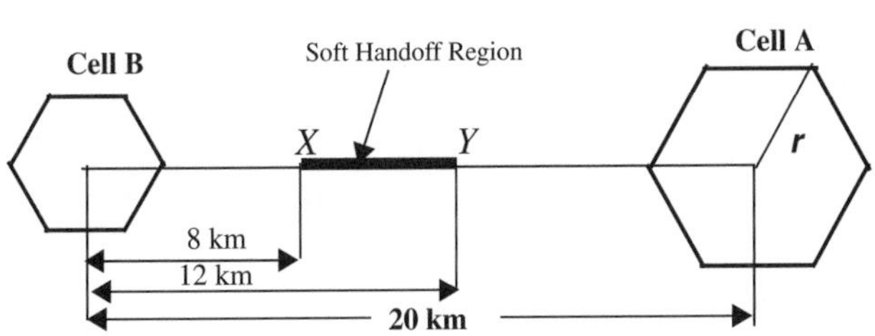

Figure 18.9 SRCH_WIN_A for Soft Handoff Between X and Y

This way, as the mobile travels from cell A to cell B, the mobile can ensure that beyond Y, the pilot from cell A drops out of the search window.

18.8.2 SRCH_WIN_N and SRCH_WIN_R

SRCH_WIN_N is the search window that is used by the mobile to monitor the neighbor set pilots. The size of this window should typically be larger than that of SRCH_WIN_A. SRCH_WIN_R is the search window that the mobile uses to track the remaining set pilots. A typical requirement for the size of this window is that it should be at least as large as SRCH_WIN_N.

The SRCH_WIN_N and SRCH_WIN_R pilots should account for the largest delay spread of the target pilot as well as the largest difference in propagation delays (i.e., difference in distance) between the reference pilot and target pilot [9].

$$T_{d,N} = T_{d,R} > \frac{2 \times D_{max}/V_c + (\tau_d)_{max}}{T_{chip}} \tag{18.42}$$

where

D_{max} = maximum difference (in miles) between (1) the mobile and cell transmitting active set pilot and (2) the mobile and the cell transmitting neighbor (or remaining) set pilot

V_c = speed of light (186,000 miles per second)

EXAMPLE 18.11

A CDMA system is located in an urban area where the maximum measured delay spread is 5 μs. The maximum distance between two neighboring base stations is 1.2 miles. Determine the active and remaining set window sizes.

$$T_{d,A} > \frac{2 \times (\tau_d)_{max}}{T_{chip}} = \frac{2 \times 5 \times 10^{-6}}{813.8 \times 10^{-9}} = 12.29 \approx 13 \text{ chips}$$

$\therefore (SRCH\_WIN\_A) = 13$ chips

$$T_{d,N} = T_{d,R} > \frac{2 \times D_{max}/V_c + (\tau_d)_{max}}{T_{chip}} = \frac{2 \times 1.2/186000 + 5 \times 10^{-6}}{813.8 \times 10^{-9}} = 22 \text{ chips}$$

$\therefore (SRCH\_WIN\_R) = 22$ chips

18.9 Summary

CDMA system planning and optimization has many more degrees of freedom than AMPS- or TDMA-based systems. Additional degrees of freedom are a blessing and a curse. Some CDMA parameter adjustments allow manipulation of cell coverage areas to cover flaws in the initial design without physically moving cells. On the other hand, traffic could create coverage gaps in city centers during peak usage periods if the system is improperly configured. A full understanding of the properties of CDMA technology is vital to the successful design and operation of CDMA networks.

Proper PN offset planning is critical to avoid alias PN sequences in a IS-95/cdma2000 networks. The PN offset planning is relatively simple for a network in which most of the cells are of same size. In a typical network, the cells in rural and highway areas are much larger compared with the cells in urban areas. For such a network, PN offset planning becomes important. The Co-PN offset alias is the major concern for the small cell cluster, whereas adjacent PN offset alias is more likely to occur in large cells.

- There are three types of PN offset confusion problems: Co-PN, adjacent PN, and confused handoff. PN confusion is more harmful than conventional interference.
- Propagation delay, search window sizes, and path-loss exponent are three key factors in PN offset planning.
- Sufficient reuse distance and PN separation, proper antenna downtilt, and proper allocation of PN offset values reduces the possibility of PN confusion.
- For nonuniform networks, reuse pattern clusters for large cells and small cells may have different shapes and may use PN offsets from two disjoint sets; small cell clusters use PN offsets with smaller values and large cell clusters use larger offset values.

Other significant issues for CDMA systems are soft handoff optimization, reverse channel noise floor management, the impact of geographical user distribution within a cell, and the criticality of power management.

In this chapter, coverage of a loaded CDMA network was studied by analyzing the uplink range. Base station multiuser detection (MUD) receiver can provide good coverage even with high system load after initial deployment. MUD decreases mobile station transmission power in a loaded network. Therefore, a network with base station MUD can be operated with a higher percentage of maximum load if the system capacity is limited by downlink. The coverage need not be sacrificed in order to use high system loads. In CDMA network planning, the performance of MUD should be considered when predicting and planning the coverage. The effect of MUD on cell range also depends on the propagation environment. Higher-data-rate services will reduce cell range in the uplink as the transmission power is limited. Therefore, in the cell design, the coverage area for low-rate services will be different than for high-rate services.

18.10 References

1. Hall, C. J., and Foose, W. A., "Practical Planning for CDMA Networks—A Design Process Overview," southcon/96, Orlando, June 25–27, 1996.
2. Garg, V. K., Smolik, K. F., Wilkes, J. E., *Applications of CDMA in Wireless Communications,* Prentice Hall: Upper Saddle River, NJ, 1997.
3. Ojanpera T., and Prasad, R., *Wideband CDMA for Third Generation Mobile Communications,* Artech House: Boston, 1998.
4. Holma, H., and Toskala, A., *WCDMA for UMTS,* John Wiley & Sons, New York, 2000.
5. Holma, H., Toskala, A., and Hottinen, A., "Performance of CDMA Multiuser Detection with Antenna Diversity and Closed Loop Power Control," Proceeding of VTC'96, Atlanta, May 1996, pp. 362–366.
6. Hamalainen, S., Holma, A., Toskala, A., and Laukkanen, M., "Analysis of CDMA Downlink Capacity Enhancements," Proceeding of PIMRC'97, Helsinki, September 1997, pp. 241–245.
7. Hamalainen, S., Holma, A., and Toskala, A., "Capacity Evaluation of a Cellular CDMA Uplink with Multiuser Detection," Proceeding of ISSSTA'96, Mainz, Germany, September 1996, pp. 339–343.
8. Chang, C. R.; Wan, J. Z.; and Yee, M. F., "PN Offset Planning Strategies for Non-Uniform CDMA Networks," Proceeding 47th IEEE Vehicular Technology Conference, Phoenix, May 4–7, 1997, pp. 1543–1547.
9. Kim, K., *Handbook of CDMA System Design, Engineering, and Optimization,* Prentice Hall: Upper Saddle River, NJ, 2000.

18.11 Problems

1. For a CDMA cellular base station, maximum transmit power is 13.2 W. The pilot power is 15% of the total transmit power, whereas the paging and sync channel powers are 35% and 10% of the pilot power, respectively. The BCR attenuation, *bcr*, is 10 dB and scale factor, *scale*, is given as 0.0017066. The forward threshold control for the traffic channel is 85% and channel activity factor (CAF) is 0.45. Find the capacity of the forward link assuming overhead due to soft handoff equal to 40% and traffic channel gain equal to 72 dgu. What are the gains for the pilot, paging, and sync channels?
2. Determine the maximum range for a handheld mobile with Rate Set 2 in Example 18.4.
3. Repeat Example 18.7 for the 8-kbps vocoder with 9.6-kbps data rate.
4. Find the size of SRCH_WIN_A for a propagation environment in a CDMA network where the signal with direct path travels 2 km to the mobile and the multipath travels 6 km before reaching the mobile.

5. Solve Example 18.11 using a maximum delay spread of 7 μs and maximum distance between neighboring base stations of 2.4 miles.
6. The maximum number of users in a cell of a WCDMA system at 12.2-kbps voice service is 100. The required E_b/N_0 in the unloaded condition is 5 dB. Assume $\bar{\mu} = 0.67$ and develop a curve to show range as a function of uplink load. Use the data given in Table 18.8. Also, develop a curve to show range as a function of uplink load for the MUD receiver at 60% efficiency.
7. Find the savings in average transmission power of a mobile station with MUD (efficiency 70%). Use the data from Table 18.9 and assume $E_b/N_0 = 1.5$ dB with 50% load (number of users, $M = 6$ at 144-kbps data rate).

CHAPTER 19

CDMA RF Optimization

19.1 Introduction

Code division multiple access (CDMA) radio frequency (RF) optimization is intended to ensure the performance and quality of the CDMA network for all users throughout the coverage area. The three main RF performance indicators are *access failures*—origination or termination, *frame error rate* (FER)—a measure of the perceived voice quality, and *dropped calls*.

A successful system optimization follows these basic steps:

1. **Planning**—Monitor hardware installation and integration status to plan clusters and drive test routes and to verify database translation parameters.
2. **Resources**—Cluster testing teams, tower crews, and available test equipment.
3. **Tools**—Mobile diagnostic monitor (MDM), RF call trace, and WatchMark Prospect.
4. **Test procedures**—Spectrum clearance to reduce external interference, sector tests for basic call processing functionality, unloaded drive tests for course adjustment, and loaded drive tests for fine tuning.
5. **Documentation**—Throughout the process, detailed records must be kept of the adjustments made during optimization (i.e., call processing parameters, antenna orientations, etc.).

The objectives of the optimization testing are

- To ensure that acceptable coverage is achieved for the pilot, paging, sync, access, and traffic channels
- To minimize the number of dropped calls, missed pages, and failed access attempts
- To control the overall percentages of soft/softer handoff
- To provide reliable hard handoff for CDMA-to-analog or CDMA f_1-to-f_2 carrier frequency

During the RF optimization process, CDMA parameters are adjusted using simulated traffic loading. The forward link loading can be implemented using an orthogonal channel noise simulator (OCNS) and reverse link loading can be achieved through the use of an attenuator and circulator at the mobile.

The RF optimization procedure includes field tests conducted to tune all aspects of CDMA air interface network performance. The RF optimization procedure is performed in two phases: (1) *cluster testing*, and (2) *complete systemwide optimization*. The initial pass of RF optimization is performed as a part of the cluster tests, and the second, more detailed tuning phase occurs after completing all cluster tests, once all cell sites in the CDMA network are activated.

Service measurements (SMs) are used to monitor overall system performance; these can indicate where system optimization is needed. Once the problem areas are identified, the performance engineer needs to further analyze the following data:

- Pilot channel E_c/I_t
- Mobile receive power
- Mobile transmit output power
- Forward/reverse link FER plots
- Call processing information
- Service measurements

The network tuning parameters that are used to mitigate system problems are neighbor list, baseband combiner and radio (BCR) attenuation and maximum output power settings, antenna orientation, pilot search window sizes, and handoff threshold parameters.

The RF performance engineer starts optimizing the CDMA system by obtaining information about the current system performance. Once the information is at hand, the performance engineer should recognize common performance problems, have a clear understanding of the available RF tuning parameters, follow a structured optimization process, and understand basic call processing and handoff methods. Key CDMA measurements show the condition of the system and RF operating environment. The measurement data is the primary gauge used to guide CDMA optimization.

In this chapter, we discuss the fundamental steps involved in the optimization of a new CDMA system. We provide a brief overview of the sources of CDMA data in the system. In addition, the tools that are necessary to extract the data are also discussed.

19.2 Cluster Testing

Cluster testing involves a series of procedures that are performed on a geographical grouping of about 19 cells each. The clusters are selected to provide a center cell with two rings of surrounding cells. The number of cells in each cluster is relatively large to provide enough forward link interference to generate realistic handoff boundaries in the vicinity of the center cell—a cluster of fewer cells would provide acceptable results over too small a geographic area. Approximately one tier of cell overlap should be provided between each cluster and the next to afford continuity across the boundaries. The goal is to complete all tests for a given cell cluster while minimizing the utilization of test equipment, personnel, and time. Cluster testing is intended to coarsely tune basic CDMA parameters and to identify, categorize, and catalog coverage problem areas. Generally, during the cluster testing, no attempt is made to resolve complex time-intensive performance problems. The location-specific, detailed refinements are deferred until the systemwide optimization phase.

19.2.1 Uplink and Downlink Interference Monitoring

The first preliminary step in cluster testing is to monitor uplink and downlink interference in the CDMA band to verify that the spectrum is clear enough for CDMA operation. For the CDMA system to properly function, the spectrum must be cleared in a sufficient guard band and guard zone. If strong in-band interference is present, then radio performance can be significantly degraded for the CDMA system. In extreme cases, it can be a time-consuming, difficult task to identify the sources of external interferences; therefore, it is important to start spectrum monitoring as early as possible. These spectral monitoring tests also provide a baseline data of measured background interference levels that can be used to optimize reverse overload control thresholds and jammer detection algorithms for the specific environment conditions.

19.2.2 Basic Call Processing Functions

The next stage of cluster testing before optimization is to exercise basic call processing functions including origination, termination, and handoff, to assure that these capabilities are operational. Quick measurements are made of CDMA signal levels to verify that each cell site is transmitting adequate power levels. These basic functional tests are intended to detect hardware, software, configuration, and translation errors for each cell site in the cluster prior to drive testing. This sector testing phase involves driving in each coverage area of each sector to ensure that installation has been completed correctly. At this stage, it is common to detect bad coaxial cables/connectors, misoriented antennas, and other similar defects.

19.2.3 Forward Link Pilot Channel Coverage

After basic cell operation has been checked, surveys of forward link pilot channel coverage are conducted with a light traffic load on the system. During the unloaded survey measurements, all cells in the cluster are simultaneously transmitting on forward link overhead channels (i.e., pilot, sync, and paging), with only a single Markov call active. In general, the drive routes should include major highways and roadways within the designed coverage area of the cluster where high levels of wireless traffic are expected. Drive routes may also be selected to explore suspected weak coverage areas and regions with multiple serving cells, as predicted by propagation modeling tool or based on the topography of the surrounding terrain. The unloaded pilot survey results identify coverage holes, handoff regions, and multiple pilot coverage areas. The pilot survey information highlights fundamental flaws in the RF design of the cluster under best-case, lightly loaded conditions. The pilot survey provides coverage maps for each sector in the cluster. These coverage maps are used to adjust system parameters during the optimization phase. Finally, measuring the pilot level without load serves as a baseline for comparison with measurements from subsequent cluster tests under loaded conditions. Characteristics of cell breathing can be compared under extremes of light and heavy traffic load.

During the unloaded coverage tests, two iterative passes of optimization are performed. The first-pass optimization provides correction of neighbor lists and adjustments to the fundamental RF environment (transmit power, antenna azimuth, height, downtilt, and antenna type). The second-pass optimization involves fine adjustments of handoff thresholds, search windows, overhead channel transmit powers, and access parameters. These two iterative optimization passes are intended to resolve problems observed by the field teams from the coverage plots and from analysis of dropped call mechanisms.

19.2.4 Coverage Drive Runs

The final measurements that are taken as a part of cluster testing are coverage drive runs performed under loaded conditions. Drive routes for the loaded coverage testing are kept to exactly the same routes as those used for the unloaded coverage surveys. The objectives are to provide coarse system tuning and to identify, categorize, and catalog coverage deficiencies so more difficult problems can be solved during later systemwide optimization tests. During loaded testing, both first-pass and second-pass tuning parameters are adjusted to fix problems observed by the field teams. At the conclusion of the loaded coverage tests, a performance validation procedure is conducted to measure system performance against cluster exit criteria.

19.3 Systemwide Optimization Testing

After testing all clusters in the CDMA network, systemwide optimization begins with all cells activated. Optimization teams will drive test each of the problem areas identified during cluster testing, using the same test conditions under which the problem was previously observed. An iterative tuning procedure is used to fix coverage problems by adjusting transmit powers and neighbor list entries. In extreme situations, handoff thresholds, search window size, or other low-level tuning parameters may have to be modified. If coverage problems cannot be resolved by the field team in a reasonable time period (~1 to 1.5 hours), the team flags the problem area for further investigation by other RF support personnel.

After attempts have been made by the site team to resolve the individual coverage problem areas, the systemwide optimization proceeds to the final phase. The final optimization step is a comprehensive drive test covering the major highways and primary roads in the defined coverage area for the CDMA network. During the systemwide drive run, simulated loading is used to model traffic on the network. Performance data is collected while a small number of active CDMA subscriber units traverse the systemwide drive route. Statistics are collected to characterize pilot, paging, traffic, and access channel coverage over the entire drive route. After the entire drive test has been completed, specific problem areas identified by the systemwide drive run are addressed on a case-by-case basis. Comprehensive statistics from the systemwide drive test are used to assess the overall performance quality of the network, including dropped call rates, handoff probabilities, and frame erasure statistics.

At the end of comprehensive systemwide drive test phase, the RF optimization procedure is considered complete and the CDMA network is ready for live traffic testing and market trials leading into commercial service. Once significant loading with live traffic is present on the CDMA network, additional tuning of system parameters is required to accommodate uneven traffic conditions (e.g., traffic hot spots, unusual traffic patterns, etc.) and other dynamic effects that cannot easily be predicted or modeled with simulated traffic loading.

19.4 CDMA System Parameters

The translation database for the CDMA system contains a large number of parameters that impact the RF performance of the network. Many of these parameters have complex interactions resulting in systemwide effects on capacity, coverage, and service quality. The main CDMA parameters can be divided into three categories: *first pass*, *second pass*, and *fixed parameters*. First-pass optimization parameters are the primary "tuning knobs" that can be used to optimize the CDMA performance. Second-pass optimization parameters are changed in unusual circumstances where problems cannot be resolved during the first-pass optimization. Fixed parameters are not changed under any circumstances during the field optimization process.

19.4.1 First-Pass Optimization Parameters

The following CDMA system parameters are used as the primary tuning controls for the RF optimization procedure during the first pass of optimization:

- Neighbor list entries
- BCR attenuation (total forward link transmit power)
- Changes in antenna configuration (azimuth orientation, antenna height, downtilt angle, antenna type)
- Hard handoff thresholds (CDMA-to-AMPS and CDMA f_1-to-f_2)

The first pass optimization parameters are the primary translations used to fix coverage deficiencies. Cell site transmit powers are adjusted with BCR attenuation to address coverage spillover, overshoot problems, and multiple pilot coverage regions. In certain cases, transmit powers are adjusted to provide fill-in coverage for weak signal strength areas. In addition, antenna azimuth, demodulating, or changing antenna patterns can be used in problem cases where transmit power adjustments are insufficient to resolve a deficiency. During adjustment of BCR attenuation, attempts are made to approximately balance the forward and reverse links.

The optimization of neighbor lists will be less of a problem during cluster tests, where only a small number of cells are active. However, with systemwide tests, in which many more sectors are simultaneously active, trade-offs are required to select neighbor list entries to minimize dropped calls due to missed handoff or handoff sequencing problems.

19.4.2 Second-Pass Optimization Parameters

These CDMA system parameters are changed only to correct performance problems at specific trouble spots during the second pass of optimization. These parameters include

- Soft handoff thresholds
- Active set and neighbor set search window sizes
- Access channel nominal and initial power settings
- Digital gain settings for the pilot, sync, and paging channels

The second-pass optimization parameters can have systemwide performance effects. These parameters should be adjusted with caution in cases where the adjusted parameters do not fully resolve a problem. For example, even small changes in soft handoff thresholds can affect overall system capacity and channel element utilization. In general, attempts should be made to maintain a consistent set of handoff thresholds for the entire CDMA network. It is not practical to change soft handoff thresholds on a sector-by-sector basis, particularly since handoff thresholds are established by the primary cell in a multiway handoff. However, a group of sectors covering a region can be changed to reflect small variations.

Search window sizes for the active and neighbor sets should initially be set based on expected cell size and multipath propagation delay spreads. If the CDMA network contains a mixture of small and large cells, then window sizes may have to be adjusted on a case-by-case basis to accommodate all handoff scenarios. For example, if there is a large variation in the antenna heights for cells in the CDMA network, the location of the search window will be skewed due to large propagation delay between the mobile and the distant cell site. In such a situation, if the search window size is not kept large enough, the mobile may fail to detect pilots from close-in neighbors due to the retarded timing reference.

19.4.3 Fixed Parameters

The following CDMA system parameters should not be adjusted during the RF optimization procedure:

- Forward power control thresholds
- Reverse power control thresholds
- Remaining set search window size

- Forward overload control set points
- Reverse overload control set points
- Minimum, maximum, and nominal traffic channel digital gains

The fixed parameters should not be adjusted during field optimization. Since power control plays such a vital role in both reverse link and forward link performance of the CDMA system, related thresholds and step sizes should be adjusted based only on simulations or lab measurements. For the optimization tests, it is recommended that reverse overload control thresholds be set to the maximum values allowed in the translation to avoid false alarms during loaded drive testing. The forward overload control parameters should be adjusted based only on lab tests and computer simulations.

19.5 Selection of Drive Routes

The following procedure should be used to select drive routes for optimization. The selection of drive routes can seriously impact overall system performance as measured during the cluster exit tests. The following steps should be taken for proper drive route selection:

1. Obtain propagation prediction tool plots for the test cluster.
2. Choose drive routes that are completely contained within the predicted coverage area of the test cluster (i.e., within the designed coverage area of the cluster).
3. Under no circumstances should be drive routes exit the designed coverage area of the cluster as determined from the prediction tool plots.
4. Choose a combination of primary and secondary roads to provide representative coverage of the interior of the cluster.
5. Avoid drive routes that border between the test cluster and adjacent clusters that have not yet been optimized, as inaccurate performance results will be obtained in these border areas until both clusters can be operated simultaneously.
6. Select drive routes primarily in the interior of the test cluster, where forward link interference can be accurately modeled.

If some of the drive routes covered areas outside of the designed coverage area of the cluster, the route segments outside the predicted coverage area should be removed from performance test results for the cluster. The data from these *out of coverage* route segments should be manually removed from the statistics collected for frame error rate (FER), dropped calls, originations, and terminations as a part of the drive test reports, cluster exit tests, and performance validation tests. At the time of the drive route selection, appropriate notation should be made to indicate which routes and segments are outside the predicted coverage of the cluster. Any coverage plots (e.g., FER, pilot E_c/I_t, mobile receive power, mobile transmit power) developed for the out-of-coverage route segments should explicitly and clearly indicate that these routes are not in the designed coverage of the cluster.

19.6 Simulation of Traffic Loading

Since all CDMA users share the same carrier frequency, the performance of the network is very much dependent on local interference levels, and hence on the traffic load of the system. The coverage areas and handoff regions are strongly affected by the amount of traffic on the network on a per-sector basis. To evaluate the CDMA system's coverage, capacity, and service quality under real-world conditions, the network must be tested under loaded conditions to match those expected during actual operation. Simulated loading must be applied to both the forward and reverse links to fully model the effects of traffic upon system performance.

19.6.1 Forward Link Loading

Orthogonal channel noise simulators (OCNS) are provided in all of the CDMA sectors to simulate forward link interference. Commands can be issued to activate forward link interference at the desired level to emulate a specified number of users; the number of equivalent users and a root mean square (rms) forward link digital gain setting can be specified on a per sector basis. OCNS is used for all forward link loading simulation during the RF optimization testing. Loading levels are determined based on estimated Erlang traffic to be served by the network.

19.6.2 Reverse Link Loading

One simple method to emulate reverse link loading is to place an attenuator in the transmit path at the CDMA subscriber unit. The mobile attenuator is used to simulate the median interference rise observed by the cell site during specified Erlang loading conditions. The attenuation simulates a static interference condition on the reverse link. No attempt is made to model the statistical variations caused by other mobiles in the system. Shadow fading effects are included for the one mobile that is involved in the test, while the effects from other mobiles are lumped into the single, static attenuation value.

For reverse link loading during the RF optimization, the main aim is to ensure that acceptable uplink coverage is maintained during loaded conditions (i.e., to check for coverage holes); for this reason, the attenuator is a simple and effective method of measuring cell shrinkage for a specified traffic load. The use of an attenuator to simulate loading is therefore proposed as a method to determine coverage holes that might develop in the CDMA system when it is loaded. The attenuator is not meant to stress the base station infrastructure equipment in a manner comparable to live traffic loading. Also, the attenuator is not proposed to simulate extreme conditions that might occur due to jamming, external interference, or other abnormal conditions.

The impact on the forward link of placing an attenuator at the mobile unit does not mimic the effects observed under actual loading of the CDMA system. Actual traffic loading on the forward link increases the amount of interference on the forward link without affecting the absolute received power of the pilot signals. In other words, as the loading is increased, the measured pilot-to-interference ratios are reduced by the presence of the additional forward traffic channels. An attenuator in the receive path of the mobile reduces the level of both the pilot channel and the interference, not just the pilot channel. As such, the attenuator distorts the pilot-to-interference levels observed by the mobile; the result is forward link performance that is not representative of the behavior observed under true loaded conditions.

To reduce the undesired impacts of the attenuator on forward link performance, microwave hardware can be used to separate the forward and reverse link paths at the mobile. A back-to-back circulator arrangement will allow attenuation to be placed only in the reverse link path, while the forward link path will only incur the circulator losses (typically less than 0.5 dB). Such an approach allows an attenuator to be used at the CDMA mobiles, without requiring increased forward link transmit powers. For the loaded RF optimization tests, a mobile attenuator with circulators will be used to simulated *blanket* background loading levels on the reverse link.

19.7 Power Allocation

Cell site output power calibration is performed before RF optimization. The purpose of power calibration is to make sure that with particular values of BCR and digital gains, the required output power is achieved at the antenna. Once the transmit path has been calibrated, power can be changed electronically by setting the values of the BCR and digital gains.

The power in Watts for a channel at connector is given as

$$P_{channel} = Scale \cdot 10^{-(bcr)/10} \cdot (G_{channel})^2 \cdot \alpha_{chan} \quad \textbf{(19.1)}$$

where

$P_{channel}$ = power in Watts for the channel at connector
$Scale$ = scale factor
bcr = BCR attenuation factor in dB
α_{chan} = channel activity factor: 1 for pilot, sync, and paging channel, and 0.4 to 0.5 for traffic channels
$G_{channel}$ = digital gain of the channel

The scale factor is calculated when the output power is calibrated for particular values of bcr and pilot digital gain. If the bcr = 10 and a pilot digital gain of 108 produces a pilot power of 2 W after calibration, then the scale factor will be

$$2 = Scale \cdot 10^{(-10)/10} \cdot (108)^2$$

$$\therefore Scale = 0.0017066$$

Assuming that the scale factor, α_{chan}, and BCR remain unchanged, powers and digital gains for a channel can be related as

$$\frac{P_1}{P_2} = \frac{(G_1)^2}{(G_2)^2} \tag{19.2}$$

where P_1 and P_2 are the powers and G_1 and G_2 are the digital gains for the P_1 and P_2 power, respectively.

The number of traffic channels to be set up on the forward link at a given time depends on the subscriber traffic at that time. Power allocated to each channel affects forward link performance. Each channel should be received by the mobile with the required strength for proper reception. If the pilot channel power is p_{pilot} percentage of the maximum power, and paging and sync channels powers are p_{paging} and p_{sync} percentage of the pilot power, respectively, then the average power, $P_{traffic/user}$, of a traffic channel is given as

$$P_{(traffic)/(user)} = \frac{(Maximum\ \ Power) - [(Pilot + Paging + Sync)Power]}{Maximum\ \ No\ \ of\ \ Traffic\ \ Channels} \tag{19.3}$$

Note that all powers are referenced to the pilot channel power. If during optimization the power in each channel is changed by changing the BCR attentuation, all channel powers change automatically by the same ratio. However, if the digital gain of any one of the channels is reduced or increased, the digital gains of other channels must be changed in the same ratio to preserve the power percentages in relation to the pilot channel and maximum power.

A suggested power allocation for the pilot channel is 15 to 20 percent of the maximum power and power allocations of 35 percent and 10 percent of the pilot power are suggested for the paging and sync channel, respectively.

The base station periodically calculates the power ratio, γ, as

$$\phi = \frac{Actual\ \ Transmit\ \ Power}{Maximum\ \ Power} \tag{19.4}$$

The amount of power reserved for handoff is upper_power_threshold minus lower_power_threshold. The purpose of the power threshold on the forward link is to limit the number of users in order to protect the amplifier from overdrive. When ϕ is equal to the lower_power_threshold, new call attempts are blocked but handoffs are allowed. When ϕ is equal to the upper_power_threshold, new call attempts and handoffs are not allowed. The recommended values for the lower_power_threshold and upper_power_threshold are 85 percent and 90 percent, respectively [1].

Forward link capacity is limited by available forward link power and forward link overload control. Forward link overload control is used to make sure that when the forward link thresholds are reached, no more users are added to the system.

If P_{pilot}/pilot_percentage < maximum transmit amplifier power, then max_power will be P_{pilot}/pilot_percentage and if P_{pilot}/pilot_percentage > maximum transmit amplifier power then pilot power is adjusted to satisfy Equation (19.1).

The total transmit power can be calculated as

$$P_{total} = Scale \cdot 10^{(-bcr)/10} \cdot \left[(G_{pilot})^2 + (G_{paging})^2 + (G_{sync})^2 + \sum_{i=1}^{M_{total}} \alpha_i \cdot (G_i)^2 \right] \quad \textbf{(19.5)}$$

where

P_{total} = total power used for all forward link channels

$Scale$ = scale factor

bcr = BCR attenuation in dB

M_{total} = total number of traffic channels including soft handoff overhead

α_i = channel activity factor for the ith traffic channel

G_{pilot}, G_{paging}, G_{sync}, G_i = digital gains for pilot, paging, sync, and *i*th traffic channels

The number of traffic channels at a given time is equal to the number of calls set up by the users. When a traffic channel is assigned to a user and communication begins, a nominal gain (nom_gain) value is assigned to traffic channel. As the forward link power control begins functioning, the gain varies between minimum (min_gain) and maximum (max_gain) gain. The min_gain avoids the *stop light effect*. This means that the forward power control will not be allowed to reduce the power so much that the mobile is not able to recover from multipath fading. The forward power control is slow and is not effective against fast fading, thus there is a need to put a lower limit on the transmitted traffic channel power. The max_gain ensures that the forward power is uniformly distributed among all traffic channels and a single channel does not become a drain on forward power. The suggested values of nom_gain, min_gain, and max_gain for Rate Set 1 and Rate Set 2 are listed in Tables 19.1 and 19.2.

Table 19.1 Traffic Channel Digital Gain—Rate Set 1

System	Nominal Gain		Minimum Gain		Maximum Gain	
	Range	Default	Range	Default	Range	Default
Cellular CDMA	34–108	57	34–50	40	50–108	80
Cellular CDMA Minicell	34–108	57	34–50	40	50–108	80
PCS CDMA Minicell	34–108	57	34–50	40	50–108	80

Table 19.2 Traffic Channel Digital Gain—Rate Set 2

System	Nominal Gain		Minimum Gain		Maximum Gain	
	Range	Default	Range	Default	Range	Default
Cellular CDMA	40–108	80	30–50	30	50–127	127
Cellular CDMA Minicel	40–108	80	30–50	30	50–127	127
PCS CDMA Minicell	40–108	80	30–50	30	50–127	127

19.8 Nominal Parameter Settings

Based on field tests, lab measurements, and computer simulations, nominal values have been developed for the most important CDMA parameters. Depending upon the local environment (terrain, cell layout, etc.), it may be desirable to adjust the nominal values (see Table 19.3) within the typical ranges.

Table 19.3 Important CDMA Parameters

Parameter	Nominal Value	Tuning Range
T_ADD	–13 dB	–12 to –14 dB
T_DROP	–15 dB	–14 to –16 dB
T_COMP	2.5 dB	2 to 3 dB
T_TDROP	2 sec	1 to 4 sec
Active Search Window	40 chips	10 to 40 chips
Neighbor Search Window	80 chips	40 to 160 chips
Pilot Digital Gain	108 dgu	108 dgu
Page Digital Gain	64 dgu	64 to 96 dgu
Sync Digital Gain	34 dgu	34 dgu

dgu = digital gain unit

EXAMPLE 19.1

Use the following data for a CDMA cellular system and calculate the capacity of the forward link:

- Scale factor = 0.0017066
- Traffic channel gain = 72 dgu
- Maximum cell site power = 13.2 W
- Pilot channel gain = 108 dgu
- Sync channel gain = 34 dgu
- Paging channel gain = 64 dgu
- Traffic channel activity factor = 0.45
- BCR attenuation (*bcr*) = 10 dB
- Forward power threshold control = 85%

$$P_{total} = Scale \times 10^{-(bcr)/10} \cdot [(G_{pilot})^2 + (G_{paging})^2 + (G_{sync})^2 + \ldots]$$

or

$$13.2 = 0.0017066 \times 10^{-1} \cdot \Phi$$

$$\therefore \Phi = 77,347 \text{ dgu}^2$$

but

$$(G_{pilot})^2 + (G_{sync})^2 + (G_{paging})^2 = (108)^2 + (34)^2 + (64)^2 = 16,916 \text{ dgu}^2$$

$$\text{Power left for traffic channels} = 0.85 \cdot [77,347 - (16,916)] = 51,366 \text{ dgu}^2$$

$$\therefore 0.45 \times M_{total} \times (72)^2 = 51366$$

$$M_{total} = 22$$

Assuming the channel overhead due to soft handoffs equal to 60%, the number of users that can be supported on the forward link will be $22/1.6 = 13.75 \approx 14$.

EXAMPLE 19.2

Use the following data for a CDMA PCS minicell and calculate the total capacity of the forward link.

- Maximum cell site power = 8 W
- Pilot channel gain = 108 dgu
- Paging channel gain = 64 dgu
- Sync channel gain = 34 dgu
- Scale = 0.0006491
- BCR attenuation (*bcr*) = 8 dB
- Traffic channel gain = 72 dgu
- Maximum power threshold = 85%
- Channel activity factor = 0.45

$$8 = 0.0006491 \times 10^{-0.8} \times \Phi$$

$$\therefore \Phi = 77,764\ \text{dgu}^2$$

but

$$(G_{pilot})^2 + (G_{sync})^2 + (G_{paging})^2 = (108)^2 + (34)^2 + (64)^2 = 16,916\ \text{dgu}^2$$

$$\text{Power left for traffic channel} = 77,764 - (16,916) = 60,848\ \text{dgu}^2$$

$$M_{total} \times 0.45 \times (72)^2 = 0.85 \times 60848$$

$$\therefore M_{total} = 22.17 \approx 22$$

Assuming the channel overhead due to soft handoff is equal to 50%, the number of users that can be supported on the forward link will be $22/1.5 = 14.67 \approx 15$.

19.9 Optional Preliminary Tests

These tests may be performed prior to the actual cluster testing. The first test is used to estimate the in-vehicle penetration loss and antenna gain differences between a roof-mounted reference antenna and a handheld antenna inside the test vehicle. If the in-vehicle penetration loss of the measurement setup has been previously characterized, then there is no need to remeasure the loss for each cluster. If a specific in-vehicle loss has been assumed in the link budget design for the market being optimized, then the loss from the link budget can be used to directly set the attenuation value. If manufacturer and service provider have previously agreed upon an acceptable attenuation value for in-vehicle loss, then there is no need to perform the test.

The second test is to characterize the forward link root mean square (rms) digital gain value for a typical CDMA user under nominal loading conditions in a particular environment. The rms digital gain value basically provides the average transmit power required for each CDMA user; therefore, rms digital gain value is useful for setting OCNS levels for use in forward link loading. In many cases, reasonable rms digital gain value can be estimated prior to cluster testing based on simulations or previous field data. If necessary, the forward digital gain test procedure can be used to verify the estimated values prior to simulated load testing.

19.10 Guide to Cell Site Engineering

In this section, we discuss initial engineering of a CDMA cell, provide information for monitoring the cell, and list possible causes and corrections for cell utilization.

- **Initial Engineering of Cell Site**
  The number of cells and traffic channels should be estimated from the expected busy hour call attempt (BHCA) and average call holding time (ACHT). The normal blocking for voice traffic is 2 percent (however, 1 percent blocking is more desirable). The Erlang B tables are used to determine how many BHCAs a radio can handle during busy hour at a certain blocking factor and a certain ACHT. The number of BHCAs per traffic channel depends upon the number of traffic channels per face. Due to RF coverage, most systems are engineered at lower blockage rate. It is critical to know what blockage rate/radio occupancy to use. The radio occupancy should be limited to 60 to 70 percent.
- **Established Calls Percentage**
  The percentage of established calls is the percentage of all seizure attempts in the system (both origination and page response [termination]) that result in an answered call. The goal should be 100 percent of the established calls. Typical results in a large metropolitan system may be 90 to 95 percent. The causes of low percentage of the established call problem are
  - Excessive seizures denied: inadequate signal, access threshold too strong, cell receive path overload, insufficient RF coverage (fringe cells), etc.
  - Excessive paging channel confirmation failures: confirmation threshold too strong, cell receive path overload, insufficient RF coverage (fringe cells), etc.
  - Excessive seizures denied (servers busy): need more voice radio and/or additional cells in the area
  - Origination/termination blocked: no voice channel, MSC says no radios available
  - Roamer service denied
  - Fraudulent mobile call attempts
  - Cell/system overload

- **Percentage of Lost Calls per Cell**
  The percentage of lost calls per cell represents the percentage of established calls that were lost as a result of either pilot pollution or any other reason. The goal should be 0 percent lost calls per cell. Typical values may be less than 3 to 4 percent. The causes of high percentage of lost calls are
  - Poor coverage, interference, and improper pilot setting. The reasons for coverage problems include cell spacing or RF shadowing (from terrain features such as mountains/hills, dense foliage, or tall building and tunnels). Fringe cells (especially if there are no handoffs to abutting systems) generally have higher lost call rates caused by mobiles driving out of the service area.
  - Interference can impair the voice radio's ability to detect the pilot and can result in lost calls.
  - Inadequate power control can cause the near-far problem, thus resulting in lost calls.

- **Percentage of Handoff Failures**
  The percentage of handoff failures represents the percentage of handoff orders that were aborted due to failure events during the handoff process. The goal should be 0 percent. Typical values may be less than 10 percent. The causes of handoff failures are
  - Call processing fails to set a timer during handoff process
  - MSC failure during handoff process
  - MSC failure on trunk associated with new cell
  - MSC shutdown
  - Transmitter activation failure at the new cell
  - Unexpected message received from the new cell
  - Time-out waiting for new cell transmitter activation confirmation
  - Time-out waiting for MSC connect acknowledgment
  - Old cell time-out
  - Time-out waiting for MSC switch path acknowledgment
  - New cell time-out
  - Time-out waiting for MSC clear acknowledgment
  - Voice channel blocking on an inter-MSC handoff

- **Percentage of Locate Requests Dropped**
  The percentage of locate requests dropped is the percentage of the received location request messages from a neighbor cell that are dropped because of too many requests in a one-second interval. The goal for percentage locate requests dropped is 0 percent. The typical value may be less than 0.1 percent. The most likely cause of a high percentage of locate request dropped is an incorrect neighbor list containing too many entries or incorrect entries.

- **Percentage of Handoff Complete for Cell**
  The percentage of handoff complete is the percentage of successful handoffs for a cell. The goal for percentage of handoff complete is 100 percent. Typical values may be greater than 95 percent. Interference and poor coverage are the possible causes of low percentage of handoff complete.

- **Percentage of Processor Occupancy**
  The percentage of processor occupancy is the percentage of time spent by the processor doing work other than fill (background) work. The maximum limit for percentage of processor occupancy is 70 to 80 percent. The causes of high percentage of processor occupancy are
  - Incorrect neighbor lists
  - Excessive number of locate requests due to a large number of neighbors and high primary thresholds
  - Large number of originations, terminations, and handoffs

- **Percentage of Cell Blocking**
  The percentage of cell blocking is the percentage of origination and page response seizures denied due to lack of resources. The goal for percentage of cell blocking is 2 percent. Typical results may range from 0 to 2 percent. On a per-system basis,
  - Percentage of cell blocking < 2% = Good system
  - Percentage of cell blocking > 2% = Bad system

  The actions to correct the problem are

  1. Cell-splitting
  2. Load balancing via software adjustment
  3. Add radios and dualize the cell

- **Percentage of Cell Trunk Group (CTG) Blocking**
  The percentage of cell trunk group blocking is the percentage of all service requests (origination, termination, handoff) for a given linear amplifier (LAF) that are denied due to lack of available servers (radios) on that LAF. The goal for percentage of CTG blocking is 2 percent. Typical results are from 0 to 2 percent. On a per-system basis,
  - Percentage of CTG blocking < 2% = Good system
  - Percentage of CTG blocking > 2% = Bad system

  The actions to correct the problem are

  1. Cell-splitting
  2. Load balancing via software adjustment
  3. Add radios and dualize the cell

- **Percentage of CTG Utilization**
  The percentage of cell trunk group utilization is the ratio of measured carried traffic load to theoretical capacity (based on an established traffic model). The goal for CTG utilization is 85 percent. Typical values are 80 to 90 percent. On a per-system basis:
  - Percentage of CTG utilization < 60% = Bad system
  - 70% < CTG utilization < 80% = Marginal system
  - 80% < CTG utilization < 90% = Good system
  - 90% < CTG utilization < 100% = Marginal system
  - Percentage of CTG utilization > 100% = Bad system

19.11 CDMA Network Performance

The overall performance of the CDMA network depends upon the area coverage probability for which the system has been designed. For example, a system designed for 90 percent area coverage would be expected to incur a higher outage probability than a system designed for 97 percent area coverage. For this reason, specific numerical exit criteria cannot be provided. The exact exit criteria used in RF optimization tests is customer- and market-specific.

For the unloaded pilot survey results, best server pilot E_c/I_t measurements should be greater than –11 dB for the designed area coverage probability of the system (typically 85 to 97 percent). The objective of the loaded coverage test is to measure the performance of the CDMA system with actual or simulated loading conditions. During the testing with traffic loading, traffic is simulated. For these tests, a CDMA mobile diagnostic monitor (MDM) is placed in a roving test vehicle and driven over the same drive routes used for the unloaded pilot channel coverage survey test. During cluster testing, the objective is to identify, categorize, and catalog the coverage problems observed during the drive testing. Any coverage problem that cannot be solved with basic parameter changes requiring less than 30 minutes of work, should be deferred until the systemwide optimization phase.

19.11.1 Data Analysis

Data analysis of the loaded coverage test data can be used to display pilot channel E_c/I_t in dB as a function of location. The resulting pilot coverage maps should be compared against predicted levels from RF planning tools. Handoff areas should be determined by mapping the handoff state zero-, one-, two-, or three-way for each geographic location. Cumulative distribution functions (CDFs) of pilot E_c/I_t can be used to determine the percentage of drive route where pilot levels exceeded a particular value. Similar statistics are useful in determining the percentages of one-, two- and three-way handoffs.

The forward and reverse link coverage maps should be used to validate the RF design for the test cluster. The measurements made under unloaded conditions represent an ideal, best-case condition. Under full loading conditions, observed pilot E_c/I_t levels could be as much as 5 dB lower than the unloaded measurement values. Any coverage holes that exist for the unloaded measurement will be enlarged once the system matures and becomes loaded. If substantial areas of poor pilot coverage exist in the measured coverage area for the test cluster, then RF design alternatives should be considered to rectify the problem areas. Possible approaches include changing pilot powers, changing antenna patterns, reorienting antenna bore sites, and adding additional cells or repeaters.

Maps of handoff activity can be useful for setting handoff thresholds, creating neighbor lists, and identifying trouble spots. For an unloaded system, one can expect significantly higher handoff percentages than for a fully loaded system. Because handoff activity is based on pilot E_c/I_t levels, the handoff areas tend to shrink as the CDMA system gets loaded and E_c/I_t levels decrease due to increased interference. The CDMA mobile unit should simultaneously receive voice signals from, at most, three sectors. For this reason, it is desirable to reduce the number of areas where four or more strong pilot signals are present because areas with large numbers of strong pilots could impact capacity of the forward link.

For the loaded pilot survey results, the following exit criteria should be used:

1. Best server pilot E_c/I_t measurements should be greater than –15 dB for the designed area coverage probability of the system (typically 85 to 97 percent).
2. Forward link FER measurements should be less than the outage threshold of 10 percent for the designed area coverage probability of the system.
3. Reverse link FER measurements should be less than the outage threshold of 10 percent for the designed area coverage probability of the system.
4. Mobile transmit power measurements should be less than 20 dBm for the designed area coverage probability of the system.
5. Typically, dropped call rates in the range of 2 to 5 percent are appropriate as exit criteria.

19.12 Causes of Poor CDMA System Performance

In this section we discuss various causes of service degradation in a CDMA system and suggest appropriate corrective actions.

19.12.1 Poor Voice Quality: Forward Link

Inadequate Traffic Channel Signal Strength. There are several reasons that could cause degraded voice quality on the forward link traffic channel. One important aspect of the forward link performance is the large variability in *S/I* ratios required for a specific forward link FER (typically 1 percent). The required *S/I* ratio is a function of the mobile speed, multipath channel conditions, and soft handoff state. The required target *S/I* ratios may change by over 16 dB depending on the mobile's conditions. The distribution of speeds of the mobile may have a large impact on the forward link performance; for example, during rush hour when many mobiles are moving at low speeds (5–15 km/hour), the forward link can be subjected to greater stress. The relative performance of the forward link is an even greater issue with the 13 kbps vocoder than with the 8 kbps vocoder. The operation of forward overload control can also be responsible for performance degradations observed on the forward link. When the forward overload control threshold is exceeded, the overload algorithm caps the forward transmit power to the mobile (i.e., it allows it to power down but no power up). In such a situation, a particular mobile may not be able to get enough transmit power to achieve the desired FER.

Intermodulation Interference. Intermodulation interference is another source of potential problems. The most severe problem observed is the forward link interference from AMPS-only cells to CDMA mobiles, and similar problems may exist for PCS when several service providers become operational.

19.12.2 Poor Voice Quality: Reverse Link

There are several possible sources of reverse link voice quality degradations. The most frequent cause is improper balance between the forward and reverse links. If the forward link has more path-loss margin than the reverse link, it is possible to get into situations where the mobile may not be able to complete the reverse link. The situation can be diagnosed by monitoring the mobile transmit power in problem area; if the mobile transmits at or near the maximum value (23 dBm = 200 mW for CDMA portables), then the problem could be due to power imbalance.

Other reasons for reverse link quality problem can be caused by corruption of the reverse closed-loop power control. If the forward link is degraded, it is possible for the mobile to misinterpret the power control commands sent by the base station. In the worst situation, the mobile would power down when it was instead requested to power up by the base station. In these conditions, the reverse link may be degraded or dropped altogether.

Another connection between forward and reverse link performance is through the mobile's forward link fade-detection mechanism. If the mobile detects 12 consecutive bad frames on the forward link, the mobile will stop transmission until it receives two consecutive good frames. In this case, bad forward link performance can cause muting of the reverse link.

19.12.3 Inadequate Pilot Signal Strength from Serving Sector

If CDMA calls cannot be originated because of inadequate pilot signal strength from a serving sector, the problem is most likely caused by a coverage hole created by excessive path loss. The excessive path loss could be due to blockages from terrain, buildings, trees, or any other radio obstacle. For CDMA systems, the maximum allowable path loss will shrink as a function of system load; therefore, coverage holes that were not evident during light load conditions may suddenly appear under a heavy traffic load.

The primary CDMA tuning parameter that can be used to address coverage holes is the BCR attenuation. By increasing the transmit power from the best serving sector in steps of approximately 2 dB, it should be possible to determine whether the coverage hole can be adequately filled. In many cases increasing transmit power may not solve the problem; for example, because of the limited power output from CDMA minicells, very little transmit power may be available for use in RF optimization.

Another technique that may be used to fix coverage holes is to change cell site antennas to higher gain varieties (narrower vertical or horizontal beamwidth) to provide more signal strength in the desired area. Increasing the antenna gain may fix the coverage problem, but at the same time may create many other problems. In some cases, reorienting the antenna pointing azimuth may be useful for filling coverage holes in particular areas. Adjusting antenna downtilt at the serving sector may also allow more energy to be radiated in the vicinity of the coverage hole. CDMA repeaters, used to rebroadcast signals from existing cell sites, may be another alternative to fill in coverage holes.

19.12.4 Reverse Overload Control False Alarm

Under some operating conditions, the reverse overload algorithm may generate false alarms to inhibit call originations or soft handoffs. The reverse overload algorithm uses estimates of the received signal strength relative to the background noise. In a real-world environment, noise contains interference from many sources including in-band signals from other services; sidebands and spurious emissions from many sources (AMPS, TDMA, paging, etc.); interference from industrial sources, machinery, microwave ovens, etc.; jamming sources; and unauthorized transmitters.

Because of the difficulty of accurately measuring both the total received power and background noise levels, it is possible for the reverse overload control algorithm to be misled into concluding the reverse link is overloaded when in fact there may be only light loading present. For the purposes of RF optimization testing, it is suggested that the reverse overload control algorithm thresholds be set to their highest allowable values. Setting high reverse overload thresholds will disable the overload control algorithm for the purpose of field testing.

19.12.5 Paging or Access Channel Message Failure

In some cases, the paging and access channel performance of the CDMA system may not exactly match the traffic channel coverage. For example, the paging channel does not benefit from forward link power control or from soft handoff combining gain. If paging channel transmit powers are not set correctly, mobiles may be able to receive acceptable pilot and sync channels, but may fail because of excessive paging message failures. In this case, the paging channel digital gain setting should be adjusted upward to allow pages to be received.

A similar situation may occur for the access channel on the reverse link. When a CDMA mobile transmits on the access channel, reverse closed-loop power control is not active; therefore, the mobile must "guess" at the appropriate transmit power to use for access. The transmit power estimate made by the mobile is based on the received power combined with offset adjustments transmitted by the cell site. In some cases, if the offset adjustments are not done properly, mobiles may not be able to access the reverse link. In a more likely situation, the mobile may require a large number of access probes before being acknowledged by the cell site. Automated call generators may be used to originate calls at periodic intervals to verify coverage of access and paging attempts.

19.12.6 Handoff Failure

Due to the complexity of the handoff messaging process combined with the large number of handoffs per CDMA call, handoff failures are probably the most common cause for dropped calls. Certain situations encountered in the field may cause the call processing responses to be too slow for the mobile to complete the required handoffs. As the cell site processors become more heavily loaded, either because of traffic load or overhead functions, the response to soft handoff requests will be slowed.

Another important aspect of the handoff call processing is the method used to swap pilot signals into and out of the mobile's active set. When a mobile in a three-way soft/softer handoff receives another strong pilot, with a signal strength greater than one of the active pilots, the mobile will request a swap operation. In the ideal situation, the mobile should simultaneously be allowed to drop the weakest active pilot and add the new stronger pilot. The handoff call processing may not perform a swap operation in the manner just described. If the swap operation is performed as a sequential drop-then-add operation, the entire process may take time to complete. During the waiting period before the operation is completed, it is possible for RF conditions to degrade to the point where the call may be dropped due to excessive FER.

19.12.7 Excessive Number of Strong Pilot Signals

If strong signals from more than three serving sectors are present (assuming implementation allows demodulating signals from a maximum of three serving sectors), the additional signals will act as interference sources. Another consequence of a large number of strong pilots is an increase in the time the mobile spends in three-way soft/softer handoff. Situations may develop where calls are dropped due to delayed handoff processing in any geographical area where three or more strong pilots are present.

The more strong plots there are in a given area, the higher the interference level will become. In other words, if many strong pilots exist in an area, the pilot-to-interference ratios for all of the pilots will be reduced. Therefore, it is always desirable to have one or two dominant servers in a given area, rather than three or more servers of equal strength.

To reduce pilot signal congestion is to reduce the transmit power at the weaker serving sectors in the area. The aim is to create one or two dominant servers to cover the problem area, or at least to shift the problem area out of major traffic locations. One should try to reduce the transmit power in 2-dB steps at the weakest sectors, and then check to be certain that the coverage area of those sectors has not degraded to an unacceptable level. If sufficient transmit power margin is available, it may be possible to increase the transmit power at the desired dominant serving sectors. Other options include changing antenna azimuths, heights, downtilt, and beam widths to control the pilot coverage.

19.12.8 Unrecognized Neighbor Sector

If a CDMA mobile drops calls in an area where a strong server exists, the cause will often be attributable to omissions in the neighbor list. In some cases, the problem may be due to search window sizes that are too narrow. Because CDMA systems operate with unity frequency reuse, it is important that the neighbor list include all strong pilots the mobile is likely to encounter in a given area. With a frequency reuse of one, there is very little margin for delayed handoff.

For the mobile to enter into soft handoff with a particular sector, the candidate sector must be in the neighbor list. The CDMA mobile may occasionally report pilot signals that are not in the neighbor list, but these handoff requests will be ignored by the base station. Tuning of neighbor lists is an iterative, trial-and-error approach. The basic approach is to observe which strong pilots are being reported by the mobile.

The traffic loading on the CDMA network will impact the observance of neighbor-list–related call drops. As the network load becomes light, interference levels will be reduced, resulting in more soft/softer handoff activity. The result is that neighbor list problems may be more evident under light loading conditions.

19.13 RF Optimization Tools

The following are the currently available tools for RF optimization:

- Service measurement processing tools
  - OMP analysis tools
  - WatchMark Prospect system tools
- Qualcomm mobile diagnostic monitor (MDM)
- Pilot scanner
- Expanded general system failure (EGSF) tool

OMP analysis tools can be used to provide an overall view of system performance. Measurements are gathered for time-scheduled or request-initiated reports in order to evaluate call processing performance. OMP analysis tools are used to schedule and output system performance reports.

The WatchMark Prospect system provides performance management and some aspects of configuration management for wireless systems. The WatchMark Prospect system includes modules to enhance the operation, maintenance, and engineering of CDMA system. The modules can be used to monitor system performance, report system configuration, and diagnose, track, and solve engineering and system problems.

The CDMA FER and power level measurements (PLM) collect statistical data on signal strength, signal-to-noise ratio, handoff periods, and frame error rates. This collecting capability allows service providers to specify cell sites and sectors where the system collects data over a given period of time. The data, collected in a histogram format and stored in binary files, can be viewed by processing the file through WatchMark Prospect. The histograms can be used to fine-tune CDMA operations at the base station and guide traffic engineering, RF optimization, and capacity planning.

The data to be collected reside predominantly in cell sites. Most of the information is computed by the CDMA power control mechanism. Service providers can schedule or define the individual FER and PLM through technician interface commands or directly though the OMP. The FER and PLM feature does not require hardware changes.

The PLM feature collects the following data:

- Reverse link FER on every traffic channel
- Forward link FER on every traffic channel
- Total output power on forward link
- Reverse link loading
- Pilot strength (E_c/I_t) of the primary cell before and after handoff
- Signal-to-noise ratio (E_b/I_t) of the primary cell when the call is lost
- Signal-to-noise ratio (E_b/I_t) when the call is lost
- Soft handoff period

The Qualcomm MDM is used to gather the following data from drive tests:

- Pilot sets—their relationship, number of members, and promotion and demotion
- Handoff parameters
- Handoff messaging

The pilot PN scanner is a comprehensive, CDMA network-independent measurement tool used to verify coverage, detect and characterize pilot pollution, and optimize neighbor lists and search window sizes. The pilot PN scanner features are scan plots, measurements of multipath delay spread around any PN offset across user-defined search windows, and display of detected pilots not in the current neighbor set.

The EGSF feature is a tool that benefits both the mobile subscriber and service provider. The tool enables the service provider to identify, and resolve faster and more accurately, call processing failures and fades resulting in dropped and uncompleted calls. This feature allows individual call processing failures to be analyzed down to the point of failure in the software. Call processing failures have a negative impact on mobile subscribers since their calls either did not complete, or were dropped while in the talk state.

During RF optimization, tools are used to generate *alerts* and *warnings*. The following alerts may be generated:

- Weak pilot alert
- Neighbor list alert

The following warnings may be generated:

- Unexpected T_COMP, T_ADD, T_DROP, or T_TDROP received by mobile
- Unexpected active, neighbor, and remaining set window sizes
- A list of sectors that were never active

- A neighbor or active search window size is too small
- Undefined PN offset detected by mobile
- Dropped call

When optimizing a CDMA system, the approach is to handle the most significant alerts first. After BCR values are changed and neighbor lists are updated, the process should be repeated and further changes should be made. The process is iterative, and the goal is to reduce the total number of alerts at each step. The figure of merit that should be minimized is the total number of alerts. Trying to optimize a CDMA system by only counting the number of dropped calls is not recommended because the number of dropped calls varies so much. By optimizing parameters so that the number of alerts is minimized, the dropped call rate will reduce automatically.

Plotting E_c/I_t for only those pilots associated with alerts, rather than for all the pilots in the cluster, allows one to quickly identify and diagnose the problem areas. It is important to collect enough data so that a large enough area can be optimized, because optimization of a single area can lead to problems elsewhere. For example, one should collect data along both highways within several miles of the intersection and optimize that area rather than optimizing one highway at a time.

Several mobiles can be used to log data simultaneously to speed up the data collection process. It is strongly suggested that alerts be collected during unloaded conditions because at that time the mobile will report more pilots. This will result in more neighbor list alerts. Also, it will be easier to diagnose multiple pilot problem areas because the sectors responsible for the interference in these areas, especially under loaded conditions, will be more apparent. The optimization of an unloaded system will automatically lead to a system which is optimized for full load.

Weak Pilot Alert. There are two conditions that may lead to a weak pilot alert: (1) there are one, two, or three weak pilots, and no other pilots, or (2) there are four or more weak pilots (multiple pilots or self-interference problem).

1. *One, two, or three weak pilots, and no other pilots*—This condition points to an RF coverage hole if the mobile receive power is low, or to external interference if the receive power is high (in PCS bands, interference is typically from microwave transmissions). The power of the serving sectors should be increased by reducing BCR attenuation, if possible, while at the same time considering the effect of increased interference in another part of the system. It may be necessary to make additions to neighbor lists after power is increased. It is preferable to increase the power of the pilot that will not appear as interference somewhere else in the system, if this option is available. In some cases, PCS antennas are not placed at the top of a tower and improved coverage may be achieved by placing the antennas higher up on the tower. Antenna orientation is also an important degree of freedom, especially in PCS. The orientation of antennas can be set so as to fill in coverage holes not predicted during the RF planning phase.

EXAMPLE 19.3

The following information was received from the weak pilot alert (WPA):

[–13.0] WPA A66γ (438), [–13.0] A73α (120), [–17.5], 20:26:40.208.

where values enclosed in [ ] are pilots' E_c/I_t and values in ( ) are the PN offsets. The weak pilot alert was generated because the mobile's strongest server pilot's E_c/I_t, at –13 dB, was less than the threshold of –12 dB at time 20:26:40.208. The mobile was in two-way handoff with the γ sector of cell 66 and the α sector of cell 73. Both the pilots are active at the time stamp given.

Viewing the individual E_c/I_t of these two pilots and verifying, it was concluded that the power of 66 γ pilot can be increased by reducing BCR attenuation without creating interference somewhere else.

2. *Four or more weak pilots*—This condition is referred to as a multiple pilot problem. As discussed earlier, it is better to have two strong pilots than three or four weak pilots because of problems with slow swapping and self-interference. The solution to this problem is to power down the less-dominant sectors while powering up the significant pilots through the use of BCR attenuation. A dominant sector may also be created by pointing the main beam of one sector so that it serves the multiple pilot area more effectively. Note that reducing and increasing the digital gain of several pilots can solve the multiple pilot problem when there is no load, but the problem will still be present under loaded conditions because the interference associated with interfering traffic signals will not be attenuated. Although there are several degrees of freedom, it is important that the number one pilot in the example above is not decreased because this will lead to a coverage hole.

EXAMPLE 19.4

The following information was received from the four or more weak pilots alert:

[–11.5] WPA A55α (126), [–13.5] A55γ (462), [–12.5] A 65α (24), [–11.5] 54β(204), [–10.5] / 200β, 53β (282) [–15.0], 53γ (450) [–12.5], 00:12:28.905.

The mobile is in three-way handoff with sectors 55α, 55γ, and 65α. "/" indicates that 54β and 200β both have the same PN offset.

The power of the dominant servers should be increased while the power of the less-dominant pilots should be decreased. If a pilot appears intermittently along a highway, then BCR for this sector should be reduced because this sector will not act like an ideal server and will be the interference source, especially if the mobile is in three-way handoff with three other pilots. An ideal server along a highway is one that has strong and contiguous E_c/I_t, not one with intermittently strong E_c/I_t.

19.14 Single-Carrier Optimization

In this section, we discuss how CDMA system parameter changes can achieve the best system performance. We present a few ideas on various optimization issues.

19.14.1 Soft Handoff Parameters Adjustments

The soft handoff parameters (T_ADD, T_DROP, TT_DROP, and T_COMP) can be used to improve isolated areas while under load. Most changes have an impact on soft handoff percentages, and thus, on system capacity. All handoff parameter changes should be implemented in regions, specifically including all sectors that could be primary in the area where improvements are required. The T_ADD allows new pilots entry into the active set. Two scenarios for changing T_ADD are pilot hovering and sudden pilot increase. Both scenarios can result in increased FER and dropped calls.

A *pilot hovering* is measured by the mobile station at relatively stable level just below the T_ADD parameter threshold (active pilots are generally weak, with 1–2 dB of T_ADD, the composite received signal is moderate, and additional pilot exists at signal levels just less than T_ADD). The pilot is never measured above T_ADD and cannot be added to the candidate set; thus, it becomes a strong forward link interferer. A pilot hovering results in high forward FER in the area and also some dropped calls. One possible solution is to reduce T_ADD to a value that will add the interfering pilot to the active set. This improves FER at the expense of soft handoff in the entire region; this can also adversely affect capacity.

A fast-rising pilot must be rapidly added into the active set before the high RF energy causes the call to drop. In case of *sudden pilot increase,* the additional pilot suddenly becomes very strong before it can be added to active set. This condition results in high forward FER and a strong likelihood of a dropped call. One possible solution is to reduce T_ADD to a value that will add any sudden pilot to the active set before the link degrades. This will increase soft handoff percentages in the entire region; this can also adversely affect capacity.

Sometimes pilots both hover and rise or fall suddenly. These situations are often called *pilot thrashing.* Generally, pilot thrashing is found in areas with weak signals and multiple pilots with high FER. Pilot thrashing is caused by a lack of pilot dominance; with the pilot changing frequently, pilot E_c/I_t typically less than –11 dB, and call processing decisions that are less than optimal. To correct such problems T_DROP/T_TDROP should be changed depending on the particular situation.

T_ADD and T_COMP work together to control a pilot addition to the active set. If the active set is full, a strong pilot is found, the base station evaluates the current active set members' pilot E_c/I_t, and swaps out an inferior pilot. Any changes to T_COMP must be evaluated carefully. The objective is to ensure that the strongest pilots in the area are included in the active set.

19.14.2 Access Parameters

Access parameters can be adjusted in areas of higher interference, soft coverage areas, and larger radii cells. Adjusting access probe power is usually a last option in areas of high interference. If reverse link path loss is somewhat greater than forward link path loss, then adjustment to NOM_PWR may also be needed.

Timing is very important to achieve optimum performance in the CDMA air interface. RF energy must first be detected by the mobile station, then collected by a correlator, and finally decoded. Mobile stations must be given a large enough time window to perform these tasks on all possible multipath signals arriving at the mobile stations antenna.

19.14.3 Traffic Growth

Operators of CDMA systems must manage traffic demand by responding to the following [2]:

- RF interference rises as traffic loading on cell/sectors increases, and follows customer location, population density, and demographics.
- Growth is usually not uniform and some cell clusters often grow faster than the system as a whole.
- The three primary concerns are channel element blocking, forward link loading (cell site amplifier exhaustion), and reverse link interference rise (coverage shrinkage).

If the receiver interference margin is exceeded (due to excessive loading on the reverse link), coverage will shrink and open coverage holes where calls cannot be made. To manage reverse link quality the following process should be used:

1. Monitor lost call volume and sources (i.e., analyze dropped calls and call processing failures).
2. Watch occurrence and sources if RF link load and messaging becomes less reliable.
3. Adjust target reverse FER and power control setpoint increments to respond more quickly to changes in RF environment.
4. Revisit the reverse link receiver interference margin in the design configuration:
   - Site location selections?
   - Sites too far apart?
   - Antennas too short for average terrain?
   - Antennas not effectively tilted?

19.15 Summary

In this chapter, we discussed the fundamental steps involved in the optimization of a new CDMA system. We examined the steps of the drive test process and presented the required exit criteria. We focused on a single-carrier optimization and pointed out that handoff parameter changes can be useful in improving isolated areas under loaded conditions, access parameters can be changed to improve both the forward and reverse links, search window sizes should be set depending on the particular system requirements, and CDMA system capacity and coverage are interdependent.

19.16 References

1. Kim., K., *Handbook of CDMA System Design, Engineering, and Optimization,* Prentice Hall: Upper Saddle River, NJ, 2000.
2. Holma, H., and Toskala, A., *WCDMA for UMTS*, John Wiley & Sons: New York, 2000.

CHAPTER 20

Network and Services Management

20.1 Introduction

Wireless networks, like other telecommunications networks, require large numbers of translations and other parameters to define their physical and logical configurations. For wireless systems, data changes are frequent because of system growth, handoffs, changes in patterns of subscriber density, "quasidynamic" frequency channel assignments, trunk assignments, and ongoing fine-tuning of the system to improve quality of service.

Management of configuration data is cumbersome and risky without the proper software tools. Configuration management of wireless systems must be quick and reliable to achieve efficient, cost-effective network management with a high quality of service (QoS). In the past, each network element of a wireless system had a corresponding operation support system to provide management capabilities. Each of these management systems had a different user interface, used a different computing platform, and typically managed only one type of network element. They made the network management task inefficient, complex, time consuming, and expensive for network elements manufactured by several vendors.

The field of network and services management (NSM) has taken on a new importance in the current highly competitive telecommunications environment. The days are gone when service providers strove above all for the greatest efficiency, as measured by the fraction of revenue spent for operations, administration, maintenance, and planning (OAM&P). Instead, a new era of customer service has arisen, with speed, predictability, and customer satisfaction at the forefront. NSM standards play an important role in defining the capabilities that network elements (NEs) and operations support systems (OSS) have for configuration, fault, accounting, performance, and security management. With network operators looking to keep their operating costs down while providing high-quality services on increasingly short schedules, and not being able to afford the cost or the time to implement custom solutions, they desperately need standards that are easy to implement.

Different management systems for each network element are no longer affordable. Therefore, the standard groups in ATIS committee T1M1, International Telecommunications Union (ITU), European Telecommunications Standard Institute (ETSI), Telecommunications Industry Association (TIA) committee TR-45, Telecommunications Information Networking Consortium (TINA-C), and the Network Management Forum (NMF) have defined the interfaces and protocols for management systems under the umbrella of the Telecommunications Management Network (TMN) [1–4, 5].

TIA committee TR-45.7 is working on the two main projects: (1) 2G transition wireless network management (NM) for North America, and (2) IMT-2000 global 3G wireless NM. Another effort to influence global IMT-2000 management standards is going on in the Third Generation Partnership Project (3GPP). 3GPP focuses on telecommunication management under the convention of TMN and specifies the requirements and the integration reference points for each of the management functional areas (MFA): fault management, configuration management, performance management, and accounting management. The mobile management team is using the Tele-Management Forum (TMF) business approach to recommend standards for wireless quality of service criteria and performance matrices. A 3G network consists of different networking technologies. A traditional approach to centralize the management functions requires a major effort to develop the new interface. Therefore, a distributed telecommunication management framework is more suitable for managing the different technologies. In this chapter we discuss the following topics:

1. Traditional approaches to network management
2. Telecommunication Management Network (TMN)
3. Characteristics of 3G network and service management
4. Telecommunication network management framework
5. Standards of 3G network and service management

20.2 Traditional Approaches to Network Management

Traditional NM practices involve a wide array of procedures, processes, and tools for configuration, fault, performance, security, accounting, and other management functions. The management functions rely on a *master-slave* relationship between operations systems (OSs) and NEs. Typically, network elements have only basic operations functionality with little or no ability to control activities or make decisions beyond call processing and information transmission tasks. The OSs perform the majority of the OAM&P tasks including processing raw data generated by individual network elements, making decisions, and instructing each individual NE to perform specific actions. The master-slave relationship contributes to operating inefficiencies in several ways. There is little sharing of logical resources such as data, because network elements and operations are designed independently. Also, each vendor's equipment has unique configuration and fault management interfaces with specific performance requirements. Network management systems characterize each NE and vendor's interface on an individual basis. This adds considerable time and complexity in introducing new services or technologies. Other factors have also compounded this complexity. NM systems are often designed to optimize the task of an individual service provider's organization or work group at a particular point in time for a particular technology. This type of development is often undertaken independently by individual organizations without paying much attention to system-level interworking. Multiple copies of data, each tied to specific system or job functions and to specific equipment vintages or implementations are used throughout the network, creating a major data synchronization problem. Thus, it becomes difficult for a service provider to evolve services, network technologies, and NM processes in a cost-effective, timely, and competitive fashion in response to the rapidly changing wireless networks.

In the traditional network management framework, an OSS is dedicated to a specific type of network element and an MFA. This framework can no longer handle a complicated 3G network. In the following sections, we introduce TMN and discuss the characteristics of a 3G network and service management (NSM), a management framework that is capable of supporting a 3G network.

20.3 Platform-Centered Management

With a platform-centered management model for a heterogeneous network, management applications are centralized in platforms, separated from managed data and control functions included in network elements. The platform-centered management systems require

- Standards to unify management-agent information access across multivendor network elements. These include standardization of managed information database structures at the agent as well as standardization of access management protocol.
- Standards to unify the meaning of managed information across multivendor systems.
- Standards to unify platform-processing environments across multivendor platforms.

Simple network management protocol (SNMP) [6] and OSI common management information protocol (CMIP) [7] address the first category standards while complementary efforts to address the last two standards are being pursued by other organizations such as NMF and TINA-C [8].

Network managers typically want to be able to determine timely information about configuration of devices and equipment in a network, the status of network devices and equipment, values of various statistical counters and usage of statistics, and notifications of changes of state of equipment configuration.

While database applications typically use query languages such as Structured Query Language (SQL) to obtain somewhat static information stored in a database, network management applications use network management protocols for transferring both static and transient information from network equipment, or intermediary systems, to network management systems.

Next, we discuss two available network management protocol standards:

- SNMP, developed by the Internet Engineering Task Force (IETF)
- CMIP, developed by ITU-T and ISO

20.3.1 Simple Network Management Protocol

The SNMP has been designed to be an easily implemented, basic management tool that can be used to meet short-term network management needs for computer networks. The SNMP standards provide a framework for definition of management information protocol for the exchange of that information between managers and agents. A *manager* is a software module that has the responsibility for managing part or all of the configuration on behalf of network management applications and users. An *agent* is also a software module in a managed network element with responsibility for maintaining local management information and delivering that information to a manager by polling, or by the agent using a trap.

The SNMP agent is uniquely identified by an Internet protocol (IP) address and user datagram protocol (UDP) port number. Thus, only a single agent is accessible at a given IP. The agent can maintain only a single management information base (MIB). Therefore, within a single IP address, only a single MIB instance can exist. This unique binding of an MIB to an IP address can limit the complexity of data than an agent can offer. When a system requires multiple MIBs to manage its different components, to access via a single agent, these MIBs need to be unified under a single MIB tree. Situations may arise where such unification cannot be achieved. In such cases, each MIB may require its own SNMP/UDP/IP stack, leading to a greater complexity in the organization of management information. In addition, the SNMP is not a standardized protocol of ITU, TIA, or ETSI.

20.3.2 OSI System Management

The OSI management is aimed at satisfying network management requirements in five functional areas specified by TMN [9]: fault management, accounting management, performance management, configuration management, and security management. These are referred to as system management functional areas (SMFAs). OSI includes a number of general purpose system management functions to support the SMFAs. Each functional area can be implemented as an application that relies on some subset of the system management functions. An example of SMFA is the event management that enables event forwarding and filtering.

The SMFAs depend upon CMIP for basic exchange of management information that is defined using the Guidelines for Definition of Managed Objects (GDMO) and ASN.1. CMIP provides a comprehensive set of features such as scooping, filtering, and multiple links replies to facilitate the exchange of management information between the manager and the agent. These features offer a distinct advantage to an OSI manager compared with an SNMP manager because the former has the capacity to retrieve bulk information and to retrieve selected information of interest from the agent.

OSI management uses an extended object-oriented model of managed information. The behaviors of interest such as noise, error, queue length, etc., are different forms of a time series. A generic managed object class (MOC) [10] is defined to describe general data attributes of the time series and to describe operations to compute functions of the time series. The MOC provides generic event notification that can indicate change in management variables due to error conditions being encountered. A management platform creates instances of MOCs within device agents' databases called managed object instances (MOI). The device agent monitors the management variables in the MOI. The platform, furthermore, enrolls with an agent to receive notification of error rates and excessive processor queue. OSI management organizes managed object instances in the management information tree (MIT), which is updated to reflect changes in the resources being managed. Manager-issued CMIP commands act upon object instances present in the MIT.

The OSI information model enables the definition of complex and flexible managed object sets. Distinguishing features of OSI management are inheritance; the ability to create and delete objects, thereby leading to a dynamic MIT; and the flexibility to define specific procedures that can be invoked by the manager by issuing an M-Action command.

20.4 SNMP and CMIP

Both CMIP and SNMP are defined to support the manager/agent paradigm, where a remote manager process obtains information by invoking protocol operations on an agent process associated with network devices and equipment. Both protocols share the following request/response protocol operation: GET/SET, which returns one or more requests to change one or more quantity values with an acknowledgment of success or failure.

The essential difference between CMIP and SNMP is in the way the information is defined and accessed. CMIP uses an object-oriented approach [7], where the network devices and equipment are represented as individually named managed objects (MOs). These are defined as managed object classes with attributes to represent the object's state. SNMP employs a remote debugging approach in which network devices and equipment are modeled using variables, with each variable representing its own type identifier.

The SNMP protocol is used to access variables in an MIB associated with a single network device. An MIB definition language is used to define GET and STRING, which are allowed for the syntax of variables.

The CMIP protocol can handle a large number of objects of the same type, since it uses globally unique names to access individual MOs. The attribute types of each MO are identified in a similar manner to SNMP variable types (using registered numbered IDs), but in addition, each MO is distinguished from others of its same type by its name. MOs are defined using templates specified in GDMO.

In addition to the GET and SET operations, both protocols have mechanisms for the agent to point out significant events to the manager. CMIP uses event reports that contain the name of the MO they originate from as well as an arbitrarily complex data structure carrying all the information associated with the event. SNMP uses traps that originate from the agent and carry a simple list of variable values, which stimulates the manager to use a sequence of GET operations to obtain the information associated with the event.

CMIP has the ability to define custom *actions* that have input and output parameters, performed by MOs (e.g., an action specified to set up a connection between two named termination points). CMIP also uses scooping and filtering mechanisms to access, through a single request, multiple attributes associated with a selected set of multiple MOs whose attribute values meet a criteria specified in a filter. SNMP has a concept of variable tables, which is used to get around some of the limitations of the single variable model. SNMP is considered most applicable for dealing with single network elements, whereas CMIP is most suitable in dealing with higher management levels where a single agent represents the information associated with several NEs.

20.5 Telecommunications Management Network

TMN principles in M.3010 [9] specify the architecture of TMN including the identification of the different types of nodes and the interfaces between them. Table 20.1 lists the various physical components of the TMN.

TMN uses the concept of functional architecture, which is defined in terms of function blocks and reference points. Functional blocks are the logical entities that can be implemented in a variety of physical configurations. Table 20.1 shows the functional blocks, which are mandatory for each TMN node. TMN reference points represent the exchange of information between two functional blocks. Table 20.2 gives the correspondence between TMN interfaces and reference points. Interfaces are denoted in uppercase, and reference points are designated in lowercase.

Table 20.1 TMN Nodes and Their Mandatory Functional Blocks

TMN Nodes	Mandatory Functional Blocks
Operation System (OS)	OSF
Mediation Device (MD)	MF
Q-Adaptor (QA)	QAF
Work Station (WS)	WSF
Network Element (NE)	NEF

Table 20.2 TMN Interfaces and Reference Points

TMN Interfaces	TMN Reference Points
Q_3	q_3
Q_x	q_x
F	f
X	x

20.5.1 TMN Layers

TMN divides management functions into logical layers to represent a hierarchy of system functionality using the logical layer architecture (LLA) concepts. Each higher layer provides a higher level of system functional abstraction. The five TMN layers in M3010 are defined as

1. Network element layer (NEL)
2. Element management layer (EML)
3. Network management layer (NML)
4. Service management layer (SML)
5. Business management layer (BML)

For resource management, only the first three layers are pertinent. The roles of these three resource management layers are discussed below.

Network Management Layer. The NML provides a management view of the network that is under one administrative domain. It has responsibility for end-to-end management of a network. Through the EML, it manages subnetworks or network elements based on the view presented by the EML. It can provide detailed views of the portions, or segments, of the connections within its domain. Basic responsibility of the NML is to provide network functions; for example, network traffic monitoring, network protection routing, and fault correlation and analysis from multiple network elements/nodes. NML provides applications and functionality in four principal areas:

- Control and coordination of network view of all network elements within its domain
- Provision, cessation, or modification of network capabilities for the support of service to customers
- Maintenance of network capabilities
- Maintenance of statistical log and other data about the network and interaction with service management layer on performance, usage, availability, etc.

Element Management Layer. The EML manages each network element on an individual or group basis and supports an abstraction of the functions provided by the network element layer. Network elements of a similar type are managed at this layer. Aggregating nodes into subnetworks for wireless systems is advantageous in some cases to manage equipment from the same manufacturer. The EML filters message traffic going to the NML (e.g., NE alarm correlation). It is responsible for operation on one or several network elements/nodes for example, remote operation and maintenance, remote hardware and software management, and local fault handling. The principal areas supported by EML are

- Control and coordination of a subset of the network on an individual network function basis
- Control and coordination of a subset of network elements on a collective basis
- Maintenance of statistical log and other data about network elements

Network Element Layer. The NEL performs basic management function for wireless equipment, such as detecting faults and counting errors. It provides the agent service functions in support of management request from the EML.

Classifying functions into NEL, EML, or NML does not imply any particular physical implementation; it simply provides a logical organization for the management functions. In the same way, an interface between two TMN layers is a logical interface, which may or may not be physically implemented. For example, two layers, such as EML and NEL, may reside in the same physical device (see Figure 20.1).

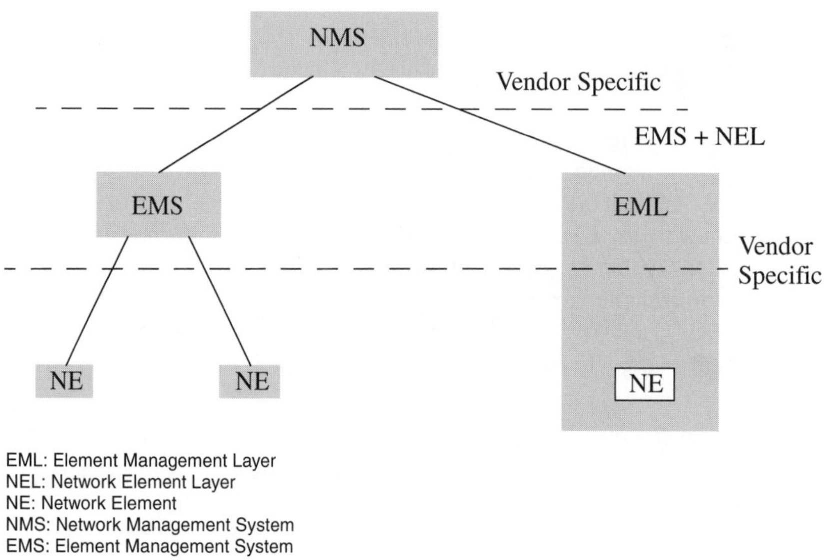

Figure 20.1 Example of Layering

20.5.2 TMN Nodes

Figure 20.2 is a simplified example of a physical TMN architecture. The OS provides the supervisory or control systems in TMN. OSs can be interconnected. Thus, OSs can form management hierarchies [11–18].

Mediation devices (MD) may provide storage, adaptation, filtering, thresholds, or condensing operations on data received from subtending equipment. They are probably the most vague component of TMN.

The Q-adapter (QA) is used to connect a TMN system to a non-TMN system. QAs may be used for integrating existing networks into TMN.

The NE is the only node that actually resides in the TMN. The primary job of an NE is to handle traffic. It is, however, the ultimate origin or destination of the management supervision and control.

The work station (WS) is an interface between a person and the TMN. It provides the presentation function to the user. It should be noted that a WS as a TMN node does not convey the same notion as the workstation of the computer world.

The nodes communicate through the data communication network (DCN), which is the transportation means used in the TMN.

Control and coordination of a subset of network elements are performed on a collective basis.

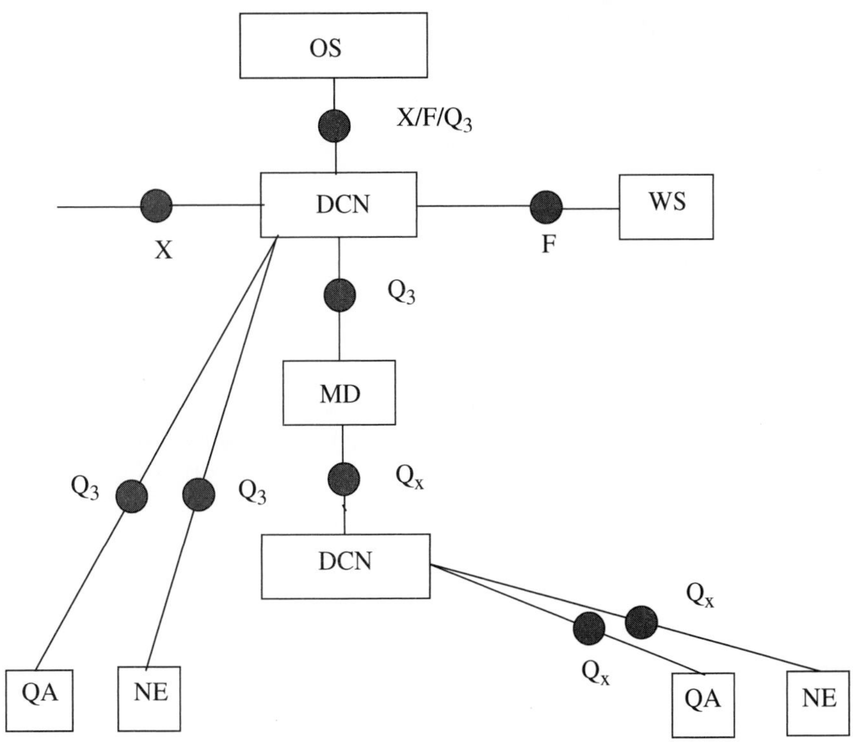

Figure 20.2 A Simplified TMN Physical Architecture

20.5.3 TMN Interfaces

The manner in which management systems interact is governed by the interfaces. Standardized interfaces allow the nodes to interwork as long as the protocols are specified to a level that allows applications to interact. The Q_3 interface connects an OS and an NE, or an OS and a QA, or an OS and an MD, or two OSs that belong to the same TMN. Q_x is like a Q_3 but with less functionality. It is intended to be used when cost or efficiency issues prevent a fully functional Q_3 interface. The X interface is used to communicate between OSs that belong to different TMNs, or between a TMN OS and a non-TMN OS that supports TMN-like interface. The F interface is used for communicating between the WS and the other nodes.

20.5.4 TMN Management Services

The Comité Consultatif International Télégraphique et Téléphonique (CCITT) divides management into five broad management functional areas to provide a framework for the determination of management service applications. The management functional areas are

- Performance management
- Fault management
- Configuration management
- Accounting management
- Security management

The management of customer access services is defined in CCITT M.3200 [19], and specifies a service related to the configuration and fault management functional areas for customer access equipment. CCITT M.3400 [20] specifies several service components for each of the relevant management functional areas.

Performance Management. Performance management is concerned with the following basic functions:

- Data collection measurement and storage
- Simple statistics showing summaries and cumulative counts
- Support of customized displays and reports for network performance information

Fault Management. Fault management deals with the following basic functions:

- GSM 12 series compliance with respect to alarm type, error ID, probable cause, severity and descriptive information
- Alarm collection and storing
- Alarm acknowledgment
- Alarm filtering and reclassification

- Alarm monitoring
- Alarm forwarding to paging and e-mail
- Scheduled loop testing

Configuration Management. The basic functions of configuration management are as follows:

- Capabilities to send files with many prespecified configuration commands to NEs
- Objects creation and deletion
- Logical interconnection schemes to show various network elements, subsystem, and functional blocks at each site
- Software management
- New features/upgrades

Security Management. Security management, the handling of access authorizations, is of real significance. A series of measures are generally used to prevent unauthorized access. This is achieved both using user- and device-related access authorization mechanisms included in the software and smart cards with coded user identifiers. Some of the tasks performed in security management include

- Password authentication of OMC and NMS users
- Logging of OMC/NMS attempts
- Automatic log-off

Accounting Management. Accounting management deals with the collection of billing information for the services provided by a network. Billing is a critical process as it delivers all the revenue to the service providers. In the GSM, there is a relationship between each public land mobile network (PLMN) and other networks to which it interconnects. These networks may be national or international carriers and may even be fixed wireline. Each PLMN has to negotiate with other networks with which they interconnect as to the basis of sharing revenue. In all cases, an accounting relationship is established between PLMNs and other networks.

The TMN has been used as a framework to organize management requirements, which should be implemented using the most efficient choice among standardized interfaces and protocols. Interfaces between equipment may use different protocols, including CMIP and SNMP, and may be structured and related in an architecture such as the Common Object Request Broker Architecture (CORBA) [21].

Figure 20.3 summarizes TMN key functions and standards, and provides one view of TMN architecture. The network element layer contains those things that carry out the actual activities of telecommunications. In the layer above are categories of functionality to monitor and control the telecommunications network.

TMN architecture concepts enable vendors and service providers to communicate their needs and expectations along several dimensions:

- Functional architecture identifies the functional blocks (OS function block, NE function block, WS function block, etc.) that collectively provide fault, configuration, accounting, performance, and security management. OS functionality is further divided into logical layers: business, service, network, and element management.
- Information architecture includes information elements and interaction models (manager/agent, client/server, etc.) and the information exchanged between them.
- Physical architecture defines the physical elements (OS, NE, WS, MD, etc.) and interfaces between them (OS-OS Q or X interfaces, OS-NE Q interfaces, etc.) that instantiate particular functional and information architecture specifications (see Figure 20.4).

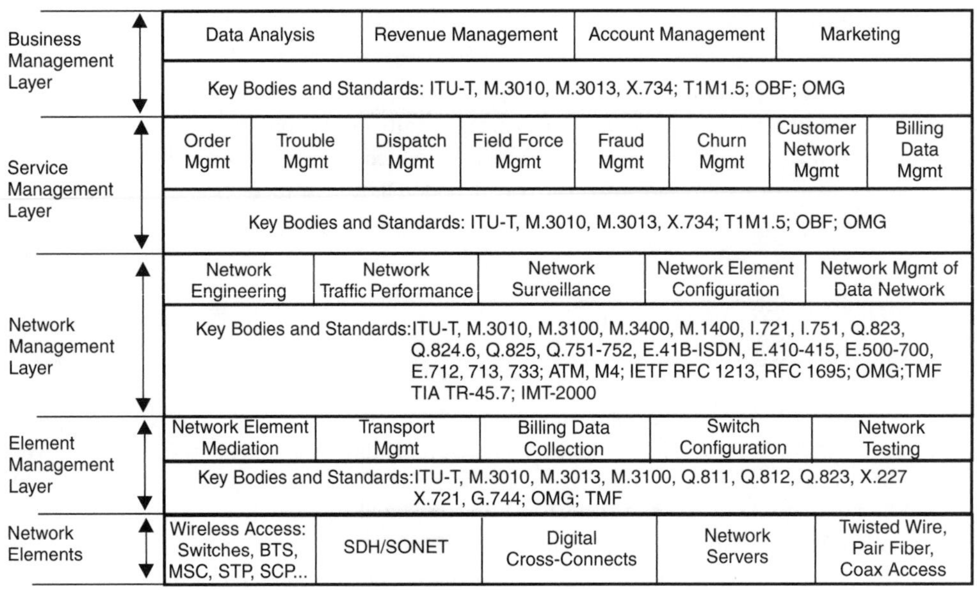

Figure 20.3 TMN Key Functions and Key Standards

M.3010 describes a TMN, and M.3013 discusses how to implement it. M.3010 and M.3013 clarify the roles of the functional and informational architecture, and how they can be realized by any number of specific physical architectures which may be distributed in various ways. These clarifications allow CORBA [19] and potentially other information technologies to develop efficient products.

The Universal Mobile Telecommunications System (UMTS) management architecture is largely based on TMN, and will reuse those functions, methods, and interfaces that are already defined and suitable to the management needs of UMTS. However, the UMTS management needs to explore other widely deployed management concepts, because UMTS incorporates other technologies with which TMN is not applicable. In addition, UMTS faces many new challenges that TMN does not address.

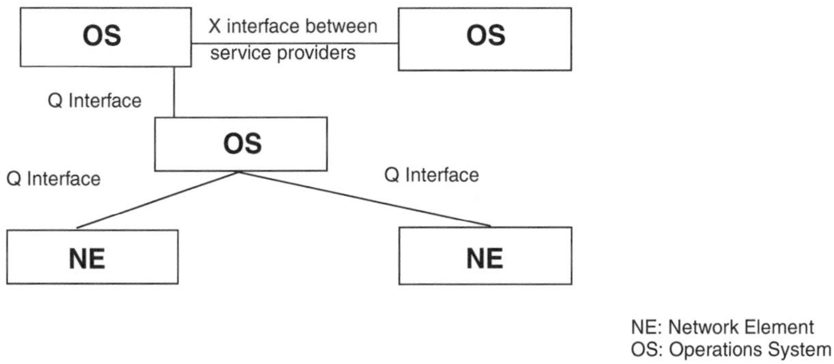

Figure 20.4 Physical Architecture for X and Q Interfaces

20.6 Common Object Request Broker Architecture (CORBA)

CORBA [19, 22, 23] is a set of standards developed and promoted by the Object Management Group (OMG). CORBA is the OMG answer to the need for interoperability among the rapidly proliferating number of hardware and software products.

CORBA is part of the TMN standard. The CORBA IIOP (Internet Inter ORB Protocol) is identified in Q.812 as one of the available protocols to use at certain interfaces. It is accepted at the X interface and is being evaluated, using a set of fitness for use criteria for use at the Q_3 interface.

CORBA provides interoperability via an object bus, which is a facility to communicate between software objects. The CORBA object bus is implemented by an ORB, which establishes the connection between the client and server. When a client invokes an operation on a server object, the ORB intercepts the request, locates an object that can carry out the request, and passes the parameters to the server and invokes the operation, then returns the results to the client. The client and the server are required to understand data only in terms of structure, which each can understand in its own language. The structure and available operations are specified in the interface definition language (IDL). CORBA objects can be invoked without knowing where they are located or what language they are implemented in.

At its core, the CORBA architecture is relatively straightforward. CORBA is a technology that offers a wide range of options and flavors. There are application-specific configurations, mechanisms for adapting CORBA to access information through foreign object systems such as the Distributed Component Object Model (DCOM), and many partially developed specifications for services and CORBA facilities. CORBA is as likely to be used as many other management standards in the telecommunications software development community, including the TMN standard GDMO.

As discussed earlier, TMN defines standardized interfaces for the exchange of information using the manager-agent approach with CMIP. This is fine when two systems that are communicating with each other have developed the application programming interface (API) and the two systems never change. Otherwise, the expenditure is quite high because the traditional approach for application integration involves the costly and time-consuming process of writing customer code to link each application to another. The systems are constantly undergoing changes and the scope of data each system has to deal with has undergone expansion in several areas including number of services, distribution, and range of users and applications. This is a problem for traditional TMN architecture, and provides an opportunity for software engineers to respond to the challenge. CORBA overcomes some of the difficulties mentioned, and is easy to use, less expensive, and more common in the commercial world.

In the most basic form, CORBA consists of a set of mechanisms for coordinating remote method invocations between clients and objects. However, there are a range of other services and architectural concepts outlined in OMG's CORBA architecture. These include

- **Basic object services**—These services include domain-independent functions for use by a wide range of programs. These services provide mechanisms that locate and categorize objects across networks of ORBs that might be used to provide required services to a user. These directory services allow users to locate objects based on either object names or object properties. These services can be found in the OMG Common Object Services Specifications.
- **Common facilities**—These facilities include a set of common application services that can be referenced from many programs. Common facilities are located in the application domain and can include high-level services such as document or data management.
- **Domain interfaces**—These are similar to the concepts applied in the common facilities and services, but are more in line with specific application domain. For example, the services and objects specified by TINA-C fall into the telecommunications domain.
- **Application interface**—CORBA is often used to develop systems in which the same vendor has developed both the client and object server side of the application. In this case, the application-specific interfaces are developed to serve specific application needs.

Much of the CORBA architecture's success is dependent on its adoption by vendors. Considerable progress has been made in the architecture's more basic areas. As various application suites are identified, constructed, and adopted by customers, more progress can be made in realizing these architectural goals.

There are a number of factors that come into play in the CORBA information and control systems' architectures. Some of the main components are those at the ends of each interaction—the clients and objects. The system's components include

- **Clients**—Clients generate requests and access object services through a variety of mechanisms provided by the underlying ORB.
- **Object implementations**—These provide the services requested by various clients in the distributed system. One benefit of CORBA architecture is that both clients and

object implementations can be written in any number of programming languages and can still provide the full range of required services. C, C++, Smalltalk, Ada, and Java have all been used to implement CORBA applications

- **The ORB**—One of the main products of CORBA, the ORB implements the mechanisms that make the object implementation services appear to the client as if the client was directly calling object services on the same machine. These services are similar to those provided by the distributed computing environment (DCE) remote procedure call (RPC) mechanisms. The ORB is able to take client requests, locate the object implementation, deliver the request, and translate and return any responses back to the original client.
- **Dynamic invocations**—Known as the dynamic invocation interface (DII), this allows the client to call the ORB's primitive request mechanisms directly. It provides added flexibility by providing clients with a set of mechanisms that permit the use of calls, avoiding some of the constraints of the blocking nature of IDL-based RPC mechanisms. Using the DII, clients can separate the request transmissions from the receipt, and issue one-way commands (for which there is no response).
- **Interface definition language (IDL) stubs**—IDL stubs provide a set of mechanisms to extract the ORB core functions into direct RPC mechanisms that can be used by the end-client applications. These stubs make the combination of the ORB and remote-object implementations appear as if they were tied to the same in-line process. In most cases, IDL compliers generate language-specific interfaces between the client and object implementations.
- **The ORB interface**—This interface is a set of generic functions that can be used by both clients and object implementations. These functions include facilities for managing some of the primitive elements of the various interactions occurring between applications and the ORBs themselves. This interface has been standardized in order to further abstract ORB vendor-specific implementations and allows easy porting between different ORB implementations.
- **Static IDL skeleton**—These components are the server-side complements to the client-side IDL stubs. They include the bindings between the ORB core and the object implementations that complete the connection between the client and object implementation.
- **Dynamic skeletons**—Similar to the relationship between IDL stubs and static IDL skeletons, the dynamic skeleton interface (DSI) completes the DII connection to the object implementation.
- **Object adapter**—These adapters provide the final mappings between the ORB and the specific objects. There are a range of object adapters that can be developed to perform particular applications, including database or remote management services.

These major functions provided within the set of CORBA implementations have a wide range of idiosyncrasies and can be constructed in any number of architectures. Each of these functions comes with a rich set of features that needs to be understood.

The challenges in implementing systems solutions using CORBA can range from trivial to complex. Developing services such as home-brewed complementary client and object implementation can be simple, particularly with one of the more mature tool sets. Larger, multivendor management systems can be constructed with CORBA, but will still require a good understanding of the system's complete dynamics.

20.6.1 TINA-C

A related network management standardization group that is considering using CORBA is the TINA-C. TINA-C is developing a set of distributed object-oriented telecommunications network management services. The TINA-C–promoted architecture (known as distributed processing environment [DPE]) contains an architecture of component-based objects that can be used to create tailored network management solutions when interconnected. By decoupling the specifics of the individual software components, changes can be made to isolated portions of the architecture without requiring extensive overhauls of the remaining portions of the system. At the DPE architecture's core, CORBA services provide mechanisms that make these component-based abstractions possible.

The TINA-C architecture is based on the ability to abstract services provided by different segments of the network infrastructure so that improvements or changes to one portion have a limited impact on other portions of the system architecture. The three major components included in TINA-C are

- **Computing architecture**—This defines the modeling concepts used in DPE. TINA-C has selected CORBA to provide these services.
- **Service architecture**—This defines the basic principles applied to the delivery of technical services within the constraints of the architecture.
- **Network architecture**—By abstracting the definition and interfaces to the system components, TINA-C provides a generic, technology-independent model for managing telecommunications networks.

TINA-C can help integrators and owners of large telecommunications networks adapt and grow as new technologies are introduced into their networks. Signs of this evolution are taking place in some of today's network control facilities. Many facilities are full of separate computing systems broken into pieces directly across vendor and technology lines. This lack of organization can present significant challenges to the coordination of configuration and fault management activities. With an integrated system, management overhead can shrink, while the ability to react to changes greatly improves.

CORBA is an architecture to define a set of services and mechanisms for the interconnection of distributed objects. Developers specify a set of external interfaces to the objects in their software components through an IDL. These IDL files define the specifics of the calling interface between client applications and the object implementations that can be accessed through the system's ORB.

With this simple architecture, separate software components (such as programs and applications) can be combined to form tailored applications suites. These newly integrated applications can then be tied together to provide a wide range of custom services that enable efficient management operation.

20.7 Management Requirements for Wireless Networks

In a wireless network where both users and terminals are mobile rather than fixed, mobility management plays an important role. Radio resources must be managed that do not exist in a wireline network. Furthermore, the wireless network can be a cellular system where radio resources and switching resources are owned and managed by one company. A second, newer model being used for a personal communication services (PCS) network has a wireline company performing the standard switching functions and a wireless company providing the radio-specific functions. With these new modes of operation, network management must include standard wireline requirements and new requirements specific to a cellular PCS system.

20.7.1 Management of Radio Resources

The cellular/PCS network allows terminals to connect via radio links. These links must be managed independently of the ownership of the access network and the switching network. The cellular/PCS network may consist of multiple service providers or one common service provider. Multiple service providers' environments will require interoperable management interfaces.

20.7.2 Personal Mobility Management

Another important aspect of a cellular/PCS network is the management of personal mobility. Users are no longer in a fixed location but may be anywhere in the world. They may be using wireline or wireless terminals to place and receive calls. In some cases, the network will determine the user's location automatically; in other cases, the user may need to report his or her location to the network. This will increase the load on management systems as subscribers manage various decision parameters about their mobility; for example, they may request different services based on the time of day or terminal busy conditions.

20.7.3 Terminal Mobility

The primary focus of a cellular/PCS network is to deliver service via wireless terminals. These terminals may appear anywhere in the worldwide wireless network. Single or multiple users may register on a wireless terminal. The terminal management functions may be integrated with the user management function or may be separate from the user management function. Different service providers may choose to operate their system in a variety of modes. The management functions must support all modes.

20.7.4 Service Mobility Management

Service mobility is the ability to use vertical features (e.g., call hold, call forwarding, etc.) in a transparent manner from remote locations or while in motion. As an example, the user should have access to the messaging service any where, any time. The ability to specify an event is provided via a user interface that is flexible enough to support a number of input formats and media. A user can specify addressing in a simple, consistent manner no matter where he or she is.

20.8 Operations, Administration Maintenance, and Planning Strategy in 3G systems

The aim of OAM&P strategy in 3G systems is to maximize the quality of service offered to the end customers of the system by

- Minimizing downtime through robust redundancy schemes and fast automatic recovery from hardware and software faults.
- Providing a set of features to maximize the efficiency of the operations staff assigned to monitor and control the system.
- Supporting a complete set of performance measurements that enables the efficient configuration and engineering of the system.
- Providing fault detection and localization facility at the network operations center or a centralized location. This is to ensure that a technician's visit to the field will replace faulty replaceable units in minimum time.

20.9 Third-Generation Partnership Project

The 3GPP produces technical specifications for a 3G mobile system based on the evolved GSM core networks and the radio access technologies supported by 3GPP partners [24]. 3GPP is not a legal entity and is not affiliated with ITU-T. 3GPP SA5 is a technical specification group (TSG) service and systems aspect (SA) working group that is responsible for defining the standard for the telecommunication management. 3GPP is defining a framework for the 3G telecommunication management. It defines the integration reference point (IRP) for protocol-neutral specification. In defining the IRPs, information models have been separated from protocol.

- **Information models** are specified with an implementation-neutral modeling language. The Unified Modeling Language (UML) has been selected.
- **Solution sets,** i.e., mapping of information models to one or several protocol (CORBA/IDL, SNMP/SMI, CMIP/GDMO etc.). Different protocols may be used for different IRPs.

The following IRPs have been defined:

- **Performance data IRP**—This is used to monitor performance information.

- **Alarm IRP**—This defines alarm surveillance aspects of fault management and an interface to allow for communicating alarm information to an IRP manager.
- **Basic configuration management IRP**—This is an interface to retrieve configuration management information and definition of a 3G management network resource model. It is needed for management of topology and logical resources in the network. It can also be used by inventory management applications to track individual equipment and related data, as well as for all types of configuration management (e.g., service activation applications, provisioning interface for frequent configuration activities). This IRP defines an IRP information model to cover both an IRP information service and a network reference model.
- **Notification IRP**—Since alarm IRP, performance IRP, and basic configuration IRP all have similar needs to use notification, the corresponding service is formalized as a notification IRP that defines a procedure to allow an IRP agent to send notifications to an IRP manager.

3GPP published a set of TSG 32 series related to telecommunication management in the 1999 release. The following specifications define a framework for 3G telecommunication management and cover four management functional areas:

- 3G TS32.101: 3G Telecom Management Principles and High Level Requirements
- 3G TS 32.102: 3G Telecom Management Architecture
- 3G TS 32.111: Fault Management, including Alarm IRP
- 3G TS 32.106: Configuration Management, including Notification IRP and Basic Configuration IRP
- 3G TS 32.104: 3G Performance Management
- 3G TS.105: 3G Charging
- 3G TS 32.005: 3G Call and Event Data for the Circuit Switch Domain
- 3G TS 32.015: 3G Call and Event Data for the Packet Switch Domain

3GPP standards is for GSM/UMTS whereas other groups are working on the standards applicable to both wired and wireless telecom networks. They are T1M1.5/ITU-T SA4 and T1M1.5/ITU-T. These organizations are defining a CORBA framework to allow CORBA to support TMN/CMIP protocol. The framework is documented in the draft ITU-T X.780 and ITU-T Q.816.

20.10 Characteristics of 3G Network and Service Management

In the 3G telecommunication management standard of 3GPP (TS 32.101-V3.3.0), the requirements and decomposition of 3G telecommunication management are not very different from those of 2G systems. The characteristics of 3G NSM can be explained with the following requirements:

- It should have capability of managing equipment supplied by different vendors.
- The management architecture should allow for easy creation of service or business management.
- The complexity of 3G NSM should be minimized.
- It should provide integrated management of MFAs (i.e., fault, configuration, accounting, performance, and security management).
- It should allow for the interoperability between network operators/service providers for the exchange of management information.
- It should reuse existing relevant standards where applicable to minimize the cost of managing a 3G network.
- It should emphasize and support new security requirements.
- It should simplify the operation and maintenance tasks.
- The platform-centered management should be used consistently across the 3G NSM infrastructure.
- It should provide and support a flexible billing and accounting administration and should support charging across UMTS and non-UMTS systems.

A 3G system consists of multiple technologies and different types of services. In a 3G system new services can be created rapidly. The architecture of a 3G system should allow for these new services to be created more easily. One approach is to provide an API across all the functional layers of a 3G network architecture. Thus, a third-party vendor can use the API to develop the services and management for the 3G system. One example of such an approach is defined in the Mobile Wireless Internet Forum (MWIF) layered functional architecture for OAM&P, in which APIs are provided across three functional layers, namely the service core layer, the control core layer, and the transport layer.

A 3G system operates under multiple networking technologies and uses products from different vendors, making 3G NSM more complicated. In order to minimize the complexities of the 3G NSM, it is obvious that the management system should follow the following rules:

1. Interface to a managed system at a network management or higher layer, leaving the element management and lower layer functions at vendor equipment and making these functions available to the network management or higher layers as needed.
2. Reuse the information model currently defined for CMIP or SNMP. CMIP has been used in the telecommunication networks and SNMP has been used to manage IP networks. Since a 3G network consists of both types of networks, the reuse of the existing information models would greatly simplify the management tasks.
3. Reuse the principles, physical architecture, and logical architecture described in the TMN standards.

The integration of 3G NSM from multiple technology environments is a big challenge. The traditional centralized NSM approach is discouraged for the following reasons:

- It is hard to scale when managing a large system.
- It is hard to achieve reliability at a lower cost.
- More effort is needed to provide the extensibility for a 3G management system, including interoperability, flexibility, and reusability of network components.

With the distributed NSM, a new system or component can easily be added to the network without the need for a centralized NSM manager to develop a new interface. Since the network management capabilities are distributed, a single point of failure is avoided and the reliability is greatly enhanced. Interoperability is required to deal with the interface between internal systems as well as with the external systems. Flexibility is needed to replace or create a new system, and reusability addresses how much software can be re-used when a system is built or replaced. These functionalities become transparent under a distributed NSM environment. Security management in the 2G networks does not have an important role because each service provider owns its private network. Since in a 3G network, the Internet will act as a backbone network, the security management is very critical as compared to the 2G systems. 3G NSM will implement the management processes based on the widely accepted telecommunication operations map from the TeleManagement Forum which maps the processes onto the service and network management layers as defined in the ITU-T recommendation M.3010 Appendix II. The management functions as defined by TMN have to be expanded in order to apply them to these processes. These management functions are

- Fault management
- Configuration and inventory management
- Performance and quality of service management
- Accounting management
- Security management
- Fraud management
- Software management
- Service management
- Roaming management
- Customer profile management

One can determine the functional architecture for certain management functions by identifying only those processes that relate to the management function through the telecommunication operation map. The description of these processes and management functions is given in the telecommunication operation map document developed by TMF [25]. They map onto the service and network management layers of ITU-T recommendation M.3010, Appendix II (see Figure 20.5).

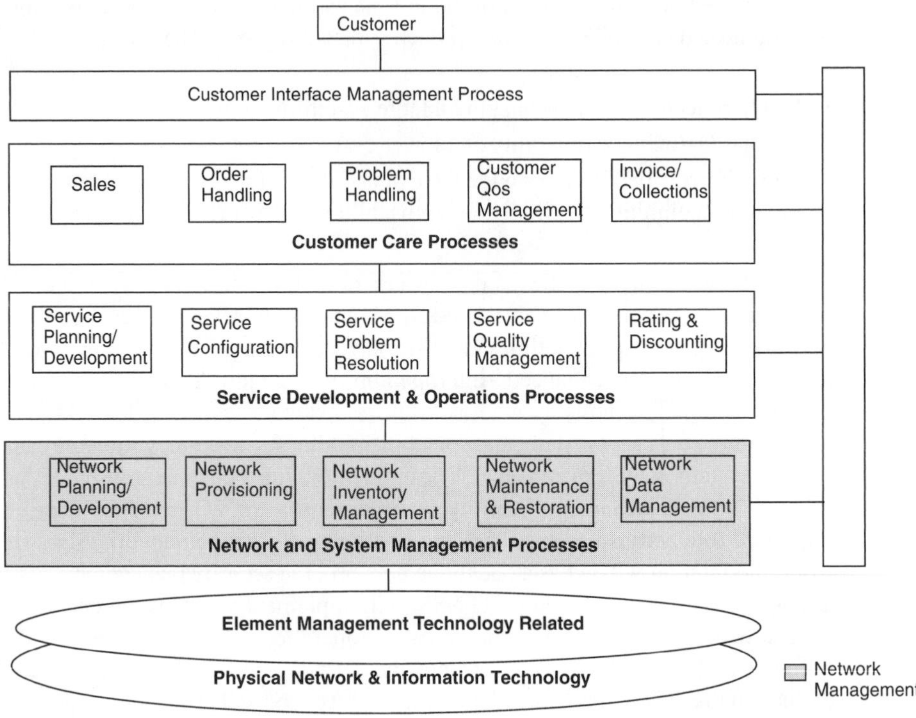

Figure 20.5 Telecommunication Operations Map Business Process Model

The management of 3G puts several new requirements on the management systems as compared with 2G. Some of the requirements that affect physical architecture are

- Capability of managing equipment supplied by different vendors
- Flexible configuration management capabilities to allow rapid deployment of services
- Accessibility to information
- Scalability and applicability to both large and small systems
- Interoperability between network operators/service providers for the exchange of management/charging information
- Events and reactions reporting capability in a common way to allow remote control
- Telecommunication management (TM) automation in a cost-efficient fashion

Many of the new requirements for the management of a 3G system can be achieved only by defining and establishing a suitable physical architecture. Because, it is not feasible to standardize one single 3G TM physical architecture, it is obvious that the success of a telecommunication management network of a 3G system will depend heavily on critical physical architectural issues.

In the UMTS, the following common NE management domains and interfaces have been identified:

- I_{tf}-N between the NE OSFs and NM/SM OSFs (see Figure 20.6). This interface could be used by the network and service management systems to transfer management messages and notification and service management requests via the NE OSF to NEs. This is an open interface with the standardized information model.

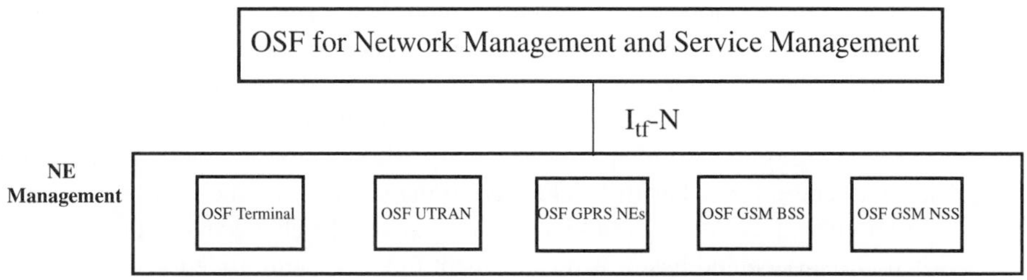

Figure 20.6 Overview of UMTS Telecommunication Management Domains and I_{tf}-N

Telecommunication management interfaces deal with the management model and the management information exchange. The management tasks vary among different NEs in the UMTS. Some NEs have high complexity (e.g., RNC), whereas others (e.g., a border gateway) are less complex. Different application protocols should be selected to satisfy the management requirements of the different NEs and technology used.

In a TMN, the necessary physical connection (e.g. circuit-switched or packet-switched) may be offered by communication paths constructed with different network components (e.g., dedicated lines, packet-switched data network, ISDN, common channel signaling network, PSTN, LAN, terminal controller, etc.). In the extreme case, the communication path provides for full connectivity, i.e., each attached system can be physically connected to all other systems. The TMN would be designed to have the capability to interface with several kinds of communication paths to ensure that a flexible framework is provided to allow an efficient communication.

Two aspects impact cost: (1) actual cost to transport data across the network between the TMN and the NE, and (2) the design of the interface including the selection of an appropriate communication protocol. Any standardized networking protocol that is capable of satisfying the functional and operational requirements of the logical and application protocol of a given UMTS management interface, is a valid protocol for that interface. The networking protocol should satisfy the following requirements:

- Capability to run over any bearer (leased lines, X.25, ATM, frame relay, etc.)
- Support of existing transport protocols and their applications, such as OSI, TCP/IP
- Widely available, cheap, and reliable

The IP ideally supports these requirements and adds flexibility to management connectivity.

20.11 Summary

In this chapter, we outlined the requirements to manage a wireless network. We discussed the SNMP- and OSI-based management schemes and presented the limitations of the SNMP in managing a wireless network. A brief discussion of the TMN architecture along with TMN layers, nodes, and interfaces was given. We also discussed the roles of five management functional areas specified in the TMN.

For years, the Internet, mobile communication, and broadband communication technologies were developed independently. With the convergence of video, voice, and data networks with wireless communication, the 3G network has been the center of various network technologies and applications. The 2G wireless network focused only on monitoring and configuring network elements. The 3G management framework should be able to handle a converged network that provides many different applications. A traditional network element-centric focus has to be changed to the service- and behavior-oriented focus. The bottom-up architecture in the TMN hierarchy should be changed to a top-down architecture where the service and business management functions will drive the network and element management functions. The new NSM framework will integrate the existing management framework at lower layers of TMN hierarchy and will provide new tools/applications at the service and business management layers.

20.12 References

1. ISO/IEC 7498-4, "Information Processing Systems—Open System Interconnection—Basic Reference Model—Part 4 Management Framework."
2. CCITT Rec. X.701/ISO/IEC 10040:1992, "Information Technology—Open Systems Interconnection—System Management Overview," 1989.
3. CCITT Rec. X.720/ISO/IEC 10165-1 "Information Technology—Open System Interconnection—Structure of Management Information: Guidelines for the Definition of Management Information Model," 1992.
4. CCITT Rec. X.721/ISO/IEC 10165-2 "Information Technology—Open System Interconnection—Structure of Management Information: Generic Management Information," 1992.
5. CCITT Rec. M.3100: "Generic Network Information Model," 1992.
6. Stallings, W., *SNMP, SNMPv2, and RMON: Practical Network Management*, 2nd ed., Addison-Wesley: Reading, MA, 1996.
7. Raman, L., "OSI Systems and Network Management," *IEEE Personal Communications*, March 1998, [Vol. 36, pp. 46–53].
8. Bloem, J., et al., "The TINA-C Connection Management Architecture," Proc. TINA'95, Melbourne, Feb. 1995, pp. 485–494.
9. CCITT Rec. M.3010: "Principles for a Telecommunications Management Network," 1992.

10. CCITT Rec. X.722/ISO/IEC 10165-4 "Information Technology—Open System Interconnection—Structure of Management Information: Guidelines for the Definition of Managed Objects," 1992.
11. CCITT Rec. X. 738: "Information Technology—Open Systems Interconnection—Systems Management: Summarization Function," 1992.
12. CCITT Rec. X.739: "Information Technology—Open Systems Interconnection—System Management: Work Load Monitoring Function," 1992.
13. CCITT Rec. Q.821: "Q_3 Interface for Alarm Surveillance," 1993.
14. CCITT Rec. X.733: "Information Technology-Open Systems Interconnection—Systems Management: Alarm Reporting Function," 1992.
15. CCITT Rec. Q.731: "Information Technology—Open Systems Interconnection—System Management: State Management Function," 1993.
16. CCITT Rec. X.734: "Information Technology—Open systems Interconnection—Systems management: Event Report Management Function," 1993.
17. GSM 12.00: "Objectives and Structure of GSM PLMN Management," 1994.
18. GSM 12.01: "Common Aspects of PLMN Network Management," 1994.
19. CCITT Rec. M.3200: "TMN Management Services: Overview," 1992.
20. CCITT Rec. M.3400: "TMN Management Functions," 1992.
21. Potonniee, O., et al., "Implementing TMN using CORBA Object Distribution," IPIP/IEEE MMNS97, Montreal, July 1997.
22. Pavon, J., et al., "CORBA for Network and Service Management in the TINA Framework," *IEEE Personal Communications*, March 1998 [Vol. 36, pp. 72–79].
23. Vinoski, S., "CORBA: Integrating Diverse Applications Within Distributed Heterogeneous Environment," *IEEE Personal Communications*, Feb. 1997, [Vol. 35, pp. 46–55].
24. 3rd Generation Partnership Project; Technical Specification Group Services and System Aspects; 3G Telecom Management Architecture (Release 1999) 3G TS 32.102 v3.2.0 (2000-07).
25. 3G TS 32.101: "3G Telecom Management Principles and High Level Requirements."

APPENDIX A

Traffic Tables

This appendix[1] provides traffic tables for a variety of blocking probabilities and channels. The blocked calls cleared (Erlang B) call model is used. In Erlang B, we assume that when traffic arrives in the system, it is either served with probability from the table, or is lost to the system. A customer attempting to place a call will therefore either see a call completion or will be blocked and abandon the call. This assumption is acceptable for low-blocking probabilities. In some cases, the call will be placed again after a short period of time. If too many calls reappear in the system after a short delay, the Erlang B model will no longer hold.

In Tables A.8 and A.9, where the number of channels is high (greater than 250 channels), linear interpolation between two table values is possible. We provide the deltas for one additional channel to assist in the interpolation.

1. The data in the tables appear through the courtesy of V. H. MacDonald.

Table A.1 Offered Loads (In Erlangs) for Various Blocking Objectives: According to the Erlang-B Model—System Capacity from 1–20 Channels

Channels P(B) =	0.01	0.015	0.02	0.03	0.05	0.07	0.1	0.2	0.5
1	0.010	0.015	0.020	0.031	0.053	0.075	0.111	0.250	1.000
2	0.153	0.190	0.223	0.282	0.381	0.471	0.595	1.000	2.732
3	0.455	0.536	0.603	0.715	0.899	1.057	1.271	1.930	4.591
4	0.870	0.992	1.092	1.259	1.526	1.748	2.045	2.944	6.501
5	1.361	1.524	1.657	1.877	2.219	2.504	2.881	4.010	8.437
6	1.913	2.114	2.277	2.544	2.961	3.305	3.758	5.108	10.389
7	2.503	2.743	2.936	3.250	3.738	4.139	4.666	6.229	12.351
8	3.129	3.405	3.627	3.987	4.543	4.999	5.597	7.369	14.318
9	3.783	4.095	4.345	4.748	5.370	5.879	6.546	8.521	16.293
10	4.462	4.808	5.084	5.529	6.216	6.776	7.511	9.684	18.271
11	5.160	5.539	5.842	6.328	7.076	7.687	8.487	10.857	20.253
12	5.876	6.287	6.615	7.141	7.950	8.610	9.477	12.036	22.237
13	6.607	7.049	7.402	7.967	8.835	9.543	10.472	13.222	24.223
14	7.352	7.824	8.200	8.803	9.730	10.485	11.475	14.412	26.211
15	8.108	8.610	9.010	9.650	10.633	11.437	12.485	15.608	28.200
16	8.875	9.406	9.828	10.505	11.544	12.393	13.501	16.807	30.190
17	9.652	10.211	10.656	11.368	12.465	13.355	14.523	18.010	32.181
18	10.450	11.024	11.491	12.245	13.389	14.323	15.549	19.215	34.173
19	11.241	11.854	12.341	13.120	14.318	15.296	16.580	20.424	36.166
20	12.041	12.680	13.188	14.002	15.252	16.273	17.614	21.635	38.159

Table A.2 Offered Loads (In Erlangs) for Various Blocking Objectives: According to the Erlang-B Model—System Capacity from 20–39 Channels

Channels P(B) =	0.005	0.01	0.015	0.02	0.03	0.05	0.07	0.1
20	11.092	12.041	12.680	13.188	14.002	15.252	16.273	17.614
21	11.860	12.848	13.514	14.042	14.890	16.191	17.255	18.652
22	12.635	13.660	14.352	14.902	15.782	17.134	18.240	19.693
23	13.429	14.479	15.196	15.766	16.679	18.082	19.229	20.737
24	14.214	15.303	16.046	16.636	17.581	19.033	20.221	21.784
25	15.007	16.132	16.900	17.509	18.486	19.987	21.216	22.834
26	15.804	16.966	17.758	18.387	19.395	20.945	22.214	23.885
27	16.607	17.804	18.621	19.269	20.308	21.905	23.214	24.939
28	17.414	18.646	19.487	20.154	21.224	22.869	24.217	25.995
29	18.226	19.493	20.357	21.043	22.143	23.835	25.222	27.053
30	19.041	20.343	21.230	21.935	23.065	24.803	26.229	28.113
31	19.861	21.196	22.107	22.830	23.989	25.774	27.239	29.174
32	20.685	22.053	22.987	23.728	24.917	26.747	28.250	30.237
33	21.512	22.913	23.869	24.629	25.846	27.722	29.263	31.302
34	22.342	23.776	24.755	25.532	26.778	28.699	30.277	32.367
35	23.175	24.642	25.643	26.438	27.712	29.678	31.294	33.435
36	24.012	25.511	26.534	27.346	28.649	30.658	32.312	34.503
37	24.852	26.382	27.427	28.256	29.587	31.641	33.331	35.572
38	25.694	27.256	28.322	29.168	30.527	32.624	34.351	36.643
39	26.539	28.132	29.219	30.083	31.469	33.610	35.373	37.715

Table A.3 Offered Loads (In Erlangs) for Various Blocking Objectives: According to the Erlang-B Model—System Capacity from 40–60 Channels

Channels P(B) =	0.005	0.01	0.015	0.02	0.03	0.05	0.07	0.1
40	27.387	29.011	30.119	30.999	32.413	34.597	36.397	38.788
41	28.237	29.891	31.021	31.918	33.359	35.585	37.421	39.861
42	29.089	30.774	31.924	32.838	34.306	36.575	38.447	40.936
43	29.944	31.659	32.830	33.760	35.255	37.565	39.473	42.012
44	30.801	32.546	33.737	34.683	36.205	38.558	40.501	43.088
45	31.660	33.435	34.646	35.609	37.156	39.551	41.530	44.165
46	32.521	34.325	35.556	36.535	38.109	40.545	42.559	45.243
47	33.385	35.217	36.468	37.463	39.063	41.541	43.590	46.322
48	34.250	36.111	37.382	38.393	40.019	42.537	44.621	47.401
49	35.116	37.007	38.297	39.324	40.976	43.535	45.654	48.481
50	35.985	37.904	39.214	40.257	41.934	44.534	46.687	49.562
51	36.856	38.802	40.132	41.190	42.893	45.533	47.721	50.644
52	37.728	39.702	41.052	42.125	43.853	46.533	48.756	51.726
53	38.601	40.604	41.972	43.061	44.814	47.535	49.791	52.808
54	39.477	41.507	42.894	43.999	45.777	48.537	50.827	53.891
55	40.354	42.411	43.817	44.937	46.740	49.540	51.864	54.975
56	41.232	43.317	44.742	45.877	47.704	50.544	52.902	56.059
57	42.112	44.224	45.667	46.817	48.669	51.548	53.940	57.144
58	42.993	45.132	46.594	47.759	49.636	52.553	54.979	58.229
59	43.875	46.041	47.522	48.701	50.603	53.559	56.018	59.315
60	44.759	46.951	48.451	49.645	51.570	54.566	57.058	60.401

Table A.4 Offered Loads (In Erlangs) for Various Blocking Objectives: According to the Erlang-B Model—System Capacity from 61–80 Channels

Channels P(B)=	0.005	0.01	0.015	0.02	0.03	0.05	0.07	0.1
61	45.644	47.863	49.381	50.590	52.539	55.573	58.099	61.488
62	46.531	48.776	50.311	51.535	53.509	56.581	59.140	62.575
63	47.418	49.689	51.243	52.482	54.479	57.590	60.181	63.663
64	48.307	50.604	52.176	53.429	55.450	58.599	61.224	64.750
65	49.197	51.520	53.110	54.377	56.422	59.609	62.266	65.839
66	50.088	52.437	54.044	55.326	57.395	60.620	63.309	66.927
67	50.980	53.355	54.980	56.276	58.368	61.631	64.353	68.016
68	51.874	54.273	55.916	57.226	59.342	62.642	65.397	69.106
69	52.768	55.193	56.853	58.178	60.316	63.654	66.442	70.196
70	53.663	56.113	57.791	59.130	61.292	64.667	67.487	71.286
71	54.560	57.035	58.730	60.083	62.268	65.680	68.532	72.376
72	55.457	57.957	59.670	61.036	63.244	66.694	69.578	73.467
73	56.356	58.880	60.610	61.991	64.222	67.708	70.624	74.558
74	57.255	59.804	61.551	62.945	65.199	68.723	71.671	75.649
75	58.155	60.729	62.493	63.901	66.178	69.738	72.718	76.741
76	59.056	61.654	63.435	64.857	67.157	70.753	73.765	77.833
77	59.958	62.581	64.379	65.814	68.136	71.769	74.813	78.925
78	60.861	63.508	65.322	66.772	69.116	72.786	75.861	80.018
79	61.765	64.435	66.267	67.730	70.097	73.803	76.909	81.110
80	62.669	65.364	67.212	68.689	71.078	74.820	77.958	82.203

Table A.5 Offered Loads (In Erlangs) for Various Blocking Objectives: According to the Erlang-B Model—System Capacity from 81–100 Channels

Channels P(B) =	0.005	0.01	0.015	0.02	0.03	0.05	0.07	0.1
81	63.574	66.293	68.158	69.648	72.059	75.838	79.007	83.297
82	64.481	67.223	69.104	70.608	73.042	76.856	80.057	84.390
83	65.387	68.153	70.051	71.568	74.024	77.874	81.107	85.484
84	66.295	69.085	70.999	72.529	75.007	78.893	82.157	86.578
85	67.204	70.016	71.947	73.491	75.991	79.912	83.207	87.672
86	68.113	70.949	72.896	74.453	76.975	80.932	84.258	88.767
87	69.023	71.882	73.846	75.416	77.959	81.952	85.309	89.861
88	69.933	72.816	74.796	76.379	78.944	82.972	86.360	90.956
89	70.844	73.750	75.746	77.342	79.929	83.993	87.411	92.051
90	71.756	74.685	76.697	78.306	80.915	85.014	88.463	93.146
91	72.669	75.621	77.649	79.271	81.901	86.035	89.515	94.242
92	73.582	76.557	78.601	80.236	82.888	87.057	90.568	95.338
93	74.496	77.493	79.553	81.202	83.875	88.079	91.620	96.434
94	75.411	78.431	80.506	82.167	84.862	89.101	92.673	97.530
95	76.326	79.368	81.460	83.134	85.850	90.123	93.726	98.626
96	77.242	80.307	82.414	84.101	86.838	91.146	94.779	99.722
97	78.158	81.245	83.368	85.068	87.827	92.169	95.833	100.819
98	79.075	82.185	84.323	86.036	88.815	93.193	96.887	101.916
99	79.993	83.125	85.279	87.004	89.805	94.217	97.941	103.013
100	80.911	84.065	86.235	87.972	90.794	95.240	98.995	104.110

Table A.6 Offered Loads (In Erlangs) for Various Blocking Objectives: According to the Erlang-B Model—System Capacity from 105–200 Channels

Channels P(B) =	0.005	0.01	0.015	0.02	0.03	0.05	0.07	0.1
105	85.518	88.822	91.030	92.823	95.747	100.371	104.270	109.598
110	90.147	93.506	95.827	97.687	100.713	105.496	109.550	115.090
115	94.768	98.238	100.631	102.552	105.680	110.632	114.833	120.585
120	99.402	102.977	105.444	107.426	110.655	115.772	120.121	126.083
125	104.047	107.725	110.265	112.307	115.636	120.918	125.413	131.583
130	108.702	112.482	115.094	117.195	120.622	126.068	130.708	137.087
135	113.366	117.247	119.930	122.089	125.615	131.222	136.007	142.593
140	118.039	122.019	124.773	126.990	130.612	136.380	141.309	148.101
145	122.720	126.798	129.622	131.896	135.614	141.542	146.613	153.611
150	127.410	131.584	134.477	136.807	140.621	146.707	151.920	159.122
155	132.106	136.377	139.337	141.724	145.632	151.875	157.230	164.636
160	136.810	141.175	144.203	146.645	150.647	157.047	162.542	170.152
165	141.520	145.979	149.074	151.571	155.665	162.221	167.856	175.668
170	146.237	150.788	153.949	156.501	160.688	167.398	173.173	181.187
175	150.959	155.602	158.829	161.435	165.713	172.577	178.491	186.706
180	155.687	160.422	163.713	166.373	170.742	177.759	183.811	192.227
185	160.421	165.246	168.602	171.315	175.774	182.943	189.133	197.750
190	165.160	170.074	173.494	176.260	180.809	188.129	194.456	203.273
195	169.905	174.906	178.390	181.209	185.847	193.318	199.781	208.797
200	174.653	179.743	183.289	186.161	190.887	198.508	205.108	214.323

Table A.7 Offered Loads (In Erlangs) for Various Blocking Objectives: According to the Erlang-B Model—System Capacity from 205–245 Channels

Channels P(B) =	0.005	0.01	0.015	0.02	0.03	0.05	0.07	0.1
205	179.407	184.584	188.192	191.116	195.930	203.700	210.436	219.849
210	184.165	189.428	193.099	196.073	200.976	208.894	215.765	225.376
215	188.927	194.276	198.008	201.034	206.023	214.089	221.096	230.904
220	193.694	199.127	202.920	205.997	211.073	219.287	226.427	236.433
225	198.464	203.981	207.836	210.963	216.125	224.485	231.760	241.963
230	203.238	208.839	212.754	215.932	221.180	229.686	237.094	247.494
235	208.016	213.700	217.675	220.902	226.236	234.887	242.430	253.025
240	212.797	218.564	222.598	225.876	231.294	240.090	247.766	258.557
245	217.582	223.430	227.524	230.851	236.354	245.295	253.103	264.089

Table A.8 Offered Loads (In Erlangs) for Various Blocking Objectives: According to the Erlang-B Model System—Capacity from 250–600 Channels

Channels P(B) =	0.005	0.01	0.015	0.02	0.03	0.05	0.07	0.1
250	222.370	228.300	232.452	235.828	241.415	250.500	258.441	269.622
delta	0.961	0.977	0.988	0.998	1.015	1.042	1.069	1.107
300	270.410	277.144	281.853	285.707	292.142	302.617	311.866	324.961
delta	0.966	0.980	0.991	1.001	1.017	1.044	1.070	1.108
350	318.698	326.155	331.424	335.738	342.995	354.836	365.359	380.384
delta	0.969	0.984	0.994	1.005	1.018	1.045	1.071	1.109
400	367.163	375.334	381.128	385.963	393.895	407.096	418.890	435.813
delta	0.972	0.989	0.998	1.004	1.020	1.046	1.071	1.109
450	415.779	424.774	431.022	436.178	444.877	459.408	472.456	491.263
delta	0.975	0.987	0.997	1.006	1.021	1.047	1.072	1.109
500	464.518	474.130	480.890	486.480	495.919	511.759	526.049	546.730
delta	0.977	0.989	0.999	1.007	1.022	1.048	1.072	1.110
550	513.361	523.600	530.843	536.846	547.012	564.142	579.663	602.208
delta	0.979	0.991	1.000	1.008	1.023	1.048	1.073	1.110
600	562.292	573.142	580.859	587.267	598.145	616.552	633.295	657.697

Table A.9 Offered Loads (In Erlangs) for Various Blocking Objectives: According to the Erlang-B Model—System Capacity from 600–1050 Channels

Channels P(B) =	0.005	0.01	0.015	0.02	0.03	0.05	0.07	0.1
600	562.292	573.142	580.859	587.267	598.145	616.552	633.295	657.697
delta	0.983	0.992	1.001	1.009	1.023	1.049	1.073	1.110
650	611.418	622.748	630.927	637.732	649.313	668.982	686.941	713.193
delta	0.981	0.993	1.002	1.010	1.024	1.049	1.073	1.110
700	660.462	672.410	681.042	688.238	700.511	721.432	740.598	768.697
delta	0.982	0.994	1.003	1.011	1.024	1.049	1.073	1.110
750	709.586	722.119	731.196	738.777	751.735	773.896	794.266	824.206
delta	0.984	0.995	1.004	1.011	1.025	1.050	1.074	1.110
800	758.762	771.872	781.386	789.346	802.981	826.375	847.943	879.719
delta	0.985	0.996	1.004	1.012	1.025	1.050	1.074	1.110
850	807.987	821.662	831.608	839.942	854.247	878.865	901.627	935.236
delta	0.985	0.996	1.005	1.012	1.026	1.050	1.074	1.110
900	857.256	871.487	881.857	890.561	905.530	931.365	955.317	990.757
delta	0.986	0.997	1.005	1.013	1.026	1.050	1.074	1.110
950	906.565	921.343	932.132	941.202	956.829	983.875	1009.013	1046.281
delta	0.987	0.998	1.006	1.013	1.026	1.050	1.074	1.111
1000	955.910	971.226	982.430	991.862	1008.142	1036.393	1062.715	1101.808
delta	0.988	0.998	1.006	1.014	1.027	1.050	1.074	1.111
1050	1005.289	1021.136	1032.748	1042.539	1059.468	1088.918	1116.420	1157.337

List of Acronyms

2G	second generation
3G	third generation
3GPP	Third-Generation Partnership Project
AAL2	ATM adaption layer 2
AAL5	ATM adaption layer 5
ABS	analysis-by-synthesis
AC	authentication center
ACELP	algebraic codebook excited linear prediction
ACH	access channel
ACHT	average call holding time
ACK	acknowledgment
ACL	asynchronous connectionless
ACL	access control list
ACSE	association control service element
ACTS	Advanced Communication Technologies and Services
ACU	analog channel unit
ADPCM	adaptive differential PCM
AI	acquisition indicator
AICH	acquisition indication channel
AIN	advanced intelligent network

ALCAP	access link control application part
AM	access manager
AMA	automatic message accounting
AM/HLR	access manager/home location register
AMPS	advanced mobile phone systems
AMR	adaptive multiple rate
ANSI	American National Standards Institute
AP	application processor
AP	access point
API	applications programming interface
APN	access point name
ARIB	Association of Radio Industries and Businesses
ARN	authentication random number
ARP	address resolution protocol
ARQ	automatic retransmission query
ARQ	automatic repeat query
AS	access slots
ASN	authentication sequence number
AT	attention
ATC	adaptive transform coding
ATIM	announcement traffic information message
ATM	asynchronous transfer mode
AWGN	additive white Gaussian noise
BCCH	broadcast control channel
BCCH-C	BCCH constant
BCCH-V	BCCH variable
BCH	broadcast channel
BCR	baseband combiner and radio
BER	bit error rate

BHCA	busy hour call attempt
BML	business management layer
BOQAM	binary offset quadrature amplitude modulation
BPSK	binary phase-shift keying
BRA	basic rate access
BRU	baseband receiver unit
BS	base station
BS/MSC	base station/mobile switching center
BSC	base station controller
BSU	base station unit
BTS	base transceiver station
BTU	baseband transmitter unit
C	codes
CAC	channel access control
CAF	channel activity factor
CAP	CAMEL Application Part
CAS	call-associated signaling
CASE	common application service element
CATU	central access and transcoding unit
CC	connection control
CC	call control
CC2	channel coding 2
CCCH	common control channel
CCIR	Comité Consultatif International des Radio Communications
CCITT	Comité Consultatif International Télégraphique et Téléphonique
CCK	complementary code keying
CCPCH	common control physical channel
CCR	commitment concurrency and recovery
CC-Tr-CH	coded composite transport channel

CDF	cumulative distribution function
CDG	CDMA Development Group
CDG-13	CDMA Development Group 13
CDL	coded digital control channel locator
CDMA	code division multiple access
CD-P	collision-detection preamble
CDPD	cellular digital packet data
CELP	code-excited linear-prediction
CEPT	Conference Européenne des Administration des Postes et des Télécommunications
CHCNT	call history count
CLIP	connectionless interworking protocol
CLNP	connectionless network protocol
CMCH	common media access control channel
CMIP	common management information protocol
CN	core network
CNB	core network bearer
CORBA	Common Object Request Broker Architecture
CPCH	common packet channel
CPICH	common pilot channel
CRC	cyclic redundancy check
CS	circuit switched
csch	common signaling channel
CSD	circuit switched data
CSMA/CA	carrier sense multiple access with collision avoidance
CSMA/CD	carrier sense multiple access with collision detection
ctch	common traffic channel
CTG	cell trunk group
CTRU	central transceiver unit

CWTS	China Wireless Telecommunications Standard
D-AMPS	Digital-Advanced Mobile Phone System
DCCH	dedicated control channel
DCN	data communications network
DCE	distributed computing environment
DCF	distributed coordination function
DCH	dedicated channel
DCHFP	DCH frame protocol
DCOM	Distributed Component Object Model
DCR	dedicated common router
DCU	digital connection unit
DECT	Digital Enhanced Cordless Telecommunications
dg	digital gains
DHCP	dynamic host configuration protocol
DII	dynamic invocation interface
DLCI	data link control identifier
dmch	dedicated media access control channel
DNS	domain name server
DPCCH	dedicated physical control channel
DPCH	dedicated physical channel
DPDCH	dedicated physical data channel
DPE	distributed processing environment
DQPSK	differential quadrature phase-shift keying
DR	direct-sequence
DRNC	drift RNC
DS	direct spread
dsch	dedicated signaling channel
DSCH	downlink shared channel
DSI	dynamic skeleton interface

DSP	digital signal processor
DSSS	direct-sequence spread-spectrum
dtch	dedicated traffic channel
DTCH	dedicated transport channel
DTMF	dual-tone multifrequency
DTX	discontinuous transmission
EACH	enhanced access channel
ECC	error control coding
ECSD	enhanced CSD
EDGE	Enhanced Data Rates for GSM Evolution
EFR	enhanced full-rate
EGC	equal-gain combining
EGPRS	enhanced GPRS
EGSF	expanded general system failure
EIB	erasure indicator bit
EIU	Ethernet interface unit
ELN	explicit loss notification
EM	electromagnetic
EMI	electromagnetic interference
EML	element management layer
Enet	Ethernet
ERIP	equivalent isotropic radiated power
ERP	effective radiated power
ESN	electronic serial number
ESS	extended service set
ESSID	extended service set identification
ETSI	European Telecommunications Standards Institute
EVRC	enhanced variable-rate codec
FA	foreign agent

FACCH	fast associated control channel
FACH	forward access channel
FACHFP	FACH frame protocol
F-APICH	forward auxiliary pilot channel
F-ATDPICH	forward auxiliary transmit diversity pilot channel
FAUSCH	fast uplink signaling channel
F-BCH	forward broadcast channel
FBI	feedback indicator
F-CACH	forward common assignment channel
FCC	Federal Communication Commission
F-CCCH	forward common control channel
FCH	fundamental channel
FCH	frame controller head
F-CPCCH	forward common power control channel
F-DAPICH	forward dedicated auxiliary pilot channel
F-DCCH	forward dedicated control channel
FDD	frequency division duplex
FDMA	frequency division multiple access
FEC	forward error control
FER	frame error rate
F-FCH	forward fundamental channel
FFPC	fast forward power control
FH	frequency hopping
FHSS	frequency-hopping spectrum spread
FLPC	forward link power control
FLPMTS	future land public mobile telephone systems
FMA	FRAMES multiple access
F-PCH	forward paging channel
F-PICH	forward pilot channel

FPLMTS	Future Public Land Mobile Telephony
f/r-csch	forward and reverse common signaling channel
F/R-DCCH	forward/reverse dedicated control channel
f/r-dmch	forward and reverse dedicated media access control channel
f/r-dsch	forward and reverse dedicated signaling channel
F/R-FCH	forward/reverse fundamental channel
F/R-PICH	forward/reverse pilot channel
F/R-SCCH	forward/reverse supplemental coded channel
F/R-SCH	forward/reverse supplementary channel
FR	frame relay
FRAMES	Future Radio Wideband Multiple Access System
F-SCCH	forward supplemental code channel
F-SCH	forward supplemental channel
FSK	frequency shift keying
F-SYNC	forward sync channel
F-SYNCH	forward sync channel
F-TDPICH	forward transmit diversity pilot channel
FTP	file transfer protocol
GBS	GPRS backbone system
GDMO	Guideline for Definition of Managed Objects
GGSN	gateway GPRS support node
GMSC	gateway MSC
GMSK	Gaussian minimum shift keying
GPRS	general packet radio service
GPS	global positioning system
GSM	Global System for Mobile Communications
GSN	GPRS support node
HA	home agent
HCI	host controller interface

HCM	handoff completion message
HDLC	high-level data link control
HDM	handoff direction message
HDR	high data rate
HLR	home location register
HS	high speed
HSD	high-speed data packet
HPSD	high-speed packet data
HSCSD	high-speed circuit-switched data
HTML	HyperText Markup Language
HTTP	HyperText Transport Protocol
I	even
IC	interference cancellation
ICMP	Internet control message protocol
IDL	interface definition language
IEEE	Institute of Electrical and Electronic Engineers
IETF	Internet Engineering Task Force
IHL	Internet header length
IIOP	Internet Inter ORB Protocol
IMA	inverse multiplexing for ATM
IMSI	International mobile subscriber identity
IMT-2000	International Mobile Telephony 2000
IN	Intelligent network
IN/CAMEL	intelligent network/customized applications for mobile enhanced logic
IP	Internet protocol
IPCP	IP control protocol
IPR	intellectual property rights
IR	infrared
IrDA	infrared data association

IrOBEX	infrared object exchange protocol
IRP	integration reference point
ISDN	integrated services digital network
ISP	Internet service provider
ITS	intelligent telephone socket
ITU	International Telecommunications Union
ITU-R	International Telecommunications Union-Radio
ITU-T	International Telecommunications Union-Telecommunications
IWF	interworking functions
JTC	Joint Technical Committee
L2CAP	logical link control and adaptation protocol
LAC	link access control
LAF	linear amplifier
LAN	local area network
LAPD	link access procedure on the D-channel
LCP	link control protocol
LD-CELP	low-delay codebook excited prediction
LIU	line interface unit
LLA	logical layer architecture
LLC	logical link control
LLC	link layer control
LMDS	local multipoint distribution system
LMP	link manager protocol
LNS	L2TP network server
LOS	line of sight
LP	linear predictor
LPAS	linear-prediction-based analysis-by-synthesis
LSPD	low-speed data packet
LT	long-term

M3UA	MTP3-user adaptation layer
MAC	media access control
MAI	multiple access interference
MAP	maximum a posterior
MASHO	mobile-assisted soft-handoff
MBS	mobile broadband systems
MC	multicarrier
MCTD	multicarrier transmit diversity
MCU	main control unit
MD	meditation device
MD-IS	mobile data intermediate system
MDM	mobile diagnostic monitor
MDR	medium data rate
MExE	mobile station execution environment
MFA	management functional area
MIB	management information base
MIN	mobile identification number
MIPS	millions of instructions per second
MIT	management information tree
MM	mobility management
MMSE	minimum mean-squared error
MNRP	mobile network registration protocol
MSE	mean-squared error
MO	managed object
MOC	managed object class
MOI	managed object instance
MOS	mean opinion score
MPT	Ministry of Posts and Telecommunications
MRC	maximal ratio-combining

MS	mobile station
MSC	mobile switching center
MS-GSN	mobile-to-GSN
MT	mobile terminal
MT0	mobile termination 0
MT2	mobile termination 2
MTP3-B	message transfer part 3-B
MUD	multiuser detection
MUX	multiplex, or multiplexer
MWIF	Mobile Wireless Internet Forum
NAK	negative acknowledgment
NBAP	Node B application part
NCAS	non-call–associated signaling
NE	network element
NEI	network entity identifier
NEL	network element layer
NIC	network interface card
NID	network identification
NIU	network interface unit
NLUM	neighbor list update message
NM	network management
NMF	Network Management Forum
NML	network management layer
NOS	network operating system
N-PDU	network layer packet data unit
NRT	non-real-time
NSM	network and services management
OAM&P	operation, administration, maintenance, and planning
OBEX	object exchange

OCCCH	ODMA common control channel
OCNS	orthogonal channel noise simulator
ODCCH	ODMA dedicated control channel
ODCH	opportunity-driven dedicated channel
ODMA	opportunity-driven multiple access
ODTCH	ODMA dedicated traffic channel
OFDMA	orthogonal frequency division multiple access
OHG	CDMA Operators Harmonization Group
OMG	Object Management Group
OQPSK	offset quadrature phase-shift keying
ORACH	ODMA random access channel
OS	operating system
OSI	open system interconnect
OSS	operations support system
OTASP	over-the-air service provisioning
OTD	orthogonal transmit diversity
OVSF	orthogonal variable spreading factor
PACCH	packet associated control channel
PACS	Personal Access Communication System
PACS-UB	Personal Access Communication System unlicensed-version
PAGCH	packet access grant channel
PBCCH	packet broadcast control channel
PC	power control
PC	personal computer
PCCCH	packet common control channel
PCCH	paging control channel
P-CCPCH	primary common control physical channel
PCG	power control group
PCH	paging channel

PCM	pulse code modulation
PC-P	power-control preamble
PCPCH	physical common packet channel
PCS	personal communication services
PCS HCA	personal communication system high compression algorithm
PCU	packet control unit
PDA	personal digital assistant
PDC	public digital cellular
PDCH	packet data channel
PDP	packet data protocol
PDSCH	physical downlink shared channel
PDSN	packet data service node
PDTCH	packet data traffic channel
PDU	packet data unit
PDU	protocol data unit
PE	polynomial expansion
PH	protocol handler
PHS	Personal Handyphone System
PHY	physical layer
PI	page indicator
PIC	parallel interference cancellation
PICH	page indication channel
PLDCF	physical layer dependent convergence function
PLICF	physical layer independent convergence function
PLM	power level measurement
PLMN	public land mobile network
PML(PN)	pair of maximal linear
PMRM	power measurement report message
PN	pseudorandom noise

PNCH	packet notification channel
POP	post office protocol
POTS	plain old telephone service
PPDN	public packet data network
PPM	pulse-position modulation
PPP	point-to-point protocol
PRACH	physical random access channel
PRAMP	power ramp
PRAT	paging channel rate
PS	packet switched
PSC	primary synchronization code
PSMM	pilot strength measurement message
PSPDN	packet-switched public data network
PSTN	public switched telephone network
PSU	packet switching unit
PTCH	packet traffic channel
PTM	point-to-multipoint
PTM-SC	point-to-multipoint service center
PTP	point-to-point
PWT	personal wireless telecommunications
PWT-E	personal wireless telecommunications-enhanced
Q	odd
QA	Q-adaptor
QIB	quality indicator bit
QNC	Quick Net Connect
QOQAM	quaternary offset quadrature amplitude modulation
QoS	quality of service
QPCH	quick paging channel
QPSK	quadrature phase-shift keying

RAB	radio access bearer
RACE	Research in Advanced Communications Equipments
RACH	random access channel
R-ACH	reverse access channel
RACHFP	RACH frame protocol
RAM	random access memory
RANAP	radio access network application part
RAND	random number
RANDSSD	random number SSD
RBC	radio bearer control
RBP	radio burst protocol
RBS	radio base station
RC	resource control
RC	radio configuration
R-CCCH	reverse common control channel
RCD	resource configuration database
RCELP	relaxed code-excited linear prediction
RCIU	RPC interface units
RCPC	rate-compatible punctured convolution
RCR	Research and Development Center for Radio Systems
R-DCCH	reverse dedicated control channel
RES	radio and equipment systems
RF	radio frequency
R-FCH	reverse fundamental channel
RFU	radio frequency unit
RILPC	reverse inner loop power control
RITT	Research Institute of Telecommunications Transmission
RIU	radio interface unit
RLAC	radio link access control

RLC	radio link control
RLP	radio link protocol
RM	resource management
rms	root mean square
RNC	radio network controller
RNCP	radio network control plane
RNG	random number generator
RNL	radio network layer
RNS	radio network subsystem
RNSAP	radio network subsystem application part
ROLPC	reverse outer loop power control
ROM	read only memory
ROPC	reverse open loop power control
ROSE	remote operation service element
RP	radio port
RPC	remote procedure call
RPC	radio port controller
RPCU	radio port controller unit
R-PICH	reverse pilot channel
RPOM	radio port operation and maintenance
RRC	radio resource control
RS	radio system
RS	Reed-Solomon
R-SCCH	reverse supplemental code channel
R-SCH	reverse supplemental channel
RSV	Reed-Solomon Viterbi
RSVD	reserved
RT	real-time
RT	random time

RTO	round trip time-out
RTS/CTS	request to send/clear to send
RTT	round trip timer
RTT	radio transmission technology
Rx	receiving
SA	systems aspect
SAAL-NNI	signaling ATM adaptation layer for network-to-network interface
SAP	service access point
SAPI	service access point identifier
SAR	segmentation and reassembly
SASE	specific application service element
SBC	subband coding
SC	selection combining
SCH	supplementary channel
SCCH	synchronization control channel
SCCP	signaling connection control part
S-CCPCH	secondary common control physical channel
SCI	synchronized capsule indicator
SCO	synchronous connection-oriented
SCP	signal control point
SCP	service control point
SCTP	simple control transmission protocol
SDB	short data bursts
SDO	Standard Development Organization
SDP	service discovery protocol
SDU	service data unit
SF	spreading factor
SGCH	signaling channel
SGSN	serving GPRS support node

S/I	signal-to-interference
SIC	successive interference cancellation
SID	system identification
SIG	special interest group
SIM	subscriber identity module
SIR	signal-to-interference ratio
SISO	soft input/soft output
SIU	serial interface unit
S-IWF	serving system IWF
SLIC	subscriber loop interface circuit
SM	service management
SME	small- to medium-sized enterprise
SMFA	system management functional area
SMG	Special Mobile Group
SML	service management layer
SMS	short message service
SMTP	Simple Mail Transfer Protocol
S/N	signal-to-noise
SNDC	subnetwork dependent convergence
SNMP	simple network management protocol
SNR	signal-to-noise ratio
SO	service option
SOM	start of message
SOVA	soft-output Viterbi algorithm
SQL	Structured Query Language
SR	spreading rate
SRBP	signaling radio burst protocol
SRLP	signaling radio link protocol
SRNC	serving RNC

SRNS	serving RNS
SRP	selective repeat protocol
SS	spread-spectrum
SS7	signaling system 7
SSC	secondary synchronization code
SSCF	service-specific coordination function
SSCF-NNI	service-specific coordination function for network-to-network interface
SSCOP	service-specific connection-oriented protocol
SSD	shared secret data
SSDT	site selection diversity
STD	silence descriptor
STRU	subscriber transceiver unit
STTD	space time block coding based transmit diversity
SU	subscriber unit
SVC	switched virtual circuit
SWAP	shared wireless access protocol
SYCH	sync channel
SYNC	synchronization
TAGs	Technical Adhoc Groups
TCAP	transactions capabilities application part
TCE	traffic channel element
TCH	traffic channel
TCP	transmission control protocol
TCS	telephony control specification
TD-CDMA	time division CDMA
TDD	time division duplexing
TDM	time division multiplexing
TDMA	time division multiple access
TE	terminal equipment

TE2	terminal equipment 2
TEI	terminal equipment identity
TeleVAS	telephony value-added services
TFCI	transport-format combination indicator
TFTP	trivial FTP
THSS	time-hop spread-spectrum
TIA	Telecommunications Industry Association
TINA-C	Telecommunications Information Networking Consortium
TLS	transport layer security
TM	telecommunication management
TMF	Tele-Management Forum
TMN	Telecommunications Management Network
TMSI	temporary mobile subscriber identifier
TNCP	transport network control plane
TNL	transport network layer
TP4	Transport Protocol Class 4
TPC	transmit power control
TRAU	transcoder rate adapter unit
TREU	transcoder and echo canceller unit
TSG	technical specifications group
TSU	telecommunications switching unit
TTA	Telecommunications Technologies Association
TTC	Telecommunications Technology Committee
TTD	time division duplex
Tx	transmitting
UCN	UMTS core network
UDP	user datagram protocol
UE	user equipment
UL	uplink

UML	Unified Modeling Language
UMTS	Universal Mobile Telecommunications System
U-NII	unlicensed National Information Infrastructure
UP	user plane
URI	uniform resource identifier
URL	uniform resource locator
USCH	uplink shared channel
USF	uplink state flag
USSD	unstructured supplementary services data
UTRA	UMTS Terrestrial Radio Access
UTRAN	UMTS Terrestrial Radio Access Network
UWCC	Universal Wireless Communication Consortium
V	Viterbi
VAD	voice activity detector
VASP	value-added service provider
VC	virtual circuit
VLR	visitor location register
VMS	voice message services
VPN	virtual private network
VPU	video provisioning unit
VRC	variable-rate codec
VSELP	vector sum-excited linear-predictor
WACS	wireless access communication system
WAE	wireless application environment
WANU	wireless access network unit
WAP	wireless application protocol
WARC	World Administration Radio Conference
WARC-92	World Administration Conference 1992
WASU	wireless access subscriber unit

WCC	WLL communication control
W-CDMA	wideband CDMA
WCMP	wireless control message protocol
WDP	wireless datagram protocol
WECA	Wireless Ethernet Compatibility Alliance
WEP	wired equivalent privacy
WG	Working Group
WIMS	wireless multimedia service
WIN	wireless intelligent networking
WLAN	wireless LAN
WLL	wireless local loop
WLT	wireless line transceiver
WML	wireless markup language
WPA	weak pilot alert
WP-CDMA	wideband packet CDMA
WPKT	WLL packet control
WPOS	wireless point-of-sale
WRRC	WLL radio resource control
WS	work station
WSP	wireless session protocol
WTA	wireless telephony application
WTAI	wireless telephony application interface
WTLS	wireless transport layer security
WTP	wireless transaction protocol
XML	Extensible Markup Language
ZF-DF	zero forcing decision feedback

Index

A

B

C

D

E

F

G

H

M

N

O

P

Q

R

M

N

O

P

Q

R

S

T

U

V

W

About the Author

Dr. Vijay Garg is a professor in the Electrical Engineering and Computer Engineering Department at the University of Illinois at Chicago, where he teaches courses in wireless communications and wireless networking. Dr. Garg worked at Bell Labs of Lucent Technologies for the last 15 years, where he performed various technical assignments dealing with wireless networks, network performance evaluations, teletraffic engineering, data communication, etc. Dr. Garg received his Ph.D. from the Illinois Institute of Technology, Chicago, in 1973 and his MS from the University of California at Berkeley, California, in 1966. Dr. Garg has co-authored several technical books, in the fields of advanced dynamics, railway vehicle dynamics, and telecommunications (*Wireless and Personal Communications Systems and Principles and Applications of GSM* with Joe Wilkes, *Application of CDMA in Wireless Communications* with Ken Smolik and Joe Wilkes, and *IS-95 CDMA and cdma2000*). Dr. Garg is a Fellow of ASCE and ASME and a senior member of IEEE. He is a registered professional engineer in the states of Maine and Illinois. Dr. Garg is an academic member of the Russian Academy of Transport. He has served on various technical committees of IEEE, ASME, and ASCE. Dr. Garg has published approximately 100 technical papers, and was a feature editor for the PCS series in *IEEE Communication Magazine* from 1996 to 2000.

About the Author

Dr. Vijay Garg is a professor in the Electrical Engineering and Computer Engineering Department at the University of Illinois at Chicago, where he teaches courses in wireless communications and wireless networking. Dr. Garg worked at Bell Labs of Lucent Technologies for the last 15 years, where he performed various technical assignments dealing with wireless networks, network performance evaluations, teletraffic engineering, data communication, etc. Dr. Garg received his Ph.D. from the Illinois Institute of Technology, Chicago, in 1973 and his MS from the University of California at Berkeley, California, in 1966. Dr. Garg has co-authored several technical books, in the fields of advanced dynamics, railway vehicle dynamics, and telecommunications (*Wireless and Personal Communications Systems and Principles and Applications of GSM* with Joe Wilkes, *Application of CDMA in Wireless Communications* with Ken Smolik and Joe Wilkes, and *IS-95 CDMA and cdma2000*). Dr. Garg is a Fellow of ASCE and ASME and a senior member of IEEE. He is a registered professional engineer in the states of Maine and Illinois. Dr. Garg is an academic member of the Russian Academy of Transport. He has served on various technical committees of IEEE, ASME, and ASCE. Dr. Garg has published approximately 100 technical papers, and was a feature editor for the PCS series in *IEEE Communication Magazine* from 1996 to 2000.

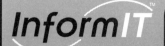

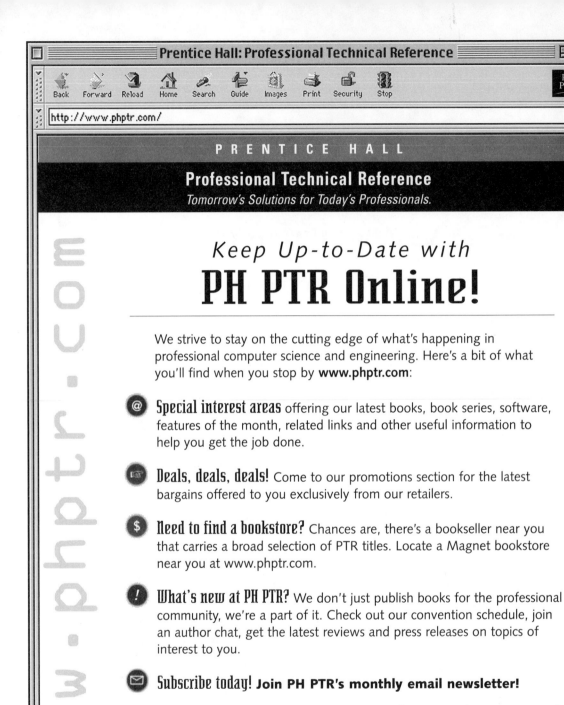